Translational Cardiology

Handbook for Designing and Conducting Clinical and Translational Research

Series Editor
Adam E. M. Eltorai, MD, PhD

Handbook for Designing and Conducting Clinical and Translational Research

Translational Cardiology

Edited by

Adam E. M. Eltorai
Brigham & Women's Hospital, Harvard Medical School, Boston, MA, United States

Jeffrey A. Bakal
Division General Internal Medicine, Faculty of Medicine and Dentistry, University of Alberta, Edmonton Alberta, Canada

C. Michael Gibson
Department of Medicine, Harvard Medical School, Boston, MA, United States

ELSEVIER

ACADEMIC PRESS
An imprint of Elsevier

Academic Press is an imprint of Elsevier
125 London Wall, London EC2Y 5AS, United Kingdom
525 B Street, Suite 1650, San Diego, CA 92101, United States
50 Hampshire Street, 5th Floor, Cambridge, MA 02139, United States

ISBN: 978-0-323-91790-2

For information on all Academic Press publications visit our website at
https://www.elsevier.com/books-and-journals

Publisher: Stacy Masucci
Acquisitions Editor: Tracey Lange
Editorial Project Manager: Barbara L. Makinster
Production Project Manager: Omer Mukthar
Cover Designer: Vicky Pearson Esser

Typeset by TNQ Technologies

Contents

Part III Clinical: Fundamentals

Part IV Statistical principles

Part V
Clinical: Study types

Dhrubajyoti Bandyopadhyay

Part VI
Clinical trials

Amr Salama

Part VIII
Regulatory basics

Aditi Nayak

Part IX
Clinical implementation

Rohin Moza

Part X
Public health

Clauden Louis

Contributors

Enoch Oluwalonimi Adewara, All Saints University School of Medicine, Roseau, Dominica

Ruchit Agrawal, School of Computer Science, University of Birmingham, Birmingham, United Kingdom

Pratik Agrawal, Division of Cardiology, WVU Heart & Vascular Institute, Morgantown, WV, United States

Fernando Aguel, Heart Failure and Circulatory Support Regulatory Affairs, MCRA, Washington, DC, United States

Monirah Albabtain, Pharmacy Department, Prince Sultan Cardiac Center, Riyadh, Saudi Arabia

Turki Al Garni, Interventional Cardiology, Division of Cardiology, Hamilton Health Sciences, McMaster University, Hamilton, ON, Canada; Department of Medicine, McMaster University, Hamilton, ON, Canada; Interventional Cardiology Department, Prince Sultan Cardiac Center, Riyadh, Saudi Arabia

Anas A. Alharbi, West Virginia University, Morgantown, WV, United States

Zaid Al Jebaje, Department of Medicine, Division of Cardiology, University of Rochester Medical Center, Rochester, NY, United States

Zakaria Almuwaqqat, Department of Medicine (Cardiology), Emory University School of Medicine, Atlanta, GA, United States

Ahmed Alnajar, The Division of Cardiothoracic Surgery, The Department of Surgery, The University of Miami Miller School of Medicine, Miami, FL, United States; Division of Biostatistics, The Department of Public Health Sciences, The University of Miami Miller School of Medicine, Miami, FL, United States

Abdulrahman Mubarak Alqahtani, Interventional Cardiology Department, King Salman Cardiac Center, King Fahd Medical City, Riyadh, Saudi Arabia

Bader Alqarni, Jeddah First Health Cluster, Alsolaimaniah, Jeddah, Saudi Arabia

Maha M. AlRasheed, Department of Clinical Pharmacy, College of Pharmacy, King Saud University, Riyadh, Saudi Arabia

Junedh M. Amrute, Center for Cardiovascular Research, Division of Cardiology, Department of Medicine, Washington University School of Medicine, Saint Louis, MO, United States

Rachel Magdalene Sabina Andrew, Dominica China Friendship Hospital, Roseau, Dominica; All Saints University School of Medicine, Roseau, Dominica; Advanced Island Care Inc., Roseau, Dominica

Precious Idorenyin Anthony, All Saints University School of Medicine, Roseau, Dominica

Amr A. Arafat, Cardiac Surgery Department, Prince Sultan Cardiac Center, Riyadh, Saudi Arabia; Cardiothoracic Surgery Department, Tanta University, Tanta, Egypt

Dhruvil Ashishkumar Patel, Wayne State University, Detroit, MI, United States

Heena Kaushal Asnani, Department of Internal Medicine, Roger Williams Medical Center, Providence, RI, United States

Elizabeth A. Babcock, University of Miami, Miami, FL, United States

Abeer Bakhsh, Heart Function and Transplant, Prince Sultan Cardiac Center, Riyadh, Saudi Arabia

Dhrubajyoti Bandyopadhyay, Cardiovascular Disease, Massachusetts General Hospital, Boston, MA, United States; Massachusetts General Hospital, Harvard Medical School, Boston, MA, United States

Basha Behrman, Department of Cardiology, Westchester Medical Center, New York Medical College, Valhalla, NY, United States

Kelley N. Benck, The University of Miami Miller School of Medicine, Miami, FL, United States; University of Southern California Keck School of Medicine, Los Angeles, CA, United States

Sanjeev P. Bhavnani, Division of Cardiology, Scripps Clinic, San Diego, CA, United States

Suryasnata Bhowmik, Critical Care Medicine, Nilratan Sircar Medical College, Kolkata, West Bengal, India

Suman Biswas, Department of Imaging, University of Rochester, Rochester, NY, United States

Lara Rodrigues Da Matta Calegari, Faculdade Evangélica Mackenzie do Paraná, Curitiba, Brazil

Luiz Campedelli, Division of Pulmonary, Allergy, Critical Care, and Sleep Medicine at the University of Pittsburgh, Pittsburgh, PA, United States

Maria Beatriz Canto, Division of Cardiovascular Medicine, Vanderbilt University Medical Center, Nashville, TN, United States

Cristiano Cardoso, Center for Preclinical Surgical & Interventional Research, The Texas Heart Institute, Houston, TX, United States

Chloe Carr, Division of Cardiovascular Medicine, Vanderbilt University Medical Center, Nashville, TN, United States

Gabriel Cavalcante Lima Chagas, Federal University of Ceará, Fortaleza, Ceará, Brazil

Shaw-Ree Chen, Oak Orchard Community Health Centers, Brockport, NY, United States

Henrique Guimarães Barbosa Coelho, Universidade Federal de Pernambuco, Recife, Brazil

Thomaz Alexandre Costa, Federal University of Ceara, School of Medicine, Fortaleza, Brazil

Ramesh Daggubati, Department of Cardiology, West Virginia, Morgantown, MV, United States

Carlos Andre Balthazar da Silveira, Bahiana School of Medicine and Public Health, Brotas, BA, Brazil

Kurt Dasse, Inspired Therapeutics LLC, Merritt Island, FL, United States

Marianna Daibes Rachid de Andrade, Spaulding Neuromodulation Center, Spaulding Rehabilitation Hospital/Harvard Medical School, Boston, MA, United States

Pedro Henrique M. Craveiro de Melo, Cardiovascular Research Foundation, New York, NY, United States

Jeremiah P. Depta, Sands-Constellation Heart Institute, Rochester Regional Health, Rochester, NY, United States

Ana Carolina do A.H. de Souza, Department of Radiology and Medicine (Cardiology), Brigham & Women's Hospital/Harvard Medical School, Boston, MA, United States

André De Souza, The Warren Alpert Medical School of Brown University, Providence, RI, United States; The Miriam Hospital, Providence, RI, United States

Kamil Detyniecki, Department of Neurology, University of Miami Miller School of Medicine, Miami, FL, United States

Maria Betzabe Diaz, Mercer University School of Medicine, Columbus, GA, United States

Sami El-Qadi, Galderma Laboratories, L.P., Dallas, TX, United States

Abdelmotagaly Elgalad, Center for Preclinical Surgical & Interventional Research, The Texas Heart Institute, Houston, TX, United States

Amari Evans, Mercer University School of Medicine, Columbus, GA, United States

Ronaldo Correa Fabiano, Department of Medicine, University of Pittsburgh Medical Center, Pittsburgh, PA, United States

Medhat Farwati, Cleveland Clinic Foundation, Cleveland Clinic, Cleveland, OH, United States

Carolina Ferraz, Universidade Santo Amaro, São Paulo, Brazil

Karley B. Fischer, Boonshoft School of Medicine, Wright State University, Dayton, OH, United States

Binita Ghosh, Internal Medicine, SSM Health St. Mary's Hospital, St. Louis, MO, United States

André Luiz Nunes Gobatto, Hospital da Cidade, Salvador, Bahia, Brazil; Hospital São Rafael, Salvador, Bahia, Brazil

Guilherme Leite B. Gonçalves, Federal University of Ceará School of Medicine, Fortaleza, Ceará, Brazil

Karthik Gonuguntla, Division of Cardiology, WVU Heart & Vascular Institute, Morgantown, WV, United States

Darbi Griffith, Boonshoft School of Medicine, Wright State University, Dayton, OH, United States

Syedah Aleena Haider, Department of Internal Medicine, Glangwili Hospital, Carmarthen, Wales, United Kingdom

Adrija Hajra, Department of Internal Medicine, Bringham and Women's Hospital/Harvard Medical School, Boston, MA, United States

Sameer Hirji, Division of Cardiac Surgery, Department of Surgery, Brigham and Women's Hospital, Harvard Medical School, Boston, MA, United States

Joel Horovitz, Nuffield Department of Surgical Sciences, University of Oxford, Oxford, United Kingdom

Mudathir Ibrahim, Department of General Surgery, Maimonides Medical Center, Brooklyn, NY, United States; Nuffield Department of Surgical Sciences, University of Oxford, Oxford, United Kingdom

Jade Iwasaka-Neder, Department of Radiology, Boston Children's Hospital/Harvard Medical School, Boston, MA, United States

Anant Jain, Department of Medicine, Westchester Medical Center, New York Medical College, Valhalla, NY, United States

Vikash Jaiswal, Larkin Community Hospital, South Miami, FL, United States

Michael John, Head of Cardiovascular Regulatory Affairs, MCRA, Washington, DC, United States

Rosaria Jordan, Wright State University, Dayton, OH, United States

Carlos Alberto Campello Jorge, Department of Medicine, Federal University of Mato Grosso, Mato Grosso, Brazil

Cedric Kamani, Faculty of Medicine, The University of British Columbia, Vancouver, BC, Canada

Tsuyoshi Kaneko, Division of Cardiothoracic Surgery, Department of Surgery, Washington University School of Medicine, Saint Louis, MO, United States

Faith Kanneh, Johns Hopkins Bloomberg School of Public Health, Baltimore, MD, United States

Aaron V. Kaplan, Dartmouth Health, Lebanon, NH, United States

João P.G. Kasakewitch, Beth Israel Deaconess Medical Center, Harvard Medical School, Boston, MA, United States

Risheek Kaul, Department of Cardiology, Westchester Medical Center and New York Medical College, Valhalla, NY, United States

Ridhima Kaul, Weill Cornell Medicine-Qatar, Education City, Qatar Foundation, Doha, Qatar

Akram Kawsara, Department of Cardiology, West Virginia, Morgantown, MV, United States

Andre P. Kengne, Department of Medicine, University of Cape Town, Cape Town, South Africa

Muhammad Zia Khan, Heart and Vascular Institute, West Virginia University, Morgantown, WV, United States

Khola Waheed Khan, Wah Medical College, National University of Medical Sciences, Rawalpindi, Pakistan

Benjamin J. Kopecky, University of Colorado Anschutz Medical Campus, Aurora, CO, United States

Yugandhar Kothapalli, Department of Pharmaceutical and Biomedical Sciences, College of Pharmacy, The University of Georgia, Athens, GA, United States

Michail Koutentakis, Department of Cardiothoracic Surgery and Transplantology, University Clinical Centre of Medical University of Warsaw, Central Teaching Clinical Hospital, Warsaw, Poland; Department of Experimental and Clinical Pharmacology, Medical University of Warsaw, Center for Preclinical Research and Technology CEPT, Warsaw, Poland

Danisha Kumar, Dow University of Health Sciences, Karachi, Pakistan

Sergio Rodriguez Labra, Department of Chemistry, The Scripps Research Institute, San Diego, CA, United States

Joseph Lamelas, Department of Cardiac Surgery, The University of Miami Miller School of Medicine, Miami, FL, United States

Ke Li, Center for Preclinical Surgical & Interventional Research, The Texas Heart Institute, Houston, TX, United States

Bruno B. Lima, Division of Cardiology, Department of Medicine, Vanderbilt University Medical Center, Nashville, TN, United States; Noninvasive Cardiovascular Imaging Program, Department of Radiology and Medicine (Cardiology), Brigham & Women's Hospital/Harvard Medical School, Boston, MA, United States; Vanderbilt University School of Medicine, Nashville, TN, United States; Department of Cardiothoracic Imaging, Vanderbilt Heart and Vascular Institute, Vanderbilt University Medical Center, Nashville, TN, United States; Division of Cardiovascular Medicine, Department of Internal Medicine, Vanderbilt University Medical Center, Nashville, TN, United States

Diego Laurentino Lima, Department of General Surgery, Montefiore Medical Center and the Albert Einstein College of Medicine, Bronx, NY, United States

Débora Moreira Lima, Walter Cantídio University Hospital, Fortaleza, Ceará, Brazil

Juan Linares, Wright State University, Dayton, OH, United States; Department of Cardiology, Kettering Health, Kettering, OH, United States

Jonathan Teng Fai Loke, Department of Medicine, Trinity College Dublin, Dublin, Ireland

Marcela de Almeida Lopes, HCor, Research Institute, São Paulo, Brazil

Clauden Louis, Structural Heart and Transcatheter Endovascular Cardiac Surgery Fellow, Cardiothoracic Surgery, Brigham and Women's Hospital, Boston, MA, United States

Dipesh Ludhwani, West Virginia University Heart and Vascular Institute, Morgantown, WV, United States

Levy Jo D. Manuntag, Department of Internal Medicine, Westchester Medical Center/New York Medical College, Valhalla, NY, United States

Samantha Eugene D. Manuntag, Department of Internal Medicine, University of Reno School of Medicine, Reno, NV, United States

Christina M. Mapp, Department of Pharmaceutical and Biomedical Sciences, College of Pharmacy, The University of Georgia, Athens, GA, United States

Alice Deberaldini Marinho, Federal University of the State of Rio de Janeiro, Rio de Janeiro, Brazil

Beatriz Marinho, School of Medicine, Universidade Estacio de Sa, Rio de Janeiro, Brazil

Peter McCulloch, Department of General Surgery, Maimonides Medical Center, Brooklyn, NY, United States

Juliet Meir, Department of Internal Medicine, Westchester Medical Center, Valhalla, NY, United States

John P. Mitchell, Division of Cardiovascular Medicine, Vanderbilt University Medical Center, Nashville, TN, United States

Angel Moctezuma-Ramirez, Center for Preclinical Surgical & Interventional Research, The Texas Heart Institute, Houston, TX, United States

Marco A. Morales-Garza, Center for Preclinical Surgical & Interventional Research, The Texas Heart Institute, Houston, TX, United States

Matheus José Barbosa Moreira, Federal University of Rio Grande do Norte, Rio Grande do Norte, Brazil

Jonathan D. Moreno, Division of Cardiology, Department of Medicine, Washington University in St. Louis, St. Louis, MO, United States

Thomaz R. Mostardeiro, Department of Medicine, Universidade Federal de Santa Maria, Santa Maria, Rio Grande do Sul, Brazil; Department of Radiology, University of Texas Southwestern, Dallas, TX, United States

Filipe A. Moura, Division of Cardiovascular Medicine, Brigham and Women's Hospital, Harvard Medical School, Boston, MA, United States; TIMI Study Group, Division of Cardiovascular Medicine, Brigham and Women's Hospital, Harvard Medical School, Boston, MA, United States

Michael G. Nanna, Department of Internal Medicine, Yale School of Medicine, New Haven, CT, United States

Zachariah B. Nealy, Cardiovascular Division, Department of Medicine, University of Virginia Health, Charlottesville, VA, United States; Riverside Regional Medical Center, Newport News, VA, United States

Vinicius Vieira Neves, Department of Behavioral Science and Health, University College London (UCL), London, United Kingdom

Guileonne Ngnitewe, Regulatory Promotional Review ProPharma Group, Madison, WI, United States

Anthony Lyonga Ngonge, Division of Cardiology, Morehouse School of Medicine, Atlanta, GA, United States

Raquel Nogueira, Department of General Surgery, Montefiore Medical Center and the Albert Einstein College of Medicine, Bronx, NY, United States; Bahiana School of Medicine and Public Health, Brotas, BA, Brazil

Brice Nouthe, Department of Medicine, The University of British Columbia, Vancouver, BC, Canada; CESSTAD-KOHEN, Cameroon

Joseph Nowatzke, Division of Cardiovascular Medicine, Vanderbilt University Medical Center, Nashville, TN, United States

Nso Nso, Division of Cardiology, University of Chicago (Northshore), Evanston, IL, United States

Favour Oludunni Ojebiyi, All Saints University School of Medicine, Roseau, Dominica

Disha B. Patel, Michigan State University College of Osteopathic Medicine, East Lansing, MI, United States

Bishow Mohan Paudel, Cardiovascular Imaging University of Virginia, Charllotsville, VA, United States

Pradipta Paul, Weill Cornell Medicine-Qatar, Education City, Qatar Foundation, Doha, Qatar

Victor Perim, Medical School, the Federal University of the State of Rio de Janeiro (UNIRIO), Rio de Janeiro, Brazil

Priscilla Petit, Inspired Therapeutics LLC, Merritt Island, FL, United States

Luka Petrovic, Mount Sinai Morningside Hospital, New York, NY, United States

William Plath, Department of General Surgery, Maimonides Medical Center, Brooklyn, NY, United States; Nuffield Department of Surgical Sciences, University of Oxford, Oxford, United Kingdom

Adam J. Prince, Dartmouth Health, Lebanon, NH, United States

Gabriel Prusch, Vanderbilt University School of Medicine, Nashville, TN, United States

Isabella S.M. Rabelo, Universidade Estadual do Ceará, Fortaleza, Ceará, Brazil

Amanda Ribeiro Rangel, Federal University of Ceará, Fortaleza, Ceará, Brazil

Ana Caroline Dias Rasador, Department of Surgery, Montefiore Medical Center, Bronx, NY, United States

Isildinha M. Reis, Division of Biostatistics, The Department of Public Health Sciences, The University of Miami Miller School of Medicine, Miami, FL, United States

Julia Tireli Rocha, Faculdade de Medicina de Jundiai, Jundiai, Brazil

Eduardo Rochitte, Centro Universitário FMABC, Faculdade de Medicina, Santo André, São Paulo, Brazil

Carlos Eduardo Rochitte, Heart Institute, InCor, University of São Paulo Medical School, São Paulo, Brazil

Amr Salama, Interventional Cardiology Department, University of Alabama at Birmingham, Birmingham, AL, United States

Gustavo Sanchez, Vanderbilt University School of Medicine, Nashville, TN, United States

João Victor S. Sanders, Universidade Federal do Ceará, Fortaleza, Ceará, Brazil

Gustavo P. de A. Saraiva, Universidade Estadual do Ceará, Fortaleza, Ceará, Brazil

Abdullah Sarkar, Cardiac Arrhythmia Institute- UCLA, Los Angeles, CA, United States

Yasar Sattar, Department of Cardiology, West Virginia, Morgantown, MV, United States

Anna Scandinaro, Cleveland Clinic Foundation, Cleveland Clinic, Cleveland, OH, United States

Luiza G. Schmitt, Department of Radiation Oncology, University of Texas Southwestern, Dallas, TX, United States; Department of Medicine, Universidade Federal de Santa Maria, Santa Maria, Rio Grande do Sul, Brazil

Mehdi Shadmand, Wright State University, Dayton, OH, United States; Department of Internal Medicine, Wright State University, Dayton, OH, United States

Ayesha Shaik, Division of Cardiology, Hartford Hospital, Hartford, CT, United States

Chen Shaw-Ree, Oak Orchard Community Health Center, Brockport, NY, United States

Christine Shen, Division of Interventional Cardiology, Scripps Clinic, La Jolla, CA, United States

Leticia Maria Tedesco Silva, Department of Neurology, University of Miami Miller School of Medicine, Miami, FL, United States

Melissa Lou Silva, Department of Radiology, University of California San Diego, San Diego, CA, United States; Department of Medicine, Universidade Estadual do Ceará, Fortaleza, Ceará, Brazil

Anderson C. Woite Silva, University of Sao Paulo, Sao Paulo, Brazil

Raisa Lomanto Silva, Department of Medicine, University of Pittsburgh Medical Center, Pittsburgh, PA, United States

Matheus Simonato, Department of Internal Medicine, Yale School of Medicine, New Haven, CT, United States

Harsimran S. Singh, Division of Cardiology, Department of Medicine, Weill Cornell Medicine - New York Presbyterian Hospital, New York, NY, United States

Uma S. Singh, Department of Pharmaceutical and Biomedical Sciences, College of Pharmacy, The University of Georgia, Athens, GA, United States

Elisa T.D. Siqueira, Universidade Federal do Ceará, Fortaleza, Ceará, Brazil

Alex Smith, The University of Miami Miller School of Medicine, Miami, FL, United States

Jayakumar Sreenivasan, Department of Cardiology, Westchester Medical Center, New York Medical College, Valhalla, NY, United States; Department of Medicine, Westchester Medical Center, New York Medical College, Valhalla, NY, United States

Samian Sulaiman, Division of Cardiology, West Virginia University, School of Medicine, Morgantown, WV, United States

J. Sunil Rao, Division of Biostatistics, The Department of Public Health Sciences, The University of Miami Miller School of Medicine, Miami, FL, United States

Badal Thakkar, Sinai Hospital of Baltimore, Baltimore, MD, United States; George Washington University School of Medicine, Washington, DC, United States; Johns Hopkins Medicine, Baltimore, MD, United States

Sherri R. Thumma, Michigan State University College of Osteopathic Medicine, East Lansing, MI, United States; Center for Healthcare Outcomes and Policy, University of Michigan, Ann Arbor, MI, United States

Aditi Ujjawal, Saint Vincent Hospital, Worcester, MA, United States

Damian Valencia, Department of Cardiovascular Medicine, Kettering Health, Kettering, OH, United States; Wright State University, Dayton, OH, United States; Department of Cardiology, Kettering Health, Kettering, OH, United States

Jef Van den Eynde, Department of Cardiovascular Sciences, KU Leuven, Leuven, Belgium

Júlia Lemos Lima Verde, Federal University of Ceará School of Medicine, Fortaleza, Ceará, Brazil

Alysa Vereen, Priovant Therapeutics, Durham, NC, United States

Dominique Vervoort, Institute of Health Policy, Management and Evaluation, University of Toronto, Toronto, ON, Canada

Adam Vohra, Division of Cardiology, Department of Medicine, Weill Cornell Medicine - New York Presbyterian Hospital, New York, NY, United States

Maham A. Waheed, Maimonides Medical Center, Brooklyn, NY, United States

Anam Akbar Waheed, Medstar Washington Hospital Center, Washington, DC, United States; Medstar Georgetown University Hospital, Washington, DC, United States

Khola Waheed Khan, Wah Medical College, National University of Medical Sciences, Rawalpindi, Pakistan

Isaiah V.N. Ware, Mercer University School of Medicine, Columbus, GA, United States

Naira Link Woite, Harvard School of Public Health, Boston, MA, United States

Emily P. Zeitler, Dartmouth Health, Lebanon, NH, United States; The Dartmouth Institute, Lebanon, NH, United States

List of section editors

I. Introduction
Rachita Navara
rnavara1@gmail.com

II. Preclinical: Discovery and Development
Samira Bahrainy
samira11@uw.edu

IV. Statistical Principles
Pratik Agrawal
pratik.agrawal@lsuhs.edu

V. Clinical: Study Types
Dhrubajyoti Bandyopadhyay
drdhrubajyoti87@gmail.com

VI. Clinical Trials
Amr Salama
amr.salama2@rochesterregional.org

VII. Clinical: Preparation
Zakaria Almuwaqet
zmwa34@gmail.com

Bruno Lima
bruno.lima@vumc.org

VIII. Regulatory Basics
Aditi Nayak
aditi.nayak.k@gmail.com

X. Public Health
Clauden Louis
claudenlouis@gmail.com

XI. Practical Resources
Amr Salama
amr.salama2@rochesterregional.org

Preface

Translational research is essential to the advancement of cardiology. Written for clinicians, scientists, students, and biotech/medtech entrepreneurs, this urology-specific guide to translational research serves as a practical, step-by-step roadmap for designing and conducting basic, translational, and clinical research, including basic regulatory considerations for new technology. *Translational Cardiology* aims to help bridge the gap between clinical problems, research, and practice. Comprehensively spanning from preclinical research, clinical research, and clinical implementation to public health, this book provides a clear process for understanding, designing, executing, and analyzing clinical and translational research.

The reader will learn how to more critically evaluate the quality of published studies with respect to measuring outcomes and to make effective use of all the types of evidence in patient care. This book covers the principles of evidence-based medicine and applies these principles to the design of translational investigations.

This book's straightforward approach aims to help the aspiring investigators navigate challenging considerations in study design and implementation. Researchers will benefit from greater confidence in their ability to initiate and execute investigations along the translational spectrum, gain valuable insights on funding, avoid common pitfalls, and know what is needed in collaborators.

Whether you are a basic scientist with a promising therapeutic agent interested in moving from laboratory to clinical experimentation, an engineer needing to complete a pilot study for a new biomedical device, a student learning the principles of statistics or epidemiology, a clinician orchestrating and preparing grants for a randomized clinical trial, or an entrepreneur seeking insight on regulatory considerations, *Translational Cardiology* is a practical guide for effectively advancing and translating a urology research question.

Adam Eltorai
Editor

Part I

Introduction

Rachita Navara

Chapter 1

Translational cardiology: Introduction

Jonathan D. Moreno
Division of Cardiology, Department of Medicine, Washington University in St. Louis, St. Louis, MO, United States

Introduction

The field of cardiology has a rich history of discovering mechanisms of disease, and the use of that knowledge to develop novel drug and device therapy. By some estimates, there has been more original investigative research within the field of cardiology than any other subspecialty in medicine. In some ways, the field has advanced at a breakneck pace because of the diversity of scientific "talent" and a broad range of disciplines (e.g., medicine, engineering, computational biology, physics, to name a few) that are attracted to translational cardiology. Not only does this make practicing cardiology exciting, but this diversity of cardiovascular research is also critical: cardiovascular disease (CVD) remains the leading cause of death globally. In 2019, nearly 18.6 million people worldwide died from CVD, representing an increase of 17% over the past decade. In addition, a staggering 523.2 million people carried a diagnosis of CVD in 2019, an increase of 26.6% from 2010.[1] With a need this great, solutions to solving this global epidemic are going to require innovative ideas from a broad range of stakeholders who are quickly able to translate both basic scientific discoveries as well as advancements in disparate fields into clinical practice.

Before jumping in and defining new research goals to pursue, it is important to take stock of where the field has come, just in the last few decades. As a heart failure cardiologist, the following list is admittedly biased and incomplete, but I would like to touch on some seminal discoveries (and some "a-ha" moments) that have inspired me in the laboratory every day. Each drug, device, diagnostic, and discovery has dramatically altered the CVD landscape and provided further impetus to push the field forward.

Research in diagnostic tools: As early as 1929, Dr. Warner Forssmann floated a catheter into his own body to measure right heart hemodynamics.[2] While initially only advanced into the right atrium, subsequent innovation allowed for directing the catheter into the pulmonary arteries; this required considerable skill and carried with it a serious risk of complications. However, in the 1970s, as the story has it, Drs. Jeremy Swan and William Gantz noticed that boats with sails seemed to move rather efficiently over the relatively calm waters of the Laguna Beach, CA shore. This "a-ha" moment inspired them to fashion a balloon-tip to the catheter to allow for easier "flotation" of the catheter through the right-sided heart chambers into the pulmonary artery. The balloon tip "democratized" the use of the PA catheter, and considerably increased its safety. While there is some conflicting data regarding the efficacy of PA catheterization, in properly selected, critically ill patients, the PA catheter has become an indispensable diagnostic tool.

Research in organ transplantation: Patients with end-stage heart failure carry an abysmal prognosis: 50% mortality within 3−5 years. After decades of preclinical research, in 1967, Dr. Christaan Barnard and his team successfully performed the first human-to-human heart transplant at the University of Capetown in South Africa. Publication of that seminal achievement remains one of the most cited articles in cardiology (I will add to that record here:[3]), and in the lay media, the story is the most publicized event in world medical history.[4] While his first patient only lived for 18 days, over the ensuing decades, remarkable progress principally within cardiac immunology and the role of immune tolerance to solid organ transplant, has made orthotopic heart transplant the most durable long-term strategy for patients afflicted with end-stage heart failure. Patients with a previous death sentence can expect a median survival of 13 years in contemporary cohorts.[5] Lastly, at the time of this writing, the first genetically modified porcine heart was transplanted successfully into man, surely ushering in an unbelievable era of cardiac xenotransplantation[6] for the years to come!

Research in devices: Owing to the limited supply of donor hearts, mechanical circulatory support (MCS) devices have revolutionized the heart failure landscape. Initial

Translational Cardiology. https://doi.org/10.1016/B978-0-323-91790-2.00032-0

ventricular assist devices (VADs) were paracorporeal, pneumatic heart pumps (e.g., Thoratec XVE) with a high device failure rate, often providing support for only 1–2 years. Over the next 40 years, pumps have become smaller, continuous flow devices (e.g., Heartmate III) allowing for surgical implantation of the pump, with only an externalized drive-line that connects to a controller and batteries. In that time frame, devices have become markedly more durable, device-related complications including stroke have plummeted, and patients can expect to attain an independent quality of life with a median survival of 5–7 years on device support.[7] The next frontier (and it is coming!) is the *fully* implanted VAD (pump, controller, batteries). The novel technologies making these devices possible span multiple disciplines: fluid mechanics, mechanical engineering (with an emphasis on blood: device interfaces), battery technology, surgical advancements, and an understanding of cardiovascular physiology to name a few.

Research in drug discovery: Among the most exciting advancements, and by far the one with the most dramatic worldwide impact on cardiovascular morbidity and mortality, has been the many seminal discoveries in heart failure pharmacology (particularly heart failure with reduced ejection fraction—HFrEF). The timeline of HFrEF therapeutics has been a steady, pedantic march of progress, starting with the introduction of vasodilator therapies (isosorbide, hydralazine), the discovery of the neurohormonal hypothesis of heart failure (beta-blockers) and the importance of blockade of the renin–angiotensin aldosterone system (ACE inhibitors, mineralocorticoid receptor blockers), and the "New Kids on the Block" (angiotensin-neprilysin inhibitors—ARNIs, and sodium glucose cotransporter type 2 inhibitors—SGLT2i). Quite simply, the HFrEF pharmacology space can now be rightly considered an "embarrassment of riches", with each of these four classes of drugs (ARNI, beta-blockade, MRA, SGLT2i), added sequentially onto the background of a previously established standard of care, and providing an additive, clinically significant morbidity and mortality benefit.

There are many lessons from this rich history of heart failure pharmacology, but three bear repeating, as instructive for future translational research. First, dare to think "outside the box," and follow the science; for many years, beta-blockers, now a cornerstone of therapy, were once considered detrimental to heart failure patients, given their negative inotropy and adrenergic blockade. However, after carefully designed clinical trials, evidence of a mortality benefit was plainly apparent and has contributed to our understanding of the role of adrenergic activation in heart failure.[8] Second, "be rigorous; " the heart failure literature is robust because investigators and the heart failure community have insisted on randomized clinical trials that incorporated the highest quality standard of care as the control group for new therapies (vs. comparison to placebo). For example, beta-blockade was tested *on the background* of proven RAAS inhibition, and ARNi therapy[9] was compared to maximally tolerated ACE inhibition, on the background of MRA and beta-blockers. It has thus become more challenging for a new therapeutic to prove significantly beneficial when added to robust cornerstone therapeutics, but as clinicians, we continue to demand this rigor for our patients. Thirdly, "follow the science" and be prepared to accept serendipity: SGLT2is were originally developed as an adjunctive oral therapy for type II diabetes. The careful clinical trials that followed, revealed, however, that the most clinically impactful benefit was a reduction in cardiovascular hospitalization and mortality *irrespective* of the presence of type II diabetes.[10] What was originally developed as a drug for diabetes, has now firmly established itself as a cornerstone therapy for patients with heart failure.

Conclusions

This list of recent cardiovascular discoveries is woefully incomplete but serves as a reminder that advancements in cardiology have heretofore been the result of tireless efforts by many physicians, scientists, and engineers (to name a few), from diverse fields, coming together to solve clinical problems. This book, and the chapters that follow, provide a framework for understanding the cardiovascular discovery process. We start with an overview of preclinical research, and the importance of defining the problem to be solved. Does the solution involve a drug, device, diagnostic, or a combination of these technologies? We then move into the clinical realm of discovery: what is clinical research, how do we define the study population, what are the data to be collected and why? We emphasize the importance of clearly and precisely defining the clinical problem of interest. From there, we briefly review key statistical principles in clinical trial design before jumping into the myriad clinical trial study types, and the pros and cons of each design strategy.

From there, we get into the "nuts and bolts" of performing the trial: how do we recruit patients, what are the budgetary and financial considerations, what types of data will be required to definitively answer the clinical question, how do we navigate regulatory agencies, and finally how do we study special populations?

Lastly, we examine special topics in cardiovascular research including clinical implementation science and public health, and end with practical resources applicable to all aspects of translational cardiology. This text is not meant to be exhaustive, rather, this book is intended to be a "toolbox," with the researcher flipping to relevant sections as the need arises during the course of their research endeavors. It is our sincere hope that the information presented herein serves as a springboard for further reading and eventual discovery.

The history of cardiovascular research and discovery is rich, but the armamentarium of new tools, techniques, computational power, and the confluence of discoveries in a broad range of disciplines all but ensures that the best is yet to come.

References

1. Virani SS, et al. Heart disease and stroke statistics-2021 update: a report from the American heart association. *Circulation*. 2021;143(8):e254–e743.
2. Chatterjee K. The Swan-Ganz catheters: past, present, and future. A viewpoint. *Circulation*. 2009;119(1):147–152.
3. Barnard CN. The operation. A human cardiac transplant: an interim report of a successful operation performed at Groote Schuur Hospital, Cape Town. *S Afr Med J*. 1967;41(48):1271–1274.
4. Brink JG, Hassoulas J. The first human heart transplant and further advances in cardiac transplantation at Groote Schuur Hospital and the University of Cape Town - with reference to : the operation. A human cardiac transplant : an interim report of a successful operation performed at Groote Schuur Hospital, Cape Town. *Cardiovasc J Afr*. 2009;20(1):31–35.
5. Moayedi Y, et al. Survival outcomes after heart transplantation: does recipient sex matter? *Circ Heart Fail*. 2019;12(10):e006218.
6. Reardon S. First pig-to-human heart transplant: what can scientists learn? *Nature*. 2022;601:305–306.
7. Zimpfer D, et al. Long-term survival of patients with advanced heart failure receiving an left ventricular assist device intended as a bridge to transplantation: the registry to evaluate the heartware left ventricular assist system. *Circ Heart Fail*. 2020;13(3):e006252.
8. Sacks CA, Jarcho JA, Curfman GD. Paradigm shifts in heart-failure therapy–a timeline. *N Engl J Med*. 2014;371(11):989–991.
9. McMurray JJ, et al. Angiotensin-neprilysin inhibition versus enalapril in heart failure. *N Engl J Med*. 2014;371(11):993–1004.
10. Packer M, et al. Cardiovascular and renal outcomes with empagliflozin in heart failure. *N Engl J Med*. 2020;383(15):1413–1424.

Chapter 2

Translational process

Bishow Mohan Paudel
Cardiovascular Imaging University of Virginia, Charllotsville, VA, United States

What is translational cardiology?

- Translational cardiology is the bridge between basic cardiac research "bench" and its clinical applications "bedside" including technological advancements and big data incorporation into more personalized patient care. It involves translating discoveries in the laboratory into viable diagnostic and therapeutic strategies or healthcare service advancement for patients suffering from cardiovascular diseases. In the modern context, translational cardiology, encompasses diverse fields, from genetics to big datasets, and global registries for both common and rare medical conditions. It also includes a wide spectrum of clinical practices, such as advanced cardiac imaging, minimally invasive surgeries, mechanical circulatory systems, and heart transplant advancements.

The importance of translational research in cardiology

Bridging the gap

- Basic science research identifies a potential new target or a diagnostic tool or intervention for treating heart diseases. Translational cardiology facilitates translating basic discoveries into appropriate clinical applications and may also bridge the gap in reverse fostering a feedback loop where clinical challenges inspire new research, driving further discoveries and in any case ultimately better patient outcomes.
  It helps accelerate drug and device development or enhance personalized treatment to individual patients based on their unique genetic makeup and other factors. Translational cardiology often involves collaboration between basic scientists, clinical researchers, and practicing cardiologists. This multidisciplinary approach ensures that all factors are considered when moving a potential treatment from the lab to the clinic.

Accelerated innovations

- Not every basic research finding has clinical potential. Translational cardiology helps prioritize the findings from the basic research that have the greatest potential to benefit patients, ensuring that resources and efforts are concentrated on the most promising avenues. It prioritizes promising discoveries, streamlines preclinical testing, and improves clinical trial design, often leading to adaptive trials. This collaborative process includes feedback loops and may expedite regulatory approvals, ensuring innovations reach the patients in a timely manner.

Personalized medicine

- Personalized medicine is possible with translational research on genetics, environmental, and lifestyle factors to guide decisions in disease prevention, diagnosis, and treatment. For example, basic science research discovered mutations in genes such as LDLR, APOB, and PCSK9 causing familial hypercholesterolemia, which is translated into the development of effective PCSK9 inhibitor treatment. Testing for specific mutations can diagnose hypertrophied cardiomyopathy, assess family risk, and tailor treatments.

Big data and AI in translational cardiology

- Big data and artificial intelligence (AI) have revolutionized translational cardiology, improving diagnostics through precise image interpretation, predicting cardiovascular risks, personalizing treatments, and expediting drug discovery. AI-powered wearables offer real-time monitoring of vital stats and cardiac rhythms, potentially enabling precise arrhythmia detection and remote

Translational Cardiology. https://doi.org/10.1016/B978-0-323-91790-2.00020-4

treatment algorithms. In addition, AI optimizes workflow in cardiac care units by predicting patient flow and resource allocation, while in clinical research, it ensures effective patient recruitment and optimized trial outcomes through rapid data analysis.

Challenges in translational cardiology

- Scientific challenges include the inherent complexity of the cardiovascular system. The variations in how diseases manifest, reliance on animal models with differences in human relevance, the need to integrate emerging knowledge, the multifactorial nature of disease interplay with social determinants of health, limitations of current imaging and diagnostic tools, the complexity of regenerative therapies, and the challenges in predicting drug effects, especially long-term outcomes. Other challenges include funding, infrastructure, and regulatory and ethical considerations.

The future of translational process

Integration with technology, collaboration, and patient-centric research

- As AI and machine learning (ML) technologies continue to advance, the future of translational cardiology has immense potential for efficiently processing vast datasets with unmatched accuracy, promising a revolution in cardiovascular research and patient care outcomes. The future powered by AI and ML is on the brink of achieving groundbreaking advancements that were once beyond imagination. Interdisciplinary collaboration is indispensable in the evolving landscape of translational cardiology as experts from diverse fields will pave the way for breakthroughs and better patient care. Translational cardiology needs to be more patient-centric, ensuring that care and treatments are tailored to individual needs, preferences, and circumstances. This approach emphasizes both advancement and a deeply human-centered focus.

Chapter 3

Scientific method

Rosaria Jordan[1], Juan Linares[1,2], Mehdi Shadmand[1,3] and Damian Valencia[1,2]

[1]*Wright State University, Dayton, OH, United States;* [2]*Department of Cardiology, Kettering Health, Kettering, OH, United States;* [3]*Department of Internal Medicine, Wright State University, Dayton, OH, United States*

Key points

- The scientific method refers to a set of steps that offer a procedural approach to answer a specific question objectively with reproducibility in mind; the steps of the scientific method are (1) observation/question, (2) research, (3) hypothesis, (4) experimentation, (5) analysis, and (6) conclusion.
- Although the scientific method is the current standard of scientific inquiry, there are several pitfalls that you should be aware of, including underpowered studies, confounding variables, bias, and misleading conclusions.
- There are thousands of real-world examples of the scientific method within the published medical literature, three of which are discussed in the chapter text.

Why it matters

The principles behind the scientific method date back thousands of years and across numerous geographical borders. Aristotle, known as one of the founders of this method, emphasized the significance of the demonstrative principle to logically explain and predict natural phenomena.[1] As the process of empirical thinking was refined, others have developed the process that we have learned to call the scientific method.[2] Today, experts believe that to successfully answer a scientific inquiry, one must methodically approach the question with observation, data collection, analysis, and retesting. Academically, these steps have been consolidated into six categories; observation/question, research, hypothesis, experimentation, analysis, and conclusion. Collectively, they offer a procedural approach to analyze a specific question empirically to support or refute a postulation.

Steps of the scientific method

Step 1—Observation/Question: It is human nature to question the world around you. From a very young age children will ask questions like "Why is the sky blue?," "How do cars move?," and "Why can't animals talk?." These are all examples of basic scientific questions. A scientific question is a question rooted in observation or prior knowledge with the intent of discovery. Scientific questions can be divided into two categories: hypothetico-deductive and hypothetico-inductive (Fig. 3.1). Hypothetico-deductive questions are those which arise from prior knowledge. Hypothetico-inductive questions are those which arise from observation.[3]

Developing a question is the first step in the scientific method. Revisions or additions to existing scientific work typically start with hypothetico-deductive questions, while novel research more commonly follows hypothetico-inductive questions. Here are two scenarios that further detail each type of question.

Scenario 1: Hypothetico-deductive question

A student reads a peer-reviewed journal reporting HMG-Co-A reductase inhibitors may decrease the risk of stroke. The student wonders if statins also lower the risk of cardiovascular events and asks, "Is there a correlation between HMG-Co-A reductase inhibitors and cardiovascular events?".

Scenario 2: Hypothetico-inductive question

A medical student on a cardiovascular rotation realizes that patients on 3-hydroxy-3-methyl-glutaryl-coenzyme A (HMG-Co-A) reductase inhibitors (also known as statins)

Translational Cardiology. https://doi.org/10.1016/B978-0-323-91790-2.00001-0

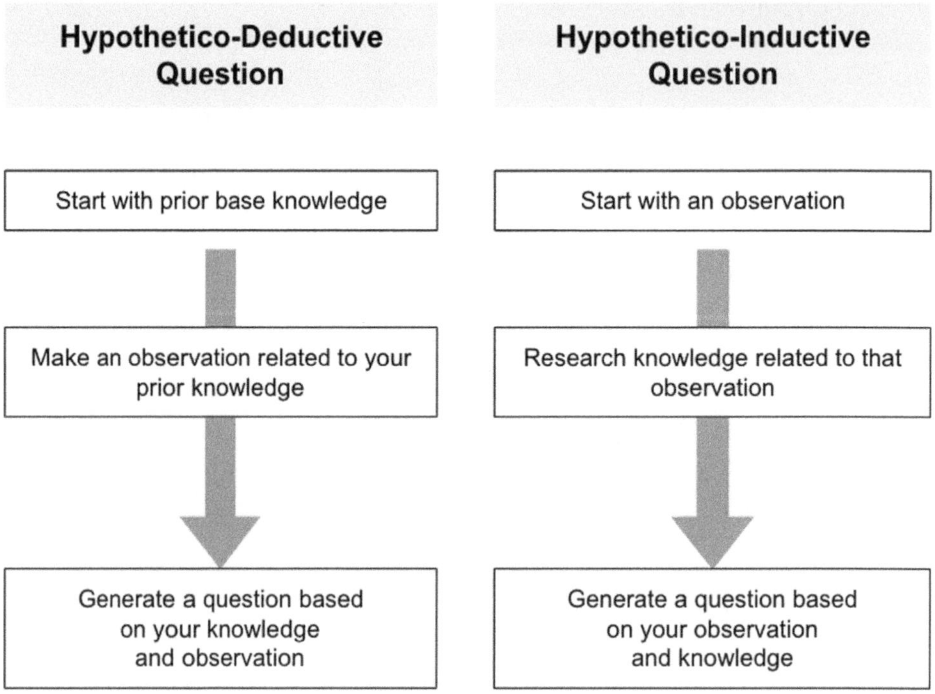

FIGURE 3.1 Types of questions.

are experiencing fewer cardiovascular events. The student asks, "Is there a correlation between HMG-Co-A reductase inhibitors and cardiovascular events?"

Step 2—Research: After a scientific question has been established, it is important that you obtain and review previous knowledge on the subject through research. This is the second step of the scientific method. In modern medicine, a wealth of knowledge already exists within the medical literature. A literature search should be performed to help guide or refine the question being asked. It is not uncommon that an answer to your question already exists. In these cases, you may change or modify your initial question. If your question changes drastically, you may need to perform a new literature search before proceeding. Of note, when changing your question due to acquired knowledge on the topic, your hypothetico-inductive question becomes a hypothetico-deductive question. During this step, you should also think about narrowing the scope of your question with the feasibility of experimentation in mind. You may also conceive follow-up questions that remain unanswered. The steps of the scientific method are not unidirectional, see Fig. 3.2 for a pictorial representation of this model.

Step 3—Hypothesis: The third step of the scientific method is to generate a hypothesis. After you have finalized your question, you must establish a conjecture or an educated guess as to what the answer of the question may be. This prediction is called a hypothesis. To form your hypothesis, you must establish your dependent and independent variables, and predict their relationship to one another. A dependent variable is a variable that depends on the values of other variables. An independent variable is a variable whose variation does not depend on the values of other variables. Hypotheses are often phrased with statistical correlation in mind. A positive correlation means one variable decreases as the other variable decreases (or as one variable increases the other increases). A negative correlation means as one variable decreases the other increases (or vice versa). Regardless of the conjecture, a null hypothesis will always exist. The null hypothesis is a term used to define a situation in which there is no statistical correlation between your conjecture and the observed phenomenon or question.

Following our previous example, a sample hypothesis would be as follows, "I suspect that use of HMG-Co-A reductase inhibitors is negatively correlated with major adverse cardiovascular events (MACE) in adults." An example of the null hypothesis would be as follows, "There is no significant correlation between HMG-Co-A reductase inhibitor use and MACE in adults."

Step 4—Experiment: The fourth step of the scientific method is to develop and run an experiment. Depending on the question and hypothesis, this step may involve prototypical experimentation, data collection, or observation. Before beginning this phase, it is critical to develop a stepwise approach, or methodology, to ensure reproducibility and elimination of bias. There are many ways to go about experimentation: retrospective, prospective

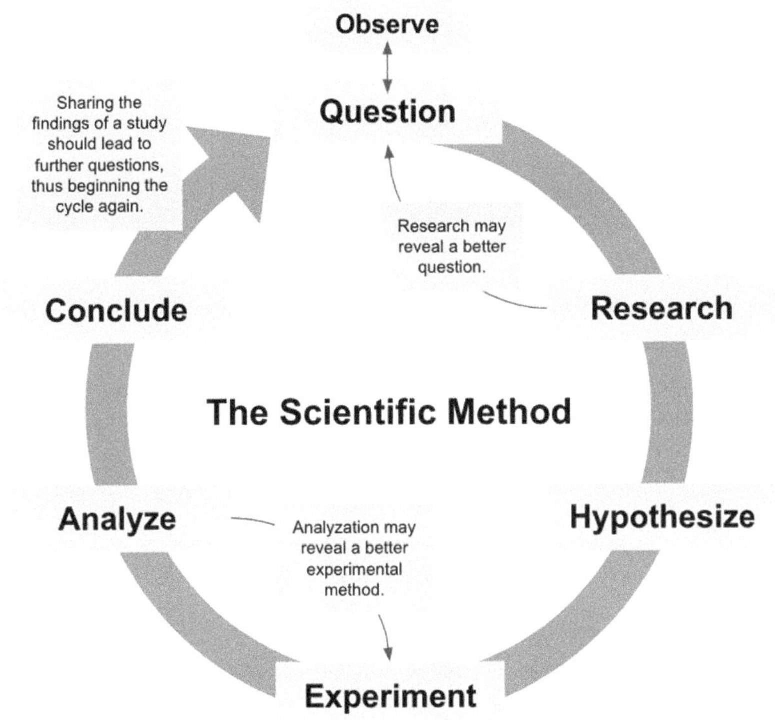

FIGURE 3.2 Dynamic steps of the scientific method.

(clinical trials), cross-sectional, and observational. Retrospective study refers to gathering and analyzing information already available (e.g., medical records) over a specific time of interest. In prospective research, the information does not yet exist but is collected as it occurs over a specified amount of time. Cross-sectional research refers to obtaining the information of interest from a single point in time (e.g., January 2022), and may be obtained by surveys or polls. Observational studies, as implied by the name, are conducted without performing any interventions on the individual or population being studied.

In medical research, experimentation is often done through clinical trials, which will be further detailed in later chapters. It is common to detail which subjects will be observed or involved in experimentation, inclusion/exclusion criteria, risks/benefits, legal/ethical concerns, monetary compensation, consent, and if any portion of the experiment is at risk of bias/manipulation. Having a set protocol for experimentation/observation will also aid future researchers in reproducing your findings. In this step, one must also decide who will be privy to data collected/observed and at which time. If none of the active participants in the study are privy to the data collected or are aware of the intervention, the protocol is termed a blinded study. If the selection of subjects for the intervention arm is random, the protocol is termed a randomized study. Throughout this step, care must be taken to safeguard any private patient information from the public or unauthorized individuals.

An example of an observational study would be to observe all adult patients taking and not taking HMG-Co-A reductase inhibitors and document any MACE with respect to time. An example of a prospective study would be to select a large number of patients and only treat half of them with HMG-Co-A reductase inhibitors and follow both arms with respect to time, documenting the occurrence of MACE. See a detailed stepwise sample of an observational study below.

Sample observational study

1. Select a sample size and patient population.
   a. Example: 1000 adult patients
2. Establish inclusion criteria.
   a. Example: ages 18–65, established with a cardiologist, able to consent
3. Establish exclusion criteria.
   a. Example: history of MACE, active malignancy, life expectancy <5 years, pregnancy, active unstable disease
4. Divide the subjects into groups based on the independent variable
   a. Example: statin use and no statin use

TABLE 3.1 Example experimentation data set.

Variables	MACE	No. MACE
Mean age		
Sex		
Male		
Female		
BMI		
Comorbidities		
Hypertension		
Hyperlipidemia		
Tobacco use		
HMG-Co-A[a] reductase inhibitor use		
Yes		
No		

[a]*3-hydroxy-3-methyl-glutaryl-coenzyme A.*

5. Follow-up period.
   a. Example: follow up every 6 months for 5 years, monitoring variables of interest (hypertension, hyperlipidemia, tobacco use, and statin use) and recording the incidence of MACE
6. Organize data for analysis.
   a. Example table below (Table 3.1)

Step 5—Analysis: After the experimental or observational phases are complete, you must organize and analyze data to determine if any correlation exists. If no correlation exists, the null hypothesis is proven to be true. If a correlation exists, it may be positive or negative but does not guarantee causation. There are many mathematical options for statistical analysis that can determine how strong this correlation is and if confounding variables are present which may skew your results. Statistical analysis will be discussed in detail in a later chapter. After data analysis is complete, it is possible to repeat the observation phase or perform further experimentation. Another approach at this step would be to modify your hypothesis and repeat testing. If the data collected is sufficient, analysis will be completed and you can proceed to develop a conclusion. In line with our previous example, statistical analysis would be performed to determine if a true difference exists between those patients taking and not taking HMG-Co-A reductase inhibitors.

Step 6—Conclusion: The conclusion is the final step of the scientific method. It is a summary of the findings and a statement regarding whether the hypothesis was proven to be true or false. In this step, you must create a means to report the results to the rest of the scientific community and to the public. An example concluding statement would be, "There is a negative correlation between HMG-Co-A reductase inhibitor use and MACE in adult patients." Depending on the strength of the data presented and the reproducibility of the findings, the authors may even change the wording of the statement to make it easier to understand and more impactful. For example, "HMG-Co-A reductase inhibitor use reduces MACE in adult patients."

Getting started

After reviewing the information contained in this chapter, the reader is encouraged to take it to the next level and apply the information learned by creating a sample research project. It is best to select a topic in a personal field of interest, this will keep the researcher motivated and engaged until completion of the project. A senior mentor can be a source of help and guidance through the process. Please review and complete the steps listed in Table 3.2. You can search through currently available research on PubMed (https://pubmed.ncbi.nlm.nih.gov) or EBSCO (https://www.ebsco.com), using keywords related to the selected topic. After you have completed a thorough literature review, refine and rewrite the original question as necessary to better fit your research objectives. Create a rough draft using the six steps of the scientific method listed above.

Pitfalls

The scientific method does not come without pitfalls. There are many factors that may influence your scientific work; common pitfalls you may encounter are detailed below.

Underpowered statistics: An important piece in the analysis step of the scientific method is statistical analysis. Underpowered statistics simply means that your sample size is too small. In most cases, it is considered unethical or misleading to publish a study with underpowered statistics.[4] The best way to combat this issue is to increase the initial recruitment size. In cases where this is not possible, multiple trials may be necessary.

Confounding variables: Confounding variables are extraneous factors that obscure the true relationship between the variables being studied, and can compromise the validity of the research.[5] You can eliminate confounding variables by the addition of strict exclusion criteria in the selection process, although this is not always possible. In cases where the elimination of confounding variables is not possible, statistical methods, such as multivariate analysis as opposed to univariate analysis, can be applied in the analysis phase to account for their effects.

Misleading conclusions: Although not a true pitfall of the scientific method, using composite endpoints may be misleading to readers. Composite endpoints are a

TABLE 3.2 List of action items.

Action items
Select a research mentor.
Select a topic of interest.
Ask a question based on a clinical suspicion or observation.
Review available literature in a scientific database using keywords.
Identify if the question being asked has been answered already.
Rephrase your question as necessary to better reflect your goals.
Formulate a hypothesis and a null hypothesis.
Select a method of research (e.g., retrospective, prospective, cross-sectional, observational study)
Design and plan a way to test your hypothesis, discuss with mentors, and verify your investigation adheres to the current standards of ethics. You may consult with a statistician about sample size for adequate power and to determine the types of statistical tests needed to analyze the data.
Obtain necessary permissions if applicable. Many studies require institutional review board (IRB) approval to access and gather data from medical records.
Conduct the research as per your prespecified design.
Enter the data on a password-protected server/computer.
Analyze the data obtained using the statistical tests selected.
Reject or accept your hypothesis.
Create a manuscript describing your research to communicate your findings to the scientific community.

combination of two or more outcomes into one concluding statement. If you must use a composite endpoint, the individual components of the composite endpoint must follow these three criteria to assure probable credibility: equal clinical importance, similar frequency of the events, and similar effect of the treatment or intervention on each component.[6] Here is an example of a misleading statement using composite endpoints: "Medication X reduces death, myocardial infarction and readmission" or "Medication X is associated with decreased death, myocardial infarction and readmission." It is possible that the statistical significance of reaching the composite endpoint of death, myocardial infarction, and readmission is driven heavily by the readmission rate alone. If each component of the endpoint is analyzed separately, it may show that Medication X does not influence death or rates of myocardial infarction, and only reduces readmission rates. Additionally, statistical significance does not always mean clinical significance. For example, a statistically significant relative risk reduction of 50% may sound impressive, but this can correspond to an absolute risk reduction of 0.0005% (risk change from 0.001% to 0.0005%), which may not be enough to make a significant impact on clinical decision-making.

Citation bias: Citation bias is the unwillingness to publish unfavorable results for fear of scientific ostracization or industry reprimand.[5] Although unfavorable or negative (null) results may be discouraging, this should not deter you from the scientific processes of research and discovery. Remember the plasticity of the scientific method (Fig. 3.2), and the ability to return to and modify prior steps. In some cases, it is best to revise your hypothesis, modify the experimental protocol, and reanalyze the data before abandoning a project. In cases where the null hypothesis is proven true, you should still share the findings as they can help others asking similar questions. For instance, the ASPREE,[7] ASCEND,[8] and ARRIVE[9] trials have all found that aspirin did not improve cardiovascular outcomes when used for primary prevention; this challenged previously held notions and guidelines. Similarly, the pragmatic TRISS[10] and TRICC[11] trials demonstrated that lower hemoglobin thresholds for blood transfusion were associated with favorable outcomes and less complications, again changing the way physicians practice. These and other trials have led to major changes throughout multiple society guidelines.

Implying causality: Oftentimes, statistics will show a significant correlation between two variables. It is important to understand that correlation does not imply causation. Here we will review a simple example where variables are correlated but no causal relationship exists. You perform data collection on several baseball players and salaries. You notice that the players with the most home runs are also the players who get paid the most. You continue to analyze the data and realize that the players who strike out the most are also the same players who get paid the most. It would be inappropriate to imply that a player can increase his salary by striking out more. The players with the highest salaries typically play the most, and such will have more strikes and home runs. Be sure to keep this in mind when formulating concluding statements of your research.

Examples of the scientific method

To further demonstrate the scientific method, three real-world examples detailing the six steps within the published literature are detailed below.

Example 1

Study name: Predictors of Aortic Valve Commissural Fusion in Patients Undergoing Left Ventricular Assist Device Therapy[12]

Question: Are there factors that influence or predict aortic valve commissural fusion (ACF) in patients supported with the HeartMate II left ventricular assist device (LVAD)?

Research: The literature review revealed several case reports of ACF in patients supported with the HeartMate II LVAD.[13,14] At the time of this study, no literature was available detailing predictive factors for ACF in this patient population.

Hypothesis: As presented at the International Society for Heart and Lung Transplant Conference, the authors hypothesized that there may be a positive correlation between higher LVAD speeds and rate of ACF. The null hypothesis would be that no association exists between LVAD speeds and rate of ACF.

Experiment: Retrospective data collection pertaining to patients who underwent LVAD implantation at their institution was performed. Patients were excluded upon heart transplantation, aortic valve suturing, pump exchange, or death. The follow-up period was 1 year with data collection at 1, 6, and 12 months. Two groups were created based on aortic valve status (open/nonfused or closed/fused).

Analysis: Each patient characteristic, echocardiographic finding, and LVAD setting were compared between groups using univariate and multivariate analysis. Patient sex, age, and LVAD indication had no association with ACF. Higher LVAD speeds, 8800 RPM as opposed to 8600 RPM, immediately postimplantation were associated with ACF at 1-month follow-up.

Conclusion: Higher LVAD speed within the first month of implantation is associated with ACF and can potentially be used as a predictor for commissural fusion.

Example 2

Study name: Risk of Cardiovascular Hospital Admission After Exposure to Fine Particulate Pollution.[15]

Question: Is there an association between heavy particulate matter air pollution and hospital admission for cardiovascular events?

Research: The literature review revealed many studies on the effects of particulate air pollution on the body, although with contradicting results.[16–20] Additional large-scale studies would be helpful to determine the effects of particulate air pollution on cardiovascular health and hospitalization.

Hypothesize: Although a hypothesis was not explicitly stated in this study, we can create one for this example. Exposure to heavy pollution will have a positive correlation with cardiovascular hospitalization rates. The null hypothesis would be that no correlation exists between air pollution levels and cardiovascular hospitalization rates.

Experiment: Data was collected from several air monitoring stations to determine daily average particulate matter levels. Criteria for what was considered heavy pollution were clearly defined. Data was collected from the Municipal Commission of Health to determine cardiovascular admissions and baseline patient characteristics. Cardiovascular admissions included myocardial infarction, chronic ischemic heart disease, arrhythmia, heart failure, angina, ischemic stroke, hemorrhagic stroke, hypertensive disease, and pulmonary embolism. Patients who had lived in the area for less than 6 months were excluded. Cardiovascular admissions were then compared to the duration of exposure to heavy particulate matter air pollution.

Analysis: Analysis found that exposure to higher levels of air pollution and duration of exposure were both associated with increased rates of hospitalization related to cardiovascular events. The authors used a time-stratified multivariate regression analysis to account for confounding variables when determining the effects of air pollution and duration of exposure to hospitalization.

Conclusion: Exposure to high particulate matter air pollution was associated with increased rates of cardiovascular admission.

Example 3

Study name: Refining clinical risk stratification for predicting stroke and thromboembolism in atrial fibrillation using a novel risk factor-based approach: the Euro Heart Survey on atrial fibrillation.[21]

Observation/Question: There are two available stroke risk prediction models for patients with atrial fibrillation (CHA2DS2-VASc and CHADS2). Which model (CHA2DS2-VASc or CHADS2) better predicts stroke risks in patients with atrial fibrillation?

Research: Previous literature suggests improved outcomes when using a risk prediction model when treating patients with anticoagulation for atrial fibrillation.[22,23] The factors included in the CHADS2 Score are congestive heart failure, hypertension, age >75, diabetes, and prior stroke/transient ischemic attack. The factors included in the CHA2DS2-VASc Score are congestive heart failure, hypertension, age >64, age >74, diabetes, prior stroke/transient ischemic attack, vascular accident, and sex category (female). Before this research, no studies have been conducted comparing the two available risk prediction models.

Hypothesis: The inclusion of additional risk factors in the CHA2DS2-VASc Score compared to the CHADS2 Score would result in a more accurate/improved risk assessment in patients with atrial fibrillation. The null hypothesis would be that there is no difference between scoring systems.

Experimentation: Patients were characterized by risk stratification scores and then grouped into low-, intermediate-, or high-risk groups. Each group was then observed for thromboembolic events. The CHA2DS2-VASc Score classified 9.2% as low risk, whereas the CHADS2 Score categorized 48.3% as low risk. Observational data showed thromboembolic events in 1.4% of low-risk patients characterized by CHADS2, while no thromboembolic events were observed in low-risk patients characterized by CHA2DS2-VASc Score.

Analysis: When expressed as a scoring system, the patient's risk stratified by CHA2DS2-VASc score experienced an increased rate of thromboembolic events with increasing scores.

Conclusion: The CHA2DS2-VASc scoring better characterizes the patient risk profile when assessing the risk of cardiac thromboembolism and determining the need for prophylactic anticoagulation.

Additional resources and readings

- PubMed Database (https://pubmed.ncbi.nlm.nih.gov)
- EBSCO (https://www.ebsco.com)
- Skinner, B. F. (1956). A case history in scientific method. *American Psychologist, 11*(5), 221–233. https://doi.org/10.1037/h0047662
- Gauch, H. G. (2003). *Scientific Method in Practice*. Cambridge University Press.
- Cohen, M. R., & Nagel, E. (2002). *An introduction to Logic and Scientific Method.* Simon Publications.
- Carey, S. S. (2012). *A Beginner's Guide to Scientific Method.* Wadsworth Cengage Learning.

References

1. Hintikka J. *Concepts of Scientific Method from Aristotle to Newton.* Netherlands: Springer; 2004.
2. Butterfield H. *The Origins of Modern Science.* Free Press; 1957.
3. Solomon E, Berg L, Martin D. *Biology.* 6th ed. Cengage Learning; 1998.
4. Halpern S, Karlawish J, Berlin J. The continuing unethical conduct of underpowered clinical trials. *JAMA.* 2002;288(3):358–362. Available from: https://doi.org/10.1001/jama.288.3.358.
5. Pannucci C, Wilkins E. Identifying and avoiding bias in research. *Plast Reconstr Surg.* 2010;126(2):619–625. Available from: https://doi.org/10.1097/PRS.0b013e3181de24bc.
6. Lia M, Dans A. Composite endpoints. *J Clin Epidemiol.* 2020;128: 157–158. Available from: https://doi.org/10.1016/j.jclinepi.2020.07.017.
7. McNeil J, Wolfe R, Woods R, et al. Effect of aspirin on cardiovascular events and bleeding in the healthy elderly. *N Engl J Med.* 2018;379(16):1509–1518. Available from: https://doi.org/10.1056/NEJMoa1805819.
8. ASCEND Study Collaborative. Effects of aspirin for primary prevention in persons with diabetes mellitus. *N Engl J Med.* 2018;379(16):1529–1539. Available from: https://doi.org/10.1056/NEJMoa1804988.
9. Gaziano J, Brotons C, Coppolecchia R, et al. Use of aspirin to reduce risk of initial vascular events in patients at moderate risk of cardiovascular disease (ARRIVE): a randomised, double-blind, placebo-controlled trial. *Lancet.* 2018;392(10152):1036–1046. Available from: https://doi.org/10.1016/S0140-6736(18)31924-X.
10. Holst L, Haase N, Wetterslev J, et al. Lower versus higher hemoglobin threshold for transfusion in septic shock. *N Engl J Med.* 2014;371(15):1381–1391. Available from: https://doi.org/10.1056/NEJMoa1406617.
11. Hébert P, Wells G, Blajchman M, et al. A multicenter, randomized, controlled clinical trial of transfusion requirements in critical care. *N Engl J Med.* 1999;340(6):409–417. Available from: https://doi.org/10.1056/NEJM199902113400601.
12. Valencia D, Kliethermas S, Bailey J, et al. Predictors of aortic valve commissural fusion in patients undergoing left ventricular assist device Therapy. *J Heart Lung Transplant.* 2015;34(4). https://doi.org/10.1016/j.healun.2015.01.554.
13. Mudd J, Cuda J, Halushka M, Soderlund K, Conte J, Russell S. Fusion of aortic valve commissures in patients supported by a continuous axial flow left ventricular assist device. *J Heart Lung Transplant.* 2008;27(12):1269–1274. Available from: https://doi.org/10.1016/j.healun.2008.05.029.
14. Gallen T, Lau W, Mehta A. Complete aortic valve fusion after HeartMate II left ventricular assist device support. *J Cardiothorac Vasc Anesth.* 2012;26(6):1060–1062. Available from: https://doi.org/10.1053/j.jvca.2012.05.010.
15. Zhang Y, Ma R, Ban J, et al. Risk of cardiovascular hospital admission after exposure to fine particulate pollution. *J Am Coll Cardiol.* 2021;78(10):1015–1024. Available from: https://doi.org/10.1016/j.jacc.2021.06.043.
16. Pope C, Renlund D, Kfoury A, May H, Horne B. Relation of heart failure hospitalization to exposure to fine particulate air pollution. *Am*

J Cardiol. 2008;102(9):1230–1234. Available from: https://doi.org/10.1016/j.amjcard.2008.06.044.

17. Lee H, Pan S, Chen B, Lo S, Guo Y. Atrial fibrillation hospitalization is associated with exposure to fine particulate air pollutants. *Environ Health.* 2019;18(1):117. Available from: https://doi.org/10.1186/s12940-019-0554-7.
18. Bunch T, Horne B, Asirvatham S, et al. Atrial fibrillation hospitalization is not increased with short-term elevations in exposure to fine particulate air pollution. *Pacing Clin Electrophysiol.* 2011;34(11):1475–1479. Available from: https://doi.org/10.1111/j.1540-8159.2011.03200.x.
19. Chiu H, Chang C, Yang C. Relationship between hemorrhagic stroke hospitalization and exposure to fine particulate air pollution in Taipei, Taiwan. *J Toxicol Environ Health.* 2014;77(19):1154–1163. Available from: https://doi.org/10.1080/15287394.2014.926801.
20. Ito K, Mathes R, Ross Z, Nádas A, Thurston G, Matte T. Fine particulate matter constituents associated with cardiovascular hospitalizations and mortality in New York City. *Environ Health Perspect.* 2011;119(4):467–473. Available from: https://doi.org/10.1289/ehp.1002667.
21. Lip G, Nieuwlaat R, Pisters R, Lane D, Crijns H. Refining clinical risk stratification for predicting stroke and thromboembolism in atrial fibrillation using a novel risk factor-based approach: the euro heart survey on atrial fibrillation. *Chest.* 2010;137(2):263–272. Available from: https://doi.org/10.1378/chest.09-1584.
22. Baruch L, Gage B, Horrow J, et al. Can patients at elevated risk of stroke treated with anticoagulants Be further risk stratified? *Stroke.* 2007;38(9):2459–2463. Available from: https://doi.org/10.1161/STROKEAHA.106.477133.
23. Camm A, Lip G, Caterina R, et al. 2012 focused update of the ESC Guidelines for the management of atrial fibrillation: an update of the 2010 ESC Guidelines for the management of atrial fibrillation. *Eur Heart J.* 2012;33(21):2719–2747. Available from: https://doi.org/10.1093/eurheartj/ehs253.

Chapter 4

Basic research

Junedh M. Amrute[1] and Benjamin J. Kopecky[2]

[1]Center for Cardiovascular Research, Division of Cardiology, Department of Medicine, Washington University School of Medicine, Saint Louis, MO, United States; [2]University of Colorado Anschutz Medical Campus, Aurora, CO, United States

Key points

- Basic research in biomedical science is complex and multivariate and demands an innovative, multidisciplinary, and systems-level approach to problem-solving.
- Basic biomedical research is ever-evolving with the emergence of novel experimental and computational tools allowing researchers to unravel complex biological networks in human health and disease.
- Here we outline some of the broader disciplines within the basic science umbrella, describe innovative tools, and discuss emerging areas of active research in basic science cardiology.

Why it matters

Clinical medicine has shifted away from a trial-and-error approach to an evidence-based paradigm grounded in fundamental scientific principles. Basic biomedical research has shown tremendous promise in delivering translatable therapeutics in cardiology—for example, the drug-eluting stent is a ground-breaking discovery motivated by an observation in cancer basic science and subsequently translated into coronary stents. As such, basic scientific research is crucial for the development of novel diagnostics, therapeutics, and interventions. However, not all basic science discoveries are translated into the next big drug, but each hypothesis-driven result acts as a building block for further scientific advancement. Basic science research is vital to furthering our understanding of the mechanistic underpinnings of cardiovascular disease and to empowering innovative solutions for patient management.

Chapter text

Objectives: Human behavior and disease are driven by underlying changes at the molecular level. To treat human health and disease at the patient level one must understand the biological mechanisms that regulate human health in a healthy individual[1,2]—for example, a seminal finding as breathing oxygen and using it for oxidative phosphorylation through complex organelles and signaling pathways in the mitochondria.[3] The pursuit of discovering the most fundamental truths is still ongoing and serves as an important foundation for understanding human disease. Human pathology arises when these fundamental processes are perturbed resulting in changes in the molecular architecture. An objective of basic biomedical research is to understand the mechanistic basis for how these perturbations at the molecular level translate into the disease phenotype.[4,5] Learning the precise biological mechanisms of disease and behavior can allow us to create tailor-specific therapeutics for individual patients while minimizing off-target effects.[6,7]

Biological systems consist of a milieu of different variables and often the number of permutations of interacting genetic networks or cells is intractable.[8,9] Consequently, traditional biomedical research has leaned on a reductionist approach whereby a complex biological system is dissected into smaller more tractable parts such as isolated genes or proteins, and these smaller systems are then studied in great depth.[10,11]

A reductionist approach to basic biomedical research allows one to meticulously dissect discrete pieces from a larger biological puzzle—by separating the specific gene or cell type of interest from its native biological environment, scientists can often ask targeted questions and perform experiments in a controlled setting.[10,12] An example of a reductionist approach is trying to address whether a specific ligand binds to a receptor in a particular cell type—by isolating the cell type and performing targeted experiments one can block out the "noise" inherent in a complex biological system. However, a reductionist approach can have limited translational utility as by decoupling the biological system into its smaller entities one ignores the complex

Translational Cardiology. https://doi.org/10.1016/B978-0-323-91790-2.00073-3

biological networks that integrate these entities[9,13] (Fig. 4.1).

By contrast, a systems biology approach to basic biomedical research occurs when one holistically studies the integrated biological networks rather than distinct pieces.[4] An important advantage of a systems-based approach is the immediate translational value as health and disease coexist in a complex biological network.[14] However, a primary drawback of a systems-based approach is the technical challenges associated with conducting meaningful and well-controlled experiments—in the pursuit of perturbing complex systems, it is often difficult to control for confounding variables[4,14,15] (Table 4.1).

With the advent of complex genetic, computational, and experimental techniques, scientists now have a unique opportunity to leverage findings from reductionist approaches to better understand how complex biological networks function in health and disease.[5]

Disciplines: Basic biomedical research consists of a plethora of subsets ranging from classical anatomy and physiology, and cell biology, to genetic and computational disciplines. We have outlined some of the major branches of basic biomedical research advancing our understanding of human health and disease (Fig. 4.2).

1. *Anatomy and physiology:* Anatomy studies the structure while physiology studies the function of the human body. The study of anatomy can be broken down into macroscopic and microscopic branches. Macroscopic anatomy or "gross anatomy" consists of that which can be studied by the eye whereas microscopic anatomy uses optical instruments to gain high-resolution views of smaller structures. Histology uses fixed tissue and dyes to stain for membranes and basic molecular structures to dissect anatomy at the cellular level. More recent advances in imaging such as magnetic resonance imaging (MRI), X-ray, and computed tomography (CT) allow real-time assessment of anatomy in the context of a physiologically active system.[16] Physiology is the study of how organisms function at a higher level as an aggregation of cellular and molecular processes to maintain homeostasis.
2. *Molecular and cell biology (MCB):* MCB focuses on cellular processes and the mechanisms that regulate

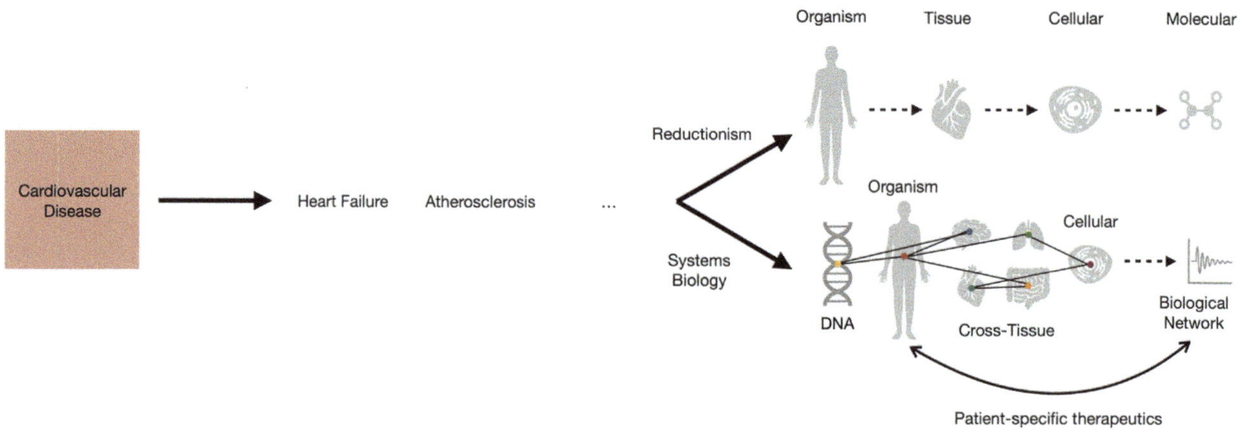

FIGURE 4.1 Overview schematic contrasting a reductionist and systems approach to biomedical research.

TABLE 4.1 Defining and comparing the pros and cons of a reductionist and systems approach.

	Reductionism	Systems biology
Definition	Dissecting a complex process into smaller pieces and studying those.	Studying the biological system as a coherent integrated network.
Pros	1. Easier to investigate 2. Fewer confounding variables 3. Easier to assess cause and effect	1. Robust framework 2. Translational to human disease
Cons	1. Less translational to human 2. Oversimplification of true biology 3. Findings are highly context dependent	1. Experimentally difficult to carry out 2. Resource limiting (too many parameters) 3. Difficult to attest cause and effect of any given experiment

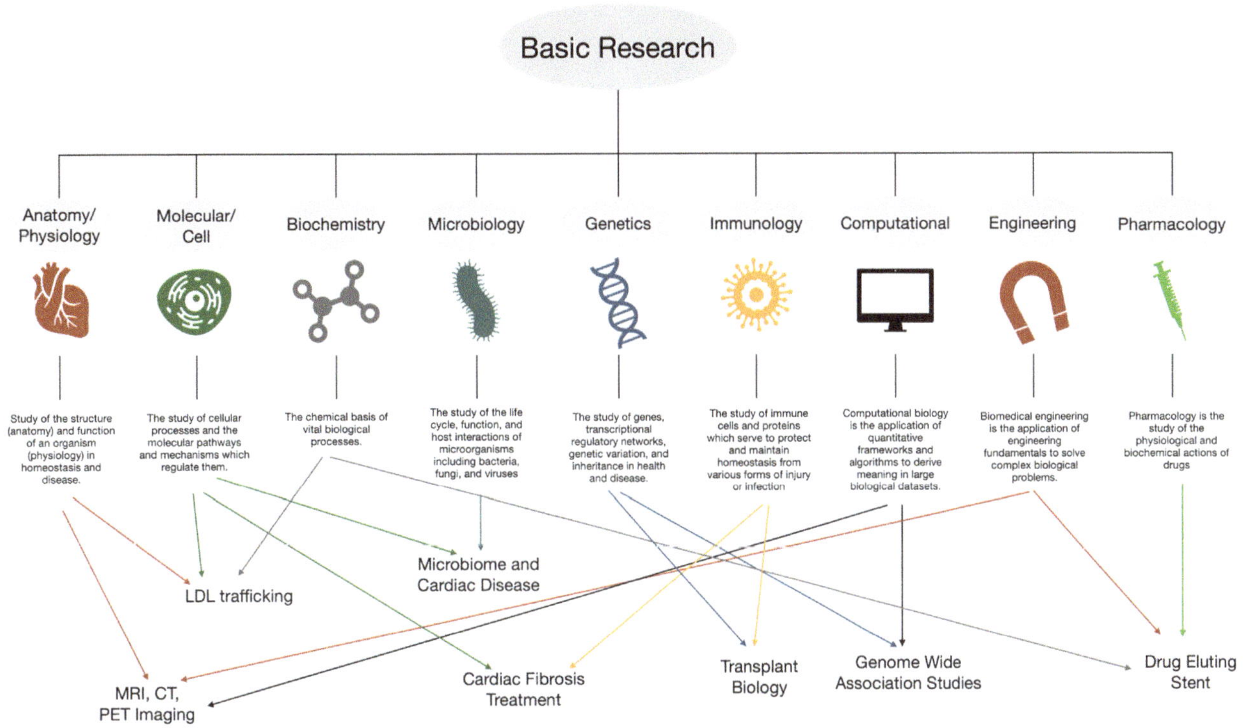

FIGURE 4.2 Different disciplines within the umbrella of basic biomedical research and examples of cross-intersection of disciplines to address problems in biomedicine today.

them. MBC includes the study of gene transcription/regulation, protein translation, cellular energetics and metabolism, autophagy, cell motility chemotaxis, trafficking of molecules in and out of cells, cell-to-cell communication, and cell growth/division/apoptosis.[17] The study of MCB is performed using cell culture models, bacterial systems, microscopy, and refined biochemical/genetic/immunological assays.[18]

3. *Biochemistry:* The study of biochemistry involves delineating the chemical basis of biological processes and vital functions inside living organisms. The key questions tackled in the biochemistry discipline revolve around the structure, function, and context of macromolecules. Biochemistry also serves a "methods" niche within biomedical research as many other biomedical disciplines utilize biochemical assays and methods. In medicine, biochemistry is closely related to drug discovery by identifying the mechanisms of macromolecule interactions in disease pathogenesis and the synthesis of active biomolecules.[19]
4. *Microbiology:* Microbiology is the study of microorganisms, including bacteria, fungi, parasites, and viruses. Microbiologists utilize biochemical, cellular, and molecular techniques to understand how microorganisms grow, function, and attack host organisms. An emerging area of interest within microbiology is the impact of the gut microbiome in regulating human health and disease.[20]
5. *Molecular genetics and genomics:* The study of genetics is concerned with genes and genetic variation. Classical genetics such as Mendelian genetics unravels the molecular inheritance of traits and certain diseases.[21] The advent of next-generation sequencing has afforded the opportunity to map genetic variation in complex polygenic diseases such as coronary artery disease.[22]
6. *Immunology:* Immunology is the study of the immune cells and proteins that serve to protect and maintain homeostasis from various forms of injury or infection. Immunology is also closely related to anatomy and cellular biology as it encompasses the study of lymphoid organs including the bone marrow, thymus, lymph nodes, spleen, tonsils, and mucous membranes.[23]
7. *Computational biology:* Computational biology is a broad term encompassing the application of tools from classical mathematics and computer science to the field of biology. It is the utilization of robust and efficient algorithms, schemas, and mathematical frameworks to concisely derive biological meaning from complex and large biological datasets. One of the most relevant applications of computer science in biology is in the analysis of next-generation sequencing data.[24]
8. *Biomedical engineering:* Biomedical engineering is the application of engineering fundamentals such as mechanics, classical physics, electronics, imaging, and

optics to the field of biology. Biomedical engineering is largely interdisciplinary and serves to bridge the gap between engineering and medicine. Some areas of exploration within the broader biomedical engineering umbrella include biomaterial design, tissue engineering, optics development, cellular and patient-level biomechanics, neural engineering, and medical device development.[25]

9. *Pharmacology:* Pharmacology is the study of the physiological and biochemical actions of drugs. Pharmacology strongly overlaps with several of the above disciplines including biochemistry (drug kinetics and chemical basis of drug interactions), physiology, and computational biology/biomedical engineering (pharmacokinetics and modeling of drug dynamics).

Common lab methods: Biomedical science research has greatly transformed from being an unimodal biology-heavy specialty to a truly multidisciplinary field with biologists, chemists, physicists, computer scientists, and engineers working together to solve biomedical problems. Consequently, the "common laboratory methods" utilized in basic biomedical research have scaled in the same vein. Here we provide examples of some of the most widely used and relevant laboratory methods relevant to basic *cardiovascular* research today. There are four broader paradigms of models used in basic cardiovascular research: in silico (experiment performed computationally), in vitro (cell culture), ex vivo (outside the living body), and in vivo (inside the living model: predominantly animal models). Many of the tools below can be used interchangeably across model systems given the scope of the question being asked (Table 4.2).

Cell biology and biochemistry

- Quantitative PCR: amplification and quantification of a target DNA sequence—can be used to detect the amount of RNA for specific genes.[26]
- Western blot: detection of a specific protein from a bulk lysate. Requires separation of proteins by size via gel electrophoresis and subsequent visualization of a target protein using a specific antibody.[27]
- Enzyme-linked immunosorbent assay (ELISA): used to quantify the amount of target ligand in a liquid solution using antibodies against the protein of interest.[28]

Sequencing

- Whole genome sequencing: a method using massively parallel sequencing to determine the DNA sequence of the entire genome of an organism. Used for unbiased profiling of the whole genome.[29,30]
- Exome sequencing: a method to sequence the exome portion (protein-coding region) of the genome. Used to assess the DNA sequence of all protein-coding sections of the genome.[31]
- Single-cell RNA sequencing: a droplet-based technique to sequence the RNA from single cells. Used as an unbiased tool to profile the transcriptome of a set of cells.[32]
- Cellular indexing of transcriptomes and epitopes by sequencing (CITE-Seq): a droplet-based technique to sequence the RNA and surface proteome from single cells. Used as an unbiased tool to profile the transcriptome and immunophenotype cells at single-cell resolution.[33]
- Assay for transposase and chromatin accessibility: an unbiased technique to profile the chromatin accessibility landscape genome-wide using transposases.[34]

TABLE 4.2 Experimental paradigms used in basic biomedical research.

	In vivo	In vitro	In silico
Definition	Inside a living organism.	Outside a living organism.	Experiment via computational simulation
Pros	1. Highly translational 2. Physiologically realistic	1. Can directly assess cause and effect 2. Can control for confounders and ask specific experimental questions	1. Can tune thousands of parameters 2. Make predictions
Cons	1. Expensive 2. Large number of variables and potential confounders 3. Difficult to carry out and test all possible parameters	1. Experimentally difficult to carry out 2. Resource limiting (too many parameters) 3. Difficult to attest cause and effect of any given experiment	1. No physiological context 2. Difficult to test mechanisms 3. No biological context
Example models	1. Small animal: mouse 2. Large animal: rabbit 3. Human	1. Organoids 2. Cell culture experimentation	1. Computational fluid dynamic simulations 2. Probabilistic models

- Chromatin immunoprecipitation (Ch-IP) sequencing: a method used to assess protein-DNA binding—specifically used to profile the binding of specific transcription factors to different parts of the genome.[35]
- Clustered regularly interspaced short palindromic repeats (CRISPR)-Cas9: used to edit target genes in an organism using guide RNAs.[36]

Immunology

- Flow cytometry: used to measure the physical and surface proteomic characteristics of cells using fluorescence-activated lasers. Cells are stained with antibodies of interest and subsequently labeled with fluorescent markers—the cell suspension is then forced into a stream of single cells which are then detected using a panel of lasers.[37]

Imaging

- Immunofluorescence: a method that utilizes fluorescently labeled antibodies to detect proteins of interest in cells and fixed sections of tissue.[38]
- RNAscope: in situ hybridization assay to detect RNA within intact cells on slides. Useful in cases where the protein product of interest may be short lived or secreted and one cannot trace back the cell origin of the protein.[39]
- Spatial transcriptomics: a broad array of methods that measures RNA on histological sections where one can map RNA expression in voxels to the corresponding histology.[40]

Animal studies: While in vitro approaches have tremendous value in elucidating interesting biology, they do not fully recapitulate the complex network of interacting biological systems present in a living organism.[41–43] Animal models (rodent, porcine, etc.) provide a useful translational bridge to study important biology associated with disease in a living species.[44] Mouse models are economical, pragmatic, and the most genetically manipulable of other animal models.[45,46] In cardiac research, mouse models have shown tremendous promise for various forms of cardiac injury.[47–49] However, given the profound differences in scale between humans and mice, they fail to mimic several aspects of human biology[50,51]—for instance, spontaneous plaque rupture is one of the most important problems in cardiovascular research but does not occur in mice.[52,53] Consequently, large animal models such as porcine overcome this limitation and are ideal for medical device development such as intravascular stents[54–56]—however, large animal models do not afford the same degree of genetic manipulation and can be cost prohibitive.[57–59]

Major areas of current investigation

Next-generation sequencing: The advent of next-generation sequencing (NGS) has revolutionized our capacity to study biology at a higher resolution with a systems-level approach in a cost-efficient manner.[60] NGS utilizes parallelized systems to sequence anywhere from one million to 43 billion short reads. Biomedical labs have used NGS to:

- Sequence whole genome: many cardiovascular diseases have strong genetic associations and NGS has afforded the opportunity to perform large-scale genome-wide association studies (GWAS). In GWAS, a large number of cases (known clinical disease phenotype) and controls (no disease) are sequenced, and computational algorithms are used to identify genetic variants associated with disease phenotype after accounting for relevant covariates.[61–63] Coronary artery disease GWAS have identified over 200 loci associated with the clinical manifestation of the disease.[64]
- Perform single-cell RNA sequencing (scRNA): the advent of droplet-based techniques coupled with NGS has allowed researchers to sequence the RNA from single cells to unbiasedly profile the transcriptome.[65] scRNA seq can be used to profile the transcriptome from freshly harvested human tissue allowing researchers to uncover cellular diversity in the human body.[66–68] Furthermore, single-*nucleus* RNA sequencing allows transcriptome profiling in frozen tissue, which has provided scientists an opportunity to leverage tissue from biobank repositories across the world.[66,69,70]
- Profile epigenomic factors such as DNA methylation and DNA–protein interactions: epigenomic profiling (e.g., ATAC-seq[34]) allows one to unravel the gene regulatory networks driving key molecular pathways. Gene regulatory information is incredibly valuable in connecting disease-associated GWAS loci (many of which span intergenic segments) to cell-specific molecular pathways.[71,72]

Cardioimmunology: The emergence of high-throughput sequencing has allowed researchers to appreciate the profound immune involvement more robustly in cardiac morphogenesis, homeostasis, and disease.[73,74] Immune cells migrate to the heart during gestation and play important roles in maintaining homeostasis. During cardiac injury, immune cells are recruited to the heart where they dedifferentiate, cross-talk with the stromal cells, and take part in cardiac remodeling.[74–79] Basic science research has shown that some immune responses can promote wound healing and be reparative while other immune signals can result in adverse cardiac remodeling and subsequent heart failure.[78,79]

Get started

(1) Solidify a strong foundation in the fundamental sciences: biology, chemistry, physics, probability, and statistics.
(2) Based on background and one's interests develop skills in R and Python to be able to computationally analyze datasets and draw biological meaning.
(3) Read cross-disciplinary review articles and seminal papers some of which are referenced in this chapter.

Pitfalls to avoid

(1) Working with everyone from the same discipline: working with individuals across fields allows for innovative solutions to complex biomedical problems
(2) Being a perfectionist when it comes to biology: nature is far too complicated, and we must make reasonable well-informed assumptions rather than trying to address every variable.
(3) The computational black box: let the underlying science drive the analysis and computational development, not the other way around.

Real-world example

(1) Cells of the adult human heart: utilization of single cell and nucleus RNA sequencing to uncover all the cell types in the adult human heart.[66,70]
(2) FAP CAR-T cells to treat cardiac fibrosis: designing engineered CAR-T cells to kill activated fibroblasts in the setting of cardiac injury and result in subsequent fibrosis.[80,81]
(3) The human heart contains distinct macrophage subsets with divergent origins and functions.[77,78]

References

1. Schor NF. Why our patients (and we) need basic science research. *Neurology*. 2013;80:2070.
2. Arias IM. Training basic scientists to bridge the gap between basic science and its application to human disease. *New Engl J Med*. 2010;321:972–974. https://doi.org/10.1056/NEJM198910053211412.
3. Prebble JN. The discovery of oxidative phosphorylation: a conceptual off-shoot from the study of glycolysis. *Stud Hist Philos Biol Biomed Sci*. 2010;41:253–262.
4. Bruggeman FJ, Westerhoff HV. The nature of systems biology. *Trends Microbiol*. 2007;15:45–50.
5. Green ED, et al. Strategic vision for improving human health at the Forefront of Genomics. *Nature*. 2020;586:683.
6. Butcher EC, Berg EL, Kunkel EJ. Systems biology in drug discovery. *Nat Biotechnol*. 2004;22:1253–1259.
7. Marshall LJ, Austin CP, Casey W, Fitzpatrick SC, Willett C. Recommendations toward a human pathway-based approach to disease research. *Drug Discov Today*. 2018;23:1824–1832.
8. Markowetz F, Spang R. Inferring cellular networks – a review. *BMC Bioinf*. 2007;8:S5.
9. Mazzocchi F. Complexity in biology. Exceeding the limits of reductionism and determinism using complexity theory. *EMBO Rep*. 2008;9:10.
10. Van Regenmortel MHV. Reductionism and complexity in molecular biology. *EMBO Rep*. 2004;5:1016.
11. Kesić S. Systems biology, emergence and antireductionism. *Saudi J Biol Sci*. 2016;23:584.
12. Studies in the Philosophy of Biology: Reduction and Related Problems; Google Books; Available at: https://books.google.co.nz/books?hl=en&lr=&id=NMAf65cDmAQC&oi=fnd&pg=PR7&ots=_DdEX_E2NF&sig=hjFxNiSub85IS_y_tGJWyi50Hqc&redir_esc=y#v=onepage&q&f=false. Accessed 10 January 2022.
13. Ahn AC, Tewari M, Poon CS, Phillips RS. The limits of reductionism in medicine: could systems biology offer an alternative? *PLoS Med*. 2006;3:0709–0713.
14. Mesarovic MD, Sreenath SN, Keene JD. Search for organising principles: understanding in systems biology. *Syst Biol*. 2004;1:19–27.
15. Strange K. The end of 'naïve reductionism': rise of systems biology or renaissance of physiology? *Am J Physiol Cell Physiol*. 2005;288.
16. Beaton L, Bandula S, Gaze MN, Sharma RA. How rapid advances in imaging are defining the future of precision radiation oncology. *Br J Cancer*. 2019;120(8):779–790.
17. Molecular Biology of the Cell; NCBI Bookshelf; Available at: https://www.ncbi.nlm.nih.gov/books/NBK21054/. Accessed 10 January 2022.
18. Kreitzer G, Jaulin F, Espenel C. *Cell Biology Assays : Proteins*. 2010:381.
19. Newsholme EA, Leech AR. *Functional Biochemistry in Health and Disease*. 2009:543.
20. Barratt MJ, Lebrilla C, Shapiro HY, Gordon JI. The gut microbiota, food science and human nutrition; a timely marriage. *Cell Host Microbe*. 2017;22:134.
21. Chakravarti A, Clark AG, Mootha VK. Distilling pathophysiology from complex disease genetics. *Cell*. 2013;155:21.
22. Kessler T, Schunkert H. Coronary artery disease genetics enlightened by genome-wide association studies. *JACC Basic Transl Sci*. 2021;6:610–623.
23. What are the organs of the immune system?; NCBI Bookshelf; Available at: https://www.ncbi.nlm.nih.gov/books/NBK279395/. Accessed 10 January 2022.
24. Pereira R, Oliveira J, Sousa M. Bioinformatics and computational tools for next-generation sequencing analysis in clinical genetics. *J Clin Med*. 2020;9.
25. Bhat S, Kumar A. Biomaterials and bioengineering tomorrow's healthcare. *Biomatter*. 2013;3.
26. Deepak S, et al. Real-time PCR: revolutionizing detection and expression analysis of genes. *Curr Genom*. 2007;8:234.
27. Mahmood T, Yang PC. Western blot: technique, theory, and trouble shooting. *N Am J Med Sci*. 2012;4:429.
28. Aydin S. A short history, principles, and types of ELISA, and our laboratory experience with peptide/protein analyses using ELISA. *Peptides*. 2015;72:4–15.
29. Ng PC, Kirkness EF. Whole genome sequencing. *Methods Mol Biol*. 2010;628:215–226.
30. Turro E, et al. Whole-genome sequencing of patients with rare diseases in a national health system. *Nature*. 2020;583(7814):96–102.

31. Van Hout CV, et al. Exome sequencing and characterization of 49,960 individuals in the UK Biobank. *Nature*. 2020;586(7831): 749–756.
32. Haque A, Engel J, Teichmann SA, Lönnberg T. A practical guide to single-cell RNA-sequencing for biomedical research and clinical applications. *Genome Med*. 2017;9:1–12.
33. Stoeckius M, et al. Simultaneous epitope and transcriptome measurement in single cells. *Nat Methods*. 2017. https://doi.org/10.1038/nmeth.4380.
34. Buenrostro JD, Wu B, Chang HY, Greenleaf WJ. ATAC-seq: a method for assaying chromatin accessibility genome-wide. *Curr Protoc Mol Biol*. 2015;109:21.29.1.
35. Park PJ. ChIP–seq: advantages and challenges of a maturing technology. *Nat Rev Genet*. 2009;10(10):669–680.
36. Ran FA, et al. Genome engineering using the CRISPR-Cas9 system. *Nat Protoc*. 2013;8(11):2281–2308.
37. McKinnon KM. Flow cytometry: an overview. *Curr Protoc Im*. 2018;120:5.1.1.
38. Im K, Mareninov S, Diaz MFP, Yong WH. An introduction to performing immunofluorescence staining. *Methods Mol Biol*. 2019;1897:299.
39. Wang F, et al. RNAscope: a novel in situ RNA analysis platform for formalin-fixed, paraffin-embedded tissues. *J Mol Diagn*. 2012;14:22.
40. Rao A, Barkley D, França GS, Yanai I. Exploring tissue architecture using spatial transcriptomics. *Nature*. 2021;596(7871):211–220.
41. Hofer M, Lutolf MP. Engineering organoids. *Nat Rev Mater*. 2021;65(6):402–420.
42. Fatehullah A, Tan SH, Barker N. Organoids as an in vitro model of human development and disease. *Nat Cell Biol*. 2016;18:246–254.
43. Clevers H. Modeling development and disease with organoids. *Cell*. 2016;165:1586–1597.
44. Baxter VK, Griffin DE. Animal models: No model is perfect, but many are useful. *Viral Pathog*. 2016;125. https://doi.org/10.1016/B978-0-12-800964-2.00010-0.
45. Frick A, Suzuki O, Butz N, Chan E, Wiltshire T. In vitro and in vivo mouse models for pharmacogenetic studies. *Methods Mol Biol*. 2013;1015:263–278.
46. Cheng Y, et al. Principles of regulatory information conservation between mouse and human. *Nature*. 2014;515:371–375.
47. Houser SR, et al. Animal models of heart failure a scientific statement from the American Heart Association. *Circ Res*. 2012;111:131–150.
48. Luther DJ, Thodeti CK, Meszaros JG. Injury models to study cardiac remodeling in the mouse: myocardial infarction and ischemia-reperfusion. *Methods Mol Biol*. 2013;1037:325–342.
49. Muthuramu I, Lox M, Jacobs F, De Geest B. Permanent ligation of the left anterior descending coronary artery in mice: a model of post-myocardial infarction remodelling and heart failure. *J Vis Exp*. 2014. https://doi.org/10.3791/52206.
50. Rydell-Törmänen K, Johnson JR. The applicability of mouse models to the study of human disease. *Methods Mol Biol*. 2019;1940:3–22.
51. Breschi A, Gingeras TR, Guigó R. Comparative transcriptomics in human and mouse. *Nat Rev Genet*. 2017;18:425.
52. Emini Veseli B, et al. Animal models of atherosclerosis. *Eur J Pharmacol*. 2017;816:3–13.
53. Schwartz SM, Galis ZS, Rosenfeld ME, Falk E. Plaque rupture in humans and mice. *Arterioscler Thromb Vasc Biol*. 2007;27:705–713.
54. Tsang HG, et al. Large animal models of cardiovascular disease. *Cell Biochem Funct*. 2016;34:113.
55. Swindle MM, Makin A, Herron AJ, Clubb FJ, Frazier KS. Swine as models in biomedical research and toxicology testing. *Vet Pathol*. 2012;49:344–356.
56. Ziegler A, Gonzalez L, Blikslager A. Large animal models: the key to translational discovery in digestive disease research. *Cell Mol Gastroenterol Hepatol*. 2016;2:716–724.
57. Shin HS, Shin HH, Shudo Y. Current status and limitations of myocardial infarction large animal models in cardiovascular translational research. *Front Bioeng Biotechnol*. 2021;9:321.
58. Ribitsch I, et al. Large animal models in regenerative medicine and tissue engineering: to do or not to do. *Front Bioeng Biotechnol*. 2020;8:972.
59. Casal M, Haskins M. Large animal models and gene therapy. *Eur J Hum Genet*. 2006;14(3):266–272.
60. Zhang J, Chiodini R, Badr A, Zhang G. The impact of next-generation sequencing on genomics. *J Genet Genomics*. 2011;38:95.
61. Visscher PM, et al. 10 Years of GWAS discovery: biology, function, and translation. *Am J Hum Genet*. 2017;101:5.
62. Tam V, et al. Benefits and limitations of genome-wide association studies. *Nat Rev Genet*. 2019;20(8):467–484.
63. Klein RJ, Xu X, Mukherjee S, Willis J, Hayes J. Successes of genome-wide association studies. *Cell*. 2010;142:350–351.
64. Nikpay M, et al. A comprehensive 1000 Genomes–based genome-wide association meta-analysis of coronary artery disease. *Nat Genet*. 2015;47(10):1121–1130.
65. Anaparthy N, Ho YJ, Martelotto L, Hammell M, Hicks J. Single-cell applications of next-generation sequencing. *Cold Spring Harb Perspect Med*. 2019;9.
66. Litviňuková M, et al. Cells of the adult human heart. *Nature*. 2020;588:466–472.
67. Williams JW, et al. Single cell RNA sequencing in atherosclerosis research. *Circ Res*. 2020:1112–1126. https://doi.org/10.1161/CIRCRESAHA.119.315940.
68. Wilk AJ, et al. A single-cell atlas of the peripheral immune response in patients with severe COVID-19. *Nat Med*. 2020. https://doi.org/10.1038/s41591-020-0944-y.
69. Tucker NR, et al. Transcriptional and cellular diversity of the human heart. *Circulation*. 2020;142:466–482.
70. Koenig AL, et al. Single cell transcriptomics reveals cell type specific diversification in human heart failure. *bioRxiv*. 2021. https://doi.org/10.1101/2021.07.06.451312.
71. Albert FW, Kruglyak L. The role of regulatory variation in complex traits and disease. *Nat Rev Genet*. 2015;16(4):197–212.
72. Farh KKH, et al. Genetic and epigenetic fine mapping of causal autoimmune disease variants. *Nature*. 2015;518:337–343.
73. Chen Y, Liu Y, Gao X. The application of single-cell technologies in cardiovascular research. *Front Cell Dev Biol*. 2021;9:751371.
74. Swirski FK, Nahrendorf M. Cardioimmunology: the immune system in cardiac homeostasis and disease. *Nat Rev Immunol*. 2018;18(12): 733–744.
75. Leid J, et al. Primitive embryonic macrophages are required for coronary development and maturation. *Circ Res*. 2016;118:1498–1511.
76. Epelman S, Lavine KJ, Randolph GJ. Origin and functions of tissue macrophages. *Immunity*. 2014;41:21–35.

77. Bajpai G, et al. Tissue resident CCR2- and CCR2+ cardiac macrophages differentially orchestrate monocyte recruitment and fate specification following myocardial injury. *Circ Res*. 2019;124:263–278.
78. Bajpai G, et al. The human heart contains distinct macrophage subsets with divergent origins and functions. *Nat Med*. 2018;24:1234–1245.
79. Svedberg FR, Guilliams M. Cellular origin of human cardiac macrophage populations. *Nat Med*. 2018;24:1091–1092.
80. Rurik JG, et al. CAR T cells produced in vivo to treat cardiac injury. *Science*. 2022;375:91–96.
81. Aghajanian H, et al. Targeting cardiac fibrosis with engineered T cells. *Nature*. 2019;573(7774):430–433.

Part II

Pre-clinical: Discovery & development

Samira Bahrainy

Chapter 5

Overview of preclinical research in cardiovascular diseases

Brice Nouthe[1,2], Cedric Kamani[3], Nso Nso[4], Anthony Lyonga Ngonge[5] and Andre P. Kengne[6]

[1]Department of Medicine, The University of British Columbia, Vancouver, BC, Canada; [2]CESSTAD-KOHEN, Cameroon; [3]Faculty of Medicine, The University of British Columbia, Vancouver, BC, Canada; [4]Division of Cardiology, University of Chicago (Northshore), Evanston, IL, United States; [5]Division of Cardiology, Morehouse School of Medicine, Atlanta, GA, United States; [6]Department of Medicine, University of Cape Town, Cape Town, South Africa

Why it matters

Preclinical development of drugs and devices in CVD research has mostly not been followed by significant advances for patients in the last decades. This calls for a reorganization of pre-human experiments and better methodology for future studies. We herein summarize the different steps and possible actionable points to consider to improve the yield of translational work in this field.

From in vitro research to animal models

In vitro research

Cell cultures are used to study disease processes, the mode of action, and the toxicologic effects of candidate drugs. The two dimensions (2D) cell culture has established comparative results useful in both hypothesis-generating and confirmatory cardiovascular studies.[1] In addition, 2D cell cultures are used to study the proarrhythmic effects of novel or repurposed medications.[2–4] Microphysiological in vitro models of 3D cell platforms have provided enough cardiac-specific systems for drug development, electrophysiological studies, myocardial functional assessment, and bidirectional interstitial fluid flow.[1,5] Those platforms have been used to study cardiac fibrosis, Chagas diseases plus other arrhythmias and are now applied in direct reprogramming but also patch-based cell therapy for cardiac regeneration.[1,6–8] The best practice guidelines for in vitro research recommend evaluating the variability patterns related to different laboratory settings, the statistical significance of the heterogeneity in assessments, and their influence on prospective predictions.[9,10] Data integrity should be maintained by monitoring: (1) sample integrity, (2) equipment and instrument, (3) result in reporting and documentation, (4) data operator's proficiency, (5) method verification and validation, (6) verification and review of raw data, (7) electronic data security, and (8) data retrieval, archive, and storage.[11] By using standard protocols in cell cultures, specific attention is paid to the assessment of cell culture coating, 3D support/scaffold (which can be tissue engineered), cell characterization, culture media (including density/medium change), risk of contamination, additive manufacturing approaches, and other methods with their reporting procedures.[12,13]

Characteristics of animals frequently used

i. **Small animals:** The advantages of using small rodents include the low cost of their maintenance, the ease of performing genetic manipulation and short gestation time, and the possibility of studying invasive parameters such as arterial pulse wave, pressure–volume loops, intracardiac electrophysiology, evaluating in situ cardiac devices or new imaging modalities.[9–12] Mouse and rat are the most used and include inbred strains, outbred stocks, spontaneous mutants, and transgenic models.[13] Models of "humanized mice" have been designed, by replacing the murine ortholog with the human version of a gene of interest, to study congenital heart diseases, heart failure/transplant (with aortic constriction overload or coronary artery ligation-induced myocardial infarction), and atherosclerosis (like in dysbetalipoproteinemia), among others.[14–17] However, significant differences in heart contraction, electrophysiology, and even molecular

Translational Cardiology. https://doi.org/10.1016/B978-0-323-91790-2.00065-4

structure of troponin in small rodents limit their reliability when studying human CVD.[14,18,19]

ii. **Large animals:** Frequently used animals are rabbits, pigs, sheep, and also canine and swine models. Humans and large mammalian hearts share the same following characteristics: lower heart rate, pericardial basic structure (although the thickness increases with the heart size), atrial/atrial appendage architecture (with differences in the number of pulmonary veins and position of venous structures), ventricular organization (but animal hearts have coarser trabeculations, widened collateral tissues), and very similar conduction system.[20–23] Relevant CVD/interventions in large animals include but are not limited to:
   1. Vascular calcification in the tunica medica of pulmonary arteries of race horses and aorta of cows[9,24];
   2. Aortic valve diseases in hypertensive rabbits and high fat-fed pigs[25,26];
   3. Marfan syndrome from spontaneous mutations of *FBN1* in cattle[27];
   4. Spontaneous familial hypercholesterolemia in pigs[9,10];
   5. Heart failure using the rabbit's heart, whose structure resembles the human heart;
   6. Heart transplant because β-MHC is expressed in canine and sheep hearts;
   7. Arrhythmias in swine due to striking similarities, with humans, in the structure and function of the excitation–contraction coupling mechanisms of the myocardium.[9]

Care of the animals: Nutrition, cages, environmental/infectious risks

The high-quality care of the animals should meet their nutritional, metabolic, and digestive needs (e.g., fibers in ruminants, nonhuman primates) while reducing the risk of infectious diseases.[28,29] When needed, medications, sterilization, changes in food structure (moist/soft diets), irradiation or increased amount of fat, and security of the environment should be modified following the predefined experimental protocol to facilitate their behavioral and physiologic requirements.[28,30] During preclinical studies, the clinical assessment of animals is necessary to identify their risk of disease exposure, distress, pain, and immune responses to environmental/infectious conditions.

Experiments in translational cardiovascular research

a. **In vitro research.** The proarrhythmogenic effects of candidate drugs can be studied using specific recombinant cell lines, including with the human ether-a-go-go related gene (hERG) assay measures a candidate drug's potency to change the function of the rapid component of the delayed rectifier potassium current (IKr) encoded by the hERG gene.[31,32] Ion channel studies and in silico modeling using a specific population model (e.g., from the ventricles) are also standardized in toxicologic studies as part of the comprehensive in vitro proarrhythmic assay.[2]

b. **Methodological study design elements: the ARRIVE (Animal Research Reporting of In Vivo Experiments) guidelines**
   Cardiovascular pharmaceutical research has seen a decline in productivity and low success rates when comparing the number of promising experimental results translated into positive clinical trials.[33] The lack of reproducibility of preclinical studies led to the development of the ARRIVE guidelines in 2010,[34] a checklist used to describe in vivo experiments to help evaluate and reproduce the work presented. In 2014, the National Institutes of Health (NIH) issued a document on the information to be included when reporting results in cardiovascular studies and both have been endorsed by many scientific societies.[33,35] The relevant study design elements (SDEs), common to the NIH report and the updated ARRIVE 2.0 guidelines, are as follows:
   i. *Exclusion and inclusion criteria:* The inclusion criteria depend on several factors, including task performance, body weight, performance levels, or exposure to a medical procedure.[34] The exclusion criteria direct the disqualification of animals from the preclinical studies based on behavioral assessments, motor impairments, and inappropriate sample collection, among other factors.
   ii. *Sample-size estimation:* The sample size for the preclinical studies is determined by the number of experimental units in the treatment and control groups and depends on the null hypothesis plus the inferential statistics.[36] Like in other well-conducted scientific analyses, the deterministic factors of the sample size include the priori data variability, biologically relevant effect size, and study power.
   iii. *Randomization and blinding, and data analysis:* The study blinding minimizes the risk of bias by concealing the data analysis, outcome assessment, experimental procedures, and allocation practices from stakeholders.[34]
   iv. *Reporting:* The reporting of outcome measures, independent/dependent variables, covariates, blocking factors, missing values, comparison adjustments, data coding procedures, and software packages improve the external and internal validity of the study outcomes.[34] The checklist has been divided into a minimal set of elements, the "ARRIVE Essential 10," to facilitate the description of the research context.

In addition to flaws in reporting preclinical studies, low-quality reagents and study design, plus incomplete transparency in data analysis, still impair their reproducibility.[37] An analysis of all preclinical research studies published in the American Heart Association journals between 2006 and 2020 showed that implementing a more rigorous checklist for all submitted articles led to a higher number of reported SDE. These could thus be scrutinized by other scientists planning to reproduce the findings.[38]

Areas of improvement

a. **The case of biological sex:** We recently commented on the sex bias in preclinical studies on left ventricular assist devices (LVADs) across different animal species.[39] This is particularly relevant when knowing that female sex is a significant risk factor for aortic insufficiency[39] in LVAD patients[40] and persisted even after the NIH report on biological sex as an important variable in preclinical studies.
b. **Pulmonary arterial hypertension (PAH):** The animal models for PAH identified different etiological processes in developing chronic hypoxia.[41] Those models, together with 3D cell cultures, are currently being used to study the processes involved in pulmonary artery proliferation/thickening, muscularization of arterioles, plexiform lesions, arteritis/fibrinoid necrosis, bronchovascular bundles, fat embolism, eccentric intimal fibrosis, and fibrosing mediastinitis.[41–43] Yet, only therapies targeting endothelial dysfunction have shown promising results.[44,45]

References

1. Zuppinger C. 3D cardiac cell culture: a critical review of current technologies and applications. *Front Cardiovasc Med.* 2019;6:87.
2. Cavero I, Holzgrefe H. Comprehensive in vitro Proarrhythmia Assay, a novel in vitro/in silico paradigm to detect ventricular proarrhythmic liability: a visionary 21st century initiative. *Expet Opin Drug Saf.* June 2014;13(6):745–758.
3. Qu Y, Gao B, Arimura Z, Fang M, Vargas HM. Comprehensive in vitro pro-arrhythmic assays demonstrate that omecamtiv mecarbil has low pro-arrhythmic risk. *Clin Transl Sci.* July 2021;14(4):1600–1610.
4. Delaunois A, Abernathy M, Anderson WD, et al. Applying the CiPA approach to evaluate cardiac proarrhythmia risk of some antimalarials used off-label in the first wave of COVID-19. *Clin Transl Sci.* May 2021;14(3):1133–1146.
5. Visone R, Talò G, Lopa S, Rasponi M, Moretti M. Enhancing all-in-one bioreactors by combining interstitial perfusion, electrical stimulation, on-line monitoring and testing within a single chamber for cardiac constructs. *Sci Rep.* November 16, 2018;8(1):16944.
6. Garzoni LR, Adesse D, Soares MJ, Rossi MID, Borojevic R, de Meirelles M de NL. Fibrosis and hypertrophy induced by Trypanosoma cruzi in a three-dimensional cardiomyocyte-culture system. *J Infect Dis.* March 15, 2008;197(6):906–915.
7. Schussler O, Falcoz PE, Chachques JC, Alifano M, Lecarpentier Y. Possible treatment of myocardial infarct based on tissue engineering using a cellularized solid collagen scaffold functionalized with Arg-Glyc-Asp (RGD) Peptide. *Int J Mol Sci.* November 22, 2021;22(22).
8. Yadid M, Oved H, Silberman E, Dvir T. Bioengineering approaches to treat the failing heart: from cell biology to 3D printing. *Nat Rev Cardiol.* February 2022;19(2):83–99.
9. Tsang HG, Rashdan NA, Whitelaw CBA, Corcoran BM, Summers KM, MacRae VE. Large animal models of cardiovascular disease. *Cell Biochem Funct.* April 2016;34(3):113–132.
10. Liao J, Huang W, Liu G. Animal models of coronary heart disease. *J Biomed Res.* August 20, 2015;30(1):3–10.
11. Tillmanns J, Schneider M, Fraccarollo D, et al. PET imaging of cardiac wound healing using a novel [68Ga]-labeled NGR probe in rat myocardial infarction. *Mol Imag Biol.* February 2015;17(1):76–86.
12. Hess A, Derlin T, Koenig T, et al. Molecular imaging-guided repair after acute myocardial infarction by targeting the chemokine receptor CXCR4. *Eur Heart J.* October 1, 2020;41(37):3564–3575.
13. Hickman DL, Johnson J, Vemulapalli TH, Crisler JR, Shepherd R. Commonly used animal models. In: *Principles of Animal Research for Graduates and Undergraduartes Students.* Elsevier Inc.; 2017:117–175.
14. Milani-Nejad N, Janssen PML. Small and large animal models in cardiac contraction research: advantages and disadvantages. *Pharmacol Ther.* March 2014;141(3):235–249.
15. Zaragoza C, Gomez-Guerrero C, Martin-Ventura JL, et al. Animal models of cardiovascular diseases. *J Biomed Biotechnol.* 2011;2011:497841.
16. Yutzey KE, Robbins J. Principles of genetic murine models for cardiac disease. *Circulation.* February 13, 2007;115(6):792–799.
17. Kreutzer FP, Meinecke A, Schmidt K, Fiedler J, Thum T. Alternative strategies in cardiac preclinical research and new clinical trial formats. *Cardiovasc Res.* March 2021;8:cvab075.
18. Slabaugh JL, Brunello L, Gyorke S, Janssen PML. Contractile parameters and occurrence of alternans in isolated rat myocardium at supra-physiological stimulation frequency. *Am J Physiol Heart Circ Physiol.* June 1, 2012;302(11):H2267–H2275.
19. Kerans V, Espinoza A, Skulstad H, Halvorsen PS, Edvardsen T, Bugge JF. Systolic left ventricular function is preserved during therapeutic hypothermia, also during increases in heart rate with impaired diastolic filling. *Intensive Care Med Exp.* December 2015;3(1):41.
20. Kent GC, Carr RK. *Comparative Anatomy of the Vertebrates.* Boston, MA: McGraw Hill; 2001.
21. Hearse DJ, Sutherland FJ. Experimental models for the study of cardiovascular function and disease. *Pharmacol Res.* June 2000;41(6):597–603.
22. Michaëlsson M, Ho SY. *Congenital Heart Malformations in Mammals: An Illustrated Text.* London: Imperial College Press; 2000.
23. Hill AJ, Iaizzo PA. Comparative cardiac anatomy. In: *Handbook of Cardiac Anatomy, Physiology, and Devices.* Humana Press; 2005:87–108.
24. Silva GT de A, Guest BB, Gomez DE, et al. Development of a technique for determination of pulmonary artery pulse wave velocity in horses. *J Appl Physiol.* May 1, 2017;122(5):1088–1094.
25. Ngo DT, Stafford I, Sverdlov AL, et al. Ramipril retards development of aortic valve stenosis in a rabbit model: mechanistic considerations. *Br J Pharmacol.* February 2011;162(3):722–732.

26. Fischer JW, Steitz SA, Johnson PY, et al. Decorin promotes aortic smooth muscle cell calcification and colocalizes to calcified regions in human atherosclerotic lesions. *Arterioscler Thromb Vasc Biol.* December 2004;24(12):2391–2396.
27. Hirano T, Matsuhashi T, Kobayashi N, Watanabe T, Sugimoto Y. Identification of an FBN1 mutation in bovine Marfan syndrome-like disease. *Anim Genet.* February 2012;43(1):11–17.
28. Kirk GW. Care in the cage: materializing moral economies of animal care in the biomedical science. In: *Animal Housing and Human–Animal Relations: Politics, Practices and Infrastructures*. Routledge; 2018.
29. Toth LA. The influence of the cage environment on rodent physiology and behavior: implications for reproducibility of pre-clinical rodent research. *Exp Neurol.* August 1, 2015;270:72–77.
30. Carter RL, Lipman NS. Feed and bedding. In: *Management of Animal Care and Use Programs in Research, Education, and Testing*. 2nd ed. CRC Press/Taylor & Francis; 2018.
31. ICH Expert Working Group. The non-clinical evaluation of the potential for delayed ventricular repolarization (QT Interval prolongation) by human pharmaceuticals. In: *S7B*; May 2005. Available from: https://database.ich.org/sites/default/files/S7B_Guideline.pdf.
32. US Food and Drug Administration C for DE and R. *Recommended Voltage Protocols to Study Drug-cardiac Ion Channel Interactions Using Recombinant Cell Lines*; September 2018. Available from: https://www.fda.gov/media/129686/download.
33. Ramirez FD, Motazedian P, Jung RG, et al. Methodological rigor in preclinical cardiovascular studies: targets to enhance reproducibility and promote research translation. *Circ Res.* June 9, 2017;120 (12):1916–1926.
34. Percie du Sert N, Hurst V, Ahluwalia A, et al. The ARRIVE guidelines 2.0: updated guidelines for reporting animal research. *BMC Vet Res.* July 14, 2020;16(1):242.
35. Collins FS, Tabak LA. Policy: NIH plans to enhance reproducibility. *Nature.* January 30, 2014;505(7485):612–613.
36. Henderson VC, Kimmelman J, Fergusson D, Grimshaw JM, Hackam DG. Threats to validity in the design and conduct of preclinical efficacy studies: a systematic review of guidelines for in vivo animal experiments. *PLoS Med.* 2013;10(7):e1001489.
37. Freedman LP, Venugopalan G, Wisman R. Reproducibility2020: progress and priorities. *F1000Res.* 2017;6:604.
38. Jung RG, Stotts C, Makwana D, et al. Methodological rigor in preclinical cardiovascular research: contemporary performance of AHA scientific publications. *Circ Res.* October 15, 2021;129(9):887–889.
39. Nouthe B, Onana A, Nouthe-Djubgang J. *Missed Opportunities to Study the Association of Female Sex With Aortic Insufficiency in Lvad Patients: A Systematic Assessment of Observational and Preclinical Studies*; 2021. Available from: https://asaio.org/conference/program/2021/EPCardiac45.cgi.
40. Truby LK, Garan AR, Givens RC, et al. Aortic insufficiency during contemporary left ventricular assist device support: analysis of the INTERMACS registry. *JACC Heart Failure.* 2018;6(11):951–960.
41. Jonigk D, Golpon H, Bockmeyer CL, et al. Plexiform lesions in pulmonary arterial hypertension composition, architecture, and microenvironment. *Am J Pathol.* July 2011;179(1):167–179.
42. Llucià-Valldeperas A, Smal R, Bekedam FT, et al. Development of a 3-dimensional model to study right heart dysfunction in pulmonary arterial hypertension: first observations. *Cells.* December 20, 2021;10(12).
43. Colvin KL, Yeager ME. Animal models of pulmonary hypertension: matching disease mechanisms to etiology of the human disease. *J Pulm Respir Med.* August 4, 2014;4(4).
44. Akashi H, Kato TS, Hashimura K, et al. Normalization of pulmonary hypertension by the use of left ventricular assist device in patients with end-stage heart failure: a possible contribution to donor pool expansion in lung transplantation. *Int J Gerontol.* 2014;8(1):42–44.
45. Vraka A, Yerly P, Aubert J-D. Comparison of risk stratification scores in pulmonary arterial hypertension: a monocentric retrospective study at lausanne university hospital. *Respiration.* January 2022;19:1–12.

Chapter 6

What problem are you solving?

Rachel Magdalene Sabina Andrew[1,2,3]

[1]*Dominica China Friendship Hospital, Roseau, Dominica;* [2]*All Saints University School of Medicine, Roseau, Dominica;* [3]*Advanced Island Care Inc., Roseau, Dominica*

Key points

- Understanding translational cardiology
- Applying a systematic approach to creating a research question
- Identifying and avoiding drawbacks to the research process

Introduction

Translational cardiology represents a branch of Translational Medicine, which is a relatively novel component of medicine as a whole. The concept has been around for more than 3 decades but has only more recently been given greater attention. In recent years, a decline in the incorporation of research methods into the daily practice at the bedside has been noted (History of Translational Medicine, 2023). Translational Medicine was therefore born out of the necessity to breach this gap. The question that many are asking is, is this great divide real or is it merely perceived? The full extent of this topic is not completely explored or even fully understood; however, the aim of its development would include improving cardiovascular diagnostics and treatment by facilitating a quick translation of experimental data into practical clinical application for everyday use.

So, what exactly is the problem? The reader-deterring style in which most scientific papers are written (O'Donnell, 2005). According to this article by Michael O'Donnell, "Why Doctors Don't Read Research Papers" Explains that the objective of the writer is not to reach clinicians but to get published. It means that the audience who truly requires this information gets "lost in translation" (Jervier Bermejo, 2009). *Only 5% of published papers reached minimum standards of scientific soundness and clinical relevance, and in most journals, the figure was less than 1%* (O'Donnell, 2005). It is, therefore, necessary to create study designs that are aimed at not simple language but a clinically applicable format for easy assimilation.

The term translational medicine was introduced in the 1990s but only gained wide usage in the early 2000s. Originally, translational medical research emerged from the concept of B2B (bench to bedside), as a class of medical research aiming to eliminate the barriers between laboratory and clinical research (History of Translational Medicine, 2023). Although the term has existed for a reasonably long time, there seems to exist some disparity or variation in the interpretation by various experts on its actual definition. *A clear and coherent definition of translational medicine is lacking because translational medicine means different things to different people* (History of Translational Medicine, 2023). It therefore seems to suggest that if a clear definition is not communicated, an objective cannot be reached. This is therefore the first objective of this handbook: understanding and defining the difficulty existing in translational medicine with specific emphasis on translational cardiology, while directing the research process to ensure achieving target outcomes.

Translational medicine is defined by the European Society for Translational Medicine (EUSTM) as "an interdisciplinary branch of the biomedical field supported by three main pillars: benchside, bedside, and community (Mesquita, 2020).

The term "translational research" *therefore, reflects today's integration of basic research (*"bench"*) findings with the clinical practice of medicine and, in a wider scope, the application of results from the individual patient (*"bedside"*) to entire populations for the improvement of public health* (Dennis V. Cokkinos MD, 2015).

Transitional Cardiology, therefore is even less common, which is therefore our focus, to provide clarity and direction, in this subset of translational medicine. With these guidelines, cardiologists and researchers in the field of cardiovascular health will be able to identify pertinent

Translational Cardiology. https://doi.org/10.1016/B978-0-323-91790-2.00060-5

cardiac related problems and select the appropriate research question that would be best suited to obtaining this much needed clinical information.

The problem

The problem refers to a certain identified difficulty, need, or challenge that requires addressing, improving, or solving, which can be resolved through research methods. *Research methods can therefore be defined as strategies, processes, or techniques utilized in the collection of data or evidence for analysis to uncover new information or create a better understanding of a topic* (Research Methods: What are research methods?, 2023). In the settling of Translational Cardiology, this problem should represent a cardiovascular issue, with data once obtained could be quickly adopted by clinicians. When seeking to formulate the perfect research problem that may satisfy the needs of translational cardiology, the PICO method is recommended. The PICO method is an acronym representing four aspects: problem, intervention, comparison, and outcome. The problem of course is as described, including identification of the specific affected population or sample. The intervention would refer to that which will be provoking a change, such as a drug or procedure. The comparison would be the control group and The outcome is that which is expected or the result (What problem are we solving?, 2023).

Types of problems

Translational cardiology, through translational research, seeks to address a variety of problems within the cardiovascular arena. Its potential can revolutionize cardiovascular disease burdens and outcomes.

New pathophysiology?

All diseases' physiopathology may be divided into three groups: understood, partially understood, or idiopathic. As we grow and develop, we notice that diseases that we thought were understood, have new components to their pathological development that would not have been known without new research. Conditions that we once considered idiopathic such as essential or primary arterial hypertension now show promise for its explanation.

Studies are demonstrating remodeling of the vascular extracellular matrix (ECM) during hypertension associated with an imbalance of matrix metalloproteinases (MMPs) and their tissue inhibitors (TIMPs) causing vascular damage and rigidity.

The innate and adaptive immune system also plays a role. Most studies correlated the diagnosis of hypertension with elevated levels of circulating inflammatory markers, antibodies, and cytokines such as IL-6, IL-1β, IL-1α, IL-18, IL-2, IL-8, TNF-α, IFN-γ, C-reactive protein (CRP), and MCP-1, and a decreased level of IL-10.

Oxidative stress and inflammation are also identified as key mediators of vascular remodeling and hypertension (Weronika Frąk, 2022).

It is expected to further build on these studies to eventually translate into hypertension prevention and treatment.

Repurposing existing agents?

Drug repurposing (DR), which is also known as drug repositioning, rescue, reprofiling, or retasking is a process of identifying new therapeutic use(s) for old/existing/available drugs. It is an effective strategy for discovering or developing drug molecules with new pharmacological/therapeutic indications (Mithun Rudrapal, 2020; Mena Abdelsayed, 2022).

This is quite a useful step in Translational medicine as we can significantly decrease the time to implementation of new therapies. Traditionally derived de novo treatments take 10 to 16 years before they can be clinically utilized. Repositioning can take as little as 3 years (Mithun Rudrapal, 2020).

Drug retasking dates back several decades. Perhaps, initially repurposed by chance, now definitely more deliberate measures are being employed as the obvious benefits have been identified. The antiinflammatory agent colchicine has been explored as a treatment for coronary artery disease and myocardial infarction, including two sizable clinical trials with mixed results (Penson, 2021). In many patients with positive responses, however, the risk for GI disturbance was also identified, including bleeding. Continued studies have been recommended.

Assessing other important features of heart pathophysiology has led to drug repurposing efforts beyond inflammation. For example, hydralazine has antioxidant effects and reduces apoptosis, in addition to antihypertensive and afterload reducing effects. Understanding that these processes have important roles in reperfusion injury after myocardial infarction led to the evaluation of hydralazine for the treatment of myocardial injury after infarction, with promising results in vivo models (Mena Abdelsayed, 2022).

Drugs previously indicated for diabetes through landmark studies of DAPA-HF (assessing dapagliflozin) and EMPEROR-reduced (assessing empagliflozin) trials showed that sodium-glucose cotransporter-2 (SGLT2) inhibition reduced the combined risk of cardiovascular death or hospitalization for heart failure in patients with heart failure with reduced ejection fraction (HFrEF) with or without diabetes (Prof Faiez Zannad, 2020). These drugs have already begun to revolutionize heart failure regimens.

New technology empowering new delivery?

A correlation has been established between the activity of AdamsTS7 and the development of atherosclerosis, and coronary artery disease (CAD) has been established. ADAMTS7 is an enzyme that in humans is encoded by the ADAMTS7 gene on chromosome 15. Research done in mice and human subjects demonstrated that subjects who had a mutation in their AdamsTS7 gene rendering it inactive had a marked reduction in atherosclerotic formation and vascular injury. Targeting the inhibition of AdamsTS7 is the focus of much research. Santamaria et al. described the first fluorescence resonance energy transfer (FRET) substrate for ADAMTS-7 called ATS7FP7. This offers a great opportunity to test inhibitors that could potentially be employed to reduce the risk of CAD and atherosclerotic related complications (Salvatore Santamaria, 2021).

The reduction in atherosclerosis via these new methods will indeed translate into new deliveries in Cardiology.

Improving existing therapies and combinations

It is very important to evaluate and improve on existing drug therapies as well as establish synergistic properties through combination therapies. This can be achieved through a process of drug refinement or reengineering. It is sometimes necessary for the drug repurposing process to undergo some refinement. The phenotypic component of the drug which leads to its reassignment may need augmentation of its potency using chemical manipulation (Mena Abdelsayed, 2022).

How to ask the right research questions

The research question begins with the identification of a problem. When selecting the problem, it should be clear and real. It should be and would be important as cardiovascular diseases continue to be the leading cause of death worldwide (Top 10 causes of death, 2020). It should be a problem that primary care physicians, Internists, and cardiologists can easily identify with; which would give it great relevance. It is likely to be a topic that is inadequately addressed. These questions should be answered when formulating the question:

Who is the target population or sample?
Where are they located geographically?
What is the exact issue with which they are faced?
What is the intervention needed to effect change?
How can we measure the problem and its response to the intervention? Is it quantifiable?
Would it be transferable or capable of translation?
Will the results apply to clinical practice?
How much will it cost to obtain this data?
Do the human, physical, and financial resources permit fulfillment of the target outcome? Is it attainable?
It can therefore be summarized that in addition to using the PICO approach, asking the right research question, also requires capturing relevance, quantifiability, and achievability (What problem are we solving?, 2023).

Potential pitfalls

1. Not formulating the right problem question can lead to an inability to measure the desired target which can lead to loss of time.
2. Not anticipating a reasonable timeline for assessing the sample and gathering statistically significant data.
3. Not factoring or securing adequate funding can lead to an incomplete or inconclusive study.
4. Consider risks versus benefits first before the commencement of the study

Resources additional reading

Translational Cardiology, Springer.

Chapter 7

Types of structural interventions

Junedh M. Amrute[1] and Tsuyoshi Kaneko[2]

[1]Center for Cardiovascular Research, Division of Cardiology, Department of Medicine, Washington University School of Medicine, Saint Louis, MO, United States; [2]Division of Cardiothoracic Surgery, Department of Surgery, Washington University School of Medicine, Saint Louis, MO, United States

Key points

- Advent of percutaneous technologies has afforded a novel way to treat valvular heart disease.
- Numerous on-going and completed clinical trials comparing the efficacy of open surgical to transcatheter valve repair/replacement.
- Transcatheter technologies are not a replacement to traditional surgical repair but rather an addition to the heart team's toolkit to better treat structural heart disease.
- Valvular heart disease is a complex multifactorial pathology and the decision for the optimal type of structural intervention warrants a collaborative multidisciplinary heart team–based patient-centric approach.

Why it matters

The incidence of valvular heart disease is growing in the context of new therapeutics in healthcare and an aging global population.[1] Given the substantial morbidity, mortality, and expanding financial costs associated with the management of valvular heart disease, there is a need for new integrated valvular treatment options. Traditional treatments for valve disease relied mainly on open heart surgery. Technological advances in percutaneous devices over the past 2 decades paved the way for new-generation transcatheter-based approaches for the treatment of valvular disease.[2,3] Classical open-heart surgery offers unprecedented anatomical access to complex valve pathologies; however, it is complicated by the plethora of risks associated with open surgery. Transcatheter devices obviate the surgical risks of traditional repair and provide an alternative treatment strategy. These seemingly disparate treatment strategies have led to an important clinical question: what is the best strategy to treat valve disease in a patient-specific manner? Numerous trials have explicated the relative pros and cons of open and transcatheter-based techniques and future studies are necessary to identify metrics and algorithms to stratify valve disease in a patient-specific manner.

Aortic valve disease

The advent of transcatheter aortic valve implantation (TAVI) has led to a paradigm change in the management of aortic valve stenosis.[4] Numerous international trials[4–11] have shown TAVI to be comparable to surgical aortic valve replacement (SAVR) in terms of clinical outcomes. A recent metaanalysis[12] showed that TAVI had a reduction in all-cause death and stroke in the first 2-year window independent of preintervention surgical risk or type of transcatheter heart valve used. These studies have spurred a growing debate on the optimal care of the aortic stenosis patient. Despite these large-scale trials, there remains profound uncertainty on the long-term durability of SAVR and TAVI. Given the nuances of the patients included and excluded in these landmark trials and the lack of long-term follow-up data, a heart team approach is necessary to provide optimal care in a patient-centric manner. In this section, we discuss some of the critical parameters in the decision-making framework for structural interventions in the aortic stenosis patient.

Anatomical

Transfemoral access and the anatomy of the native aortic valve are a critical deciding factor for the feasibility for TAVI. If transfemoral access is adequate and aortic anatomy is conducive to TAVI (as in the patients from the various trials[5–12]), TAVI has at least a comparable clinical outcomes to SAVR. However, the following anatomical considerations need to be considered as the heart team

Translational Cardiology. https://doi.org/10.1016/B978-0-323-91790-2.00110-6

decides on the suitability for TAVI as the landmark trials had numerous crucial exclusion criteria[4] (Table 7.1):

- Transfemoral access: The umbrella of unfavorable transfemoral access includes iliofemoral vessel tortuosity, calcification, complete occlusion, or status-post endovascular aortic repair. For patients who fall in this category, there is limited evidence of the efficacy of TAVI versus SAVR and current studies are limited to subgroup analysis, case reports, and observational studies.[11–13] In such patients, alternative strategies such as lithotripsy-enabled vascular access,[14–17] surgical-cut down,[18] and nonfemoral vascular access routes, such as transaxillary, transcarotid, transaortic, and transapical approach,[19–22] are necessary.
- Nature of calcification: for TAVI, symmetric valve calcification is preferred. In younger patients with minimal calcification, there is a risk for valve migration post-TAVI. There are on-going efforts for novel device development for optimal valve sizing in patients with aortic regurgitation with minimal calcification in early-stage trials[23–28] and small observational studies.[29–31] In patients with heavy asymmetric calcification on the leaflets or raphe, TAVI can lead to paravalvular leaks and the potential risk of annular rupture. Numerous efforts in transcatheter valve design have utilized a textured polyethylene terephthalate outer sealing skirt to obviate PVL concerns.[4] Despite such advances, the clinical risk of adverse events associated with TAVI in severe valvular and aortic calcification remains present.[32–34]
- Risk of collateral conduction defects: In the setting of potential calcification of the conduction system and the anatomical proximity to the aortic root, a low computer-simulated valve contact pressure and a long membranous septum are favorable features for TAVI, while a high contact pressure, short membranous septum, and preexisting right bundle branch block are unfavorable parameters for TAVI.[4] Notably, holistic evaluation of collateral damage on the conduction system is necessary in deciding between TAVI and SAVR especially in younger patients. At present, for patients who have unfavorable conduction parameters (contact pressure, septum length, prior RBBB), SAVR is the preferred intervention.
- Annulus dimensions: Valve sizing is critical for effective TAVI—annulus sizes that are too small or large are not conducive to TAVI and these patients should instead undergo an SAVR. Patients with a small annulus may benefit from an SAVR with a root enlargement procedure instead of standard TAVI implants. The Small Annuli Randomized to Evolut or Sapien trial is an ongoing multicenter study comparing TAVI devices in symptomatic severe AS patients with a small aortic valve annulus (<430 mm^2).[35] In patients with a large annulus area (>683 mm^2), a recent trial showed that TAVI with the Edwards SAPIEN3 29-mm with overexpansion is safe with acceptable PVL and pacemaker implant rates.[36] Most recently, the MyVal-1 (covering areas up to 840 mm^2) study demonstrated safety with no pacemaker rates up to 12-month follow-up[37]; however, further clinical trials are necessary to characterize the PVL rates in these patients and long-term follow-up.
- Bicuspid aortic valve: Notably, the landmark SAVR/TAVI trials have excluded BAV patients yet account for 5%–10%. TAVI in BAV stenosis is favorable in patients where there is no aortopathy and minimal calcification in the raphe and valve leaflets. In patients with BAV stenosis and concomitant aortopathy (aortic root or ascending aortic dilation) necessitating prophylactic repair, SAVR is preferred to obviate the risk of

TABLE 7.1 Landmark clinical trials comparing SAVR to TAVI with most relevant clinical outcomes.[2]

	TAVI versus SAVR
30-day stroke	
PARTNER 1A	$P = .2$
CoreValve US HR	$P = .55$
PARTNER 2A	$P = .2$
SURTAVI (TAVI − SAVR) with Bayesian 95% CI	−2.3 [−2.6, 0.6]
NOTION	$P = .37$
PARTNER3	NA
Evolut LR (TAVI − SAVR) with Bayesian 95% CI	−1.2 [−2.4, −0.2]
30-day mortality	
PARTNER1A	$P = .07$
CoreValve US HR	$P = .43$
PARTNER 2A	$P = .78$
SURTAVI (TAVI − SAVR) with Bayesian 95% CI	0.5 [−0.9, 1.8]
NOTION	$P = .43$
PARTNER3 (hazard ratio with 95% CI)	0.37 [−0.07, 1.88]
Evolut LR (TAVI − SAVR) with Bayesian 95% CI	−0.8 [−1.9, 0.2]
2-year mortality	
PARTNER3	$P = .47$
Evolut LR (TAVI − SAVR) with Bayesian 95% CI	0.0 [−3.2, 3.2]
5-year mortality	
PARTNER1A	$P = .76$
CoreValve US HR	$P = .50$
PARTNER 2A	1.02 [0.85, 1.23]
SURTAVI	$P = .55$
NOTION	$P = .75$

dissection or rupture.[38] There is no RCT characterizing the efficacy of TAVI in BAV patients; however, numerous observational studies have compared TAVI in BAV stenosis to SAVR in BAV stenosis and TAVI in tricuspid AS.[39–43] Analysis from the STS registry showed similar clinical outcomes between TAVI in BAV and tricuspid AS except for greater PVL in the BAV cohort. In low-risk patients who underwent a SAPIEN3 TAVI, there was no difference in mortality or stroke at the 30-day and 1-year time points.[40] Notably, this analysis did not include control patients with BAV stenosis who were treated surgically warranting additional trials evaluating TAVI in BAV patients. Another study queried the National Inpatient Sample Database for a propensity-matched comparison between TAVI and SAVR in BAV stenosis patients[39] and found no difference in in-hospital mortality rates between the groups. Given the profound impact of BAV morphology on intervention, a recent observational study used CT analysis to correlate BAV calcification to outcomes and found that raphe calcification and eccentric leaflet calcification were associated with increased risk of procedural complications and midterm mortality.[44] Given the prevalence of BAV patients and the heterogeneity of the valve anatomy, especially in younger low-risk patients, larger clinical trials are necessary to compare newer generation TAVI devices to SAVR across a broad array of risk profiles.

- Aortic anatomy: Aortic angulation is defined as the angle between the horizontal plane and the plane of the aortic annulus. A high aortic angulation angle >48° is associated with adverse outcomes following TAVI, likely because of the difficulty in valve deployment. A recent observational study looked at the effect of aortic angulation on outcomes after TAVI using the newer generation of expandable valves—they found no impact on procedural success or clinical outcomes.[45] However, patients with an aortic angulation >70° were excluded from the landmark TAVI trials, and in this high-risk cohort, there are little data supporting the use of TAVI.[5–12] As such, patients with severe aortic angulation (>70°) should undergo SAVR.
- Valve-in-valve: In the landmark trials, patients who had a prior prosthesis were excluded; however, as the cardiac patient population ages, the durability of SAVR bio prosthesis diminishes and these patients require an additional valve intervention (at a much older age). Given the high-risk nature of redo cardiac surgery in elderly patients, there is a growing body of data advocating for the use of valve-in-valve (VinV) TAVI procedures.[46–50] Compared to TAVI in a native aortic valve, VinV is associated with greater rates of prosthesis malposition, coronary obstruction, and residual gradient flows.[51] The risk of coronary ostia obstruction is greatest in bioprosthesis with an externally mounted stent compared to stentless valves.[4,51] To evaluate the risk of coronary obstruction in stented prosthesis, the effective or virtual valve height is calculated and if this is < 3–4 mm, it is considered high risk.[52,53] In stentless prosthesis, the coronary take-off position and length of the sinus of Valsalva (as described above) are the critical factors to gauge the risk of coronary obstruction. The risk of a high residual gradient post-VinV is greater in patients with a small root—alternatives like bioprosthetic valve fracture with a high-pressure balloon have shown promise to obviate flow gradient concerns in simulated and preclinical settings.[54,55] Given the dearth of clinical data in small prosthesis VinV cases and the abundance of surgical options to enlarge the root, SAVR should be considered in low-risk patients. Furthermore, clinical metaanalysis has shown that surgical aortic root enlargement is not associated with adverse outcomes compared to standard AVR.[56] In patients with clinically significant PVL, there is a Class I recommendation for SAVR in the operable candidates and a Class IIa for TAVI in the nonsurgical candidates.[57,58] For failed bioprosthesis valves, a recent propensity score analysis compared redo-SAVR with a VinV approach and found similar 30-day and 1-year outcomes.[59] A metaanalysis of four observational studies corroborated these findings; however, there is a paucity of data in large patient cohorts with long-term follow-up.[60] For degenerated prosthesis, a 3-year outcome analysis from the PARTNER2 Registry found favorable survival, hemodynamics, and quality of life after a VinV.[61]
- Coronary anatomy: In patients where the coronary ostia take-off close to the aortic root, there is a concern for obstruction of flow post-TAVI causing catastrophic complications. This is significant issue in VinV TAVI or repeat TAVI. Various techniques such as the chimney stenting for coronary occlusion during TAVI have shown efficacy.[62–65] BASILICA is novel method which utilizes electrosurgical leaflet laceration to avoid coronary artery obstruction during TAVI[66] and a recent trial showed favorable 1-year outcomes with no new evidence of coronary occlusion following the BASILICA procedure in TAVI.[66] These studies have small patient populations with limited long-term follow-up and are performed at select clinical centers. There are new devices that are currently being developed to overcome this problem. The ShortCut study is an on-going clinical trial (NCT04952909) investigating the safety and efficacy of the ShortCut device for VinV in patients at risk for coronary artery ostium obstruction (Table 7.2).[67]

TABLE 7.2 Exclusion criteria form landmark randomized clinical trials comparing TAVI to SAVR.

Anatomical criteria	Clinical/hemodynamic criteria
- Poor transfemoral access	- Mixed valvular disease
- Bicuspid AV	- Multivessel or left main coronary artery disease.
- Excessive calcification on AV or outflow tract	- LV ejection fraction <20%
- Noncalcified AV	- Hypertrophic obstructive cardiomyopathy
- Porcelain aorta	- Blood dyscrasias
- Aortic root angulation >70 degrees	- Hemodynamic instability
- Small sinus of Valsalva	- Contraindication to antithrombotic agents
- Prior AV surgery/implant	- Prior MI or cerebrovascular event
- Aortopathy requiring intervention.	- Chronic disease comorbidities (CKD, pulmonary disease, hepatic insufficiency)
- Intracardiac vegetations or masses	- Pulmonary artery hypertension
	- Life expectancy <24 months

Adapted from Windecker S, et al. Which patients with aortic stenosis should be referred to surgery rather than transcatheter aortic valve implantation?. *Eur Heart J.* 2022; 43: 2729–2750.

Coronary artery disease

Clinically significant coronary artery disease (CAD) is prevalent in severe AS patients. The key parameter in treating the AS + CAD patient is age, CAD severity (vessel involvement and SYNTAX score), and coronary access for transcatheter approach (Table 7.3).

In patients with left main disease or an SYNTAX >32, CABG is a Class I indication and, in these patients, SAVR + CABG is indicated. In patients who are high surgical risk but have LM disease + SYNTAX <32 OR 3-vessel disease and SYNTAX <22, TAVI + PCI can be considered—if low surgical risk, then SAVR + CABG is still indicated. In patients with low CAD complexity (1–2 vessel disease + SYNTAX <22) who are poor surgical candidates, then TAVI + PCI is appropriate. Moreover, a recent study showed that in patients with complex CAD (LM disease or SYNTAX >22) who are poor surgical candidates, TAVI + PCI and SAVR + CABG were associated with similar rates of major adverse cardiac and cerebrovascular events with a 3-year follow-up, but notably, TAVI + PCI had a higher rate of coronary revascularization.[68]

Long-term management

Transcatheter aortic valves are still in their infancy and there is a lack of long-term follow-up data available. The PARTNER2 trial did a subgroup analysis with a follow-up to 5 years and found that TAVI led to a higher rate of all-cause death and stroke as compared to SAVR.[61,69] Given the expansion of TAVI indications for low-risk and young patients, there is a need for more long-term studies to effectively compare the efficacy of SAVR versus TAVI and guide management. TAVI is associated with higher rates of PVR compared to SAVR, and the impact of this on chronic LV remodeling and clinical outcomes and some preliminary studies has shown that mild PVR post-TAVI is associated with increased mortality after 5 years.[70–72] Similarly, TAVI is associated with greater rates of AV and left bundle branch blocks as compared to SAVR, whereas

TABLE 7.3 Management for severe AS and clinically significant CAD.

Age	<70	70–80	>80
CAD severity	LM disease + SYNTAX >32 OR 3 vessel disease + SYNTAX >22	LM disease + SYNTAX <32 OR 3 vessel disease + SYNTAX <22	1–2 vessel disease + SYNTAX <22
Coronary access	Poor	Moderate	Good
Guideline	SAVR + CABG	SAVR + CABG OR TAVI + PCI	TAVI + PCI

Adapted from Windecker S, et al. Which patients with aortic stenosis should be referred to surgery rather than transcatheter aortic valve implantation?. *Eur Heart J.* 2022; 43: 2729–2750.

SAVR is associated with greater rates of postop atrial fibrillation. A metaanalysis found that greater rates of LBBB and AV block post-TAVI were associated with increased risk of all-cause mortality and HF hospitalizations at 1-year follow-up.[73] On the contrary, postop Afib after SAVR was not independently associated with adverse outcomes in the PARTNER3 trial.[74] Recent data from the Evolut Low-risk trial 4 year follow-up showed a 26% relative reduction in hazard for all-cause mortality or disabling stroke in TAVI versus SAVR.[75] Current durability data are available for TAVI versus SAVR up-to 10 years in older patients from the NOTION trial[76]; however, the TAVR valve was an old generation valves, and on the SAVR side, the some of the used valves are already out of the market. 10-year data from PARTNER3 and Evolut Low-risk trial will provide further information to guide us in understanding the valve durability.

The other critical information in the decision-making between a mechanical valve is the choice of surgical bioprosthesis, including the Ross procedure. Several RCTs and observational studies have compared mechanical valves to bioprosthesis in SAVR and have found that in younger patients who are tolerate anticoagulation, mechanical valves are recommended.[77,78] Bioprosthetic valves were associated with greater reoperation rates; however, at 15-year follow-up, a propensity-matched analysis showed that there were no differences in survival or stroke.[79] An alternative to bioprosthesis is the Ross, where the surgeon implants the pulmonary valve into the aortic position and the pulmonic valve is replaced with a homograft. A metaanalysis showed that the Ross is associated with lower all-cause mortality but higher rates of reoperation compared to mechanical aortic valves after a 5.8-year median follow-up (Fig. 7.1).[80]

Mitral valve disease

There is a growing incidence of patients with primary and secondary mitral valve disease and a concomitant emergence of transcatheter mitral technologies. Given the traditional downsides of open-heart surgery, especially in older patients, there are growing clinical trials comparing transcatheter mitral devices to surgical repair. The etiology of mitral valve disease (primary vs. secondary) is an important factor in the clinical decision-making process.

Primary MR

Primary mitral valve regurgitation is driven by an abnormality within the mitral apparatus (leaflets, chordae, papillary muscles, or annulus).[82] In the United States, mitral valve surgery is recommended for patients with primary MR who are surgical candidates.[57,58] The first percutaneous mitral repair trial (EVEREST I) showed safety and mid-term durability with the MitraClip system

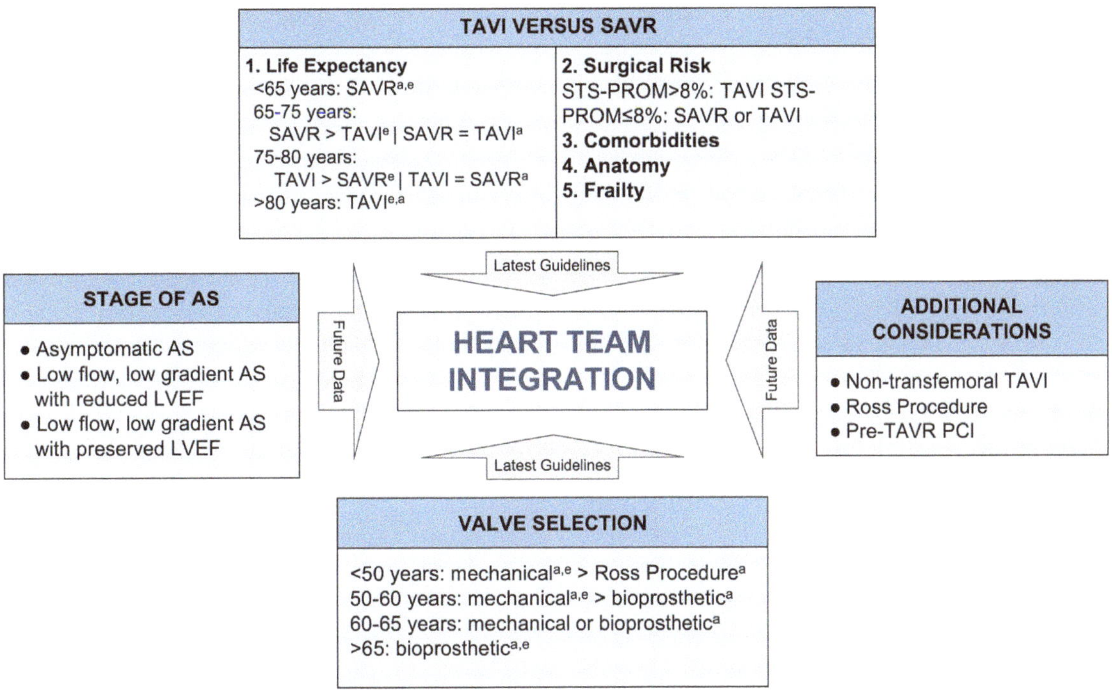

a Recommended by the 2020 ACC/AHA Guidelines
e Recommended by the 2021 ESC/EACTS Guidelines

FIGURE 7.1 Structural intervention strategy in aortic stenosis—figure directly from ESC/EACTS versus AHA/ACC guideline management.[81]

(Abbott structural heart, Minneapolis, MN).[83] The subsequent EVEREST II trial compared percutaneous repair to surgery in patients with grade 3+ to 4+ MR and found that percutaneous repair was less effective at reducing MR, but was associated with greater safety and similar clinical outcomes.[84] The 5-year follow-up analysis of the EVEREST II trial showed that patients with percutaneous repair had higher rates of redo repair for MR during the first year of treatment, but the 1- and 5-year rates for redo surgery and MV dysfunction were comparable in percutaneous and surgical cohorts.[85] A notable subanalysis within the EVEREST II trial showed that most failures were within the first 6 months attributed to lack of technical experience.[83–85] Additionally, the etiology of the primary MR in the EVEREST trials was diverse and approximately 30% of patients in EVEREST II had complex Barlow's.[86] Current recommendations for complex degenerative MR such as Barlow's are managed surgically.[57,58] The EVEREST II high-risk registry showed that percutaneous treatment reduced MR and LV dimensions and improved clinical outcomes at 12 months.[87] A retrospective analysis from this high-risk registry showed that for patients with degenerative severe MR who are not surgical candidates, percutaneous repair led to fewer hospitalizations and functional improvements with adaptive LV remodeling.[88] A metaanalysis showed that percutaneous MV repair is safe and efficacious in primary MR.[89] The first human trial (CLASP study) for the Edwards PASCAL TVMr system (Edwards Lifesciences, Irvine, CA) showed feasibility and a high technical success rate and reduction of MR severity at the 1-month time point in 2017.[90] The 1-year follow from the CLASP study showed safety, low complication rates, sustained reduction in MR, and improvement in functional status.[91] The CLASP IID trial[92] which compared the Edwards PASCAL TVMr system with the MitraClip showed at the 1-year end point noninferiority of the Edwards PASCAL TVMr system to MitraClip, high survival, significant reduction in MR and functional status, and improvements in quality of life in patients with primary degenerative MR who are not surgical candidates.[92,93]

A recent registry-based cohort study showed that in patients with isolated moderate-severe MR or degenerative MR undergoing transcatheter mitral valve repair, the procedure led to successful repair in 88.9% of patients. Mortality was inversely correlated with associated with MR burden.[94] Another retrospective cohort study compared surgical to transcatheter mitral repair in primary MR and found that transcatheter repair had better short-term outcomes, whereas surgical repair had better medium-term MACE scores.[95] A recent French study compared surgical to transcatheter repair in patients with severe MR and found that at 3-year follow-up transcatheter repair was associated with no differences in all-cause mortality between the two techniques.[96]

There are two clinical trials that will compare the effectiveness of M-TEER and mitral valve repair. The REPAIR MR trial was designed to compare the MitraClip (a transcatheter device for mitral valve repair) vs. surgical repair for primary MR in older or moderate risk patients.[97] This study has two primary endpoints measured at 2 years: (1) all-cause mortality, stroke, hospitalization, and AKI requiring renal replacement therapy and (2) recurrent MR, MV replacement, r other MV intervention (NCT04198870). Another on-going trial, the PRIMARY trial, is comparing percutaneous or surgical repair in mitral prolapse and regurgitation for individuals >60 years of age (NCT05051033)—this is an international trial comparing any device used within the county for primary degenerative MR. The primary outcome in the PRIMARY trial is all-cause mortality, valve reintervention, hospitalization for HF, or >3+ MR composite score.

Chordal replacement: Transapical chordal replacement using the NeoChord device in patients with severe primary MR with flail or prolapsed leaflets mid-term outcomes showed good clinical outcomes and echocardiographic features up to 3-year follow-up.[98,99] Notably, in this analysis, patients with complex leaflet anatomy had worse outcomes therefore optimal patient selection for such procedures is of utmost importance. Midterm results of transventricular mitral valve repair for severe MR showed that P2 prolapse repair was possible but found durable in only 82% at 3-year follow-up and complex initial leaflet degeneration was associated with greater risk of MR recurrence.[100] A randomized trial comparing the Neochord DS100 system to open surgical repair (NCT02803957) is currently ongoing. The primary outcomes in this study are subjects free of major adverse events and MR burden or MV reintervention.[101] There are currently investigational efforts into a new transseptal TMVR device (Innovalve TMVR).[102] Future trials are necessary to compare this technique to open mitral surgery with long-term follow-up.

Transcatheter mitral valve repair is utilized in select patient populations in primary MR, albeit is not the first-line treatment. Complex degenerative disease such as Barlow's disease, heavy calcification of leaflets, rheumatic etiology, small annular size, and short posterior leaflets are anatomical barriers to effective repair with transcatheter approaches.[103]

Secondary MR

Secondary MR is caused by LV or LA remodeling or subvalvular apparatus with structurally normal mitral valve leaflets.[104] Secondary MR can be further classified into ischemic (post-MI)[105] or nonischemic (secondary to atrial fibrillation or dilated cardiomyopathy).[106] Management of functional MR is challenging as it is a by-product of a primary cause, and in many cases, fixing the primary problem can alleviate the MR.[104,105] Therefore, patients with severe MR should be on maximal goal-directed medical therapy with appropriate interventions (coronary revascularization, CRT) measures taken to address the underlying problem. Mitral valve surgery currently has a class IIb indication for severe MR with NYHA class III-IV heart failure.[57,58,104] The guidelines also state that MV surgery can be performed in patients undergoing concomitant CABG or AVR.[57,58,104] The key decision points for mitral surgery in functional MR include repair versus replacement and surgery versus percutaneous approaches.

Mitral valve surgery for functional MR: A landmark randomized clinical trial compared mitral valve repair to replacement in patients with ischemic MR and found no significant differences in LV remodeling or survival at 1-year follow-up, although replacement provided a durable repair of the valve.[107–109] Two-year outcomes for the surgical treatment of severe ischemic MR from this patient cohort found no differences in LV remodeling or survival—recurrent MR and associated adverse HF events were more frequent in the repair group.[110] Given the profound impact of recurring MR, a subgroup analysis used baseline clinical and echocardiographic parameters to predict MR recurrence (AUC = 0.82). This analysis underscores the importance of a Heart Team approach to mitral surgery given the patient-specific propensity for recurrent MR but included a small patient cohort and needs further validation.[111] Another study compared found that in patients with severe MR due to ischemic and dilated cardiomyopathy, mitral valve replacement is associated with higher mortality than repair.[112]

Transcatheter MV repair (TMVr): The COAPT trial examined the benefit of TMVr in patients with heart failure + moderate-severe or severe MR and found that TMVr reduced HF hospitalization and all-cause mortality in a 24-month follow-up.[113] The MITRA-FR trial in patients with HFrEF and severe MR found that the rate of death or HF hospitalization was unchanged at 1-year follow-up.[114] These two trials targeted the same broad clinical population with the same clinical intervention, yet had diametrically opposite results. There are several important differences between these trial designs as well as the clinical state of the trial participants which must be entangled before making conclusions[113–115]: (1) Sample size: the COAPT trial had a larger patient population (by twofold); (2) Endpoint: COAPT measured heart failure hospitalizations at 2-year follow-up while MIITRA-FR looked at the composite of all-cause death and heart failure–related hospitalization at 1-year follow-up; (3) Disease burden: MITRA-FR patients had more severe baseline MR and advanced LV disease compared to COAPT. The COAPT trial also excluded patients with severe LV dysfunction, whereas the MITRA-FR did not. Furthermore, baseline MR in the MITRA-FR trial was more severe compared to that in the COAPT trial (EROA 31 vs. 41 mm^2) respectively; (4) Concomitant medical management: both trials required persistent HF even with GDMT treatment; however, patients in the COAPT trial had greater GDMT optimization at baseline, whereas patients in the MITRA-FR trial had a softer inclusion criteria; (5) Valve repair: the COAPT-trial employed a more aggressive repair strategy (more average clips) which also manifested as a more sustained reduction in MR at 1-year (5% with moderate-severe MR in COAPT vs. 17% in MITRA-FR). Collectively, these findings highlight that the disparate results of these trials could be explained by less severe baseline MR/LV dysfunction, greater GDMT optimization, and more aggressive repair approach in COAPT versus the MITRA-FR trial. The upcoming ReshapeHF2 trial (NCT02444338) has the same inclusion criteria as COAPT for MR severity, but intermediate inclusion criteria for baseline LV dysfunction.[116] This trial will help guide the suitability of the MitraClip in patients with more severe baseline LV dysfunction.

Transcatheter MV replacement: Early experience for transcatheter MVR has been with the Edwards Sapien valves with anchors (valve, ring, or calcium) and these studies showed the feasibility of transcatheter MVR.[117] Mitral valve surgery via transcatheter techniques is complicated by the profound complexity of the mitral valve including the asymmetry, subvalvular apparatus, annular calcification, and absence of a single valvular plane.[104] Numerous devices and international trials have emerged to treat functional MR using transcatheter techniques:

- Tendyne: The first feasibility study with the Tendyne transcatheter MVR system from 2019 in the first 100 patients with severe MR who are not surgical candidates lowered MR burden with an appropriate safety profile.[118] The recent 2-year follow-up of this cohort showed reduced MR burden, fewer HF-related hospitalizations, and improved symptoms.[119] Notably, the greatest HF-related hospitalizations and all-cause mortality occurred within the first 3 months highlighting the need for optimal patient selection (from an anatomical standpoint) and more operative experience. A European study found in patients with severe MR and annular calcification, transcatheter MVR with the Tendyne valve had positive acute MR reduction, midterm durability, and clinical improvement.[120] A metaanalysis found that the Tendyne valve had a 96.9% technical success across 309 patients.[121]
- Intrepid: The Intrepid system is a self-expanding, trileaflet bovine pericardial prosthesis with a nitinol frame.[122] A feasibility study from 2015 showed feasibility in a high-severe risk cohort of 50 patients. The study established feasibility, 96% valve deployment rate, 14% 30-day mortality, and a median follow-up of 143 days that confirmed sustained low MR rates

and symptomatic improvement.[123] Ongoing trials will evaluate the efficacy of the Intrepid system compared to other transcatheter devices and open surgery.[122]

On-going trials for functional MR:

- SUMMIT: The SUMIT trial is designed to look at the feasibility of Tendyne valve system in symptomatic patients with MR with a randomized nonrandomized and a MAC cohort. The randomized arm will compare Tendyne to the MitraClip, the nonrandomized group will consist of nonreparable MR, and the MAC cohort contains patients with MR and severe annular calcification. The primary endpoint in all arms is survival free of HF hospitalizations at 12 months.[124]
- APOLLO: The APOLLO trial is designed to evaluate the Medtronic Intrepid transcatheter valve system.[125] The APOLLO trial has two arms: primary/secondary MR and MAC. Unlike the SUMMIT trial, in the APOLLO trial all patients will undergo transcatheter mitral replacement with the Intrepid system.[104,124,125] The primary endpoint in the primary cohort is all cause mortality or HF hospitalization post-30 days or KCCQ improvement <10 composite and only all-cause mortality in the secondary cohort.[125]
- ENCIRCLE: An on-going trial designed to establish safety and efficacy of the SAPIEN M3 (Edwards Lifesciences) in patients with 3+ symptomatic MR who are not candidates for currently available transcatheter or surgical devices (NCT04153292).

Annuloplasty: annuloplasty devices have primarily been used in secondary MR. There are interesting reports of using annuloplasty devices in conjunction with transcatheter repair.[103] One report used a Carillon (Cardiac Dimensions) coronary sinus annuloplasty device with MitraClip using the Alferie technique in an 84-year-old with degenerative MR with a EuroSCORE II of 9.64%—the surgery resulted in an adequate reduction in MR and with no postoperative complications suggesting that such a technique could be used in patients with annular dilation who are not surgical candidates.[103,126] Another report used ring annuloplasty and neochord leaflet implantation in degenerative MR with prolapse in a 63-year-old with an STS predicted mortality of 8%.[103,127] A clinical trial using CardioBand (Edwards Lifesciences) a percutaneous device for direct mitral annuloplasty showed safety and efficacy and is currently approved for clinical use in the European Union.[128]

Tricuspid valve

Tricuspid valve disease affects 4% of patients above 75 years of age.[129] TR can be driven by left heart failure and many undergoing surgeries for left heart disease undergo concomitant TV surgical repair.[130] Isolated TV is rarer and more isolated surgical repair of the TV has shown to carry up to 10% in-hospital mortality.[131] The advent of transcatheter technologies has given rise to numerous opportunities for minimally invasive repair or replacement of the tricuspid valve.

Tricuspid valve repair: A recent prospective randomized clinical trial compared transcatheter repair with the TriClip (Abbott) decide for severe TR to medical therapy and found reduced TR burden and improvement in quality of life.[132,133] These new devices and on-going clinical trials exploring transcatheter treatment of TR hold promise; however, further studies are needed to isolate the optimal patient population who would most benefit from these interventions. Notably, transcatheter TV devices fix the TR but potentiate an afterload mismatch between a remodeled RV which may lead to postoperative compromise.[134] Additionally, the long-term durability of these devices is not well characterized.

Orthotopic valves leverage a bioprosthetic valve that is implanted on an existing tricuspid anulus.

- EVOQUE: The first-in-human application of the EVOQUE system (FDA approved) in 27 patients who were deemed high surgical risk established feasibility and safety with a 92% success rate, reduction in TR at 1-year with improvement in HF symptoms, and 7% all-cause mortality.[135] The recent results from the TRISCEND trial[135–137] in 176 patients showed TR reduction, increase in SV and CO, high survival and low HF related hospitalization rates, and improved functional quality of life at 1-year.[137,138]
- Lux-Valve: The Lux-Valve is deployed via a minimally invasive right thoracotomy transatrial approach.[134] An initial study established feasibility and safety of the Lux-Valve for patients with a 100% technical success rate, reduction in TR, and improved functional quality of life at 1-year.[139] There is an on-going trial to investigate the second-generation Lux-Valve which is deployed via a transjugular access with primary endpoint a composite score of any major adverse event at 30-day postprocedure.[140]
- GATE: The GATE device is also deployed via a minimally invasive right thoracotomy transatrial approach.[134] An early study showed feasibility, improved RV hemodynamics, increased CO, and improvement in HF associated symptoms.[141]
- Cardiovalve: The Cardiovalve technology is deployed via transfemoral access. An early study showed compassionate use I 15 patients with no perioperative mortality and a reduction in TR at 30-days.[134] There is an ongoing European clinical trial evaluating the efficacy of this device further with a primary end point of all-cause mortality and any major adverse event.[142]
- Intrepid: The Intrepid device has been applied in mitral transcatheter replacement previously.[122,125] An early

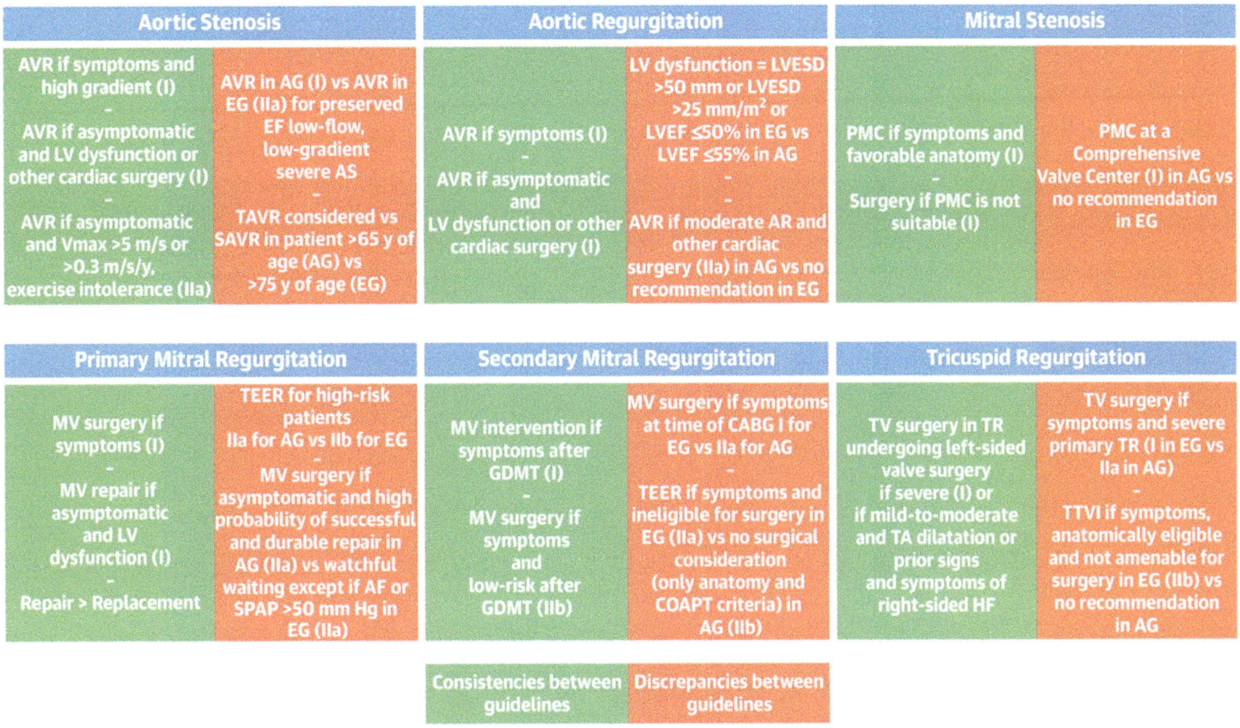

FIGURE 7.2 ACC/AHA and ESC/EACTS guidelines for the structural management of valvular heart disease. Figure directly from Coisne et al.[150]

feasibility safety showed 100% technical success and reduction in TR from 15 patients.[134,143,144]

Heterotopic valves allow for bicaval valve implantation enabling a reduction in venous regurgitation and alleviating right HF symptoms. Unlike orthotopic valves, bicaval valve implantation is able to accommodate for eccentric device angles and large annular sizes.

- Sapien: An off-label study deployed the balloon-expandable Spaien valve in the IVC after an anchoring stent was deployed.[134] A randomized controlled trial was prematurely terminated as 4/28 patients required open heart surgery because the device dislodged.[145,146]
- TricValve: The TricValve is deployed via transfemoral access. The TRICUS EURO study (nonrandomized feasibility trial with 35 patients) showed a high procedural success rate, improvement in quality of life, and functional classification at 6 months follow-up.[147] A recent report also showed successful application of this device in an LVAD patient with severe TR.[148]
- TRICENTO: The TRICENTO is deployed via transfemoral access into the SVC and IVC with a bicuspid pericardial valve facing into the right atrium.[134] A recent multicenter registry analysis of the TRICENTO device showed feasibility, functional improvement, and RV recovery.[149] Further studies are needed to dissect eligibility criteria given incidence of stent fractures (Fig. 7.2).[134]

Importance of the heart team

Given the growing patient population with complex valvular disease and the expansion of emerging percutaneous devices a multidisciplinary Heart Team is necessary for the care of patients with structural heart disease. A multidisciplinary heart team-based approach a class I indication for the management of complex valvular disease.[151] Such a team needs to consist of cardiac surgeons, interventional cardiologists, cardiologists, imaging experts, and translational scientists. Each patient needs to be carefully evaluated based on clinical and nonclinical comorbidities to identify the optimal structural intervention. Integrating translational scientists can allow for better characterization of valvular disease, development of computational hemodynamic models, integration of AI alongside imaging to build patient-specific strategies, characterization of molecular changes driving valvular dysfunction, and development of newer technologies.

References

1. Coffey S, et al. Global epidemiology of valvular heart disease. *Nat Rev Cardiol.* 2021;18(12):853–864. https://doi.org/10.1038/s41569-021-00570-z.
2. Cahill TJ, et al. Transcatheter aortic valve implantation: current status and future perspectives. *Eur Heart J.* 2018;39:2625–2634.
3. Cahill TJ, Terre JA, George I. Over 15 years: the advancement of transcatheter aortic valve replacement. *Ann Cardiothorac Surg.* 2020;9:442.

4. Windecker S, et al. Which patients with aortic stenosis should be referred to surgery rather than transcatheter aortic valve implantation? *Eur Heart J.* 2022;43:2729–2750.
5. Smith CR, et al. Transcatheter versus surgical aortic-valve replacement in high-risk patients. *N Engl J Med.* 2011;364:2187–2198.
6. Adams DH, et al. Transcatheter aortic-valve replacement with a self-expanding prosthesis. *N Engl J Med.* 2014;370:1790–1798.
7. Leon MB, et al. Transcatheter or surgical aortic-valve replacement in intermediate-risk patients. *N Engl J Med.* 2016;374:1609–1620.
8. Reardon MJ, et al. Surgical or transcatheter aortic-valve replacement in intermediate-risk patients. *N Engl J Med.* 2017;376:1321–1331.
9. Thyregod HGH, et al. Transcatheter versus surgical aortic valve replacement in patients with severe aortic valve stenosis: 1-year results from the all-comers NOTION randomized clinical trial. *J Am Coll Cardiol.* 2015;65:2184–2194.
10. Mack MJ, et al. Transcatheter aortic-valve replacement with a balloon-expandable valve in low-risk patients. *N Engl J Med.* 2019;380:1695–1705.
11. Popma JJ, et al. Transcatheter aortic-valve replacement with a self-expanding valve in low-risk patients. *N Engl J Med.* 2019;380:1706–1715.
12. Siontis GCM, et al. Transcatheter aortic valve implantation vs. surgical aortic valve replacement for treatment of symptomatic severe aortic stenosis: an updated meta-analysis. *Eur Heart J.* 2019;40:3143–3153.
13. Toggweiler S, et al. Management of vascular access in transcatheter aortic valve replacement: part 1: basic anatomy, imaging, sheaths, wires, and access routes. *JACC Cardiovasc Interv.* 2013;6:643–653.
14. Di Mario C, et al. A prospective registry of intravascular lithotripsy-enabled vascular access for transfemoral transcatheter aortic valve replacement. *JACC Cardiovasc Interv.* 2019;12:502–504.
15. Cruz-González I, et al. Facilitated transfemoral access by shockwave lithoplasty for transcatheter aortic valve replacement. *JACC Cardiovasc Interv.* 2019;12:e35–e38.
16. Gorla R, Cannone GS, Bedogni F, De Marco F. Transfemoral aortic valve implantation following lithoplasty of iliac artery in a patient with poor vascular access. *Cathet Cardiovasc Interv.* 2019;93:E140–E142.
17. Sanghvi K, Singh M. Intervention of prohibitive aortic and iliac atherosclerotic stenosis to allow successful transfemoral transcatheter aortic valve replacement. *JACC Cardiovasc Interv.* 2018;11:e137–e140.
18. Drafts BC, et al. Comparison of outcomes with surgical cut-down versus percutaneous transfemoral transcatheter aortic valve replacement: TAVR transfemoral access comparisons between surgical cut-down and percutaneous approach. *Cathet Cardiovasc Interv.* 2018;91:1354–1362.
19. Damluji AA, et al. Alternative access for transcatheter aortic valve replacement in older adults: a collaborative study from France and United States. *Cathet Cardiovasc Interv.* 2018;92:1182–1193.
20. Gleason TG, et al. Subclavian/axillary access for self-expanding transcatheter aortic valve replacement renders equivalent outcomes as transfemoral. *Ann Thorac Surg.* 2018;105:477–483.
21. Mohammadi S, Rodés-Cabau J, Kalavrouziotis D. TAVR access: fitting the approach to the patient and not the patient to the approach. *J Am Coll Cardiol.* 2019;74:2740–2742.
22. Beurtheret S, et al. Femoral versus nonfemoral peripheral access for transcatheter aortic valve replacement. *J Am Coll Cardiol.* 2019;74:2728–2739.
23. Chen X, et al. Improved transcatheter aortic valve implantation for aortic regurgitation using a snare loop-assisted device: the first preclinical experience. *Eur J Clin Invest.* 2016;46:714–720.
24. Scherman J, et al. Preclinical evaluation of a transcatheter aortic valve replacement system for patients with rheumatic heart disease. *EuroIntervention.* 2019;15:E975–E982.
25. Babaliaros V, Cribier A. The expansion of transcatheter technology to treat aortic insufficiency: everything old becomes new again. *JACC Cardiovasc Interv.* 2014;7:1175–1176.
26. Seiffert M, et al. Initial German experience with transapical implantation of a second-generation transcatheter heart valve for the treatment of aortic regurgitation. *JACC Cardiovasc Interv.* 2014;7:1168–1174.
27. Schlingloff F, Schäfer U, Frerker C, Schmoeckel M, Bader R. Transcatheter aortic valve implantation of a second-generation valve for pure aortic regurgitation: procedural outcome, haemodynamic data and follow-up. *Interact Cardiovasc Thorac Surg.* 2014;19:388–393.
28. Zhu D, Chen W, Peng L, Guo Y. Valve sizing for pure aortic regurgitation during transcatheter aortic valve replacement: deformation dynamic of the aortic annulus in different valve pathology may be different. *JACC Cardiovasc Interv.* 2015;8:372–373.
29. Mentias A, et al. Transcatheter versus surgical aortic valve replacement in patients with rheumatic aortic stenosis. *J Am Coll Cardiol.* 2021;77:1703–1713.
30. Abramowitz Y, et al. Severe aortic stenosis with low aortic valve calcification: characteristics and outcome following transcatheter aortic valve implantation. *Eur Heart J Cardiovasc Imag.* 2017;18:639–647.
31. Xiong TY, et al. Transcatheter aortic valve replacement in patients with non-calcific aortic stenosis. *EuroIntervention.* 2018;13:e1756–e1763.
32. Farhan S, et al. Impact of moderate or severe left ventricular outflow tract calcification on clinical outcomes of patients with severe aortic stenosis undergoing transcatheter aortic valve implantation with self- and balloon-expandable valves: a post hoc analysis from the SOLVE-TAVI trial. *EuroIntervention.* 2022;18:759–768.
33. Okuno T, et al. Impact of left ventricular outflow tract calcification on procedural outcomes after transcatheter aortic valve replacement. *JACC Cardiovasc Interv.* 2020;13:1789–1799.
34. Pasic M. Annular rupture during TAVR: despite it, Dr. Alain Cribier should receive a Nobel Prize. *JACC Cardiovasc Interv.* 2020;13:1800–1802.
35. Herrmann HC, et al. Rationale and design of the SMall Annuli randomized to Evolut or SAPIEN trial (SMART trial). *Am Heart J.* 2022;243:92–102.
36. Tang GHL, et al. Impact of aortic root anatomy and geometry on paravalvular leak in transcatheter aortic valve replacement with extremely large Annuli using the Edwards SAPIEN 3 valve. *JACC Cardiovasc Interv.* 2018;11:1377–1387.
37. Sharma SK, et al. First-in-human evaluation of a novel balloon-expandable transcatheter heart valve in patients with severe symptomatic native aortic stenosis: the MyVal-1 study. *EuroIntervention.* 2020;16:421–429.
38. Vincent F, et al. Transcatheter aortic valve replacement in bicuspid aortic valve stenosis. *Circulation.* 2021;143:1043–1061.
39. Elbadawi A, et al. Temporal trends and outcomes of transcatheter versus surgical aortic valve replacement for bicuspid aortic valve stenosis. *JACC Cardiovasc Interv.* 2019;12:1811–1822.

40. Makkar RR, et al. Association between transcatheter aortic valve replacement for bicuspid vs tricuspid aortic stenosis and mortality or stroke among patients at low surgical risk. *JAMA*. 2021;326:1034–1044.
41. Forrest JK, et al. Transcatheter aortic valve replacement in bicuspid versus tricuspid aortic valves from the STS/ACC TVT registry. *JACC Cardiovasc Interv*. 2020;13:1749–1759.
42. Makkar RR, et al. Association between transcatheter aortic valve replacement for bicuspid vs tricuspid aortic stenosis and mortality or stroke. *JAMA*. 2019;321:2193–2202.
43. Halim SA, et al. Outcomes of transcatheter aortic valve replacement in patients with bicuspid aortic valve disease: a report from the society of thoracic surgeons/American college of cardiology transcatheter valve therapy registry. *Circulation*. 2020;141:1071–1079.
44. Yoon SH, et al. Bicuspid aortic valve morphology and outcomes after transcatheter aortic valve replacement. *J Am Coll Cardiol*. 2020;76:1018–1030.
45. Medranda GA, et al. The impact of aortic angulation on contemporary transcatheter aortic valve replacement outcomes. *JACC Cardiovasc Interv*. 2021;14:1209–1215.
46. Dvir D, et al. Transcatheter aortic valve implantation in failed bioprosthetic surgical valves. *JAMA*. 2014;312:162–170.
47. Tuzcu EM, et al. Transcatheter aortic valve replacement of failed surgically implanted bioprostheses: the STS/ACC registry. *J Am Coll Cardiol*. 2018;72:370–382.
48. Webb JG, et al. Transcatheter aortic valve implantation within degenerated aortic surgical bioprostheses: PARTNER 2 valve-in-valve registry. *J Am Coll Cardiol*. 2017;69:2253–2262.
49. Deeb GM, et al. 1-year results in patients undergoing transcatheter aortic valve replacement with failed surgical bioprostheses. *JACC Cardiovasc Interv*. 2017;10:1034–1044.
50. Mahmoud AN, et al. Systematic review and meta-analysis of valve-in-valve transcatheter aortic valve replacement in patients with failed bioprosthetic aortic valves. *EuroIntervention*. 2020;16:539–548.
51. Webb JG, Dvir D. Transcatheter aortic valve replacement for bioprosthetic aortic valve failure. *Circulation*. 2013;127:2542–2550.
52. Dvir D, et al. Coronary obstruction in transcatheter aortic valve-in-valve implantation: preprocedural evaluation, device selection, protection, and treatment. *Circ Cardiovasc Interv*. 2015;8.
53. Blanke P, et al. Computed tomography imaging in the context of transcatheter aortic valve implantation (TAVI)/Transcatheter aortic valve replacement (TAVR): an expert consensus document of the society of cardiovascular computed tomography. *JACC Cardiovasc Imaging*. 2019;12:1–24.
54. Allen KB, et al. Bioprosthetic valve fracture to facilitate transcatheter valve-in-valve implantation. *Ann Thorac Surg*. 2017;104:1501–1508.
55. Chhatriwalla AK, et al. Bioprosthetic valve fracture improves the hemodynamic results of valve-in-valve transcatheter aortic valve replacement. *Circ Cardiovasc Interv*. 2017;10.
56. Rocha RV, et al. Surgical enlargement of the aortic root does not increase the operative risk of aortic valve replacement. *Circulation*. 2018;137:1585–1594.
57. Vahanian A, et al. 2021 ESC/EACTS Guidelines for the management of valvular heart disease. *Eur Heart J*. 2022;43:561–632.
58. Otto CM, et al. 2020 ACC/AHA guideline for the management of patients with valvular heart disease: a report of the American college of cardiology/American heart association joint committee on clinical practice guidelines. *Circulation*. 2021;143:E72–E227.
59. Spaziano M, et al. Transcatheter aortic valve implantation versus redo surgery for failing surgical aortic bioprostheses: a multicentre propensity score analysis. *EuroIntervention*. 2017;13:1149–1156.
60. Neupane S, et al. Meta-analysis of transcatheter valve-in-valve implantation versus redo aortic valve surgery for bioprosthetic aortic valve dysfunction. *Am J Cardiol*. 2018;121:1593–1600.
61. Webb JG, et al. 3-year outcomes after valve-in-valve transcatheter aortic valve replacement for degenerated bioprostheses: the PARTNER 2 registry. *J Am Coll Cardiol*. 2019;73:2647–2655.
62. Palmerini T, et al. Coronary protection to prevent coronary obstruction during TAVR: a multicenter international registry. *JACC Cardiovasc Interv*. 2020;13:739–747.
63. Mercanti F, et al. Chimney stenting for coronary occlusion during TAVR: insights from the chimney registry. *JACC Cardiovasc Interv*. 2020;13:751–761.
64. Li H, et al. Delayed coronary artery occlusion after transcatheter aortic valve replacement and chimney stenting: a case report. *BMC Cardiovasc Disord*. 2021;21.
65. Kandzari DE, Yadav PK. Coronary occlusion during TAVR: if you think about it, protect it. *JACC Cardiovasc Interv*. 2020;13:762–764.
66. Khan JM, et al. BASILICA trial: one-year outcomes of transcatheter electrosurgical leaflet laceration to prevent TAVR coronary obstruction. *Circ Cardiovasc Interv*. 2021;14:E010238.
67. Study Details. *The ShortCut^{TM} Study Protocol*. ClinicalTrials.gov; 2024. https://www.clinicaltrials.gov/study/NCT04952909.
68. Alperi A, et al. Transcatheter versus surgical aortic valve replacement in patients with complex coronary artery disease. *JACC Cardiovasc Interv*. 2021;14:2490–2499.
69. Makkar RR, et al. Five-year outcomes of transcatheter or surgical aortic-valve replacement. *N Engl J Med*. 2020;382:799–809.
70. Okuno T, et al. Five-year outcomes of mild paravalvular regurgitation after transcatheter aortic valve implantation. *EuroIntervention*. 2022;18:33–42.
71. Pibarot P, et al. Association of paravalvular regurgitation with 1-year outcomes after transcatheter aortic valve replacement with the SAPIEN 3 valve. *JAMA Cardiol*. 2017;2:1208–1216.
72. Athappan G, et al. Incidence, predictors, and outcomes of aortic regurgitation after transcatheter aortic valve replacement: meta-analysis and systematic review of literature. *J Am Coll Cardiol*. 2013;61:1585–1595.
73. Faroux L, et al. Clinical impact of conduction disturbances in transcatheter aortic valve replacement recipients: a systematic review and meta-analysis. *Eur Heart J*. 2020;41:2771–2781.
74. Shahim B, et al. Postoperative atrial fibrillation or flutter following transcatheter or surgical aortic valve replacement: PARTNER 3 trial. *JACC Cardiovasc Interv*. 2021;14:1565–1574.
75. Forrest JK, et al. 4-year outcomes of patients with aortic stenosis in the evolut low risk trial. *J Am Coll Cardiol*. 2023;82:2163–2165.
76. Thyregod HGH, et al. Transcatheter or surgical aortic valve implantation: 10-year outcomes of the NOTION trial. *Eur Heart J*. 2024. https://doi.org/10.1093/EURHEARTJ/EHAE043.
77. Glaser N, Jackson V, Holzmann MJ, Franco-Cereceda A, Sartipy U. Aortic valve replacement with mechanical vs. biological prostheses in patients aged 50-69 years. *Eur Heart J*. 2016;37:2658–2667.

78. Goldstone AB, et al. Mechanical or biologic prostheses for aortic-valve and mitral-valve replacement. *N Engl J Med.* 2017;377:1847–1857.
79. Chiang YP, et al. Survival and long-term outcomes following bioprosthetic vs mechanical aortic valve replacement in patients aged 50 to 69 years. *JAMA.* 2014;312:1323–1329.
80. Mazine A, et al. Ross procedure vs mechanical aortic valve replacement in adults: a systematic review and meta-analysis. *JAMA Cardiol.* 2018;3:978–987.
81. Lee G, et al. ESC/EACTS vs. ACC/AHA guidelines for the management of severe aortic stenosis. *Eur Heart J.* 2023;44:796–812.
82. Jolobe O. Recognition of the distinction between primary and secondary mitral regurgitation is also important. *QJM: Int J Med.* 2016;109.
83. Feldman T, et al. Percutaneous mitral repair with the MitraClip system. Safety and midterm durability in the initial EVEREST (endovascular valve edge-to-edge REpair study) cohort. *J Am Coll Cardiol.* 2009;54:686–694.
84. Feldman T, et al. Percutaneous repair or surgery for mitral regurgitation. *N Engl J Med.* 2011;364:1395–1406.
85. Feldman T, et al. Randomized comparison of percutaneous repair and surgery for mitral regurgitation: 5-year results of EVEREST II. *J Am Coll Cardiol.* 2015;66:2844–2854.
86. Boekstegers P, et al. Percutaneous interventional mitral regurgitation treatment using the mitra-clip system. *Clin Res Cardiol.* 2014;103:85–96.
87. Glower DD, et al. Percutaneous mitral valve repair for mitral regurgitation in high-risk patients: results of the EVEREST II study. *J Am Coll Cardiol.* 2014;64:172–181.
88. Lim DS, et al. Improved functional status and quality of life in prohibitive surgical risk patients with degenerative mitral regurgitation after transcatheter mitral valve repair. *J Am Coll Cardiol.* 2014;64:182–192.
89. Chiarito M, et al. Outcome after percutaneous edge-to-edge mitral repair for functional and degenerative mitral regurgitation: a systematic review and meta-analysis. *Heart.* 2018;104:306–312.
90. Praz F, et al. Compassionate use of the PASCAL transcatheter mitral valve repair system for patients with severe mitral regurgitation: a multicentre, prospective, observational, first-in-man study. *Lancet.* 2017;390:773–780.
91. Webb JG, et al. 1-year outcomes for transcatheter repair in patients with mitral regurgitation from the CLASP study. *JACC Cardiovasc Interv.* 2020;13:2344–2357.
92. Lim DS, et al. Randomized comparison of transcatheter edge-to-edge repair for degenerative mitral regurgitation in prohibitive surgical risk patients. *JACC Cardiovasc Interv.* 2022;15:2523–2536.
93. Zahr F, et al. 1-year outcomes from the CLASP IID randomized trial for degenerative mitral regurgitation. *Cardiovasc Interv.* 2023;16:2803–2816.
94. Makkar RR, et al. Transcatheter mitral valve repair for degenerative mitral regurgitation. *JAMA.* 2023;329:1778–1788.
95. Majmundar M, et al. Transcatheter versus surgical mitral valve repair in patients with mitral regurgitation. *Eur J Cardio Thorac Surg.* 2024;65.
96. Deharo P, et al. Mitral transcatheter edge to edge repair versus isolated mitral surgery for severe mitral regurgitation: a French nationwide study. *Eur Heart J.* 2024. https://doi.org/10.1093/EURHEARTJ/EHAE046.
97. McCarthy PM, et al. Percutaneous mitraclip device or surgical mitral valve repair in patients with primary mitral regurgitation who are candidates for surgery: design and rationale of the REPAIR MR trial. *J Am Heart Assoc.* 2023;12:27504.
98. Colli A, et al. An early European experience with transapical off-pump mitral valve repair with NeoChord implantation. *Eur J Cardio Thorac Surg.* 2018;54:460–466.
99. Gerosa G, et al. Transapical off-pump echo-guided mitral valve repair with neochordae implantation mid-term outcomes. *Ann Cardiothorac Surg.* 2021;10:131–140.
100. Budra M, et al. Midterm results of transventricular mitral valve repair: single-center experience. *J Thorac Cardiovasc Surg.* 2022;164:1820–1828.
101. Study Details. *Randomized Trial of the Neochord DS1000 System Versus Open Surgical Repair.* ClinicalTrials.gov; 2019. https://clinicaltrials.gov/study/NCT02803957#publications.
102. Avner A, Borenstein N, Orlov B, Raanani E. Innovalve transcatheter mitral valve replacement system: short-term pre-clinical study in pigs. *Appl Eng Sci.* 2023;15:100144.
103. Feldman T, Fernandes E, Levisay JP. Transcatheter mitral valve repair/replacement for primary mitral regurgitation. *Ann Cardiothorac Surg.* 2018;7:75563–75763.
104. Scott EJ, Rotar EP, Charles EJ, Lim DS, Ailawadi G. Surgical versus transcatheter mitral valve replacement in functional mitral valve regurgitation. *Ann Cardiothorac Surg.* 2021;10:75.
105. Chehab O, et al. Secondary mitral regurgitation: pathophysiology, proportionality and prognosis. *Heart.* 2020;106:716–723.
106. Krawczyk-Ożóg A, et al. Anatomy of the mitral subvalvular apparatus. *J Thorac Cardiovasc Surg.* 2018;155:2002–2010.
107. Mearns BM. Valvular disease: mitral valve repair and replacement compared. *Nat Rev Cardiol.* 2014;11:4.
108. Acker MA, et al. Mitral-valve repair versus replacement for severe ischemic mitral regurgitation. *N Engl J Med.* 2014;370:23–32.
109. Kron IL, et al. 2016 update to the American association for thoracic surgery consensus guidelines: ischemic mitral valve regurgitation. *J Thorac Cardiovasc Surg.* 2017;153:1076–1079.
110. Goldstein D, et al. Two-year outcomes of surgical treatment of severe ischemic mitral regurgitation. *N Engl J Med.* 2016;374:344–353.
111. Kron IL, et al. Predicting recurrent mitral regurgitation after mitral valve repair for severe ischemic mitral regurgitation. *J Thorac Cardiovasc Surg.* 2015;149:752–761.e1.
112. De Bonis M, et al. Mitral replacement or repair for functional mitral regurgitation in dilated and ischemic cardiomyopathy: is it really the same? *Ann Thorac Surg.* 2012;94:44–51.
113. Stone GW, et al. Transcatheter mitral-valve repair in patients with heart failure. *N Engl J Med.* 2018;379:2307–2318.
114. Obadia J-F, et al. Percutaneous repair or medical treatment for secondary mitral regurgitation. *N Engl J Med.* 2018;379:2297–2306.
115. Pibarot P, Delgado V, Bax JJ. MITRA-FR vs. COAPT: lessons from two trials with diametrically opposed results. *Eur Heart J Cardiovasc Imag.* 2019;20:620.
116. Sorajja P, Ukaigwe AC. Edge-to-edge repair: past challenge, current case selection and future advances. *Ann Cardiothorac Surg.* 2021;10:43.
117. Del Val D, et al. Early experience with transcatheter mitral valve replacement: a systematic review. *J Am Heart Assoc.* 2019;8.

118. Sorajja P, et al. Initial feasibility study of a new transcatheter mitral prosthesis: the first 100 patients. *J Am Coll Cardiol.* 2019;73:1250–1260.
119. Muller DWM, et al. 2-Year outcomes of transcatheter mitral valve replacement in patients with severe symptomatic mitral regurgitation. *J Am Coll Cardiol.* 2021;78:1847–1859.
120. Gössl M, et al. Early outcomes of transcatheter mitral valve replacement with the tendyne system in severe mitral annular calcification. *EuroIntervention.* 2022;17:1523–1531.
121. Ahmed A, Aziz TAA, AlAsaad MMR, Majthoob M, Toema A. Transcatheter mitral valve implantation with tendyne system ten years since the first in-human implant a systematic review. *J Cardiothorac Surg.* 2023;18.
122. Sorajja P, Bapat V. Early experience with the Intrepid system for transcatheter mitral valve replacement. *Ann Cardiothorac Surg.* 2018;7:792.
123. Bapat V, et al. Early experience with new transcatheter mitral valve replacement. *J Am Coll Cardiol.* 2018;71:12–21.
124. Study Details. *Clinical Trial to Evaluate the Safety and Effectiveness of Using the Tendyne Transcatheter Mitral Valve System for the Treatment of Symptomatic Mitral Regurgitation.* ClinicalTrials.gov; 2024. https://clinicaltrials.gov/study/NCT03433274.
125. Study Details. *Transcatheter Mitral Valve Replacement With the Medtronic IntrepidTM TMVR System in Patients With Severe Symptomatic Mitral Regurgitation.* ClinicalTrials.gov; 2024. https://clinicaltrials.gov/study/NCT03242642.
126. Schueler R, Werner N, Nickenig G, Hammerstingl C. Catheter-based complete 'Alfieri-Stich' via interventional annuloplasty and edge-to-edge repair for degenerative mitral regurgitation. *Eur Heart J.* 2016;37:2201.
127. Von Bardeleben RS, et al. First in human transcatheter COMBO mitral valve repair with direct ring annuloplasty and neochord leaflet implantation to treat degenerative mitral regurgitation: feasibility of the simultaneous toolbox concept guided by 3D echo and computed tomography fusion imaging. *Eur Heart J.* 2018;39:1314–1315.
128. Maisano F, et al. Cardioband, a transcatheter surgical-like direct mitral valve annuloplasty system: early results of the feasibility trial. *Eur Heart J.* 2016;37:817–825.
129. Topilsky Y, et al. Burden of tricuspid regurgitation in patients diagnosed in the community setting. *JACC Cardiovasc Imag.* 2019;12:433–442.
130. Taramasso M, et al. Transcatheter versus medical treatment of patients with symptomatic severe tricuspid regurgitation. *J Am Coll Cardiol.* 2019;74:2998–3008.
131. Dreyfus J, et al. Isolated tricuspid valve surgery: impact of aetiology and clinical presentation on outcomes. *Eur Heart J.* 2020;41:4304–4317.
132. Study Details. *TRILUMINATE Pivotal Trial.* ClinicalTrials.gov; 2024. https://clinicaltrials.gov/study/NCT03904147.
133. Sorajja P, et al. Transcatheter repair for patients with tricuspid regurgitation. *N Engl J Med.* 2023;388:1833–1842.
134. *Cutting-Edge Structural Interventions | Transcatheter Tricuspid Valve Replacement: Useful or Not?* American College of Cardiology; 2022. https://www.acc.org/Latest-in-Cardiology/Articles/2022/10/01/01/42/Cutting-Edge-Structural-Interventions-Transcatheter-Tricuspid-Valve-Replacement-Useful-or-Not#.
135. Webb JG, et al. Transcatheter tricuspid valve replacement with the EVOQUE system: 1-year outcomes of a multicenter, first-in-human experience. *JACC Cardiovasc Interv.* 2022;15:481–491.
136. Kodali S, et al. Transfemoral tricuspid valve replacement in patients with tricuspid regurgitation: TRISCEND study 30-day results. *JACC Cardiovasc Interv.* 2022;15:471–480.
137. Study Details. *2019-06 TRISCEND Study.* ClinicalTrials.gov; 2024. https://clinicaltrials.gov/study/NCT04221490.
138. Kodali S, et al. Transfemoral tricuspid valve replacement and one-year outcomes: the TRISCEND study. *Eur Heart J.* 2023;44:4862–4873.
139. Sun Z, et al. Twelve-month outcomes of the LuX-Valve for transcatheter treatment of severe tricuspid regurgitation. *EuroIntervention.* 2021;17:818–826.
140. Study Details. *A Study to Evaluate the Safety and Performance of LuX-Valve Plus System for Tricuspid Replacement.* ClinicalTrials.gov; 2024. https://clinicaltrials.gov/study/NCT05436028.
141. Hahn RT, et al. Early single-site experience with transcatheter tricuspid valve replacement. *JACC Cardiovasc Imag.* 2019;12:416–429.
142. Study Details. *AHEAD: European Feasibility Study of the Cardiovalve Transfemoral Mitral Valve System.* ClinicalTrials.gov; 2024. https://clinicaltrials.gov/study/NCT03339115.
143. Study Details. *TTVR Early Feasibility Study.* ClinicalTrials.gov; 2024. https://clinicaltrials.gov/study/NCT04433065.
144. Blusztein DI, Hahn RT, Godoy Rivas C, George I, Kodali SK. Transcatheter tricuspid valve replacement with novel self-expanding valve: secure fixation in insecure anatomy. *Case Rep.* 2023;12.
145. Lauten A, et al. Interventional treatment of severe tricuspid regurgitation: early clinical experience in a multicenter, observational, first-in-man study. *Circ Cardiovasc Interv.* 2018;11.
146. Dreger H, et al. Treatment of severe TRIcuspid regurgitation in patients with advanced heart failure with CAval vein implantation of the Edwards Sapien XT VALve (TRICAVAL): a randomised controlled trial. *EuroIntervention.* 2020;15:1506–1513.
147. Estévez-Loureiro R, et al. 6-Month outcomes of the TricValve system in patients with tricuspid regurgitation: the TRICUS EURO study. *JACC Cardiovasc Interv.* 2022;15:1366–1377.
148. Chandran K, et al. First in-man experience with TricValve transcatheter bicaval valve system in left ventricular assist device heartmate II patient for high-risk tricuspid regurgitation. *Circ Heart Fail.* 2023;16:E010027.
149. Wild MG, et al. Early clinical experience with the TRICENTO bicaval valved stent for treatment of symptomatic severe tricuspid regurgitation: a multicenter registry. *Circ Cardiovasc Interv.* 2022;15:E011302.
150. Coisne A, et al. ACC/AHA and ESC/EACTS guidelines for the management of valvular heart diseases: JACC guideline comparison. *J Am Coll Cardiol.* 2023;82:721–734.
151. Batchelor WB, et al. The multidisciplinary heart team in cardiovascular medicine: current role and future challenges. *JACC (J Am Coll Cardiol): Adv.* 2023;2.

Chapter 8

Drug discovery

Sami El-Qadi
Galderma Laboratories, L.P., Dallas, TX, United States

Key points

- The process of drug discovery involves five mains steps: target identification, target validation, lead discovery, lead optimization, and drug development. This process may take up to 15 years and over 2 billion dollars per novel drug.
- The success rates of clinical drug development are close to 10%−15% with most drug candidates failing due to poor target validation, target optimization, and low therapeutic efficacy.
- Clinical trials are split into three different trials. Phase 1 clinical trials are primarily used to establish new drug safety and dose range, Phase 2 clinical trials determine the effectiveness of the new drug on the disease, and Phase 3 clinical trials are conducted at multiple centers with several hundred or thousand patients to continue generation of data of drug safety and effectiveness.

Introductory section

Drug discovery is a lengthy and arduous process demanding significant time and financial resources often exceeding 15 years and $2 billion in costs to bring a new drug to market. Typically originating in academic settings through identification of disease pathways or proteins that the inhibition or activation of leads to a therapeutic effect. Following identification, there is frequently, although not always, funding from pharmaceutical industries to pinpoint molecules or biologics that could potentially offer a therapeutic solution for the targeted disease. After intensive testing and optimization of the development candidate, the chosen molecule will progress into preclinical testing in laboratory animals to evaluate toxicity and safety, followed by human testing through clinical trials and eventually commercialization of the drug. The aim of this chapter is to provide a detailed process of drug discovery from the initial set of target identification to the commercialization of the pharmaceutical product.

Target identification

The process of drug discovery is often initiated from the aspect of necessity and lack of effective treatments for a specific disease state. In the case of many disease states, the drug discovery process relies heavily on academia to discover proteins or pathways that when inhibited or activated result in a therapeutic effect. After this discovery, target identification and validation are the first step in this long and costly process. Good target identification is incredibly crucial because it focuses on the key elements of a successful drug, which are efficacy, safety, meeting a need for a disease state or population, and in the case of small molecule drugs, follows Lipinski's Rule of Five (molecular weight less than 500 Da, Log $P \leq 5$, H-Bond donors ≤ 5, H-Bond acceptors ≤ 10). Following these general guidelines is important as it generally improves potency and selectivity and may also improve safety and tolerability.

Successful target identification significantly enhances the probability of developing effective treatments for a targeted disease state. Targets often fall into two distinct categories: established and new. Established targets have a history of scientific validation and a pharmaceutical understanding of their roles in normal physiology and human pathology. On the other hand, new targets are frequently selected for specific diseases from vast libraries of chemicals using high-throughput screening (HTS). HTS involves running hundreds to thousands of assays to assess the likelihood of a small molecule interacting with or influencing the target protein or pathways, HTS can evaluate up to 100,000 samples in a single day. Typical cellular targets for drug discovery are G-protein-coupled receptors (GPCRs) and disease-associated protein kinases. Traditionally, many drugs and other chemicals with biological activity are derived from plants, animals, and bacteria, and those drugs do not follow the same HTS discovery process.

However, this process is protracted, as the "ideal" drug candidate often does not emerge in the initial screenings.

Translational Cardiology. https://doi.org/10.1016/B978-0-323-91790-2.00047-2

When a reaction occurs, the identified molecule is cataloged and progresses to the subsequent stage of drug discovery. Following HTS, promising compounds undergo hit-to-lead optimization and lead optimization phases. During hit-to-lead optimization, identified molecules are refined to enhance their binding affinity (improved activity toward the chosen target) and specificity (reduced for unrelated targets). The lead optimization also improved pharmacokinetic properties such as absorption, distribution, metabolism, and extraction along with safety and effectiveness. This process is usually done by attempting to improve the structure–activity relationships often increasing the molecular weight of the compound through the addition of specific functional groups which in turn may increase safety and tolerability.

Target optimization and preclinical phase

The lead optimization stage typically yields around 250 potential drug candidates deemed feasible for further progression in the drug development pipeline. An exhaustive analysis of each lead candidate's drug profile, emphasizing aspects like potency, selectivity, and favorable drug-like properties, results in a significant reduction to approximately 2–6 top candidates. These potential candidates exhibit crucial characteristics such as minimal exposure to healthy tissues while showing affinity for disease-targeted tissues. They boast high bioavailability, moderate half-life (usually within the range of 4–6 h), and efficient hepatic clearance. Molecules that align with these criteria prove challenging and might not be achievable for every case. Nonetheless, the consensus in the pharmaceutical space indicated that the more optimized a lead candidate is in meeting these parameters, the higher the likelihood of its successful progression through clinical trials and eventual commercialization.

Following the identification of lead candidates, the drug development process progresses to the preclinical phase, where adherence to FDA regulations necessitates adequate animal testing to ensure initial safety for human trials. This crucial step aims to evaluate the potential impacts on various organs, determine suitable administration methods, exposure durations, and frequencies for individuals, and assess any inherent toxicity. Animal testing serves as a pivotal stage to gauge the drug's safety profile before human trials commence. Subsequently, some drug sponsors run an additional relatively new practice known as phase 0 human studies after animal testing. In this phase, a minute dosage (typically 1/100th) of the pharmacologically active agent is administered to evaluate its toxicity in humans at significantly lower doses. The preclinical phase is quite costly and only 5 out of 5000 lead candidates make it to human testing due to issues with toxicity or efficacy.

Clinical studies

Following the rigorous preclinical phases, the drug sponsor proceeds by submitting an Investigational New Drug application (IND) to the FDA, marking the initiation of human testing and the onset of the clinical trials phase in drug discovery. Upon receiving IND approval, Phase 1 clinical trials commence. These initial trials primarily focus on evaluating the treatment's safety profile and typically involve a small number of participants over several months. In Phase I trials, the first cohort receives a very low dose of the treatment, and their responses are closely observed for any potential side effects. If no adverse effects or only minor side effects are observed, subsequent groups of participants receive progressively higher doses. Dose escalation continues until investigators determine the dose that is most likely to be effective while maintaining an acceptable level of side effects. Throughout Phase 1 clinical trials, the investigational drug is assessed for its pharmacokinetic properties (how the body processes the drug), pharmacodynamic effects (how the drug affects the body), and overall safety profile. This phase helps determine the optimal dosage and safety parameters for the unapproved drug and helps guide the subsequent clinical studies.

After the completion of Phase 1 clinical trials, Phase 2 clinical trials which are primarily aimed at assessing the investigational drug's efficacy are initiated. Drawing insights from the Phase 1 findings, Phase 2 trials focus on appropriately dosing the investigational drug and typically involve enrolling approximately 100–300 volunteers diagnosed with the specific disease or condition under investigation. These trials typically extend over several months to up to 2 years, concentrating on evaluating baseline effectiveness and identifying any additional side effects. Phase 2 clinical trials also mark the introduction of randomization and the inclusion of a placebo group. This stage also introduces randomization in which subjects either receive the investigational drug or a placebo. This randomization and setup allow researchers to draw more definitive conclusions about the drug's efficacy by comparing its effects against a placebo. While Phase 2 clinical studies may not be large enough to definitively determine a drug's overall benefit, they serve as pivotal stages to refine research inquiries, establish research methodologies, and design new protocols for Phase 3 investigations.

The final step before marketing and commercialization is the Phase 3 clinical studies. These trials serve as pivotal demonstrations to ascertain whether a product provides a treatment benefit to a specific population. Spanning an average duration of 1–4 years, Phase 3 trials enroll a larger cohort, typically involving a range of 300 to 3000 participants. This large cohort of subjects allows for a more comprehensive assessment of the treatment's effectiveness, safety profile, and the identification of long-term or rare side effects. These trials are conducted across multiple locations, sometimes on a global scale, and are often carried out in community hospitals and doctor's offices to ensure diverse representation within the studied population. After completion of Phase 3 studies, the drug sponsor submits a new drug application to the FDA. The NDA includes comprehensive data and analyses from the clinical trials, aiming to demonstrate the investigational drug's superior efficacy or safety compared to existing treatments. If the FDA determines that the new investigational drug presents significant improvements in efficacy or safety over current treatments, it grants approval, allowing for commercialization and making the drug available for medical use.

Conclusion

The process of drug discovery, from target identification to eventual commercialization, represents a long, complex, and expensive process with the end goal of improving patient care and disease state management. The stages involve the target screening, optimization, and testing of potential drug candidates through preclinical and clinical trials. These trials, spanning from Phase 1 to Phase 3, assess safety, efficacy, and overall effectiveness of investigational drugs. This process is crucial for patient well-being, ensuring that any drug entering the market has undergone stringent evaluation to guarantee its safety and efficacy. Ultimately, drug discovery can significantly contribute to managing disease and enhancing the quality of life for patients worldwide.

Get started

Embarking on the exploration of drug discovery and development involves several key steps to grasp the intricacies of this field and its impact on healthcare. Begin by understanding the basics and familiarize yourself with the fundamental concepts of drug discovery. Dive into understanding how researchers and academics identify potential therapeutic targets. Learn about the different phases of preclinical drug development and the different phases of clinical trials. Understand the steps in place to ensure that patients receive safe and effective treatments while improving quality of life and healthcare outcomes.

Potential pitfalls

It is important to remember that not only does this process take more than 10 years to complete, but also that there are multiple different drug sponsors spending billions of dollars to find a treatment for a specific disease making the space very competitive.

This is not the only way drugs are approved, but it is the most common. Drugs may also receive approval as orphan drugs if they are treating an unmet medical need for a rare condition. Drug sponsors may receive tax credits, grants, and streamlined regulatory pathways.

The vast majority of molecules tested do not progress to clinical trials. Of those that reach Phase 1, 70% move to the next phase. However, only 33% of those that reach Phase 2 progress to Phase 3. Moreover, among those that reach Phase 3, merely 25% receive approval for a New Drug Application (NDA).

The majority of currently approved medications do not follow the HTS process and mainly come from animal, plant, or bacteria samples.

Examples of HTS drug discovery

In a paper published in the American Journal of Translational Research, the use of HTS to discover small-molecule drugs was described. For instance, the discovery of MCL-1 inhibitors resulted from these predictive assays and the outlined drug discovery pathway.

There have been many medications that were discovered using HTS including but not limited to the following: Gefitinib, Tipranavir, Sitagliptin, and Ambrisentan. These medications are being used for different disease states and are incredibly effective in their respective therapeutic areas.

In a paper published in Nature Reviews Drug Discovery, the utilization of HTS by GlaxoSmithKline led to the discovery and eventual commercialization of Promacta/Revolade, a thrombopoietin receptor agonist. This stands as an excellent example of a lead target initially possessing poor physiochemical properties, which were substantially improved and optimized to ultimately yield a marketable drug.

Resources

1. https://www.fda.gov/drugs/science-and-research-drugs/research-tools-and-resources is a valuable resources that provides necessary tools and information on drug discover.

Further reading

1. Hughes JP, Rees S, Kalindjian SB, et al. Principles of early drug discovery. *Br J Pharmacol.* 2011;162(6):1239–1249.
2. Li Q, Kang C. Mechanisms of action for small molecules revealed by structural biology in drug discovery. *Int J Mol Sci.* 2020;21(15):5262.
3. Lipinski CA, Lombardo F, Dominy BW, et al. Experimental and computational approaches to estimate solubility and permeability in drug discovery and development settings. *Adv Drug Deliv Rev.* 2001;46(1–3):3–26.
4. Ma YS, Xin R, Yang XL, et al. Paving the way for small-molecule drug discovery. *Am J Transl Res.* 2021;13(3):853–870.
5. Macarron R, Banks MN, Bojanic D, et al. Impact of high-throughput screening in biomedical research. *Nat Rev Drug Discov.* 2011;10(3):188–195.
6. Sun D, Gao W, Hu H, et al. Why 90% of clinical drug development fails and how to improve it? *Acta Pharm Sin B.* 2022;12(7):3049–3062.
7. U.S. Food and Drug Administration. Clinical Research. FDA.gov. Available from: https://www.fda.gov/. Accessed 26 November 2023.

Chapter 9

Drug testing

João Victor S. Sanders[1], Carolina Ferraz[2], Elisa T.D. Siqueira[1], Isabella S.M. Rabelo[3], Gustavo P. de A. Saraiva[3] and Bruno B. Lima[4]
[1]*Universidade Federal do Ceará, Fortaleza, Ceará, Brazil;* [2]*Universidade Santo Amaro, São Paulo, Brazil;* [3]*Universidade Estadual do Ceará, Fortaleza, Ceará, Brazil;* [4]*Vanderbilt University School of Medicine, Nashville, TN, United States*

Key points

- Drug development is a lengthy, time-consuming, and high-cost process that is divided into preclinical and clinical phases.
- Preclinical phase focuses on providing drug pharmacologic, pharmacokinetic, and toxicologic data using in vivo and in vitro models.
- Clinical phase focuses on drug toxicity and efficacy in human beings.

Introduction

The "bench to bedside" process illustrates how preclinical laboratory findings may be developed into a clinical treatment. This is a challenging task due to its complexity and high rates of failure; more than 90% of drugs investigated that enter human trials fail. Since the drug development process cost may be estimated in millions of dollars, this is a high-risk investment that often fails to succeed. This scenario has urged that novel drug development has to be clinically focused from the early beginning with clinician and scientists' teams. Furthermore, the traditional in vivo and in vitro models have demanded new approaches that focus on providing more predictable clinical results, instead of only providing pharmacologic, pharmacokinetic, and toxicologic data.

Get started

Before being considered a treatment option, a novel drug must undergo extensive testing to prove its efficacy and safety to regulatory institutions. In the United States, drug approval lies within the scope of the FDA. Drug testing is divided into preclinical and clinical trials, the latter uses human beings to access data. Preclinical data comprises a drug's pharmacodynamics, pharmacokinetics, and toxicity.

1. **Pharmacokinetics**
   Pharmacokinetics studies the movement of drugs within the body, encompassing their absorption, distribution, metabolism, and excretion. A drug's pharmacokinetic profile is often described using several parameters, such as half-life, clearance, volume of distribution, bioavailability, and time to peak concentration. These factors are critical in characterizing a drug's behavior in the body, determining dosing regimens, and predicting the responses in different patient populations. Analysis of drug absorption will determine how a drug enters the systemic circulation. Once in the body, a drug's distribution pattern is analyzed. After, drug metabolism evaluates biochemical transformation, which often results in metabolites with varying activity and toxicity. Excretion concludes the route drugs and metabolites exit the body.
2. **Pharmacodynamics**
   Pharmacodynamics (PD) describes the biochemical and physiological effects of drugs, including the mechanisms of action, the relationship between drug concentration and effect, and any side effects. Understanding the PD of drugs allows researchers to:
   - Determine the dose-response relationship, which elucidates the therapeutic window.
   - Examine receptor interactions, signaling pathways, and the duration of drug action to optimize therapeutic effects.
   - Identify the potential for adverse effects and interactions with other medications.
3. **In vitro testing**
   In vitro testing refers to several types of experiments that are conducted outside of a living organism in an artificial environment, such as in test tubes, petri dishes, or other laboratory containers. The focus of in vitro testing is on the drug's direct effects. At this phase,

Translational Cardiology. https://doi.org/10.1016/B978-0-323-91790-2.00077-0

using clinical-grade materials with a molecular structure matching the final product is vital because it meets regulatory requirements and pinpoints Phase I trial dosages. Adherence to the International Council on Harmonization Requirements for Registration of Pharmaceutical for Human Use (ICH) guideline is mandatory and avoids repeating experiments, aiming for global consistency in drug evaluation. It is suggested to assess the drug in various experimental settings to gauge its clinical relevance accurately. Using different models to mimic various biological states gives a better view of the drug's therapeutic potential and safety.

4. **In vivo Testing**
   In vivo testing refers to experimentation using a whole, living organism. It allows confirmation of drug targets and evaluation of biodistribution including difficult-to-access compartments, toxicity, and efficacy. It is important to choose a model that accurately reflects the human environment. Employing various in vivo models can strengthen the justification for pursuing clinical trials and securing funding.
5. **Toxicology, Dosing and Harm**
   Toxicology studies establish safe and appropriate dosing parameters, including the initial dose for Human studies, the rate at which doses can be escalated during trials, and the maximum tolerable dose. These studies are designed to use the same drug formulation and route of administration that are intended for the eventual clinical trials. A primary objective of toxicology studies is to identify the dose at which no adverse events occur. They thoroughly assess the severity and reversibility of acute toxicities and the effects of short-term (acute), medium-term (subacute), and long-term (chronic) drug exposure. Tests are performed on both rodent and non-rodent species to provide a safety profile, and all studies are conducted in accordance with Good Laboratory Practice (GLP) regulations to ensure the quality and reliability of the results.
6. **Good Laboratory Practices (GLP)**
   These are principles made to confirm the quality and dependability of nonclinical lab experiments, relevant to preclinical testing of drugs and chemicals. GLP provides guidelines for conducting studies and managing data, promoting consistent, reliable results.

Major areas of current investigation

- Beside the classic in vivo and in vitro models, there are new methodologies that focus on predicting clinical outcomes. Two examples of these innovative approaches include in *chemico* and *silico* testing. Other areas of investigation are developing improved in vitro and in vivo models to reduce animal use and improve clinical predictability. An innovative example is the in vitro proarrhythmic assay, which combines in vitro and in *silico* assays. This innovative technique consists of testing how drug structure would react with multiple ion channels, to determine its potential cardiovascular proarrhythmic risks.

Budgetary considerations and how to fund this stage

- Drug development is a lengthy and expensive process. Its cost is estimated from millions to billions of dollars, which is often not feasible to be done by a single research group, requiring funding. Investors vary on each stage of development. The government and philanthropic organizations engage in early development and capitalist ventures participate in later stages. Since investing in this sector is very risky, especially during mid-development, the US Department of Health and National Institutes of Health have programs providing financial assistance for various stages. Secure funding to afford the preclinical and Phase I clinical trials is a prudent primary goal.

Potential pitfalls

After properly determining the molecule target, there are some potential pitfalls to have in mind to prevent time consuming and increased costs in drug development.

1. **Proper preparation**
   The first potential problem is underestimating the cost, effort, and time required during the preclinical phase. Drug development sometimes fails early in the process due to a lack of proper preparation.
2. **Pharmacodynamics and pharmacokinetics**
   Understand the new drug properties early in development. This can lead to early molecule structure optimization.
3. **Early focus on clinical translation**
   Preclinical drug testing is guided by clinical translation steps, such as using identical preclinical and clinical drug constructs; conducting preclinical studies with a production process yielding a clinically compatible drug; and developing batch manufacturing processes.
4. **Good laboratory practices regulations**
   Understand that these international regulations prevent mismanagement. Its focus is on procedure standardization, not on scientific validity.

Examples of …

1. **Leqvio's formulation: An in vitro testing**
   - Leqvio is a drug approved by the FDA that has inclisiran as its active ingredient, which is a double-stranded small interfering RNA (siRNA) that acts against PCSK9 mRNA. During the preclinical stage of the drug testing, one of the in vitro experiments altered siRNAs, which have as target the human PCSK9 transcript and the ortholog mRNA sequence in cynomolgus monkeys, to contain different patterns of chemical modifications in the DNA backbone.
2. **Camzyos development: An in vivo experiment**
   - Camzyos is an FDA-approved drug that has mavacamten as its active ingredient, acting as a reversible inhibitor of cardiac myosin. During the in vivo trials of this drug, mice, rats, monkeys, and dogs were evaluated, allowing to evaluation of the pharmacokinetics (PK) across varied species and also, based on the data, to assume a model to predict the human PK and estimate the drug volume and distribution for humans.
3. **Entresto's production: A phase 1 trial**
   - Entresto (LCZ696) is an FDA-approved drug that has as active ingredients sacubitril and valsartan, promoting the inhibition of angiotensin and neprilysin receptors, contributing to patients with heart failure. During the testing phase, this trial evaluated and compared the effects of pharmacodynamics and pharmacokinetics between placebo and 400 mg of LCZ696 in healthy volunteers.

Resources

1. Drug Development Tools Research Grants (U01) Clinical Trial Optional. https://grants.nih.gov/grants/guide/pa-files/PAR-21-178.html.
2. NIH Research Project Grant Program (R01). https://grants.nih.gov/grants/funding/r01.htm.
3. Simoens S., Huys I. R&D costs of new medicines: a landscape analysis. Front Med. 2021;8.
4. Sertkaya A., Wong H.H., Jessup A., Beleche T. Key cost drivers of pharmaceutical clinical trials in the United States. Clin Trials. 2016;13:117e126.

Further reading

1. Barh D, Chaitankar V, Yiannakopoulou EC, et al. In silico models. In: *Animal Biotechnology*. 2014:385–404.
2. Chow SC. Bioavailability and bioequivalence in drug development. *Wiley Interdiscip Rev Comput Stat*. 2014;6(4):304–312.
3. Cronin MTD, Bajot F, Enoch SJ, Madden JC, Roberts DW, Schwöbel J. The in chemico–in silico interface: challenges for integrating experimental and computational chemistry to identify toxicity. *Altern Lab Anim*. November 2009;37(5):513–521.
4. David S, Kim PY. Drug Trials [Updated 2023 June 20]. In: *StatPearls*. Treasure Island (FL): StatPearls Publishing; January 2023. Available from: https://www.ncbi.nlm.nih.gov/books/NBK546595/.
5. Singh K, Sampson JH, Gedeon PC. Drug testing: translating a novel immunotherapeutic from bench to bedside. In: *Translational Surgery*. 2023:43–51.
6. Grillo MP, Erve JCL, Dick R, et al. In vitro and in vivo pharmacokinetic characterization of mavacamten, a first-in-class small molecule allosteric modulator of beta cardiac myosin. *Xenobiotica*. October 1, 2018;49(6):718–733.
7. Hallare J, Gerriets V. *Half Life*. PubMed. Treasure Island (FL): StatPearls Publishing; 2022. Available from: https://www.ncbi.nlm.nih.gov/books/NBK554498/.
8. Kandi V, Vadakedath S. Clinical trials and clinical research: a comprehensive review. *Cureus*. February 16, 2023;15(2). Available from: https://assets.cureus.com/uploads/review_article/pdf/128436/20230317-1378-1d3buy1.pdf.
9. Langenickel TH, Tsubouchi C, Ayalasomayajula S, et al. The effect of LCZ696 (sacubitril/valsartan) on amyloid-β concentrations in cerebrospinal fluid in healthy subjects. *Br J Clin Pharmacol*. March 8, 2016;81(5):878–890.
10. Nambisan P. Laboratory biosafety and good laboratory practices. In: *An Introduction to Ethical, Safety and Intellectual Property Rights Issues in Biotechnology*; 2017:253–271. Available from: https://www.sciencedirect.com/science/article/pii/B9780128092316000119?via%3Dihub.
11. Nishikido T, Ray KK. Non-antibody approaches to proprotein convertase subtilisin kexin 9 inhibition: siRNA, antisense oligonucleotides, adnectins, vaccination, and new attempts at small-molecule inhibitors based on new discoveries. *Front Cardiovasc Med*. January 29, 2019;5.
12. Shen J, Swift B, Mamelok R, Pine S, Sinclair J, Attar M. Design and conduct considerations for first-in-human trials. *Clin Transl Sci*. January 1, 2019;12(1):6–19. Available from: https://www.ncbi.nlm.nih.gov/pmc/articles/PMC6342261/.
13. Sontag JM, Page NP, Saffiott U. *Guidelines for Carcinogen Bioassay in Small Rodents. Technical Report Series*. National Cancer Institute Carcinogenesis; 1976. No 1. February.

Chapter 10

Device testing

Marco A. Morales-Garza[1], Cristiano Cardoso[1], Abdelmotagaly Elgalad[1], Joseph Lamelas[2] and Ahmed Alnajar[3]

[1]Center for Preclinical Surgical & Interventional Research, The Texas Heart Institute, Houston, TX, United States; [2]Department of Cardiac Surgery, The University of Miami Miller School of Medicine, Miami, FL, United States; [3]The Division of Cardiothoracic Surgery, The Department of Surgery, The University of Miami Miller School of Medicine, Miami, FL, United States

Key points

- Medical devices are used to cure, treat, mitigate, or prevent diseases.
- Medical devices must go through a rigorous regulatory process to establish safety and efficacy before marketing to the public.
- Because medical device development is a complex and expensive process, a variety of funding possibilities are available.

Why does it matter?

Medical devices are important tools to cure, treat, mitigate, or prevent diseases in the medical field. However, having a device approved and ready to market can be time consuming and expensive. This chapter provides essential information about the regulatory process and key steps when applying for medical device approval by the US Food and Drug Administration. Additionally, we provide available opportunities to fund the development of a medical device.

Introduction

The explosive growth of medical technology since the mid-1960s has paved the way for fields as disparate as heart transplantation and circulatory support for patients with end-stage heart failure.[1] As the oldest comprehensive consumer protection agency in the United States, the US Food and Drug Administration (FDA) is charged with protecting and promoting the development of human and veterinary drugs, biological products, radiation-emitting products, human and animal food, cosmetics, and medical devices.[2] The FDA defines a medical device as:

An instrument, apparatus, implement, machine, contrivance, implant, in vitro reagent, or other similar or related article, including a part or accessory which is: recognized in the official National Formulary, or the United States Pharmacopeia, or any supplement to them, intended for use in the diagnosis of disease or other conditions, or in the cure, mitigation, treatment, or prevention of disease, in man or other animals, or intended to affect the structure or any function of the body of man or other animals, and which does not achieve its primary intended purposes through chemical action within or on the body of man or other animals and which is not dependent upon being metabolized for the achievement of any of its primary intended purposes.[3]

Given the fast pace of emerging medical technology, few physicians and patients appreciate the rigorous process by which new medical devices are evaluated and approved by the FDA. Although the FDA's device approval processes are as rigorous as its drug approval processes, the controversial and well-publicized problems[4,5] that led to vaccine hesitation during the COVID-19 pandemic[6] highlight the need to enhance our scrutiny of medical devices.

Medical device safety, efficacy, and regulation in the United States

Federal oversight of food and drugs began in 1906 when President Theodore Roosevelt signed the Pure Food and Drugs Act, which established the FDA's precursor organization and prohibited interstate trade in misbranded or adulterated food and drugs. Since then, the FDA's role has been expanded considerably. In the 1960s and 1970s, Congress responded to the public's desire for more oversight over medical devices by passing the Medical Device

Translational Cardiology. https://doi.org/10.1016/B978-0-323-91790-2.00011-3

Amendments to the 1938 Federal Food, Drug, and Cosmetic Act. In 1982, the organizational units that regulated medical devices and radiation-emitting products at the FDA merged to form the Center for Devices and Radiological Health (CDRH). Table 10.1 details individual legislative acts and milestones in the history of medical device legislation in the United States.[2]

Safety and effectiveness

Safety and effectiveness are the two pillars of medical device regulation. A product is considered safe if it can be used safely under the conditions prescribed, recommended, or suggested in its labeling. Effectiveness is determined according to whether a product achieves its intended purpose when used according to the label's directions.[3]

The FDA has established standards that are designed to ensure that all medical devices are safe and effective before they can be marketed in the United States. The FDA regulates safety and effectiveness by requiring premarket approval (PMA) for new devices and postmarket surveillance to monitor the safety of existing devices. The FDA analyzes all available data to establish reasonable assurance that a device is safe and effective; PMA approval is based on a risk/benefit evaluation.[3]

Medical device reporting

Medical device reporting is one of the FDA's postmarket surveillance techniques for monitoring device performance, detecting potential device-related safety concerns, and contributing to risk/benefit analyses of these products. Every year, the FDA receives hundreds of thousands of complaints of suspected medical device-related deaths, serious injuries, and malfunctions. Manufacturers, facilities that use medical devices (e.g., hospitals, nursing homes), and importers are obliged to submit to the FDA specific types of reports on adverse occurrences and product problems involving medical devices. In addition, the FDA encourages healthcare professionals, patients, caregivers, and consumers to voluntarily report major adverse events, as well as use errors, product quality issues, and treatment failures.[3]

The FDA maintains a record of these reports of adverse events and other issues with medical devices and notifies health professionals and the public as necessary to promote proper device use and ensure patient health and safety. Congress passed the Food and Drug Administration Safety and Innovation Act on July 9, 2012. Under Section 608(c) of this Act, the FDA is required to publish annually the number and type of medical devices reclassified in the previous calendar year.[2]

Medical device classification

The FDA has classified more than 1700 generic device types into 19 medical specialties known as *panels*. Classification is defined by a device's intended application and its indications for use. Each panel is then assigned to one of three regulatory classes (Class I, II, or III),[7] depending on their hazards and the regulatory controls required to reasonably ensure their safety and effectiveness (Table 10.2). Class I devices are the safest for the patient or user, whereas Class III devices are those that present the highest risk. A product's classification and other descriptive data can be found in the FDA's Product Classification Database.[8]

If the panel to which a device belongs is known, its classification can be determined by browsing through the list of classified devices in the US Code of Federal Regulations (CFR); otherwise, the Product Classification database's keyword directory can be searched.[8] See Table 10.3.[9] The CFR starts each classification panel with a list of the devices that are classified in that panel. A seven-digit number and a three-letter product code are assigned to each classified device—for example, 21 CFR 880.2920 and FLK for a clinical mercury thermometer. Similarly, the Product Classification database has several entries for various types of thermometers under the keyword "thermometer." The three-letter product code is also used as the classification number on the Medical Device Listing form.[7]

The Food, Drug, and Cosmetic Act of 1938 (FD&C Act) provides minimum regulations (general controls) for all three medical device classes (Table 10.2).[7] Premarket notification and good manufacturing practice standards are not required for approximately three-fourths of Class I devices. These exemptions are detailed in the 21 CFR classification regulations and the Medical Device Exemptions document.[9]

Beyond classification based on overall intended use, a subset of intended uses arises when a more particular indication is added to the device's labeling. For example, the purpose of a scalpel is to cut tissue; a more specific use could be "for generating incisions in the cornea." The device's instructions for use are written in the labeling and may also be delivered verbally during the product's sale. The notion of intended use is explained in the 510(k) program Evaluating Substantial Equivalence.[7,9]

How to choose the proper classification

To determine the classification of a new medical device and whether any exemptions apply, search the Product Classification database for a part of the device name, or browse the device panel (medical specialty) listing to which the

TABLE 10.1 History of the FDA: legislative acts and events.

1906	Pure Food and Drugs Act (also called the Federal Food and Drugs Act)	Established the FDA's authority to regulate and oversee medical items.
1938	Federal Food, Drug, and Cosmetic Act (FD&C Act)	Extended the reach of the 1906 Pure Food and Drugs Act to include cosmetics and therapeutic medical devices; authorized factory inspections.
1944	Public Health Service Act	Provided for laboratory certification and increased scrutiny of biologics.
1968	Radiation Control for Health and Safety Act	Created performance criteria for reducing exposure to radiation from magnetic fields and electronic products (e.g., diagnostic X-ray and magnetic resonance imaging machines; microwave, ultrasound, ultraviolet, and diathermy devices; lasers).
1970	Cooper Committee recommendations	Determined that medical device-specific legislation was needed, given that devices and pharmaceuticals present different challenges; instituted the concept of risk-based categorization for medical devices.
1976	Medical Device Amendments to the FD&C Act	Established a risk-based, three-level classification system for all medical devices; created new premarket and postmarket regulatory procedures for new or significantly modified medical devices; established a regulatory avenue for new investigational medical devices to be researched in patients; authorized the FDA to ban devices.
1990	Safe Medical Devices Act	Enhanced postmarket surveillance: required facilities using medical devices to report adverse events; required manufacturers to monitor permanently implanted devices if device failure could result in permanent damage or death; empowered the FDA to issue device recalls and levy civil fines for FD&C Act violations; altered procedures for establishing, amending, or revoking performance criteria; established humanitarian use programs for patients with rare diseases.
1992	Mammography Quality Standards Act	Required accreditation and government certification of mammography facilities and annual federal or state inspections.
1997	Food and Drug Administration Modernization Act	Created "least burdensome" premarket review provisions and widened access to investigational devices; allowed new low-to-moderate risk devices to be categorized as Class I or II rather than Class III.
2002	Medical Device User Fee and Modernization Act	Permitted the FDA to charge user fees for specified medical device premarket submissions; set lower premarket approval fees for qualifying small firms; set FDA performance goals for some premarket submission decisions and new regulatory requirements for reprocessed devices; required electronic registration for medical device companies; established the Office of Combination Products.
2007	Food and Drug Administration Amendments Act	Enhanced premarket review times; required unique device identifiers and electronic registration and listing for medical devices.
2012	Food and Drug Administration Safety and Innovation Act	Improved premarket review times and expanded use of "least burdensome" criteria in premarket evaluations; established industry-shared outcomes goals; permitted FDA to collaborate with other governments to unify regulatory standards; allowed novel, low-to-moderate risk devices to be classified into Class I or II (rather than Class III) without first submitting a 510(k) application; altered the standards for IDE disapproval.
2016	21st Century Cures Act	Facilitated patient access to new medical equipment: enacted expedited review procedure for breakthrough devices, extended use of "least burdensome" concepts, and streamlined exemption processes in premarket evaluations; increased the annual population estimate required for humanitarian use; allowed the use of central institutional review boards for IDE and HDE activities; regulated medical software and digital health goods.
2017	Food and Drug Administration Reauthorization Act	Reauthorized the medical device user fee program; enhanced premarket review timelines; invested in strategic programs such as the National Evaluation System for Health Technology and patient involvement; improved inspection processes for device manufacturers; separated accessories from parent device classification; required the FDA to investigate how real-world evidence can help with postmarket surveillance.

FDA, US Food and Drug Administration; *HDE*, humanitarian device exemption; *IDE*, investigational device exemption.
US Food and Drug Administration. A History of Medical Device Regulation & Oversight in the United States; 2019. https://www.fda.gov/medical-devices/overview-device-regulation/history-medical-device-regulation-oversight-united-states.

TABLE 10.2 Medical device classifications.

	Risk	Required control	Examples
Class I	Low risk for illness or injury	General control	Gauze, toothbrushes, bandages, manual stethoscopes
Class II	Moderate risk for illness or injury	General control + Specific regulatory control	Sutures, needles, catheters, contact lenses
Class III	Significant risk for illness or injury	General control + Specific regulatory control + Premarket approval	Pacemaker, cardiac valves, breast implants, ventilators, defibrillators

US Food and Drug Administration.

TABLE 10.3 Locations of device classifications and specific regulations.

Medical specialty		
Panel	Specialty	Regulation citation (21 CFR)
73	Anesthesiology	Part 868
74	Cardiovascular	Part 870
75	Chemistry	Part 862
76	Dental	Part 872
77	Ear, nose, and throat	Part 874
78	Gastroenterology and urology	Part 876
79	General and plastic surgery	Part 878
80	General hospital	Part 880
81	Hematology	Part 864
82	Immunology	Part 866
83	Microbiology	Part 866
84	Neurology	Part 882
85	Obstetrical and gynecological	Part 884
86	Ophthalmic	Part 886
87	Orthopedic	Part 888
88	Pathology	Part 864
89	Physical medicine	Part 890
90	Radiology	Part 892
91	Toxicology	Part 862

CFR, US Code of Federal Regulations.
US Food and Drug Administration. Device Classification Panels; 2018. https://www.fda.gov/medical-devices/classify-your-medical-device/device-classification-panels.

device belongs to locate the relevant regulation.[10] Even though medical device classification seems to be straightforward, doctors and medical devices companies may have doubts about which classification their device should receive. The FDA has guidance to help investigators classify the medical device correctly: Medical device companies can submit a 513(g) request to obtain information about the FDA's views regarding the classification of a new or upgraded device. The 513(g) request guarantees formal device determination or categorization, avoiding the waste of time and money.[10]

Safety and efficacy data for medical device approval

Generally, Class I and II devices undergo a less-restrictive regulatory procedure. The approval process for Classes I and II is focused on registration, manufacturing, and labeling and often does not require extensive preclinical or clinical data. Class III devices with only minor differences from already-approved, so-called predicate devices may be reclassified as Class I or II and, if so, are subject to less-stringent testing requirements than most Class III devices without predicates. Because Class III devices may confer more significant risks to patients, they are subject to more rigorous processes and will require the collection of safety data before PMA or premarketing notification (PMN) is granted.[7]

Pathways to medical device approval

Once a medical device is classified, the investigator must request medical device approval from the FDA before testing it on human patients. There are three different pathways for obtaining FDA approval and permission to market the medical device:

(1) Premarket approval: A PMA application requests FDA approval for a new medical device that has no equivalent in the market. To facilitate the PMA submission, the FDA can host a presubmission meeting to review the file and provide guidance to manufacturers before submission.

(2) Premarketing notification: A PMN (510(k)) application is used for medical devices substantially equivalent to an existing device that is already approved for marketing.

(3) Humanitarian device exemption (HDE): An HDE classification is requested for medical devices intended to treat or diagnose a disease or condition that affects no more than 8000 individuals in the United States. Because of the rarity of the disease or condition, HDE approval requires data for device safety but not for efficacy. Additionally, local institutional review board (IRB) approval is mandatory before submission for HDE classification.[7,11]

The appropriate premarketing pathway is determined according to the medical device's class designation (Table 10.4). A 510(k) submission is required for marketing

TABLE 10.4 Medical device approval pathways.

	Premarket approval	Premarketing notification	Humanitarian device exemption
Classification	Class III	Class I or II	N/A
The purpose of the device	Completely new devices, nothing similar on the market	Devices that are substantially similar to a predicate device	Intended to treat or diagnose a disease or condition with no more than 8000 individuals in the United States
Need safety data before application submission	Yes	May not require clinical data	Yes
Need efficacy data before application submission	Yes	May not require clinical data	No
Need local IRB approval before application submission	Yes	No	Yes
Time to application answer	180 days	30–90 days	45 days
Example	Pulmonary valve prosthesis delivered percutaneously	Cardiac pacemaker	Stent graft for coronary perforation

IRB, institutional review board.
US Food and Drug Administration. Overview of Medical Device Classification and Reclassification; 2018. https://www.fda.gov/about-fda/cdrh-transparency/overview-medical-device-classification-and-reclassification; US Food and Drug Administration. Device Classification Panels; 2018. https://www.fda.gov/medical-devices/classify-your-medical-device/device-classification-panels.

if the device is designated as Class I or II and is not exempt from 510(k) requirements. Device exemption limitations are outlined in 21 CFR xxx.9, where xxx refers to Parts 862–892. A PMA is necessary for Class III devices unless the device is an amendments device (i.e., on the market before the Medical Device Amendments to the FD&C Act was passed in 1976, or substantially equivalent to such a device) and PMAs were not originally requested. In such an instance, a 510(k) submission is the best and easiest way to get the product to market.[11]

Investigational device exemption

If no therapy is available to treat patients, the FDA allows an investigational and unapproved device to be classified with an investigational device exemption (IDE) (Fig. 10.1). However, this pathway is restricted to emergency use, emergency research, and compassionate use. In the case of an ongoing clinical trial, the sponsor may request an IDE for treatment use when data suggest the device is effective and the investigator wants to expand the number of enrollees. In the case of a concluded trial, an IDE might be granted to allow enrollees to continue to receive treatment while the approval process is underway. Requirements for IDE devices include an investigational plan approved by an IRB, informed consent from all patients, labeling stating that the device is for investigational use only, monitoring of the study, and required records and reports. Also, if the device involves significant risk, the IDE must be approved by the FDA.[12,13]

Early feasibility studies

The FDA frequently requires that pivotal research be preceded by an early feasibility study to establish basic safety and efficacy prospects. An early feasibility study will include a small number of patients, usually 10 or fewer, and will collect comprehensive data on device use and performance. The FDA's review is primarily concerned with safety and whether the data support that the device's potential benefit and value outweigh any risk.[12]

Pivotal studies

A pivotal study provides the primary clinical support for a marketing application in most cases. Pivotal studies were developed to show "reasonable confidence of safety and effectiveness," based on statistically driven endpoints and sample size. The FDA review process is substantially more complex in a pivotal study than in an early feasibility study.[12]

Budget considerations

Because medical devices are regulated by the FDA, medical device manufacturers are required to pay fees to the federal government to cover the cost of regulatory review (Table 10.5). These fees support the regulatory activities of the FDA and enhance its efficiency, thereby shortening the time required to bring safe and effective medical devices to the market in the United States. The fees for medical devices are established in legislation and updated every year.

The FDA fee is one of the most important factors that should be considered when budgeting for medical devices, because the fee is based on a percentage of the device's cost, which can range from 1% to 20%. In addition, FDA approval costs more for high-risk devices, and fees will vary depending on company size, ranging from $2500 to $1 million per product application. It is also essential to consider the costs of production, manufacturing, distribution, and even human capital and consultants, because these costs can vary widely depending on the type of device being made.[14]

Small businesses with an approved Small Business Determination and gross receipts or sales of less than $30 million per year can reduce some of these fees by submitting a Request for Small Business Certification. The

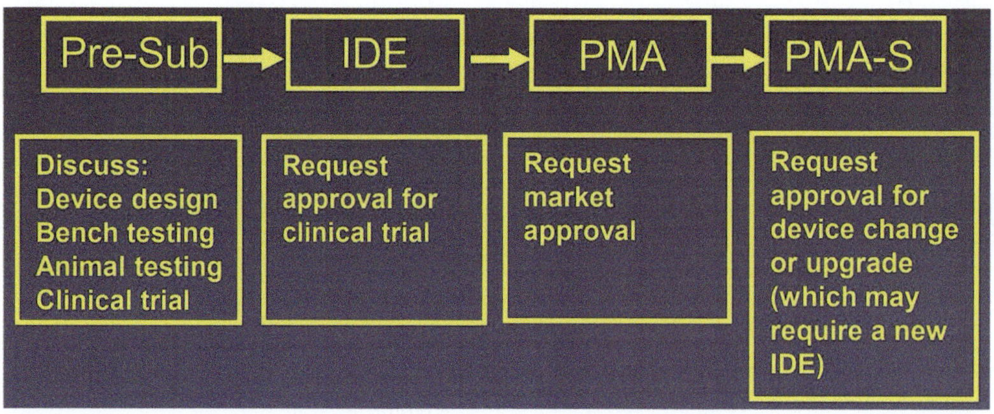

FIGURE 10.1 Approval pathway from presubmission to premarket approval.[12] *US Food and Drug Administration. Faris O.* Clinical Trials for Medical Devices*:* FDA and the IDE Process*; 2018. https://www.fda.gov/media/87603/download.*

TABLE 10.5 FDA fees for medical device submission (as of 2022).

Application type	Standard fee	Small business fee[a]
510(k)[b]	$12,745	$3186
513(g)	$5061	$2530
PMA, PDP, PMR, BLA[c]	$374,858	$93,714
De novo classification request	$112,457	$28,114
Panel-track supplement	$281,143	$70,286
180-day supplement	$56,229	$14,057
Real-time supplement	$26,240	$6560
BLA efficacy supplement	$374,858	$93,714
30-day notice	$5998	$2999
Annual fee for periodic reporting on a Class III device (PMA, PDP, PMR)[d,e]	$13,120	$3280

BLA, biologics license application; *CDRH*, Center for Devices and Radiological Health; *FDA*, US Food and Drug Administration; *PDP*, product development protocol; *PMA*, premarket approval; *PMR*, premarket report.
[a]*Small business fee: For businesses that have been designated as small businesses by the CDRH.*
[b]*510(k) fees: The user charge applies to all types of 510(k) submissions (traditional, abbreviated, and special). No user charge is assessed for 510(k) applications submitted to the FDA on behalf of an FDA-accredited third-party reviewer.*
[c]*Small business waivers: On their first PMA, PDP, PMR, or BLA, small firms with an approved Small Business Determination and gross receipts or sales of $30 million per year or less are eligible to have the charge waived.*
[d]*Submitting a small business certification request: The CDRH's Small Business Program assesses whether a company qualifies and is recognized as a small business, which would make it eligible for a charge reduction on some CDRH submissions that require a user fee.*
[e]*Reduced medical device user fees: The Small Business Determination Program has more information on how to submit a certification request. Guidance: https://www.fda.gov/media/93354/download.*
US Food and Drug Administration. Medical Device User Fee Amendments (MDUFA); 2022. https://www.fda.gov/industry/fda-user-fee-programs/medical-device-user-fee-amendments-mdufa.

CDRH's Small Business Program assesses whether a business qualifies and can be recognized as a small business, making it eligible for lower prices for certain types of CDRH submissions that require user fees.[15]

A survey of more than 200 medical technology companies attempted to quantify the costs of bringing a medical device to market. Participants included members of the Medical Device and Manufacturers Association and the National Venture Capital Association networks. Price Waterhouse Coopers LLP conducted an independent review and verification of the data. The overall cost to bring a 510(k) device from concept to clearance was approximately $31 million, $24 million of which was for FDA-related expenses; for a Class III device, these costs were $94 million and $75 million, respectively.[14]

Medical device manufacturers pay FDA user fees when they register their establishments and list their devices with the agency, when they submit an application or notification to sell a new medical device in the United States, and for certain other types of submissions. These fees are not refunded if the FDA rejects an application or if the applicant withdraws their application.

Funding opportunities for device approval

Federal agencies support small enterprises through the Small Business Innovation Research (SBIR) and Small Business Technology Transfer (STTR) programs, the US National Institutes of Health (NIH), the Centers for Disease Control, the FDA, and the Administration for Community Living, all of which are part of the US Department of Health and Human Services. Scientists and entrepreneurs working to deliver their discoveries to patients are funded by 24 NIH institutes and centers.[16] The NIH Matchmaker tool can help locate the best match in a given research area. Entering publication abstracts, study bios, or other scientific content into the NIH Matchmaker will list similar studies and find institutes and centers that have previously sponsored research in that area.[17]

Small business entrepreneurs can apply for SBIR and STTR grants to receive nondilutive funding for early-stage research and development; applications are accepted three times a year,[16,18] although SBIR contract solicitations are issued only once a year and are based on specific research requirements at NIH's numerous institutes and centers.[16,18] The SBIR and STTR award bids are used to submit ideas proposed by researchers. These funding opportunity announcements do not focus on a single topic, but they link to selected areas of interest for each awarding component. These solicitations are where most small company applications to the NIH are submitted.[18]

Conclusion

Medical science and technology development is a complex process. The development cycle of a medical device is seldom a straightforward, linear progression from basic research to the patient's bedside. The process involved in bringing the medical device to market includes research and development, satisfying regulations and requirements, marketing, manufacturing, and legal considerations. This process is likely to prompt multiple iterations of the medical device as concepts evolve; redesigns typically change the initial idea or pivot into a better one, which brings new challenges. The FDA is an innovative organization that must adapt constantly to the new technologies and the medical applications for these technologies.

Much of the cutting-edge research in academia is funded by industry. Although this is a beneficial influence, journals and academia should be transparent about the involvement of private financing in the research and development process.

Researchers also require more efficient procedures for creating reimbursement rules for medical devices. The primary question here is: What approach should be used to assess the costs, risks and benefits, and values of new medical devices, which are constantly evolving?

Get started

- When developing a medical device, keep in mind the essential concepts outlined in this review before delving into the details of FDA regulations.
- Ensure that your device is classified properly and then prepare your plan and paperwork accordingly.
- Start writing funding proposals for your project.

Pitfalls to avoid

- The FDA can help you classify your device and avoid costly errors: Beginning with the wrong classification can cost your company a significant amount of money and could lead to a failed go-to-market strategy.
- A Regulatory Affairs Certification Professional can assist companies and startups applying for FDA approval and serve as a consultant for such companies.
- Because FDA regulatory pathways are expensive, it is advisable to speak with the FDA in the pre-IDE phase to understand the pathway for your device; meetings can generally be scheduled within 90–120 days of your request.

Acknowledgments

Jeanie F. Woodruff, BS, ELS, of the Scientific Publications & Grants Department at The Texas Heart Institute, contributed to the editing of the manuscript.

References

1. Alnajar A, Frazier OH. The state of artificial heart therapy. *Tex Heart Inst J*. 2019;46(1):77–79. https://doi.org/10.14503/THIJ-18-6756.
2. US Food and Drug Administration. *A History of Medical Device Regulation & Oversight in the United States*; 2019, 10.03.22 https://www.fda.gov/medical-devices/overview-device-regulation/history-medical-device-regulation-oversight-united-states.
3. US Food and Drug Administration. *Medical Device Overview*; 2018, 10.03.22 https://www.fda.gov/industry/regulated-products/medical-device-overview.
4. Berenson A, Harris G, Meier B, Pollack A. Despite warnings, drug giant took long path to Vioxx recall. *N Y Times Web*. 2004; November 14;A1:A32.
5. Krause PR, Gruber MF. Emergency use authorization of covid vaccines - safety and efficacy follow-up considerations. *N Engl J Med*. 2020;383(19):e107. https://doi.org/10.1056/NEJMp2031373.
6. Rosenbaum L. Escaping Catch-22 - overcoming Covid vaccine hesitancy. *N Engl J Med*. 2021;384(14):1367–1371. https://doi.org/10.1056/NEJMms2101220.
7. US Food and Drug Administration. *Overview of Medical Device Classification and Reclassification*; 2018, 10.03.22 https://www.fda.gov/about-fda/cdrh-transparency/overview-medical-device-classification-and-reclassification.
8. US Food and Drug Administration. *Product Classification Database*; 2022, 10.03.22 https://www.accessdata.fda.gov/scripts/cdrh/cfdocs/cfPCD/classification.cfm.
9. US Food and Drug Administration. *Device Classification Panels*; 2018, 10.03.22 https://www.fda.gov/medical-devices/classify-your-medical-device/device-classification-panels.
10. US Food and Drug Administration. *FDA and Industry Procedures for Section 513(g) Requests for Information Under the Federal Food, Drug, and Cosmetic Act*; 2021, 10.03.22 https://www.fda.gov/regulatory-information/search-fda-guidance-documents/fda-and-industry-procedures-section-513g-requests-information-under-federal-food-drug-and-cosmetic.
11. US Food and Drug Administration. *Premarket Approval (PMA)*; 2022, 10.03.22 https://www.accessdata.fda.gov/scripts/cdrh/cfdocs/cfpma/pma.cfm?id=P200015S017.
12. Faris O. *Clinical Trials for Medical Devices: FDA and the IDE Process*; 2018, 10.03.22 https://www.fda.gov/media/87603/download.

13. US Food and Drug Administration. *Investigational Device Exemption (IDE)*; 2019, 10.03.22 https://www.fda.gov/medical-devices/premarket-submissions-selecting-and-preparing-correct-submission/investigational-device-exemption-ide.
14. StarFish Medical. *How Much Does it Cost to Develop a Medical Device?*; 2020, 10.03.22 https://starfishmedical.com/assets/StarFish-Whitepaper-Cost-to-Develop-Medical-Devices-July-2020.pdf.
15. US Food and Drug Administration. *Medical Device User Fee Amendments (MDUFA)*; 2022, 10.03.22 https://www.fda.gov/industry/fda-user-fee-programs/medical-device-user-fee-amendments-mdufa.
16. National Institutes of Health. *Small Business Funding: Find Funding*; 2022, 10.03.22 https://seed.nih.gov/small-business-funding/find-funding.
17. Mohan K. Summary and conclusions. In: Hanna KE, Manning FJ, Bouxsein P, Pope A, eds. *Innovation and Invention in Medical Devices: Workshop Summary*. Washington DC: National Academies Press; 2001.
18. National Science Foundation. *America's Seed Fund: Medical Devices (MD)*; 2022, 10.03.22 https://seedfund.nsf.gov/topics/medical-devices/.

Chapter 11

Diagnostic discovery

Chloe Carr[1] and Bruno B. Lima[2]

[1]*Division of Cardiovascular Medicine, Vanderbilt University Medical Center, Nashville, TN, United States;* [2]*Division of Cardiovascular Medicine, Department of Internal Medicine, Vanderbilt University Medical Center, Nashville, TN, United States*

Key points

- **Parameter selection in cardiac tests:** Choosing parameters that best describe heart diseases to positively impact patient outcomes is crucial.
- **Choosing test type and data:** Initial steps involve selecting the proper test and identifying critical data for effective clinical management.
- **Clinical utility evaluation:** Assess the test's utility based on reproducibility, accuracy, and feasibility across four phases.
- **Balancing statistical and clinical significance**: Researchers should analyze statistical results comprehensively to ensure the test's real-world effectiveness.

Introductory section/why it matters

Diagnostic discovery is an essential component of healthcare innovation, focusing on identifying and resolving unmet clinical needs through the development of cutting-edge diagnostic tools and methodologies. This chapter provides an in-depth exploration of the diagnostic discovery process, emphasizing the significance of creating diagnostics that are not only technologically advanced but also clinically relevant and effective in improving patient outcomes.

In the realm of cardiology, unmet clinical needs are glaringly evident and underscore the urgency for diagnostic discovery. One prominent example is the challenge of detecting coronary artery disease (CAD) at its earliest stages. Current diagnostic methods often miss subtle signs of CAD, leaving patients at risk of heart attacks. These missed opportunities for early intervention directly impact patient outcomes, emphasizing the pressing need for innovative diagnostic tools to provide more accurate and timely detection.

Another critical area within cardiology is the management of atrial fibrillation (AF). Many cases of AF remain undiagnosed until patients experience complications, such as strokes. The unmet need here lies in developing more accessible and efficient diagnostic solutions that can detect AF earlier in its course, allowing for preventative measures to reduce the risk of stroke. These examples illustrate the tangible consequences of unmet clinical needs in cardiology and underscore the pivotal role that diagnostic discovery plays in improving patient care and outcomes. Addressing these needs not only enhances the quality of cardiology practice but also contributes to reducing the burden of cardiovascular diseases on healthcare systems and society.

Get started

In the field of cardiology, there is a critical decision to be made regarding the path forward in diagnostic technology: whether to embrace cutting-edge innovations or to refine and enhance existing methods. Choosing between these approaches has significant implications for the field and patient care.

Embracing innovation allows for the development of new, noninvasive diagnostic methods like coronary computed tomography angiography (CCTA), which can revolutionize the way cardiac conditions are identified and managed. Imagine having the ability to assess coronary arteries without invasive procedures, providing unparalleled precision and comfort for patients. These innovations have the potential to reshape the entire landscape of cardiovascular diagnostics, pushing the boundaries of what is achievable in the field.

On the other hand, refinement involves leveraging the wisdom and data accumulated through established diagnostic tools to enhance their accuracy and effectiveness. Consider electrocardiography (ECG), a tried-and-true method. By integrating digital technology and continuous improvement, we can now detect irregular heartbeats with greater accuracy, all while preserving the reliability of the familiar tools we have relied upon for years.

The choice between these two paths depends on the specific clinical need and each approach's potential impact,

Translational Cardiology. https://doi.org/10.1016/B978-0-323-91790-2.00074-5

feasibility, and resource requirements. It is a decision that requires careful evaluation to ensure that the trajectory of diagnostic discovery in cardiology continues to evolve, ultimately leading to better outcomes for patients and the field as a whole. By choosing wisely, we can create a future where innovation and refinement work in harmony, driving us toward higher standards in diagnosing and treating heart disease.

At the end of this chapter, the reader should:

1. Comprehend the role of diagnostic discovery in Cardiology.
2. Examine the areas of investigation in discovering diagnostic tools and devices, such as serum biomarkers, multimodality imaging techniques, and invasive methods.
3. Learn how to create and utilize multidisciplinary frameworks for discovering diagnostic tools.

Potential pitfalls

1. Sole reliance on a single diagnostic tool for cardiovascular condition diagnosis can lead to inaccuracies; hence, a combination of modalities supported by the clinical presentation is essential.
2. Researchers must be mindful of the cost and accessibility of diagnostic tests for patients from diverse socioeconomic backgrounds when designing diagnostic tools.
3. Instead of adopting a broad approach to establishing best practice guidelines, research should employ structured frameworks that incorporate a standardized methodology.
4. Staying up-to-date with current FDA regulatory guidelines for various diagnostic technologies is crucial.

Prototyping

Prototyping is an essential aspect of diagnostic tool and methodology development, involving the creation of functional models that progress from low-fidelity conceptualizations to high-fidelity prototypes resembling the final product.

The initial stages involve the development of low-fidelity prototypes, such as paper sketches or computer simulations, which serve as a canvas for exploring diverse design concepts and gathering stakeholder feedback. As concepts mature, high-fidelity prototypes are created, which may include functional software prototypes, 3D-printed models, or the selective deployment of new biomarkers in pilot studies.

The iterative process of refinement and optimization involves evaluating each prototype through technical assessments, feasibility studies, and user feedback, which helps identify areas for improvement and ensures that the final product aligns with the dynamic landscape of modern healthcare. This commitment to perpetual learning and improvement propels diagnostic discovery toward dependable and impactful solutions.

Disciplines and common lab methods/techniques

In the field of cardiology, an expansive array of laboratory methods and techniques serves as the backbone of diagnostic discovery:

- Cardiovascular magnetic resonance (CMR): provides detailed, noninvasive images of the heart's structure and function, offering invaluable insights into cardiac health.
- Coronary computed tomography angiography (CCTA): enables the visualization of coronary arteries, aiding in the diagnosis of coronary artery disease with reduced invasiveness.
- Nuclear cardiology: leverages radiotracers to assess myocardial perfusion, helping identify areas of reduced blood flow indicative of coronary artery disease.
- Echocardiography: uses ultrasound to visualize the heart's structure and function, providing real-time data for diagnostic purposes.
- Interventional cardiology: uses X-ray to directly visualize the coronaries, measure intracardiac pressures, and perform cardiac biopsy, enabling a microscopic examination of cardiac tissue.

Major areas of current investigation

- Early detection and risk assessment: Newer studies are exploring the use of genetic markers and biomarkers to identify individuals at high risk of developing cardiovascular diseases, such as heart failure and coronary artery disease.
- Noninvasive assessment: Speckle tracking echocardiography is being refined to provide more accurate assessments of myocardial function. Advanced imaging techniques, CMR and CCTA, are also being developed to improve early-stage disease detection.
- Improved diagnosis of specific cardiac conditions: Researchers are working on developing more accurate and accessible methods for diagnosing cardiac amyloidosis, particularly transthyretin cardiac amyloidosis (ATTR-CA), has been significantly improved by the use of technetium-99m pyrophosphate (PYP) scintigraphy, a noninvasive method for detecting amyloid deposits. PYP scans have shown high sensitivity and specificity for the diagnosis of ATTR-CA, making it a standard of care for identifying this condition.

- Artificial intelligence and machine learning: AI algorithms are being utilized to analyze and interpret cardiac images, improving the accuracy and efficiency of diagnosis and identifying subtle patterns missed by human eyes. AI-powered risk prediction models are also being developed to predict individual risk of developing cardiovascular diseases or complications, enabling personalized preventive strategies.

Examples of ...

New pan-amyloid detector

In a recent study, Wall et al. introduced a novel amyloid-reactive peptide labeled with iodine-124 (124I) to detect cardiac amyloidosis and systemic amyloidosis. This phase 1/2 study assessed the safety, biodistribution, and sensitivity of 124I-evuzamitide (the labeled peptide) in patients with systemic amyloidosis, asymptomatic transthyretin sequence variant carriers, and healthy volunteers. The results showed that 124I-evuzamitide could effectively detect amyloid deposits in the heart and other abdominal-thoracic organs, with high sensitivity in patients with light chain-associated amyloidosis. The radiotracer's uptake correlated significantly with serum N-terminal pro-B-type natriuretic peptide measurements in patients with this form of amyloidosis. Importantly, 124I-evuzamitide was well-tolerated, with minimal adverse events reported. This novel amyloid-binding radiotracer holds promise for the sensitive detection of cardiac amyloidosis, representing a significant advancement in the field of amyloidosis diagnosis.

Transient endothelial dysfunction induced by mental stress

Lima et al. investigated a new biomarker to assess the effects of mental stress on the heart. The study found a graded positive association between transient endothelial dysfunction caused by mental stress and major adverse cardiovascular events. Patients who experienced a decrease in flow-mediated vasodilation after a mental stress task had a higher incidence of cardiovascular events. This result suggests that impairment in endothelium-dependent relaxation induced by mental stress could serve as an essential risk biomarker for individuals with stable coronary artery disease, highlighting the potential role of psychological stress in affecting cardiovascular outcomes in this patient population.

Computed tomography-based fractional flow reserve

In 2019, Tang et al. compared the diagnostic performance of computed tomography-based fractional flow reserve (CT-FFR) to traditional invasive FFR for detecting lesion-specific ischemia in a multicenter study with meta-analysis. This study used a prototype for the computation of FFR derived from CCTA based on deep machine learning, compared to the traditional FFR, and analyzed the combined diagnostic performance of several different CT-FFR algorithms from other published studies. Overall, this study found that CT-FFR with machine learning had high diagnostic accuracy (sensitivity and specificity of 0.84 and 0.94) in determining lesion-specific ischemia compared to CCTA alone. Secondly, they found no significant difference in the diagnostic performance of the CT-FFR algorithms tested. This multicenter study shows how noninvasive CT-FFR testing can accurately determine clinically significant ischemia in individual vessels without the need for invasive FFR testing.

Resources

- Government grants: Explore grants from organizations like the National Heart, Lung, and Blood Institute (NHLBI). Collaborate with experienced researchers and institutions.
- Nonprofit organizations: Partner with organizations like the American Heart Association (AHA) for funding and support.
- Pilot grants: Avail yourself of pilot grants from academic institutions and medical centers for proof-of-concept studies.
- Industry collaborations: Partner with pharmaceutical companies, medical device manufacturers, and healthcare organizations with a vested interest in cardiology.
- Clinical trials and commercialization: Incorporate clinical trials into your development process and develop a clear strategy for commercialization.
- Public engagement and crowdfunding: Engage with the public and cardiac patient communities to raise awareness and funding for cardiology diagnostic research.
- International funding: Explore international funding opportunities for cardiovascular diagnostics research.
- Continual networking: Attend cardiology conferences, seminars, and networking events to connect with potential funders, collaborators, and investors.

Further reading

1. Monaghan PJ, Lord SJ, St John A, et al. Biomarker development targeting unmet clinical needs. *Clin Chim Acta.* 2016;460:211−219.
2. Sharma K, Kass DA. Heart failure with preserved ejection fraction: mechanisms, clinical features, and therapies. *Circ Res.* 2014;115(1): 79−96.
3. Ioannidis JP. Personalized genetic prediction: too limited, too expensive, or too soon? *Ann Intern Med.* 2009;150(2):139−141.
4. Abdelrahman KM, Chen MY, Dey AK, et al. Coronary computed tomography angiography: from clinical uses to emerging technologies - JACC state-of-the-art review. *J Am Coll Cardiol.* 2020;76(10):1226−1243.
5. Wall JS, Martin EB, Lands R, et al. Cardiac amyloid detection by PET/CT imaging of iodine (124I) evuzamitide (124I-p5+14): a phase 1/2 study. *JACC Cardiovasc Imag.* 2023;16(11):1433−1448.
6. Lima BB, Hammadah M, Kim JH, et al. Association of transient endothelial dysfunction induced by mental stress with major adverse cardiovascular events in men and women with coronary artery disease. *JAMA Cardiol.* 2019;4(10):988−996.
7. Tang CX, Wang YN, Zhou F, et al. Diagnostic performance of fractional flow reserve derived from coronary CT angiography for detection of lesion-specific ischemia: a multi-center study and meta-analysis. *Eur J Radiol.* 2019;116:90−97.

Chapter 12

Diagnostic testing

John P. Mitchell[1], Joseph Nowatzke[1] and Bruno B. Lima[2]

[1]Division of Cardiovascular Medicine, Vanderbilt University Medical Center, Nashville, TN, United States; [2]Division of Cardiovascular Medicine, Department of Internal Medicine, Vanderbilt University Medical Center, Nashville, TN, United States

Key points

- Diagnostic tests are tools clinicians use to evaluate for the presence or absence of a disease.
- The accuracy of diagnostic tests is measured via the test's sensitivity and specificity
- The FDA governs the approval of new diagnostic tests in the United States via a scientific review process to establish the tests' safety and efficacy.

Introduction

Diagnostic testing is a cornerstone of adequate medical care clinicians use to determine whether a patient has a particular disease. These tests vary widely, including imaging, lab tests, biopsies, and physical examinations. Ideally, a diagnostic test would deliver absolute certainty about the presence of a disease, but all tests have limitations and must be interpreted with care.

Cardiologists are challenged to decide which patients require further intervention based on test results. They must also select the most appropriate testing method based on each patient's clinical situation. As medical technology advances, with new imaging techniques and laboratory tests developing rapidly, cardiologists must be adept at customizing these diagnostics to individual patient needs. They also need to understand the intricacies of different diseases and how these nuances influence test interpretation. For instance, the choice of an imaging test may differ for a patient with suspected acute plaque rupture compared to one with new, mild angina symptoms.

This chapter will explore the key metrics used to evaluate diagnostic tests, illustrated with practical examples. It will delve into the role of the FDA in assessing and approving new diagnostic methods and conclude with an analysis of the financial aspects involved in funding advancements in diagnostic testing.

Evaluation of diagnostic tests

The foundational metrics for evaluating the accuracy of a diagnostic test are sensitivity and specificity. Sensitivity quantifies the likelihood that individuals with a disease will obtain a positive result from the test. Conversely, specificity calculates the probability that disease-free individuals will receive a negative result. Two other critical metrics are the positive predictive value (PPV) and negative predictive value (NPV), which measure the likelihood that positive and negative test results accurately reflect the disease status (Fig. 12.1).

A diagnostic test with high sensitivity is pivotal for exclusion diagnosis—a negative result confers a high level of assurance that a patient is disease-free. In contrast, a test that demonstrates high specificity is crucial for confirmatory diagnosis, as a positive result strongly indicates the presence of the disease. Many diagnostic tests yield quantitative data, and the thresholds for positive or negative results are established considering their impact on sensitivity and specificity. Additionally, the clinician's assessment of a patient's pretest probability of disease influences test selection. For instance, a patient with a low likelihood of acute coronary syndrome might undergo noninvasive evaluation such as EKG and cardiac biomarkers analysis. In contrast, one with a high pretest likelihood could be expedited to invasive cardiac catheterization.

Beyond the performance parameters, the ideal diagnostic test would be cost-effective, deliver rapid results, and pose minimal risk to the patient. A diagnostic test boasting perfect sensitivity and specificity has limited clinical utility if it is prohibitively costly, yields results too slowly, or entails significant risk. Therefore, the practical application of a diagnostic test must balance accuracy with safety, efficiency, and cost considerations for optimal patient care.

Translational Cardiology. https://doi.org/10.1016/B978-0-323-91790-2.00102-7

		Disease		Predictive Value	
		(+)	(-)		
Test	(+)	A True Positive (TP)	B False Positive (FP)	Positive Predictive Value (PPV) $\frac{TP}{TP + FP} = \frac{A}{A + B}$	Total Positive Results (A + B)
	(-)	C False Negative (FN)	D True Negative (TN)	Negative Predictive Value (NPV) $\frac{TN}{FN + TN} = \frac{D}{C + D}$	Total Negative Results (C + D)
Sensitivity & Specificity		Sensitivity $\frac{TP}{TP + FN} = \frac{A}{A + C}$	Specificity $\frac{TN}{FP + TN} = \frac{D}{B + D}$		
		All diseased patients (A + C)	All non-diseased patients (B + D)		

FIGURE 12.1 Table for calculating sensitivity, specificity, PPV, and NPV of diagnostic tests.

Real-world examples

Coronary artery calcium score

Coronary artery calcium (CAC) scoring has emerged as a significant prognostic tool for predicting coronary artery disease (CAD) based on extensive evidence from large, long-term, population-based observational studies. These include:

- The multiethnic study of atherosclerosis (MESA) was a prospective, multicenter study involving 6814 individuals aged 45–84 who underwent CAC scoring at baseline. Participants were monitored for major cardiac events, and the study confirmed that CAC is a robust predictor of CAD across four racial and ethnic groups in the United States.
- The Early Identification of Subclinical Atherosclerosis by Noninvasive Imaging Research (EISNER) trial, a prospective randomized study of 2137 volunteers, evaluated the additional benefits of CAC scoring over traditional risk factor modification. The findings indicated that CAC scoring led to better control of CAD risk factors without increasing downstream medical testing.
- The CAC consortium, a multicenter cohort study of 66,636 patients who underwent clinically indicated CAC scoring from 1991 to 2010, demonstrated that CAC predicts adverse outcomes consistently across all race/ethnicity subgroups over an average follow-up of 12 years.

These findings are integrated into the clinical context where, for adults aged 40–75, routine assessment of cardiovascular risk factors and calculation of 10-year ASCVD risk is recommended. While CAC scoring is highly informative, particularly for asymptomatic patients with intermediate ASCVD risk, it does not detect noncalcified plaques that may pose a significant risk of rupture and ACS.

The Agatston score, part of the CAC scoring system, is used to quantify the calcium burden of coronary arteries and is calculated by summing the lesion scores with a calcium density over 130 HU. Despite its correlation with plaque burden, it is essential to note that CAC scoring is not strongly associated with the presence and severity of angiographic stenosis, and a high CAC score does not necessitate an automatic angiogram in asymptomatic individuals.

In clinical practice, a CAC score of 0 can justify delaying statin therapy, barring other risk factors. In contrast, a score of 1–99 may suggest initiating such treatment, especially in those over 55 years. A score of 100 or above or at or above the 75th percentile for the patient's demographic strongly indicates the need for statin therapy.

The evidence from MESA, EISNER, and the CAC consortium reinforces using CAC scoring in risk stratification and guiding individualized patient care, considering the statistical risk and the patient's overall clinical profile.

Natriuretic peptides

Diagnosing acute decompensated heart failure (ADHF) presents a significant clinical challenge. It typically requires a combination of thorough patient history, meticulous physical examination, and the clinician's acumen to assess volume status. Despite the clinician's expertise, traditional physical examination findings like lower extremity edema, jugular venous distension, and rales have been notoriously unreliable for estimating pulmonary capillary wedge pressure. Indeed, studies have demonstrated that the sensitivity of these physical signs can be as low as 58%, although they are highly specific when present.

To improve diagnostic accuracy, natriuretic peptides such as B-type natriuretic peptide (BNP) and N-terminal pro-b-type natriuretic peptide (NT-proBNP) have been introduced as diagnostic markers. Ventricular myocytes release these peptides in response to increased wall stress, often due to raised intravascular volume, a hallmark of heart failure. The assays for these biomarkers have high sensitivities, approximately 95%, and a high negative predictive value of around 94% when lower threshold levels are used. However, their specificity is more variable, ranging from 26% to 90%, and the confidence intervals around these estimates are typically wide, reflecting variability across populations and clinical settings.

The clinical utility of natriuretic peptide testing lies primarily in its ability to rule out ADHF, particularly when lower cut-off values are applied. The high sensitivity and negative predictive value mean that a low level of natriuretic peptides makes the diagnosis of ADHF unlikely. However, the wide range of specificity suggests that elevated levels of natriuretic peptides are less definitive and should be used in collaboration to confirm a diagnosis of ADHF. Instead, high levels should prompt further investigation, possibly including echocardiography or other imaging modalities, to understand the context of the elevated biomarker levels.

Various studies and guidelines have supported the integration of natriuretic peptide testing into clinical practice. For instance, the 2017 ACC/AHA/HFSA heart failure guidelines recommend the use of natriuretic peptide levels for the diagnosis and prognosis of heart failure and to guide therapy in chronic management. These biomarkers' role in acute and chronic settings underlines their significance in modern cardiology, providing clinicians with an objective measure to complement their clinical judgment in diagnosing and managing heart failure.

Potential pitfalls

1. Utilizing the coronary artery calcium (CAC) scoring system for intermediate-to-high-risk symptomatic patients, such as those with suspected angina, is inappropriate due to its low specificity for obstructive coronary artery disease (CAD). This misapplication could lead to further unnecessary testing and increased radiation exposure.
2. Developing a diagnostic test without a central question to be answered by the test.
3. Comparing a new diagnostic test against an inadequate reference test.

Regulatory process

The Food and Drug Administration (FDA), a crucial branch of the Department of Health and Human Services, governs the approval of diagnostic tests, also known as "in vitro diagnostic devices" (IVD), ensuring they meet safety and efficacy standards before reaching the market. These devices are categorized into three risk-based classes. Class I devices, like bandages, are considered low-risk and are typically exempt from rigorous premarket processes. Class II devices, including CT scanners and EKG electrodes, must undergo premarket notification or a 510(k) submission, demonstrating their effectiveness is on par with existing devices. Class III devices, such as pacemakers and prosthetic valves, face the highest level of scrutiny. They require premarket approval through a rigorous process that includes a comprehensive application and a thorough scientific review to validate their safety and effectiveness. This tiered system considers the direct risk associated with using the device and the potential consequences of inaccurate diagnostic results, thereby protecting patient safety at every level.

Get started

To develop a reliable diagnostic test, it is essential to establish its purpose and identify the appropriate reference test for comparison. If the diagnostic test involves a scale, conducting a Receiver Operating Characteristic (ROC) analysis can significantly enhance its sensitivity and specificity by determining the optimal cutoff point for positive and negative values. It is imperative that the investigator clearly defines the intended use of the diagnostic test and ensures full compliance with the relevant regulatory guidelines to obtain regulatory approval.

Resources

1. Clinical Lab Consulting. https://clinicallabconsulting.com/.
2. DxQI Seed Grant. https://www.improvediagnosis.org/dxqi/.

3. Early Career Research Support From the American Heart Association: to the Second Century and Beyond https://doi.org/10.1161/CIRCRESAHA.123.323060.

Further reading

1. Taron J, Lyass A, Mahoney TF, et al. Coronary artery calcium score—directed primary prevention with statins on the basis of the 2018 American College of Cardiology/American Heart Association/Multisociety Cholesterol Guidelines. *J Am Heart Assoc*. 2021;10(1):e018342. https://doi.org/10.1161/JAHA.120.018342.
2. Carr JJ, Jacobs Jr DR, Terry JG, et al. Association of coronary artery calcium in adults aged 32 to 46 years with incident coronary heart disease and death. *JAMA Cardiol*. 2017;2(4):391—399. https://doi.org/10.1001/jamacardio.2016.5493.
3. Ferencik M, Pencina KM, Liu T, et al. Coronary artery calcium distribution is an independent predictor of incident major coronary heart disease events. *Circ Cardiovasc Imag*. 2017;10(10):e006592. https://doi.org/10.1161/CIRCIMAGING.117.006592.
4. Stevenson LW. The limited reliability of physical signs for estimating hemodynamics in chronic heart failure. *JAMA*. 1989;261(6):884. https://doi.org/10.1001/jama.1989.03420060100040.
5. Roberts E, Ludman AJ, Dworzynski K, et al. The diagnostic accuracy of the natriuretic peptides in heart failure: systematic review and diagnostic meta-analysis in the acute care setting. *BMJ*. 2015;350:h910. https://doi.org/10.1136/bmj.h910.
6. Greenland P, Blaha MJ, Budoff MJ, Erbel R, Watson KE. Coronary calcium score and cardiovascular risk. *J Am Coll Cardiol*. 2018;72(4):434—447. https://doi.org/10.1016/j.jacc.2018.05.027.
7. Budoff MJ, Nasir K, McClelland RL, et al. Coronary calcium predicts events better with absolute calcium scores than age-sex-race/ethnicity percentiles: MESA (Multi-Ethnic Study of Atherosclerosis). *J Am Coll Cardiol*. 2009;53(4):345—352. https://doi.org/10.1016/j.jacc.2008.07.072.
8. Rozanski A, Gransar H, Shaw LJ, et al. Impact of coronary artery calcium scanning on coronary risk factors and downstream testing: the EISNER (Early Identification of Subclinical Atherosclerosis by Noninvasive Imaging Research) prospective randomized trial. *J Am Coll Cardiol*. 2011;57(15):1622—1632. https://doi.org/10.1016/j.jacc.2011.01.019.
9. Peng AW, Mirbolouk M, Orimoloye OA, et al. Long-term all-cause and cause-specific mortality in asymptomatic patients with CAC ≥1000: results from the CAC consortium. *JACC Cardiovasc Imag*. 2020;13(1 Pt 1):83—93. https://doi.org/10.1016/j.jcmg.2019.02.005.

Chapter 13

Other product types

Ronaldo Correa Fabiano[1], Maria Beatriz Canto[2], Bruno B. Lima[3] and Raisa Lomanto Silva[1]

[1]*Department of Medicine, University of Pittsburgh Medical Center, Pittsburgh, PA, United States;* [2]*Division of Cardiovascular Medicine, Vanderbilt University Medical Center, Nashville, TN, United States;* [3]*Division of Cardiovascular Medicine, Department of Internal Medicine, Vanderbilt University Medical Center, Nashville, TN, United States*

Key points

- **FDA Regulatory Scope:** The Food and Drug Administration (FDA) regulates a variety of categories, such as food, drugs, biologics, medical devices, radiation-emitting devices, cosmetics, veterinary products, and tobacco.
- **Medical Device Classification:** Medical devices are categorized into three classes (I to III) based on risk, with Class I being the lowest and Class III the highest.
- **Product Category Understanding:** Identifying your product's specific FDA category is crucial for accessing the proper guidelines and achieving market success.

Introduction

The FDA, a central part of the Department of Health and Human Services, is critical in ensuring public health safety in the United States. As a regulatory body, the FDA oversees a wide range of products, including human and veterinary drugs, biologics, medical devices, foods, cosmetics, and tobacco products. This diverse portfolio signifies its extensive responsibility in safeguarding public health through rigorous regulation and oversight (Table 13.1).[1] A review of the products regulated and not regulated by the FDA is the focus of this chapter.

FDA overview

The FDA's regulatory principle to protect public health traces back to the Pure Food and Drug Act of 1906, which prohibited interstate transport of unlawful (adulterated and misbranded) food and drugs.[2] Over the decades, the FDA has had social, economic, and political changes to serve its purposes better, which has led to a significant expansion of the range of products under its regulation.[2] The FDA regulation also includes products not produced in the United States but sold in all 50 states, including the District of Columbia, Puerto Rico, Guam, the Virgin Islands, American Samoa, and other U.S. territories.[3] Therefore, the FDA is a highly organized institution that uses scientific evidence to ensure safety and security in its product evaluation. The institution is compartmentalized into 16 departments with intrinsic cross-collaboration.[4] The departments are the Center for Biologics Evaluation and Research; Center for Device and Radiological Health; Center for Drug Evaluation and Research; Center for Food Safety and Applied Nutrition; Center for Tobacco Products; Center for Veterinary Medicine; Oncology Center of Excellence; Office of Regulatory Affairs; Office of Clinical Policy and Programs; Office of External Affairs; Office of Food Policy and Response; Office of Minority Health and Health Equity; Office of Operations; Office of Policy, Legislation and International Affairs; Office of the Chief Scientist; and Office of Women's Health.[4]

Given the breadth of FDA regulation and its importance, it is essential to state its role to avoid semantic confusion or misinterpretation. Briefly, we highlight the below FDA myths and truths[1–4]:

- Approve tobacco products (myth) →. FDA regulates tobacco's (and similar) manufacturers' compliance with formulation, selling, and distribution (true).
- Approve products' pricing (myths) → A product cost or price is blinded from and is not analyzed or determined by the FDA (true)
- Approve product manufacturers, laboratories, or healthcare facilities (myth) → FDA site inspections' goal is to guarantee the adherence to regulatory requirements of maintaining, handling, manufacturing, storing, or providing services (e.g., computerized tomography) (true)
- Test or develop new drugs, devices, biologics, or vaccines (myths) → FDA experts analyze the data (studies/tests) done by the manufacturers (true)

Translational Cardiology. https://doi.org/10.1016/B978-0-323-91790-2.00111-8

- It requires clinical trials for all drugs and devices for market authorization (myth). The proof of "substance equivalence" may be applied to a drug, precluding the need for clinical trials. Clinical data and trial are required for devices class III (true).

Regulated and not regulated products

Human food and supplements

The FDA establishes requirements to minimize food safety hazards, including preventive measures in food processing, packing, and storing.[1] The safety assessment of new dietary ingredients has been regulated since the 1958 Food Additives Amendment.[2] Dietary supplements are not considered a conventional meal or solely used as a complete meal and are only intended for oral ingestion.[5] Regulations prohibit disease prevention or treatment claims on supplement labels, but statements about nutrient content, structure, and health are encouraged and required for some supplements.[5]

Human drugs

An official pharmacopeia substance used for preventing, diagnosing, mitigating, treating, or curing a health condition meets the FDA's drug definition.[6] As mentioned, dietary supplements, including vitamins and herbal products, do not meet the drug definition by the FDA.[5]

From creation to the market (drugs and devices)

A new successful drug creation requires a deep understanding of the target condition's pathological disease process and pharmacology dynamics and mechanisms.[7] Many steps are required, from the drug idea, design, and conception to its approval for market release, in which many factors are considered in the FDA's final decision (Fig. 13.1).[6] Given that it can take many years from the drug creation to its market release, depending on the disease circumstances, the FDA has created pathways to speed up drug creation, development, and market release approval.[8] Typically, once a new drug has passed the preliminary animal toxicities studies and the next step is to start testing in humans, the manufacturer will need to apply for one of the three types of Investigational New Drug (IND) application to transport across states and to be used in human tests.[9]

The following are the IND types:

- Investigational/research IND is submitted by a physician who is directly involved in studying a new/unapproved drug or an approved drug for a further indication or in a new population of patients.
- Emergency Use IND is generally used in emergencies, allowing the experimental drug to be used in regulated circumstances or for patients in new without alternatives who do not meet existing inclusion experimental protocols.

TABLE 13.1 General summary view of FDA's regulatory products categories.

FDA's category	Examples of included items		
Foods	Dietary supplements	Bottled water	Food additives
	Infant formulas	Foods	
Drugs	Prescription (Brand-name and generic) drugs		
	Nonprescription (over-the-counter) drugs		
Biologics	Vaccines for humans	Blood and blood products	Tissue and tissue products
	Allergenics	Cellular and gene therapy products	
Medical devices	Simple devices: Tongue depressor, blood pressure cuff		Dental devices
	Complex devices: Pacemakers		Surgical implants/prosthetics
Electronic radiation products	Microwave ovens	Medical imaging equipment	Laser products
	Ultrasonic therapies	Mercury vapor lamps	Sunlamps
Cosmetics	Color addictive of cosmetic products, including skin and nail care products		
Veterinary products	Pet foods	Livestock feeds	Veterinary drugs and devices
Tobacco products	Cigarettes, cigars	Tobacco (cigarette, roll your own, smokeless)	
	Hookah	E-cigarettes	

- Treatment IND is used when there is preliminary evidence of promising clinical testing results for life-threatening conditions while the final clinical FDA review occurs.

Once the new or unapproved drug starts to be used in human clinical tests, its final approval depends on several aspects, and the target condition severity, available treatments, and assessment of clinical data benefits and risks highly determine the timeline for its support.[6] These processes are "Accelerated Approval", "Priority Review", "Fast Track", and "Breakthrough Therapy"; and they need to be requested by the manufacturers on drug application (not meant to be comprehensive)[6]:

- **Accelerated approval:** Clinical surrogate endpoints can approve drugs for severe conditions without better treatment options. This is meant for conditions where the observed outcomes can take years but a meaningful clinical benefit for the patient can be observed earlier—a drug that substantially reduces cancer mass on imaging.
- **Priority Review:** The FDA review experts' decision is made within 6 months instead of the standard 10 months. It is essential to notice that Priority Review only changes the FDA review decision time and not other processes during drug approval. Once again, only drugs with significant impact in severe conditions are eligible. This might be suitable for established drugs used for new purposes (indications) with compelling clinical evidence.
- **Fast track:** Development of drugs that treat severe conditions for which animal or human data is promising. Critical new drugs were getting earlier to patients in need.
- **Breakthrough Therapy:** The goal is to expedite the new drug development intended to treat severe conditions for which preliminary clinical data suggest potential significant improvement compared to available therapy on clinically important endpoints.

Medical devices

The FDA process for device market approval is also summarized in Fig. 13.1. In brief, "Medical Devices" encompass various healthcare products, from essential tools like thermometers and tongue depressors to cutting-edge technologies such as pacemakers and MRI machines. These devices serve a multitude of functions within the field of cardiology, where precise diagnosis, prevention, and treatment of cardiovascular diseases are paramount.

1. Class I (Low Risk): In cardiology, Class I devices can include simple instruments like basic stethoscopes used for routine patient examinations. These devices are considered low risk and require minimal regulatory oversight, as they are unlikely to cause harm when used correctly.
2. Class II (Moderate Risk): Cardiology encompasses a wide array of Class II devices, including electrocardiogram (ECG or EKG) machines, which monitor the heart's electrical activity. These devices play a crucial role in diagnosing and monitoring heart conditions. Manufacturers of ECG machines must demonstrate substantial equivalence to existing devices through the 510(k) premarket notification process.
3. Class III (High Risk): Cardiology is home to numerous Class III devices, especially implantable medical devices such as cardiac pacemakers, implantable cardioverter-defibrillators, and left ventricular assist devices. These life-saving devices are used to treat severe cardiac conditions. Due to the potential for serious harm if they malfunction, Class III devices undergo rigorous scrutiny and clinical trials, often requiring premarket approval.

The "De Novo" and Humanitarian Device Exemption (HDE) are less common device class applications. "De Novo" is a "type" of device that has never been marketed in the United States, but profile safety and technology are well known. HDE is a device for rare cases (<4000 U S. cases per year).[10]

- **Step 1.** Discovery, concept, and development (Laboratory environment)
  - ➢ Deep disease mechanism understanding, and identification of appropriate disease modify target
- **Step 2.** Preclinical Research/Prototype
  - ➢ Laboratory and animal testing to basic safety questions
- **Step 3.** Clinical Research/Pathway to Approval
  - ➢ Devices* depends on its potential harm risk class (I, II, III): ↑ class, ↑ risk, ↑↑ safety and efficacy necessary
- **Step 4.** FDA review
  - ➢ Experts review in detail the inventors submitted data to decide to approve market release or not.
- **Step 5.** FDA Post-Market Safety Monitoring
  - ➢ FDA monitors products' safety once they are available for public use

FIGURE 13.1 Summary of drug and devices FDA development process.

Biologics

"Biologics" refers to a category of medical products derived from living organisms used for diagnosing, preventing, treating, or curing diseases and medical conditions in humans. These include various products such as vaccines, blood and blood components, allergenics, somatic cells, gene therapy, tissues, and recombinant therapeutic proteins. Biologics are distinct from traditional pharmaceutical drugs in their complexity and method of manufacture. They are often more sensitive to changes in manufacturing processes and require stringent regulatory oversight to ensure safety, purity, and effectiveness. The FDA's regulation of biologics is comprehensive, covering aspects from product development and manufacturing to postmarket surveillance.[11]

Electronic radiation products

The Radiation Control Provision oversees electronic products that emit radiation during operation. This category includes various devices, such as X-ray machines and CT scanners. The regulation's primary objective is to enforce safety standards and reduce radiation exposure to patients and healthcare professionals, ultimately ensuring public health protection.[12]

Cosmetics

Cosmetic products and their ingredients are not subject to FDA approval, except for color additives. However, regulations and laws are in place to oversee the cosmetic industry, ensuring the safety of products when they are in the hands of consumers. The Federal Food, Drug, and Cosmetic Act and Fair Packaging and Labeling Act grant the FDA authority over cosmetics. Cosmetics are products intended for application to the human body for purposes such as cleansing, beautification, enhancing attractiveness, or altering appearance. Examples include facial creams, lotions, and makeup products.[13]

Veterinary products

Products regulated are veterinary devices, animal drugs, animal food and feed, pet food, and medicated feed.[14]

Tobacco products

The FDA regulates many aspects of it which was provided by the Tobacco Control Act in 2009, establishing restrictive FDA capabilities in the tobacco being sold in the United States. Some of the most important regulatory aspects are regarding restrictions on marketing and sales to the youth population, warm health potential consequences of tobacco consumption on labels, and disclosure of tobacco ingredients.[15]

Get started

- A good understanding of FDA regulations and regulatory products and requirements are essential for basic and translational research. Take advantage of this widely available resource and be ahead in your next project involving regulated FDA products.
- The practical FDA regulations implications (required steps) can help you planning your project feasibility and preclinical phase.

Potential pitfalls

- Knowledge gap on the FDA-regulated products and its regulations, including inspection factors and storing of laboratory products.
- Inappropriate understanding of the submitting process of a new or unapproved drug or device to FDA.
- Underestimating the amount that you can learn simply by accessing the FDA website.

Cardiovascular real-world examples of products regulated by the FDA

Drugs: Ticagrelor, mavacamten, sacubitril, and valsartan.

Medical Devices: Implantable pacemakers and defibrillators, Impella 5.5.

Electronic Products: Positron emission tomography (PET)

Resources: Additional reading, consultants, contractors

- US. Food and Drug website: www.fda.gov
- US. Food and Drug website: FDA Authority Over Cosmetics: How Cosmetics Are Not FDA-Approved, but Are FDA-Regulated: www.fda.gov/cosmetics/cosmetics-laws-regulations/fda-authority-over-cosmetics-how-cosmetics-are-not-fda-approved-are-fda-regulated
- US. Food and Drug website: Supplement Your Knowledge: www.fda.gov/food/information-consumers-using-dietary-supplements/supplement-your-knowledge
- US. Food and Drug website: FDA Fundamentals: www.fda.gov/about-fda/fda-basics/fda-fundamentals

References

1. FDA Fundamentals. Published on. FDA; January 8, 2021. Available at: https://www.fda.gov/about-fda/what-we-do/what-does-fda-regulate. Accessed January 16, 2024.

2. Published on. *Changes in Science, Law and Regulatory Authorities.* FDA; January 31 2018. Available at: https://www.fda.gov/about-fda/fda-history/changes-science-law-and-regulatory-authorities. Accessed January 16, 2024.
3. Published on. *What does FDA do?* FDA; June 28, 2021. Available at: https://www.fda.gov/about-fda/fda-basics/what-does-fda-do. Accessed January 16, 2024.
4. Published on. *FDA Organization Charts.* FDA; March 21, 2019. Available at: https://www.fda.gov/about-fda/fda-organization/fda-organization-charts. Accessed January 16, 2024.
5. Published on. *Supplement Your Knowledge.* FDA; May 21, 2022. Available at: https://www.fda.gov/food/information-consumers-using-dietary-supplements/supplement-your-knowledge. Accessed January 16, 2024.
6. Published on. *Development & Approval Process Drugs.* FDA; August 8, 2022. Available at: https://www.fda.gov/drugs/development-approval-process-drugs. Accessed January 16, 2024.
7. Published on. *The Device Development Process.* FDA; January 4, 2018. Available at: https://www.fda.gov/patients/learn-about-drug-and-device-approvals/device-development-process. Accessed January 16, 2024.
8. Published on. *How Drugs are Developed and Approved.* FDA; October 24, 2022. Available at: https://www.fda.gov/drugs/development-approval-process-drugs/how-drugs-are-developed-and-approved. Accessed January 16, 2024.
9. Published on. *Investigational New Drug (IND) Application.* FDA; July 20, 2022. Available at: https://www.fda.gov/drugs/types-applications/investigational-new-drug-ind-application. Accessed January 16, 2024.
10. Published on. *Step 3: Pathway to Approval.* FDA; February 9, 2018. Available at: https://www.fda.gov/patients/device-development-process/step-3-pathway-approval. Accessed January 16, 2024.
11. Published on. *Guidance, Compliance & Regulatory Information (Biologics).* FDA; December 14, 2022. Available at: https://www.fda.gov/vaccines-blood-biologics/guidance-compliance-regulatory-information-biologics. Accessed January 16, 2024.
12. Published on. *Electronic Product Radiation Control Program.* FDA; February 23, 2023. Available at: https://www.fda.gov/radiation-emitting-products/electronic-product-radiation-control-program. Accessed January 16, 2024.
13. Published on. *FDA Authority Over Cosmetics: How Cosmetics Are Not FDA-Approved, but Are FDA-Regulated.* FDA; March 2, 2022. Available at: https://www.fda.gov/cosmetics/cosmetics-laws-regulations/fda-authority-over-cosmetics-how-cosmetics-are-not-fda-approved-are-fda-regulated. Accessed January 16, 2024.
14. Published on. *FDA Regulation of Animal Drugs.* FDA; December 14, 2023. Available at: https://www.fda.gov/animal-veterinary/resources-you/fda-regulation-animal-drugs. Accessed January 16, 2024.
15. Published on. *Family Smoking Prevention and Tobacco Control Act - An Overview.* FDA; June 3, 2020. Available at: https://www.fda.gov/tobacco-products/rules-regulations-and-guidance/family-smoking-prevention-and-tobacco-control-act-overview. Accessed January 16, 2024.

Further reading

1. Published on. *Overview of Device Regulation.* FDA; September 20, 2020. Available at: https://www.fda.gov/medical-devices/device-advice-comprehensive-regulatory-assistance/overview-device-regulation. Accessed January 16, 2024.

Chapter 14

Procedural technique development

Angel Moctezuma-Ramirez[1], Ke Li[1], Ahmed Alnajar[2] and Abdelmotagaly Elgalad[1]
[1]*Center for Preclinical Surgical & Interventional Research, The Texas Heart Institute, Houston, TX, United States;* [2]*The Division of Cardiothoracic Surgery, The Department of Surgery, The University of Miami Miller School of Medicine, Miami, FL, United States*

Key points

- A fundamental aspect of any successful preclinical study is the detailed preparation of its protocol.
- Refinement of a procedural technique is achieved by a continuous cycle of PDCA: plan, do, check, and act.
- No variable is too small to consider when planning a preclinical study. Choosing the animal model, deciding on an appropriate protocol for necropsy, and everything in between are all important.
- The success of clinical procedures depends on the physician's clinical experience and involvement in technical innovations.

Why does it matter?

Preclinical studies serve as the bridge between basic science and clinical practice. They include many types of experiments, ranging from in vitro computational simulations to in vivo evaluation of new therapies or innovative devices in animal models before clinical application. Because the data provided by these experiments are pivotal for advancing new technologies and for determining whether a new therapy is safe for use in clinical trials, study design is critical. A poorly designed preclinical study can delay the introduction of new therapies and treatments and waste resources, potentially affecting product or protocol approval and funding. In this chapter, we discuss the general principles of in vivo preclinical study design, including general concepts, hypothesis testing, and common mistakes made during the planning and testing stages (as well as how to avoid them).

Introduction

Early preclinical research studies were heavily focused on innovation, an approach that proved fruitful by some measures but incredibly wasteful by others. After a much-needed reexamination of the methodological approaches, heavy yet flexible regulation and supervision of study design was implemented, which changed the way we conduct preclinical research to this day. The research and development stage in the preclinical field is full of ideas, new approaches, curiosity, and experimentation that lead to the generation and testing of hypotheses. Depending on what kind of information is yielded, the experiments conducted during this stage can be classified into two broad categories: hypothesis-generating experiments, also called exploratory, and hypothesis-testing experiments, known as confirmatory research.[1]

Regardless of the experiment category, all in vivo studies need a clearly outlined protocol. An up-to-date literature search should be performed to avoid duplicating previous research and to identify alternatives for replacing and refining experimental approaches. A description of the test and control articles (if applicable), administration routes, justification, and any special handling instructions should also be included. Identifying the appropriate animal species is of the utmost importance because different species are especially useful for certain procedures. For example, swine are an optimal animal model for the study of acute myocardial infarction because their hearts have similar dimensions, structure, and physiologic properties to human hearts.

Providing appropriate justification for the number of animals to be used in the study is essential when proposing the protocol to the Institutional Animal Care and Use Committee (IACUC). Exploratory, proof-of-concept experiments may use only a small number of animals. In contrast, confirmatory studies require a larger sample size for the findings to achieve statistical significance. Randomization is critical; even when randomization is not possible because of a small sample size, it is essential to create different study groups to ensure that every aspect of care is the same except for the test article or therapy to be evaluated.

Translational Cardiology. https://doi.org/10.1016/B978-0-323-91790-2.00036-8

When it comes to procedure preparation, a detailed description of preprocedural care (e.g., use of paralytic agents or muscle relaxants), surgical preparation, anesthesia, critical steps in the surgical procedure, and postoperative care must be included in the protocol.

General approach to procedural technique development

Planning is essential; every critical step of the procedure must be planned in detail before the experimentation phase can begin. In preclinical research, the experimental model's limitations often present certain challenges. For example, ex vivo simulations may produce extraordinary projected results that do not correlate with real results from an in vivo animal model. Functionalized anatomic models used in simulated preclinical studies provide an excellent opportunity for training. However, functional computational models for therapy or device evaluation are still in early development. A thorough and up-to-date literature search must be conducted to determine the best animal model, surgical technique, or interventional technique to use.

The following steps are the most common ones to consider when conducting in vivo testing. Table 14.1 provides a brief description of each step in procedural technique development.

TABLE 14.1 General approach to procedural technique development.

Step	What does it entail?
Preprocedural medications and care	Determining the prerequisites that must be met before the start of the study. Example: Initiation of antiplatelet therapy with aspirin 3 days before the procedure.
Surgical preparation and anesthesia	Selecting the agents and methods to be used for anesthetic induction.
Surgical procedure	Determining aspects of the "Day 0" portion of the experiment, such as study type (survival vs. nonsurvival), complexity (major vs. minor), and phases (multiple vs. single-stage).
Follow-up procedures	Planning any follow-up studies that will be done after the initial procedure. Example: Echocardiography performed 7, 14, and 21 days after induction of acute myocardial infarction.
Recovery and postoperative medications	Detailing any specific care, monitoring, or medications that will be needed for adequate recovery. This includes any treatment necessary during or immediately after surgery. Example: Anticoagulation with heparin and eventual transition to warfarin. Scheduling of vital signs and medications is a priority. How long the animal will need to recover in intensive care should also be discussed.
Euthanasia	Selecting an appropriate method of euthanasia from the AVMA guidelines. Animals that appear in poor condition should be euthanized according to institutional policies of humane care of animals.
Necropsy and tissue harvest	Deciding between gross necropsy versus directed necropsy, which tissues are to be harvested, and what methods will be used to preserve and transport them.

AVMA, American Veterinary Medical Association.

Preprocedural medications and care

Certain experiments require specific preparations for the procedure, such as fasting or medication administration during a particular time frame (e.g., chemotherapy to ensure an immunocompromised state).

Surgical preparation and anesthesia

All agents and methods to be used during anesthesia and surgical preparation must be included in the protocol. Knowing in advance the areas to be worked on determines how specific anatomical sites should be prepped or cleaned. The type of experiment may determine which anesthetic agents should or should not be used (e.g., administering atropine during anesthesia induction can alter the results of some electrophysiology studies). Dosages of anesthetic agents should be chosen that will achieve adequate levels of anesthesia without exposing the animal to unnecessary pharmacological effects.

Surgical procedure

Planning in this area includes choosing survival versus nonsurvival, major versus minor, and single-stage versus multistage procedures. The phases that are used when developing a surgical technique in the clinical setting may serve as an example for the initial approach to be used in a preclinical experiment. Just as in the clinical surgical scenario, preclinical surgical techniques need to undergo an initial proof-of-concept experiment followed by constant refinement until a systematic approach is established.[2] A proof-of-principle experiment may be beneficial when the best approach has not been established in the literature; these sorts of exploratory experiments tend to be nonsurvival procedures intended to allow maximum anatomical and physiological analysis without compromising animal recovery. This step may use a small sample because its main objective is to identify the best technique and establish its safety and feasibility.

The next phase, "refinement and definition," involves making changes to the procedural technique based on evidence collected in the previous phase; this step generally requires a slightly larger sample than the proof-of-principle step. The PDCA (Plan, Do, Check, and Act) cycle has proven extremely useful for continuous improvement in preclinical studies. Going back to the drawing board and identifying even small variables that can affect the results, such as the animal's position on the table, can be crucial to the study's success.

Listing all critical steps—as is done in the clinical setting—aids in the description of the procedure. Any effort to make the procedure reproducible must include a detailed checklist of the critical steps to follow during the process.

Additionally, any perioperative medications should be selected before the procedure is performed, as should any cardioprotective or resuscitative drugs that may be needed intraoperatively. Any medications requested by the attending veterinarian or surgeon to minimize animal suffering or to promote animal survival without interfering with the scientific integrity of the study must be documented in the appropriate animal or study records.

As mentioned before, an initial series of studies may be performed as a proof of concept or to determine the most efficient way to establish a disease model. A study conducted by Li et al. in a swine model of myocardial infarction showed how a systematic approach and an adequate antiarrhythmic treatment might considerably reduce mortality in the pigs.[3] On the other hand, a refined study design includes an initial procedure—most often for inducing a disease model—followed by a test of a new therapy. As an example of synergy between these two types of experiments, the pivotal study conducted by Liu et al. benefited from the refined model developed by Li et al. The group first induced myocardial infarction in adult pigs. They then used gene therapy to knock down the endogenous Hippo signaling pathway in the pigs' cardiomyocytes.[4]

Follow-up procedures

Any follow-up studies after the initial procedure must be worked out in detail, including any necessary sedation or anesthesia. Other details may include blood draws for laboratory testing, echocardiography to evaluate ejection fraction, and therapy administration.

Postoperative medications and care

Any specific care, monitoring, or medications required while the animal is recovering must be included in the surgical protocol. Because there is no standardized scheme for managing recovering animals, special care must be taken, and decisions must be based on both evidence and the center's experience.

Postoperative care should be tailored to the animal used and the treatments tested. Some medications are problematic in certain species, such as anticoagulation drugs in sheep. To provide more clarity in this specific scenario, Weigand et al. tested the efficacy of different antiplatelet (aspirin, clopidogrel, ticagrelor) and anticoagulant (enoxaparin, dabigatran) drugs in sheep of different ages.[5] Their findings suggest that anticoagulation is most efficient with sodium enoxaparin, rather than dabigatran or antiplatelet therapy. In addition to medications, health screening and lab tests may be required in this step.

Euthanasia

Once the study's endpoints have been met, euthanasia must be performed in a manner consistent with current American Veterinary Medical Association (AVMA) guidelines.[6] If an animal appears in poor condition or in extremis before the endpoints have been met, it can be euthanized as per institutional policies on the humane care of animals. All animals that undergo unscheduled euthanasia should be necropsied immediately; if an immediate necropsy is not possible, the animal should be refrigerated to minimize autolysis, and the necropsy should be done within 24–48 h.

Necropsy and tissue harvest

If the study requires it, a necropsy must be performed by an independent, experienced veterinary pathologist. Successful necropsy requires that the methods, analyses to be outsourced, and unique materials be determined beforehand.

A step-based approach

Once the protocol outline is complete, the following steps can be done. Hereby, we propose a step-based approach to developing a procedural technique, which is summarized in Table 14.2.

Step 1: Do the research

- What type of study will be done? Acute or chronic? Major versus minor?
- Have any previous studies or other research been done?
- Is any pathological model needed?
- What is the best species for this study?
- What is the best methodology to use in this study?

TABLE 14.2 Steps for preclinical procedure development.

Steps	Questions to be answered
1: Do the research	What type of study are you going to do? Are any data available from previous studies? Is any pathological model needed? What is the best animal to use in the study? What is the best methodology to use? Critical step: determine the best animal model.
2: Refine the procedure	How will the therapy be delivered? What are the primary endpoints? How many procedures are needed? Is the group able to perform all operations? Does the study need any special supplies or equipment? How many animals are required? Critical steps: developing the procedure design, methods, and timeline.
3: Perform the pilot study	How should proof of concept be established before a larger study is begun?
4: Perform the complete study	How can accurate and precise data be collected? As more data are collected, they should be used to design a more robust study. The design is constantly refined.

Many new therapy ideas for ischemic heart disease require preclinical testing every year. The test article, whether a drug or a device, is first evaluated in an animal model. For testing a new drug in a large-animal model, swine may be the best choice because their anatomy and physiology are similar to those of humans. But for testing specific types of devices, such as a left ventricular assist device, a bovine or ovine model is better because it is difficult to restrain swine without violating animal welfare guidelines. Anything implanted inside the pig that needs a connection to an outside controller or monitor may be compromised by the pig's movement.

As for the ischemia itself, it can be induced temporarily by percutaneous balloon inflation-reperfusion or permanently by surgical occlusion or microsphere injection. To choose the best method, researchers should consider whether the study needs a functional change that can mimic the actual clinical scenario (for which balloon occlusion is preferred) or a well-developed scar that can cause a significant left ventricular remodeling (for which permanent occlusion is better).

Step 2: Refine the procedure

- How will the therapy be delivered?
- What are the primary endpoints?
- How many procedures are needed?
- Is the group able to perform all procedures?
- Are any special supplies or equipment needed?
- How many animals are needed for further testing?

In step 1, the researchers determine the animal model the study needs. In step 2, the important questions to address are how best to deliver the test article and to evaluate the treatment effect. For example, there are many methods of delivering drugs to treat heart disease. Finding the best route for delivering a new drug requires examining previous data regarding the efficacy and technical feasibility of the delivery and corresponding mortality. Intravenous injection is easy but has low efficacy. Catheter-based endomyocardial, intracoronary arterial, and intracoronary sinus injection are less-invasive procedures (preferred in the clinical setting) but need a professional team with surgical skills. The open-chest epicardial injection is more straightforward, but its invasiveness may limit the drug's future clinical application.

Once the delivery method is selected, researchers should decide how to evaluate the treatment effect by choosing and defining the primary endpoints. Ejection fraction (EF) is the most common assessment parameter for regenerative heart therapies and is measurable by echocardiography. However, EF change may not be a meaningful endpoint because it correlates more to global response than to regional functional change.[7] Some researchers choose scarring as a proxy for heart dysfunction or use other parameters that need different imaging modalities, such as magnetic resonance imaging (MRI), which is relatively expensive and not widely available.

Once the primary endpoints are confirmed, an initial evaluation analysis of the data obtained should be performed. Also, explicit inclusion and exclusion criteria need to be set. For example, the researchers may decide that only animals whose EF declines by 30% within 14 days after myocardial infarction induction will be used to test the experimental therapy. Finally, a power calculation to estimate the needed sample size should be made from previous data. If no such data are available, researchers usually use a small sample size for a feasibility study, or clinical judgment. Still, we recommend performing statistical simulation studies of different sample sizes to improve the power of the pilot study (as we discuss in Chapters 20 and 35).

Step 3: Pilot study

A pilot study is characterized by the use of a small sample to test the feasibility of a new research idea. We recommend that this step begin with a series of acute studies. These acute studies aid researchers in answering all questions raised in step 2. The pilot study helps fill the gap between Plan and Act by enabling researchers to practice the procedure and evaluate the results. If simulation data are available, a sample size power calculation can be performed, which would increase the reliability of the results. This, too, is consistent with the PDCA cycle of plan, do, check, and act.

Step 4: Performing the complete study

After all of the previous preparation, researchers can confidently perform the entire study at this stage. However, new issues can still arise, so researchers should continue to follow the PDCA cycle to improve the whole procedure. Being able to summarize graphically (e.g., in a flowchart), the study design suggests that the protocol is well written, concise, and to the point. Fig. 14.1 shows an example of a refined protocol with different time points.

Clinical procedures

The development process for clinical procedures is different from that for drugs and medical devices. Nonetheless, it is essential to introduce new technology or techniques that can change clinical practice. Clinical

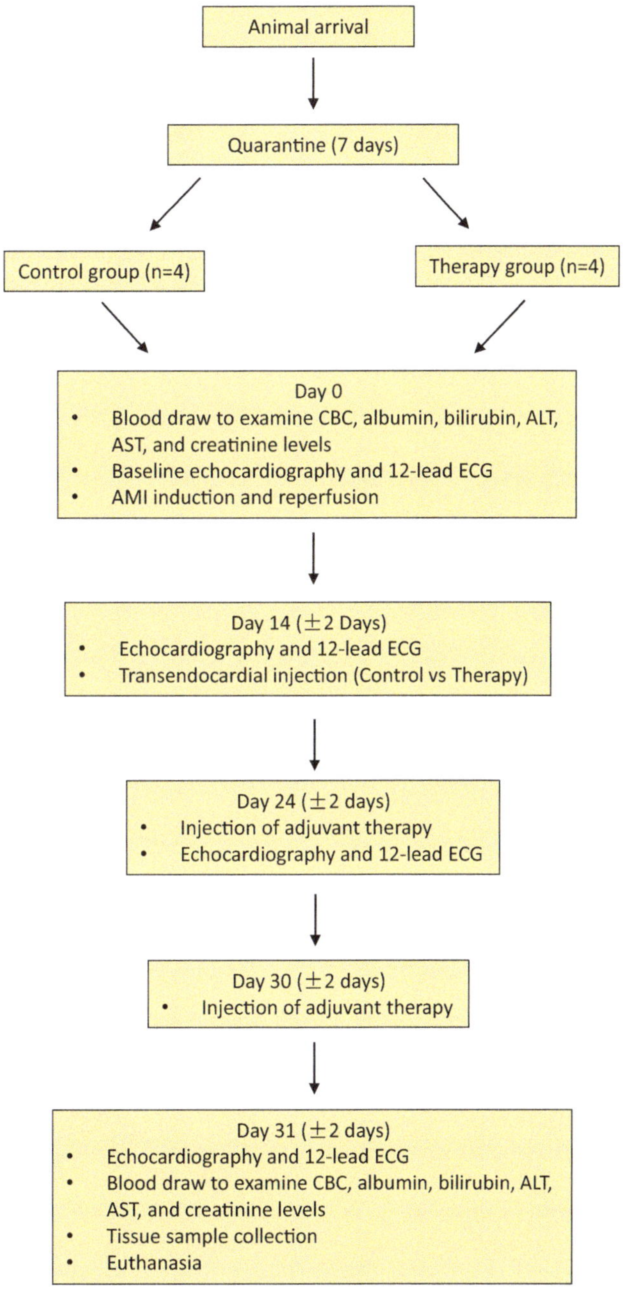

FIGURE 14.1 Flowchart of a refined study design in a porcine model of acute myocardial infarction. *ALT*, alanine transaminase; *AMI*, acute myocardial infarction; *AST*, aspartate transaminase; *CBC*, complete blood count; *ECG*, electrocardiogram.

procedures have no standard governmental regulatory system for development and evaluation, unlike drugs and devices. In fact, the development process has traditionally depended on physicians' ability to make reasonable decisions and patients' trust in those physicians. This process bridges the gap between engineering and clinical research, such as when the tools were developed for the Miami Method of aortic valve replacement.[8] A biomedical engineer had to undergo an open-heart valve-replacement procedure because no instruments were available for replacing the aortic valve in a minimally invasive fashion. After recovering from the procedure, the engineer devoted his expertise and resources to helping his surgeon create the tools the surgeon envisioned to facilitate a mini-thoracotomy approach for minimally invasive cardiac surgery that would produce outstanding outcomes.[9,10]

Because the process of developing procedures seldom includes adequate preclinical experimentation, the first human applications of clinical procedures are often premature. Yet, with experience, clinicians can refine their new techniques. For example, mitral valve operations were performed on animals before their first application to humans in 1923. Still, because of the high mortality rate associated with the procedure, it was suspended in 1928. After a series of drug, device, and clinical advances were made in other disciplines—including cardiac imaging, anesthetic technique, and antibiotics—the clinical development of mitral valve surgery resumed in 1945, which was followed by declining mortality rates as experience increased and the technique became more established.[11]

Pitfalls to avoid

- *Copying an animal model without proper adaptation to the research endpoints*
  Each model has its advantages and limitations. In the design stage, the research group should consider all factors, including animal species, procedures, and primary endpoint, to decide which model is best, rather than basing the selection solely on published data.
- *Designing a study that surpasses the research center's capabilities*
  Researchers always want to use the newest available techniques in their studies. For example, MRI is a valuable tool in cardiovascular research, but it is not widely available in the preclinical area. Using MRI in a preclinical study can increase the budget significantly; additional expenses include animal transfer and procedure costs, especially if you need to cooperate with other facilities to perform the MRI procedures. In this situation, carefully review your study design and determine whether a cheaper option could meet your study requirements. For example, echocardiography can provide reliable data regarding cardiovascular functional change; even some of the mechanical quantifications used with echocardiography, such as strain analysis, can produce data similar to those obtainable by MRI.
- *Not prioritizing animal welfare*
  Protocols with large numbers of animals and heavy procedure schedules are virtually always rejected by the IACUC. In any preclinical study, the one critical consideration apart from evidence-based decision-making is animal welfare. Researchers must consistently implement the "Three Rs" principle of animal welfare: replacement, reduction, and refinement. Replacement means using methods that replace or otherwise avoid the use of animals; reduction refers to minimizing the number of animals used per experiment; and refinement deals with minimizing animal suffering and improving animal welfare. These 3 Rs are essential factors when the IACUC reviews a study. A result-oriented approach is appealing to most researchers, especially since the faster results are obtained, the faster the research may advance. However, we must never forget that animals sacrifice their lives to help humans improve their health; decreasing their suffering should also be our priority during animal studies.

Getting started

- The critical procedural aspects of the study should be identified.
- The choice of animal model should be evidence-based. The model should be chosen once the group has outlined the phases of the protocol.
- The surgical procedure itself should not be the sole focus of planning; other preoperative, intraoperative, and postoperative aspects of the treatment (e.g., medications, follow-up procedures) must be considered, as well.
- No variable is too small to consider when it comes to developing a procedural technique.
- A thoroughly planned surgical procedure follows appropriate preoperative preparation. An adequate recovery and tailored postoperative care need to be planned out, as well.

Examples of additional resources

Alving B, Dai K, Chan S. *Translational Medicine – What, Why and How: An International Perspective.* Vol 3. Karger; 2012. https://doi.org/10.1159/isbn.978-3-318-02285-8.

de Caestecker M, Humphreys BD, Liu KD, ASN AKI Advisory Group. Bridging translation by improving preclinical study design in AKI. *J Am Soc Nephrol.* 2015;26(12):2905–2916. https://doi.org/10.1681/ASN.2015070832. Epub 2015 Nov 4. PMID: 26538634; PMCID: PMC4657852.

Fish R, Brown M, Danneman P, Karas, A. *Anesthesia and Analgesia in Laborator Animals.* Academic Press; 2008. https://doi.org/10.1016/B978-0-12-373898-1.X5001-3.

Girard P. *Design and Analysis of Animal Studies in Pharmaceutical Development.* Vol 19. Wiley; 2000. https://doi.org/10.1002/1097-0258(20000730)19:14<1849::AID-SIM506>3.0.CO;2-1.

Kiani AK, Naureen Z, Pheby D, et al. International Bioethics Study Group. Methodology for clinical research. *J Prev Med Hyg.* 2022;63(2 Suppl 3):E267–E278. https://doi.org/10.15167/2421-4248/jpmh2022.63.2S3.2769. PMID: 36479476; PMCID: PMC9710407.

Kim B, Trounson A, Design preclinical studies for reproducibility. *Nat Biomed Eng.* 2018;2:789–790. https://doi.org/10.1038/s41551-018-0322-y.

Acknowledgments

Stephen N. Palmer, PhD, ELS, and Nicole Stancel, PhD, ELS(D), contributed to the editing of the manuscript.

References

1. Huang W, Percie du Sert N, Vollert J, et al. General principles of preclinical study design. In: Bespalov A, Michel MC, Steckler T, eds. *Good Research Practice in Non-clinical Pharmacology and Biomedicine.* Cham: Springer International Publishing; 2020:55–69.
2. McCulloch P. Developing appropriate methodology for the study of surgical techniques. *J R Soc Med.* 2009;102:51–55.
3. Li K, Wagner L, Moctezuma-Ramirez A, et al. A robust percutaneous myocardial infarction model in pigs and its effect on left ventricular function. *J Cardiovasc Transl Res.* 2021;14:1075–1084.
4. Liu S, Li K, Wagner Florencio L, et al. Gene therapy knockdown of Hippo signaling induces cardiomyocyte renewal in pigs after myocardial infarction. *Sci Transl Med.* 2021;13.
5. Weigand A, Boos AM, Ringwald J, et al. New aspects on efficient anticoagulation and antiplatelet strategies in sheep. *BMC Vet Res.* 2013;9:192.
6. American Veterinary Medical Association. *AVMA Guidelines for the Euthanasia of Animals*; March 13, 2022. https://www.avma.org/resources-tools/avma-policies/avma-guidelines-euthanasia-animals, 2022.
7. Kosaraju A, Goyal A, Grigorova Y, et al. *Left Ventricular Ejection Fraction*; March 13, 2021. https://www.ncbi.nlm.nih.gov/books/NBK459131/, 2022.
8. Lamelas J. Minimally invasive aortic valve replacement: the "Miami Method". *Ann Cardiothorac Surg.* 2015;4:71–77.
9. Mihos CG, Santana O, Pineda AM, et al. Right anterior mini-thoracotomy versus median sternotomy surgery for native mitral valve infective endocarditis. *J Heart Valve Dis.* 2014;23:343–349.
10. Pineda AM, Santana O, Reyna J, et al. Outcomes of reoperative aortic valve replacement via right mini-thoracotomy versus median sternotomy. *J Heart Valve Dis.* 2013;22:50–55.
11. Gelijns AC. The development of clinical procedures. In: *Institute of Medicine (US) Committee on Technological Innovation in Medicine. Technological Innovation: Comparing Development of Drugs, Devices, and Procedures in Medicine.* Washington DC: National Academies Press; 1989.

Chapter 15

Behavioral intervention

Favour Oludunni Ojebiyi and Enoch Oluwalonimi Adewara
All Saints University School of Medicine, Roseau, Dominica

Key points

- Behavioral interventions aim to modify, strengthen, or eliminate targeted health behaviors to achieve improved health outcomes.
- Preclinical considerations are crucial to ensure translatable results addressing factors such as feasibility, fidelity, and cost-effectiveness among others. Guidelines such as PREPARE (Planning Research and Experimental Procedures on Animals) and ARRIVE (The Animal Research: Reporting of In Vivo Experiments) provide comprehensive checklists to improve validity and reproducibility in research involving animals.
- The National Institute of Health (NIH) stage model for behavioral intervention development consists of six stages: basic research, intervention generation/refinement, efficacy testing with research providers, efficacy testing in community settings, effectiveness and implementation, and dissemination.

Introduction

Behavioral interventions seek to change, reinforce, or eradicate specific health behaviors that result in better health outcomes for a patient. Many major health issues such as obesity, smoking, and mental illness are primarily influenced by behavioral and lifestyle factors making the development and testing of behavioral interventions to address these challenges vital.[1] Behavioral interventions begin with the identification of specific problems, gaps, disparities, and issues and require various techniques that involve modifying various biosocioeconomic factors at the individual, societal, or governmental level, or a combination of all levels, to improve results for a specific demographic.[1,2]

Developing an intervention often involves designing and implementing studies on nonhuman subjects (e.g., animal models) to understand the effect of interventions on behavioral changes and provide insight into the underlying mechanism of behavior change. Several important preclinical considerations need to be considered to ensure that the chosen model system yields reproducible and translatable results. Such considerations include questions regarding the type of animal model selection, feasibility, fidelity, safety, ethical appropriateness, study designs, cost effectiveness, and statistical analysis. Guidelines such as the Planning Research and Experimental Procedures on Animals (PREPARE) guidelines, a 15 module checklist designed to improve scientific validity, reproducibility, and animal welfare[3] in animal research, provide a comprehensive checklist of items that should be considered and addressed when planning and conducting animal experiments. Another guideline, The Animal Research: Reporting of In Vivo Experiments (ARRIVE) guidelines, designed to enhance the reproducibility and quality of reporting in animal research, provides a checklist of essential information that should be included in research publications involving animal experiments.[4]

Methodologies

Animal models

Animal models are essential because they allow researchers to comprehend the complex biological mechanisms behind behavior, development, and health.[5] Since the pathophysiology of the cardiovascular system is significantly influenced by environmental and genetic factors, no absolute model precisely replicates the complexity of the human cardiac disease,[6] that is, a single animal model should not be used in studies to provide answers to all questions as there is no perfect animal model of the human cardiovascular system.[7] Therefore, factors that should be considered when choosing an appropriate animal model include the cost, infrastructure, reproducibility, and close similarity of the model to the disease being studied.[6,8]

Translational Cardiology. https://doi.org/10.1016/B978-0-323-91790-2.00088-5

Computational models

Computational models refer to mathematical or computational representations of biological systems that are used to simulate and predict behaviors, interactions, and outcomes under different conditions. The degree to which a computational model's predictions are tested against experimental data that was not employed in the model's formulation is the gold standard for assessing the model's quality[9]

Stages of behavioral intervention

Various concepts have been developed over the years through research on the stages or phases of intervention development. Most research models arrive at a consensus that the efficacy and effectiveness of intervention development and implementation vary based on optimal accuracy and precision. Some research models used a numbering system for the stages which was utilized in creating a schematic representation of the stages of behavioral intervention adapted from Onken LS et al.[10] shown in Fig. 15.1.

The NIH stage model for behavioral intervention development provides a systematic framework for developing maximally potent and implementable interventions that can be translated into clinical practice, emphasizing the importance of an iterative, recursive, and multidirectional approach. By following this model, researchers and practitioners can enhance the effectiveness and relevance of behavioral interventions, ultimately improving health outcomes for individuals and communities.

- **Stage 0: Basic Science Research**
  Stage 0 can serve as the precursor and foundation for the creation of a new intervention or the modification of an existing one. At this stage, basic science research is conducted, encompassing various fields such as behavioral, cognitive, affective, social science, and neuroscience, with the primary objective of gathering knowledge that will inform the development of behavioral interventions, that is, conducting basic science research to lay the groundwork for intervention development. Basic science involves asking various specific questions to behavioral intervention research even before it is developed or completed. This stage of research is carried out to know and understand how the intervention could work on its target, why the

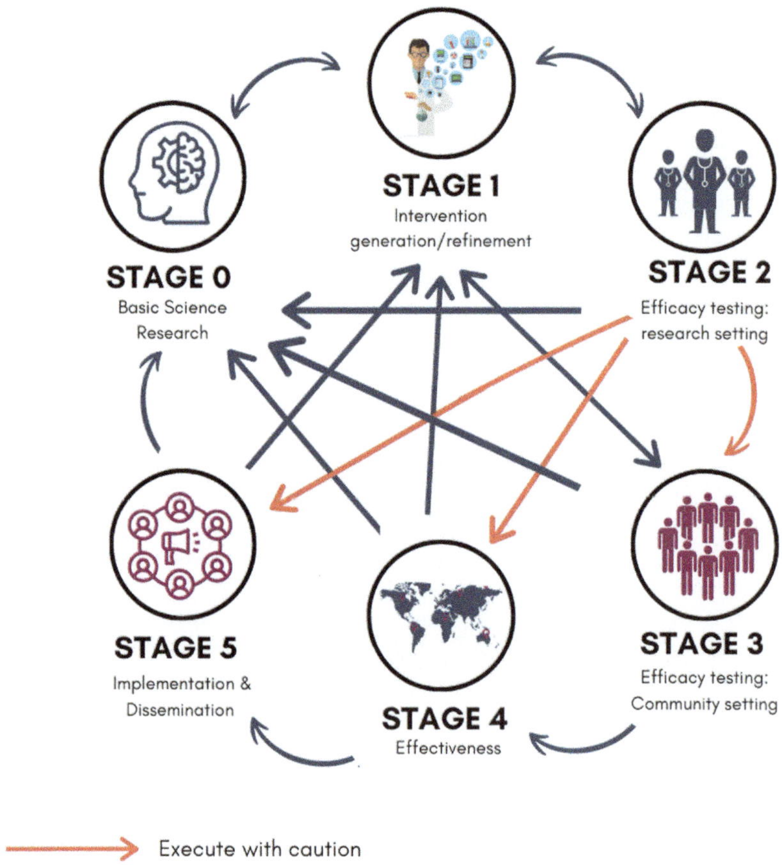

FIGURE 15.1 NIH stage model.

researcher believes it could work and the process involved in executing it.
Another type of basic science research—research on the mechanisms of behavior change—permeates all stages of intervention development and involves posing fundamental questions rooted in basic science disciplines, aiming to understand how interventions work and for whom they are most effective. Basic science research is not limited to Stage 0 but can be integrated into all other stages of intervention development. With an understanding of the step by step process for an intervention, it could now be used optimally to influence behavior and could benefit the researcher and the target in various ways, some of which are; reducing costs, improving efficacy, and strengthening how well it is implemented.

- **Stage I: Intervention Generation/Refinement**
Stage I encompasses activities related to designing, adapting, or modifying (Stage IA), as well as the feasibility and pilot testing of an intervention (Stage IB) to evaluate its efficacy concerning the intervention's mechanism of behavior change.
Stage 1 involves a lot of experimentation. 1A encompasses the development of the intervention coupled with all the errors and experiments that modify the result and also provides a leeway for changes to be made to ensure improvement of the final result. 1B then evaluates the intervention's feasibility; if the intervention could work and if it can be administered with internal validity remaining intact.
Stage 1 studies can occur in any setting, whether in research settings with research providers and subjects or in community settings with community providers or caregivers. Since training materials are integral to correctly administering the intervention, Stage 1 also involves developing, modifying, and pilot testing training materials and fidelity procedures and interventions for community providers. Since modification to enhance implementation ability in real world settings and adaptation can only occur in Stage 1 when making any modification or planned deviation to an intervention, the study is considered to be in Stage 1.
- **Stage 2: Efficacy Testing: Research Setting and Provider**
Stage 2 assesses the efficacy of interventions through the experimental testing of behavioral interventions in research settings with research based providers. Stage 2 allows for the highest possible degree of control in an efficacy trial as well as the highest internal validity possible, thereby giving researchers the best environment to test their hypotheses or principles underlying an intervention.
Unlike Stage 1, this stage is highly controlled using well trained researchers and randomized targets or participants so that findings from the intervention used can be credited to the research without the influence of any extraneous factors. This stage maximizes the efficacy and internal validity of the said intervention.
- **Stage 3: Efficacy Testing: Community Setting and Provider**
Stage 2 assesses the efficacy of interventions through the experimental testing of behavioral interventions in community settings with community based providers or caregivers under highly controlled conditions. Stage 3 is sometimes referred to as a hybrid between efficacy and effectiveness testing.
It is important to note that this stage does not test for effectiveness. Like Stage 2, it maximizes the internal accuracy of the intervention so that the anticipated effect in a target community setting or real world population can be reliable. But unlike Stage 2, intervention administration is specific to a real world target population and it is not tightly controlled.
- **Stage 4: Effectiveness**
The goal of Stage 4 is to demonstrate the effectiveness of the intervention before moving forward with implementation and dissemination in the real world. At this level, the emphasis is on maximizing external validity as behavioral therapies are assessed in real world environments by local therapists and clinicians. Unlike Stage 3, the population reach is widened and the circumstances associated with the intervention are broader.
- **Stage 5: Implementation and Dissemination**
This stage involves the development and testing of strategies for the successful implementation and dissemination of interventions. In this stage, the potential for use of the current tentatively supported intervention is maximized by figuring out the optimal design for propagation and administration through research.
In this stage, there are challenges when it comes to the context in which the intervention is administered. Administration of the intervention at this stage cuts across the individual, community, and even national levels, and sometimes the community meant to receive the intervention poorly matches it. Therefore, acceptance is decreasing, affecting the implementation and dissemination of behavioral intervention.
This stage model has multiple directions as shown in Fig. 15.1. This allows researchers to complete each intervention in multiple different but still similar ways, especially on an individual basis. It also gives the chance of doing and redoing the steps multiple times to make the intervention maximally effective and useful for the population it was created for.

TABLE 15.1 Summary of NIH'S stages of behavioral intervention.

Stages	Title	Summary
0	Basic science research	Asking questions and understanding the intervention development
1	Intervention generation/refinement	Experimentation, modification, feasibility assessment, and pilot testing of the intervention
2	Efficacy testing: research setting and provider	Tightly controlled assessment of intervention to maximize efficacy and internal validity
3	Efficacy testing: community setting and provider	Less controlled administration of intervention to a real world target population
4	Effectiveness	Administration of intervention to different population targets to maximize external validity
5	Implementation and dissemination	Development and testing of strategies for optimal propagation and administration of intervention

From the arrows shown in Fig. 15.1, a researcher could decide to follow the pattern: $1 \rightarrow 2 \rightarrow 1 \rightarrow 2 \rightarrow 3 \rightarrow 4 \rightarrow 5$ or another more thorough pattern: $1 \rightarrow 2 \rightarrow 1 \rightarrow 3 \rightarrow 1 \rightarrow 3 \rightarrow 4 \rightarrow 5$. It has been observed that along the development of any intervention, it is better to revert to Stage 1 multiple times before continuing down Stages 3, 4, and 5. Also, the red arrows indicate steps that should be executed with caution because they bypass the regeneration and refinement phase done in Stage 1. The cautionary steps include $2 \rightarrow 3$, $2 \rightarrow 4$, and $2 \rightarrow 5$; going from Stage 2 directly to Stages 3, 4, or 5 deprives the intervention of the potential to attain maximum efficacy and internal accuracy.[10,11]

This allows continuous correction and modification of the intervention development to contribute optimally to the success of the end result. The text in Table 15.1 adapted from[1,12] summarizes the importance of each of the stages involved in behavioral interventional development according to the NIH.

Get started

- Understand the importance of behavioral intervention
- Learn the Preclinical considerations and methodologies
- Know and understand the six stages of the NIH stages of behavioral interventions
- Observe real world examples of behavior interventions

Potential pitfalls

- When using animal models to understand the effect of interventions on behavioral changes, it is important to ensure that the chosen model system yields reproducible and translatable results.
- Since no single animal model perfectly replicates the complexity of human behavior and health, relying solely on one animal model may lead to limited insights and less reliable results.
- It is essential to ensure that the predictions made by computational models align with empirical observations and are not solely based on theoretical assumptions
- It is also vital to utilize the multidirectional stages of interventional development to maximize its effectiveness. Skipping steps or not going through stage 1 multiple times before testing on a real world population could lead to errors and therefore decrease acceptance of the intervention.

Real world examples

- The Multiple Risk Factor Intervention Trial (MRFIT), an individual intervention, was a large scale clinical trial conducted between 1973 and 1982 aimed at assessing the impact of interventions targeting multiple cardiovascular disease risk factors on mortality from coronary heart disease (CHD). Over 12,000 high risk middle aged men were divided into Special Intervention (SI) and Usual Care (UC) groups. The SI group received intensive interventions to modify risk factors, such as smoking, blood pressure, cholesterol, and weight. While the trial showed a modest reduction in CHD mortality in the SI group, the overall mortality rates did not differ significantly between the two groups.[2]
- The Stanford Five City Project was one of the community level cardiovascular disease intervention trials that occurred between 1980 and 1987. The intervention which included two treatment and two control cities was delivered through mass media (TV, radio, and newspapers) and direct education (classes, pamphlets, and kits) with a focus on dietary changes, physical activity, and medication use. At the end of the 6 year period, there was an increase in knowledge of coronary heart disease risk factors

in treatment cities compared to the control cities, positive changes in the risk factors for both treatment and control cities, and a modest difference in blood pressure reduction and obesity in treatment cities in some years. However, there was no difference in cholesterol, smoking, and mortality in treatment cities.[2]

- The smoking cessation movement is one of the single most successful health interventions. Before 1964, surveys revealed that while more than half percent of individuals recognized that cigarettes were harmful, they believed they did not smoke enough to hurt themselves. This came to an end in 1964 when the Surgeon General published a study declaring that smoking cigarettes is the primary cause of chronic bronchitis, a cause of lung and laryngeal cancer in men, and a potential cause of lung cancer in women. The report was widely publicized and was listed among the top stories of 1964. Other organizations sent similar messages in response to the findings, and by 2000, cigarette smoking had dropped by half from its peak in 1964.[2]

Additional resources for the reader

Gitlin LN, Czaja SJ. Behavioral intervention research designing, evaluating, and implementing. New York: Springer Publishing Company; 2016.

References

1. Gitlin L, Czaja S. *Behavioral Intervention Research: Designing, Evaluating, and Implementing*. Springer Publishing Company; 2016.
2. Norman B,A, Rodolfo A,B, Barney C. *Behavioral Health Interventions: What Works and Why?*. 2003.
3. Smith AJ, Clutton RE, Lilley E, Hansen KEA, Brattelid T. PREPARE: guidelines for planning animal research and testing. *Lab Anim*. April 2018;52(2):135−141.
4. Percie Du Sert N, Ahluwalia A, Alam S, et al. Reporting animal research: explanation and elaboration for the ARRIVE guidelines 2.0. *PLoS Biol*. July 14, 2020;18(7):e3000411.
5. Phillips NLH, Roth TL. Animal models and their contribution to our understanding of the relationship between environments, epigenetic modifications, and behavior. *Genes*. January 15, 2019;10(1):47.
6. Zaragoza C, Gomez-Guerrero C, Martin-Ventura JL, et al. Animal models of cardiovascular diseases. *J Biomed Biotechnol*. 2011;2011:1−13.
7. Camacho P, Fan H, Liu Z, He JQ. Large mammalian animal models of heart disease. *JCDD*. October 5, 2016;3(4):30.
8. Bianco RW, Gallegos RP, Rivard AL, Voight J, Dalmasso AP. Animal models for cardiac research [cited 2023 July 18]. In: Iaizzo PA, ed. *Handbook of Cardiac Anatomy, Physiology, and Devices*. Totowa, NJ: Humana Press; 2009:393−410. Available from: http://link.springer.com/10.1007/978-1-60327-372-5_25.
9. Yang JH, Saucerman JJ. Computational models reduce complexity and accelerate insight into cardiac signaling networksWinslow R, ed. *Circ Res*. January 7, 2011;108(1):85−97.
10. Onken LS, Carroll KM, Shoham V, Cuthbert BN, Riddle M. Reenvisioning clinical science: unifying the discipline to improve the public health. *Clin Psychol Sci*. January 2014;2(1):22−34.
11. Adam EM, Jeffery AB, Paige CN, Adena JO. *Handbook for Designing and Conducting Clinical and Translational Research*. London: Elsevier; 2023.
12. NIH Stage Model for Behavioral Intervention Development. National Institute on Aging; 2022. Available from: https://www.nia.nih.gov/research/dbsr/nih-stage-model-behavioral-intervention-development.

Part III

Clinical: Fundamentals

Constantine Monzidelis

Chapter 16

Introduction to clinical research: What is it? Why is it needed?

Ahmed Alnajar
The Division of Cardiothoracic Surgery, The Department of Surgery, The University of Miami Miller School of Medicine, Miami, FL, United States

Key points

- Clinical research is a fundamental step for evaluating the effectiveness of an intervention in human beings and requires careful design and ethical consideration.
- Many regulatory agencies are involved in the early steps of conducting clinical research, so investigators should be familiar with their institution's research oversight board and follow the rules.
- The results of evidence-based research are critical for improving clinical practice; while randomized controlled trials are the gold standard for research, even observational research can provide important insights.

Why it matters?

Clinical practice has been influenced by the results of randomized controlled trials. Up to four in 10 previously accepted medical practices did not pass the test when subjected to proper trials. The reversal of established medical practice is common and occurs across all classes of medical practice, including cardiology. Thus, in our rapidly moving field, even more than others, cardiology researchers and physicians should focus on obtaining and disseminating high-quality research. When the primary objective of a research design is to provide clear evidence that allows us to answer a specific research question, we need to understand the basic steps involved in clinical research, including ethical considerations. These steps will allow cardiologists to reevaluate our practices, produce new knowledge, and critically appraise the results obtained from other studies.

Introduction

The ultimate goal of clinical research is to generate knowledge capable of improving patients' health outcomes. When a new drug or surgical technique is developed and used on patients, many ethical issues may occur because of the possibility of causing harm rather than good. By studying animals, however, it is possible to obtain information that cannot be learned in any other way. However, animal models may fall far short of being able to predict human responses and may present ethical issues of their own. Thus, there are many attempts to translate basic scientific findings from a laboratory setting into human applications and potential treatments or biomarkers for diseases. The level of evidence required for adoption by the community of cardiovascular medicine depends on the strength of the evidence provided in relation to the type of disease and therapy. For that reason, clinical research has many types and involves many methods. The initial stages of such research in humans are similar and the success of a study depends upon how many issues are thought out in advance. Most similarities lie in following regulatory approvals and proper conduct of ethical research once a human subject is involved.

Regulatory approval

Once there's an intent to conduct human-subject research, whether phase I trials (Table 16.1) or retrospective research involving human participants, researchers seek protocol approval from their Institutional Review Board (IRB) or the Human Studies Review Committee that oversees the institutional clinical research in their institution. In the United States, IRB ensures that the research is conducted in compliance with the Health Insurance Portability and Accountability Act (HIPAA). The US government has mandated that clinicians should protect any confidential or sensitive information of a patient. HIPAA guidelines ensure that researchers avoid collecting patients' identification unnecessarily when they conduct research, and certainly

Translational Cardiology. https://doi.org/10.1016/B978-0-323-91790-2.00008-3

TABLE 16.1 Types and phases of clinical trials.

Phases of a trial		Brief description
Nonclinical	Preclinical phase	Test an intervention's harmful or toxicity potential through in vitro and in vivo studies before involving human subjects
	Phase 0	Quantify and detect any harmful effect on a very small number of people
Clinical trials	Phase I	Rule out harm by testing whether a new intervention is safe and look for maximally tolerated dose and the best route to deliver the treatment
	Phase II	Study efficacy of new treatment in a few subjects to determine risk/benefit ratio
	Phase III	Test whether a new treatment is better than a standard treatment.
	Phase IV	Find more information about long-term benefits and side effects in a large population following marketing.

when they publish. IRB also determines what qualifies to be considered human-subject research in the first place. For example, the analysis of deidentified nationwide Healthcare Cost and Utilization Project databases or deidentified United Network for Organ Sharing datasets is considered data-driven research but not human subject research, despite being clinical research, since the analysis is conducted on retrospective data not governed by the researchers' institution directly. However, the agencies providing the data will still require the researchers to participate in training and agree to conduct their research ethically. IRB determines who can qualify to be a primary investigator, regulates collaboration between different centers, and may have prerequisite courses to ensure that the research team is familiar with current guidelines for research conduct and ethical consideration (e.g., IRB usually requires the Collaborative Institutional Training Initiative certificates, also known as CITI Program). The Human Research Protection Office (HRPO) audits and mandates professionalism within clinical research to protect patients from fraud and improper clinical research and lends legitimacy to the clinical research being performed by an academic institution. Fleshman's summary of all agencies involved in clinical research more than a decade ago remains comprehensive.[1] Regulators from different countries have their own criteria to evaluate a new intervention, and many of them—from the United States, European Union, Japan, and China—have issued guiding documents to allow globalized and multiregional clinical trials to improve patient access to new and innovative treatments.[2] This granular effort to support ethical consideration and research meticulous development in clinical settings significantly influenced empirical therapy (based on clinicians' traditions, beliefs, or biased opinions).[3] However, this new era of practice has not yet completely transformed cardiology to be an evidence-based field, where therapies are based on the results of well-performed and informative clinical trials.

Observational or interventional research

Quantitative study design is broadly subdivided into observational or interventional research. Observational design may employ groups and control variables to answer a research question with a hypothesis, resulting in a type of analytical research designed to increase the quality of yielded evidence (Fig. 16.1). Analytical research seeks to find a significant correlation between exposure and outcome in observational studies. Otherwise, observational studies without groups will be descriptive research. The descriptive type of research describes an individual case or a cohort of patients in their natural settings and explores the characteristics of patient participants. Most observational studies that explore the likelihood of death, recovery, complication rates, or quality of life outcomes from a single intervention fall under this category. On the other hand, interventional research (also called experimental research), prospectively applies an intervention in a group, which will then get measured and compared to a control group, with or without randomization. In a randomized design, the participants are randomly assigned to two or more groups using a known randomization technique (e.g., random number table). Nonrandomized design is used when it is neither feasible nor ethical to randomize participants into groups for comparison. It is sometimes referred to as a quasiexperimental design, which involves the use of prospective or retrospective data from the same or different participants as the control group (Fig. 16.1).

Randomized controlled trials (RCTs) provide the highest level of evidence in research, and among RCTs, multicenter studies are—usually—better than single-center ones. However, RCTs are not feasible for rare or high-risk events such as end-stage heart failure. In 1968, Dr. Cooley performed the first successful heart transplantation and in 1969, he took a total artificial heart from Dr. DeBakey's lab to perform the first clinical implantation of the device.[4] The

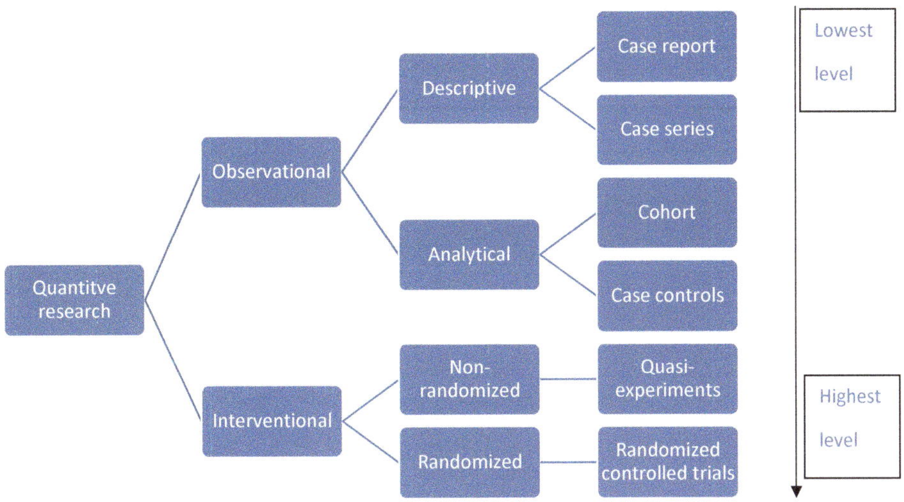

FIGURE 16.1 Types of research and the level of evidence (*arrow*).

dire need at the time provided a level of evidence that now allows us to perform about 2000–3000 transplants each year.[5,6] For more than half a century, the clinical evidence of the benefit of heart survival accumulated in this field from observational studies, since an RCT is ethically unacceptable. One of the principles of such trials in these vulnerable patients with rare diseases is that there is genuine uncertainty about the comparative therapeutic merits of each arm in a clinical trial.[7] This indicates the importance of observational studies—and sometimes expert opinion—in providing reliable evidence. Furthermore, RCTs cannot be truly free of bias.[8,9] Nevertheless, experimental clinical research—whether in the context of RCTs or adjusted data-driven analyses—shares many similarities besides the paperwork process, and we will discuss throughout this section "fundamentals of clinical research" how to choose and refine a research question, select a study population, design a study, and specify an outcome, as well as discuss the impact of different statistical tests and sample sizes on the study results. In the rest of this chapter, instead of discussing similarities, we will focus on the differences between the two ends of experimental research: RCTs and observational studies (Table 16.2).

Randomized controlled trials

When performing clinical trials, researchers compare groups of patients receiving the experimental intervention to standard therapy (or no therapy) to see if the intervention is more effective or has fewer side effects. Randomization helps prevent bias that may occur when a trial's results are affected by human choices or other factors not related to the treatment being tested. The Food and Drug Administration (FDA) works to protect participants in clinical trials and inform them adequately. FDA oversees the regulations and guidelines for each new intervention labeled as either an investigational drug or device. An investigational new drug application (IND) or investigational device exemption (IDE) through the FDA allows investigators to administer a drug or use a device in a clinical study. Following considerable efforts by institutional and national regulatory bodies, a rigorous process takes place to account for medical ethics[10] before an intervention gets tested on a human subject.

RCTs are high-stakes research, as they could identify medical practices that offer no net benefits or even do harm. They can result in the reversal of established medical

TABLE 16.2 Difference between observational and interventional studies.

	Interventional studies	Observational studies
Design	Assign exposures—observe outcomes	Observe both exposures and outcomes
Advantage	Yield a high level of evidence with more accuracy and certainty	Less expensive since the data could be available or patients are already undergoing the intervention
Disadvantage	Tailor-made data require complex tools and a considerable budget	Difficult to make casual inferences between exposure and outcome

practice,[11] as in a review of trials in the New England Journal of Medicine over a decade (2001–2010). This review found that, of 363 trials that tested an accepted medical practice, 146 (40%) were ineffective compared with a previous standard or its omission.[12] However, clinicians should consider two important concepts while using medical evidence from RCTs:

1. The nature of medical illness and its progress pattern may not be predicted precisely with a trial. Real-world data may reveal a different pattern.
2. Trials give us the average effects of a treatment on a select group of patients; however, the way that average effect applies to each patient could be unpredictable and variable.

Observational studies

Since there is still a great demand for research to support evidence-based practices for different clinical application sizes, scopes, and complexity, many investigators analyze their own data to enhance their practice model. These investigators try to generate satisfactory evidence from observational data based on trend data and correlation significance or mimic RCTs with sophisticated research design and statistical methods.[13,14] Others combine evidence obtained from multiple clinical trials, observational studies, or both to fill in scientific gaps.[15,16] The results of these studies will influence the evidence obtained from future RCTs—especially those that reevaluate old practices against state-of-the-art therapies and interventions. The way forward seems clear in an era of evidence-based medicine. Both observational studies and RCTs will allow better and more reliable results to be the basis for research-based decisions going forward. An appropriate selection of study designs will help ensure that researchers will achieve their research objectives and address their research questions.

Get started

- Research goals should be clear, and the tools required to answer a specific research question should be predetermined to complete a research project's objectives.
- Multiple regulatory committees and organizations have rules and protocols that dictate and facilitate or hinder the ability to conduct research timely and efficiently. Being familiar with and following the rules and protocols early on will reduce unnecessary delays
- Collaborating with a qualified team in data management and analysis will ensure that a study is designed properly and sufficient and accurate data are obtained to answer the research question.

Pitfalls to avoid

- Researchers should be aware of the bias and confounding associated with studies that are nonrandomized, uncontrolled, and without proper statistical adjustment.
- Single-center and/or small sample-sized RCTs may not be reliable for changing practices or adopting a new intervention.
- Levels of evidence provide a guide, and researchers need to be cautious when building their research based on interpreting results on the low end of the evidence hierarchy.

References

1. Fleshman Jr JW. Getting started in clinical research. *Clin Colon Rectal Surg*. 2011;24(2):100–105.
2. Shenoy P. Multi-regional clinical trials and global drug development. *Perspect Clin Res*. 2016;7(2):62–67.
3. Lamas GA, Pfeffer MA, Hamm P, Wertheimer J, Rouleau JL, Braunwald E. Do the results of randomized clinical trials of cardiovascular drugs influence medical practice? The save investigators. *N Engl J Med*. 1992;327(4):241–247.
4. Cooley DA. Feuds: social and medical. *Tex Heart Inst J*. 2010;37(6):649–651.
5. Alnajar A, Frazier OH. The state of artificial heart therapy. *Tex Heart Inst J*. 2019;46(1):77–79.
6. Everly MJ. Cardiac transplantation in the United States: an analysis of the UNOS registry. *Clin Transpl*. 2008:35–43.
7. Freedman B. Equipoise and the ethics of clinical research. *N Engl J Med*. 1987;317(3):141–145.
8. Byar DP, Simon RM, Friedewald WT, et al. Randomized clinical trials. Perspectives on some recent ideas. *N Engl J Med*. 1976;295(2):74–80.
9. Jadad AR, Rennie D. The randomized controlled trial gets a middle-aged checkup. *JAMA*. 1998;279(4):319–320.
10. Van Norman GA. Drugs, devices, and the FDA: Part 2: an overview of approval processes: FDA approval of medical devices. *JACC Basic Transl Sci*. 2016;1(4):277–287.
11. Prasad V, Gall V, Cifu A. The frequency of medical reversal. *Arch Intern Med*. 2011;171(18):1675–1676.
12. Prasad V, Vandross A, Toomey C, et al. A decade of reversal: an analysis of 146 contradicted medical practices. *Mayo Clin Proc*. 2013;88(8):790–798.
13. Gay HC, Baldridge AS, Huffman MD. Feasibility, process, and outcomes of cardiovascular clinical trial data sharing: a reproduction analysis of the SMART-AF trial. *JAMA Cardiol*. 2017;2(12):1375–1379.
14. Bartlett VL, Dhruva SS, Shah ND, Ryan P, Ross JS. Feasibility of using real-world data to replicate clinical trial evidence. *JAMA Netw Open*. 2019;2(10):e1912869.
15. Hoffman SN, TenBrook JA, Wolf MP, Pauker SG, Salem DN, Wong JB. A meta-analysis of randomized controlled trials comparing coronary artery bypass graft with percutaneous transluminal coronary angioplasty: one- to eight-year outcomes. *J Am Coll Cardiol*. 2003;41(8):1293–1304.
16. Lopes RD, Hong H, Harskamp RE, et al. Safety and efficacy of antithrombotic strategies in patients with atrial fibrillation undergoing percutaneous coronary intervention: a Network meta-analysis of randomized controlled trials. *JAMA Cardiol*. 2019;4(8):747–755.

Chapter 17

The question: Types of research questions and how to develop them

Medhat Farwati[1], Anna Scandinaro[1], Elizabeth A. Babcock[2] and Ahmed Alnajar[3]

[1]Cleveland Clinic Foundation, Cleveland Clinic, Cleveland, OH, United States; [2]University of Miami, Miami, FL, United States; [3]The Division of Cardiothoracic Surgery, The Department of Surgery, The University of Miami Miller School of Medicine, Miami, FL, United States

Key points

- The correct research question must be posed to identify the appropriate study design and statistical analysis to be employed in a research study.
- If the appropriate study design and analysis are not used, data can easily be misunderstood in the absence of well-defined questions.
- Correlation is not equal to causation. Research questions that aim at establishing that a certain exposure/ intervention is on the causal pathway for a specific outcome are best addressed with a randomized clinical trial.
- It is critical to decide on an appropriate statistical analysis at the planning stage before initiating the study, taking into consideration the desired sample size and the type of data that will be collected and analyzed.

Why it matters?

The overall conclusion of a study and the interpretation of the results, both depend on the specific question one seeks to answer and the study designed to provide that answer. Incorrect conclusions can be presented when a study is interpreted outside of the scope of the research question and the study design. Therefore, posing the appropriate study questions and choosing the proper study designs are imperative to correctly interpret the results of a study and communicate the findings with the scientific community as well as the general public.[1]

For example, a systematic review of 93 randomized clinical trials (RCTs) of cardiovascular research in high-impact journals showed that manipulative language or "spin" was used in 57% of abstracts and 67% of main texts. Whether intentional or not, this misrepresentation can lead to commonly accepted falsehoods among scholars and the general public at large.[2] As a researcher, it is critical to develop a well-thought-out study question and design to interpret data effectively and prevent common reporting mistakes. As a literature reviewer, it is essential to grasp the value of assessing the research question to properly analyze the presented data and proposed conclusions. As a trialist, it is invaluable to consider prior knowledge when formulating a question to understand whether a proper analysis will need an adaptive design for Bayesian statistics or not.

Selecting the appropriate research question

To determine the study design and statistical analysis to be used in a research study, the right question must be asked. Without a well-defined question, data can easily be misinterpreted if the appropriate corresponding study design and analysis are not applied. Understanding the specific type of question being addressed in a given research endeavor allows for data to be reproducible—a key feature of a successful research study. It is important to evaluate the data that is available to determine what questions a study can answer. There are six types of questions researchers commonly consider when designing a research study: descriptive, exploratory, inferential, predictive, causal, and mechanistic.[3,4] These types are summarized in Table 17.1. The qualities of datasets that the researcher has access to, as well as the general purpose of the study, are frequently used to select what type of question to utilize (Fig. 17.1).

A *descriptive* question asks "what" instead of "how." For example: what proportion of people born with atrial septal defects (ASDs) living in the Kentucky area have family members who were also born with ASDs? In a study by Caputo et al., the authors set out to investigate the familial recurrence of ostium secundum ASDs by examining the family history of 538 patients with this condition

Translational Cardiology. https://doi.org/10.1016/B978-0-323-91790-2.00061-7

TABLE 17.1 Types of questions asked in research studies.

Question type	Goal	Example
Descriptive	**Describe** or **summarize** a set of data.	What proportion of people born with atrial septal defects (ASDs) living in the Kentucky area have family members who were also born with ASDs?
Exploratory	**Examine** or **explore** data and find **relationships** between variables that were previously unknown.	What is the familial distribution of people born with an ASD?
Inferential	*Using a **small** sample of data to **infer** about a **larger** population.*	What proportion of people born with atrial septal defects (ASDs) living in the United States have family members who were also born with ASDs?
Predictive	Using historical or current **data** to find patterns to make **predictions** about the **future.**	What are the characteristics of a person born with ASD?
Causal	Looks at the **cause and effect** of relationships between variables, focused on **finding the cause** of a correlation.	What are the rates of ASDs in children born to mothers with ASDs compared to those without?
Mechanistic	Understand **exact** changes in variables that lead to other changes in other variables.	What gene or genes are responsible for the familial inheritance of ASD?

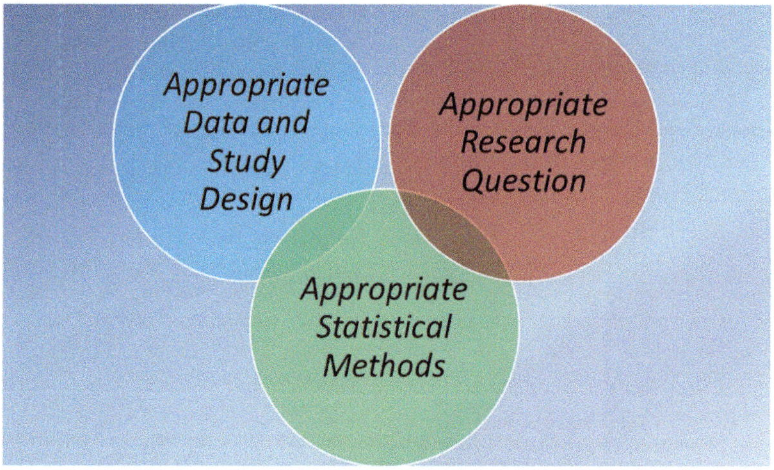

FIGURE 17.1 Key components of successful research study proposal.

compared to 408 healthy patients.[5] This is an example of a descriptive study asking: what is the familial recurrence of ASD? Another example is describing trends of admissions and readmissions for the specific patient population. In this regard, many researchers utilize nationwide databases for this purpose, such as the National Inpatient Sample (NIS) and/or Nationwide Readmissions Database (NRD) provided by the Healthcare Cost and Utilization Project. Agarwal et al. assessed the national trends in heart failure (HF) hospitalizations and readmissions spanning 7 years and including 8,273,270 patients admitted for HF in the United States.[6] They found that postdischarge HF readmissions declined between 2010 and 2014, followed by an increase between 2014 and 2017.

In contrast, an *exploratory* question attempts to generate a hypothesis by looking for specific trends or patterns seen in a data set. For example, by examining the familial distribution of people born with ASD, you could propose there is a genetic component to being born with this congenital anomaly. An important component of an exploratory question is that it can generate hypotheses but does not seek to understand them. These questions are used to propose directions for further research. An example from the literature is the exploratory question posed by Vaidya et al. looking at trends in cardiovascular implanted electronic devices.[7] This study investigated a particular population, patients in Olmstead County who received an implanted electronic device between 1988 and 2018, and is an example of an exploratory question aimed to evaluate trends and generate possible hypotheses for further research. Comparison studies are another example, they explore the outcomes of one treatment or approach over the other. In such a case, this type of question would answer whether an emerging treatment, such as Edoxaban, would be noninferior to Warfarin in patients with venous thromboembolism,[8,9] or whether minimally invasive approaches, such as open minimally invasive approach versus traditional sternotomy or transcatheter approach for aortic valve replacement versus traditional open approach,[10,11] result in better (superior) outcomes. When percutaneous left atrial appendage closure emerged as a minimally invasive option to prevent stroke, Vuddanda et al. determined to explore the outcomes of this option and compare both epicardial and endocardial approaches to explore which approach has better outcomes.[12] The authors used the NIS data for in-hospital outcomes and NRD data for readmission outcomes to answer their research question. They found that the epicardial approach has higher complications and readmissions; however, given recent improvements in this approach's technique and postprocedural care, it demonstrated a significant reduction in pericardial complications.

An *inferential* question is used to determine if the proposed hypothesis from an exploratory question holds true in another data set. For example, the hypotheses generated from the previous exploratory question, suggest that a person born with ASD has a genetic predisposition. An inferential question can be utilized to establish if this conclusion continues to be valid in a different setting or population, for example looking at those born with ASDs in the whole United States. In an example from the recent cardiology literature, Whitlock et al. asked an inferential question where they aimed to determine whether left atrial appendage occlusion will reduce ischemic strokes in a subgroup of patients with atrial fibrillation (AF) undergoing cardiac surgery, similar to the general AF population.[13]

A *predictive* question asks if there are factors that are associated with the proposed given outcome. For example: what are the features of a person born with ASD? This question is not intended to ascertain whether the individual born with an ASD has a family member who also has an ASD; rather, it is intended to elicit additional information about other factors that may influence whether a person is born with an ASD.

Additionally, this question can be used to demonstrate that a variable predicts a particular event but does not explain why it occurs. Callahan et al. proposed a predictive study question when looking at factors associated with death or transplant after cavopulmonary shunt compared to those who underwent Fontan completion in patients with tricuspid atresia.[14] This study was used to find information about the events surrounding a particular outcome: death, transplant, or Fontan completion, in a specific population, patients with tricuspid atresia. In other words, this question was used to design research that identifies characteristics associated with a certain result but does not test for a causal link. Referring back to Vuddanda et al.'s study for epi- versus endocardial percutaneous left atrial appendage closure technique, the first author in his published thesis[15] attempted to identify the predictors of unplanned readmission within 30 days. Using the same data, the author used a machine learning approach, however, since the study did not consider the survey design of the data, this introduced a bias into the analysis. For prediction, Bayesian models can add value from expert knowledge and clinical data to predict critical events early and accurately. For example, Ruiz et al. collected data from 93 patients with a single ventricle and used naïve Bayesian models for prediction.[16] They found that these models may enhance clinicians' ability to identify infants with single-ventricle physiology at high risk of critical events with high sensitivity (84%) and specificity (81%). This leveraged a researcher's predictive question to detect events early on and allow for timely interventions; thus, reducing morbidity, mortality, and healthcare costs.

A *causal* question is used to determine the relationship between two variables. If you change one variable, does it directly impact the outcome? It can also measure the size and direction of the relationship. For example, if you have two mothers, one was born with an ASD and one who was not, what are the rates of ASDs in children born to mothers with ASDs compared to those without. Cha et al. aimed to identify if a difference in symptom-to-door time was associated with worse outcomes in a patient hospitalized with non-ST-elevation myocardial infarction (NSTEMI), by stratifying the group into those whose symptom-to-door time was $\geq$24 h compared with less than 24 h.[17] In this example, answering the study question aids in establishing a causal relationship between time from symptom onset to presentation and mortality.

Finally, a *mechanistic* question seeks to answer "how." In our previous example, let us assume we have concluded that there appears to be a genetic component to the development of an ASD but we do not know how this occurs. A mechanistic question would seek to answer the genetic basis for the proposed relationship. Yeh et al. studied atrial ERK1/2 activation in a mouse model and its role in developing ASD. They found that enhanced atrial MEK1-ERK1/2 signaling during fetal development disrupts normal atrial septation, possibly regulated by the balance of ERK1/2 phosphorylation.[18] Another example is the work of Devarakonda et al. who examined how a specific cardiac gene, relaxin family peptide receptor 1 (RXFP1), protects against acute myocardial infarction in a mouse model.[19] Mechanistic questions are commonly asked in basic science as an exact mechanism often requires molecular and animal studies.

Appropriate data and study design

After selecting an appropriate research question, it is equally important to identify the study design that is best suited to answer the question at hand. Frequently, the research question determines the study design and type of data used in the study. Researchers may worry about whether or not the question they are attempting to answer is the correct question. However, it is equally important to have a valid procedure to answer such research questions and to verify the statistical validity of the results. A relatively simple approach to ascertain the appropriate study design is to review the language used in drafting the research question. For example, descriptive questions usually attempt to answer "how often" or "how many" to determine the prevalence or occurrence of a phenomenon. One example would be "how many individuals in the US with coronary heart disease (CHD) continued to smoke cigarettes in the past year?" In this scenario, an epidemiological study with a cross-sectional design and data collected via a postal survey would be acceptable. If the inquiry begins with "why" or "how," it is typically a query aimed at learning more about phenomena. For instance, "how did internal medicine residents deal with burnout during the COVID-19 pandemic?" The data to answer this question would be best collected utilizing a qualitative research design that includes in-depth interviews or focus groups. When the wording "is A better than B as it relates to a certain outcome?" is used, this usually necessitates obtaining randomized data that directly compare both interventions/strategies. An example would be "is torsemide better than furosemide at reducing readmissions in patients with heart HF?" A randomized controlled trial that compares these two medications head-to-head in HF patients would probably provide the greatest insight into this question.

Researchers should be mindful of how their research questions and hypotheses tie together. In general, we assess averages from our sample in which we utilize for the experiment. For example, a clinical trial experiment question will sound like "Is treatment X_1 superior to treatment X_2 on condition Y, on average?" To be specific, since Shim et al. proved Edoxaban noninferior to Warfarin,[20] let us generate a research question for a superiority trial "Does Edoxaban, on average, reduce the risk of stroke following heart valve repair or replacement more than Warfarin?" Assuming our experiment does show a positive effect, the statistical significance tests will answer other questions, such as: what are the chances that Edoxaban was an ineffective treatment, yet it could produce the results that we saw? How likely is it that another researcher could validate our results and be able to replicate our study with the same size trial? We will be asking when facing a real patient in our practice "If I prescribe Warfarin to this patient—the one that just recovered from a valve replacement procedure and waiting to discharge home—will his/her stroke risk decrease?" Since there is only one patient, there is no use for an "average." In this case, we want to incorporate the prior knowledge into the probability of preventing stroke for this patient, and that is a simple Bayesian question. Bayes rule allows us to reverse probabilities and calculate the probability of a hypothesis given the data instead of just calculating the probability of data given a hypothesis (the basis of frequentist statistics including likelihood). In many scenarios, the probability of the hypothesis could be more useful. While the results of the analyses can be similar, Bayesian and Frequentist methods are based on fundamentally different theories about probability. Thus, at the research question phase, the major difference between the Bayesian's and Frequentist's approaches to conduct research should be understood early on. In Table 17.2, we summarize the major differences between Bayesian and Frequentist statistics.

When considering the appropriate study design for a research question, it is important to avoid some common pitfalls. One of the most widespread errors in the research environment is to draw a causal link between exposure and results from observational data. Ideally, research questions that aim at establishing that a certain exposure/intervention is on the causal pathway for a given outcome are best addressed with a randomized clinical trial. However, the medical literature is replete with examples where mere correlations are misconstrued as causative on the basis of observational data. In a recent preprint study from the Vaccine Adverse Event Reporting System (VAERS) retrospective database looking into adverse events post-vaccination with the mRNA COVID-19 vaccine, cardiac adverse events including myocarditis were found to be occurring at a rate of 162.2 per million among males 12−15 years old.[21] This incidence was reported to be higher than the expected 120-day COVID-19 hospitalization rate for males in this age group. Although it has not been published in a peer-reviewed journal yet, this study has sparked intense debate in an era of vaccine hesitancy[22,23] and its findings are frequently quoted to hint toward a causal relationship between COVID-19 mRNA vaccine and the incidence of myocarditis.[24,25] Despite statistically sound methods and subject-expert authors of this preprint, the message of this work was heavily criticized due to the numerous limitations inherent to the VAERS database.[22] Therefore, it is critical to thoroughly analyze the study questions and the data employed to answer them before reaching any conclusions; particularly for research whose objective is to establish a causal link between variables.

TABLE 17.2 Bayesian versus Frequentist statistics.

Frequentist approach	Bayesian approach
Probability is the fraction of times that an event will occur in many trials.	Probability is a degree of belief that something will occur.
Calculate the probability of getting the observed data or something more extreme given the null hypothesis.	Calculate the probability that each hypothesis is true given the data.
True parameter values are unknown constants.	Parameter values have probability density functions representing their uncertainty.
The observed data set is a random draw from a population (a variable).	The data are a constant.
If an experiment is repeated many times, the true value will be included in the 95% confidence interval 95% of the time.	Parameter values have probability density functions that reflect uncertainty about the value of the parameter. There is a 95% chance that the true mean is in the 95% credible interval.
Examples	
• The probability of getting a second heart attack before discharge is 0.1%. • If we repeat the experiment many times, there will be a type I error 5% of the time.	• Your probability of surviving a heart attack is 55%. • There is a 67% chance that the median length of stay is greater than 2 days following ASD occlusion.

Appropriate statistical methods

Following the formulation of a research question and selecting the most appropriate study design to address the question, consideration should be given to the statistical methods that will be utilized to evaluate the data. As with choosing the appropriate study design, a few key concepts may assist in selecting the proper statistical test for hypothesis testing. Similarly, it is crucial to decide on an appropriate statistical analysis at the planning stage before initiating the study and to choose an acceptable sample size. These cannot be determined randomly following the conclusion of the investigation and the collection of data. An important factor to consider is the type of data that will be analyzed (numerical vs. categorical, paired vs. unpaired, 2 groups vs. > 2 groups). For example, it is necessary when dealing with numeric data to determine if they conform to the parameters of the normal distribution in which case parametric tests are used. On the other hand, it is preferable to employ nonparametric testing if the data distribution is not normal or if the distribution is unknown.[26]

As demonstrated in the study design examples above, it should be noted that using inappropriate statistical models might result in misleading conclusions and inaccurate interpretations of study data. The list of potential statistical blunders is exhaustive, but some of the most frequently encountered ones include incorrect interpretation of *P*-values, insufficient sample size and power estimates to support study findings, and deceptive data visualizations.[27,28] Having these pitfalls in mind when formulating the question will lead to selecting proper statistical analyses and rigorous study design, which is critical for an accurate answer to the research question, as well as understanding and interpretation of the study's findings by readers from different backgrounds.

Get started

1. Determine the type of question you look for when searching for evidence. This will determine the type of question, the study design, and the type of data needed to answer the question.
2. Develop a question. Consider a thorough question while avoiding spending precious free time aimlessly searching through journals. The question must be clinically relevant, well thought out, and answerable. As we stated in the examples above, treatment modalities and techniques could be associated with better outcomes and less complication rates, in such cases, impacting the outcomes will make your study timely and significant.
3. Utilize PICO to refine your question and SMART for your objectives. PICO is an acronym for patient/problem, intervention, comparison intervention, and outcomes. This tool can help make your question more focused. SMART is an acronym that can help outline a strategy for reaching the study objectives. SMART goals are specific, measurable, achievable, realistic, and anchored within a time frame. However, when discussing acronyms for medical research questions, one should not forget about the importance of ethics, which is incorporated into the FINER acronym, which formulates the question to be feasible, interesting, novel, ethical, and relevant.

4. **Involve a statistician early on.** Since the analysis is crucial to answering a question, the study design has already been determined when the question is developed. Late collaboration may lead to a heavily disliked postmortem statistical analysis.

Pitfalls to avoid

1. Failure to critically assess the qualities of datasets that the researcher has access to, as well as the general purpose of the study in the process of deciding what type of research question to investigate.
2. Drawing a causal link between exposure and results from observational data is one of the most common pitfalls when interpreting research findings. Making valid causal inferences is challenging because it requires high-quality data and adequate statistical methods. While causal inference from observational data can be valid, it is an attempt to emulate a (hypothetical) randomized trial. Nevertheless, an actual randomized clinical trial is still the best technique to answer research questions that try to prove that a specific exposure/intervention is on the causal pathway for a specific outcome. However, there are numerous examples in the medical literature, as discussed above, where correlations may be misinterpreted as causative based on observational data.
3. Failure to select an acceptable sample size due to a lack of adequate statistical planning before starting the investigation. As previously stated, deciding on a suitable statistical analysis plan a priori is critical to allow for meaningful analysis and interpretation of study findings.

References

1. Moyé L. Statistical methods for cardiovascular researchers. *Circ Res.* 2016;118(3):439–453.
2. Khan MS, Lateef N, Siddiqi TJ, et al. Level and prevalence of spin in published cardiovascular randomized clinical trial reports with statistically nonsignificant primary outcomes: a systematic review. *JAMA Netw Open.* 2019;2(5). e192622–e192622.
3. Peng RD, Matsui E. *The Art of Data Science: A Guide for Anyone Who Works with Data.* Skybrude consulting LLC; 2016.
4. Leek JT, Peng RD. What is the question? *Science.* 2015;347(6228): 1314–1315.
5. Caputo S, Capozzi G, Russo MG, et al. Familial recurrence of congenital heart disease in patients with ostium secundum atrial septal defect. *Eur Heart J.* 2005;26(20):2179–2184.
6. Agarwal MA, Fonarow GC, Ziaeian B. National trends in heart failure hospitalizations and readmissions from 2010 to 2017. *JAMA Cardiol.* 2021;6(8):952–956.
7. Vaidya VR, Asirvatham R, Kowlgi GN, et al. Trends in cardiovascular implantable electronic device insertion between 1988 and 2018 in olmsted county. *Clin Electrophysiol.* 2022;8(1):88–100.
8. Büller HR, Décousus H, Grosso MA, et al. Edoxaban versus warfarin for the treatment of symptomatic venous thromboembolism. *N Engl J Med.* 2013;369(15):1406–1415.
9. Alnajar A, Rao JS, Cayetano SM, Lamelas J. Commentary: battle of the bioprosthetic valve blood thinners. *J Thorac Cardiovasc Surg.* 2021;165(1):69–70.
10. Lamelas J, Sarria A, Santana O, Pineda AM, Lamas GA. Outcomes of minimally invasive valve surgery versus median sternotomy in patients age 75 years or greater. *Ann Thorac Surg.* 2011;91(1):79–84.
11. Popma JJ, Deeb GM, Yakubov SJ, et al. Transcatheter aortic-valve replacement with a self-expanding valve in low-risk patients. *N Engl J Med.* 2019;380(18):1706–1715.
12. Vuddanda VLK, Turagam MK, Umale NA, et al. Incidence and causes of in-hospital outcomes and 30-day readmissions after percutaneous left atrial appendage closure: a US nationwide retrospective cohort study using claims data. *Heart Rhythm.* 2020;17(3):374–382.
13. Whitlock RP, Belley-Cote EP, Paparella D, et al. Left atrial appendage occlusion during cardiac surgery to prevent stroke. *N Engl J Med.* 2021;384(22):2081–2091.
14. Callahan CP, Jegatheeswaran A, Barron DJ, et al. Factors associated with mortality or transplantation versus Fontan completion after cavopulmonary shunt for patients with tricuspid atresia. *J Thorac Cardiovasc Surg.* 2021;163(2):399–409.
15. Vuddanda VLK. *Percutaneous Left AtrialAppendage Closure: Incidence, Causes and Predictors of InHospital Outcomes and Readmissions a United States Nationwide Retrospective Cohort Study Using Claims Data.* Master's thesis. Harvard Medical School; 2019.
16. Ruiz VM, Saenz L, Lopez-Magallon A, et al. Early prediction of critical events for infants with single-ventricle physiology in critical care using routinely collected data. *J Thorac Cardiovasc Surg.* 2019;158(1):234–243.e3.
17. Cha J-J, Bae S, Park D-W, et al. Clinical outcomes in patients with delayed hospitalization for non–ST-segment elevation myocardial infarction. *J Am Coll Cardiol.* 2022;79(4):311–323.
18. Yeh CC, Fan Y, Yang YL, Mann MJ. Atrial ERK1/2 activation in the embryo leads to incomplete Septal closure: a novel mouse model of atrial Septal defect. *J Biomed Sci.* 2017;24(1):89.
19. Devarakonda T, Mauro AG, Cain C, Das A, Salloum FN. Cardiac gene therapy with relaxin receptor 1 overexpression protects against acute myocardial infarction. *JACC Basic to Transl Sci.* 2021;7(1): 53–63.
20. Shim CY, Seo J, Kim YJ, Lee SH, De Caterina R, Lee S, et al. Efficacy and safety of edoxaban in patients early after surgical bioprosthetic valve implantation or valve repair: a randomized clinical trial. *J Thorac Cardiovasc Surg.* 2021;165(1):58–67.
21. Høeg TB, Krug A, Stevenson J, Mandrola J. SARS-CoV-2 mRNA vaccination-associated myocarditis in children ages 12-17: a stratified national database analysis. *medRxiv.* 2021:21262866. https://doi.org/10.1101/2021.08.30.21262866.
22. Munro C. Covid-19: study that claimed boys are at increased risk of myocarditis after vaccination is deeply flawed, say critics. *British Med J Publish Group.* 2021;374:n2251.
23. Verd S, Fernández-Bernabeu M, Cardo E. The controversy surrounding vaccination of young people against COVID-19. *Acta Paediatr.* 2021;111(1):187.
24. Liester MB. A review of the use of persuasion and coercion to overcome COVID-19 vaccine hesitancy. *J PeerScientist.* 2021;4(2): e1000037.

25. Martin D. *Learning to Think from Both Sides—and to Make Your Own Decision with Regard to Vaccinating 5-11 Year-Olds against Covid-19*. 2021.
26. Nayak BK, Hazra A. How to choose the right statistical test? *Indian J Ophthalmol*. 2011;59(2):85.
27. Engledowl C, Weiland T. Data (Mis) representation and COVID-19: Leveraging misleading data visualizations for developing statistical literacy across Grades 6−16. *J Stat Data Sci Educ*. 2021:1−5.
28. Slutsky DJ. Statistical errors in clinical studies. *J Wrist Surg*. 2013; 2(04):285−287.

Suggested readings

1. Peng RD, Matsui E. The Art of Data Science: a guide for anyone who works with Data. Skybrude consulting LLC; 2016. Nayak BK, Hazra A. How to choose the right statistical test? *Indian J Ophthalmol*. 2011;59(2):85.
2. Slutsky DJ. Statistical errors in clinical studies. *J Wrist Surg*. 2013;2(04):285−287.

Chapter 18

Study population: Who and why them?

Kelley N. Benck[1], J. Sunil Rao[2] and Ahmed Alnajar[2]
[1]University of Southern California Keck School of Medicine, Los Angeles, CA, United States; [2]Division of Biostatistics, The Department of Public Health Sciences, The University of Miami Miller School of Medicine, Miami, FL, United States

Key points

- Being familiar with the study populations and sampling methods is a fundamental step before the study design.
- Study population size and demographics are important components of a study's generalizability.
- Inclusion of underrepresented populations can decrease the likelihood of having erroneous conclusions stemming from the study results.

Why it matters?

The results of both randomized clinical trials (RCTs) and real-world observational studies can be compromised if their samples are not representative of the general population. This may occur especially when RCTs are conducted under idealized and rigorously controlled conditions, compromising the study's external validity. Thus, it is important to utilize representative samples, in which the relevant characteristics of the study participants are generally the same as those of the population, especially in a critical and quickly advancing field such as cardiology.

Introduction

The study population is one of the most important components of a research project, providing the foundation with which to assert generalizability. In well-composed studies, the study population is a representative sample of the target population. However, it is often difficult to achieve truly representative study populations, and this failing is frequently listed as a limitation. Such a weakness will result in poor representativeness, which limits a study's generalizability and may require varying case-by-case correction to draw meaningful conclusions about the target population. Furthermore, any sort of bias at the early stages will impact the reproducibility and external validity of the work. In this chapter, an introduction to the key sampling techniques will be provided along with examples from cardiovascular research. Since understanding the differences in sampling techniques may aid researchers in effectively appraising literature and designing sound projects, this chapter aims to describe how to draw a representative study population using robust inclusion and exclusion criteria. We then offer methods to overcome challenges in obtaining a representative sample to provide a reference point for researchers as well as students, residents, and physicians who engage in cardiovascular research.

Size and power

One of the first considerations researchers must make when creating a robust study is generating a sample size representative of the target population. As one would expect, the larger the study population, the greater the generalizability; however, researchers must be realistic with a size that is both manageable and attainable. Power and sample size calculations can be used to find that balance.

The power of a study is derived from its ability to detect a deviation from the null hypothesis—no difference between the two groups being studied—if one exists. Power calculations are important in helping avoid two main types of statistical errors: type I (falsely concluding there is a difference) or type II (falsely concluding there is no difference). As the relationship between power and sample size is directly proportional, power calculations tell researchers the size needed to avoid those two errors.

It is difficult for studies without sufficient sample size or power to derive adequate and generalizable conclusions from their data. For example, Shim et al. reported that the anticoagulant edoxaban was a noninferior alternative to warfarin in 220 patients with postsurgical bioprosthetic valve implantation or repair.[1] Though such a conclusion has significant implications for clinicians,[2] the study was

Translational Cardiology. https://doi.org/10.1016/B978-0-323-91790-2.00068-X

limited by the small number of patients and events in each group. In contrast with a prior guidelines-changing study, Giugliano et al. included 21,105 patients with atrial fibrillation to demonstrate edoxaban noninferiority to warfarin.[3] Thus, researchers should foresee limitations that detract from their study's applicability in clinical practice upon calculating study power with adequate noninferiority margins (as we will discuss in the next chapters). Inadequate sample sizes also have the ability to affect data. Arora et al.[4] and Velu et al.[4] both aimed to study MitraClip device outcomes in patients with and without atrial fibrillation; however, the former study had almost five times the population as the latter. While Velu's study with its small sample size found no difference in 1-year outcomes between patients with and without atrial fibrillation after MitraClip therapy, Arora's group—with the arguably more generalizable study population based on size—found patients with atrial fibrillation had statistically significant increases in mortality and rehospitalization after 1 year. These examples demonstrate the impact of sample size on data quality and the study's outcomes.

Sampling

Sampling is the method by which participants are selected for inclusion into a study population. This process must be established before the sampling begins and should be comprehensive to avoid the omission of an integral part of the sample that will result in selection bias. There are two broad categories of sampling—nonprobability and probability—for researchers to select their population.[5,6] The main difference in study design between probability and nonprobability samples is the use of randomization. Some researchers propose using methods to link the two,[7,8] as this could improve the efficiency and accuracy by reducing the mean squared errors for the estimates. In this section, we will discuss these sampling methods and their strengths, limitations, and applicability.

Nonprobability sampling involves subjective, rather than random selection of participants, therefore a representation of the target population is not achieved. The lack of generalizability makes this category less popular; however, it can be useful in scenarios where there are ethical concerns in subjecting participants to an intervention, or the population of interest is too small and niche for random selection techniques. There are five types of nonprobability sampling: convenience, purposive, quota, snowball, and self-selection.

- *Convenience sampling*, as the name implies, involves selecting participants based on the ease of access for the researcher. The definition of "ease" can vary with the researcher, examples include geographic proximity and choosing consecutive patients from a list.
- *Purposive sampling*, also known as judgment sampling, is often used in qualitative studies and involves the researchers using their own judgment to select participants. It is often classically seen in studies aiming to report expert opinions on a subject of interest.
- *Quota sampling* aims to have proportional groups included in the study. The group's identity (e.g., age, sex, race, disease type) must be established first for researchers to attempt to seek proportional representation.
- *Snowball sampling* uses its first participant(s) as a mechanism to gain more participants that also meet the study's inclusion criteria. This technique is especially useful for unique populations, which may require referrals to other relevant potential participants, such as when global surgery initiatives require identifying stakeholders in distant countries,[9] or when studies require interviews with participants with substance abuse and stigmatized lifestyles and/or diseases.
- *Self-selection sampling* involves participants volunteering to take part in the study.

Probability sampling uses the theory that all members of the target population have an equal probability of being selected. Through this category's random selection of participants, researchers can draw meaningful statistical conclusions from the data. Well-composed probability sampling has the ability to produce generalizable study populations. There are four types: simple random, systematic random, stratified, and clustered.

- *Simple random sampling* is when participants are numbered and picked based on randomly selecting numbers (e.g., random number generator).
- *Systematic random sampling* is similar to simple random sampling; however, the numbered participants are picked through a system of predefined, fixed intervals (e.g., every fifth participant).
- *Stratified sampling* involves dividing the overall target population into subgroups (strata) based on an identifying factor (e.g., income bracket, age groups) and randomly selecting participants from each subgroup. Selections can be done equally among the strata or proportionally depending on the variability in the size of the subgroups.
- *Cluster sampling* is similar to stratified sampling; this method subdivides the population, but each subgroup is equal in characteristics. Examples include dividing by institutions such as hospitals or schools or separating by cities within a county.

There are many sampling techniques available to researchers; however, the selection of the appropriate method depends on the goals of the study. A technique utilizing randomization is often seen as the best approach for achieving generalizability.

Inclusion and exclusion criteria

Diversity within study populations is perhaps one of the most important components to producing generalizable data. Investigators need to be alert to the possibility that their inclusion and exclusion criteria reflect sex, racial, ethnic, economic, or information disparities biases. Cardiology has a long history of studying homogenous groups, a limitation that has only begun to be recognized within the last decade. Representation has been lacking for women, Black, Indigenous, and other People of Color (BIPOC), older adults, and patients with comorbidities.[10]

Sex

Cardiovascular disease (CVD) has remained the leading cause of death for women of all ages, with some reports indicating 1 in 16 women have coronary heart disease alone.[11] Even though there is a higher reported CVD incidence in men, mortality from acute cardiovascular events is worse in women.[12] One contributing factor to this disparity is the lack of inclusion of women in cardiovascular research in areas of screening, diagnosis, and management of CVDs. Two large-scale systematic reviews[13,14] investigating study population diversity in cardiovascular studies reported an underrepresentation of women, who only made up 28% of participants in heart failure (HF) trials and 24.9% of participants in acute coronary syndrome (ACS) trials. This sex-based disparity is supported by other CVD studies,[15,16] including stroke trials[17] and lipid-lowering therapy trials.[18] More equitable inclusion of women in study populations is important, as CVD has such a high disease burden in both sexes. In addition, studies have shown that women present differently than men for diseases such as coronary artery disease (CAD), stroke, and HF. Women have been shown to present older for CAD and first-stroke events[12] and have greater mortality rates from HF even though they are more likely to have preserved ejection fraction, which, in men, has been linked to lower mortality.[19] One example that highlights how study population generalizability can better improve management guidelines is the work done by Skibba et al. investigating gender differences with abdominal aortic aneurysms (AAAs). Even though women only made up 23% of the study, they found AAAs ruptured at smaller diameters in women than men, and more frequently at diameters <5.5 cm.[20] This was significant because previous gender-neutral guidelines suggested repair only for AAAs >5.5 cm. The generalizability of this study contributed to changes in published AAA screening and treatment guidelines.[21]

In addition, studies have found women-specific risk factors[22] such as complications of pregnancy,[23] age of menarche,[24] and onset of menopause.[25,26] These differences in presentation, onset, and risk factors demonstrate the importance of female representation in study populations aiming to be generalizable.

Lastly, the difference in the presentation of women may affect their inclusion. For example, the later onset of stroke and CAD—as described above—as well as the lower rates of interventions[27,28] and treatment[29,30] provided to women may exclude them from certain trials.

Race and ethnicity

Racial and ethnic minorities have comparable, and in some cases worse, rates of mortality from CVD as white patients[31–34]; however, they have been disproportionately underrepresented in cardiovascular research. One systematic review reported that less than half of the 118 HF trials being investigated reported race/ethnicity, and of those that reported, only 37% stratified the primary outcome by race/ethnicity.[13] Although the inclusion of certain minority groups such as Hispanic and Asian patients has increased over the years, BIPOC groups remain underrepresented.[13,14,16] This disparity is most pronounced among Black patients as they consistently endure the highest mortality rates among all racial and ethnic groups.[35] This degree of underrepresentation has continued to be reported in clinical trials of cardiovascular medications.[16,36] Reporting and inclusion of these groups are important not only due to the higher burden of disease but also because of the greater incidence of comorbidities leading to an earlier and differing onset of CVD.[37–39]

One example of the impact of diverse study populations can be seen with research done on breast arterial calcifications (BAC), a potential indicator of CVD for women, detected on mammography.[40] Studies looking into the prevalence of BAC reported widely different results that varied with the inclusion of race. Literature including only white or >90% white participants[41,42] reported BAC rates between 11%–14%, while studies with more equal distributions of race or non-White majorities[43,44] reported prevalence closer to 29%–34%. Another demonstration of the need for generalizable study populations comes from the African American Study of Kidney and Hypertension (AASK) trial that investigated the benefit of blood pressure (BP) management in African Americans with chronic kidney disease (CKD). Previous research with minimal inclusion of African American participants[45] indicated that effective BP management slowed the progression of renal failure; however, Jackson et al. from the AASK trial demonstrated in this population, no slowing of CKD progression was found in patients with better controlled BP.[46]

Researchers should aim to include underrepresented groups in the analysis of their primary outcome but should be mindful of potential obstacles in enrollment. Such barriers include socioeconomic, cultural, occupational, familial, and transportation factors.[47–49] In addition, some populations exhibit more fear and distrust of research and/or the healthcare industry.[50] Literature reviews have shown that BIPOC patients report high rates of biased treatment[51] and lack of satisfaction in the treatment they are receiving leading to mistrust and perceived bias of care.[52]

Elderly and comorbidities

Age and comorbidities such as obesity, diabetes mellitus, and renal disease are established risk factors for the development of CVD[53–55]; however, they are underrepresented.[10,15,56,57] Studies have shown limited reporting of age, but of those that did report, less than one-quarter of patients were over age 76 years.[13,14,18] Older patients often possess a greater amount of comorbidities, which can lead to exclusion from trials.[58,59] The inclusion of these populations is necessary for the generalizability of studies as many patients with CVD are of older age and possess comorbidities. In addition, as the elderly continue to possess a greater proportion of the population,[60] inclusion of these patients in studies is pertinent to effective screening, diagnosis, and management.

Similar to the exclusion of women and BIPOC, obstacles researchers may face in recruiting elderly patients or those with comorbidities include socioeconomic, cultural, and transportation factors.

Pitfalls to avoid

- Timing and geographical limitations: This includes forgetting that a small portion of patients will be affected by the climate, supportive relatives, and access to healthcare and communication technologies.
- Failing to perform appropriate power analysis: Accounting for the likelihood of missing patients early on by involving an expert's opinion will mitigate the risk of running the study short. In the case of having a pilot study with a population without an expert involved with them before, accounting for 10% missingness may suffice; however, there is a need for simulation analysis to generate an appropriate study sample number instead of convenience sampling.
- Missing patients during follow-up: missing patients will impact the completeness of the study and the quality of inference, thus appropriate referral from the primary doctor, involvement of patients' supporting family, and network become especially important. To minimize nonresponse and omission of participants, see Table 18.1.

TABLE 18.1 Challenges and solutions aimed to decrease omission.

Challenge	Suggested solutions
Language barrier	1. Knowing the languages of the target population will help with translation; however, it is equally important to keep the instruction simple for better communication.
Sampling from rural/hard-to-reach areas or communities	1. Consider technology to help expand recruitment. Some tools will facilitate patient engagement, monitoring and tracking them, and conduct decentralized trials virtually and remotely. 2. Build community relations and have community partners to get information and support isolated groups of eligible study participants (homeless, cultural, or religious groups).
Lack of follow-up	1. Upon recruitment, set the expectation for follow-up and its timing very clearly. 2. Increase the follow-up window to give patients adequate time to respond. 3. Circle back reminders for the patients as well as the team interacting with patients. 4. Use incentive programs to reimburse patients' time following successful follow-up.

Getting started

- *Generating a good research question*
  The first step in composing any study is to define the research question (as discussed in other chapters within this section). Once inclusion and exclusion criteria are established, we work on selecting the group of interest that represents a larger group of interest for inference. Early on, a population-specific approach strategy should be established and refined. Fig. 18.1 indicates some questions to consider at this stage.
- *Diversifying your team*
  Awareness of the general, race, and/or ethnic representation within one's team has the ability to reflect the representation of one's study populations. For example, Reza et al. found a direct correlation between the proportion of female authors in HF clinical trials and the percentage

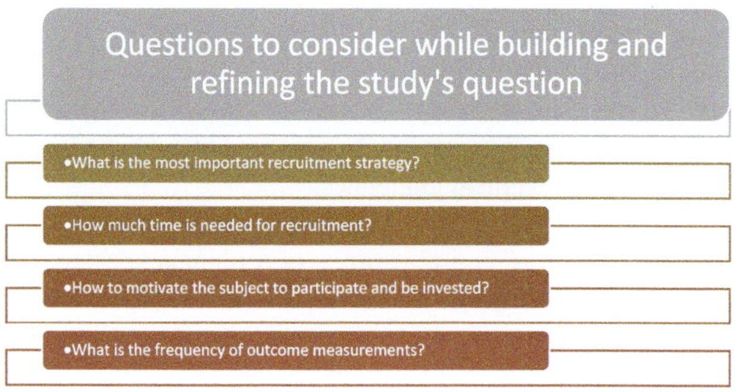

FIGURE 18.1 Some questions to consider at the early stage of research.

of female participation in those trials.[61] The etiology of this trend is not known but it may originate from the author's mindfulness of including those similar to themselves or the greater willingness of certain demographics of patients to participate. Researchers belonging to understudied groups may be more intentional in recruiting or including such populations. In addition, participants may be more trustful of a study when they see increased representation. Regardless, a diverse team has the ability to form a more generalizable study population, as well as bring varying perspectives into the work.

- *Collaborate early with a statistician*
The inclusion of a statistician is a key step to bolstering one's study. Statisticians are not only helpful—if not critical—in the management and interpretation of data but can be intrinsic to the study design. They can provide guidance on choosing study outcomes, calculating sample size and power, randomization, or sampling strategies, and minimizing biases in study design. Therefore, the earlier inclusion of a statistician (i.e., while forming one's research question) is an important component in developing a strong, generalizable study.

References

1. Shim CY, Seo J, Kim YJ, et al. Efficacy and safety of edoxaban in patients early after surgical bioprosthetic valve implantation or valve repair: a randomized clinical trial. *J Thorac Cardiovasc Surg*. 2021;165(1):58–67.
2. Alnajar A, Rao JS, Cayetano SM, Lamelas J. Commentary: battle of the bioprosthetic valve blood thinners. *J Thorac Cardiovasc Surg*. 2021;165(1):69–70.
3. Giugliano RP, Ruff CT, Braunwald E, et al. Edoxaban versus warfarin in patients with atrial fibrillation. *N Engl J Med*. 2013;369(22):2093–2104.
4. Arora S, Vemulapalli S, Stebbins A, et al. The prevalence and impact of atrial fibrillation on 1-year outcomes in patients undergoing transcatheter mitral valve repair: results from the society of thoracic surgeons/American college of cardiology transcatheter valve therapy registry. *JACC Cardiovasc Interv*. 2019;12(6):569–578.
5. Kish L. *Survey Sampling*. New York: John Wiley & Sons; 1965.
6. Cochran WG. *Sampling Techniques*. 3rd ed. Wiley; 1977.
7. Wiśniowski A, Sakshaug JW, Perez Ruiz DA, Blom AG. Integrating probability and nonprobability samples for survey inference. *J Surv Statist Methodol*. 2020;8(1):120–147.
8. Elliot MR. Data from probability and non- probability samples using pseudo-weights. *Surv Practice*. 2009;2(6).
9. Forcillo J, Watkins DA, Brooks A, et al. Making cardiac surgery feasible in African countries: experience from Namibia, Uganda, and Zambia. *J Thorac Cardiovasc Surg*. 2019;158(5):1384–1393.
10. Bourgeois FT, Orenstein L, Ballakur S, Mandl KD, Ioannidis JPA. Exclusion of elderly People from randomized clinical trials of drugs for ischemic heart disease. *J Am Geriatr Soc*. 2017;65(11):2354–2361.
11. Benjamin EJ, Muntner P, Alonso A, et al. Heart disease and stroke statistics-2019 update: a report from the American heart association. *Circulation*. 2019;139(10):e56–e528.
12. Gao Z, Chen Z, Sun A, Deng X. Gender differences in cardiovascular disease. *Med Novel Technol Dev*. 2019;4:100025.
13. Tahhan AS, Vaduganathan M, Greene SJ, et al. Enrollment of older patients, women, and racial and ethnic minorities in contemporary heart failure clinical trials: a systematic review. *JAMA Cardiol*. 2018;3(10):1011–1019.
14. Tahhan AS, Vaduganathan M, Greene SJ, et al. Enrollment of older patients, women, and racial/ethnic minority groups in contemporary acute coronary syndrome clinical trials: a systematic review. *JAMA Cardiol*. 2020;5(6):714–722.
15. Nguyen QD, Peters E, Wassef A, Desmarais P, Remillard-Labrosse D, Tremblay-Gravel M. Evolution of age and female representation in the most-cited randomized controlled trials of cardiology of the last 20 years. *Circ Cardiovasc Qual Outcomes*. 2018;11(6):e004713.
16. Khan MS, Shahid I, Siddiqi TJ, et al. Ten-year trends in enrollment of women and minorities in pivotal trials supporting recent US food and drug administration approval of novel cardiometabolic drugs. *J Am Heart Assoc*. 2020;9(11):e015594.
17. Khan SU, Michos ED. Women in stroke trials-A tale of perpetual inequity in cardiovascular research. *JAMA Neurol*. 2021;78(6):654–656.
18. Khan SU, Khan MZ, Raghu Subramanian C, et al. Participation of women and older participants in randomized clinical trials of lipid-lowering therapies: a systematic review. *JAMA Netw Open*. 2020;3(5):e205202.
19. Virani SS, Alonso A, Benjamin EJ, et al. Heart disease and stroke statistics-2020 update: a report from the American heart association. *Circulation*. 2020;141(9):e139–e596.
20. Skibba AA, Evans JR, Hopkins SP, et al. Reconsidering gender relative to risk of rupture in the contemporary management of abdominal aortic aneurysms. *J Vasc Surg*. 2015;62(6):1429–1436.
21. Guirguis-Blake JM, Beil TL, Senger CA, Coppola EL. Primary care screening for abdominal aortic aneurysm: updated evidence report and systematic review for the US preventive services task force. *JAMA*. 2019;322(22):2219–2238.
22. Appelman Y, van Rijn BB, Ten Haaf ME, Boersma E, Peters SA. Sex differences in cardiovascular risk factors and disease prevention. *Atherosclerosis*. 2015;241(1):211–218.
23. Poorthuis MH, Algra AM, Algra A, Kappelle LJ, Klijn CJ. Female- and male-specific risk factors for stroke: a systematic review and meta-analysis. *JAMA Neurol*. 2017;74(1):75–81.
24. Canoy D, Beral V, Balkwill A, et al. Age at menarche and risks of coronary heart and other vascular diseases in a large UK cohort. *Circulation*. 2015;131(3):237–244.
25. Lisabeth LD, Beiser AS, Brown DL, Murabito JM, Kelly-Hayes M, Wolf PA. Age at natural menopause and risk of ischemic stroke: the Framingham heart study. *Stroke*. 2009;40(4):1044–1049.
26. Muka T, Oliver-Williams C, Kunutsor S, et al. Association of age at onset of menopause and time since onset of menopause with cardiovascular outcomes, intermediate vascular traits, and all-cause mortality: a systematic review and meta-analysis. *JAMA Cardiol*. 2016;1(7):767–776.
27. Arora S, Stouffer GA, Kucharska-Newton AM, et al. Twenty year trends and sex differences in young adults hospitalized with acute myocardial infarction. *Circulation*. 2019;139(8):1047–1056.
28. Singh JA, Lu X, Ibrahim S, Cram P. Trends in and disparities for acute myocardial infarction: an analysis of Medicare claims data from 1992 to 2010. *BMC Med*. 2014;12:190.
29. Ballo P, Balzi D, Barchielli A, Turco L, Franconi F, Zuppiroli A. Gender differences in statin prescription rates, adequacy of dosing,

and association of statin therapy with outcome after heart failure hospitalization: a retrospective analysis in a community setting. *Eur J Clin Pharmacol*. 2016;72(3):311−319.

30. Karalis DG, Wild RA, Maki KC, et al. Gender differences in side effects and attitudes regarding statin use in the Understanding Statin Use in America and Gaps in Patient Education (USAGE) study. *J Clin Lipidol*. 2016;10(4):833−841.
31. Mensah GA, Mokdad AH, Ford ES, Greenlund KJ, Croft JB. State of disparities in cardiovascular health in the United States. *Circulation*. 2005;111(10):1233−1241.
32. Heron M. Deaths: leading causes for 2017. *Natl Vital Stat Rep*. 2019;68(6):1−77.
33. Wadhera RK, Figueroa JF, Rodriguez F, et al. Racial and ethnic disparities in heart and cerebrovascular disease deaths during the COVID-19 pandemic in the United States. *Circulation*. 2021;143(24):2346−2354.
34. Breathett K, Sims M, Gross M, et al. Cardiovascular health in American Indians and Alaska natives: a scientific statement from the American heart association. *Circulation*. 2020;141(25):e948−e959.
35. Ritchey MD, Wall HK, Owens PL, Wright JS. Vital signs: state-level variation in nonfatal and fatal cardiovascular events targeted for prevention by million hearts 2022. *MMWR Morb Mortal Wkly Rep*. 2018;67(35):974−982.
36. Chen S, Li J. Participation of Black US residents in clinical trials of 24 cardiovascular drugs granted FDA approval, 2006-2020. *JAMA Netw Open*. 2021;4(3):e212640.
37. Vivo RP, Krim SR, Krim NR, et al. Care and outcomes of Hispanic patients admitted with heart failure with preserved or reduced ejection fraction: findings from get with the guidelines-heart failure. *Circ Heart Fail*. 2012;5(2):167−175.
38. Loehr LR, Rosamond WD, Chang PP, Folsom AR, Chambless LE. Heart failure incidence and survival (from the Atherosclerosis Risk in Communities study). *Am J Cardiol*. 2008;101(7):1016−1022.
39. Bajaj NS, Gutierrez OM, Arora G, et al. Racial differences in plasma levels of N-terminal pro-B-type natriuretic peptide and outcomes: the reasons for geographic and racial differences in stroke (REGARDS) study. *JAMA Cardiol*. 2018;3(1):11−17.
40. Yoon YE, Kim KM, Han JS, et al. Prediction of subclinical coronary artery disease with breast arterial calcification and low bone mass in asymptomatic women: registry for the women health cohort for the BBC study. *JACC Cardiovasc Imaging*. 2019;12(7 Pt 1):1202−1211.
41. Rotter MA, Schnatz PF, Currier Jr AA, O'Sullivan DM. Breast arterial calcifications (BACs) found on screening mammography and their association with cardiovascular disease. *Menopause*. 2008;15(2):276−281.
42. Maas AH, van der Schouw YT, Beijerinck D, Deurenberg JJ, Mali WP, van der Graaf Y. Arterial calcifications seen on mammograms: cardiovascular risk factors, pregnancy, and lactation. *Radiology*. 2006;240(1):33−38.
43. Gennarelli M, Jedynak A, Forman L, et al. The potential impact of mammographic breast arterial calcification on physician practices in a primary care setting. *Future Cardiol*. 2021;17(7):1241−1248.
44. Reddy J, Son H, Smith SJ, Paultre F, Mosca L. Prevalence of breast arterial calcifications in an ethnically diverse population of women. *Ann Epidemiol*. 2005;15(5):344−350.
45. 1995 Update of the Working Group Reports on Chronic Renal Failure and Renovascular Hypertension. National high blood pressure education program working group. *Arch Intern Med*. 1996;156(17):1938−1947.
46. Wright Jr JT, Bakris G, Greene T, et al. Effect of blood pressure lowering and antihypertensive drug class on progression of hypertensive kidney disease: results from the AASK trial. *JAMA*. 2002;288(19):2421−2431.
47. Oh SS, Galanter J, Thakur N, et al. Diversity in clinical and biomedical research: a promise yet to Be fulfilled. *PLoS Med*. 2015;12(12):e1001918.
48. Virani SS, Alonso A, Aparicio HJ, et al. Heart disease and stroke statistics-2021 update: a report from the American heart association. *Circulation*. 2021;143(8):e254−e743.
49. Clark LT, Watkins L, Pina IL, et al. Increasing diversity in clinical trials: overcoming critical barriers. *Curr Probl Cardiol*. 2019;44(5):148−172.
50. Kim ES, Menon V. Status of women in cardiovascular clinical trials. *Arterioscler Thromb Vasc Biol*. 2009;29(3):279−283.
51. Shavers VL, Fagan P, Jones D, et al. The state of research on racial/ethnic discrimination in the receipt of health care. *Am J Public Health*. 2012;102(5):953−966.
52. LaVeist TA, Nickerson KJ, Bowie JV. Attitudes about racism, medical mistrust, and satisfaction with care among African American and white cardiac patients. *Med Care Res Rev*. 2000;57(Suppl 1):146−161.
53. Xiao Q, Berrigan D, Powell-Wiley TM, Matthews CE. Ten-year change in neighborhood socioeconomic deprivation and rates of total, cardiovascular disease, and cancer mortality in older US adults. *Am J Epidemiol*. 2018;187(12):2642−2650.
54. Forti P, Maioli F, Procaccianti G, et al. Independent predictors of ischemic stroke in the elderly: prospective data from a stroke unit. *Neurology*. 2013;80(1):29−38.
55. Howard G, Goff DC. Population shifts and the future of stroke: forecasts of the future burden of stroke. *Ann N Y Acad Sci*. 2012;1268:14−20.
56. Nanna MG, Chen ST, Nelson AJ, Navar AM, Peterson ED. Representation of older adults in cardiovascular disease trials since the inclusion across the lifespan policy. *JAMA Intern Med*. 2020;180(11):1531−1533.
57. Caughey GE, Inacio MC, Bell JS, Vitry AI, Shakib S. Inclusion of older People reflective of real-world clinical practice in cardiovascular drug trials. *J Am Heart Assoc*. 2020;9(21):e016936.
58. Masoudi FA, Havranek EP, Wolfe P, et al. Most hospitalized older persons do not meet the enrollment criteria for clinical trials in heart failure. *Am Heart J*. 2003;146(2):250−257.
59. Cherubini A, Oristrell J, Pla X, et al. The persistent exclusion of older patients from ongoing clinical trials regarding heart failure. *Arch Intern Med*. 2011;171(6):550−556.
60. Jennifer M, Ortman VAV, Hogan H. *An Aging Nation: The Older Population in the United States*. United States Census Burea; 2014.
61. Reza N, Tahhan AS, Mahmud N, et al. Representation of women authors in international heart failure guidelines and contemporary clinical trials. *Circ Heart Fail*. 2020;13(8):e006605.

Chapter 19

Introduction to clinical research: What data are being collected and why?

Abdullah Sarkar[1] and Ahmed Alnajar[2]

[1]*Cardiac Arrhythmia Institute- UCLA, Los Angeles, CA, United States;* [2]*The Division of Cardiothoracic Surgery, The Department of Surgery, The University of Miami Miller School of Medicine, Miami, FL, United States*

Key points

- Outcome measurements are the study components that can affect the study's real-life applicability.
- The source of data and the study population need to be critically appraised for appropriateness—in relation to the intended clinical question and outcome of interest.
- Clearly defining appropriate study outcomes is important and should be predetermined when possible.

Why it matters?

The effective use of outcome measures is an important aspect of study design and subsequently clinical care to decide which interventions are relevant to which specific patient population. This is especially relevant given the multitude of therapeutic options in modern-day medicine and in the treatment of cardiovascular disease (CVD). Hence, selecting appropriate measures of study outcomes requires an understanding of the clinical significance of the variables assessed in the research study and its analysis. To function well, these tools must be used with skill and understanding.

Introduction

> *If we knew what we were doing, it would not be called research, would it?*
>
> Albert Einstein.

As research investigates what is unknown, it is vital for researchers to critically evaluate their study frameworks. The significant items to evaluate include components that can affect study outcomes—these can be referred to as outcome measurements. This chapter touches on outcome measurements that can impact a study's results divided into three categories: choosing appropriate variables, data sources and study populations, and data analysis methods (Fig. 19.1). Subsequent chapters cover specifics of study types and design, data analysis, and resources for further reading.

Choosing appropriate variables

Variable selection is a strong influencer of outcome measurement. It is best to choose variables systematically. For one, variables should be pertinent to the subject matter and easily sourceable from the data. It is best to review existing literature to identify potential variables, especially identifying those that have been associated with independent effects and clinical significance.[1] It is also important to be practical in data selection, both in number and in definition. Choosing many variables will make data collection more challenging (as not all the variables may be available for each patient) and more costly. It can also make data interpretation more difficult and less generalizable to the target population. We will discuss some examples to better review the importance of these choices on research outcomes and their applicability in real-world settings.

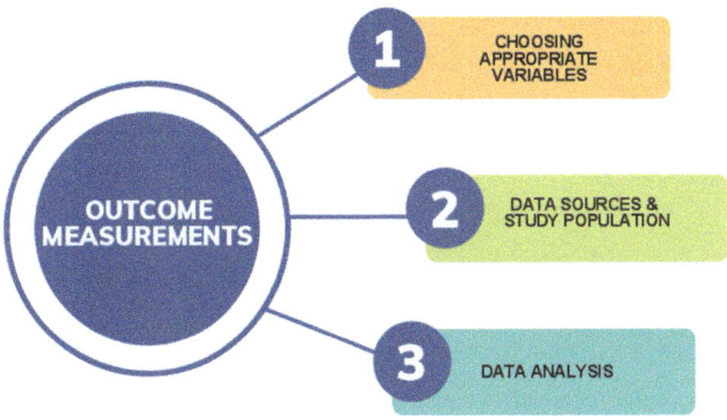

FIGURE 19.1 Outcome measurements.

Translational Cardiology. https://doi.org/10.1016/B978-0-323-91790-2.00016-2

Clinical evaluation of chest pain is a daily practice among physicians in all medical care settings. In 1979, Drs. Diamond and Forrester laid the foundation for estimating the probability of disease in patients with chest pain using three variables: age, sex, and angina typicality.[2] However with the advancement of disease assessment using imaging, functional testing, and angiography, further variables are available to accurately estimate pretest probability (PTP) of disease. This was shown in several studies using multiple modalities including coronary computed tomographic angiography (CCTA)[3] and myocardial perfusion imaging (MPI). Specifically, Soman et al. demonstrated the poor performance by overestimation of "traditional" PTP estimation in stable patients referred for MPI.[4] In 2019, the European Society of Cardiology (ESC) issued guidelines for the diagnosis and management of chronic coronary syndromes, adding additional variables for PTP to the previous three—risk factors, EKG changes, and other test findings if available.[5] This was then studied again by Soman et al., and they showed a close association of expected prevalence and PTP estimation using the additional determinants of clinical likelihood of obstructive coronary artery disease.[6] Thus, future studies looking to predict disease will need to use additional variables than were originally recommended, just as clinical practice too has adopted.

In addition to obtaining appropriate variables, defining variables of interest is important as well. The Rotterdam study was established in 1990 to evaluate cardiovascular disease determinants and prognosis among other organ system diseases.[7] Two decades later, the research team published updated definitions of cardiac outcomes, such as myocardial infarction (MI), revascularization, coronary heart disease (CHD) mortality, heart failure (HF), atrial arrhythmia, and sudden cardiac death (SCD).[8] The authors recognized the evolving understanding of these variables and the need for identification to facilitate uniform data collection among study sites and improve readers' understanding.

Data sources and study populations

The amount of data available in the 21st century has been unprecedented and overwhelming. Rumsfeld et al. summarized examples of data sources for big data applications[9] (Table 19.1). Datasets commonly used for published articles can have several limitations. A widely used dataset and an example of this, is the Nationwide Inpatient Sample (NIS)[10] database which is compiled by the Agency for Healthcare Research and Quality (AHRQ). Several papers and editorials[11,12] have discussed the limitations of studies using this database and the most common include: (1) sampling design—since the inception of the database, there have been significant changes to sampling design, including a change from including 100% of hospital patients to surveying 20% of hospitals in the United States, (2) each data point, rather than representing an individual patient, represents a hospital admission, possibly leading to overestimations in incidence, (3) identification of diagnosis based on claim codes, as those that are not used for reimbursement may be inconsistent.[13] This is not to suggest that datasets with limitations should be avoided, but rather, it is incumbent on the researcher to understand the data source, for example, reviewing the dataset documentation,[14] to ensure the appropriate outcome measurements are used for the specific outcome of interest. Essentially, one must learn and understand the limitations of a dataset and ensure that the variables of interest necessary to determine the desired outcome measurement are included.

TABLE 19.1 Major data sources in clinical cardiology research.

Data sources
Administrative
Registry
Metanalysis (MA)
Randomized control trial (RCT)
Electronic health record (EHR)
Biometric
Patient reported
Genomics/Proteomics

When considering which datasets to use, researchers should consider the target population and overall generalizability. A dataset's demographic distribution should be determined before utilization. An example of the importance of understanding the study population can be found in studies of Aortic Stenosis (AS) therapies such as Transcatheter Aortic Valve Replacement (TAVR), where only one of the pivotal clinical trials[15] even included race breakdown in their publication.[16] It is difficult to conclude that the benefit of TAVR was applicable to all races that met the inclusion criteria. Fortunately, subsequent registry studies have explored racial outcomes, supporting result validity in minority populations.[17]

Data analysis

The way in which obtained outcome measurements are statistically analyzed, impacts the outcome. There are many examples of published data that have suboptimal analysis, leading to conclusions that are not clinically useful. For example, Charig et al. published their study comparing treatment modalities on renal calculi and reported higher success rates with the closed surgical approach versus the open approach.[18] However, when accounting for stone size (small vs. large), the open approach had a higher success ratio in each subgroup contrary to the data combining all groups, under each treatment modality. This discrepancy is due to what is now known as Simpson's Paradox, where a trend reverses when data is combined and does not account for confounding variables.[19] Hence, it is important to understand the inability to dichotomize continuous variables and their impact on the outcome.

Again, independent of study design, variables, population, and data source, the methodology of analysis can impact study outcomes and results. Nosek et al. gathered 21 teams comprising 61 analysts and distributed the same data with one study question: whether soccer referees are more likely to give red cards to dark-skin-toned players versus light-skin-toned players. The researchers reported 29 different analytic methods that resulted in 69% of teams finding a statistically significant positive effect while the remaining teams did not.[20] The study demonstrates how analytic choices influence study results.

The choice of outcome type

In addition to discussing outcome measurements, choosing an appropriate outcome of interest is crucial. The selection of the latter is necessary to properly utilize the former. A study's goal to report Health-Related Quality of Life (HRQL) among postoperative patients[21] can be used to understand limitations of outcome measurements on outcomes. The tool for outcome assessment was the 36-item Short Form of the Medical Outcomes Study (SF-36) questionnaire, which explores the functional status of patients with some questions on emotions and general perception of well-being—it provides an understanding of health status rather than "Quality of Life."[22] Additionally, the validity is questionable given the lack of understanding of the patients' values.[23] Hence this tool is limited to explore the intended outcome.

Finally, it is advisable to incorporate early statistician support. This can provide an evaluation of study feasibility, power, and sample size estimation, techniques to avoid bias, the aim of the study (superiority or noninferiority), and outcomes (primary and secondary, with or without surrogate endpoints, and use of composite endpoints).[24] This will be discussed in the next chapter, Optimizing the question: Balancing significance and feasibility.

Get started

- Incorporation of statistician support early in the research design is well advised to use appropriate methods to answer the study question and fit the data sampling method, and plan—in advance—how to report the outcomes measured in a clear and reproducible fashion.
- While contemplating a study design, defining study variables, methods, and outcomes—before beginning with the research and following a careful review of prior studies/literature—is prudent for reproducibility and overall clinical adoption.

Pitfalls to avoid

Given that there are multiple factors to consider when designing a study, some considerations are highlighted:

1. The risk of relying solely on statistically significant variables to measure outcomes. This approach can overlook the importance of clinically relevant variables that are informed by expert opinions or prior studies. *P*-values may fail to detect the predictive power of variables due to correlation effects and other issues.
2. Not considering weighted analysis methods for survey data (e.g., NIS). Researchers should ensure the study population/data source can appropriately assess the intended outcome with the correct methods.
3. Chasing significance. When researchers do not predetermine their methodology of data analysis and study outcomes, they tend to underreport null results, often without adequately considering the original research question.

Resources/Further reading

1. https://training.cochrane.org/handbook/current/chapter-05.
2. https://ccnmtl.columbia.edu/projects/rcr/rcr_data/foundation/.

References

1. Chowdhury MZI, Turin TC. Variable selection strategies and its importance in clinical prediction modelling. *Fam Med community Heal.* 2020;8(1). https://doi.org/10.1136/fmch-2019-000262. e000262-e000262.
2. Diamond GA, Forrester JS. Analysis of probability as an aid in the clinical diagnosis of coronary-artery disease. *N Engl J Med.* 1979;300(24):1350–1358. https://doi.org/10.1056/NEJM197906143002402.
3. Cheng VY, Berman DS, Rozanski A, et al. Imaging performance of the traditional age, sex, and angina typicality-based approach for estimating pretest probability of angiographically significant coronary artery disease in patients undergoing coronary computed tomographic angiography results from the multinational coronary CT angiography evaluation for clinical outcomes: an international multicenter registry (CONFIRM). *Circulation.* 2011. https://doi.org/10.1161/CIRCULATIONAHA.111.

4. Batal O, Malhotra S, Harinstein M, et al. Performance of traditional pretest probability estimates in stable patients undergoing myocardial perfusion imaging. *Circ Cardiovasc Imaging*. 2019;12(10):e008473. https://doi.org/10.1161/CIRCIMAGING.118.008473.
5. Knuuti J, Wijns W, Saraste A, et al. 2019 ESC Guidelines for the diagnosis and management of chronic coronary syndromes. *Eur Heart J*. 2020;41(3):407−477. https://doi.org/10.1093/eurheartj/ehz425.
6. Malhotra S, Batal O, Douglas P, Soman P. Accurate prediction of myocardial perfusion abnormality by the European society of cardiology pretest probability estimates of coronary artery disease. *Circ Cardiovasc Imaging*. 2020;13(10):e011342. https://doi.org/10.1161/CIRCIMAGING.120.011342.
7. Hofman A, Grobbee DE, de Jong PT, van den Ouweland FA. Determinants of disease and disability in the elderly: the rotterdam elderly study. *Eur J Epidemiol*. 1991;7(4):403−422. https://doi.org/10.1007/BF00145007.
8. Leening MJ, Kavousi M, Heeringa J, et al. Methods of data collection and definitions of cardiac outcomes in the Rotterdam Study. *Eur J Epidemiol*. 2012. https://doi.org/10.1007/s10654-012-9668-8.
9. Rumsfeld JS, Joynt KE, Maddox TM. Big data analytics to improve cardiovascular care: promise and challenges. *Nat Rev Cardiol*. 2016;13(6):350−359. https://doi.org/10.1038/nrcardio.2016.42.
10. HCUP. Healthcare Cost and Utilization Project (HCUP) Overview of the National (Nationwide) Inpatient Sample (NIS).
11. Memtsoudis SG. Limitations associated with the analysis of data from administrative databases. *Anesthesiology*. 2009;111(2):449. https://doi.org/10.1097/ALN.0b013e3181adf739.
12. Alnajar A, Ibrahim W, Mendoza CE. Are the outcomes of TAVR significantly riskier for solid organ transplant recipients than for the general population? *J Card Surg*. 2022;37(3):608−609. https://doi.org/10.1111/jocs.16207.
13. Khera R, Krumholz HM. With great power comes great responsibility: big data research from the national inpatient sample. *Circ Cardiovasc Qual Outcomes*. 2017;10(7):e003846. https://doi.org/10.1161/CIRCOUTCOMES.117.003846.
14. Sample NI. *Healthcare Cost and Utilization in Project (HCUP)*. Rockville, MD: Agency Healthc Res Qual; 2003. Published online.
15. Mack MJ, Leon MB, Thourani VH, et al. Transcatheter aortic-valve replacement with a balloon-expandable Valve in low-risk patients. *N Engl J Med*. 2019;380(18):1695−1705. https://doi.org/10.1056/NEJMoa1814052.
16. Edelman JJ, Thourani VH. Racial disparity in the treatment of aortic Stenosis. *J Am Heart Assoc*. 2021;10(14):e019875. https://doi.org/10.1161/JAHA.120.019875.
17. Alkhouli M, Holmes DRJ, Carroll JD, et al. Racial disparities in the utilization and outcomes of TAVR: TVT registry report. *JACC Cardiovasc Interv*. 2019;12(10):936−948. https://doi.org/10.1016/j.jcin.2019.03.007.
18. Charig CR, Webb DR, Payne SR, Wickham JE. Comparison of treatment of renal calculi by open surgery, percutaneous nephrolithotomy, and extracorporeal shockwave lithotripsy. *Br Med J*. 1986;292(6524):879−882. https://doi.org/10.1136/bmj.292.6524.879.
19. Julious SA, Mullee MA. Confounding and Simpson{\textquoteright}s paradox. *BMJ*. 1994;309(6967):1480−1481. https://doi.org/10.1136/bmj.309.6967.1480.
20. Silberzahn R, Uhlmann EL, Martin DP, et al. Many analysts, one data set: making transparent how variations in analytic choices affect results. *Adv Methods Pract Psychol Sci*. 2018;1(3):337−356. https://doi.org/10.1177/2515245917747646.
21. Mangione CM, Goldman L, Orav EJ, et al. Health-related quality of life after elective surgery: measurement of longitudinal changes. *J Gen Intern Med*. 1997;12(11):686−697. https://doi.org/10.1046/j.1525-1497.1997.07142.x.
22. Guyatt G. Insights and limitations from health-related quality-of-life research. *J Gen Intern Med*. 1997;12(11):720−721. https://doi.org/10.1046/j.1525-1497.1997.07149.x.
23. Gill TM, Feinstein AR. A critical appraisal of the quality of quality-of-life measurements. *JAMA*. 1994;272(8):619−626.
24. Zapf A, Rauch G, Kieser M. Why do you need a biostatistician? *BMC Med Res Methodol*. 2020;20(1):23. https://doi.org/10.1186/s12874-020-0916-4.

Chapter 20

Optimizing the question: Balancing significance and feasibility

Mudathir Ibrahim[1,2] and Ahmed Alnajar[3]

[1]*Department of General Surgery, Maimonides Medical Center, Brooklyn, NY, United States;* [2]*Nuffield Department of Surgical Sciences, University of Oxford, Oxford, United Kingdom;* [3]*The Division of Cardiothoracic Surgery, The Department of Surgery, The University of Miami Miller School of Medicine, Miami, FL, United States*

Key points

1. Research questions serve an important role in forming the foundation and backbone of a research study. Therefore, the more optimized the research questions are, the more likely the study will be successful from conception to completion and impactful after dissemination.
2. Well-optimized research questions share the characteristics of being empirical and foregrounded, clear, simple, focused, well-structured, and most importantly, feasible and relevant.
3. The PICO and FINER criteria are useful tools for creating well-optimized and high-quality research questions.

Why it matters?

A research question is an important aspect of any research study. It serves not to only inform about the aim and purpose of the study, but also as a guide for the researcher from conception through the conduction, completion, and dissemination of the study. If the question is not optimized to pass the test of feasibility and significance, the study may fail to complete. If it succeeds to completion, it may not have any meaningful impact on the community, therefore, contributing to research waste.

Introduction

Every research study, regardless of the field, begins with a problem the researcher intends to solve. For the problem to be successfully answered through research, it must be amenable to study through the collection and analysis of data.[1] While some problems may be interesting, they are not answerable through research. These are problems relating to morals and are known as "ethical problems." For example, a question like "Should health coverage be nationalized in the United States?" is an ethical problem. The true answer is "it depends on the personal values and view of who you ask,"—meaning there is no one right or wrong answer. However, this is not to say that problems regarding health coverage cannot be studied. For example, one could study how different health coverage systems impact the treatment or prevention of a particular disease or in a particular population. These are termed "empirical problems"—these are problems that can be answered by real experiences in the real world, meaning they are researchable.[2] To solve a problem in clinical research, the problem should be empirical. Researchable problems are endless and often identified through multiple sources, such as personal experiences as a physician and/or clinical researcher or consumer of healthcare, from interaction with colleagues through conference presentations and proceedings, and reading of the literature.[3]

Once a researchable problem is identified, the researcher has a broad idea of what to compare, understand, or change. However, what is the next step? An important preliminary and basic step is developing a clear research question. Creating a research question is an iterative process, with one version leading to another. In the previous chapter, we discussed different types of research questions and how to formulate them. This chapter focuses on how to refine and optimize a research question to produce one that is well structured, considering issues of answerability and relevance.

Research questions

What is the research question under investigation? This is often one of the first queries raised in any research study.

Translational Cardiology. https://doi.org/10.1016/B978-0-323-91790-2.00042-3

This information is not only important to the researcher, but also to statisticians, collaborators, funding agencies, journal editors, and reviewers, and even to the frontline clinicians, patients, and policymakers who may read and use the published study. Wood et al. defined a research question as "an uncertainty about a problem that can be challenged, examined, and analyzed to provide useful information."[4] According to Eldredge et al., questions are the driving force behind evidence-based practice (EBP). If there were no questions, EBP would not be necessary.[5] Richardson et al. proposed two kinds of research questions, termed background and foreground questions.[6] Background questions are typically asked by researchers who are relatively new to a subject area and are often used to address a general knowledge gap. For example, "What is atrial fibrillation?" or "How does warfarin work?" are background questions. In contrast, a more experienced researcher with background knowledge is likely to need to address a specific question like, "In patients with new-onset atrial fibrillation, is warfarin more effective than apixaban at reducing the risk of long-term ischemic stroke?"—this is termed a foreground question. Compared to background questions, the level of knowledge required to pose such a question is greater and the target knowledge is very specific.[6,7] Mayo et al. report that the findings of a study may only acquire relevance if it provides an unbiased answer to a question that is specific and clear.[8] Therefore, defining a specific and clear foreground question that will be studied is an important first step in any clinical research.

Research questions may also take different forms; they may be interrogatory, like "Does ligation of mitral appendage reduce the risk of ischemic stroke in patients with atrial fibrillation undergoing mitral valve surgery?" or in statement form, like "Ligation of the mitral appendage reduces the risk of long-term ischemic stroke in patients with baseline atrial fibrillation undergoing mitral valve surgery." In either case, it is important that the question clearly defines the purpose of the study.[3] To stress the significance of research questions, Riva et al. suggest that up to one-third of the time spent in the research process (from conception to publication) could be invested in finding the right primary study question.[9] However, while identifying a clear and specific question that will be studied is important, it remains one of the most challenging aspects of a research process because it affects the direction of subsequent planning and analysis and strategies employed to locate relevant information to answer the question. In fact, Mayo et al. report that 30% of published articles would have needed a major rewording of the study question.[8]

John Ruskin eloquently said, "To be able to ask a question clearly is two-thirds of the way to getting it answered" According to the Center for Evidence-based Medicine (CEBM) at the University of Oxford, "One of the fundamental requirements of EBM is asking a well-built clinical question. To benefit patients and clinicians, such questions need to be directly relevant to patients' problems and phrased in ways that direct your search to relevant and precise answers"[10] A successful research study relies heavily on a well-structured research question, based on the problems faced in day-to-day research activities and clinical practice. Also, the underlying research question should provide important information about the answerability and relevance of the study, while serving as a building block for the implementation of the study. Furthermore, a research question should be well-focused and precise. A well-focused question helps to tune out irrelevant information and keep the researcher on track with the study's aim and goal. Several authors have linked precise questions to efficient search for evidence and finding of precise answers.[5,11] The delineation of a precise question is an analytic and iterative process, requiring reexamination and redefinition until, finally, a well-structured and focused question is established.

Developing an optimized research question

Optimized clinical research questions are empirical and foregrounded, clear, focused, and specific, and, most importantly, feasible and relevant (Fig. 20.1). How can this be achieved? Here, we will be discussing two well-established frameworks to guide the process of refining research questions to finally develop a high-quality question for a high-quality research study. They are **PICO** (population, intervention, comparison, outcome) and **FINER** (feasible, interesting, novel, ethical, relevant).

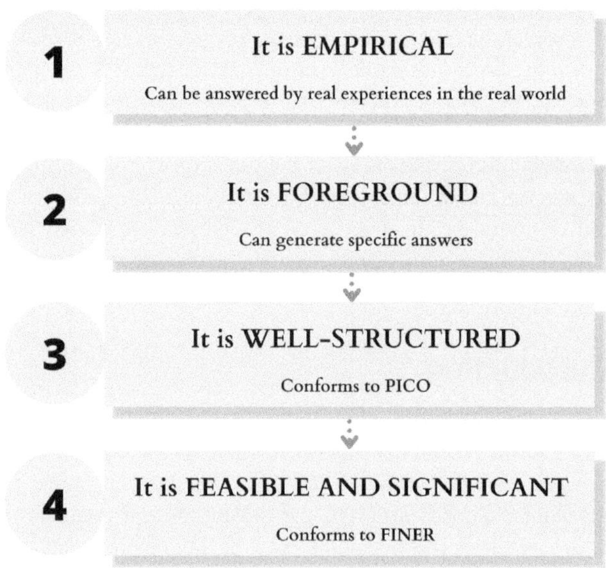

FIGURE 20.1 Characteristics of an optimized research question.

PICO

Richardson et al. introduced the concept of PICO to break down clinical research questions into four smaller components (population, intervention, comparison, and outcome), with the aim of facilitating clarity of research questions to help generate suitable and relevant answers.[12] (Fig. 20.2). While some studies would not fulfill all four components of PICO because of their structure and type, all studies have at least two of the components, the population and outcome being measured.[6] Also, there are variations of the framework with more than four components, such as PICOT (population, intervention, comparison, outcome, and time) and PICOS (population, intervention, comparison, outcome, and study design).[13] In all cases, the underlying aim of the framework remains the same, which is to provide a systematic way to identify the crucial components of a research question and methodology of a research study.

P—population, patient, or problem: refers to the individuals or problem being studied. Asking and addressing questions such as, who is the patient/population? what characteristics make them important for the study? what is the problem or disease of interest? sets a strong basis for a high-quality research question and fosters successful study design and implementation. Also, it elucidates study eligibility criteria, suggests recruitment strategy, and potentially informs on the generalizability of the study.

I—intervention: refers to the planned intervention. The Intervention could be a treatment (medical or surgical), diagnostic test, or prognostic factor. Also, in some studies, the "I" in PICO is replaced with E (PECO), which stands for "Exposure." Asking questions such as, what is the intervention or exposure? What are available alternatives (standard of care)? will help the researcher gain a better understanding of what the proposed change is or will be, what its potential positive or negative effects might be, who it might benefit or harm, and overall provide a valuable guide for the study design and conduction.

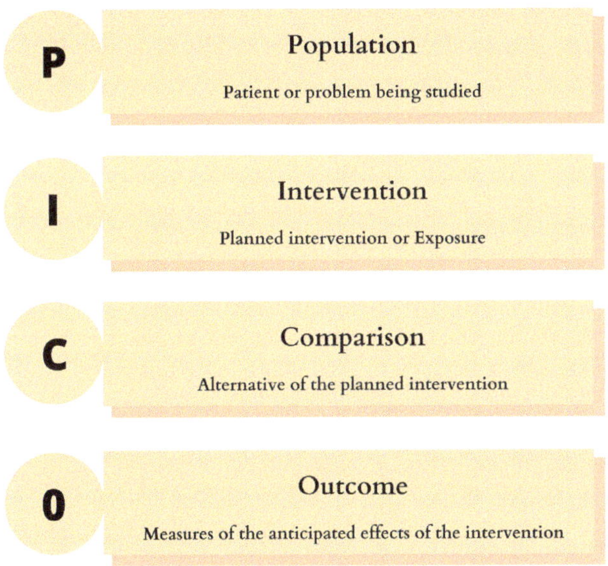

FIGURE 20.2 PICO criteria.

C—comparison: refers to the alternative of the intervention being tested. This is what is being compared to the intervention of interest. It could be standard of care (an established therapy or method) or placebo (a sham therapy or method). Where there is no intervention, it could be a population with different factors (disease, risk, or prognosis) to the interested population. In defining a research question, it is often advisable to avoid using multiple comparators, unless there is a large enough sample to generate robust statistical power for the outcomes of interest and multiple testing can be adjusted for in the statistical analyses. Clearly asking and appreciating available comparators (therapies and method) helps clarify the aim and potential impact of the intervention, identify and address any ethical concerns, especially relating to the use of placebo or sham therapy, and select appropriate study design and statistical method.

O—outcome: refers to what will be measured to identify the end result of the intervention. This is the anticipated effects of the intervention or exposure. Questions such as "what is the aim of the intervention? what do I want to improve, eliminate, or decrease?" provide valuable insight into how to generate and measure important outcomes for the main effect of the proposed intervention, termed the primary outcome. Having a clear primary outcome is of vital importance because it guides numerous methodological aspects of the research study, such as sample size calculation, data collection, and statistical analysis, while also providing stakeholders with valuable information, like the relevance of the intervention. It is often advisable to keep the primary outcome very simple and limited to one. A good primary outcome should also be easily quantifiable, validated (backed by strong evidence of being associated with the exposure or intervention of interest), relevant, and reproducible.

While the primary outcome remains the main focus of the study, it is equally important to appreciate other effects of the intervention that are not captured in the primary outcome. These are termed secondary outcomes. Noting secondary outcomes can be useful in capturing potential side effects or safety issues of an intervention, and they are particularly helpful when used to lend supporting evidence for the primary outcome.

We will wrap this section up with a research question formulated using the PICO framework.

"In adult patients with an abdominal aortic aneurysm, does endovascular repair reduce mortality better than open repair?"

Population—adult patients with an abdominal aortic aneurysm.

Intervention—endovascular repair.
Comparison—open repair.
Outcome—Mortality reduction.

FINER

Having a clearly defined question is half the optimization process of a research question. The other half is assessing how possible it is to conduct the study and answer the well-built question (feasibility), and whether the generated answer will be useful to important stakeholders (relevance). Hulley et al. introduced the concept of FINER, which produces a mind map for researchers to ensure that a study idea is Feasible, Interesting, Novel, Ethical, and Relevant.[14] (Fig. 20.3).

F—feasible: refers to whether the study can be conducted and the question answered sufficiently within a reasonable time and with the available resources. Many clinical trials are terminated because of futility. Twenty-five percent of the trials approved by six research ethics committees between 2000 and 2003 in Canada, Germany, and Switzerland were discontinued and the discontinuation rate was 43% for a cohort of surgical trials registered on ClinicalTrials.gov between 2008 and 2009.[15–17] For other types of research, feasibility problems are less accurately known but probably even more common. Numerous factors influence the feasibility of a research study, most of which should be considered very early in the process to avoid waste of effort and resources. Important factors include but are not limited to, project cost, time and skills of the researcher(s), access to participants and cooperative collaborators (biostatistician, colleagues, advisors, managers, and administrators) with similar interests and goals, data availability, facilities, and equipment. Although it can be difficult to predict some of these factors upfront, controlling unwarranted optimism among investigators and funders and being realistic are valuable considerations. Pilot and feasibility studies can also be used to "test-drive" the methodology of a study, identifying any issues or areas of weakness before launching a major study.[18]

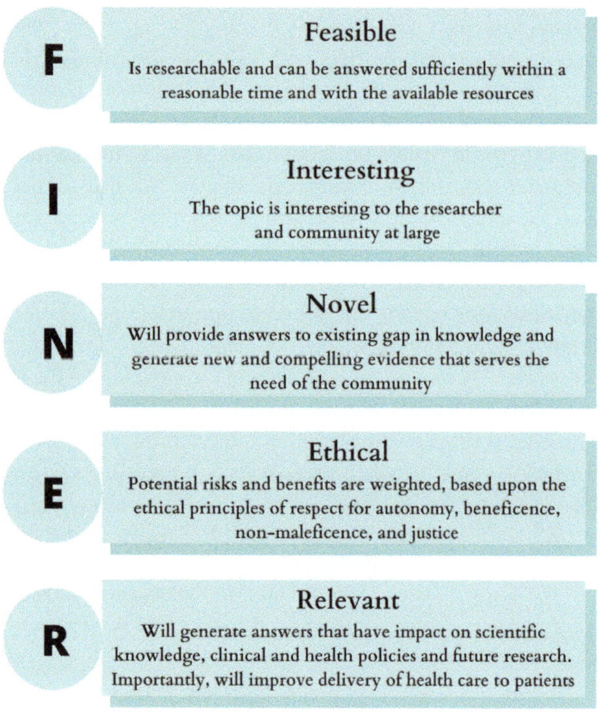

FIGURE 20.3 FINER criteria.

I—interesting: this is subjective and open to multiple interpretations. First, it is important that the topic and research question be interesting to the researcher. While conducting research can be rewarding, the process can be laborious, and without the ongoing enthusiasm and motivation of the researcher throughout the duration of the study, it risks failure. However, what interests the researcher might not necessarily interest the rest of the community. Therefore, it is very important that in the question development phase, there is communication and collaboration with colleagues in similar research or clinical fields, funders, and most importantly, patients to identify areas of interest that will contribute positively to the community at large. For example, to understand the effects of sociodemographic factors and social determinants of health (SDH) on outcomes, behavioral researchers, public health leaders, and clinicians have collaborated to construct SDH scores and employed them in studying heart transplant candidates and recipients' survival.[19,20] There is also usually high utility and interest in questioning and solving problems with higher disease burdens, an example is, "examining the prevalence of asymptomatic COVID-19 infection in cancer patients in New York City."[21] Context is always key, however. Solving problems with low prevalence but grave consequences can be equally interesting and valuable,[15] an example is, "examining the consequences and management of postoperative open chest in patients after cardiac surgery and lung transplantation."[22,23]

N—novel: this is defined as "the provision of answer to an existing gap in knowledge" A novel research question and study should generate new and compelling evidence that serves the needs of the community. Novelty ensures that the answers generated from the question will add new insight to the scientific literature and may serve as a foundation for future explorations and knowledge. Editors and peer-reviewers of high-impact journals are usually reluctant to publish studies that lack inherent novelty and relevant results although this does not mean that researchers cannot replicate previous work.[24] Doing so in a different population or setting, with a different method or variables could add novelty to a question, and the resulting answer could confirm and validate or refute established knowledge or create new facts about existing knowledge. A useful strategy for assessing the novelty of a question at an early stage involves reflective evaluation of what is known and unknown about the topic of interest through a search of the literature.[25] Doing so will help researchers to identify what is already published about similar questions and may also provide a guide in choosing the appropriate design and methodology.

E—ethical: in empirical research, developing an ethical research question is a MUST. The research question is the foundation upon which all other research decisions are based; it is therefore very important to consider the ethical aspects of a research question very early on in the process to inform the ethical obligations and processes throughout the research project. Potential risks and benefits should be carefully weighed, based on the ethical principles of respect for autonomy, beneficence, nonmaleficence, and justice.[26] Ethical issues could be related to the population (how and which patients were selected or what eligibility criteria were used), intervention (side effects and complications of the intervention), comparison (use of sham or placebo therapy), or outcome (longer exposure to intervention, need for repetitive or invasive testing).[27] Considering these issues and embedding respect for them in the research question helps to ensure that the study fulfills its ethical obligations, both institutionally and professionally.

R—relevant: Research relevancy can be established by considering the impact of its findings on scientific knowledge, clinical and health policy, and future research.[14] While it is important to keep a research study simple and direct, it should not be so simple as to make no difference whatsoever. Useful clinical research is that in which results can lead to a favorable change in decision-making (when changes in benefits, harm, or cost are considered) either by itself or when integrated with other evidence in systematic reviews, meta-analyses, and guidelines.[15] The goal of clinical research is to improve the delivery of health care to patients. Therefore, researchers should address questions that are prioritized by the healthcare community and study outcomes that are important to relevant stakeholders, especially patients. Otherwise, the entire study may be considered "waste."[28] Interest is usually linked to the novelty and relevance of a research question. It is important to note that while they may be related, they remain distinct. It is reflected in a recent survey by Johnson et al. that demonstrated only a third of respondents who downloaded articles declared to have read it thoroughly, and at least half of those manuscripts remained unread. Also, up to one-third of papers remained uncited 5 years after publication, and only 20% of papers accounted for 80% of the citations.[29] A relevant research question should be patient-centered (be in alignment with patient priorities),[15] and asked with the aim of providing answers that preserve health and enhance wellness instead of academic achievement and career advancement. Furthermore, it should aim to contribute to the body of knowledge in the field and fill any identified gap. Putting these characteristics into consideration when developing research questions ensures appropriate justification for the study and minimizes waste.

Pitfalls to avoid

1. Failure to modify and optimize a research question very early in the process of a research study. This could result in unwarranted, even unethical iterations to the study protocol and design and affect the overall implementation process, quality, and success of the study. This could result in a waste instead of achieving the aim of clinical research which is "to bring meaningful change to practice."
2. Generating complex questions with multiple focus and aims, instead of specific, simple, and clear ones. These are likely to require extensive resources which could in turn affect the feasibility of the study without necessarily improving the relevance.
3. Narrowed focus on developing a well-structured question without giving consideration to whether it can be answered within a reasonable time and with the available resources (feasibility) and whether the answer would be useful (relevance).
4. Narrowed focus on the researcher's needs without giving consideration to what is best for the community at large.

Acknowledgments

We thank April Mann, Director of the Writing Center at the University of Miami, for her writing assistance and proofreading of the chapter.

References

1. Trice LB, Bloom KC. *Chapter 5: PICOT, Problem Statement, Research Question, Hypothesis*. Jones & Bartlett Learning, LLC. Available from: https://samples.jbpub.com/9781449695071/81968_CH05_121_146.pdf. Accessed 5 March 2022.
2. DeCarlo M. *Scientific inquiry in social work. SWK 340: Social Work Research Methods OER Collection*. 2018.
3. Portney LG, Watkins MP, eds. *Asking the Research Question. Foundations of Clinical Research: Applications to Practice*. 3rd ed. McGraw Hill; 2017. https://fadavispt.mhmedical.com/content.aspx?bookid=1987&sectionid=149503178.
4. Wood MJ, Ross-Kerr JC. From question to proposal. In: *Basic Steps in Planning Nursing Research*. 7th ed. Jones and Bartlett Publishers; 2006.
5. Eldredge JD. Evidence-based librarianship: an overview. *Bull Med Libr Assoc*. 2000;88(4):289–302.

6. Richardson WS, Wilson MC. *On Questions, Background and Foreground*. vol. 17. Evidence Based Healthcare Newsletter; 1997:8–9.
7. Booth A, Noyes J, Flemming K, et al. Formulating questions to explore complex interventions within qualitative evidence synthesis. *BMJ Glob Health*. 2019;4:e001107.
8. Mayo NE, Asano M, Pamela Barbic S. When is a research question not a research question? *J Rehabil Med*. 2013;45:513–518.
9. Riva JJ, Malik KM, Burnie SJ, Endicott AR, Busse JW. What is your research question? An introduction to the PICOT format for clinicians. *J Can Chiropr Assoc*. 2012;56:167–171.
10. https://www.cebm.ox.ac.uk/. Accessed 5 March 5 2022.
11. Snowball R. Using the clinical question to teach search strategy: fostering transferable conceptual skills in user education by active learning. *Health Libr Rev*. 1997;14(3):167–172.
12. Richardson WS, Wilson MC, Nishikawa J, Hayward RSA. The well-built clinical question: a key to evidence-based decisions. *ACP J Club*. 1995;123:A12–A13.
13. Davies KS. Formulating the evidence based practice question: a review of the frameworks. *Evid Base Libr Inf Pract*. 2011;6(2):75–80. https://doi.org/10.18438/B8WS5N.
14. Hulley SB, Cummings SR, Browner WS, et al. *Designing Clinical Research*. 3rd ed. Philadelphia, PA: Lippincott Williams and Wilkins; 2007.
15. Ioannidis JP. Why most clinical research is not useful. *PLoS Med*. 2016;13(6):e1002049. https://doi.org/10.1371/journal.pmed.1002049.
16. Kasenda B, von Elm E, You J, Blümle A, Tomonaga Y, et al. Prevalence, characteristics, and publication of discontinued randomized trials. *JAMA*. 2014;311(10):1045–1051. https://doi.org/10.1001/jama.2014.1361.
17. Chapman SJ, Shelton B, Mahmood H, Fitzgerald JE, Harrison EM, et al. Discontinuation and non-publication of surgical randomised controlled trials: observational study. *BMJ*. 2014;349:g6870. https://doi.org/10.1136/bmj.g6870.
18. Blatch-Jones AJ, Pek W, Kirkpatrick E, et al. Role of feasibility and pilot studies in randomised controlled trials: a cross-sectional study. *BMJ Open*. 2018;8:e022233.
19. Alnajar A. *Identifying Factors Associated With Long-Term Graft Survival Following Heart Transplantation Using Statistics and Machine Learning Methods*. University of Miami; 2021.
20. Clerkin KJ, Garan AR, Wayda B, et al. Impact of socioeconomic status on patients supported with a left ventricular assist device: an analysis of the UNOS database (united network for organ sharing). *Circ Heart Fail*. 2016;9(10).
21. Ibrahim M, Natarajan V, Murthy P, Meghal T, Xu Y, Wiesel O. The prevalence of asymptomatic COVID-19 infection in cancer patients. A cross-sectional study at a tertiary cancer center in New York City. *Cancer Treat Res Commun*. 2021;27:100346. https://doi.org/10.1016/j.ctarc.2021.100346.
22. Rafiroiu S, Hassouna H, Ahmad U, et al. Consequences of delayed chest closure during lung transplantation. *Ann Thorac Surg*. 2020;109(1):277–284. https://doi.org/10.1016/j.athoracsur.2019.08.016.
23. Bakaeen FG, Haddad O, Ibrahim M, et al. Advances in managing the noninfected open chest after cardiac surgery: negative-pressure wound therapy. *J Thorac Cardiovasc Surg*. 2019;157(5):1891–1903.e9. https://doi.org/10.1016/j.jtcvs.2018.10.152.
24. Davidson A, Delbridge E. How to write a research paper. *Paediatr Child Health*. 2012;22:61–65.
25. Kim YJ, Mack SJ, Chung KC. Articulating the "so, what?" in clinical research: insight from the M-CHOIR group. *Plast Reconstr Surg Glob Open*. 2020;8(5):e2848. https://doi.org/10.1097/GOX.0000000000002848.
26. Hall R, McKnight D, Cox R, Coonan T. Guidelines on the ethics of clinical research in anesthesia. *Can J Anesth*. 2011;58:1115–1124.
27. Fandino W. Formulating a good research question: pearls and pitfalls. *Indian J Anaesth*. 2019;63(8):611–616. https://doi.org/10.4103/ija.IJA_198_19.
28. Chalmers I, Bracken MB, Djulbegovic B, et al. How to increase value and reduce waste when research priorities are set. *Lancet*. 2014;383(9912):156–165. https://doi.org/10.1016/S0140-6736(13)62229-1.
29. Johnson R, Watkinson A, Mabe M. *The STM Report: An Overview of Scientific and Scholarly Journals Publishing*. International Association of Scientific, Technical and Medical Publishers; 2018. Available from: https://www.stm-assoc.org/2018_10_04_STM_Report_2018.pdf. Accessed March 5, 2022.

Suggested readings

1. Portney LG, Watkins MP, eds. *Asking the Research Question. Foundations of Clinical Research: Applications to Practice*. 3rd ed. McGraw Hill; 2017.
2. Wood MJ, Ross-Kerr JC. From question to proposal. In: *Basic Steps in Planning Nursing Research*. 7th ed. Jones and Bartlett Publishers; 2006.
3. Riva JJ, Malik KM, Burnie SJ, Endicott AR, Busse JW. What is your research question? An introduction to the PICOT format for clinicians. *J Can Chiropr Assoc*. 2012;56:167–171.
4. Hulley SB, Cummings SR, Browner WS, et al. *Designing Clinical Research*. 3rd ed. Philadelphia, PA: Lippincott Williams and Wilkins; 2007.

Part IV

Statistical principles

Pratik Agrawal

Chapter 21

Common issues in analysis

Michail Koutentakis[1,2]

[1]*Department of Cardiothoracic Surgery and Transplantology, University Clinical Centre of Medical University of Warsaw, Central Teaching Clinical Hospital, Warsaw, Poland;* [2]*Department of Experimental and Clinical Pharmacology, Medical University of Warsaw, Center for Preclinical Research and Technology CEPT, Warsaw, Poland*

Key points

- Pinpointing analytical errors: This chapter focuses on the frequent analysis mistakes and misconceptions in translational cardiology, aiming not only to identify but also to design strategies to avoid them.
- Mitigating data analysis pitfalls: Prioritizing misunderstood concepts, limitations imposed by study designs, and the fine line between valid and excessive conclusions, this chapter provides the reader with actionable insights to navigate obstacles within translational cardiology research.
- Enhancing clinical validity: The primary objective is the precise, articulate, and reliable interpretation of data, underlining its pivotal function in augmenting clinical validity and its substantial effect on the progressively advancing discipline of translational cardiology.

Introductory section/why it matters

In the intricate and multifaceted field of translational cardiology, a wide assortment of analytical methodologies are enlisted. These styles span from conventionally employed statistical analysis to contemporary and progressively advanced tools like machine learning algorithms and integrational bioinformatics. These analytical tools bear the formidable task of decoding the static condition of raw data into dynamic, significant, and interpretable facets which can provide groundbreaking insights into patient care.[1–3]

Notwithstanding, these potent analytical resources often pose their own unique set of potential hurdles. Misapplications or misconceptions in the usage of these analytical methodologies are not rare. Such misunderstandings, hovering from the principle of a method to its practical application, can lead to distorted conclusions, impeding the progress of vital research and its subsequent application in improving patient prognosis.[4] Moreover, the design of each study, while providing a structure and guiding path for the research, could also impose its own set of constraints, thus limiting the scope of results and their interpretations.[5–7]

Interpreting results and drawing conclusions from complicated analyses in the realm of cardiology is a specialized skill that demands a refined balance between substantiated assertions based on empirical evidence and innovative hypotheses that push the boundaries of knowledge.[8–10] The assessment of this delicate equilibrium presents significant challenges, particularly related to excessive interpretation and conjecture, which have the potential to distort the understanding and utilization of research findings in the field of cardiology.[11,12]

The "Common Issues in Analysis" chapter attempts to shed light on these complex and often underappreciated elements. The objective is to clarify common misconceptions surrounding the utilization of analytical methodologies,[4] focusing on their purpose, accuracy, and areas where practitioners may encounter challenges. Additionally, it explores the limitations imposed by study design, highlighting potential pitfalls and providing guidance on navigating them effectively.

Finally, this chapter delves into the meticulous process of deriving conclusions from abundant data output, distinguishing well-supported conclusions from premature and overly optimistic speculations. By uncovering these issues, the chapter aims to facilitate and equip healthcare professionals and researchers in effectively navigating the diverse range of analytical challenges that pervade the field of translational cardiology.[13]

Translational Cardiology. https://doi.org/10.1016/B978-0-323-91790-2.00054-X

Misunderstood concepts in statistical analysis

Examples of common misconceptions in translational cardiology

Frequently in the field of translational cardiology research, there exist certain concepts in statistical analysis that often fail to be understood or interpreted correctly. Misunderstandings of statistical methodology can precede from the foundation of conceptual knowledge to the practical application of these methods.[14–16]

P-values

In translational cardiology research, as in many other scientific fields, a common misconception is the overemphasis and misinterpretation of the *P*-value—a problem so widespread that it can be considered a chief culprit of statistical misconception. The misunderstood interpretation essentially lies within viewing *P*-values as direct measures of effect size or strength of evidence regarding an observed effect.[17–20]

Though, this interpretation is not accurate. Contrarily, a *P*-value is a conditional probability. It measures the likelihood of obtaining the observed data (or data more extreme) given that the null hypothesis is true. It does not measure the likeliness or size of an effect.[21–23]

One critical misunderstanding is the "*P*-value fallacy," where a *P*-value less than .05 is considered as strong evidence against the null hypothesis and indicates a genuine effect, while a *P*-value greater than .05 is viewed as evidence for the null hypothesis, rendering the effect nonexistent. This is not correct because a *P*-value is not a proof of truth, but rather a tool to assist in forming conclusions.[17,18,24]

Another erroneous assumption is that a lower *P*-value represents a stronger effect. A smaller *P*-value simply indicates that the observed data is less likely to occur by chance under the null hypothesis and does not quantify the size or importance of an effect.[17,18,24]

There is also a misconception about the interpretation of the *P*-value with respect to replication. A common belief is that a *P*-value less than .05 ($P < .05$) ensures that the results will be replicated 95% of the time. This interpretation does not account for the variations inherent in experimental reproduction and misapprehends the meaning of probability involved in *P*-values.[25,26]

Effectively, the misuse and overemphasis on *P*-values in translational cardiology research can lead to questionable results and conclusions. It does not signify the presence of an effect, rather it informs that if an effect were absent, what the probabilities of observing the collected data or even more extreme data would be. Understanding the real meaning and application of *P*-values can enhance the scientific rigor and reliability of studies in the field.[18,27]

Regression analysis

Regression analysis is a powerful statistical tool often used in translational cardiology research. It allows researchers to understand the relationships between different variables and to predict one variable based on the knowledge of others.[28–31] Still, there are several common misconceptions and inappropriate practices surrounding its use that can compromise the reliability and interpretability of research findings.[32,33]

One common misunderstanding pertains to the assumptions underlying regression models. These models make significant assumptions about the nature of the data, including the linearity of relationships, independence of observations, homoscedasticity of residuals, and normality of error terms. Violation of these assumptions can result in biased or incorrect conclusions. For instance, a typical misuse involves applying linear regression to data that violate the linearity and equal variance (homoscedasticity) assumptions. This can lead to inefficient, biased, or inconsistent estimates of the model parameters.[33] In translational cardiology research, checks for these assumptions are often overlooked or inadequately reported, with researchers failing to appreciate the implications these violations have on the outcomes of their analysis. This lack of understanding compromises the validity of the statistical inferences drawn from the analyses.[34]

The misuse or overuse of multivariable models without clear justifications is another problematic area in regression analysis. Multivariable modeling is a valuable tool that allows the simultaneous analysis of multiple predictors. However, its inappropriate overuse has taken root in the translational cardiology field. It is common to see a multitude of variables included in models without any methodological justification, leading to issues like multicollinearity, overfitting (creating models that too closely align with particular dataset nuances), and escalating model complexity, affecting the comprehensibility and generalizability of research.[35]

Particularly, overfitting models with too many variables relative to the number of outcomes can lead to exaggerated effect estimates and narrow confidence intervals, providing an unreliable impression of precision. These overly complex, "overfit" models also decrease the model's ability to predict outcomes in new or different populations, impairing their generalizability.[36]

To prevent these misconceptions and misuses, understanding and application of regression analysis should be promoted. Applying the assumptions checks, using sensible variable selection methods, avoiding overfitting, and ensuring the comprehensibility of models would go a long way in enhancing the accuracy and reliability of the outcomes in translational cardiology research.[37]

Overfitting and underfitting

The application of machine learning techniques in translational cardiology research has dramatically increased, offering potent tools for predictive modeling, diagnosis and prognosis, and clustering, among other purposes.[38,39] Despite that, along with the advantages comes the susceptibility to issues of overfitting and underfitting, which are often misunderstood and improperly handled, thus compromising research validity and reliability.[40,41]

Overfitting is a phenomenon where the model becomes overly complex, capturing the noise and inconsistencies in the data rather than the underlying pattern. This predicament happens when the model is excessively reliant on the training data used, leading it to perform excellently in the training phase but poorly when exposed to new or unseen data.[40,41] In the context of translational cardiology, overfitted models may predict heart disease with superlative accuracy on the training data but fail to generalize to new patients, calling into question the utility of the research.[38,42]

On the contrary, underfitting refers to a scenario where the model is overly simplistic, failing to capture the underlying complexities in the data. An underfitted model, in essence, is characterized by high bias and leads to inaccurate predictions and conclusions as it ignores relevant variables or relationships in the data.[40,41] This is just as problematic. For instance, in predicting cardiac events, an underfitted model might overlook critical risk factors like hypertension or previous cardiac history, producing predictions that are woefully inadequate and thus jeopardizing patient care.[38,42]

In translational cardiology research employing machine learning techniques, both overfitting and underfitting weaken out-of-sample predictive performance—that is, the model's ability to predict outcomes accurately in new patients or datasets different from the ones on which the model was trained.[43]

To prevent these pitfalls, a careful balance between model complexity and simplicity, often referred to as the bias-variance trade-off, must be struck.[1,44–47] Additionally, appropriate resampling techniques (cross-validation or bootstrapping),[48–51] model validation on independent data,[52,53] and regularization methods[54–56] should be used to guard against overfitting and underfitting.

Confidence intervals

In translational cardiology research, the concept of confidence intervals (CI) is often misunderstood or misinterpreted, leading to inaccurate conclusions being drawn from the data.[57–59] One common misconception is that a 95% confidence interval refers to a range that will contain the true population parameter 95% of the time.[59–61] This is incorrect. The correct interpretation is that if we were to repeat the study many times, we would expect that 95% of those confidence intervals would contain the true parameter.[62–64] This misinterpretation often leads to the erroneous conclusion that, with a 95% confidence interval, there is a 95% chance the true population parameter lies within the given interval, creating false certainty about the findings.[59,65,66]

Another common misunderstanding is equating a smaller confidence interval with a more precise estimate. While the width of the confidence interval is related to precision, it is also affected by variability in the data and sample size. While the interval may be narrow, it is important to note that a smaller interval does not inherently imply greater precision in the estimate. Similarly, a wide interval can indicate greater uncertainty, but could also be due to a large sample size, leading to falsely perceived less precision.[62,64,67]

Lastly, if a confidence interval overlaps with the value representing no effect (like 0 or 1), it is often incorrectly concluded that there is "no effect." Nonetheless, failure to reject a null hypothesis (i.e., the confidence interval contains the null value) does not prove the null is true. A study might simply lack the power to detect the true effect. Therefore, framing confidence intervals as an indicator of effect size is a more accurate interpretation.[63,68,69]

To mitigate these misconceptions, it is necessary to promote proper understanding and interpretation of confidence intervals in translational cardiology, emphasizing their real meaning and factors affecting their width. Clarifying that no definitive conclusions can be made about the null hypothesis from a single study will also help avoid misconceptions.[57–59]

Multivariate analysis

In translational cardiology research, the application and interpretation of multivariate analysis (MVA) require careful consideration. A common misconception is that MVA is a fully automated procedure, simply yielding definitive results.[70,71] Yet, it requires meticulous selection of variables by the researcher, guarding against the incorporation of unrelated factors that may skew the analysis.[72–74]

It is also crucial to understand that while MVA aids in adjusting for confounding factors, this adjustment is limited to the variables accounted for in the model.[75–77] Overlooking key confounders can undermine the validity of the analysis, demonstrating the importance of comprehensive and thoughtful variable selection.[72–74]

Furthermore, it is easy to misinterpret associations identified by MVA as causal relationships. While MVA effectively identifies associations and adjusts for confounding variables, it cannot conclusively establish causation due to potential unmeasured or unrecognized confounders.[70]

Last but not least, more discerning use of MVA is required to avoid the misconception that adding more variables improves the analysis. Overfitting the model with excessive variables can diminish its predictive efficacy.[78–80]

Collectively, the adept application of MVA in translational cardiology necessitates both an understanding of its limitations and refined discernment in selecting variables and interpreting associations. This ensures research findings are both statistically sound and clinically meaningful.[70,71]

Multiple comparison analysis

Understanding and accurately applying multiple comparison analysis (MCA) is of central importance to translational cardiology research, wherein it allows for robust and reliable results. Misinterpretations can guide clinicians and researchers toward incorrect conclusions, affecting both study outcomes and clinical decision-making.[81–83]

Firstly, it is often forgotten that MCA's purpose is to control the Type I error rate, which increases as the number of comparisons grows.[84–86] MCA does not intend to discover all relationships within the data, thus overlooking significant associations is to be expected.[81–83]

Secondly, there is a misconception about the "family-wise error rate (FWER)." It is mistakenly believed that FWER is the probability of making one or more Type I errors in the group of comparisons. However, FWER is the proportion of similar groups that would produce at least one Type I error.[87–89]

Moreover, MCA should not be invoked indiscriminately. While it can control the potential for false positives, it raises the potential for false negatives due to its conservative nature. Therefore, it is critical to balance the necessity of decreasing Type I errors with potential increases in Type II errors.[83,86,90]

Finally, it is important not to misinterpret nonsignificant *P*-values in MCA. Nonsignificant results in MCA do not necessarily signify no effect. They instead indicate that the data does not provide enough evidence to reject the null hypothesis.[91,92]

Proper application and interpretation of MCA enhance the validity and reliability of translational cardiology research, leading to more accurate clinical decisions and improved patient outcomes.[81–83]

Endogeneity

Within translational cardiology, navigating endogeneity is integral to generating meaningful results. Misconceptions around the concept can significantly impact study findings and subsequently influence clinical practice.[93,94]

To begin with, endogeneity does not imply the simultaneous determination of variables, and the dependent variable influences the independent variable. Endogeneity indicates an unmeasured correlation of the independent variable that influences the dependent variable, which can lead to biased estimation of effect sizes.[95,96]

Within this framework, a common misunderstanding is that using observational data automatically results in endogeneity. Although, while endogeneity is more likely in observational studies, appropriate model specification, measurement of all relevant variables, and correct methodological techniques can manage endogeneity.[93,97]

In addition, there is a common misconception that endogeneity arises predominantly from omitted variables. It is indeed true that omitted variables, which are relevant predictors not included in the model, represent a significant source of endogeneity. Their exclusion can lead to biased estimates of the relationships under study, as the influence they have on the outcome variable gets wrongly attributed to the included predictor variables.[96]

Nevertheless, it is critical to recognize that omitted variables aren't the sole contributors to endogeneity.[95] Simultaneity and measurement errors are other significant sources.[98,99] Simultaneity refers to a situation where cause-and-effect relationships run in both directions between the outcome and predictor variables, making it difficult to disentangle their respective influences. This circularity can introduce bias in the estimated relationships.[100–102] Measurement error contributes to endogeneity as well, and it occurs when the predictor variables are measured inaccurately. If such errors are systematically associated with the error term in the model, they may create a false association between the predictor and outcome variables, leading to erroneous conclusions.[98,99] Thus endogeneity is a multifaceted issue, and addressing it requires a thorough understanding of all these contributing elements.

Overcoming these misconceptions and understanding endogeneity ensures the appropriate modeling of relationships and parameters, leading to more accurate interpretations of cardiovascular disease mechanisms.[93,94]

Limitations by study design

In the context of translational cardiology, different study designs come with their unique set of limitations.[103–105] Understanding these can help in mitigating the limitations and enhancing experimental precision.

Observational studies

Observational studies, though commonly selected within this field due to ethical and pragmatic considerations,[106] have their limitations statistically.[107–110]

The primary concern lies with confounding, where extraneous variables exert influence on both the independent and dependent variables, thereby skewing results. For instance, while evaluating the influence of a novel intervention on cardiac outcomes, lifestyle variables such as dietary habits or physical exercise levels could also impact cardiac outcomes, thereby acting as confounders.[111–113]

Furthermore, these studies may also be susceptible to selection bias, where the chosen sample does not accurately epitomize the broader population, distorting statistical interpretations.[107,108,114] Selection bias arises when the chosen participants in a study possess distinct characteristics from those eligible individuals who are not included in the study.[114]

Randomized controlled trials

Randomized controlled trials (RCTs), considered the gold standard due to the random assignment of interventions, also entail certain statistical limitations.[115,116]

In light of this, attrition bias is a frequent concern, where an unequal dropout rate among participants can lead to inaccurate interpretations. For instance, a higher dropout rate among nonresponders or those with adverse outcomes in a cardiology trial could overstate the efficacy of an intervention. The substantial costs involved and the requirement for long follow-up periods in RCTs often restrict the sample size, consequently affecting the statistical power of the study.[117–119]

In a more refined context, consider a cardiological trial designed to evaluate a specific intervention purposed to lower blood pressure.[120–123] Subsequent to this study, an elevated attrition rate among nonrespondents or individuals encountering adverse events could artificially enhance the perceived potency of the intervention.[117–119] The financial implications associated with conducting RCTs, along with the necessity for extended follow-up durations, frequently impose limitations on the sample size. This could, consequently, jeopardize the amiability of the study's overall statistical power.[115,116]

Case-control studies

Case-control studies offer speed and cost-effectiveness but are not devoid of limitations when it comes to statistical analysis.[108,124–126] Recall bias is often an issue, as retrospective queries about factors influencing cardiovascular health depend on the accuracy of participants' memories, thereby affecting the reliability of data.[127,128] Selection bias could also occur when choosing control subjects, consequently influencing the findings.[108,124–126]

For example, a case-control study might be carried out to investigate whether high cholesterol levels could be a risk factor for heart attacks.[129–132] Here, the participants diagnosed with heart attacks (cases), and those without such a diagnosis (controls), would need to recall their past cholesterol levels or related dietary habits—potentially introducing recall bias if they cannot remember accurately.[127,128] Similarly, if the control group is not chosen from the same population, or if health-conscious individuals are more likely to participate, it could lead to a selection bias, thereby influencing the outcome of the study.[124–126]

Cohort studies

Despite the strength of cohort studies in providing superior evidence of causality as opposed to case-control studies and their popularity in cardiology research due to the ability to scrutinize multiple outcomes from a single exposure, they present significant constraints.[133–135] The handling of large sample sizes and extended durations, particularly in the context of rare outcomes, can pose a significant challenge.[134,136] This challenge is exacerbated in the study of chronic diseases such as cardiac conditions, where the necessity for long follow-up periods intensifies these difficulties.[133–135] Compounding the issue, the protracted durations could elevate the costs and increase the participant drop-out rates, thereby potentially diminishing the study power and introducing bias into the findings.[137–139]

Consider a cohort study designed to examine the association between a sedentary lifestyle and the risk of developing cardiovascular diseases. Such a study would require long-term monitoring of a large participant group, assessing factors such as lifestyle habits, health status alterations, and emergent cardiovascular symptoms.[133–135]

In executing this study, the researchers may encounter potential stumbling blocks. Participant attrition over the extensive follow-up period can create bias in the study. Participants may relocate, lose interest, or worse, suffer mortality from unrelated causes.[137–139] The remaining cohort may therefore be skewed toward those with better health conditions or more health-conscious behaviors, creating a potential source of bias.[140–142]

Furthermore, the extensive amount of data requiring management, including changes in lifestyle habits and health status over time, amplifies the complexity of data analysis.[133–135] Coordinating frequent health examinations, maintaining administrative processes, and managing long-term data collection can challenge resource allocation, leading to increased costs.[134,136] Cumulatively, these factors could compromise the study's power and the precision of the estimated association between sedentary behavior and the risk of cardiovascular disease.[133–135]

Recognizing these limitations allows researchers in translational cardiology to take appropriate measures to reduce bias, improve accuracy, and therefore, the reliability of their findings, ultimately advancing cardiology practices.

Appropriate and overstepping conclusions

In the complex and evolving sphere of translational cardiology research, careful attention must be paid to the conclusions that we draw from our studies. In particular, the themes of "Appropriate Conclusions" and "Overstepping Conclusions" are crucial to emphasize and understand to maintain the highest levels of scientific integrity and accuracy.

When considering the idea of "Appropriate Conclusions," we are referring to the importance of having a conclusion that is directly backed up by the data obtained in the study.[143,144] For instance, a study published in 2018 on a large cohort showed that the drug Canakinumab resulted in a significant reduction in incidents of cardiovascular events in postmyocardial infarction (post-MI) patients. The researchers accurately concluded that Canakinumab could be a potential therapeutic addition for high-risk patients post-MI.[145,146] This conclusion was directly supported by the empirical evidence that the study yielded, and so represents a model of an "Appropriate Conclusion."

On the other hand, "Overstepping Conclusions" arise when the researchers make wider claims that are not sufficiently supported by the actual study results.[143,144] An example of this comes from a study published in 2022, which looked at the global burden of cardiovascular disease and found that hypertension, hypercholesterolemia, dietary risks, and air pollution were the leading causes of cardiovascular disease worldwide.[147] Nevertheless, it would be an overstep to conclude that these are the only factors contributing to cardiovascular disease based on this data alone. There could be other factors not accounted for in the study that also contribute to cardiovascular disease.

Maintaining the balance between "Appropriate" and "Overstepping" conclusions is crucial in the field of translational cardiology research. It requires meticulous attention to detail, a solid understanding of the data and its limitations, as well as clear, precise communication of the findings. Moreover, peer review and the consistent application of scientific methods also play significant roles in preventing overreaching conclusions and ensuring the research's proper contribution to translational cardiology.

Examples

1. **Confounding variables:** A recently published study explored the potential impact of hormone replacements as part of gender affirmation therapy on the risk of serious cardiac events, including stroke, heart attack, and pulmonary embolism The researchers found that hormone replacements were associated with higher rates of cardiac events, mostly related to dangerous blood clots. However, they were not associated with higher rates of death. The study highlighted the importance of considering confounding variables such as age, gender, and comorbidities, which can influence both the exposure (medication) and the outcome (cardiovascular events), leading to biased results.[148] To address this challenge, researchers can use techniques like multivariable regression analysis to adjust for confounding factors and obtain more accurate estimates of the medication's effect on cardiovascular events.
2. **Time-to-event analysis:** A second study explored the potential impact of competing risks on time-to-event endpoints in cardiology trials. They found that standard survival analysis techniques, such as the Kaplan–Meier estimator, can yield seriously biased results in the presence of competing risks. Although methods to account for competing risks are well known in the statistical literature, they are rarely applied in clinical trials.[149] Researchers can employ survival analysis techniques like Kaplan–Meier curves and Cox proportional hazards models to appropriately handle censoring and estimate the probability of events over time.
3. **Longitudinal Data Analysis:** Another study investigated the value of longitudinal data over and above the use of cross-sectional data via six distinct modeling strategies from statistics, machine learning, and deep learning that incorporate repeated measures for survival analysis of the time-to-cardiovascular event in the Coronary Artery Risk Development in Young Adults (CARDIA) cohort. They found that longitudinal information from 35 variables that were repeatedly collected in six exam visits over 15 years improved subsequent long-term (17 years after) risk prediction by up to 8.3% in C-index compared to using baseline data (0.78 vs. 0.72), and up to approximately 4% compared to using the last observed cross-sectional data (0.75).[150] Researchers can utilize methods like mixed-effects models or generalized estimating equations to properly model longitudinal data, consider the correlation structure, and address missing data through appropriate imputation techniques.

Potential pitfalls

1. **Misunderstood concepts in statistical analysis:** Without a thorough understanding of statistical concepts, there is the potential pitfall of misinterpreting data, leading to flawed conclusions and deterring the advancement of translational cardiology.
2. **Limitations by study design:** Ignoring or overlooking the limitations of a study design can result in erroneous interpretations of results, potentially misleading future research and clinical practices in translational cardiology.
3. **Overstepping conclusions in translational cardiology:** Overreaching or overgeneralizing based on the

study data can lead to miscommunication of research findings, ultimately diluting the validity and application of translational cardiology research.

Get started

To enhance the quality and impact of translational cardiology research, researchers must address some focal points related to statistical analysis, study design, and conclusions.

1. Gain a detailed understanding of statistical concepts.
2. Diligently acknowledge and articulate the limitations inherent to their study design.
3. Steadfastly adhere to the data when drawing conclusions.

Consultants/cocontractors

The composition of the chapters stands as the sole effort of the author, with no participation from external individuals or organizations during the writing process.

References

1. Shameer K, Johnson KW, Glicksberg BS, Dudley JT, Sengupta PP. Machine learning in cardiovascular medicine: are we there yet? *Heart*. 2018;104(14):1156–1164. https://doi.org/10.1136/HEARTJNL-2017-311198.
2. Meder B, Katus HA, Keller A. Computational cardiology — a new discipline of translational research. *Dev Reprod Biol*. 2016;14(4):177. https://doi.org/10.1016/J.GPB.2016.08.001.
3. Khomtchouk BB, Tran DT, Vand KA, Might M, Gozani O, Assimes TL. Cardioinformatics: the nexus of bioinformatics and precision cardiology. *Brief Bioinform*. 2020;21(6):2031–2051. https://doi.org/10.1093/BIB/BBZ119.
4. Austin CP. Translational misconceptions. *Nat Rev Drug Discov*. 2021;20(7):489–490. https://doi.org/10.1038/d41573-021-00008-8.
5. Hyder A, May AA. Translational data analytics in exposure science and environmental health: a citizen science approach with high school students. *Environ Health*. 2020;19(1):1–12. https://doi.org/10.1186/S12940-020-00627-5/TABLES/2.
6. Rumsfeld JS, Alexander KP, Goff DC, et al. Cardiovascular health: the importance of measuring patient-reported health status. *Circulation*. 2013;127(22):2233–2249. https://doi.org/10.1161/CIR.0B013E3182949A2E.
7. Weant KA, Bailey AM, Baker SN. Strategies for reducing medication errors in the emergency department. *Open Access Emerg Med*. 2014;6:45. https://doi.org/10.2147/OAEM.S64174.
8. Manrai AK, Patel CJ, Ioannidis JPA. In the era of precision medicine and big data, who is normal? *JAMA*. 2018;319(19):1981–1982. https://doi.org/10.1001/JAMA.2018.2009.
9. Bossuyt P, Davenport C, Deeks J, Hyde C, Leeflang M, Scholten R. *Cochrane Handbook for Systematic Reviews of Diagnostic Test Accuracy Chapter 11 Interpreting results and drawing conclusions*; 2023. http://srdta.cochrane.org/. Accessed December 3, 2023.
10. *Chapter 15: Interpreting Results and Drawing Conclusions*: Cochrane Training. https://training.cochrane.org/handbook/current/chapter-15. Accessed 3 December 2023.
11. Fihn SD. Combating misrepresentation of research findings. *JAMA Netw Open*. 2019;2(5). https://doi.org/10.1001/JAMANETWORKOPEN.2019.2553.
12. Rafie N, Kashou AH, Noseworthy PA. ECG interpretation: clinical relevance, challenges, and advances. *Hearts*. 2021;2(4):505–513. https://doi.org/10.3390/HEARTS2040039.
13. Dai H, Younis A, Kong JD, et al. Big data in cardiology: state-of-art and future prospects. *Front Cardiovasc Med*. 2022;9:844296. https://doi.org/10.3389/FCVM.2022.844296/BIBTEX.
14. Wasserstein RL, Lazar NA. The ASA statement on p-values: context, process, and purpose. *Am Stat*. 2016;70(2):129–133. https://doi.org/10.1080/00031305.2016.1154108.
15. Altman N, Krzywinski M. Points of significance: P values and the search for significance. *Nat Methods*. 2016;14(1):3–4. https://doi.org/10.1038/NMETH.4120.
16. Althouse AD, Below JE, Claggett BL, et al. Recommendations for statistical reporting in cardiovascular medicine: a special report from the American heart association. *Circulation*. 2021;144(4):E70–E91. https://doi.org/10.1161/CIRCULATIONAHA.121.055393.
17. Mark DB, Lee KL, Harrell FE. Understanding the role of P values and hypothesis tests in clinical research. *JAMA Cardiol*. 2016;1(9):1048–1054. https://doi.org/10.1001/JAMACARDIO.2016.3312.
18. Thomas LE, Pencina MJ. Do not over (P) value your research article. *JAMA Cardiol*. 2016;1(9). https://doi.org/10.1001/JAMACARDIO.2016.3827.
19. Fletcher J. Misinterpretation of p values. *BMJ*. 2023;337. https://doi.org/10.1136/BMJ.A201.
20. Stovitz SD, Verhagen E, Shrier I. Misinterpretations of the 'P value': a brief primer for academic sports medicine. *Br J Sports Med*. 2017;51(16):1176–1177. https://doi.org/10.1136/BJSPORTS-2016-097072.
21. Ferguson J. Bayesian interpretation of P values in clinical trials. *BMJ Evid Based Med*. 2022;27(5):313–316. https://doi.org/10.1136/BMJEBM-2020-111603.
22. López-Astorga M, Ragni M, Johnson-Laird PN. The probability of conditionals: a review. *Psychon Bull Rev*. 2022;29(1):1–20. https://doi.org/10.3758/S13423-021-01938-5/TABLES/16.
23. Sedgwick P. Re: understanding P values. *BMJ (Online)*. 2023;349. https://doi.org/10.1136/BMJ.G4550.
24. Goodman SN. Toward evidence-based medical statistics. 1: the P value fallacy. *Ann Intern Med*. 1999;130(12):995–1004. https://doi.org/10.7326/0003-4819-130-12-199906150-00008.
25. Lyu XK, Xu Y, Zhao XF, Zuo XN, Hu CP. Beyond psychology: prevalence of P value and confidence interval misinterpretation across different fields. *J Pac Rim Psychol*. 2020;14:e6. https://doi.org/10.1017/PRP.2019.28.
26. Lyu Z, Peng K, Hu CP. P-value, confidence intervals, and statistical inference: a new dataset of misinterpretation. *Front Psychol*. 2018;9(JUN). https://doi.org/10.3389/FPSYG.2018.00868.
27. Wang M, Long Q. Addressing common misuses and pitfalls of P values in biomedical research. *Cancer Res*. 2022;82(15):2674–2677. https://doi.org/10.1158/0008-5472.CAN-21-2978.
28. Regression Model Assumptions. *Introduction to Statistics*: JMP. https://www.jmp.com/en_us/statistics-knowledge-portal/what-is-reg

ression/simple-linear-regression-assumptions.html. Accessed 3 December 2023.
29. Moyé L. Statistical methods for cardiovascular researchers. *Circ Res.* 2016;118(3):439. https://doi.org/10.1161/CIRCRESAHA.115.306305.
30. Zhang J, Bolli R, Garry DJ, et al. Basic and translational research in cardiac repair and regeneration: JACC state-of-the-art review. *J Am Coll Cardiol.* 2021;78(21):2092–2105. https://doi.org/10.1016/J.JACC.2021.09.019.
31. Sánchez de la Nava AM, Gómez-Cid L, Ríos-Muñoz GR, et al. Cardiovascular diseases in the digital health era: a translational approach from the lab to the clinic. *BioTech.* 2022;11(3). https://doi.org/10.3390/BIOTECH11030023.
32. *6 Assumptions of Linear Regression With Plots and Solutions*; 2024. https://www.analyticsvidhya.com/blog/2016/07/deeper-regression-analysis-assumptions-plots-solutions/. Accessed December 3, 2023.
33. *The Four Assumptions of Linear Regression - Statology*; 2020. https://www.statology.org/linear-regression-assumptions/. Accessed December 3, 2023.
34. Hickey GL, Kontopantelis E, Takkenberg JJM, Beyersdorf F. Statistical primer: checking model assumptions with regression diagnostics. *Interact Cardiovasc Thorac Surg.* 2019;28(1):1–8. https://doi.org/10.1093/ICVTS/IVY207.
35. Grant SW, Hickey GL, Head SJ. Statistical primer: multivariable regression considerations and pitfalls. *Eur J Cardio Thorac Surg.* 2019;55(2):179–185. https://doi.org/10.1093/EJCTS/EZY403.
36. Grant SW, Collins GS, Nashef SAM. Statistical primer: developing and validating a risk prediction model. *Eur J Cardio Thorac Surg.* 2018;54(2):203–208. https://doi.org/10.1093/EJCTS/EZY180.
37. Atkinson A, Riani M. *Robust Diagnostic Regression Analysis.* 2000. https://doi.org/10.1007/978-1-4612-1160-0.
38. Kresoja KP, Unterhuber M, Wachter R, Thiele H, Lurz P. A cardiologist's guide to machine learning in cardiovascular disease prognosis prediction. *Basic Res Cardiol.* 2023;118(1):1–12. https://doi.org/10.1007/S00395-023-00982-7/TABLES/2.
39. Couckuyt A, Seurinck R, Emmaneel A, et al. Challenges in translational machine learning. *Hum Genet.* 2022;141(9):1451–1466. https://doi.org/10.1007/S00439-022-02439-8.
40. *Overfitting and Underfitting in Machine Learning - Javatpoint.* https://www.javatpoint.com/overfitting-and-underfitting-in-machine-learning. Accessed 3 December 2023.
41. *ML Underfitting and Overfitting.* GeeksforGeeks; 2024. https://www.geeksforgeeks.org/underfitting-and-overfitting-in-machine-learning/. Accessed December 3, 2023.
42. Terranova N, Venkatakrishnan K, Benincosa LJ. Application of machine learning in translational medicine: current status and future opportunities. *AAPS J.* 2021;23(4):1–10. https://doi.org/10.1208/S12248-021-00593-X/FIGURES/2.
43. Quer G, Arnaout R, Henne M, Arnaout R. Machine learning and the future of cardiovascular care: JACC state-of-the-art review. *J Am Coll Cardiol.* 2021;77(3):300–313. https://doi.org/10.1016/J.JACC.2020.11.030.
44. Guan X, Burton H. Bias-variance tradeoff in machine learning: theoretical formulation and implications to structural engineering applications. *Structures.* 2022;46:17–30. https://doi.org/10.1016/J.ISTRUC.2022.10.004.
45. Doroudi S. The bias-variance tradeoff: how data science can inform educational debates. *AERA Open.* 2020;6(4):233285842097720. https://doi.org/10.1177/2332858420977208.
46. Arfat Y, Mittone G, Esposito R, Cantalupo B, Deferrari GM, Aldinucci M. Machine learning for cardiology. *Minerva Cardiol Angiol.* 2023;70(1):1–75. https://doi.org/10.23736/S2724-5683.21.05709-4.
47. Ben Ali W, Pesaranghader A, Avram R, et al. Implementing machine learning in interventional cardiology: the benefits are worth the trouble. *Front Cardiovasc Med.* 2021;8:711401. https://doi.org/10.3389/FCVM.2021.711401/BIBTEX.
48. Singh V, Pencina M, Einstein AJ, Liang JX, Berman DS, Slomka P. Impact of train/test sample regimen on performance estimate stability of machine learning in cardiovascular imaging. *Sci Rep.* 2021;11(1):1–8. https://doi.org/10.1038/s41598-021-93651-5.
49. Alonso-Atienza F, Rojo-Álvarez JL, Rosado-Muñoz A, Vinagre JJ, García-Alberola A, Camps-Valls G. Feature selection using support vector machines and bootstrap methods for ventricular fibrillation detection. *Expert Syst Appl.* 2012;39(2):1956–1967. https://doi.org/10.1016/J.ESWA.2011.08.051.
50. Garg R, Dong S, Shah S, Jonnalagadda SR. *A Bootstrap Machine Learning Approach to Identify Rare Disease Patients From Electronic Health Records*; 2016. Published online September 6, 2016 https://arxiv.org/abs/1609.01586v1. Accessed December 3, 2023.
51. Isler Y, Narin A, Ozer M. Comparison of the effects of cross-validation methods on determining performances of classifiers used in diagnosing congestive heart failure. *Meas Sci Rev.* 2015;15(4):196–201. https://doi.org/10.1515/MSR-2015-0027.
52. *Machine Learning Model Validation.* The Data-Centric Approach; 2022. https://appen.com/blog/machine-learning-model-validation/. Accessed December 3, 2023.
53. Wang H, Zheng H. Model validation, machine learning. *Encycl Syst Biol.* 2013:1406–1407. https://doi.org/10.1007/978-1-4419-9863-7_233.
54. Albahr A, Albahar M, Thanoon M, Binsawad M. Computational learning model for prediction of heart disease using machine learning based on a new regularizer. *Comput Intell Neurosci.* 2021;2021. https://doi.org/10.1155/2021/8628335.
55. Chen C, Qin C, Qiu H, et al. Deep learning for cardiac image segmentation: a review. *Front Cardiovasc Med.* 2020;7:508599. https://doi.org/10.3389/FCVM.2020.00025/BIBTEX.
56. Zhang B, Chai H, Yang Z, Liang Y, Chu G, Liu X. Application of L 1/2 regularization logistic method in heart disease diagnosis. *Bio Med Mater Eng.* 2014;24(6):3447–3454. https://doi.org/10.3233/BME-141169.
57. Tan SH, Tan SB. The correct interpretation of confidence intervals. *Proc Singapore Healthc.* 2010;19(3):276.
58. Hoekstra R, Morey RD, Rouder JN, Wagenmakers EJ. Robust misinterpretation of confidence intervals. *Psychon Bull Rev.* 2014;21(5):1157–1164. https://doi.org/10.3758/S13423-013-0572-3/FIGURES/2.
59. *Misconceptions About Confidence Intervals*: Statistics Solutions. https://www.statisticssolutions.com/misconceptions-about-confidence-intervals/. Accessed 3 December 2023.
60. McCormack J, Vandermeer B, Allan GM. How confidence intervals become confusion intervals. *BMC Med Res Methodol.* 2013;13(1):1–6. https://doi.org/10.1186/1471-2288-13-134/FIGURES/3.

61. Dekkers OM. Meta-analysis: key features, potentials and misunderstandings. *Res Pract Thromb Haemost*. 2018;2(4):658–663. https://doi.org/10.1002/RTH2.12153.
62. *6: Confidence Intervals and Sample Size*. Statistics LibreTexts; 2021. https://stats.libretexts.org/Bookshelves/Introductory_Statistics/Foundations_in_Statistical_Reasoning_%28Kaslik%29/06%3A_Confidence_Intervals_and_Sample_Size. Accessed December 3, 2023.
63. *6.6 - Confidence Intervals & Hypothesis Testing*: STAT. https://online.stat.psu.edu/stat200/lesson/6/6.6. Accessed 3 December 2023.
64. *Understanding Confidence Intervals*. Easy Examples & Formulas; 2020. https://www.scribbr.com/statistics/confidence-interval/. Accessed December 3, 2023.
65. Angus N, Patience F, Maclean E, et al. Cardiac misconceptions in healthcare workers. *Eur J Cardiovasc Nurs*. 2012;11(4):396–401. https://doi.org/10.1016/J.EJCNURSE.2011.01.004.
66. Stang A, Deckert M, Stolpe S. Statistical inference in abstracts published in cardiovascular journals. *J Am Coll Cardiol*. 2021;77(12):1554–1561. https://doi.org/10.1016/J.JACC.2021.01.031.
67. *4.6 - Impact of Sample Size on Confidence Intervals*: STAT. https://online.stat.psu.edu/stat200/lesson/4/4.7. Accessed 3 December 2023.
68. *Understanding Hypothesis Tests: Confidence Intervals and Confidence Levels*; 2015. https://blog.minitab.com/en/adventures-in-statistics-2/understanding-hypothesis-tests-confidence-intervals-and-confidence-levels. Accessed December 3, 2023.
69. Perdices M. Null hypothesis significance testing, p-values, effects sizes and confidence intervals. *Brain Impair*. 2018;19(1):70–80. https://doi.org/10.1017/BRIMP.2017.28.
70. Tessler M. *Multivariate Analysis: Causation, Control, and Conditionality*. 2022:87–133. https://doi.org/10.1007/978-3-031-13838-6_4. Published online 2023.
71. *Multivariate Analysis: An Overview*. Students 4 Best Evidence; 2021. https://s4be.cochrane.org/blog/2021/09/09/multivariate-analysis-an-overview/. Accessed December 3, 2023.
72. Variyath AM, Brobbey A. Variable selection in multivariate multiple regression. *PLoS One*. 2020;15(7):e0236067. https://doi.org/10.1371/JOURNAL.PONE.0236067.
73. Variable Selection in Multiple Regression. *Introduction to Statistics*: JMP. https://www.jmp.com/en_us/statistics-knowledge-portal/what-is-multiple-regression/variable-selection.html. Accessed 3 December 2023.
74. Chowdhury MZI, Turin TC. Variable selection strategies and its importance in clinical prediction modelling. *Fam Med Community Health*. 2020;8(1):e000262. https://doi.org/10.1136/FMCH-2019-000262.
75. *3.5 - Bias, Confounding and Effect Modification*: STAT. https://online.stat.psu.edu/stat507/lesson/3/3.5. Accessed 3 December 2023.
76. *Confounding and Multivariate Models — Winter Applied Data Analysis*. http://adatawinter.site.wesleyan.edu/schedule-2/confounding-and-multivariate-models/. Accessed 3 December 2023.
77. *A Beginner's Guide to Confounding*. Students 4 Best Evidence; 2018. https://s4be.cochrane.org/blog/2018/10/01/a-beginners-guide-to-confounding/. Accessed December 3, 2023.
78. Bartlett PL, Long PM, Lugosi G, Tsigler A. Benign overfitting in linear regression. *Proc. Natl. Acad. Sci*. 2020;117(48):30063–30070.
79. *Overfitting Regression Models: Problems, Detection, and Avoidance*: Statistics By Jim. https://statisticsbyjim.com/regression/overfitting-regression-models/. Accessed 3 December 2023.
80. *The Danger of Overfitting Regression Models*; 2015. https://blog.minitab.com/en/adventures-in-statistics-2/the-danger-of-overfitting-regression-models. Accessed December 3, 2023.
81. Kaler AS, Gillman JD, Beissinger T, Purcell LC. Comparing different statistical models and multiple testing corrections for association mapping in soybean and maize. *Front Plant Sci*. 2020;10:486047. https://doi.org/10.3389/FPLS.2019.01794/BIBTEX.
82. Sullivan GM, Feinn RS. Facts and fictions about handling multiple comparisons. *J Grad Med Educ*. 2021;13(4):457–460. https://doi.org/10.4300/JGME-D-21-00599.1.
83. *6.1: Multiple Comparisons*. Statistics LibreTexts; 2024. https://stats.libretexts.org/Bookshelves/Applied_Statistics/Biological_Statistics_%28McDonald%29/06%3A_Multiple_Tests/6.01%3A_Multiple_Comparisons. Accessed December 3, 2023.
84. Bernhardson CS. 375: Type I error rates when multiple comparison procedures follow a significant F test of ANOVA. *Biometrics*. 1975;31(1):229. https://doi.org/10.2307/2529724.
85. *Multiple Comparisons - Overall Type I Error When Repeatedly Testing Accumulating Data*: Cross Validated. https://stats.stackexchange.com/questions/7780/overall-type-i-error-when-repeatedly-testing-accumulating-data. Accessed 3 December 2023.
86. *Type I and Type II Errors*: Wikipedia. https://en.wikipedia.org/wiki/Type_I_and_type_II_errors. Accessed 3 December 2023.
87. *What is the Family-Wise Error Rate?* Statology; 2020. https://www.statology.org/family-wise-error-rate/. Accessed December 3, 2023.
88. *Family-Wise Error Rate*: Wikipedia. https://en.wikipedia.org/wiki/Family-wise_error_rate. Accessed 3 December 2023.
89. *What is the Family-Wise Error Rate?*: Online Statistics library StatisticalPoint.com. https://statisticalpoint.com/family-wise-error-rate/. Accessed 3 December 2023.
90. *What is the Bonferroni Correction and How to Use It*: Statistics By Jim. https://statisticsbyjim.com/hypothesis-testing/bonferroni-correction/. Accessed 3 December 2023.
91. *A Guide to Using Post Hoc Tests with ANOVA*. Statology; 2019. https://www.statology.org/anova-post-hoc-tests/. Accessed December 3, 2023.
92. *How to Interpret Contradictory Results Between ANOVA and Multiple Pairwise Comparisons?*: XLSTAT Help Center. https://help.xlstat.com/6741-how-interpret-contradictory-results-between-anova-and. Accessed 3 December 2023.
93. Hill AD, Johnson SG, Greco LM, O'boyle EH, Walter SL. Endogeneity: a review and agenda for the methodology-practice divide affecting micro and macro research. *J Manag*. 2021;47(1):105–143. https://doi.org/10.1177/0149206320960533.
94. Chen S, Li Z, Li H, Zeng X, Yuan H, Li Y. RNA sequencing of whole blood in premature coronary Artery disease: identification of novel biomarkers and involvement of T cell imbalance. *J Cardiovasc Transl Res*. 2023. https://doi.org/10.1007/S12265-023-10465-8.
95. *Endogeneity (econometrics)*: Wikipedia. https://en.wikipedia.org/wiki/Endogeneity_(econometrics). Accessed 3 December 2023.
96. *What is Endogeneity?*: Statistics Solutions. https://www.statisticssolutions.com/what-is-endogeneity/. Accessed 3 December 2023.
97. Athey, S., Chetty, R., Imbens, G. Combining experimental and observational data to estimate treatment effects on long term outcomes. 2020. arXiv preprint arXiv:2006.09676.
98. *Chapter 30 Endogeneity A Guide on Data Analysis*. https://bookdown.org/mike/data_analysis/endogeneity.html. Accessed 3 December 2023.

99. Bjørn E, Krishnakumar J. Measurement errors and simultaneity. *Adv Stud Theor Appl Econom.* 2008;46:323–367. https://doi.org/10.1007/978-3-540-75892-1_10/COVER.
100. Le Poidevin R. Causation and simultaneity. In: *Change, Cause and Contradiction.* 1991:83–98. https://doi.org/10.1007/978-1-349-21146-3_7. Published online 1991.
101. *Simultaneity Bias: Simple Definition*: Statistics How To. https://www.statisticshowto.com/simultaneity-bias/. Accessed 3 December 2023.
102. Leszczensky L, Wolbring T. How to deal with reverse causality using panel data? Recommendations for researchers based on a simulation study. *Sociol. Methods Res.* 2022;51(2):837–865.
103. Shin HS, Shin HH, Shudo Y. Current status and limitations of myocardial infarction large animal models in cardiovascular translational research. *Front Bioeng Biotechnol.* 2021;9:673683. https://doi.org/10.3389/FBIOE.2021.673683/BIBTEX.
104. Odening KE, Baczko I, Brunner M, Mechanisms KG, de Medeiros RA, Hausen ZA. Animals in cardiovascular research: important role of rabbit models in cardiac electrophysiology. *Eur Heart J.* 2020;41(21). https://doi.org/10.1093/EURHEARTJ/EHAA251.
105. *9 Study Design Limitations*: Scientific Research and Methodology. https://bookdown.org/pkaldunn/SRM-Textbook/Interpretation.html. Accessed 3 December 2023.
106. Nallamothu BK, Hayward RA, Bates ER. Beyond the randomized clinical trial. *Circulation.* 2008;118(12):1294–1303. https://doi.org/10.1161/CIRCULATIONAHA.107.703579.
107. Beck CA. Selection bias in observational studies. *Neurology.* 2009;72(2):108–109. https://doi.org/10.1212/01.WNL.0000339046.10288.FB.
108. *What Is Selection Bias?*: Definition & Examples. https://www.scribbr.com/research-bias/selection-bias/. Accessed December 3 2023.
109. *What Is an Observational Study?*: Guide & Examples. https://www.scribbr.com/methodology/observational-study/. Accessed 3 December 2023.
110. *Observational Studies, Confounders, and Stratification*: Data Science Discovery. https://discovery.cs.illinois.edu/learn/Basics-of-Data-Science-with-Python/Observational-Studies-Confounders-and-Stratification/. Accessed 3 December 2023.
111. Smith CJ. Methods to account for confounding in observational studies. *Phlebology.* 2011;26. https://doi.org/10.1258/phleb.2011.011j01.
112. *What is Confounding?*. https://sphweb.bumc.bu.edu/otlt/MPH-Modules/BS/BS704-EP713_Confounding-EM/BS704-EP713_Confounding-EM2.html Accessed 3 December 2023.
113. Lamorte WW, Sullivan L. Confounding and effect measure modification, weblearn program. *Eur J Vasc Endovasc Surg.* 2018;55:737. https://doi.org/10.1016/j.ejvs.2018.02.028.
114. Pandis N. Bias in observational studies. *Am J Orthod Dentofacial Orthop.* 2014;145. https://doi.org/10.1016/j.ajodo.2014.01.008.
115. Greene T. Randomized controlled trials 6: determining the sample size and power for clinical trials and cohort studies. *Methods Mol Biol.* 2021;2249:281–305. https://doi.org/10.1007/978-1-0716-1138-8_16/COVER.
116. Moher D, Wells GA, Dulberg CS. Statistical power, sample size, and their reporting in randomized controlled trials. *JAMA.* 1994;272(2):122–124. https://doi.org/10.1001/JAMA.1994.03520020048013.
117. *Types of Bias in Randomized Controlled Trials: A Refresher for Military Mental Health Providers*: Health Mil. https://health.mil/Military-Health-Topics/Centers-of-Excellence/Psychological-Health-Center-of-Excellence/Clinicians-Corner-Blog/Types-of-Bias-in-Randomized-Controlled-Trials-A-Refresher-for-Military-Mental-Health-Providers. Accessed 3 December 2023.
118. Dumville JC, Torgerson DJ, Hewitt CE. Reporting attrition in randomised controlled trials. *BMJ.* 2006;332(7547):969–971. https://doi.org/10.1136/BMJ.332.7547.969.
119. Hewitt CE, Kumaravel B, Dumville JC, Torgerson DJ. Assessing the impact of attrition in randomized controlled trials. *J Clin Epidemiol.* 2010;63(11):1264–1270. https://doi.org/10.1016/J.JCLINEPI.2010.01.010.
120. Zhang J, Saju C. A systematic review of randomised controlled trials with adaptive and traditional group sequential designs – applications in cardiovascular clinical trials. *BMC Med Res Methodol.* 2023;23(1):1–9. https://doi.org/10.1186/S12874-023-02024-1/FIGURES/2.
121. Gonzalez Garcia M, Fatehi F, Bashi N, et al. A review of randomized controlled trials utilizing telemedicine for improving heart failure readmission: can a realist approach bridge the translational divide? *Clin Med Insight Cardiol.* 2019;13. https://doi.org/10.1177/1179546819861396.
122. *The Utility of CAC in Efficient Randomized Control Trial Design.* American College of Cardiology; 2021. https://www.acc.org/latest-in-cardiology/articles/2021/10/26/18/59/the-utility-of-cac-in-efficient-randomized-control-trial-design. Accessed December 3, 2023.
123. Bowman L, Weidinger F, Albert MA, et al. Randomized trials fit for the 21st century. A joint opinion from the European society of cardiology, American heart association, American college of cardiology, and the world heart federation. *Eur Heart J.* 2023;44(11):931–934. https://doi.org/10.1093/EURHEARTJ/EHAC633.
124. *Self-Selection Bias in a Case-Control Study.* https://sphweb.bumc.bu.edu/otlt/MPH-Modules/PH717-QuantCore/PH717-Module10-Bias/PH717-Module10-Bias3.html. Accessed 3 December 2023.
125. Savitz DA, Wellenius GA. Selection bias in case-control studies. In: *Interpreting Epidemiologic Evidence.* 2016:93–112. https://doi.org/10.1093/acprof:oso/9780190243777.003.0008. Published online November 17, 2016.
126. *Control Selection Bias in a Case-Control Study.* https://sphweb.bumc.bu.edu/otlt/MPH-Modules/PH717-QuantCore/PH717-Module10-Bias/PH717-Module10-Bias2.html. Accessed 3 December 2023.
127. Raphael K. Recall bias: a proposal for assessment and control. *Int J Epidemiol.* 1987;16(2):167–170. https://doi.org/10.1093/IJE/16.2.167.
128. Kopec JA, Esdaile JM. Bias in case-control studies. A review. *J Epidemiol Community Health.* 1990;44(3):179–186. https://doi.org/10.1136/jech.44.3.179.
129. Irony TZ. Case-control studies: using "real-world" evidence to assess association. *JAMA.* 2018;320(10):1027–1028. https://doi.org/10.1001/JAMA.2018.12115.
130. Herman JM, Slawson DC. Case-control studies. *Fam Med.* 1990;22(1):52–56. https://doi.org/10.1007/978-3-030-05120-4_17/COVER.
131. Roever L. Critical analysis of case-control studies in cardiovascular research. *Health Prim Care.* 2020;4(3). https://doi.org/10.15761/HPC.1000195.

132. Zhang C, Neuzil P, Petru J, et al. Coronary Artery spasm during pulsed field vs radiofrequency catheter ablation of the mitral isthmus. *JAMA Cardiol.* 2024;9(1). https://doi.org/10.1001/JAMACARDIO.2023.4405.
133. *Advantages and Disadvantages of Cohort Studies.* https://sphweb.bumc.bu.edu/otlt/MPH-Modules/EP/EP713_CohortStudies/EP713_CohortStudies5.html. Accessed 3 December 2023.
134. Wang X, Kattan MW. Cohort studies: design, analysis, and reporting. *Chest.* 2020;158(1S):S72–S78. https://doi.org/10.1016/J.CHEST.2020.03.014.
135. Ransohoff DF. Cultivating cohort studies for observational translational research. *Cancer Epidemiol Biomarkers Prev.* 2013;22(4):481–484. https://doi.org/10.1158/1055-9965.EPI-13-0140/67654/AM/CULTIVATING-COHORT-STUDIES-FOR-OBSERVATIONAL.
136. *9.2 - Comparison of Cohort to Case/Control Study Designs With Regard to Sample Size*: STAT. https://online.stat.psu.edu/stat507/lesson/9/9.2. Accessed 3 December 2023.
137. Xue X, Agalliu I, Kim MY, et al. New methods for estimating follow-up rates in cohort studies. *BMC Med Res Methodol.* 2017;17(1). https://doi.org/10.1186/s12874-017-0436-z.
138. Sindhu KN, Srinivasan M, Subramaniam S, et al. Why do participants drop-out: findings from a prospective pediatric cohort for fever surveillance established at Vellore, southern India. *BMC Med Res Methodol.* 2019;19(1):1–12. https://doi.org/10.1186/S12874-019-0881-Y/TABLES/5.
139. Bell ML, Kenward MG, Fairclough DL, Horton NJ. Differential dropout and bias in randomised controlled trials: when it matters and when it may not. *BMJ.* 2013;346. https://doi.org/10.1136/BMJ.E8668.
140. Ramirez-Santana M, Ramirez-Santana M. Limitations and biases in cohort studies. In: *Cohort Studies in Health Sciences.* 2018. https://doi.org/10.5772/intechopen.74324.
141. *Chapter 25: Assessing Risk of Bias in A Non-Randomized Study.* Cochrane Training; 2023. https://training.cochrane.org/handbook/current/chapter-25. Accessed December 3, 2023.
142. *Introduction to Study Designs - Cohort Studies*: Health Knowledge. https://www.healthknowledge.org.uk/e-learning/epidemiology/practitioners/introduction-study-design-cs. Accessed 3 December 2023.
143. Frati G, Benedetto U, Biondi-Zoccai G, Sciarretta S. Bridging the gap between translational and outcome research in cardiovascular disease. *BioMed Res Int.* 2015;2015. https://doi.org/10.1155/2015/454680.
144. Bermejo J, Heras M, Segovia J, Alfonso F. Translational cardiovascular medicine. Now or never. *Rev Española Cardiol (English Ed).* 2009;62(1):66–68. https://doi.org/10.1016/S1885-5857(09)71515-4.
145. Ridker PM, Everett BM, Thuren T, et al. Antiinflammatory therapy with canakinumab for atherosclerotic disease. *N Engl J Med.* 2017;377(12):1119–1131. https://doi.org/10.1056/NEJMOA1707914.
146. Everett BM, Cornel JH, Lainscak M, et al. Anti-inflammatory therapy with canakinumab for the prevention of hospitalization for heart failure. *Circulation.* 2019;139(10):1289–1299. https://doi.org/10.1161/CIRCULATIONAHA.118.038010.
147. Savage P, Cox B, Linden K, Coburn J, Shahmohammadi M, Menown I. Advances in clinical cardiology 2021: a summary of key clinical trials. *Adv Ther.* 2022;39(6):2398–2437. https://doi.org/10.1007/S12325-022-02136-Y/FIGURES/3.
148. *Hormone Therapy for Gender Dysphoria May Raise Cardiovascular Risks.* American College of Cardiology; 2023. https://www.acc.org/About-ACC/Press-Releases/2023/02/22/20/29/Hormone-Therapy-for-Gender-Dysphoria-May-Raise-Cardiovascular-Risks. Accessed December 17, 2023.
149. Rauch G, Kieser M, Ulrich S, et al. Competing time-to-event endpoints in cardiology trials: a simulation study to illustrate the importance of an adequate statistical analysis. *Eur J Prev Cardiol.* 2014;21(1):74–80. https://doi.org/10.1177/2047487312460518.
150. Nguyen HT, Vasconcellos HD, Keck K, et al. Multivariate longitudinal data for survival analysis of cardiovascular event prediction in young adults: insights from a comparative explainable study. *BMC Med Res Methodol.* 2023;23(1):1–19. https://doi.org/10.1186/S12874-023-01845-4/FIGURES/5.

Chapter 22

Basic statistical principles

Ruchit Agrawal
School of Computer Science, University of Birmingham, Birmingham, United Kingdom

Key points

- Confidence intervals and *P*-values are essential in interpreting statistical results, offering insights into the reliability of estimates and the strength of evidence.
- Confusion matrices and their derived metrics (sensitivity, specificity, precision, and recall) are crucial in evaluating classification models, especially in medical tests or machine learning.
- Accuracy, incidence, and prevalence help provide context for understanding how well a model or diagnostic test is likely to perform in real-world scenarios.
- Quantifying risk, using measures like relative risk and odds ratios, is essential in assessing potential impacts and guiding decision-making.

Introduction: Why it matters

Statistical principles form the backbone of decision-making in various fields such as medicine, public health, social sciences, and machine learning. Whether you're evaluating a clinical trial's effectiveness, assessing a diagnostic test, or building a classification model, understanding key concepts like confidence intervals, *P*-values, sensitivity, and specificity is critical. These tools not only help quantify uncertainty but also guide evidence-based decisions. Furthermore, quantifying risk and understanding the prevalence of conditions in populations are essential for resource allocation and managing public health interventions. This chapter offers a foundation in these principles, ensuring you can both interpret and apply statistical data correctly. By mastering the statistical principles in this chapter, readers will be well-equipped to understand research results, evaluate models, and apply statistical reasoning in both academic and practical contexts.

Confidence intervals

The confidence interval refers to the probability that the value of a parameter will fall between a pair of values around the mean. Confidence intervals are a measure of the *degree of uncertainty* in a sampling method, and indicate a range (upper and lower bound) of the estimated value for a particular parameter at a designated *confidence level.*

For example, let us assume that we have 100 students whose IQ levels have been noted. The 101th student misses the IQ test but wants to know his IQ level. In such a scenario, a range could be provided (for instance 80−120) with a confidence level. Most random processes in the world tend to be Gaussian, so let us try to see the confidence interval calculation in that distribution.

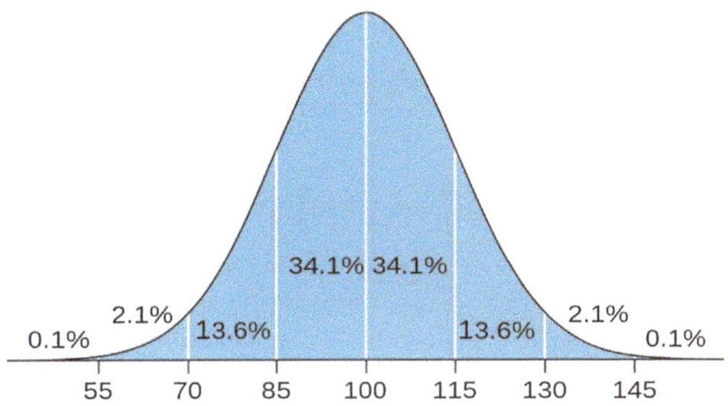

The above distribution suggests that the average IQ level of all humans is 100, and that we can say that a randomly sampled human being will have an IQ level in the range of 85−115 with a confidence of 68.2%.

The confidence interval is computed as follows:

$$CI = \bar{x} \pm z\frac{s}{\sqrt{n}}$$

where $\bar{x}$ is the sample mean, s is the standard deviation, n is the number of samples, and z is the confidence level value.

Translational Cardiology. https://doi.org/10.1016/B978-0-323-91790-2.00096-4

P-values

The *P*-value is the probability that a particular statistical measure for our sample will be greater than or equal to that of the observed results. Put simply, a *P*-value is a measure of the probability that an observed difference could have occurred just by random chance. The *P*-value suggests how significant are the results *statistically speaking*. The lower the *P*-value, the higher the statistical significance of the observed difference.

Confusion matrix

Consider the following confusion matrix that will be helpful to understand the subsequent concepts. Assuming we conduct a test for a specific disease, there are four possible outcomes of this test,[1] depending upon the disease state (disease present or not) and the test prediction (test positive or not). These are depicted in the following table:

	Test result		
True condition		Positive	Negative
	Disease	True positive	False negative
	No disease	False positive	True negative

True positive cases are those wherein a test is positive for a person having the disease. False negative cases are those wherein a test is negative for a person having the disease. False positive cases are those wherein a test is positive for a person not having the disease. True negative cases are those wherein the test is negative for a person not having the disease.

Sensitivity

The sensitivity of a medical test refers to its ability to correctly detect ill patients who do have the condition.[2] It is therefore used to calculate the effectiveness of a medical test.

Sensitivity is also referred to as the "True positive rate" and is given by the following ratio:

$$\textbf{sensitivity} = \frac{\text{tp}}{\text{tp} + \text{fn}}$$

False negatives are typically highly undesirable for medical tests since they result in ill patients not getting the treatment due to the incorrect test result. A high sensitivity value maximizes true positives and minimizes false negatives, and is therefore a desirable trait for medical tests.

Specificity

Specificity denotes the probability of a negative test result when the disease is absent. It is calculated as the percentage of individuals without the condition correctly categorized as not having the disease.

Specificity is also referred to as the "True negative rate" and is derived by taking the following ratio:

$$\textbf{specificity} = \frac{\text{tn}}{\text{tn} + \text{fp}}$$

A high specificity value maximizes true negatives and minimizes false positives, which would mean that the test is very good at diagnosing negative results for patients who are not ill.

Positive and negative predictive values

The positive predictive value is the probability that a subject that gets tested positive for a condition truly has the condition.[3] It is computed as follows:

$$\text{PPV} = \frac{\text{Number of true positives}}{\text{Number of true positives} + \text{Number of false positives}} = \frac{\text{Number of true positives}}{\text{Number of positive calls}}$$

Similarly, the negative predictive value is the probability that a subject that gets a negative test outcome for a condition truly does not have the condition, computed as

$$\text{NPV} = \frac{\text{Number of true negatives}}{\text{Number of true negatives} + \text{Number of false negatives}} = \frac{\text{Number of true negatives}}{\text{Number of negatives calls}}$$

Precision and recall

Precision and recall are the terms used in informatics for positive predictive value and sensitivity, respectively, which are defined in the medical context previously.

$$\textbf{precision} = \frac{\text{tp}}{\text{tp} + \text{fp}} \quad \textbf{recall} = \frac{\text{tp}}{\text{tp} + \text{fn}}$$

Accuracy

The accuracy of the medical test or any prediction is a basic measure of the efficiency of the test, and calculates the percentage of the test results that are correct. Accuracy is derived as follows:

$$\text{Accuracy} = \frac{(\text{TP} + \text{TN})}{(\text{TP} + \text{FP} + \text{TN} + \text{FN})}$$

Incidence

Incidence is a measure of a disease that can determine the likelihood that a person will be diagnosed with the disease within a certain period.

For example: In a study, 100 people with an age of 60 or more who did not have cancer were selected and within a year, two of them were diagnosed with cancer. The incidence of cancer in this case for humans aged 60+ is 2 per 100 in a year.

Prevalence

Prevalence is the proportion of a population with a specific attribute during a given period. It is very similar to Incidence, but the essential difference is in Incidence; we look for a total number of new cases that appeared in a period, while in Prevalence, we look at the number of people already having a disease over a period of time.

Quantifying risk

Unlike risk in everyday language, which is frequently linked with a negative event, risk in statistics refers to the likelihood that an event will occur. Let us assume that we are testing a vaccine for COVID-19, and a possible event is a person getting infected with COVID-19. In order to test the vaccine, we would have two groups, namely the control group and the treatment group. As the name suggests, the treatment group will get the vaccine, and the control group will get a placebo.

The risk level can then be assessed for both the groups by examining the total number of people in the control and treatment groups that get infected with COVID-19. The reduction in risk can thereby be calculated with the vaccine in place.

Potential pitfalls

- **Misinterpreting Confidence Intervals**: A common mistake is assuming that a 95% confidence interval means there is a 95% probability that the true parameter is within that interval. In fact, it means that if you repeated the study 100 times, 95 of those intervals would contain the true parameter.
- **P-Value Misconception**: *P*-values do not measure the probability that the null hypothesis is true, nor do they offer the magnitude of an effect. *P*-values merely indicate whether the data observed would be unusual under the null hypothesis.
- **Misclassification in the Confusion Matrix**: Without considering the balance between sensitivity and specificity, it's easy to over-interpret accuracy without understanding the true performance of a model, especially in cases of imbalanced classes.
- **Confusing Precision and Recall**: Precision and recall are not interchangeable. Depending on the application, one may be more important than the other. In medical diagnostics, for example, high recall (sensitivity) might be prioritized to catch as many true positives as possible.
- **Ignoring Incidence and Prevalence**: When working with diagnostic tests, neglecting prevalence can lead to overconfidence in positive results, as the context of a rare disease can skew the real-world applicability of test results.
- **Risk Misinterpretation**: Misunderstanding relative risk or odds ratio can lead to incorrect conclusions about the impact of exposure, leading to misguided recommendations or decisions.

Examples of terms

- **Confidence Interval Example**: In a clinical trial of a new drug, a 95% confidence interval for the drug's effect on reducing symptoms might range from 10% to 20%. This suggests the true reduction in symptoms is likely within that range, though the study is not certain of the exact value.
- **P-Value Example**: In testing the hypothesis that a new medication lowers blood pressure, researchers find a *P*-value of 0.03. This means there's a 3% chance that the results they observed would occur if the medication had no effect.
- **Confusion Matrix Example**: A model to predict whether patients have a disease produces the following confusion matrix: 80 true positives, 20 false positives, 50 true negatives, and 10 false negatives. Sensitivity, specificity, precision, and accuracy can all be derived from this matrix.
- **Incidence Example**: If 1000 people in a town are followed for a year and 50 develop a certain disease, the incidence rate of the disease is 50 per 1000 people per year.
- **Prevalence Example**: At a given time, 200 people in a town of 10,000 have diabetes, meaning the prevalence of diabetes is 2%.
- **Quantifying Risk Example**: A study shows that smokers have a relative risk of three for developing lung cancer compared to nonsmokers, meaning smokers are three times more likely to develop lung cancer.

Resources

1. **Books and Texts**:
   - "Statistical Methods for the Social Sciences" by Alan Agresti

- "Introduction to the Practice of Statistics" by David S. Moore and George P. McCabe
- "Principles of Biostatistics" by Marcello Pagano and Kimberlee Gauvreau

2. **Online Resources**:
   - **Khan Academy**: Provides tutorials on confidence intervals, *P*-values, and other statistical concepts.
   - **Coursera**: Offers courses on data analysis, biostatistics, and machine learning which include many of the statistical principles outlined here.
   - **CDC and WHO Statistics Resources**: For real-world applications of incidence, prevalence, and risk quantification in public health.
3. **Software Tools**:
   - **R and Python**: Both have extensive libraries (e.g., `statsmodels` in Python or `tidyverse` in R) for statistical analysis, including calculating confidence intervals, performing hypothesis tests, and building classification models.
   - **SPSS or SAS**: Statistical software often used in academic and professional research for in-depth data analysis and reporting.

References

1. Fletcher RH, Fletcher SW. *Clinical Epidemiology: The Essentials*. 4th ed. Baltimore, Md: Lippincott Williams & Wilkins; 2005:45.
2. Altman DG, Bland JM. Diagnostic tests. 1: sensitivity and specificity. *BMJ Br Med J (Clin Res Ed)*. 1994;308(6943):1552.
3. Altman DG, Bland JM. Diagnostic tests 2: predictive values. *BMJ*. 1994;309(6947):102. https://doi.org/10.1136/bmj.309.6947.102. PMC 2540558. PMID 8038641.

Chapter 23

Distributions

Michail Koutentakis[1,2]

[1]*Department of Cardiothoracic Surgery and Transplantology, University Clinical Centre of Medical University of Warsaw, Central Teaching Clinical Hospital, Warsaw, Poland;* [2]*Department of Experimental and Clinical Pharmacology, Medical University of Warsaw, Center for Preclinical Research and Technology CEPT, Warsaw, Poland*

Key points

- The essential principles of statistical distributions, which encompass the normal, binomial, Poisson, and exponential types, enable a comprehensive understanding of the dispersion, variability, and inherent patterns in data sets.
- These classifications of distributions provide suitable methodologies for modeling and analyzing various types of data arising in research.
- Statistical distributions are imperative components in translational cardiology research. They offer strong, mathematical apparatus for constructing patient outcome models and in-depth analyses of clinical trial data, thereby facilitating refined clinical decision-making processes and the optimization of patient healthcare.

Introductory section/why it matters

In the field of translational cardiology research, understanding distributions and their significance in statistical principles is crucial. Distributions play a vital role in analyzing data and drawing meaningful conclusions from it. By providing a framework for describing variability and uncertainty, they allow researchers to accurately interpret study results and make informed decisions regarding patient care.[1]

An important reason why distributions matter in translational cardiology research is that they reflect the inherent variation present within biological systems. Cardiac parameters such as heart rate (HR), blood pressure (BP), or biomarker levels can vary significantly between individuals due to factors like age, sex, genetics, lifestyle choices, or underlying health conditions. Understanding these variations helps us identify trends and patterns that could potentially lead to better diagnostic tools or more efficient treatments.[1–3]

Moreover, distributions assist researchers by quantifying probabilities associated with certain outcomes or events. This enables them to assess risk factors more accurately. To put it into context, in cardiology research, such knowledge turns out key when evaluating treatment effectiveness, tailoring therapies according to patients' needs, and predicting possible complications.[4]

Therefore, introducing the importance of distributions sets the stage for understanding how statistical principles underpin translational cardiology studies. It serves as a foundation for extracting meaning from complex datasets, resulting ultimately in improved clinical management.[5]

This chapter yields an extensive appraisal of the vital frequently employed distributions utilized in the field of translational cardiology. The analysis accentuates their unique attributes, their importance in the discipline, and their specific implementations within the context of this scientific domain.[6]

The normal distribution

Characteristics and properties

The normal distribution, often referred to as the Gaussian distribution, is a continuous probability distribution[7] that is frequently utilized in translational cardiology research and takes center stage due to its fundamental characteristics and properties. It is symmetrical in nature and demonstrates a bell-shaped curve when graphically depicted, where the mean, median, and mode all coincide at the center of the distribution.[7,8] This distribution presents infinite tails that approach but never touch the horizontal axis, always maintaining nonnegative values.[8]

The key parameters defining the shape of the normal distribution are the mean, which represents the location of

Translational Cardiology. https://doi.org/10.1016/B978-0-323-91790-2.00051-4

the peak of the distribution, and the standard deviation, denoting the degree of dispersion of the data from the mean or, simply put, the "width" of the distribution.[7,8] It is the seamless adjustability of these parameters that allows the normal distribution to represent a plethora of continuous variables in cardiology like blood pressure or cholesterol levels.[9,10]

One prominent attribute of the normal distribution is the empirical rule, often called the "68-95-99.7" rule. It depicts the proportion of data falling within one, two, and three standard deviations from the mean, respectively. This rule provides an intuitive understanding of interpreting clinical data.[11,12]

Finally, the normal distribution enjoys a unique relationship with probability. It links to several probability concepts, such as the central limit theorem, which states that the sum of a large number of independent, identically distributed variables will tend toward a normal distribution, irrespective of the original distribution shape.[13] This theorem is a crucial property empowering the wide application of the normal distribution in statistical analysis.

These properties collectively contribute to the universal usage and vital role the normal distribution plays, serving as a powerful tool for data interpretation in translational cardiology research.

Standardization and Z-scores

Standardization is a frequently employed method in statistical evaluation designed to reconfigure data to conform to a standard normal distribution, characterized by a mean of 0 and a standard deviation of 1. It is a procedure dedicated to maintaining data uniformity across various origins. Specifically, this involves deducting the mean of the data from each individual data point and subsequently dividing the result by the standard deviation. The culminating values, referred to as Z-scores, denote the number (quantity) of standard deviations a data point is distanced from the mean. In translational cardiology, standardization is important for comparing data from different studies.[14]

On the other hand, Z-scores are a means of expressing the deviation of a given anatomic or physical measurement from a size- or age-specific population mean. Z-scores can be applied to echocardiographic measurements, height, weight, and blood pressure, and thus may assist in clinical assessment and decision-making.[15]

Within translational cardiology, standardization and Z-scores are applied to juxtapose observations from disparate populations or to estimate the probability of an observation based on an understood distribution. For example, given a population of patients with hypertension and the knowledge that their blood pressure metrics follow a normal distribution, we can leverage Z-scores to determine the probability of a patient's blood pressure reading falling above or below a designated threshold.[15] This can assist in identifying patients who may potentially be at an elevated risk, thus necessitating a more comprehensive assessment or monitoring.

Applications in translational cardiology research

As aforementioned, in translational cardiology research, the normal distribution is often used to model several variables, including heart rate, blood pressure, or cholesterol levels.[9] To identify the population, mean, and standard deviation, a study might, for instance, utilize normal distribution to examine the distribution of blood pressure readings in a sizable cohort of hypertensive patients. The risk associated with cardiac disease can then be ascertained utilizing such information, thereby enabling the creation of treatment plans tailored to each patient. This distribution type facilitates the calculation of probabilities and confidence intervals,[16] thus allowing researchers to make well-informed determinations based on the probability of specific results.

The Gaussian distribution also forms the basis for statistical inference, where sample means are premised to follow a normal distribution. When these sample means are postulated as normally distributed, researchers can employ more advanced statistical tests, including *t*-tests, analysis of variance (ANOVA), linear regression, and hypothesis testing, to compare groups and evaluate the statistical significance of differences between them.[17,18] This methodology can prove beneficial for the identification of cardiovascular disease risk factors, or for assessing the effectiveness of various treatments.

Overall, normal distribution is a powerful tool in translational cardiology that allows researchers to analyze continuous variables, assess risk, and make statistical inferences. Its properties and applications are well understood, making it a widely used and valuable statistical tool to better understand, diagnose, and manage cardiovascular diseases, ultimately leading to improved patient outcomes.

The binomial distribution

Definition, components, and parameters

Acquainting ourselves with the binomial distribution becomes fundamental when dealing with binary outcomes in statistical research.[19] By definition, the binomial distribution is a discrete probability distribution that counts the number of successes in a series of n-independent experiments. It represents a scenario where an isolated event generates one of two outcomes—success (often assigned 1) or failure (typically ascribed 0).[20,21] For instance, within cardiology, this could denote the presence (success) or the absence (failure) of a specific cardiac condition.

The binomial distribution primarily consists of two parameters: the number of trials (n) and the success probability on each trial (p). This distribution's key trait is that each trial is independent, and the success probability remains consistent throughout all experiments.[22,23]

The representation of the binomial distribution is not always symmetrical, it is reliant on the success probability. If the probability (P) is .5, the distribution attains symmetry while, if P is either less than or more than 0.5, the graph veers right ($P < .5$) or left ($P > .5$). The distribution mean (μ), given by the formula $\mu = np$, depicts the expected number of successes, while the standard deviation (σ), which signifies dispersion, is defined by the equation $\sigma = \sqrt{npq}$, with "q" symbolizing the probability of failure $(1 - p)$.[22,24]

A concept closely linked with the binomial distribution is the Bernoulli trials. When an experiment has exactly two possible outcomes and is repeated numerous times, each individual attempt is known as a Bernoulli trial.[25–27]

In terms of statistical analysis, the binomial distribution provides a firm foundation for assessing the likelihood of observation. The ability of the distribution to model binary outcomes makes it exceptionally apt for a plethora of applications, including the analysis of clinical data in cardiology research.[28–30]

Implementation in translational cardiology research

As mentioned previously, the binomial distribution serves as a valuable tool in translational cardiology research, particularly when investigating binary outcomes.[19] Its utility resides fundamentally in the modeling and analysis of clinical events that have two possible outcomes. For instance, it is used to establish the occurrence (denoted as a success) or nonoccurrence (marked as failure) of a particular heart condition within a specific population group.[31]

One application of binomial distribution in translational cardiology might be to evaluate the efficacy of a novel cardiac drug.[32–35] Consider a study where a new cardiological drug is administered to a sample of patients. Success can be considered as improvement or cure, and failure can be no change or the worsening of the condition. If you administer it to "n" patients, and "X" number of patients shows improvement, this event will follow a binomial distribution. By modeling this as a binomial distribution, research can generate substantial predictions about the drug's effectiveness.

Similarly, this form of distribution can be utilized to assess the likelihood of risk factors linked to cardiac disease in a demographic group.[36,37] This can more specifically be extended to binary variables such as the existence/nonexistence of hypertension, elevated cholesterol, obesity, or other modifiable factors.[38,39] All these can be examined through this statistical tool. Consider a scenario whereby one is investigating whether obesity (represented by a Yes or No response) poses a significant cardiac disease risk factor. Here, again, if one has "n" subjects and "X" among them are observed to be obese and have heart disease, this event is a depiction of the binomial distribution. These studies can provide insights into the likelihood of developing cardiac complications related to risk factors within a specific population.

Last but not least, the binomial distribution may be used even in studies on the presence or absence of a specific cardiac condition, such as myocardial infarction (MI) in a sample population.[40]

The Poisson distribution

Attributes

The Poisson distribution is a discrete probability distribution, named after the French mathematician Siméon Denis Poisson, that characterizes the occurrence of independent events within a fixed interval of time or space.[41] It is vital to fields such as physics, engineering, economics, and significantly, the areas of health and medical sciences, inclusive of cardiology.[41–44]

Several attributes distinctly characterize the Poisson distribution. Firstly, the events are independent, suggesting that the occurrence of one event does not influence or modify the outcome of another. Secondly, events transpire at a steady mean rate with no change in frequency throughout the observed interval. Lastly, the simultaneous occurrence of two events at an identical point in time is not accounted for, although this principle may be relaxed for continuous monitoring scenarios.[45,46] Consequently, researchers and practitioners routinely employ the Poisson distribution to model data, particularly when counting occurrences over a specific time or spatial range.[21,45–47]

Nonetheless, the constant mean rate assumption of the Poisson distribution may not accurately encapsulate real-world situations, where the mean rate of occurrences might likely fluctuate.[48,49] Moreover, the distribution assumes events to be independent and singular, an assumption that may not hold true in specific disciplines like biology.[41,50] Despite these inherent challenges, the Poisson distribution remains a fundamental instrument in many areas such as ecological inference, machine learning, and medicine due to its conceptual simplicity and pragmatic predictive capacity.[44,51]

The probability mass function in the Poisson distribution

The probability mass function (PMF) of the Poisson distribution is key to delineating its functionality and properties. It provides the probability that a particular number of events will occur within a given interval, provided we know the average number of events that usually occur per interval.[45,46,52]

Formally, the PMF of a Poisson distribution is defined by the parameter λ (lambda), where $\lambda > 0$ (λ greater than 0) signifies that λ represents the average event rate. This distribution is expressed in relation to a discrete random variable denoted as k, which indicates a nonnegative integer and represents the actual number of events. This can be formulated in an equation, presented as follows:

$$P(X = k) = \lambda^k \frac{(e^{-\lambda})}{k!}$$

In this equation, "λ" is the mean number of occurrences in an interval, "e" is the base of the natural logarithm (approximately 2.71828), X is the random variable that takes the value "k," which is the number of occurrences, and "k!" is the factorial of k. Essentially, this formula calculates the likelihood of observing exactly k events in a fixed interval, given that the average number of events is λ.[45,53–56]

The PMF of the Poisson distribution brings to light the fact that, with the assumption of a fixed interval, the distribution only depends on "λ," the average rate of an event's occurrence.[56]

It is important to emphasize that in the PMF, the "k" events are independent, meaning the outcome of one does not determine or alter the outcome of another event. The distribution also presumes that events occur at a constant mean rate and two events cannot occur at exactly the same instant. This PMF assumption is instrumental in practical applications, including several realms such as physics, engineering, finance, and more notably, epidemiology and translational cardiology.[56]

The Poisson distribution can be used in cardiology for studying and modeling the occurrence of certain discrete events like arrhythmias in heart patients over a specific period of time.[57,58] For instance, suppose we know that a patient has an average of two arrhythmic events per day ($\lambda = 2$). In this case, if we want to determine the probability that this particular patient will have exactly four arrhythmic events on a given day, we can use the Poisson distribution's PMF.

Let $k = 4$ (the number of arrhythmic events).

Then:

$$P(X = 4) = \lambda^k \frac{(e^{-\lambda})}{k!} = 2^4 \frac{(e^{-2})}{4!} = 16 \frac{(e^{-2})}{24} = 0.0902$$

This implies there is approximately a 9.02% probability that a patient who averages two arrhythmic events per day will have exactly four such events on a given day.

Its use in translational cardiology research

In the realm of translational cardiology, the Poisson distribution has been found to be indispensable. As abovementioned, this statistical method models the probability of a given number of events occurring in a fixed interval of time or space and is particularly adept at managing rare events, which are common in medical research and practice.

One prime example of the application of the Poisson distribution is in the prediction of the incidence of pacemaker replacements. The incidence of pacemaker implantation after valve surgery, for example, depends on the type of procedure being performed and on risk factors such as age and a baseline conduction defect, for example, bundle branch block. The prevalence of pacemaker implantation was lowest after MVr (4.5%) and highest after AVR plus MVR (13.3%).[59] This statistical tool allows hospitals and researchers to estimate the number of replacements that could occur in a given time frame based on the average rate of pacemaker replacements per year, informing resource planning and allocation.

On top of that, the Poisson distribution has proved instrumental in evaluating the efficacy of cardiological drugs. For instance, in a clinical study investigating a drug designed to reduce heart attack incidents,[60] the Poisson distribution can be utilized to compare the rate parameter λ (average number of incidents) before and after drug intake, providing a robust measure of the drug's effectiveness.

In the same vein, in patients at high risk of heart failure, the Poisson distribution could serve as a predictor of the number of heart failure incidences over a specific period, offering clinicians a predictive tool that can aid in timely interventions.

Importantly, the Poisson distribution is also employed in arrhythmia studies.[61] By leveraging this method, researchers can quantify the probability of a patient experiencing a defined number of abnormal episodes within a fixed time window thereby influencing management and treatment strategies.

The application of the Poisson distribution in translational cardiology extends to the quantification of stenosis events, epitomizing narrowing within blood vessels. A study employed a variant of this statistical model, the truncated zero-inflated double Poisson, to pinpoint determinants impacting the incidence of stenoses within patients suffering from obstructive coronary artery disease (OCAD). This statistical modeling facilitated the identification of risk factors like age, gender, specific lifestyle behaviors, and preexisting conditions.[62] By identifying these risk factors, healthcare professionals can better predict which patients might be at higher risk of having multiple narrowed or blocked vessels. This information can then guide the development of individualized treatment plans for these patients.

Before employing the Poisson distribution in any given context, it is imperative to first ascertain that all the necessary conditions of the distribution are met. This validation process is a crucial preliminary step that ensures the appropriateness and accuracy of the distribution's application.

The exponential distribution

Features and elements

The exponential distribution, also known as the negative exponential distribution, is commonly used in the field of statistical analysis and probability theory. It is uniquely discernible by its singular defining parameter, λ, which is technically referred to as the rate parameter. Its formulation stems from the Poisson process, typically used to model the number of times an event occurs in a fixed interval of time or space.[63–66]

One of the striking characteristics of the exponential distribution includes its constant hazard function.[66–69] A hazard function, alternatively known as the failure rate, is a characteristic of a probability distribution used to model failure time in reliability studies, such as those involving mechanical or electronic components that fail independently of age.[65,70] Meanwhile, in the field of medical and life sciences, this function plays a pivotal role in understanding and predicting events where the risk does not necessarily increase over time, such as spontaneous radioactive decay or subsequent memoryless events.[71] The capacity to model such scenarios depicts the flexible and diverse applicability of the exponential distribution, extending its reach beyond the conventional realms of statistical analyses.

In light of the above-mentioned, another key feature of the exponential distribution is that it has a "lack of memory." This means that the waiting time until a certain event does not depend on how much time has already passed without the event. Adopting the lack of memory property for statistical modeling can simplify the complexity and enhance the computational feasibility of the model, although it may not fully mirror the actual behavior or performance of a certain subject or system.[72]

In addition to that, the exponential distribution is characterized by skewness and kurtosis. The exponential distribution exhibits positive skewness, meaning its curve is skewed to the right with the mass of the distribution concentrated on the left. This characteristic makes it particularly useful for studying phenomena that have a definite skew in their occurrence. It also possesses excess kurtosis which suggests a "peakedness" or flatness in relation to the normal distribution, further highlighting its ability to model rare events with notable precision.[63,73]

The forward-looking features of the exponential distribution make it a core component of many areas of study, including translational cardiology. Despite its simplicity, its characteristics and elements offer a profound understanding of the data at hand, and its manipulation can yield insightful conclusions.

Probability density function in the exponential distribution

The probability density function (PDF) is a significant concept in the field of statistics and probability, delicately defining the probability of a random variable lying within a certain range.[63,74]

For an exponential distribution, the PDF can be represented as follows: if X is an exponential random variable with parameter λ, the PDF of X for $X > 0$ is given by: $f(X;\lambda) = \lambda e^{(-\lambda X)}$.[68,75]

Here, e is the base of the natural logarithm (~2.71828), and λ is the rate parameter, often interpreted as the number of events per unit of time or space. This framework insinuates that for $X \geq 0$, the function reflects the likelihood of a random variable X taking the value "X."[74]

In many instances, the exponential distribution model serves to depict the temporal duration between events in a Poisson point process. This process is a mathematical framework, designed to capture specific spatial or temporal random phenomena. An inherent correlation is therefore established between the exponential and Poisson distributions. Subsequent analysis will delve deeper into this relationship.

Relationship between the Poisson and exponential distribution

To be more specific, an intriguing connection exists between the exponential and Poisson distributions.[76] Imagine a situation where the period between two consecutive events adheres to an exponential distribution, with the average time interval being μ units of time.

The assumption is that these time durations are independent, meaning the time span between events is not influenced by the time durations between preceding events. If these conditions are satisfied, then the count of events that occur per unit time adheres to a Poisson distribution with a mean of $\lambda = \frac{1}{\mu}$. As aforementioned, if a given variable X follows a Poisson distribution with mean λ, the probability of observing k events is given by the equation: $P(X = k) = \lambda^k \frac{(e^{-\lambda})}{k!}$ (k! signifies the factorial of k, that is, the product of all positive integers up to k).[56,74]

Conversely, if the count of events per unit time adheres to a Poisson distribution, the time lapse between these events follows an exponential distribution.

Applications in translational cardiology research

Within the scope of translational cardiology research, the exponential distribution finds immense applicability, exhibiting noteworthy potential in dealing with time-to-event data, often termed as "survival times."[77,78]

For instance, the exponential distribution is frequently employed in the analysis of time-to-failure of cardiac devices and interventions. In scenarios where the expected length of time until a specific event (like device failure) is continuously decreasing, the exponential distribution proves to be a suitable statistical model.[79]

Moreover, it is utilized in MI and cerebrovascular accidents following noncardiac surgery. In a specific study conducted by Kikura et al., the exponential model was used to compare the preoperative characteristics and postoperative daily distribution of postoperative myocardial infarction and stroke.[80]

Likewise, the exponential distribution is integral in queuing models, which simulate patient flow in hospital settings.[81] These models assist in making informed decisions about resource allocation, patient scheduling, and operational efficiency, indirectly contributing to improved patient outcomes.

Apart from that, this kind of distribution contributes to heart rate variability (HRV) studies. The intervals between successive heartbeats exhibit considerable variability, which can be a crucial physiological marker of cardiovascular health and the autonomic nervous system's function. The exponential distribution provides a suitable model for the distribution of these interbeat intervals, assisting in heart rate variability studies.[82,83]

its significance extends to the realm of arrhythmia episodes. Within this context, the exponential distribution serves as a valuable tool for modeling the temporal intervals between occurrences of arrhythmia episodes, characterized by irregular heartbeats. This approach aids in predicting the likelihood of future arrhythmia episodes, drawing upon historical data for predictive insights.[84]

In summary, the exponential distribution stands as a pivotal statistical tool in translational cardiology research. Its application facilitates invaluable insights into diverse facets, including cardiac device reliability, patient survival analysis, healthcare operational efficiency, and the development of risk prediction models.

Each statistical distribution presents its unique assortment of predefined assumptions and inherent restrictions. Deciding on a suitable statistical distribution needs to emerge from a careful consideration of the relevance of its assumptions to the particular dataset being used and the research question being investigated. This systematic and rigorous consideration ensures a coherent match between the selected statistical distribution and both, the empirical data realities and the investigative objectives.

Get started

For a proficient application of statistical distributions in the field of translational cardiology research, it is recommended for researchers to:

1. Acquaint themselves with the features and uses of different statistical distributions.
2. Select an apt distribution that aligns with the unique attributes of their dataset as well as the pertinent research inquiry.
3. Select advanced statistical software or programming languages to expedite the calculation and analysis of statistics relevant to the chosen distribution.

Potential pitfalls

1. **Misinterpretation of data:** One pitfall in utilizing statistical distributions in translational cardiology research is the potential for data misinterpretation due to inappropriate application of the chosen statistical principle. This can lead to incorrect conclusions and later issues during the application or translation of the findings into practice.
2. **Incorrect distribution assumptions:** Many statistical principles assume that data follows a specific distribution, such as normal or exponential. The derived conclusions may be inaccurate or misleading if these assumptions are incorrect. In cardiology research, certain variables (e.g., heart rate, blood pressure) may not follow the assumed distribution due to various influencing factors like medication or lifestyle differences.
3. **Overreliance on statistical significance:** Another potential pitfall is the overemphasis on statistical significance without considering clinical significance. A statistical model might find that a certain variable has a statistically significant effect, but the actual impact on patient outcomes may be minimal. This may divert

resources away from interventions that could have larger clinical effects but did not reach statistical significance in the study.

Additional reading

1. Pilling M. Handbook of Statistical Distributions with Applications. J R Stat Soc Ser A Stat Soc. 2017; 180(3):941−941. https://doi.org/10.1111/RSSA.12292
2. Bakouch HS. Using the Weibull distribution: Reliability, modeling and inference. J Appl Stat. 2014; 41(4):913−914. https://doi.org/10.1080/02664763.2013.853927
3. Patterson C, Willis MS. Translational Cardiology: Molecular Basis of Cardiac Metabolism, Cardiac Remodeling, Translational Therapies and Imaging Techniques. Translational Cardiology: Molecular Basis of Cardiac Metabolism, Cardiac Remodeling, Translational Therapies and Imaging Techniques. Published online January 1, 2012:1−543. https://doi.org/10.1007/978-1-61779-891-7/COVER
4. Cleophas TJ, Zwinderman AH, Cleophas TF. Relationship Among Statistical Distributions. Statistics Applied to Clinical Trials: Self-Assessment Book. Published online 2002:175−188. https://doi.org/10.1007/978-94-010-0285-1_11

Consultants/contractors

The composition of the chapters stands as the sole effort of the author, with no participation from external individuals or organizations during the writing process.

References

1. Samuels ML, Witmer JA, Schaffner AA. *Statistics for the Life Sciences*. Pearson College Division; 2016.
2. Kusuoka H, Hoffman JIE. Advice on statistical analysis for circulation research. *Circ Res*. 2002;91(8):662−671. https://doi.org/10.1161/01.RES.0000037427.73184.C1.
3. Dietrich DF, Schwartz J, Schindler C, et al. Effects of passive smoking on heart rate variability, heart rate and blood pressure: an observational study. *Int J Epidemiol*. 2007;36(4):834−840. https://doi.org/10.1093/IJE/DYM031.
4. Vetter TR. Fundamentals of research data and variables: the devil is in the details. *Anesth Analg*. 2017;125(4):1375−1380. https://doi.org/10.1213/ANE.0000000000002370.
5. Gottwik M, Zahn R, Schiele R, et al. Differences in treatment and outcome of patients with acute myocardial infarction admitted to hospitals with compared to without departments of cardiology; results from the pooled data of the Maximal Individual Therapy in Acute Myocardial Infarction (MITRA 1+2) Registries and the Myocardial Infarction Registry (MIR). *Eur Heart J*. 2001;22(19):1794−1801. https://doi.org/10.1053/EUHJ.2001.2630.
6. Thrun MC, Gehlert T, Ultsch A. Analyzing the fine structure of distributions. *PLoS One*. 2020;15(10). https://doi.org/10.1371/JOURNAL.PONE.0238835.
7. Tsokos C, Wooten R. Normal probability. *Joy Finite Math*. 2016:231−263. https://doi.org/10.1016/B978-0-12-802967-1.00007-3.
8. Wolfram MathWorld. The web's most extensive mathematics resource. Normal Distribution. https://mathworld.wolfram.com/. Accessed 4 September 2023.
9. Mishra P, Pandey CM, Singh U, Gupta A, Sahu C, Keshri A. Descriptive statistics and normality tests for statistical data. *Ann Card Anaesth*. 2019;22(1):67. https://doi.org/10.4103/ACA.ACA_157_18.
10. Saely CH, Risch L, Frey F, et al. Body-mass-index, blood pressure, and serum cholesterol in young Swiss men: an analysis on 56784 army conscripts. *Swiss Med Wkly*. 2009;139(3536). https://doi.org/10.4414/SMW.2009.12653.
11. Wooditch A, Johnson NJ, Solymosi R, Medina Ariza J, Langton S. The normal distribution and single-sample significance tests. In: *A Beginner's Guide to Statistics for Criminology and Criminal Justice Using R*. 2021:155−168. https://doi.org/10.1007/978-3-030-50625-4_10. Published online 2021.
12. Ferrare JJ. Normal distributions. http://www.josephferrare.com/. Accessed 14 August 2023.
13. Kwak SG, Kim JH. Central limit theorem: the cornerstone of modern statistics. *Korean J Anesthesiol*. 2017;70(2):144−156. https://doi.org/10.4097/KJAE.2017.70.2.144.
14. Kranzusch R, Aus Dem Siepen F, Wiesemann S, et al. Z-score mapping for standardized analysis and reporting of cardiovascular magnetic resonance modified Look-Locker inversion recovery (MOLLI) T1 data: normal behavior and validation in patients with amyloidosis. *J Cardiovasc Magn Reson*. 2020;22(1). https://doi.org/10.1186/S12968-019-0595-7.
15. Dallaire F, Bigras JL, Prsa M, Dahdah N. Bias related to body mass index in pediatric echocardiographic Z scores. *Pediatr Cardiol*. 2015;36(3):667−676. https://doi.org/10.1007/S00246-014-1063-7.
16. Dunn OJ. Confidence intervals for the means of dependent, normally distributed variables. *J Am Stat Assoc*. 1959;54(287):613−621. https://doi.org/10.1080/01621459.1959.10501524.
17. Kim H-Y. Statistical notes for clinical researchers: assessing normal distribution (1). *Restor Dent Endod*. 2012;37(4):245. https://doi.org/10.5395/rde.2012.37.4.245.
18. Kim H-Y. Statistical notes for clinical researchers: assessing normal distribution (2) using skewness and kurtosis. *Restor Dent Endod*. 2013;38(1):52. https://doi.org/10.5395/rde.2013.38.1.52.
19. Abdi H. Binomial distribution/binomial and sign tests. *Encycl Meas Stat*. 2007. https://doi.org/10.4135/9781412952644.n54.
20. Edwards AWF. The meaning of binomial distribution. *Nature*. 1960;186(4730):1074. https://doi.org/10.1038/1861074A0.
21. Ross SM. Introduction to probability models: eleventh edition. In: *Introduction to Probability Models*. 11th ed. 2014:1−767. https://doi.org/10.1016/C2012-0-03564-8. Published online 2014.
22. Johnson RA, Miller I, Freund JE. *Miller & Freund's Probability and Statistics for Engineers*. Pearson Education, Inc.; 2019.
23. Wolfram MathWorld. The web's most extensive mathematics resource. *Binomial Distribution*. https://mathworld.wolfram.com/. Accessed 4 September 2023.

24. Wallis S. Binomial confidence intervals and contingency tests: mathematical fundamentals and the evaluation of alternative methods. *J Quant Ling.* 2013;20(3):178–208. https://doi.org/10.1080/09296174.2013.799918.
25. Arellano G. Calculation of narrower confidence intervals for tree mortality rates when we know nothing but the location of the death/survival events. *Ecol Evol.* 2019;9(17):9644–9653. https://doi.org/10.1002/ECE3.5495.
26. Zhang NF. The use of correlated binomial distribution in estimating error rates for firearm evidence identification. *J Res Natl Inst Stand Technol.* 2019;124:1. https://doi.org/10.6028/JRES.124.026.
27. Fernández Coronado JI, Arredondo NA, Rivera I, et al. The binomial distribution: historical origin and evolution of its problem situations. *Mathematics.* 2022;10(15):2680. https://doi.org/10.3390/MATH10152680.
28. Hu C, Zhou H, Sharma A. Application of beta-distribution and combined uniform and binomial methods in longitudinal modeling of bounded outcome score data. *AAPS J.* 2020;22(5):1–9. https://doi.org/10.1208/S12248-020-00478-5/FIGURES/6.
29. Dai H, Younis A, Kong JD, et al. Big data in cardiology: state-of-art and future prospects. *Front Cardiovasc Med.* 2022;9:844296. https://doi.org/10.3389/FCVM.2022.844296/BIBTEX.
30. Wesołowski S, Lemmon G, Hernandez EJ, et al. An explainable artificial intelligence approach for predicting cardiovascular outcomes using electronic health records. *PLOS Digit Health.* 2022;1(1):e0000004. https://doi.org/10.1371/JOURNAL.PDIG.0000004.
31. Peacock JL, Peacock PJ. *Probability and distributions.* Oxford Handbook of Medical Statistics; 2011:243–280. https://doi.org/10.1093/MED/9780198743583.003.0007. Published online May 2020.
32. Kosiborod M, Arnold SV, Spertus JA, et al. Evaluation of ranolazine in patients with type 2 diabetes mellitus and chronic stable angina: results from the TERISA randomized clinical trial (type 2 diabetes evaluation of ranolazine in subjects with chronic stable angina). *J Am Coll Cardiol.* 2013;61(20):2038–2045. https://doi.org/10.1016/J.JACC.2013.02.011.
33. Pan X, Kamble S, Burns L, et al. What do real world data say about safety and resource use of oral antagonists? Early analysis of newly anticoagulated NON-valvular atrial fibrillation patients using either apixaban, dabigatran, rivaroxaban or warfarin. *J Am Coll Cardiol.* 2016;67(13):894. https://doi.org/10.1016/S0735-1097(16)30895-6.
34. Ageno W, Garcia D, Silingardi M, Galli M, Crowther M. A randomized trial comparing 1 mg of oral vitamin K with no treatment in the management of warfarin-associated coagulopathy in patients with mechanical heart valves. *J Am Coll Cardiol.* 2005;46(4):732–733. https://doi.org/10.1016/J.JACC.2005.05.022.
35. Kandzari DE, Kini AS, Karmpaliotis D, et al. Safety and effectiveness of everolimus-eluting stents in chronic total coronary occlusion revascularization: results from the EXPERT CTO multicenter trial (evaluation of the XIENCE coronary stent, performance, and technique in chronic total occlusions). *Cardiovasc Interv.* 2015;8(6):761–769. https://doi.org/10.1016/J.JCIN.2014.12.238.
36. Harandi SA, Sarrafzadegan N, Sadeghi M, et al. Do cardiometabolic risk factors relative risks differ for the occurrence of ischemic heart disease and stroke? *Res Cardiovasc Med.* 2016;5(1):30619. https://doi.org/10.5812/CARDIOVASCMED.30619.
37. Sprecher DL, Pearce GL. How deadly is the "deadly quartet"?: a post-CABG evaluation. *J Am Coll Cardiol.* 2000;36(4):1159–1165. https://doi.org/10.1016/S0735-1097(00)00867-6.
38. Chandra A, Neeland IJ, Berry JD, et al. The relationship of body mass and fat distribution with incident hypertension: observations from the Dallas Heart Study. *J Am Coll Cardiol.* 2014;64(10):997–1002. https://doi.org/10.1016/J.JACC.2014.05.057.
39. Neizel M, Korosoglou G, Lossnitzer D, et al. Impact of systolic and diastolic deformation indexes assessed by strain-encoded imaging to predict persistent severe myocardial dysfunction in patients after acute myocardial infarction at follow-up. *J Am Coll Cardiol.* 2010;56(13):1056–1062. https://doi.org/10.1016/J.JACC.2010.02.070.
40. Martin TN, Groenning BA, Murray HM, et al. ST-segment deviation analysis of the admission 12-lead electrocardiogram as an aid to early diagnosis of acute myocardial infarction with a cardiac magnetic resonance imaging gold standard. *J Am Coll Cardiol.* 2007;50(11):1021–1028. https://doi.org/10.1016/J.JACC.2007.04.090.
41. Hu H. *Poisson Distribution and Application.* 2008.
42. Zrelak PA. Use of the Poisson distribution is a helpful tool that is underused in nursing practice. *J Nurs Care Qual.* 2022;37(3):E54–E57. https://doi.org/10.1097/NCQ.0000000000000612.
43. Sheldon R, Opie-Moran M. The placebo effect in cardiology: understanding and using it. *Can J Cardiol.* 2017;33(12):1535–1542. https://doi.org/10.1016/J.CJCA.2017.09.017.
44. El-Dawoody M, Eliwa MS, El-Morshedy M. An extension of the Poisson distribution: features and application for medical data modeling. *Processes.* 2023;11(4):1195. https://doi.org/10.3390/PR11041195.
45. Kissell R, Poserina J, Advanced Math and Statistics. *Optimal Sports Math, Statistics, and Fantasy.* 2017:103–135. https://doi.org/10.1016/B978-0-12-805163-4.00004-9. Published online.
46. Forthofer RN, Lee ES, Hernandez M. Probability distributions. *Biostatistics.* 2007:103–133. https://doi.org/10.1016/B978-0-12-369492-8.50010-8.
47. Fokianos K. Count time series models. *Handb Stat.* 2012;30:315–347. https://doi.org/10.1016/B978-0-444-53858-1.00012-0.
48. Stigler SM. Poisson on the Poisson distribution. *Stat Probab Lett.* 1982;1(1):33–35. https://doi.org/10.1016/0167-7152(82)90010-4.
49. Koehrsen W. The Poisson distribution and Poisson process explained. *Towar. Data Sci.* 2019;(13):1–13. https://towardsdatascience.com/the-poisson-distribution-and-poisson-process-explained-4e2cb17d459. Accessed August 29, 2023.
50. Jolicoeur P. The Poisson distribution. *Introd Biometry.* 1999:124–133. https://doi.org/10.1007/978-1-4615-4777-8_19.
51. Lin Z, Wang Y, Hong Y. The computing of the Poisson multinomial distribution and applications in ecological inference and machine learning. *Comput Stat.* 2022:1–27. https://doi.org/10.1007/S00180-022-01299-0/FIGURES/7.
52. Sharma DK, Singh B, Raja M, Regin R, Rajest SS. An efficient python approach for simulation of Poisson distribution. In: *2021 7th International Conference on Advanced Computing and Communication Systems, ICACCS.* 2021:2011–2014. https://doi.org/10.1109/ICACCS51430.2021.9441895. Published online March 19, 2021.
53. Teerapabolarn K, Wongkasem P. Poisson approximation for independent geometric random variables. *Int Math Forum.* 2007;2(65):3211–3218.

54. Fox H, Bolton B. Probability and statistics. In: *Mathematics for Engineers and Technologists*. 2022:291–320. https://doi.org/10.1016/B978-075065544-6/50010-2.
55. Dunn WL, Shultis JK. Some common probability distributions. *Explor Monte Carlo Methods*. 2022:447–495. https://doi.org/10.1016/B978-0-12-819739-4.00019-6.
56. *Poisson Distribution*: Wikipedia. https://en.wikipedia.org/wiki/Poisson_distribution.
57. Jehangir Q, Lee Y, Latack K, et al. Incidence, mortality, and imaging outcomes of atrial arrhythmias in COVID-19. *Am J Cardiol*. 2022;173:64–72. https://doi.org/10.1016/J.AMJCARD.2022.02.051.
58. Akoum N, Zelnick LR, De Boer IH, et al. Rates of cardiac rhythm abnormalities in patients with CKD and diabetes. *Clin J Am Soc Nephrol*. 2019;14(4):549–556. https://doi.org/10.2215/CJN.09420818.
59. Moskowitz G, Hong KN, Giustino G, et al. Incidence and risk factors for permanent pacemaker implantation following mitral or aortic valve surgery. *J Am Coll Cardiol*. 2019;74(21):2607–2620. https://doi.org/10.1016/J.JACC.2019.08.1064.
60. Egermann H, Frank P, Krumphuber A. Evaluation of the Poisson distribution for estimating the quality of drug/diluent random powder mixtures. I. High particle size of the drug constituent. *Pharmaceut Res: An Off J Am Assoc Pharmaceut Sci*. 1992;9(3):385–389. https://doi.org/10.1023/A:1015851204770/METRICS.
61. Quah J, Dharmaprani D, Lahiri A, et al. Prospective cross-sectional study using Poisson renewal theory to study phase singularity formation and destruction rates in atrial fibrillation (RENEWAL-AF): study design. *J Arrhythm*. 2020;36(4):660. https://doi.org/10.1002/JOA3.12363.
62. Orooji A, Sahranavard T, Shakeri MT, Tajfard M, Saffari SE. Application of the truncated zero-inflated Double Poisson for determining of the effecting factors on the number of coronary artery stenosis. *Comput Math Methods Med*. 2022;2022. https://doi.org/10.1155/2022/5353539.
63. Exponential distribution - an overview. ScienceDirect Top. https://www.sciencedirect.com/topics/mathematics/exponential-distribution.
64. Ristić MM, Kundu D. Marshall-Olkin generalized exponential distribution. *Metron*. 2015;73(3):317–333. https://doi.org/10.1007/S40300-014-0056-X/TABLES/2.
65. Yadav AS, Altun E, Yousof HM. Burr–hatke exponential distribution: a decreasing failure rate model, statistical inference and applications. *Ann Data Sci*. 2021;8(2):241–260. https://doi.org/10.1007/S40745-019-00213-8/FIGURES/6.
66. Rossello X, González-Del-Hoyo M. Survival analyses in cardiovascular research, part I: the essentials. *Rev Española Cardiol (English Ed)*. 2022;75(1):67–76. https://doi.org/10.1016/J.REC.2021.06.003.
67. Ristić MM, Balakrishnan N. The gamma-exponentiated exponential distribution. *J Stat Comput Simul*. 2012;82(8):1191–1206. https://doi.org/10.1080/00949655.2011.574633.
68. Mahdavi A, Kundu D. A new method for generating distributions with an application to exponential distribution. *Commun Stat Theory Methods*. 2017;46(13):6543–6557. https://doi.org/10.1080/03610926.2015.1130839.
69. Kartsonaki C. Survival analysis. *Diagn Histopathol*. 2016;22(7):263–270. https://doi.org/10.1016/J.MPDHP.2016.06.005.
70. Oguntunde PE, Adejumo AO. The transmuted inverse exponential distribution. *Int J Adv Stat Probab*. 2015;3(1):1–7. https://doi.org/10.14419/ijasp.v3i1.3684.
71. Kim L. Survival analysis for epidemiologic and medical research. Steve selvin. *Int J Epidemiol*. 2009;38(5):1420–1421. https://doi.org/10.1093/IJE/DYN190.
72. Infante Lopez GG, Hermanns H, Katoen JP. Beyond memoryless distributions: model checking semi-Markov chains. *Lect Notes Comput Sci*. 2001;2165:57–70. https://doi.org/10.1007/3-540-44804-7_4/COVER.
73. Nadarajah S, Kotz S. The beta exponential distribution. *Reliab Eng Syst Saf*. 2006;91(6):689–697. https://doi.org/10.1016/J.RESS.2005.05.008.
74. *Exponential Distribution*: Wikipedia. https://en.wikipedia.org/wiki/Exponential_distribution.
75. Jdid M, Alhabib R, Salama AA. The basics of neutrosophic simulation for converting random numbers associated with a uniform probability distribution into random variables follow an exponential distribution. *Neutrosophic Sets Syst*. 2023;53:358–366. https://doi.org/10.5281/ZENODO.7536049.
76. Skogvoll E, Lindqvist BH. Modeling the occurrence of cardiac arrest as a Poisson process. *Ann Emerg Med*. 1999;33(4):409–417. https://doi.org/10.1016/S0196-0644(99)70305-7.
77. Kantidakis G, Hazewinkel AD, Fiocco M. Neural networks for survival prediction in medicine using prognostic factors: a review and critical appraisal. *Comput Math Methods Med*. 2022;2022. https://doi.org/10.1155/2022/1176060.
78. Meng C, Esserman D, Li F, et al. Simulating time-to-event data subject to competing risks and clustering: a review and synthesis. *Stat Methods Med Res*. 2023;32(2):305–333. https://doi.org/10.1177/09622802221136067.
79. Lindsey ML, Gray GA, Wood SK, Curran-Everett D. Statistical considerations in reporting cardiovascular research. *Am J Physiol Heart Circ Physiol*. 2018;315(2):H303–H313. https://doi.org/10.1152/AJPHEART.00309.2018.
80. Kikura M, Oikawa F, Yamamoto K, et al. Myocardial infarction and cerebrovascular accident following non-cardiac surgery: differences in postoperative temporal distribution and risk factors. *J Thromb Haemost*. 2008;6(5):742–748. https://doi.org/10.1111/J.1538-7836.2008.02948.X.
81. Kolker A. Process modeling of emergency department patient flow: effect of patient length of stay on ED diversion. *J Med Syst*. 2008;32(5):389–401. https://doi.org/10.1007/S10916-008-9144-X/FIGURES/12.
82. Van Steenis HG, Tulen JHM. The exponential distribution applied to nonequidistantly sampled cardiovascular time series. *Comput Biomed Res*. 1996;29(3):174–193. https://doi.org/10.1006/cbmr.1996.0015.
83. Bento Silva LR, Viana Gentil PR, Beltrame T, et al. Exponential model for analysis of heart rate responses and autonomic cardiac modulation during different intensities of physical exercise. *R Soc Open Sci*. 2019;6(10). https://doi.org/10.1098/RSOS.190639.
84. Hennig T, Maass P, Hayano J, Heinrichs S. Exponential distribution of long heart beat intervals during atrial fibrillation and their relevance for white noise behaviour in power spectrum. *J Biol Phys*. 2006;32(5):383–392. https://doi.org/10.1007/S10867-006-9022-Z.

Chapter 24

Hypothesis and error types

Dipesh Ludhwani[1] and Pratik Agrawal[2]

[1]*West Virginia University Heart and Vascular Institute, Morgantown, WV, United States;* [2]*Division of Cardiology, WVU Heart & Vascular Institute, Morgantown, WV, United States*

Introduction

Modern medicine and biostatistics go hand in hand, especially given the current emphasis on evidence-based medicine. Besides forming a hypothesis, the certainty with which we can accept or reject a hypothesis constitutes the framework for our medical endeavors. One of the most critical steps in answering a research question is forming a well-worked-up hypothesis. The observations and statistics are then integrated to verify or falsify that hypothesis [1]. It is important to remember that falsifying or rejecting a hypothesis is equally essential.

At the scaffold of each clinical trial, inferences drawn from a selected data sample are generalized to a broader population. To confidently generalize the trial findings, we need to minimize the errors. Similar recurrent observations in the sample may not truly represent the population. The presence of a single observation can be enough to falsify the hypothesis. For example, a single observation of a lion without a mane can falsify the hypothesis that all lions have a mane despite having repeated observations of the latter.

Hypothesis

The hypothesis is based on a good research question and should be simple, specific, and stated in advance.[1] Hypothesis, in general, tries, to establish an association between the predictor (risk factor) and the outcome variable (disease). By default, the assumption is made that there is no statistically significant difference or association between the two variables in question, also known as the null hypothesis (H_0). For example, when studying the association between diabetes mellitus (DM) and chronic kidney disease (CKD), the null hypothesis infers no association between the two. The results are due to chance alone and are not statistically significant. The alternate hypothesis (H_1) is extrapolated after the initial null hypothesis is rejected. H_1 states that there is a statistically significant relationship between the two variables; in other words, whatever you are trying to prove happened. The alternate hypothesis can either establish causality between predictor and outcomes (referred to as one-tailed hypothesis) or establish the association without causality (two-tailed hypothesis).

P-value approach of hypothesis testing

Predictive value or *P*-value describes how likely the observed difference is related to chance alone and that the null hypothesis is valid. In other words, *P*-value is the probability of reaching the observed findings assuming the null hypothesis is correct. A *P*-value of .035 equates to a 3.5% probability that the results are accidental and related to chance. Lower the *P*-value, lower the chance and stronger the evidence (Fig. 24.1).

On the probability distribution curve, the *P*-value can be demonstrated toward the tail end of the graph. Depending on the type of hypothesis testing, it can be one-sided or two-sided (single- vs. two-tailed test). The area to the right of the *P*-value is the probability of obtaining that observed value.

Alpha-value (α) is used in hypothesis testing to define the probability of making an error. It is a surrogate for confidence level and assists with accepting or rejecting the null hypothesis. The investigator establishes the α before the start of the study. Once you run the test and obtain the

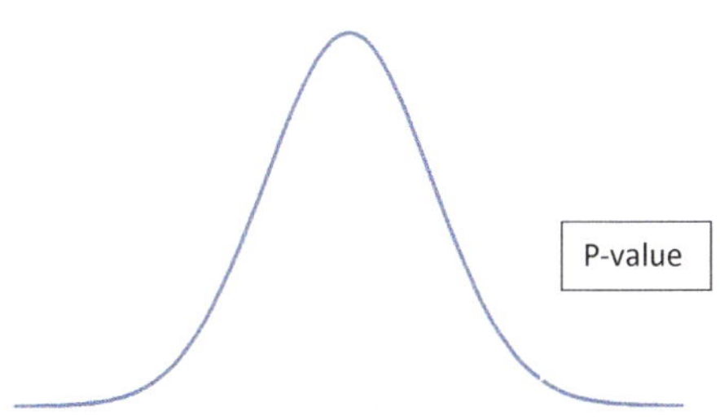

FIGURE 24.1 Graphical representation of *P*-value on distribution curve.

Translational Cardiology. https://doi.org/10.1016/B978-0-323-91790-2.00048-4

P-value, this is compared with the preset α. For example, if you decide to run the test with a 95% confidence level, your α would be 0.05% or 5% (100−95). Now, if the calculated *P*-value is .02 or 2%, less than the preset alpha level of 0.05, H_0 can be rejected, and H_1 can be favored. On the contrary, if the calculated *P*-value is .10 or 10%, then the evidence to favor H_1 is weak, and, in such cases, H_0 should not be rejected.

Types of error

Now that we have some basic understanding of *P*-value and alpha-level, we can talk about the two predominant types of statistical error. It is important to understand that you can never prove H_1 as valid, but you can reject H_0 as being unlikely.

Type I error (α)

In type I error, the H_0 is wrongly rejected in favor of H_1. Type I error is also called a false-positive error. Wrongfully convicting an innocent person is an example of a type I error if the H_0 is defined as the person being innocent. Alpha (α) is the probability of making a type I error. If α is set as 0.05, there is a 5% chance that an association or effect is established when there is none.

Type II error (β)

In type II error, H_0 is wrongly accepted when there is an effect or difference. Type II error is also called a false negative. In the above example, letting a guilty man go would be a type II error. An entity similar to α is beta (β). Beta is the probability of committing a type II error. 1-β is the power defined as the probability of observing an effect or significance when one exists. If β is set at 0.05, then there is a 5% chance of missing an association and failing to reject H_0 when it is false. The power in such a scenario would be 0.95% or 95% (1−0.05). There would be a 95% chance of finding a true association. By increasing the power, we can decrease type II errors. One way of increasing power is to have a large sample size.

In an ideal scenario, we would always like to set α and β as zero to avoid types I and II errors. However, this is usually not possible as this would need a large sample size. Therefore, both α and β are set as small as possible.

Reference

1. Banerjee A, Chitnis UB, Jadhav SL, Bhawalkar JS, Chaudhury S. Hypothesis testing, type I and type II errors. *Ind Psychiatry J.* 2009;18(2):127−131. https://doi.org/10.4103/0972-6748.62274.

Chapter 25

Power of the study

Yasar Sattar[1], Dhruvil Ashishkumar Patel[2], Khola Waheed Khan[3], Akram Kawsara[1] and Ramesh Daggubati[1]

[1]*Department of Cardiology, West Virginia, Morgantown, MV, United States;* [2]*Wayne State University, Detroit, MI, United States;* [3]*Wah Medical College, National University of Medical Sciences, Rawalpindi, Pakistan*

Key points

- The concept of power simply permits an inference of reliability of the conclusions obtained from a study.
- The power is interwoven with the hypothesis testing as the prevention of a type II error is the main aim of increasing statistical power

The importance of power in clinical and translational research

The concept of statistical power is imperative to producing clinical and translational research that has a high confidence of having the desired effect in the real-world setting. The concept of power simply permits an inference of the reliability of the conclusions obtained from a study. In a study there are multiple controlled factors that are not controlled for in a clinical setting. It is therefore imperative for a clinical investigator to have both an intuitive understanding and awareness of the fundamental process of statistical power and what important thresholds are necessary to permit findings that are clinically relevant and translatable.

To evaluate this phenomenon as a clinical investigator, this chapter will define the opportunities during study planning, conducting of a study, and analysis of data where statistical power can be optimized to prevent erroneous conclusions from being supported that are not related to the actual nature of the relationship. Furthermore, we will delve into what the effect of these decisions will be on statistical power to emerge an intuitive understanding for how both study planning and analyses of data can be modulated to maximize the power while describing the consequences of each. This will also allow one to estimate the value of their statistical power both intrinsically in variables within a single study and comparing the value of their results with other study types as well—informing real-world translatability of findings.

Throughout this chapter, the concept of power is interwoven with hypothesis testing as the prevention of a type II error is the main aim of increasing statistical power. The concept of a type II error emerges in hypothesis testing when the threshold for accepting a hypothesis that is incorrect is large, thereby increasing the probability that no relationship between a variable exists even though it does. Such conclusions prevent new discoveries and findings to emerge and thus become important for translatable research to be validated across new paradigms. Therefore, optimizing for power is in the benefit of scientific research as the potential for new discoveries that have a positive impact and the identification of relationships that have a negative impact on patient outcomes are vital for real-world applicability. What is important to understand is that, unlike many other concepts in statistics, increasing the power of a study will only have positive implications for findings as the confidence for translatability will only increase.

Establishing a context for power by introducing hypothesis testing

Hypothesis testing is a ubiquitous methodology within translation research to identify clear utilizable relationships based on the study question. Yet there is always room for error in these contexts, which indicate that the parameters that inform the statistic being studied must be optimized to minimize the likelihood of these errors occurring. There are two main kinds of errors in hypothesis testing that are represented based on their reference to the null hypothesis.

Specifically, these are as follows:

Translational Cardiology. https://doi.org/10.1016/B978-0-323-91790-2.00086-1

A. Type I error—where the null hypothesis is rejected despite it being true
B. Type II error—where the null hypothesis is not rejected despite it being false

In the context of clinical and translational research, either of these errors can have negative consequences and must be optimized. This optimization is done both at the stage of study planning and analysis of data. Furthermore, minimizing the presence of these errors is important to establish the reproducibility of findings. Minimizing both type I and type II error likelihood is important to prevent conclusions that can have clinical consequences if implemented.

A high-level intuition of how these errors can be committed

To establish an intuitive understanding of minimizing both errors, creating a context for how they are represented across analyses is important. The error, in both the type I and type II errors, is in reference to the null hypothesis—or the assumed status quo. In the paradigms of clinical research, that status quo is in the general form of "no relationship exists" between the variables being studied. If a relationship does exist and has a high likelihood, then the null hypothesis is rejected—and this rejection of the null hypothesis is referred to as the alternative hypothesis. Therefore, the frame of reference in hypothesis testing is always relative to the null hypothesis.

In the context of clinical research, assessing for a relationship between two variables is commonly done using regression inferential statistics. That is either a linear or logistics regression analysis. To use an example between two clinical variables—let us say mortality from mechanical ventilation was compared with a patient's age.

(1) A type I error in this context would be that there is no relationship, either positive or negative, between mortality from mechanical ventilation and a patient's age but our hypothesis testing suggested there is.
(2) A type II error in this context would be that there is a vital relationship between mortality from mechanical ventilation and the patient's age, but our analyses suggested that the null hypothesis (i.e., no relationship exists) occurred.

The final paradigm to consider is the value that is being used to compare to the null hypothesis mean. Essentially, where does the value that is testing against the null hypothesis come from? The source of this comparator value is from a sample of the data collected during the study. If the source—for example, a sample mean, has been taken from the actual data and is representative of the mean of the entire population data, then testing its likelihood against the null hypothesis will lead to the highest accuracy of results.

Understanding that the likelihood of committing a type I or type II error is relative to the null hypothesis means that any modulation done to the parameters to minimize these errors will create a sample mean that is most accurate and the hypothesis testing result will be as close to the truthful relationship as possible. Therefore, modulation to optimize for type I and type II is done in the following two paradigms:

1. Setting up hypothesis testing thresholds that confirm the actual truth of the relationship
2. Using test values that are most representative of the truth as possible

When optimizing for type I errors, only modulation of paradigm #1 can be done while for type II errors, modulation of both paradigm #1 and #2 can be done.

Exploring type I errors to understand type II errors and power

The probability of having a type I error is denoted by α. When the test statistical probability used to evaluate for a relationship has a probability more than the threshold that is set for α in the test, then the alternative hypothesis is accepted while the null hypothesis is correctly rejected as we are now sure that there is very minimal probability of committing a type I error.

If the opposite were to occur whereby the test statistical probability for a relationship falls outside the threshold set for α, then setting a threshold extremely small will guarantee that if a test statistics probability of a type I error falls under the α value, then the alternative hypothesis is most likely to be accepted. Therefore, the likelihood of the relationship being falsely significant is based on what threshold of confidence is being used. In the context of biostatistical studies, the most common α threshold, referred to as the likelihood of making a type I error, is 5% (0.05). If a test statistic value of a type I error: defined as p, falls underneath this value ($P < .05$), then the conclusion is most likely not influenced by type I error. Therefore, the investigators are taking a 5% chance that they will support a conclusion that a significant relationship is present in the total population between the variables studied when in fact that is not true.

A type II error can also be communicated as a probability. The probability of committing a type II error for a study is denoted as β. Unlike a type I error, there is no threshold set whereby if the sample analyses probability falls within this range, we can be sure that a type II error has not been committed. Instead, ensuring that the mean from the sample—that was gained from the study—is as close to the truthful value as possible becomes the most necessary way. The probability in this scenario—where the null hypothesis is correctly rejected when it is false—is referred to as the statistical power of the study. Furthermore, if the probability that a type II error being committed is β, then the probability of not committing a type II error can also be denoted as β subtracted from 1 (which is the total probability), which equates to the statistical power of the study (1-β).

What is statistical power and how can we optimize for it?

To ensure a type II error does not occur, the value that is used to compare against the null hypothesis in hypothesis testing must be as close to the truthful value as possible. In the context of clinical research, obtaining a comparator value with the highest accuracy (referring to the comparator value being as close to the true value) as possible and precision (referring to the comparator values over multiple samples from the mean being as consistently as the true value_ as possible is aimed for.

Therefore, the power of a study provides confidence that the most truthful value has been used to reject the null hypothesis when it is false, which would affirm the alternative hypothesis. Optimizing for power is crucial for actual valuable conclusions to emerge, but the question now becomes: what can be done when conducting a study to optimize for the greatest power? This optimization occurs during the planning and analysis phase of the study and can be done using the following strategies, each with its own levels of control that investigators have over them to manage outcomes:

(1) What can be in control of the investigators

A. *Sample size increase:*

The one metric that is in most control of the investigators is the size of the study population. Essentially, increasing the number of participants in the study will allow for findings to be as related to the actual phenomenon in the overall population being studied as possible. Therefore, the accuracy of the study findings relative to the population is directly proportional to the sample size of the actual study.

Framing this concept relative to committing a type II error: Due to a better representation of the actual population being elicited with a greater study size, a sample taken from the study population would also have this same property. This would narrow down the range of values that would otherwise not correspond to the truthful relation when comparing toward the null hypothesis in hypothesis testing.

Reduction in variance:

It is also important to mention that the increase in sample size will also collect metrics that are close in value to each other. This will decrease the variance of the dataset, reducing the proportion of erroneous or outlier calculations as a true average emerges from more of a population being studied. This will reduce variability in data points which reduces the variance of the values of the variables collected. This reduction in variance, as a value of the sample size increase, will thereby also increase the power of the study.

B. *Level of significance used for analysis:*

Increasing the level of significance, meaning the α, will increase the probability for the sample test value to reject the null hypothesis. This occurs because the level of significance defines the probability that a given relationship tested is not due to chance. In clinical research, often the threshold significance can be increased from 5% (where $\alpha = 0.05$) to 10% (where $\alpha = 0.1$). As mentioned earlier, this increase in significance consequently also increases the risk of a type I error.

(2) What is not in control of the investigators

C. *The effect size:*

a. The effect size refers to the extent of the difference between the value of the null hypothesis and the alternative hypothesis. As mentioned earlier, the value of the alternative hypothesis used for testing against the null hypothesis emerges from the sample taken of the data obtained from the overall study population. If the difference between the sample mean and the null hypothesis is substantive, then the presence of difference will have a higher probability of being significant.

It is also important to understand that this difference in effect size can be a difference between variables within the study cohort. This is the principle responsible for associations between one outcome and other factors—which is a common study paradigm done in clinical and translation research. For example, mortality in an arbitrary population can be associated with certain factors including an ICU admission, an in-hospital myocardial infarction, and the need for mechanical ventilation, but could not be associated with the presence of acute kidney injury. Furthermore, the effect size between mortality and these variables can be different across different cohorts when studies are conducted for different questions, different patient factors, and various study hypotheses. Therefore, the effect size is not a factor of how the study can be set up or data can be analyzed but rather is solely dependent on the data collection during the study and the effect size difference magnitude measured within them. As a result, it cannot be controlled by investigators and is instead an intrinsic property of the results collected.

Due to the establishment of these differences, the larger the effect size (meaning the greater the difference between the null hypothesis and alternative hypothesis), the higher the likelihood that a true difference occurs. This difference would allow confidence that the null hypothesis is correctly rejected because it is false—reducing the probability that a type II error has occurred and therefore maximizing the power of the study.

Standards for power in clinical and translational research: Introducing sensitivity

While till now, statistical power has been referred to as a conceptual metric, assigning a value to it will allow for comparative power between studies to be determined is important. As mentioned before, the power of a study is 1-β, where β is defined as the probability of a type II error when undergoing hypothesis testing. Since a type II error is defined as the probability of not rejecting the null hypothesis when it is false, the statistical power is the likelihood of rejecting the null hypothesis when it is in fact false, thereby accepting the alternative hypothesis.

Another way to frame this conceptually is that the power then becomes accepting a relationship when it is in fact true. This is also referred to as the sensitivity of the test. Sensitivity is a post hoc analysis value assigned to confirm relationships if they are there. The higher the sensitivity of a test, the greater the number of points that represent a true relationship are confirmed.

A common threshold for power used in designing clinical and translation research is 80%—meaning that there is an 80% likelihood that the test will have a true significant relationship when in fact there is one. This is the exact definition of sensitivity and therefore a sensitivity of a test being 80% is generally acceptable. Yet it has been noted that the statistical power can vary in published studies, often at a range of 0%–20%.[1] Investigation of research methodology, population levels, and complexity of the study embarked are all other factors that can also affect power, contributing to this elevated variability. Therefore, while striving for an 80% power that is considered conventional, understanding the influence of the complexity of the study being done can also limit the power that is attained—but using the principles mentioned in this chapter, an intuitive understanding of power can still help to develop a study that provides results as significant and close to actual truth as possible.

For many teachers of introductory statistics, power is a concept that is often not used. In many cases, it's avoided altogether. In fact, many Advanced Placement (AP) teachers stay away from the topic when they teach tests of significance, according to Floyd Bullard in "Power in Tests of Significance." However, power is an important concept to understand as a consumer of research, no matter what field or profession a student may enter as an adult. Hence, a discussion of power should be included in an introductory course.

To discuss and understand power, one must be clear on the concepts of type I and type II errors. Doug Rush provides a refresher on type I and type II errors (including power and effect size) in the Spring 2015 issue of the *Statistics Teacher Network*, but, briefly, a type I Error is rejecting the null hypothesis in favor of a false alternative hypothesis, and a type II Error is failing to reject a false null hypothesis in favor of a true alternative hypothesis. The probability of a type I error is typically known as Alpha, while the probability of a type II error is typically known as Beta.

Now on to power. Many learners need to be exposed to a variety of perspectives on the definition of power. Bullard describes multiple ways to interpret power correctly:

- Power is the probability of rejecting the null hypothesis when, in fact, it is false.
- Power is the probability of making a correct decision (to reject the null hypothesis) when the null hypothesis is false.
- Power is the probability that a test of significance will pick up on an effect that is present.
- Power is the probability that a test of significance will detect a deviation from the null hypothesis, should such a deviation exist.
- Power is the probability of avoiding a type II error.

Simply put, power is the probability of not making a type II error, according to Neil Weiss in *Introductory Statistics*.

Mathematically, power is 1-beta. The power of a hypothesis test is between 0 and 1; if the power is close to 1, the hypothesis test is very good at detecting a false null hypothesis. Beta is commonly set at 0.2, but may be set by the researchers to be smaller.

Consequently, power may be as low as 0.8, but may be higher. Powers lower than 0.8, while not impossible, would typically be considered too low for most areas of research.

Bullard also states there are the following four primary factors affecting power:

1. Significance level (or alpha)
2. Sample size
3. Variability, or variance, in the measured response variable

4. Magnitude of the effect of the variable

Power is increased when a researcher increases sample size, as well as when a researcher increases effect sizes and significance levels. There are other variables that also influence power, including variance ($\sigma 2$), but we will limit our conversation to the relationships among power, sample size, effect size, and alpha for this discussion.

In reality, a researcher wants both type I and type II errors to be small. In terms of significance level and power, Weiss says this means we want a small significance level (close to 0) and a large power (close to 1).

Having stated a little bit about the concept of power, the authors have found it is most important for students to understand the importance of power as related to sample size when analyzing a study or research article versus actually calculating power. We have found students generally understand the concepts of sampling, study design, and basic statistical tests, but sometimes struggle with the importance of power and necessary sample size.

Example

1. Research is carried out to examine whether a new app called StatMaster can help students learn statistical concepts better than traditional methods. The study was run with high-school students, comparing the morning AP Statistics class (35 students) that incorporated the StatMaster app to the afternoon AP Statistics class (35 students) that did not use the StatMaster app. After the application of the statistical analysis and test of significance the *P*-value of 0.034 was obtained. Since the alpha was set at 0.05, the results are statistically significant. While the study is still at risk of making a type I error, this result does not leave open the possibility of a type II error. Said another way, the power is adequate to detect a difference because they did detect a difference that was statistically significant. It does not matter that there is no power or sample size calculation when the *P*-value is less than alpha.
2. A biotechnology company wants to compare how effective two of its drugs are for treating diabetes. The null hypothesis states the two medications are equally effective. A null hypothesis, H_0, is the claim that the company hopes to reject using the one-tailed test. The alternative hypothesis, H_a, states the two drugs are not equally effective. The alternative hypothesis, H_a, is the state of nature that is supported by rejecting the null hypothesis. The biotech company implements a large clinical trial of 3000 patients with diabetes to compare the treatments. The company randomly divides the 3000 patients into two equally sized groups, giving one group one of the treatments and the other group the other treatment. It selects a significance level of 0.05, which indicates it is willing to accept a 5% chance it may reject the null hypothesis when it is true or a 5% chance of committing a type I error.
   Assume the *P* value is calculated to be 0.025. Therefore, the probability of committing a type II error is 97.5%. If the two medications are not equal, the null hypothesis should be rejected. However, if the biotech company does not reject the null hypothesis when the drugs are not equally effective, a type II error occurs.
3. Essential oils are gaining popularity nowadays due to their various benefits. You assume that the lavender essential oil has the ability to reduce stress and anxiety. To check this assumption you may conduct the hypothesis testing by restating the hypothesis as follows,
   Null hypothesis (Ho): Lavender essential has no effect on reducing anxiety.
   Alternative hypothesis (Ha): Lavender oil helps in reducing anxiety.
   In this experiment, group A, that is, the experimental group is provided with lavender oil, while group B, that is, the control group is provided with a placebo. The data is then collected using various statistical tools and the stress levels of both the groups, that is, the experimental and the control groups are then analyzed. After the calculation, the *P*-value is found to be 0.05. The *P* values are less than the significance values; hence, the null hypothesis is rejected, and it can be concluded that lavender oil helps in reducing stress among people.

Reference

1. Dumas-Mallet E, Button KS, Boraud T, Gonon F, Munafò MR. Low statistical power in biomedical science: a review of three human research domains. *R Soc Open Sci*. February 1, 2017;4(2):160254. https://doi.org/10.1098/rsos.160254.

Chapter 26

Regression

Samian Sulaiman
Division of Cardiology, West Virginia University, School of Medicine, Morgantown, WV, United States

Key points

- **Definition of Regression:** Regression is a statistical method that evaluates the relationship between a dependent variable and independent variables.
- **Types of Regression**:
  - **Univariable Regression**: Involves one independent variable.
  - **Multivariable Regression**: Involves multiple independent variables.
- **Linear Regression**: Used for continuous dependent variables. The outcome variable depends on the sum of the regression coefficients.
- **Logistic Regression**: Used for binary dependent variables.
- **Poisson Regression**: Used for discrete dependent variables.
- **Purpose of Regression**: Estimates the association between variables but does not indicate causality.

Definition and types of regression models

Regression is a statistical method that evaluates the relationship between a dependent variable (also known as outcome) and independent variables (also known as predictors). The regression models are broadly divided into linear and nonlinear. However, the distinction between linear and nonlinear models is not as straightforward as expected. Linear regression is by far the most common type. The main difference is that the outcome variable is dependent on the sum of the regression coefficients for linear models.

Linear regression has multiple subtypes: straight-line (commonly known as linear regression), logistic regression, and Poisson regression. The numerical characteristics of the outcome variable define the correct regression type to use. Fig. 26.1 shows an algorithm for choosing the appropriate regression type.

Mathematical basics

Since straight-line regression (hereafter "linear") is the easiest to understand, we will use this type to explain the mathematical basics.

Linear regression takes the following form:

$$Y = \beta * X + \varepsilon$$

Y is the outcome variable, X represents independent variables, β represents coefficients, and ε is residual errors. X includes a single predictor in simple or univariable regression and multiple predictors in multiple or multivariable regression. To understand the concepts, we built fictitious data of 100 patients. Our goal was to investigate the relationship between daily sodium intake and systolic blood pressure (SBP). Fig. 26.2 demonstrates the relationship between actual SBP, predicted SBP, and residual errors. The mathematical details of fitting the regression line are complicated; however, the main idea is that the best-fitted line is found by minimizing the residual errors (the least-squares method).

Statistical assumptions

To be able to generalize regression results, multiple assumptions must be fulfilled. We will use the above fictitious data to exemplify these concepts:

1. Independence of outcome events: it means one subject's outcome is not related to another's outcome. For example, the fact that a random patient has an SBP of 160 does not affect the SBP of another patient. However, most medical data are clustered within patients (such as repeated measurements) or hospitals or clinics. For example, the National

Translational Cardiology. https://doi.org/10.1016/B978-0-323-91790-2.00045-9

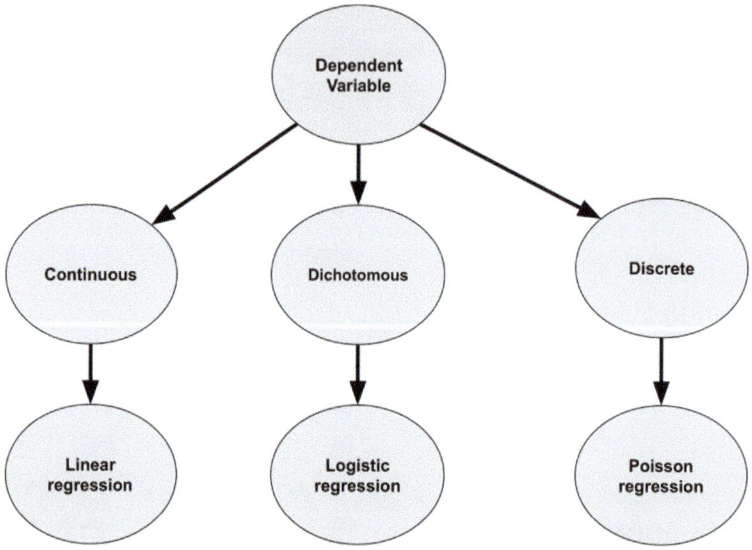

FIGURE 26.1 Algorithm for choosing the correct regression model depending on the dependent variable type.

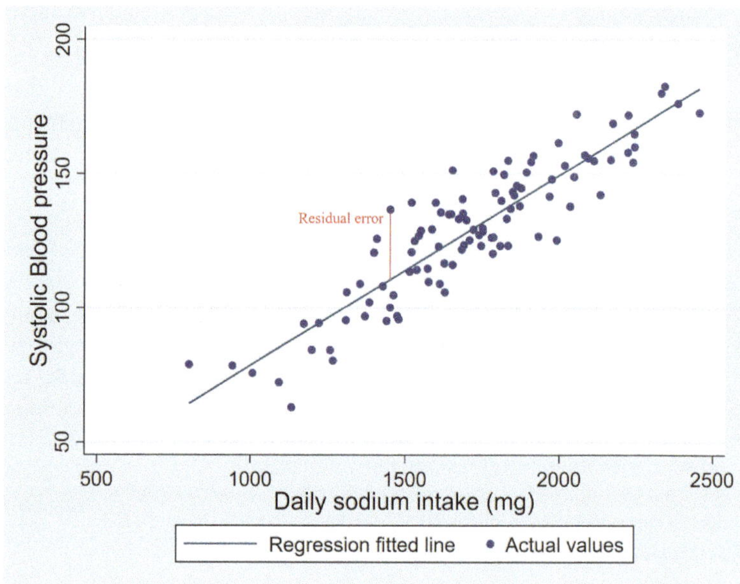

FIGURE 26.2 Scatter plot of the daily sodium intake and systolic blood pressures of the fictitious data.

Inpatient Database[1] is a national administrative database that includes about 20% of hospital stays in the United States. The patients within each hospital are treated in the same clinical setting and are more likely to share a socioeconomic background. Therefore a correlation between them cannot be excluded. As a result, hierarchical or mixed-effect regression models should be used in these cases. Linear, logistic, and Poisson regressions can be used along with hierarchical models.

2. Linearity: the means of the outcome variable at any value of the predictor variable lie along a line. A scatter plot can be used to explore this assumption. After plotting, ask yourself the following question: is it possible to fit a line in the cloud of points? If the answer is yes, then it is plausible to assume that the relationship is linear.
3. Homoscedasticity: It means the variance of the residual errors is similar in every level of the predictor variable. The variance of errors in predicting SBP should be the same at 1500 and 2000 mg/d sodium intake in the example above. After running regression models, statistical software allows diagnostic tests, including the evaluation of residual error distribution.
4. Independent errors: the residual error of a patient is not related to another's error.

5. Normal distribution of errors: After running the regression model, the investigator can plot a histogram of residual errors to evaluate normality.
6. Avoiding perfect multicollinearity: Imagine putting sodium intake in grams and milligrams in the same regression model. Both measurements are perfectly correlated, and the use of both is redundant. In real life, multicollinearity is more subtle. For instance, diabetes mellitus status and glycohemoglobin levels are related, so using one of them to adjust for diabetes in a regression model would be enough. We recommend evaluating pairwise correlations for all the predictors planned to be included in the regression model before running the regression model.

When to use regression

Regression analysis is used for different purposes:

1. Estimate the direction and extent of association between variables
2. Predict outcomes
3. Adjust for confounders
4. Assess interactions between variables

It should be emphasized that regression coefficients represent an association between the predictors and the outcome and not a causal effect. Regression models tell us the amount of change in the outcome variable associated with a change in the independent variable.

Examples

Linear regression for continuous outcomes

Continuous variables are measurable and can take any value between two intervals. Variables such as age and BMI are continuous. For example, BMI can be 30, 30.5, or any infinite possibility between 30 and 31.

Jin et al. Investigated the impact of coronary plaque calcification on plaque volume change.[2] Some statistical concepts used in their study are relevant to this section. First, since the outcome (plaque volume change) was a continuous variable, the authors used linear regression to evaluate the relationship between the outcome and coronary plaque calcification, a predictor variable. Initially, they used univariable regression, which included one predictor variable, coronary plaque calcification. Next, to adjust for different confounders, they used a multivariable regression including multiple predictors in addition to coronary plaque calcification. The relationship between plaque calcification and plaque volume change was reversed after adjustment, underscoring the importance of regression as a tool for adjustment. Finally, it seems that initially, after running the regression models, the authors noted that the residual errors were not normally distributed, so they did a logarithmic transformation of the outcome variable and reran the model.

Logistic regression for dichotomous/binary outcomes

Binary outcomes have only two values such as death or survival. Transforming continuous variables into binary variables is called dichotomization. Although it is a common practice, it is discouraged because it causes information loss.[3] Occasionally, dichotomization might be useful if it has a specific biological meaning, such as dichotomizing glycohemoglobin (HbA1c) into diabetes ($\geq$ 6.5%) and nondiabetes (<6.5%) ranges.

Castro-Dominguez et al. used logistic regression to build a model for predicting in-hospital mortality after PCI.[4] The outcome was binary (died or survived), so the appropriate regression model was logistic. Here the logistic regression was used for the purpose of prediction. The PCI registry that included 706,263 PCIs between 2018 and 2019 was split into a development subset (70%) and a validation subset (30%). Logistic regression achieved excellent results in prediction with a C-index of 0.943.

Poisson regression for discrete outcomes

Discrete outcomes are countable whole numbers that take specific values and are not dividable. For example, the number of patients admitted to the hospital on a particular day might be 10 or 11; however, it cannot be 10.5 because a patient cannot be divided into half.

Bueno et al. evaluated the relationship between the hospital's coronary revascularization rate and 2-year mortality rate in 11,931 non-ST elevation acute coronary syndrome patients.[5] Since the outcome variable, the number of deaths within 2 years postdischarge is discrete; the authors used univariable and multivariable Poisson regression. For instance, the incidence rate ratio for those who had in-hospital cardiac complications versus those who did not have any complications was 2.86 on univariate analysis and 1.39 after adjustment.

Get started

Regression analysis evaluates the relationship between a dependent variable and independent variables. Key types include linear regression for continuous outcomes, logistic regression for binary outcomes, and Poisson regression for discrete outcomes. Regression helps estimate associations, make predictions, and adjust for risk, though they don't indicate causality.

Potential pitfalls

1. **Overfitting**: Creating a model that is too complex and fits the training data too closely, which can reduce its ability to generalize to new data.
2. **Multicollinearity**: When independent variables are highly correlated with each other, it can make it difficult to determine the individual effect of each variable.
3. **Heteroscedasticity**: When the variance of the errors is not constant across all levels of the independent variables, it can affect the reliability of the regression coefficients.
4. **Autocorrelation**: When residuals (errors) are correlated with each other, it can indicate that important variables are missing from the model.

Resources

1. **Omitted Variable Bias**: Failing to include relevant variables in the model can lead to biased and inconsistent estimates.
2. **Sample Size**: Small sample sizes can lead to unreliable estimates and reduce the power of the regression analysis.
3. **Assumption Violations**: Regression models rely on certain assumptions (e.g., linearity, independence, normality of errors). Violating these assumptions can lead to incorrect conclusions.

References

1. HCUP. HCUP NIS Database Documentation. *Healthcare Cost and Utilization Project (HCUP)*. Rockville, MD: Agency for Healthcare Research and Quality; December 2020. www.hcup-us.ahrq.gov. Accessed January 2, 2021.
2. Jin H-Y, Weir-McCall JR, Leipsic JA, et al. The relationship between coronary calcification and the natural history of coronary artery disease. *JACC Cardiovasc Imag*. 2021/01/01;14(1):233–242. https://doi.org/10.1016/j.jcmg.2020.08.036.
3. Fedorov V, Mannino F, Zhang R. Consequences of dichotomization. Article. *Pharmaceut Stat*. 2009;8(1):50–61. https://doi.org/10.1002/pst.331.
4. Castro-Dominguez YS, Wang Y, Minges KE, et al. Predicting in-hospital mortality in patients undergoing percutaneous coronary intervention. *J Am Coll Cardiol*. 2021/07/20;78(3):216–229. https://doi.org/10.1016/j.jacc.2021.04.067.
5. Bueno H, Rossello X, Pocock Stuart J, et al. In-hospital coronary revascularization rates and post-discharge mortality risk in non–ST-segment elevation acute coronary syndrome. *J Am Coll Cardiol*. 2019/09/17;74(11):1454–1461. https://doi.org/10.1016/j.jacc.2019.06.068.

Further reading

1. Kleinbaum DG, Kupper LL, Nizam A, Rosenberg ES. *Applied Regression Analysis and Other Multivariable Methods*. Cengage Learning; 2013.
2. Field A. *Discovering Statistics Using Spss*. 3rd ed. 2009.

Chapter 27

Continuous variable analyses: Student's *t*-test, Mann–Whitney *U* test, Wilcoxon signed-rank test

Badal Thakkar[1,2,3]

[1]*Sinai Hospital of Baltimore, Baltimore, MD, United States;* [2]*George Washington University School of Medicine, Washington, DC, United States;* [3]*Johns Hopkins Medicine, Baltimore, MD, United States*

Key points

- Continuous variables are commonly encountered in clinical research.
- There are various tests that can be employed to analyze continuous variables, most common among them are Student's *t*-test, Mann–Whitney *U* test (also known as the Wilcoxon rank-sum test), and Wilcoxon signed-rank test.
- The choice of which test to employ depends on whether the data is normally distributed or non-normally distributed and if it is paired data or not. There are also certain assumptions that need to be met for each test.

Chapter introduction

In this chapter, we will review some of the common tests used to compare continuous variables between two groups. We will specifically discuss about Student's *t*-test, Mann–Whitney *U* test (also known as the Wilcoxon rank-sum test), and Wilcoxon signed-rank test. The choice of which test to employ depends on whether the data is normally distributed or non-normally distributed and if it is paired data or not. This chapter not only discusses which test to use based on the data, but it also discusses various assumptions that need to be met for a particular test. Further, each of the three tests is explained in simple language with the help of an example.

Comparing continuous variables when data is normally distributed

The *t*-test (also known as Student's *t*-test) is used to determine if there is a statistically significant difference in the means of a continuous variable between two independent groups, assuming the data is normally distributed. Let's review how this test can be used with an example.

Example 1: A researcher is conducting a study to determine if a drug X is more effective in reducing LDL-C levels than a placebo. The researcher recruits 400 patients for the study and randomly assigns them to either receive drug X (treatment group of 200 patients) or receive a placebo (control group of 200 patients) for a period of 3 months. After 3 months, the mean LDL-C levels were reduced by 45 mg/dL in the treatment group, whereas it was reduced by 20 mg/dL in the control group. The researcher would like to determine if the reduction in mean LDL-C levels is significantly different in the treatment group versus the control group. With the help of a statistician or statistical software, the researcher obtains a t-statistic and a *P*-value. If the *P*-value is less than their chosen significant level (e.g., $\alpha = 0.05$), then they are able to conclude that the mean reduction in the LDL-C levels for the treatment group is statistically significant (*P*-value 0 < .05).

As can be seen in the above example, the *t*-test is used commonly in clinical studies given its utility in comparing means of continuous variables like LDL-C

Translational Cardiology. https://doi.org/10.1016/B978-0-323-91790-2.00052-6

levels, blood pressure readings, and heart rate readings between two independent groups. However, the *t*-test can only be used when certain assumptions are met, those are as follows:

- the data is normally distributed (assumption of normality),
- the variances of the two groups being compared are approximately equal (assumption of homogeneity of variances).

If these assumptions are not met, then *t*-test is not an appropriate test, and instead, nonparametric tests should be used; we will discuss those later in the chapter.

Apart from comparing the means of a continuous variables between two independent groups, the *t*-test can also be used in some other scenarios:

- Paired *t*-test: used when we are dealing with paired data such as measuring a parameter before and after an intervention (e.g., blood pressure readings before and after treatment). In this scenario, a paired or dependent samples *t*-test is used, which helps assess if there is a significant difference between the means of the paired measurements.
- One-sample *t*-test: used when we want to determine if the mean of a single sample is significantly different from a known or hypothesized population mean. For example, we might use this test to assess if the mean blood pressure in our patient population differs significantly from a known national average.
- Post hoc *t*-test: in cases with more than two groups, we can use one-way analysis of variance (ANOVA) followed by post hoc *t*-tests to identify which specific groups differ significantly from each other. For example, comparing LDL-C levels among three groups receiving different doses of Drug X.

Comparing continuous variables when data is not normally distributed

The Mann–Whitney *U* test, also known as the Wilcoxon rank-sum test, is a nonparametric statistical test used to compare continuous variables among two independent groups when the data is not normally distributed. Let us explore this test using an example.

Example 2: Let us consider a study that examines the effectiveness of two different cholesterol-lowering drugs, Drug A and Drug B, in a group of patients with hypercholesterolemia. The researchers have collected data on the reduction in total cholesterol levels (measured in mg/dL) in two independent groups:

- Group A: Patients receiving Drug A ($n = 30$).
- Group B: Patients receiving Drug B ($n = 25$).

Unlike Example 1, in this scenario, the data for cholesterol reduction in both groups may not meet the assumption of normality. Therefore, the researchers opt for the Mann–Whitney *U* test to determine if there is a statistically significant difference in the effectiveness of the two drugs.

After performing the Mann–Whitney *U* test using appropriate statistical software, they obtain a U-statistic and a *P*-value. If the *P*-value is less than their chosen significance level (e.g., $\alpha = 0.05$), they conclude that there is a statistically significant difference between the two drugs in terms of their effectiveness in reducing cholesterol levels.

The Mann–Whitney *U* test offers several advantages, particularly when dealing with non-normally distributed data or ordinal variables. Unlike the *t*-test, it does not assume normality or equal variances. However, it does assume that the two groups being compared have similar shapes of distributions, except for possible differences in central tendencies (medians).

Considerations for the Mann–Whitney *U* Test:

i. Suitable for comparing two independent groups.
ii. While it does not assume normality, the test assumes that the shapes of the distributions for the two groups are similar, except for possible differences in central tendencies (medians).
iii. Suitable for both ordinal and continuous data.
iv. Can be applied to various sample sizes, from small to large.

The Wilcoxon signed-rank test is a nonparametric statistical test used to compare paired continuous variables when the data may not meet the assumption of normality. It assesses whether there is a statistically significant difference between two measurements taken on the same subjects, typically before and after an intervention.

Example 3: Consider a study aimed at assessing the impact of a new exercise program on the exercise capacity of cardiac patients. In this research, a group of 20 patients undergoes a 6-week exercise program. Their exercise capacity, measured in metabolic equivalents (METs), is recorded both before and after the intervention. After applying the Wilcoxon signed-rank test using appropriate statistical software, researchers obtain a test statistic (often denoted as W) and a *P*-value. If the *P*-value is less than their chosen significance level (e.g., $\alpha = 0.05$), they conclude that there is a statistically significant improvement in exercise capacity following the program.

Considerations for the Wilcoxon signed-rank test:

i. Used for paired data when the measurements in each pair are dependent or related.
ii. Suitable for analyzing continuous or ordinal data.
iii. The test assumes that the distribution of the differences between paired observations is symmetric around a

central value (usually zero). This means that differences can be positive or negative.

iv. While it can be applied to various sample sizes, it is typically used for relatively small sample sizes (e.g., less than 30 pairs).

In summary, these statistical tests offer valuable tools for analyzing continuous variables in clinical research, whether data follows a normal distribution or not. Choosing the appropriate test depends on the specific characteristics of the data and the research question at hand.

Additional note: If there are more than two groups, then the test used is analysis of variance (ANOVA) for normally distributed data or nonparametric ANOVA for non-normally distributed data. However, ANOVA tests are not covered in this chapter.

Chapter 28

Categorical variable analyses: Chi-square test, Fisher's exact test, Mantel-Haenszel test

Badal Thakkar[1,2,3]
[1]Sinai Hospital of Baltimore, Baltimore, MD, United States; [2]George Washington University School of Medicine, Washington, DC, United States; [3]Johns Hopkins Medicine, Baltimore, MD, United States

Key points

1. Categorical variables are commonly encountered in clinical research.
2. There are various tests that can be employed to analyze categorical variables, the most common among them are: Chi-Square test, Fisher's Exact test, and Mantel-Haenszel test.
3. The choice of which test to employ depends on the sample size and whether there is a confounding third variable present or not.

Introduction

In this chapter, we will review some of the common tests used to compare categorical variables between two groups. We will discuss three tests: chi-square test, Fisher's exact test, and Mantel–Haenszel test. The choice of which test to employ depends on the sample size and whether there is a confounding third variable present or not. This chapter not only discusses which test to be use based on the data, but it also discusses various assumptions that need to be met for a particular test. Further, each of the three tests are explained in simple language with the help of an example.

Chi-square test

The chi-square test is among the most used tests to examine the association between two categorical variables. It utilizes a contingency table to present the distribution of observations across various categories. Let us explore it with the help of an example.

Example 1. A resident physician wants to know if hospitalized COVID-19 patients with diabetes mellitus are at a higher risk of requiring mechanical ventilation from COVID-19-induced respiratory failure compared to those without diabetes mellitus. He collects data on 100 hospitalized COVID-19 patients. He finds out that 40 of those patients eventually required mechanical ventilation. He notices that 30 among those had diabetes, whereas among those who did not require ventilation, 25 had diabetes. He wants to know if having diabetes is significantly associated with a higher risk of requiring mechanical ventilation. He prepares a contingency table.

	Requires mechanical ventilation	
	Yes	No
Patients with diabetes mellitus	30	25
Patients without diabetes mellitus	10	35

The null hypothesis is that there is no association between diabetes mellitus and the need for mechanical ventilation in hospitalized COVID-19 patients. He would need to perform a chi-square test to accept or reject the null hypothesis. He proceeds to perform the chi-square test by calculating the expected frequencies (E) of each cell in the above table. The formula for calculating E is:

$$E = \frac{(\text{Row Total}) \times (\text{Column Total})}{\text{Grand Total}}$$

Translational Cardiology. https://doi.org/10.1016/B978-0-323-91790-2.00072-1

After obtaining the expected frequencies of each cell, the next step is to calculate a chi-square statistic by summing $\frac{(Observed - Expected)^2}{Expected}$ for each cell. With this chi-square statistic, a P-value is obtained, and if it less than a chosen significance level (e.g., $\alpha = 0.05$), he rejects the null hypothesis and concludes that there is a significant association between diabetes mellitus and need for mechanical ventilation in hospitalized COVID-19 patients.

There are certain assumptions that need to be met while conducting a chi-square test. The first assumption is that the observations in the contingency table are independent of each other. The second assumption is that the data is collected through a random sampling process. The chi-square test is usually a robust test when the sample size is large. The expected frequency for each cell in the contingency table should be greater than 5. If the expected frequencies are not greater than 5, then the chi-square test is not reliable, and Fisher's exact test becomes the test of choice in such a situation.

Fisher's exact test

Just like the chi-square test, Fisher's exact test is utilized for analyzing categorical variables, which can be tabulated in a 2 × 2 contingency table. Unlike the chi-square test which is not reliable with a smaller sample size where expected frequencies are not greater than 5, the Fisher's exact test is a reliable alternative in such a situation. Let us explore it with the help of an example.

Example 2. A researcher is conducting a case-control study on experimental drug X. Few participants in the study have a rare side effect. The researcher wants to know if the experimental drug X is significantly associated with this rare side effect. He tabulates the contingency table as below.

	Occurrence of rare side effects	
	Yes	No
Cases (receiving experimental drug X)	3	12
Controls (receiving a placebo)	1	14

The null hypothesis is that there is no significant association between experimental drug X and the rare side effect. As the expected frequencies in some of the cells above are not greater than 5, the researcher is unable to utilize the chi-square test to study the association. Thankfully, the researcher has at his disposal a reliable test for this situation—Fisher's exact test. With the help of statistical software, he is able to run Fisher's exact test and obtain a P-value. If the obtained P-value is less than a chosen significance level (e.g., $\alpha = 0.05$), then the researcher rejects the null hypothesis and concludes that the association between experimental drug X and the rare side effect is significant.

Like the chi-square test, Fisher's exact test assumes independence of observations. It is used for smaller sample sizes where expected frequencies are not greater than 5. For larger sample sizes, the exact calculation of the P-value becomes computationally intensive, so a chi-square test is usually more appropriate in such a situation.

Mantel—Haenszel test

The Mantel—Haenszel test is used to study the association between two categorical variables after controlling the effect of a potential confounding third variable. Let us explore it with the help of an example.

Example 3. A researcher is conducting a study to assess the relationship between smoking and lung cancer. He would like to control for gender as a potential confounding variable. He collects the data and tabulates it in a contingency table stratified by gender.

		Lung cancer	
		Yes	No
Smoking: Yes	Male	14	86
	Female	11	89
Smoking: No	Male	5	95
	Female	6	94

The null hypothesis is that there is no association between smoking and lung cancer after controlling for gender. The Mantel—Haenszel test is conducted using appropriate statistical software to obtain a test statistic and its associated P-value. If the obtained P-value is less than a chosen significance level (e.g., $\alpha = 0.05$), then the researcher rejects the null hypothesis and concludes that smoking is significantly associated with lung cancer even after controlling for gender.

While both the Mantel—Haenszel test and the chi-square test assess associations between categorical variables, the advantage of the Mantel—Haenszel test is its ability to control for a potential confounding variable.

Chapter 29

ANOVA in statistics

Yasar Sattar[1], Dhruvil Ashishkumar Patel[2], Khola Waheed Khan[3], Akram Kawsara[1] and Ramesh Daggubati[1]
[1]Department of Cardiology, West Virginia, Morgantown, MV, United States; [2]Wayne State University, Detroit, MI, United States; [3]Wah Medical College, National University of Medical Sciences, Rawalpindi, Pakistan

Key points

- When the variance of values is compared and contrasted between MULTIPLE GROUPS, these potential differences in the context of clinical research may provide a surrogate for the efficacy of one intervened group compared to another
- The analysis of variance, termed ANOVA, is the general methodology to be used in such instances whereby multiple groups (more than 2) that are affected by independent variables are compared with each other due to having the same primary outcome (dependent variables).

The aim of clinical and translation research is to provide better patient care. To do this effectively, the clinical status of a patient should be understood since the health of a patient is dynamic and interrelated with many variables. Complex set of data points must be analyzed together to determine the overall effect. Drawing conclusion from a single relationship without placing it into the context of the overall participant status can be detrimental to predicting the effects of an intervention. Moreover, as the complexity of a study increases, so does the analysis of the data also increases. Up until now, we have mainly discussed the relationship between two groups to determine if significant differences exist. In other words, we compare an intervention in one group to another group—which determine whether the intervention had an affect or not. Extending this paradigm further, what if there are multiple interventions that need to be tested on the same target? How can this relationship be confirmed to be significant and true?

These forms of analysis are quite easily advantageous. Studies that are able to see the effects of multiple interventions aid in reducing the cost of conducting these studies individually, mitigate any temporal or extraneous variables that may impact comparative results if these studies were done separately, and reduce resource load on analysis and presentation of the findings with the strength of comparing within the same overall study group and investigating team. A very common form of this kind of investigation in clinical and translational research is multidrug testing for a particular pathology. Defining the same marker outcomes and administering multiple separate drugs, each to one group that is usually in a single center as a part of a multicenter trial, during the same study period, can be easily seen to ease the comparisons emerging after the trial ends.

In this paradigm, the question now becomes, in what way can a significant relationship between these multiple interventions be demonstrated? In other words, what is the statistical analysis that can be done which provides evidence that multiple interventions on the same pathology may present a different effect and thereby indicate the superiority or inferiority of an intervention? As compared to analyzing two groups together (prototypically an intervention and a control group), when three or more groups are involved, then the variance of values IN BETWEEN those groups and the variance of values INSIDE each of these groups can be compared and contrasted. Conceptually, the variance is a way of illustrating the distribution of values within a SINGLE GROUP of measurements. Now when the variance of values is compared and contrasted between MULTIPLE GROUPS, these potential differences in the context of clinical research may provide a surrogate for the efficacy of one intervened group compared to another.[1] Analyzing this difference using inferential statistics can provide evidence for a legitimate effect that can be clinically applicable and thereby translationally possible.

Translational Cardiology. https://doi.org/10.1016/B978-0-323-91790-2.00101-5

Analysis of variance

The analysis of variance, termed ANOVA, is the general methodology to be used in such instances whereby multiple groups (more than 2) that are affected by independent variables are compared with each other due to having the same primary outcome (dependent variables). As referred to earlier, the greater the variance between each of the independent variable groups, the larger the possibility that a true difference exists at which point investigators can be sure that there might be a positive or negative effect of an intervention—the information for which comes through taking the data through the ANOVA process.[1]

The question now becomes, what is the ANOVA process, and what must occur to lead to the conclusion that there is an effect of different interventions on the same outcome?

The aim of this process is twofold:

(1) To analyze the difference in variance between groups and within the groups;
(2) To then compare these differences with each other through inferential statistics to confirm or deny there being a significant difference. To observe these properties in a dataset with multiple arms, such as the example of multiple pharmaceutical interventions used for the same disease, an overall test statistic is derived: through ANOVA, this is the F-statistic.

The F-statistics is a ratio between the variance of each of the test arms. It is defined as the "Sum of Squares Between/Degrees of freedom BETWEEN each group" ratio with the variance WITHIN each of the independent arms: defined as the "Sum of Squares Within/Degrees of freedom within each group."

The grand mean and the Sum of Squares Between

The value of the "Sum of Squares Between" defines the variance between each arm of the dataset. It is derived through the process of first calculating the overall mean of the dataset (addition of all values in the dataset/total number of data points), which is also referred to as the grand mean of the dataset. Simultaneously, the mean from the values of each of the independent arms is then also derived. The Sum of Squares Between is derived by first calculating the mean of the values of each of the independent arms. Now the average of each independent arm of the study is representative of the difference between the independent arm mean and the grand mean, and the Sum of Squares Between is calculated by differencing the independent arm from the grand mean and squaring that difference for every single value within the independent arm. This is then done for all values within all arms separately as each arm has its own mean. The accumulation of these values derives the Sum of Squares Within with the degrees of freedom being the total number of independent arms subtracted by 1.

The Sum of Squares Within

The value of the "Sum of Squares Within" is defined as the variance within each of the groups. Essentially taking the value of each unit in an independent arm and finding its difference from the independent arm's mean and squaring that value. This is done for each independent arm separately and the accumulation provides the value of the Sum of Squares Within, with the degrees of freedom being the total number of all individual values subtracted by 1.[2]

With both these values obtained F-statistics can now be derived. The F-statistic is a representation of the variation that there is between each independent arm compared to the variation within each independent arm. In the context of clinical and translational research, the greater the value of the F-statistic, with larger degrees of freedom the higher the Sum of Square Between is relative to the Sum of Squares Within. This inferentially indicates that there do exist marked differences among each of the independent variables arms and as such likely represents a different effect per intervention for the same outcome. What needs to be clearly understood though is that the derived value affirms that a global difference is present within the data but does not indicate that there is an intergroup difference or the magnitude of difference between independent arms compared to each other. There are subsequent analyses that can be done for this process.

Calculating the F-statistics can now be used in inferential statistics to determine if a potential relationship exists. Relationships that are significant through this process can affirm that there is a statistically significant chance that a true difference does exist between intervention arms. Specifically, the fact that there is equality between all the intervention groups (which is the null hypothesis) is not true and thereby there is in fact a difference present (the alternative hypothesis). Yet, this process does not affirm which intervention specifically the most different, as well as if there is a specific value that potentially was not different from another (which would mean the null hypothesis could be true for this subset of groups). Comparing each intervention arm with each other is done through a separate set of subsequent analyses.

ANOVA in clinical and translation studies—The basic principles and caveats

While we have discussed till now, the basic principles of ANOVA, placing ANOVA in the context of clinical and translational studies sturcturingfor such an analysis or determining if an ANOVA is required for a study. Clinical studies can often include a variety of variables that are measured in each intervention arm. This can include bio-clinical status, biomarkers, demographic variables, and even associated temporality. The complexity can drastically increase, especially in studies assessing dynamics of biological agents. Yet, applying ANOVA still determine if there is an overall difference across an aggregate of values.[1–6]

One-way versus two-way ANOVA

When there is only one variable being observed within each intervention group, then the pooled analysis is referred to as a one-way ANOVA. For example, this includes only a single variable—such as a different drug type between each independent arm whose effect on the outcome is compared together.[3]

A two-way ANOVA can be performed when two variables within each study arm are compared with the outcomes (such as the type of drug and a patient's systolic vitals) to the outcome. In theory, multiple independent variables can be included and can be pooled within each study arm for ANOVA analysis to ascertain an effect on the outcome (dependent variable), but this can increase the risk of inferential statistical errors.

Inflation of type I errors

As defined in earlier chapters, a Type I error occurs from rejecting the null hypothesis when it in fact is true, essentially meaning that there was in fact no difference between the groups in the context of ANOVA, but the contrary was instead accepted. In ANOVA, where three or more groups are compared, the likelihood that one of these groups in fact does not have a varying difference with the other will increase as there are more groups to compare. This risk intuitively emerges for each independent arm as the complexity of the datasets increases, such as in clinical research studies where variables do have complexity. Thereby the more groups that are in the ANOVA, the greater the likelihood of a type I error as intragroup differences are not calculated, rather than overall variance is assessed—as noted by the calculation of the F-statistic.[1]

Post hoc analysis

The conclusion that we can derive from the ANOVA has limitations. It only tells us that the means of the groups may differ and that any one group is different. The answer to this, that is, which of the groups differ is provided by the post hoc test. Post hoc tests that can be utilized are Bonferroni's correction, Turkey, Schéffe, and Holm methods[1,4]

Potential pitfalls

- The statistical conclusion arrived at after ANOVA can determine that a difference does exist between each of the independent arms of a study in aggregate, but it does not conclude the magnitude of this difference when comparing specific groups together as well as determining what the implications of this difference can be. Thereby what can only be confirmed is that the values in each group are either closely related to each other or they are not on the scale of the global dataset on which an ANOVA was performed.
- Increasing the number of groups for an ANOVA analysis may elevate the ability to discern a difference but this simultaneously continues to elevate the risk of a type I error as well.

Example 1[3]

Data of the 20 patients.

Sr. No.	1	2	3	4	5	6	7	8	9	10	11	12	13	14	15	16	17	18	19	20
Gender	M	M	M	M	M	M	M	M	M	M	F	F	F	F	F	F	F	F	F	F
Age groups	1	2	3	1	1	3	3	2	1	3	3	2	2	2	3	1	1	3	3	2
BMI	23	25	26	24	23	27	27	24	21	28	26	26	23	24	25	22	19	25	26	25
DBP (B/L)	68	80	85	75	77	82	80	82	80	74	75	86	85	87	83	79	73	77	82	81
DBP (at 30 min)	75	79	90	80	81	89	92	90	80	80	89	88	87	91	88	78	74	84	83	80
DBP (at 60 min)	70	78	85	75	72	72	75	85	80	75	85	80	82	82	85	74	87	86	85	72

Age groups: 1 (<30 years), 2 (30–50 years), 3 (>50 years). *B/L*, baseline; *BMI*, body mass index; *DBP*, diastolic blood pressure; *Gender: M*, male; *F*, female; *min*, minute.

A one-way ANOVA test was used to compare the mean DBP in three age groups (independent variable), which was found statistically significant ($P = .002$). Levene test for homogeneity was insignificant ($P = .231$); as a result, the Bonferroni test was used for multiple comparisons, which showed that DBP was significantly different between two pairs that is, age groups of <30 to 30–50 and <30 to >50 ($P < .05$) but insignificant between one pair, that is, 30–50 to >50 ($P > .05$).

Example 2[5]

Medical researchers want to know if four different medications lead to different mean blood pressure reductions in patients. They randomly assigned 20 patients to use each medication for 1 month, then measured the blood pressure both before and after the patient started using the medication to find the mean blood pressure reduction for each medication.

To understand whether there is a statistically significant difference in the mean blood pressure reduction that results from these medications, researchers can conduct a one-way ANOVA, using "type of medication" as the factor and "blood pressure reduction" as the response.

If the overall *P*-value of the ANOVA is lower than our significance level, then we can conclude that there is a statistically significant difference in mean blood pressure reduction between the four medications. We can then conduct post hoc tests to determine exactly which medications lead to significantly different results.

Example 3[6]

A study is designed to test whether there is a difference in mean daily calcium intake in adults with normal bone density, adults with osteopenia (a low bone density that may lead to osteoporosis), and adults with osteoporosis. Adults 60 years of age with normal bone density, osteopenia, and osteoporosis are selected at random from hospital records and invited to participate in the study. Each participant's daily calcium intake is measured based on reported food intake and supplements. The data are shown below.

Normal bone density	Osteopenia	Osteoporosis
1200	1000	890
1000	1100	650
980	700	1100
900	800	900
750	500	400
800	700	350

Source of variation	Sums of squares (SS)	Degrees of freedom (df)	Mean squares (MS)	F
Between treatments	152,477.7	2	76,238.6	1.395
Error or residual	819,833.3	15	54,655.5	
Total	972,311.0	17		

We do not reject H_0 because $1.395 < 3.68$. We do not have statistically significant evidence at $a = 0.05$ to show that there is a difference in mean calcium intake in patients with normal bone density as compared to osteopenia and osteoporosis. Are the differences in mean calcium intake clinically meaningful? If so, what might account for the lack of statistical significance?

Resources

- Essential Medical Statistics by Betty R. Kirkwood, Jonathan A. C. Sterne

References

1. Kim TK. Understanding one-way ANOVA using conceptual figures. *Korean J Anesthesiol.* 2017;70(1):22–26. https://doi.org/10.4097/kjae.2017.70.1.22.
2. Rouder JN, Schnuerch M, Haaf JM, Morey RD. Principles of model specification in ANOVA designs. *Comput Brain Behav.* 2023;6:50–63.
3. Mishra P, Singh U, Pandey CM, Mishra P, Pandey G. Application of Student's *t*-test, analysis of variance, and covariance. *Ann Card Anaesth.* 2019;22(4):407–411. https://doi.org/10.4103/aca.ACA_94_19.
4. McHugh ML. Multiple comparison analysis testing in ANOVA. *Biochem Med.* 2011;21(3).
5. https://www.statology.org/anova-real-life-examples/.
6. https://sphweb.bumc.bu.edu/otlt/MPH-Modules/BS/BS704_HypothesisTesting-ANOVA/BS704_HypothesisTesting-Anova4.html.

Chapter 30

Correlation

Júlia Lemos Lima Verde[1], Guilherme Leite B. Gonçalves[1], Gabriel Prusch[2], Gustavo Sanchez[2] and Bruno B. Lima[2]

[1]*Federal University of Ceará School of Medicine, Fortaleza, Ceará, Brazil;* [2]*Vanderbilt University School of Medicine, Nashville, TN, United States*

Key points

- The correlation coefficient (r) allows for the interpretation of the scientific concept of correlation, offering an objective alternative for the former subjective judgment of the term, and mitigates the possibility of dubious interpretation between different individuals and research centers.
- The correlation coefficient reproduces the strength of the relationship between two variables on a linear scale from -1 to 1, but it cannot be used as a measure of causality.
- Pearson's correlation coefficient can be used to measure the strength of association between two variables where there is a normal (linear) distribution of data, while Spearman's rank correlation coefficient is frequently used to determine the strength of association between two variables when data is ordinal or when it does not follow a linear pattern.

Introduction

When writing up scientific productions, it is common to use terms such as strong, weak, great, and negligible to describe the association between two variables. However, these terms, when used by themselves, result in a lack of objectivity and reproducibility that is needed for further interpretation of manuscripts in a standardized manner between different research centers and individuals who may interpret the paper. To standardize an international method for assessing the strength of association between variables, the mathematical concept of correlation is used.

In medical and health sciences, correlation is vital for understanding the relationships between different health factors. For instance, correlating lifestyle habits with cardiovascular outcomes or identifying correlations between genetic markers and diseases.

To calculate, interpret, and report correlations correctly, one must be familiar with the concept of correlation, the different correlation coefficients, how and where to apply them, and the pitfalls to avoid while doing so.

Get started

For cardiologists applying correlations in their research, particularly in observational studies, it is essential to design these studies well and obtain institutional review board approval. Utilize available online tools, whether free or paid, to calculate the correlation coefficient. It is recommended to consult with a statistician before interpreting and reporting the correlation coefficient for a more accurate understanding of the results.

Correlation

Correlation is a statistical measure that helps us visualize how two different variables are related to each other. It demonstrates that, as one variable changes in value, the other either has a tendency to vary in a specific direction or to not vary at all in a predictable way.

Therefore, it makes it possible to pinpoint correlations such as rainfall amount and plants growth and mean temperature and ice cream sales, bearing in mind the maxim that correlation does imply causation. Otherwise, we would have to believe that the divorce rate in Maine and the per capita consumption of margarine had something to do with each other, just because of its coefficient of correlation (r) of 0.9925.

Every correlation has a sign and a form, the sign could be positive, negative, or neutral, which allows the evidence of strength and the direction in which these relations occur. This mathematical entity is represented by values that range from -1 to $+1$.

Translational Cardiology. https://doi.org/10.1016/B978-0-323-91790-2.00030-7

- Positive ($0 < x \leq 1$): The two variables grow simultaneously in the same direction (either both increase or decrease).
- Neutral: No relationship is identified between the variables.
- Negative ($-1 \leq X < 0$): The two variables grow in opposite directions (when one increases, the other decreases, and vice versa)

Concerning the form of a correlation, it can be monotonic (linear or not) and nonmonotonic:

Monotonic correlation: The variables tend to move in the same relative direction, but are not considered linear unless they move at a constant rate. Therefore, all linear correlations are monotonic but the opposite is not always true.

Nonmonotonic correlation: A correlation is nonmonotonic when two variables do not change at a constant rate and do not always trend in the same direction. In these cases, the relationship between the variables does not graph as a straight line, but as a curved pattern, such as parabola or hyperbola.

Pearson's correlation coefficient

This coefficient is broadly employed when the variables have a normal, linear distribution, due to its value being highly prone to disruption caused by outliers.

When applied to a sample, the calculation for Pearson's correlation coefficient is as follows:

$$r = \left[\Sigma\left(\mathrm{Xi} - \overline{\mathrm{X}}\right)\left(\mathrm{Yi} - \overline{\mathrm{Y}}\right)\right] / \left\{\sqrt{\left[\Sigma\left(\mathrm{Xi} - \overline{\mathrm{X}}\right)^{2}\right]} \sqrt{\left[\Sigma\left(\mathrm{Yi} - \overline{\mathrm{Y}}\right)^{2}\right]}\right\}$$

where:

- r is the Pearson correlation coefficient.
- Xi and Yi are the individual data points for the two variables.
- $\overline{X}$ and $\overline{Y}$ and are the means (averages) of the respective variables. are the means (averages) of the respective variables.
- Σ represents the sum of the calculations over all data points.

This formula calculates the correlation coefficient, denoted as "*r*" in the equation, by calculating the covariance between the two variables and then dividing it by the product of the standard deviations of the two variables. The result, r, which ranges from -1 to $+1$, can be interpreted considering that -1 value indicates a perfect negative linear relationship, $+1$ indicates a perfect positive linear relationship, and 0 indicates no linear relationship between the two variables.

Spearman's rank correlation coefficient

The Spearman's rank correlation coefficient becomes highly useful in cases when data is still monotonic, but nonparametric, meaning that one of the variables is ordinal or not normally distributed, or even when there are outlier values. Despite being calculated from the ranks of the data, and not from their actual values, it is less sensitive to extreme values.

The formula for calculating Spearman's rank correlation coefficient, often denoted as "ρ" (rho), is used to assess the strength and direction of a monotonic relationship (which is a nonlinear but consistent relationship) between two variables.

When applied to a sample, the calculation for Spearman's rank correlation coefficient is as follows:

$$\rho = 1 - \left\{\left(6\Sigma d^{2}\right) \ / \ \left[n\left(n^{2} - 1\right)\right]\right\}$$

where:

- ρ (rho) is Spearman's rank correlation coefficient.
- Σd^2 is the sum of the squared differences between the ranks of corresponding data points in the two variables.
- n is the number of data points.

The formula determines the strength and direction of the relationship between the two variables based on the differences in their ranks. The result ρ can range from -1 to $+1$, where -1 indicates a perfect inverse relationship, $+1$ indicates a perfect direct relationship, and 0 means no monotonic relationship between ranks.

Other notable correlation coefficients

- Kendall's Tau (τ): Similar to Spearman's rank correlation, Kendall's Tau is a nonparametric measure of correlation based on the ranks of the data. It is usually used when the same rank is repeated too many times in a small dataset, which according to some authors makes it the most accurate when analyzing data for population.
- Point-biserial correlation coefficient: This coefficient is used when one variable is dichotomous (e.g., binary) and the other variable is continuous. It is a special case of Pearson's correlation coefficient.
- Phi coefficient (φ): The Phi coefficient is used for assessing the association between two dichotomous

variables. It is essentially also a special case of the Pearson correlation coefficient.

Correlation interpretation

Table interpretation of the strength of the relationship between variables based on correlation coefficients[1]

Correlation value	Interpretation
$0.7 < \|r\| < 1$:	Moderate to strong (positive or negative) correlation
$0.3 < \|r\| < .7$	Weak to moderate (positive or negative) correlation
$\|r\| < .3$	Weak (positive or negative) correlation
$r = 0$	Absence of linear correlation

Potential pitfalls

- **Correlating facts that are derived from each other**
  It is inadequate to infer that there is a correlation between variables that are derived from each other, since unavoidably one changes when the other changes. If we consider the income of doctors in their first year after training and the average income of doctors over all their years of work, the average income of the first year makes up the overall average, which would create a bias in the study.
- **Correlation is different from causation**
  Oftentimes, the term correlation can give the impression that there is a cause and consequence relationship due to the strong correlation that exists (a high value of $|r|$). However, correlations are established by mathematical models that seek to visualize the direction of behavior of each variable, requiring controlled and more detailed experiments to prove causal relationships.
- **Ignoring a *P* value > .05 just because of an r close to |1|**
  The *P* value focuses on the statistical relevance of the null hypothesis, while r describes the strength and direction between two variables, without evaluating the statistical significance of this relationship. Thus, even if there is a correlation coefficient equal to -1 or $+1$, if there is a $P > .05$, that is, the null hypothesis cannot be rejected, the correlation should be discarded due to its statistical insignificance.
- **Qualify a correlation without adequate caption and parametrization of "r"**
  Many studies use terms such as "strong correlation" without establishing the value of r or tabulating what each range of r values represents. This often leads to studies being interpreted inappropriately, valuing correlations that are often insignificant.
- **Not contextualizing**
  It is extremely important not to forget the context in which that correlation coefficient was established. In 1897, Karl Pearson described the concept of "spurious correlation," where he described that it is possible to obtain extremely significant r values between two extremely random variables. For instance, if we took the number of measles cases in the United States each year and the marriage rate, we would find that these data are correlated, but this does not mean that reducing measles cases would reduce the number of marriages because these two variables are independent. Modern medicine has managed to progressively reduce the number of measles cases and, coincidentally, people are getting married less these days for a number of other reasons.

Real-world examples

1. Sleep quality, duration, and consistency are associated with better academic performance in college students
   Kana Okano et al.[2] sought to analyze the association between sleep and academic performance with more objective data. Activity trackers were distributed among 100 students in an introductory chemistry class (88 completed the study), which made it possible to evaluate sleep variables and performance in tests applied throughout the study.
   A negative association was found between the average bedtime and the test scores ($r(86) = -0.45$, $P < .0001$), so an earlier bedtime was associated with higher average test scores.
   In addition, it was observed that an average wake-up time had a negative association with the average scores $(86) = -0.35$, $P < .001$) so that an earlier average wake-up time was associated with a higher overall score.
   Sleep inconsistency was defined based on the standard deviation, so those who obtained a higher standard deviation, that is, less homogeneity between the number of hours slept each day, had greater sleep inconsistency. Thus, it was noted that greater sleep inconsistency was correlated with lower average grades ($r(86) = -0.36$, $P < .001$).
2. Negative correlation between bilirubin levels and the severity of atherosclerosis
   Nancy Gupta et al.[3] presented a literature review to explore the correlation between bilirubin levels and the risk of coronary artery disease (CAD). Traditionally, bilirubin has been understood as a cytotoxic element, however, it has now been linked by several studies to

its antiinflammatory, antioxidant, and vasodilator potential, properties useful for protecting against CAD. In 11 studies, a negative correlation was observed between bilirubin levels and the severity of atherosclerosis, a key element in the consolidation of CAD (spearman rank coefficient of $r = -0.31$; $P < .0001$).

3. Correlation of duodenal histology with tissue transglutaminase and endomysial antibody levels in pediatric celiac disease
Matthew R Donaldson et al.[4] aimed to analyze the proportions of IgA antibodies against tissue transglutaminase (tTg-IgA) and endomysium (EMA) and the classifications of histological findings in pediatric patients with Crohn's disease. The study analyzed 117 patients with serological tests for IgA antibodies against tissue transglutaminase (tTg-IgA) and endomysium (EMA) and duodenal biopsies graded according to the modified Marsh classification.
tTg-IgA ($r = 0.704$, $P < .001$) and IgA EMA ($r = 0.740$, $P < .001$) were shown to correlate with intestinal villous atrophy in pediatric CD patients using Spearman's rank correlation coefficient.
tTg-IgA titers >100 units or EMA >1:1280 were highly specific (>98%) for Marsh 3a lesions or greater. tTg-IgA antibody levels >100 or EMA >1:1280 were almost always associated with CD histopathology, which showed that these titers may be sufficient for diagnosing Crohn's disease in this population.

Resources

1. Aggarwal R, Ranganathan P. Common pitfalls in statistical analysis: the use of correlation techniques. *Perspectives in Clinical Research*. 2016;7(4):187–190. https://doi.org/10.4103/2229-3485.192046

2. Axmedovna MM, Muminjonovich TS, Akramaliyevich AD. Correlation coefficient as a mathematical solution of economic issues. *International Journal of Research in Commerce, it, Engineering and Social Sciences*. 2022;16(06):72–75. ISSN: 2349–7793 Impact Factor: 6.876.

3. Li G, Zhang A, Zhang Q, Wu D, Zhan C. Pearson correlation coefficient-based performance enhancement of broad learning system for stock price prediction. *IEEE Transactions on Circuits and Systems II: Express Briefs*. 2022;69(5):2413–2417.

4. Ali Abd Al-Hameed K. Spearman's correlation coefficient in statistical analysis. *International Journal of Nonlinear Analysis and Applications*. 2022;13(1): 3249–3255.

References

1. Witte RS, Witte JS. *Statistics*. 11th ed. Wiley; 2017.
2. Okano K, Kaczmarzyk JR, Dave N, Gabrieli JD, Grossman JC. Sleep quality, duration, and consistency are associated with better academic performance in college students. *NPJ Sci Learn*. 2019;4(1):16.
3. Gupta N, Singh T, Chaudhary R, et al. Bilirubin in coronary artery disease: cytotoxic or protective? *World J Gastrointest Pharmacol Therapeut*. 2016;7(4):469.
4. Donaldson MR, Firth SD, Wimpee H, et al. Correlation of duodenal histology with tissue transglutaminase and endomysial antibody levels in pediatric celiac disease. *Clin Gastroenterol Hepatol*. 2007;5(5):567–573.

Chapter 31

Statistical bias

Karthik Gonuguntla[1], Ayesha Shaik[2] and Pratik Agrawal[1]
[1]*Division of Cardiology, WVU Heart & Vascular Institute, Morgantown, WV, United States;* [2]*Division of Cardiology, Hartford Hospital, Hartford, CT, United States*

Selection bias

Selection bias occurs when a specific study population is included more often in the study leading to distorted data that might not represent the target population. It can occur at any phase of the study: at the initial stage of patient enrollment when a study is designed, or at randomization where patients with the disease might be enrolled in the diagnostic or treatment procedure.[1,2] Examples of selection bias include Attrition bias, Neyman bias, Berkson bias, nonresponse bias, volunteer bias, and late look bias.

- *Attrition bias* occurs when the participants in the study are lost to follow-up. Naturally, the sample lost to follow-up can have a different prognosis compared to those who complete the study.
- *Neyman bias* aka *selective survival bias* occurs when the exposure/risk factor is associated with mortality of the disease. In a sample of patients with chronic obstructive pulmonary disease (COPD) and study its relation to smoking, smokers with COPD might die sooner leading to a patient population with less prevalence of smoking as smoking is associated with mortality of the disease.
- *Berkson bias* occurs when the sample does not include the general population but a subset of it. For example, cases and controls in a hospitalized patient sample differ in exposures and have a higher likelihood of illness compared to a general population.
- *Nonresponse bias* occurs when participants who respond have less severe illness compared to those who respond. For example, in a study looking at the association between smoking and myocardial infarction (MI), nonresponse could be higher among MI patients who smoke.
- *Volunteer bias* occurs when participants who volunteer are significantly different from those who do not volunteer.
- *Late look bias* occurs when patients either die early due to the severity of the illness or are cured from the mild illness before they participate in the study.

Strategies to avoid selection bias include randomization, inclusion of patients within different settings such as inpatient, intensive care, and outpatient—and designing a longitudinal study rather than a cross-sectional study.[3] For example, *Attrition bias can be overcome by including more participants to compensate for potential withdrawal during the study period or using Intention to treat analysis where in all participant are included in the final analysis irrespective of them being noncompliant or lost to follow up.*

Recall bias

Recall bias occurs when participants do not recall events accurately, that is, participants with disease outcomes of interest report exposures to risk factors differently than participants without the disease. For example, patients with COPD or cancer most likely remember that they smoke. One way to avoid recall bias is to use hospital records rather than rely on patient's report and/or to decrease the time from exposure to follow-up. Measurement bias leads to information obtained, which is not accurate. For example, the *Hawthorne effect* where patients change their behavior when they are aware of being studied.[4] Methods to avoid it would be to use standardized data collection methods and use placebo groups.

Translational Cardiology. https://doi.org/10.1016/B978-0-323-91790-2.00039-3

Confounding bias

Confounding bias occurs when a variable is associated with both the exposure and outcome. For example, a study can show an association between coffee intake and lung cancer. However, it could be due to the fact that patients who smoke tend to drink more coffee and their lung cancer could be secondary to smoking. It can be prevented by matching, cross-over studies or randomization to ensure similar characteristics between cases and controls, and restriction of enrollment (e.g., If smoking is a confounding variable, a study can be conducted on either smokers or nonsmokers as the study population). It is important to keep in mind that effect modification is not a bias; it occurs when the third variable is related to the exposure and outcome. For example, smoking can lead to cancer, and smoking is frequently related to alcohol use. Both smoking and alcohol are risk factors for cancer.

Lead time bias

Lead time bias occurs when the screened group has a longer duration of disease compared to the nonscreened group because of early detection. However, the overall health outcome is the same in both the groups but early diagnosis gives a false sense of prolonging survival. This happens when the disease is diagnosed at a latent stage, which leads to a longer duration but in reality, there is no change in the duration of illness. For example, a cancer screening test might seem to prolong survival, but it only helps to detect cancer at an earlier stage. This can be avoided by getting an estimate of the lead time and comparing survival in the two groups [(screened group) versus (unscreened groups plus estimated lead time)]. For example, lung cancer screening in two patients A and B. Patient A was screened at age 70 without any obvious symptoms and patient B had symptoms at age 75. As we followed their clinical course both patient's A and B died at age 80. This gives us a lead time bias of 5 years between the two groups where it appears that screening leads to longer survival. To avoid the bias in survival time, add estimated lead time to patient B survival time, that is, patient A $[80 - 70 = 10]$ versus patient B $[(80 - 75) + 5 = 10]$. This shows disease onset was similar in both groups with no difference in survival time. Another solution would be to compare the reduction in the mortality between the two groups rather than survival time, which is not subject to lead time bias.

Other biases

- Design bias is when the control group is significantly different from the intervention group. It can be avoided by randomization or blinding.
- Surveillance bias occurs when certain outcomes are detected in patients who are more closely monitored.
- Accumulation effect occurs when patients must be exposed to a risk factor for a long period to develop an event, for example, sun exposure for a long period to develop skin cancer. It can be prevented by following the patients for as long as possible.

References

1. Labos C, Thanassoulis G. Selection bias in cardiology research: another thing to worry about (and how to correct for it). *Can J Cardiol*. 2018;34(6):705–708. https://doi.org/10.1016/J.CJCA.2018.03.010.
2. Hernán MA, Hernández-Díaz S, Robins JM. A structural approach to selection bias. *Epidemiology*. 2004;15(5):615–625. https://doi.org/10.1097/01.EDE.0000135174.63482.43.
3. Sj D, Dg A. Statistics notes: blinding in clinical trials and other studies. *BMJ*. 2000;321(7259). https://doi.org/10.1136/BMJ.321.7259.504.
4. Wickstrom G, Bendix T. The "Hawthorne effect" - what did the original Hawthorne studies actually show? *Scand J Work Environ Health*. 2000;26(4):363–367. https://doi.org/10.5271/sjweh.555.

Chapter 32

Basic science statistics

Anas A. Alharbi
West Virginia University, Morgantown, WV, United States

Introduction

Statistics is the science that deals with developing methods for collecting as well as interpreting, analyzing, and presenting data. In this chapter, we will go over the basic principles of statistics that are used in basic science including types of data, descriptive and inferential statistics, central tendency, and dispersion.

Basic concepts

Types of data

There are two main types of data in statistics: qualitative and quantitative data. Qualitative data can be subdivided into nominal and ordinal data. Data that is used to label variables and does not provide any quantitative values to them is called nominal data, for example, gender (male, female) or race (Whites, Blacks, etc.) are nominal data. Ordinal data on the other hand is used to describe variables that are occurring in naturally ordered categories, for example, level of education (primary, junior high, high school, and college degree). Quantitative data can be also subdivided into discrete and continuous data. Discrete data is the data that has a finite value like the number of people in a classroom, whereas continuous data is the data that has an infinite number of values like length or weight.

Descriptive statistics

Descriptive statistics are means or methods to describe or summarize a given data set. Examples of descriptive statistics are measures of central tendency like the mean or median or measures of variability like standard deviation or the variance.

Inferential statistics

The main objective of inferential statistics is to assess the generalizability of the sample data to the population of interest. Many types of research use models like the generalized linear model, which includes a *t*-test or analysis of variance among other tests in an attempt to answer this question.

Central tendency and dispersion

Measures of central tendency

Central tendency is usually described using one of the three most used measures of central tendency and these are the mean, median, and mode with the main objective of identifying the central position within a set of observations. The most used measure is the arithmetic mean or simply mean (sometimes also described as the average) to distinguish it from the different types of means like the weighted mean and geometric mean among other types (outside the scope of this book). It is computed by adding all the values (or observations) and dividing by the number of observations. The mean is an excellent measure for a roughly symmetrical set of values. The median on the other hand is defined as the middle value when the values are ranked in ascending or descending order, for odd samples, the median is the central value in an orderly ranked set of values, and in an even sample, it is the average of the two central values. Unlike the mean, the median is not influenced by outliers and thus is a useful measure of central tendency when the data is skewed. Lastly, the mode is defined as the most occurring value among a set of values and is useful in ordinal and categorical variables.

$$\text{Mean} = \overline{\text{X}} = S\text{x}/n$$

where $S\text{x}$ refers to the summation of individual values and n to the number of values.

Translational Cardiology. https://doi.org/10.1016/B978-0-323-91790-2.00080-0

Measures of dispersion

Measures of dispersion attempt to describe the spread of the data (or values). The most used measure of dispersion is the standard deviation (SD) but also variance, interquartile range, and the range can be used. Standard deviation is the statistic that measures the spread of a set of values relative to its mean and can be calculated by finding the square root of the variance.

$$SD = s = \sqrt{S(c_i - m)^2 / N}$$

where σ is the standard deviation, N is the size of the population, χi is each value from the population, and μ is the population mean. $\sigma 2$ is usually used to describe the variance and it estimates the average of the squared differences from the mean. The simplest measure of dispersion is the range, which is basically the difference between the highest and the lowest values and can be significantly influenced by outliers. Lastly, the interquartile range (sometimes called the midspread) is used to describe the values that lie between the 25th and 75th percentile when the values are ordered from lowest to highest and are least influenced by outliers.

Hypothesis testing

Hypothesis testing is a method used to make a possible conclusion based on two different and usually conflicting hypotheses. These two conflicting hypotheses are usually referred to as the null hypothesis and the alternative hypothesis. The null hypothesis in statistics is the hypothesis that assumes no statistical significance exists between the groups, populations, or observations of interest. Whereas the alternative hypothesis is the one we conclude when we are able to reject the null hypothesis and it assumes that there is a statistical difference between the groups, populations, or observations of interest. In basic science statistics, researchers test their observations against the null hypothesis to make a conclusion. If they are able to reject the null hypothesis, the alternative hypothesis will be concluded and if they fail to reject the null hypothesis, then the null hypothesis will be concluded. In doing so, the null hypothesis might be rejected when it is in fact true, and this is called a type I error or false positive. Also, failure to reject the null hypothesis when it is in fact false results in type II error or false negative. The probability of committing a type I error is called the significance level or the size of the test and is referred to as α and (1-α) is the specificity of the test. On the other hand, the probability of committing type II error is referred to as b and (1-b) is the power of the test (Table 32.1).

Based on what was mentioned earlier we can define the following concepts. Specificity (Sp) in statistics is the true negative rate where Sp = true negative/true negative + false positive. Sensitivity (Sn) is defined as the true positive rate where Sn = true positive/true positive + false negative. In other words, specificity of a test is defined as the proportion of those who test negative and do not have the disease while sensitivity of the test is defined as the proportion of those who test positive and have the disease. Both sensitivity and specificity are characteristics of the tests and are not influenced by the prevalence of a condition, unlike positive and negative predictive values which can be influenced by the prevalence of a condition and are defined as the proportions of positive and negative results that are true positive and true negative, respectively.

Another important concept that we need to define is the probability level (a level) or *P*-value which defines when a sample result is strong enough to reject the null hypothesis and it expresses the probability that the outcome is due to a random chance and in most cases, a *P*-value of less than .05 (sometimes 0.01 is used) is considered an acceptable level to reject the null hypothesis (in this example it means that the probability that the observed results are due to chance alone is less than 0.05). Lastly, the power of the test is affected by four main factors and the most important one of them is the sample size as the power increases as the sample size increases, the power also increases if we increase the a level (changing the probability level from .01 to .05 for example). Also, the power of the test is increased if the variability of the observations is decreased or the difference in outcomes increases.[1–3]

TABLE 32.1 Statistical outcomes in hypothesis testing.

Decision about the null hypothesis	Null hypothesis		
		True	False
	Not rejected	Correct Probability = 1-a	Type II error Probability = b
	Rejected	Type I error Probability = a	Correct Probability = 1-b

References

1. Lane DM, Scott D, Hebl M, Guerra R, Osherson D, Zimmer H. *Introduction to Statistics: An Interactive E-Book*. University of Houston; 2013.
2. Trochim WM, Donnelly JP. *Research Methods Knowledge Base*. vol. 2. New York: Atomic Dog Pub: Macmillan Publishing Company; 2001.
3. Ross SM. *Introductory Statistics*. Academic Press; 2017.

Chapter 33

Sample size

Luka Petrovic
Mount Sinai Morningside Hospital, New York, NY, United States

Key points

- Adequate estimation of sample size is critical in planning any research study and assures sufficient power to detect the desired effect for a specified significance level.
- Power analysis is the most common approach in sample size estimation.
- Collaboration with a statistician is essential for successful sample size estimation.

Get started

Understanding the concept of sample size and its practical application is essential in conducting virtually any biomedical research and in understanding scientific literature. In general, research usually aims to draw inferences about the characteristics of a population by examining a sample drawn from the population of interest. One of the main reasons for such logistical workflow is that sample studies impose much less economic and time costs compared to population studies. For the inference to be successful the selected sample needs to be representative of the population that the researcher wishes to examine. Important inference issues can arise due to a sample being too large as well as with the samples that are too small.[1] A very large sample size enhances the power of a statistical test and can demonstrate that there is a statistical significance even for very small differences with questionable clinical significance. A sample size that is too small can hinder the ability of a statistical test to detect a difference, when in fact there is a meaningful difference that can be demonstrated with a larger sample size. Many considerations are considered to determine what is the optimal sample size for a particular study, but one of the most important ones is the statistical test that is being used, the precision of the measures, and the design of the study. Sample size calculation is an unavoidable step in the planning phase of any research study. It assures that the study will have the power to detect the desired effect for a specified significance level. It provides economic and practical insights into planning a research study and it is important from an ethical standpoint to minimize the number of subjects needed for a particular research question.

Sample size calculation

Sample size calculation usually happens in the planning phase of the research study and/or while transitioning into the step of designing the study.[2] During the planning stage of a project, two particularly important questions need to be answered. First, what is the number of study subjects necessary to have a certain desired power to detect a minimal clinically meaningful difference (i.e., sample size estimation/determination)? Second, if the researcher has limitations in the number of study subjects available (e.g., budget constraints, other medical considerations, etc.), how will that affect the power to detect the minimal clinically meaningful difference (i.e., sample size justification).[3,4]

Common approaches to estimating a sample size are *precision analysis* and *power analysis*.[5] Precision analysis refers to an approach where a sample size is estimated in a way that there is the desired precision at a fixed confidence level (i.e., fixed type I error). In other words, in precision analysis, we estimate the sample size needed in a study to ensure that the confidence interval width of a desired measured outcome is within a prespecified limit. This approach is usually used in studies where confidence interval is the method of statistical inference, instead of hypothesis testing. Power analysis is the most widely used approach for sample size estimation in studies where hypothesis testing is the method of statistical inference. The main principle of power analysis is to select the required

Translational Cardiology. https://doi.org/10.1016/B978-0-323-91790-2.00056-3

sample size for achieving the desired power for detecting a clinically or scientifically meaningful difference at a fixed type I error rate.

To perform a valid sample size calculation, the researcher must select an appropriate statistical test for the hypothesis of interest that accurately reflects the study objectives under a valid study design. This means that the critical step before sample size calculation is to clearly state the hypothesis of interest and the null hypothesis. This will facilitate every next step in the process—including valid study design and selection of appropriate statistical tests for the hypothesis of interest and later making sure that sample size is determined based on a primary endpoint of a study and a clinically meaningful difference of the primary endpoint that the study is intended to detect.

To put this into perspective, the following is a brief overview of the basic principles of hypothesis formulation, study design, choice of statistical test, the primary endpoint, choice of clinically meaningful difference, and how these affect the sample size calculation.

Study design

To achieve the objective of a research study, different study designs could be used. As we already mentioned, it is necessary to choose a valid study design that will lead to appropriate data collection needed to answer the hypothesis of interest. Common examples in clinical trials include parallel-group design or crossover design, while some of the usual examples of design in observational studies are case-control, cohort, or cross-sectional.[3] Sample size calculation is affected by the design strategy because statistical methods or tests are derived under the hypotheses and study design.[5]

Hypotheses

The standard sequence of testing the hypothesis of interest involves drawing a random sample from the population of interest, then a statistical test is used to determine whether the null hypothesis will be rejected at some prespecified significance level.[6] Depending on the test result, a certain conclusion about the hypothesis of interest can be made. In hypothesis testing research, the focus of interest can be to test the population mean or proportion to a predefined value, a difference of two or multiple groups' means, equality of variances, significance of correlation coefficient and regression coefficient, etc.

A common approach for clinical research is to have a point hypothesis that will test for equality or an interval hypothesis that will evaluate noninferiority or superiority.[3]

For equality, one can formulate the hypothesis as follows:

$$H_0: \mu_e = \mu_c \text{ versus } H_a: \mu_e \neq \mu_c$$

where μ_e and μ_c are the mean response of the outcome variable for the experimental group and the control group, respectively. In this formulation of the hypothesis, the rejection of the null hypothesis demonstrates that there is a statistically significant difference.

For superiority, hypothesis formulation can be presented as:

$$H_0: \mu_e - \mu_c \leq \delta \text{ versus } H_a: \mu_e - \mu_c > \delta$$

μ_c is the mean for the control group and μ_e is the mean for the experimental group, while δ represents the difference in clinical importance. In this type of hypothesis formulation, if the null hypothesis is rejected, the difference between the mean response of the outcome variable in the experimental group and the mean response of the outcome variable in the control group is larger (i.e., superior) than the minimal difference of clinical importance. This same concept can be used for the formulation of a noninferiority hypothesis, which would be the following:

$$H_0: \mu_e - \mu_c \geq \delta \text{ versus } H_a: \mu_e - \mu_c < \delta$$

For hypothesis as a test for equivalence, we can formulate:

$$H_0: |\mu_e - \mu_c| \geq \delta \text{ versus } H_a: |\mu_e - \mu_c| < \delta$$

Each one of these different formulations of hypotheses will have different requirements for the sample size to achieve the desired power or precision for the corresponding test.

Primary study endpoint

The influence of the choice of primary study endpoint on sample size is most evident in clinical trials. A primary clinical endpoint is a study variable that is considered by the researcher to be the most relevant to represent the objective of the study. There can be multiple primary endpoints in a study, and usually, for an endpoint to be primary, certain criteria should be satisfied: (1) biological and/or clinical importance; (2) form the basis of the objective of the trial; (3) not highly correlated; (4) should have sufficient power for the statistical hypotheses formulated from the objectives of the trial; (5) should be relatively few.[3]

It is recommended that sample size calculation should be done based on a single primary endpoint.[3] The reason for such a recommendation is if the sample size calculation is based on detecting a difference in multiple or all primary endpoints, this may result in false-positive or false-negative results. Also, note that sample size may be affected by the necessity of achieving sufficient power for a particular chosen primary endpoint.

Effect size and clinically meaningful difference

The choice of a clinically meaningful difference may affect the sample size calculation and alter the conclusion of clinical results. The determination of clinically meaningful differences (usually denoted as δ) is most important in clinical trials of equivalence and noninferiority. In equivalence trials, δ is usually referred to as the equivalence limit, while in noninferiority trials, δ is referred to as the noninferiority margin. From a practical aspect, there are no golden rules for determining the clinically meaningful difference in clinical trials. Some recommendations suggest that the noninferiority margin should not be greater than the smallest effect size that the tested intervention would be reliably expected to have compared with the control. At the same time, recommendations do allow that a combination of statistical calculation and clinical judgment should be the basis for the determination of δ. Certain regulatory bodies provide recommendations for choosing the clinically meaningful difference in particular study designs.[7] In practice, determination of effect size is performed by either using information from prior studies that examined a similar research question or if such information is unavailable, conducting a smaller pilot study first by the researcher. If none of the prior possibilities are available, researchers sometimes use Cohen guidelines[8] for specifying the effect size when determining sample size. These guidelines define small (0.2), medium (0.5), and large (0.8) effects, but with the important caveat that these conventional recommendations may have different significance in different types of research. For example, in biological studies, a more conservative approach is considered, and the definition of small, medium, and large effect sizes is 0.5, 1, and 1.5, respectively.

Power analysis

As mentioned before, the sample size may be determined by precision analysis and power analysis. A precision analysis is performed by controlling type I error (or confidence level), while power analysis is performed by controlling type II error (or power). Since a type I error is usually considered more important, the most common approach in hypothesis testing is to choose a fixed alpha level that is acceptable and then try to minimize beta by determining the optimal sample size. This concept for determining sample size is referred to as power analysis for sample size determination. General.

When an investigator wants to perform a power analysis, specific information needs to be specified. First, a significance level needs to be selected. This means what will be the tolerable probability of concluding that a difference exists when in fact there is no real difference. The most common selection for alpha level is 0.05. Second, the investigator specifies what is the desired power. This means to specify what is the chance of correctly detecting a difference when that difference truly exists. Usually, this value is chosen to be at least 0.8. Next, the researcher specifies clinically meaningful differences (also referred to as minimal detectable difference or hypothetical size of the effect desired to achieve—δ). Finally, it is necessary to have the knowledge of the standard deviation (σ) of the measured outcome that will be used in the study. Depending on how precisely the standard deviation is measured, it will impact the sample size necessary to detect a given difference. Precise measurement (i.e., smaller σ) will allow for detection of a given difference with a smaller sample size compared to measurement with lower precision (i.e., larger σ).

Statistical power can be defined as the probability of observing a statistically significant effect, given sample size, alpha level (i.e., probability of type I error), and a hypothetical effect size of interest. The definition illustrates that if we know three out of four mentioned elements—we can calculate the remaining one. In other words, power is directly dependent on type I and type II errors, the estimated difference between groups that should be shown (estimated effect size), the variance of the data (assumed standard deviation), and sample size.

Type I and Type II errors represent statistical errors that can occur in the hypothesis testing process. **Type I error** is the probability of rejecting the null hypothesis when, in fact, it is true. This probability is represented by α. Keeping this in mind, we can define the term confidence level (1-α). A confidence level is the probability of not rejecting the true null hypothesis (i.e., probability that one will correctly identify that there is no effect to be detected, in the case where a true effect really does not exist). Since implications of Type I error are considered more deleterious than Type II error, the usual alpha level that is fixed in research studies is at least 0.05. This means that there is only a 5% chance that the researcher will detect an effect, when in fact one does not exist.

Type II error is the probability of not rejecting the null hypothesis when it, in fact, is false. Type II error is represented by β. The power of a test is the probability of correctly not rejecting the null hypothesis (i.e., probability of detecting an effect, in the case there is a true effect present to detect). When it comes to the sample size estimation, one must choose which power level will be sufficient for the research question that will be examined. Most commonly in contemporary medical research, researchers choose to have power no less than 0.8, meaning that there is at least an 80% chance that the true effect will be identified if such does exist. Fig. 33.1 demonstrates the possible outcomes of hypothesis testing including Type I and II errors.

It is important to understand the relationship between α (type I error), β (type II error), and power. Fig. 33.2 depicts

	Null hypothesis should not be rejected	Null hypothesis should be rejected
Reject null hypothesis	"False positive" = Type I error (α)	True negative = power (1-β)
Do not reject null hypothesis	True positive = confidence level (1-α)	"False negative" = Type II error (β)

FIGURE 33.1 Possible outcomes of hypothesis testing and potential errors associated with it.

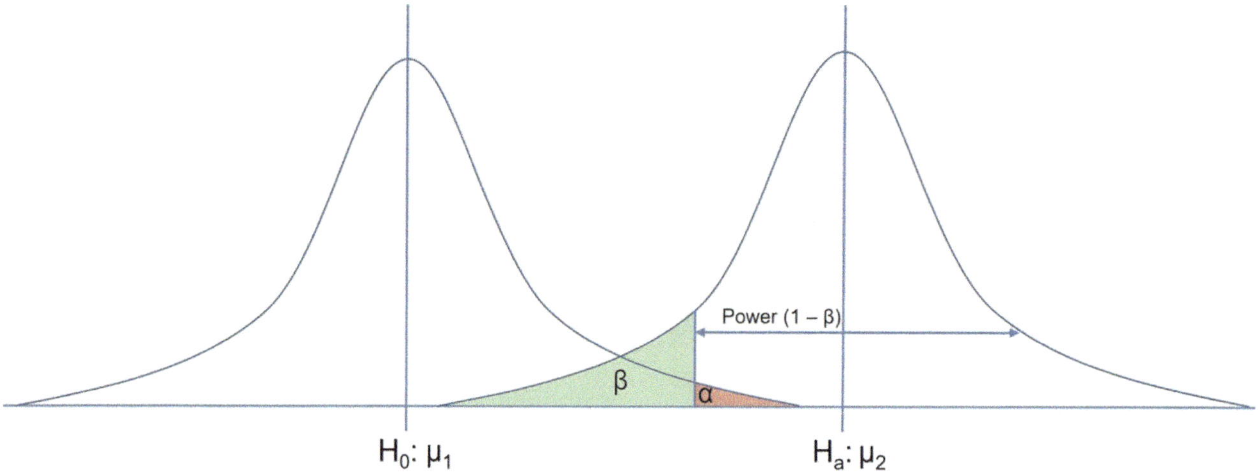

FIGURE 33.2 Relationship between α (type I error) and β (type II error). H_a—alternative hypothesis; H_0—null hypothesis.

two distributions of sample means under the null hypothesis (H_0) and alternative hypothesis (H_a). In this illustration, the area that is marked as α represents the probability of making a type I error and β represents the probability of making a type II error when testing the null hypothesis against the alternative hypothesis. Note that the area to the right of the β represents the power (1—β). When planning a research study, one needs to specify the α and β errors that are appropriate depending on the context of the research. Hence, knowing how the adjustment of the probability of type I error will affect the probability of type II error and vice versa is of critical importance in the planning phase of research. Note in Fig. 33.3 that if we wish to decrease α, then β will increase and the power of the study will decrease. The opposite is true if we wish to increase the power of the study just by decreasing the β, then α will increase, making our statistical test more prone to making type I errors. How is this relevant when it comes to the sample size? Increasing the sample size is the only way to increase the power of the study without increasing the probability of type I error (i.e., increasing α). Fig. 33.3 demonstrates that increasing sample size leads to a narrower sampling distribution which essentially means that there is less overlap between the sampling distribution for the null hypothesis and the sampling distribution for the alternative hypothesis.

Numerous formulas for estimating sample size exist.[3,9,10] One of the common scenarios is the situation where one wants to determine sample size in a two-sample *t*-test for a desired power to detect a given difference between two population means. In this scenario, the null hypothesis can be formulated as $H_0 : \mu_1 - \mu_2 = 0$ against the alternative hypothesis $H_1 : \mu_1 - \mu_2 > 0$. After specifying the alpha level, power, and minimal desired difference $d(\mu_1 - \mu_2)$, if the population variance is known, the sample size can be calculated as:

$$n = \frac{2\sigma^2}{d^2}\left(Z_\alpha + Z_\beta\right)^2$$

Note that if a two-tailed test is necessary to be used, then $Z_{\alpha/2}$ should be utilized.

$$n = \frac{2S^2}{d^2}\left(t_{\alpha,v} + t_{\beta,v}\right)^2$$

Note that if a two-tailed test is necessary to be used, then $t_{\alpha/2,v}$ should be utilized.

Time-to-event endpoints

In clinical research, it is very common that the endpoint of interest is time-to-event. If a researcher is interested in the occurrence of an undesirable event (e.g., disease

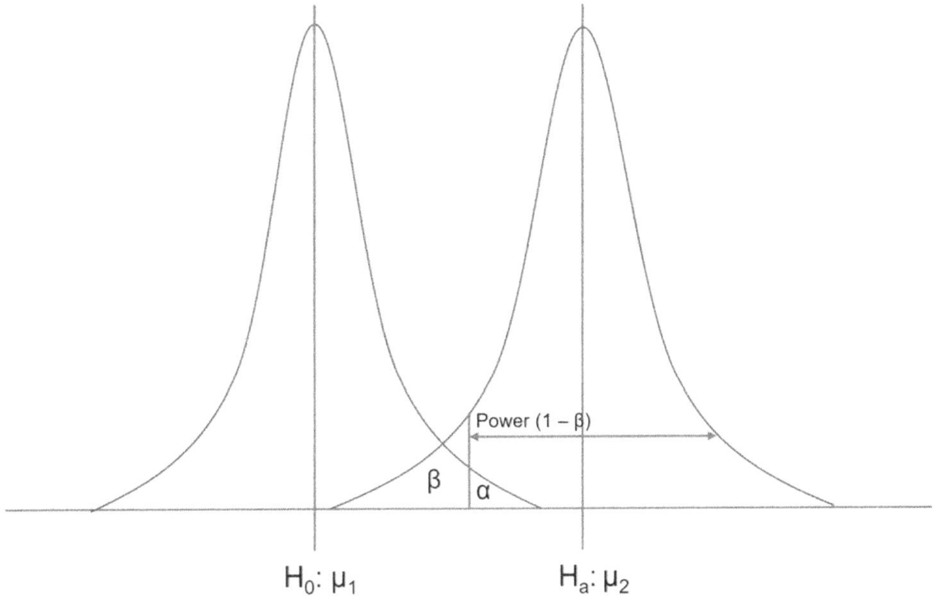

FIGURE 33.3 Effect of increased sample size on α (type I error) and β (type II error). H_a—alternative hypothesis; H_0—null hypothesis.

progression, hospitalization, myocardial infarction, death), one of the objectives is to evaluate whether an intervention will have an effect on delaying such an event. From a practical standpoint, statistical methods in survival analysis are different compared to analysis of continuous variables.[2,11] Some of the reasons include censoring and highly skewed data, which violates the normality assumption. The most common statistical models used for time-to-event data include the exponential model, Cox's proportional hazards model, and the weighted log-rank test.[2]

When the null hypothesis that is tested is expressed by hazard ratio or relative risk, the following variables are necessary for sample size calculation. In such a situation, the researcher needs to specify the median survival time on control treatment m_1 and the hazard ratio R (or relative risk) of the control treatment relative to the experimental treatment. Similarly, as in any type of hypothesis testing, the alpha level is specified and a particular desired power for detecting a difference in the hazard ratio (relative risk) of the experimental group relative to the control group if such exists. Accrual time during which patients are recruited and additional follow-up time after the end of recruitment should be specified. Finally, the ratio of control to experimental patients should be specified too. Inputting the mentioned variables in some of the widely available sample size calculators can provide a general idea of the sample size needed for the research question. Sole reliance on online calculators for the purpose of sample size calculation may be misleading and early collaboration with a statistician during the planning phase of the study is necessary.

Other important practical issues to consider that can modify the sample size necessary for the desired research question are the possibility of confounding or interaction factors, unequal treatment allocation, adjustment for dropouts, mixed-up randomization schedules, treatment study center imbalance, multiplicity, and sample size calculation based on the very low incidence rate of events. All these considerations require a statistical method of modifying the initially determined sample size. For detailed statistical formulas and their derivation, the reader is referred to specialized textbooks on sample size.[3,5,9] The two most common considerations—uneven allocation to the study groups and accounting for the dropout rate.

Examples of sample size calculations from clinical trials

The examples from the following clinical trials can illustrate which parameters were necessary to determine for computation of sample size to be successful.[12] Sample sizes were computed through a specific software sample size calculator. Hence, the goal is not to present complex statistical formulas but to demonstrate real-world examples of published studies and what input parameters researchers need to specify to accurately determine sample size by using statistical software.

Granger et al.[13] in a landmark clinical trial investigated apixaban compared with warfarin in patients with atrial fibrillation. The trial was designed as a noninferiority trial. Their primary endpoints were ischemic or hemorrhagic stroke or systemic embolism.

The authors decided to calculate which sample size they would need to have a power of 90%, at a significance level of 0.005, for a noninferiority margin of 0.005. Note here that the noninferiority margin is analogous to the concept of clinically meaningful difference or effect size (δ) that we addressed in the chapter. Also, note that this study design will use survival analysis as a statistical test for the interpretation of outcomes, thus authors needed to estimate the incidence rate of the primary outcome, which was deemed to be 0.012 (based on available literature). In the sample size calculation, the authors also adjusted for an expected dropout rate of 1%. The average follow-up was planned to be 2.1 years. With the above parameters, researchers calculated that they would need 448 events to detect the desired effect with the above-mentioned power and alpha level. Given the incidence rate of 0.012, the sample size can be calculated as follows:

$$N = \frac{\#\text{Events}}{(\text{Incidence rate} \times \text{Follow up time})} = \frac{448}{(0.012 \times 2.\ 1)} = 17778$$

Once adjusted for the 1% dropout rate, the sample size is the following:

$$N_{\text{adjusted}} = \frac{N_{\text{unadjusted}}}{\left(1 - p_{\text{dropout}}\right)} = \frac{17778}{1 - 0.01} = 17957$$

Hence, in the original paper, the authors chose to recruit 18,000 patients.

In a clinical trial by Ghofrani et al.[14] researchers wanted to compare riociguat to placebo in patients with symptomatic pulmonary arterial hypertension. The primary endpoint was a change in the 6-minute walk test distance. In this study, the researchers specified power and alpha levels as usual ($1\text{-}\beta = 0.9$, $\alpha = 0.05$), and the standard deviation was assumed to be 70. The estimated clinically meaningful effect was 25 m in the 6-min walking test distance. Since the randomization ratio was 4:2, the sample size estimated was 250 and 125 patients in the experimental group and control group, respectively. Note that in the study, the statistical analysis is a two-sample student *T*-test, given that there is a comparison of means between two groups, with a modification due to an unequal number of subjects in each group ($n_1/n_2 \neq 1$).

In a study by Zamorano et al.[15] where a goal was to investigate whether a proactive multifactorial risk factor intervention strategy using single-pill amlodipine/atorvastatin in addition to other antihypertensive and lipid-lowering therapy, as required, resulted in a greater reduction in calculated Framingham 10-year cardiovascular risk compared with usual care after 52-week treatment. The investigator calculated that to detect a 10% relative difference in the primary endpoint between the intervention and control group, with 90% power, at a significance level of 0.05, they would need at least 132 sites and 1378 patients. The investigators estimated the population standard deviation to be 0.350 from previous studies. Due to the particular design of this trial and the use of cluster samples, the intraclass correlation coefficient (ICC) needs to be specified for sample size calculation to be performed. This was also estimated from prior similar studies and an ICC 0.2 was used. Finally, researchers wanted to account for a potential dropout rate of 20% for investigators and 30% for patients, leading to a final determination that the sample needed for the study would be 164 sites and 1968 patients.

In conclusion, sample size calculation is a necessary step in the planning phase of any research study. It provides the assurance that the study will have the power to detect the desired effect for a specified significance level. It provides economic and practical insights into planning a research study and it is important from an ethical standpoint to minimize the number of subjects needed for a particular research question. To perform a valid sample size calculation, the researcher must select an appropriate statistical test for the hypothesis of interest that accurately reflects the study objectives under a valid study design. Sample size determination is most commonly done by precision analysis or power analysis. Precision analysis is utilized in survey studies and is based on the theory of estimation, while power analysis is used in hypothesis testing experiments. In practice, necessary variables for the sample size calculation formula will depend on the hypothesis formulation and study design. Sample size calculation must include adjustments for certain practical issues such as unequal treatment allocation, adjustment for dropouts, and endpoints with very low event rates. Close communication with a statistician during the planning phase of research is of critical importance to successfully determine the correct sample size for a contemplated research study. The use of software for sample size calculation is standard practice. Some of the available software for sample size calculation are G*Power, PASS (Power Analysis and Sample Size Software), and nQuery.[16–18]

References

1. Faber J, Fonseca LM. How sample size influences research outcomes. *Dental Press J Orthod.* 2014;19(4):27–29. https://doi.org/10.1590/2176-9451.19.4.027-029.ebo.
2. Altman DG. *Practical Statistics for Medical Research.* Chapman and Hall/CRC; 1990. https://doi.org/10.1201/9780429258589.
3. Chow SC, Shao J, Wang H, Lokhnygina Y. *Sample Size Calculations in Clinical Research.* 3rd ed. Chapman and Hall/CRC; 2017. https://doi.org/10.1201/9781315183084.
4. Lakens D. Sample size justification. *Collabra: Psychol.* 2022;8(1):33267. https://doi.org/10.1525/collabra.33267.
5. Verma JP, Verma P. *Determining Sample Size and Power in Research Studies: A Manual for Researchers.* Springer; 2020. https://doi.org/10.1007/978-981-15-5204-5.

6. Biau DJ, Jolles BM, Porcher R. P value and the theory of hypothesis testing: an explanation for new researchers. *Clin Orthop Relat Res.* 2010;468(3):885–892. https://doi.org/10.1007/s11999-009-1164-4.
7. Keefe RSE, Kraemer HC, Epstein RS, et al. Defining a clinically meaningful effect for the design and interpretation of randomized controlled trials. *Innov Clin Neurosci.* 2013;10(5–6 Suppl A):4S–19S.
8. Cohen J. *Statistical Power Analysis for the Behavioral Sciences.* 2nd ed. Routledge; 1988. https://doi.org/10.4324/9780203771587.
9. Sozu T, Sugimoto T, Hamasaki T, Evans SR. *Sample Size Determination in Clinical Trials with Multiple Endpoints.* Springer International Publishing; 2015. https://doi.org/10.1007/978-3-319-22005-5.
10. Nanjundeswaraswamy TS, Divakar S. Determination of sample size and sampling methods in applied research. *PES.* 2021;3(1):25–32. https://doi.org/10.24874/PES03.01.003.
11. Etikan İ, Bukirova K, Yuvalı M. Choosing statistical tests for survival analysis. *Biometrics Biostat Int J.* 2018;7(5). https://doi.org/10.15406/bbij.2018.07.00249.
12. Statsols. *Real Sample Size Examples From Clinical Trials With nQuery.* https://www.statsols.com/hypothesis-testing-example. Accessed 18 November 2023.
13. Granger CB, Alexander JH, McMurray JJV, et al. Apixaban versus warfarin in patients with atrial fibrillation. *N Engl J Med.* 2011;365(11):981–992. https://doi.org/10.1056/NEJMoa1107039.
14. Ghofrani HA, Galiè N, Grimminger F, et al. Riociguat for the treatment of pulmonary arterial hypertension. *N Engl J Med.* 2013;369(4):330–340. https://doi.org/10.1056/NEJMoa1209655.
15. Zamorano J, Erdine S, Pavia A, et al. Proactive multiple cardiovascular risk factor management compared with usual care in patients with hypertension and additional risk factors: the CRUCIAL trial. *Curr Med Res Opin.* 2011;27(4):821–833. https://doi.org/10.1185/03007995.2011.555754.
16. *Universität Düsseldorf: G*Power.* https://www.psychologie.hhu.de/arbeitsgruppen/allgemeine-psychologie-und-arbeitspsychologie/gpower. Accessed 18 November 2023.
17. Statsols. *nQuery Platform for optimizing Trial Design.* https://www.statsols.com. Accessed 18 November 2023.
18. *PowerSampleSize Main Vanderbilt Biostatistics Wiki*; 2018. https://biostat.app.vumc.org/wiki/Main/PowerSampleSize. Accessed November 18, 2023.

Chapter 34

Statistical software

Muhammad Zia Khan
Heart and Vascular Institute, West Virginia University, Morgantown, WV, United States

Introduction

Statistical software is a computing tool that assists in the collection, modification, and analysis of information. We have seen a revolution in statistics in the last 5 decades driven by new statistical techniques and an exponential increase in computing power. The development and availability of fast personal computers to collect, store, and analyze large amounts of data have created the need for new statistical techniques, as older techniques are deemed to be insufficient to tackle the statistical problems of large databases. This ever-increasing complexity of data has led to the constant development and evolution of statistical software. We have seen the increasing popularity of both open-source programming language-based software (such as R) and enterprise software with graphical interfaces (such as SPSS).

One of the earliest techniques for hypothesis testing was the *t*-test. Data analysis in that era was relatively straightforward and easily reproducible, though cumbersome given the lack of computing aids. The size of the data was limited, and sampling was random. This is no longer the case. With the increase in computing power, size, and complexity of datasets, more robust forms of data analysis techniques are needed.

Ideally, statistical software should have certain characteristics. These include:

i. Data management: It should have ease of data entry, data manipulation, merging datasets, the ability to import and transform data created through different resources, ability to organize, mutate, and transform data. Creating subsets should be easy. It should also be able to identify and address missing data.

ii. Replication of data analysis: Replication is a very important part of science. At a minimum, two scientists analyzing the same data should arrive at the same conclusion if similar statistical techniques are employed. Therefore, statistical software should have a log that correctly records each step of data analysis, this allows for replication studies. Although there has not been much emphasis on sharing data and output logs (code) in the past, code sharing for transparency and replication is the way forward. Many journals and the scientific community are asking for greater transparency via sharing data and data analysis logs.

iii. Flexibility: It must be flexible in data analysis and should allow the use of newer techniques through packages or updates. It must be reasonably available to all scientists (at a reasonable cost) who want to replicate the work. For this reason, open-source software will be preferred. However, the validity of any new package or script especially if not published in a peer-reviewed journal, is a challenge since vendor support is lacking.

iv. High performance: Ability to handle large datasets (sufficient memory) with reasonable computing speed (these are also dependent upon computing apparatus). Ability to test the basic and complex hypothesis.

v. Visualization of dataset results: Must be able to create and customize charts, tables, and graphs that are brief yet easy to comprehend.

Given the wide variety of statistical software, a complete overview of all available statistical software is beyond the scope of this chapter. We will focus on software that are popular and relevant to biostatistics in this chapter.

Statistical Package for the Social Sciences

Statistical Package for the Social Sciences (SPSS) was developed by social scientists at the University of Chicago, starting around 1968. Around 1985 support was added for personal computers, and eventually a graphical interface. IBM bought SPSS in 2009. This led to the emphasis on business analytics, a shift from social science data analysis. SPSS is currently available as a subscription model from

Translational Cardiology. https://doi.org/10.1016/B978-0-323-91790-2.00033-2

IBM, many websites provide low-cost discounted versions for students and social and health scientists.

Many features of SPSS Statistics are accessible via pull-down menus for simple and most common tasks. This graphical interface setup has led to the popularity of SPSS in social sciences. However, advanced analysis or complex data manipulation can only be done by using command syntax language. Command syntax programming has the benefit of simplifying repetitive and complex tasks. Using the graphical interface for analysis will also create appropriate syntax commands in the output log for that analysis. The output can be pasted into the syntax file to run the previous analysis again. The command syntax generated can be saved for review and shared with colleagues/other scientists for replication. Syntax commands can be run interactively or unattended. Output can be easily exported into MS Word or other document files. SPSS can also be driven externally through programmable extensions from Python or R. All data processing occurs sequentially case-by-case through the dataset.

SPSS can import a variety of data files including CSV, Stata, Excel, and SAS. SPSS datasets have a two-dimensional table structure with a data and variables tab that can be toggled with a single click. In the data tab, rows typically represent cases, and the columns represent variables. In the variable tab, the characteristics of variables are displayed and can be changed easily with a pull-down interface. Menu items are simple and self-explanatory with a data menu for data sorting, a transform menu for data transformation, and an analyze menu for data analysis. A separate graph menu exists to create customizable graphs though these can be created at the time of analysis. Add-ons and extensions can be added to the base SPSS package.

SPSS continues to have statistics modules, but the focus has shifted to more business-oriented solutions, including analytics, forecasting, and marketing.

R and R studio

At Bell Inc., John Chambers and his group started developing the S program language in the late seventies. In the 1990s, R was developed by Ross and Robert. R was written as an alternative implementation of the S language, using most of the same functions (macros). This was later released under GNU public license making R a free software. This was a revolutionary step as it was one of the first statistical software under GNU. This led to the rapid popularity of R, especially among health scientists. R is a programming language with a minimum graphical interface. This may seem forbidding to noncoders; however, beginners can load software packages developed by other scientists for data analysis and data visualization. These packages create a middle ground between pure command line and graphical interfaced software. With online open-source public support, new statistical techniques can be easily loaded in R as packages and tested, R acts like a rapid prototyping environment for testing new statistical techniques. This makes R independent of the need for constant updates that are required with commercial software, which in turn makes the user independent of vendors. However, any new scripts or packages introduced by an individual data scientist or research group need independent validation, and this may be a challenge if the package or script has not been published in a peer-reviewed statistical journal. The relative ease of introducing a variety of new statistical techniques through packages has led to the popularity of R.

Learning R can be time consuming, but it is not as intimidating as other programming languages like Python or C. This may be a put-off for some scientists who are interested in one type of analysis that can be done through a graphical interface software, however, with ever-evolving statistical techniques and data complexity, the use of R is becoming more popular. Many statisticians learn R as their first programming language, instead of learning more basic programming languages such as Python or C. This may be counterintuitive as R is not a basic programming language; however, the popularity of R means that it does not need to be one. This open architecture has made R the de facto software for teaching biostatistics in colleges and universities.

Several integrated development environments have been developed for R, the most popular version is RStudio. RStudio is an integrated development environment for R (therefore, it requires R for it to function properly). RStudio comes with a minimal graphical interface. This makes simple tasks like saving, sharing, or creating new files easy while preserving the versatility and flexibility of R. Typical layout of RStudio consists of the console window, workspace window, and files/plots/packages windows. There is a menu tab on top for basic functions.

The console window is used for typing the command line. Through the command line packages can be installed using *install.packages* command line while packages can be recalled for use during a session via *library* function. The most popular nonstandard package is *Tidyverse* for data manipulation and visualization (*ggplot2*). This is especially popular among health scientists. *Tidyverse* is based upon the concept of "piping" or chains of command lines. Using multiple subsets of commands makes data manipulation easier in *Tidyverse*. Given its popularity, *Tidyverse* can be considered as a special dialect of R. *Tidyverse* has free extensive online support and is relatively easy to learn.

A working directory can be set using a command line or graphical interface. Data can be imported using the command line, almost any form of data can be imported once the packages required are installed. You can either type code directly into the Console window, or you can type it into a text file, and then run the code you have written in the Console window. All codes can appear in the workspace window under R history, which can be saved as an *R history* file for later review. This means that codes can be shared, and analysis can be easily replicated. All graphs are displayed in the files/plots/packages window, from where they can be easily exported into multiple formats using the graphical interface.

Stata

The development of Stata began in 1984, by William Gould and later by Sean Becketti. Stata was written, in the C programming language, initially for PCs running the DOS operating system. Since then, versions of Stata have been released for systems running Linux, Windows, and MacOS. Stata is a powerful statistical package with data-management facilities, a wide array of statistical techniques, and the ability to produce good-quality graphs. Stata is fast, with multicore support, and easy to use due to its hybrid command and graphical interface. Recent editions have allowed users to use Python and R scripts using special commands. Stata has a good online presence and online community support with a wide range of statistical packages.

Stata can import a wide variety of data files including CSV, Excel, and SAS. Stata is widely available at a reasonable price and with discounted versions for students and nonprofit organizations. It also offers a perpetual license in addition to a subscription-based license.

SAS and JMP

SAS was developed at North Carolina State University from 1966 until 1976 and later at SAS Institute in the 1980s and 1990s. SAS is available as an enterprise license, for students a limited free version, SAS university edition is available. SAS's strong vendor support has made it popular among large institutions.

SAS basic interface has an editor for code creation and can be saved as *.SAS* program, an output window where all reports are created, and a log window where logs are created for the analysis. Data can be viewed from the library window. SAS can read a variety of RAW data to create SAS data which it can intuitively use to perform a wide variety of analyses. SAS uses syntax, with a single-line command called statements, each statement ends with a semicolon, and multiple statements combine to form a step. Each step is executed by using the "run; " command.

JMP was developed by the JMP business unit of SAS Institute in 1989 with an intuitive and easy-to-use graphical user interface. JMP began integrating with SAS in version 7.0 in 2007 and has strengthened this integration ever since. Users can write SAS code in JMP, connect to SAS servers, and retrieve and use data from SAS. JMP has some users and admirers but never became as popular as SPSS or SATA in either statistical research or statistical applications.

Other software

Recently, a new generation of open-access software are being developed. These are mostly task-specific, easy to use, and are based on a graphical user interface. These also focus on the reproducibility of analysis by integrating analysis settings into results. The most popular and arguably the easiest to use is *JASP, released by the* University of Amsterdam. It has extensive online community support. It produces high-quality graphics and has the most popular kind of analysis built into a graphical user interface. JASP can read a variety of data types. JASP is getting popular especially in social sciences as an open-source alternative to SPSS.

Further reading

1. George D, Mallery P. *An Overview of IBM® SPSS® Statistics*. 2021:1–7. https://doi.org/10.4324/9781003205333-1.
2. Ihaka R, Robert GR. A language for data analysis and graphics. *J Comput Graph Stat*. 1996;5(3):299–314.
3. Ismay C, Kim AY. *Statistical Inference via Data Science: A ModernDive Into R and the Tidyverse*. Chapman and Hall/CRC; 2019. Published December 13.
4. Baum CF. *An Introduction to Stata Programming*. College Station: College Station, TX, USA: Stata Press; 2015.
5. Schechter C. Review of Christopher F. Baum's an introduction to stata programming. In: *The Stata Journal*. 2nd ed. vol. 16. 2016:511–516.
6. Barr AJ, Goodnight JH, Sall JP, Helwig JT. *A User's Guide to SAS 76*. Raleigh, North Carolina: SAS Institute, Inc.; 1976.
7. Han H, Dawson K. *JASP (Software)*. 2020. https://doi.org/10.31234/osf.io/67dcb.

Part V

Clinical: Study types

Dhrubajyoti Bandyopadhyay

Chapter 35

Design principles: Hierarchy of study types

Basha Behrman and Jayakumar Sreenivasan
Department of Cardiology, Westchester Medical Center, New York Medical College, Valhalla, NY, United States

Key points

- There is a hierarchy of evidence widely accepted, which reflects the likelihood of elimination of bias and validity of results in each particular study design.
- In general, systematic reviews provide a higher level of evidence than individual trials.
- There is guidance as to which study designs are the most suitable for individual research question categories.

Introduction

Once a well-designed research question has been developed, the next step is to determine the optimal study design to provide the highest quality data for that question. This is especially relevant for translational science, as the grade of clinical recommendation is contingent upon the level of evidence provided by researchers. This chapter will discuss the principles of study design and introduce the various study types and the strength and generalizability of the evidence they provide. The hierarchy of evidence is a foundational concept in evidence-based medicine, and it is necessary for prospective authors and audiences to understand that not all types of studies carry the same weight, referred to as the level of evidence, as well as the fact that not all studies within a category are designed to be equally as strong in the evidence they produce. In general, systematic reviews and meta-analyses are considered the strongest form of evidence, followed by experimental studies, and then observational studies, to be discussed in detail below.

Observational and experimental studies

Before detailing levels of particular types of evidence, it is important to introduce two broad categories of research: "observational" and "experimental" studies.[1] In observational studies, data is collected without intervention or interference in the exposures of the study population. These studies can be descriptive, providing statistical information about the study group without data on causation, or analytical, in which hypotheses can be tested and causal relationships can be evaluated. Examples of observational studies include surveys, controlled cohort studies, case-control studies, and case series or studies.[1,2]

In experimental studies, there is a control group and an intervention group, and unlike descriptive studies, researchers do manipulate variables to study. An example of this study type includes randomized controlled trials (RCTs).

The data generated by these study types have different strengths and utility. By the nature of their design, the experimental RCTs, if well designed, provide data that can answer many clinical questions with a high level of confidence. Observational studies, such as case reports and surveys, yield interesting data but are neither generalizable nor reliably free of confounding variables. Our current schema for ranking evidence incorporates an assessment of the likelihood of bias influencing the outcomes of the study as well as a probability that the conclusion of causality the study generates is valid.[2,3] This leads the data from an RCT to be accepted as a superior level of evidence as compared to a case report.

Translational Cardiology. https://doi.org/10.1016/B978-0-323-91790-2.00025-3

Hierarchy of evidence

Since the original discussion of levels of evidence in 1979, there have been many international and professional groups who have discussed this topic and each has generated their own hierarchy. The list of these groups includes, but is not limited to, the American College of Chest Physicians (ACCP), Australian National Health and Medical Research Council (ANHMRC), Grading of Recommendations Assessment, Development and Evaluation (GRADE) Working Group, Oxford Center for Evidence-Based Medicine (OCEBM), Scottish Intercollegiate Guidelines Network (SIGN), US Preventive Services Task Force (USPSTF), and US Task Force on Community Preventive Services (USTFCPS). While they do not all agree on the exact system of ranking, one group of researchers, the GRADE working group, published a comparison and evaluation of the other six well-known frameworks for grading levels of evidence in 2004.[4] The authors analyzed the ACCP, ANHMRC, OCEBM, SIGN, USPSTF, and USTFCPS guidelines for ranking evidence and, importantly, addressed issues with reproducibility of grading evidence, applicability to different types of clinical questions, ability for widespread use by professionals, patients, and policymakers, among other criteria. While this study had some notable issues, such as the fact that this was an informal assessment by a group of authors, all of whom had experience with at least one system, and many had helped develop one of the systems included in the appraisal, it does highlight the need for continued development of systems to rank evidence more effectively. Unfortunately, there was little consensus about which system was most well suited to grading the quality of evidence. At that time, the conclusion of the study reported that all the systems had meaningful limitations in utility.

While we are in search of a more perfect system for grading evidence to address the minutiae of studies, the tools we have at present do still have value in assessing the strength of evidence and the utility and applicability of the conclusions from the resulting data. Understanding the shortcomings of our current systems and pitfalls in their applicability should caution users to investigate which system is best for their uses (for example, the OCEBM system was unanimously thought to be applicable to questions of effectiveness, diagnosis, prognosis, and harm).

The widely agreed upon general schematic for the hierarchy of evidence can be seen in Fig. 35.1, which ranks studies by decreasing the strength of evidence, going from top to bottom.

As can be seen by this figure, systematic reviews and meta-analyses they generate are considered to be the highest level of evidence, followed by the experimental RCTs and nonRCTs, and then followed by the observational studies.

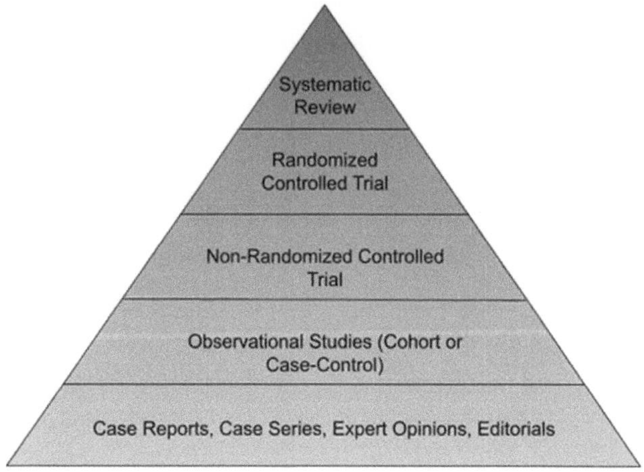

FIGURE 35.1 Hierarchy of evidence, from the strongest evidence at the top of the pyramid to the weakest evidence at the bottom of the pyramid.

This hierarchy was originally built for interventional studies, however, when thinking about other categories of research questions, adaptations were made to address the strength of evidence in studies as applied to each category and can be useful in assisting researchers with selecting a study design which is most appropriate for their question, as can be seen in Table 35.1.[5] It must be kept in mind, however, that the level of evidence will also depend on the individual study quality, effect size, and precision.

Systematic review and meta-analysis

In general, a systematic review of all relevant studies on a topic provides a higher level of evidence than a single study alone. This is supported by the fact that these studies utilize data from numerous individual studies to arrive at their conclusions. A systematic review uses specified search criteria to aggregate data from a broad spectrum of sources to answer the research question. To minimize bias in either the review process or the bias carried over from the design of the included studies for review, systematic reviews employ published and transparent protocols to determine which evidence is eligible for inclusion into the review, as well as "risk of bias" tools to assess the included studies. These steps result in a reproducible summary of the state of current research in a particular area.[6]

Once the eligible studies have been aggregated, the data from these sources can be pooled and synthesized to generate novel or stronger conclusions than those from an individual study. This employs the technique of meta-analysis, which can be used to analyze data from multiple studies, theoretically improve the precision of evidence, and provide the power to allow for subgroup analyses not found in the original study.

There are, however, limitations to the systematic review and meta-analysis. As alluded to above, one major pitfall

TABLE 35.1 Level of evidence according to research question category.

	Research question categories				
Level of evidence	**Treatment Effects**	**Diagnostic test accuracy**	**Prognosis**	**Screening**	**Prevalence**
1	Systematic review of RCTs	Systematic review of cross-sectional studies	Systematic review of (inception) cohort studies	Systematic review of RCTs	Local random sample survey
2	RCT; observational study with exceptionally evident effects	Individual cross-sectional studios with blinding and a consistently applied reference standards	Individual inception cohort study	RCT	Systematic review of surveys that allow matching to local circumstances
3	Cohort study; follow-up study	Nonconsecutive studies; studies without consistently applied reference standards	Noninception cohort study; control group of RCT	Cohort study; follow-up study	Local nonrandom sample
4	Case-control study; historically controlled study; case series	Case-control studies; studies where poor-quality reference standards used	Poor quality cohort studies; case-control studies; case series	Case-control study; historically controlled study; case series	Case series
5	Mechanism-based reasoning	Mechanism-based reasoning	N/A	Mechanism-based reasoning	N/A

described in numerous writings is the fact that data from systematic reviews are only as strong as the quality of the evidence they include and the factors employed to minimize error and bias in each study included.[7,8] While the risk of bias tools are used, there are those who would argue that despite these measures, there is no way to escape the inherent subjectivity of a technique that selects which studies to include in the analysis,[9] as well as those who cite the fact that published trials available for inclusion are more likely to be those with favorable outcomes, versus those with inconclusive or negative results, referred to as "publication bias"[10] and highlight the lack of inclusion of so-called "gray literature," including results from unpublished studies and those printed in non-English language publications among others, in the majority of meta-analyses.[11]

Despite these apparent limitations, systematic reviews and meta-analyses are widely regarded as the highest level of evidence, and some such studies have been used to change the landscape of current knowledge in a field.

Randomized control trial

RCTs are considered to be the ideal individual study to assess the benefits and harms of an intervention. In these studies, participants are assigned to either the intervention group(s) or the control group by a purely random process, minimizing the chance of a confounding factor being heavily present in one group and maximizing the likelihood that the intervention in question will be the only significant difference between arms of the study, allowing for a strong conclusion on causality and impact of intervention to be drawn if the study is adequately sized.[2] RCTs can be limited by practical or ethical concerns relating to the feasibility of recruiting participants or of administering an intervention.

Some of the most influential studies in medicine have been RCTs, including studies that showed that beta blockers, previously thought to be contraindicated in patients with heart failure, were actually beneficial in improving mortality and clinical outcomes in these patients.[12–14]

Cohort study

Cohort studies are designed to assess changes over time in a specific population. These studies utilize a cohort of disease-free subjects with defined statistical similarity and known exposure status and follow them over time to assess the outcome of interest and its association with the exposure.[15,16] These studies provide the strongest evidence regarding prognosis, as the question is what the natural course of a condition is without intervention.

Case-control study

Case-control studies are observational studies that break down study participants into groups of individuals with the disease/outcome of interest and those without it and then study the relationship with exposures. This study design has intrinsic drawbacks, including the obligatorily retrospective nature and recall bias regarding exposures; however, it has increased utility in cases of rare diseases where it can be helpful to identify those with a disease before the study.[2,15]

Cross-sectional study

A cross-sectional study provides a static observation where exposure and outcome are measured simultaneously. Since the status of exposure and outcome are assessed at the same time, researchers are often unable to draw causal conclusions; however, this type of study is useful for assessing the burden of a disease among a specific population.[1,15,16]

Survey

The exception to the systematic review of RCT as the strongest form of data would be in the case where the clinical question focuses on an assessment of the burden and extent of an issue among a population. In these cases, as shown in Table 35.1 above, a survey or questionnaire would be the most appropriate study. These studies are also useful in measuring public opinion, identifying opinions/beliefs that impact public health, and self-perception of changes in clinical status, for example, the article published in JAMA Pediatrics in 2021 by Chido-Amajuoyi et al., which identified that concerns over safety and adverse events were increasingly being reported as the main reason for HPV vaccine nonacceptance over the study period.[17] These findings can impact patient education discussions and public health campaigns. Limitations of these studies include the fact that these studies depend heavily on the participation of the surveyed group and are subject to nonresponse bias and recall bias. As the researcher is unable to verify responses, the data may be subjective.

Case series

Case series, which are multiple reports of cases of a disease, are retrospective, uncontrolled studies and hold one of the lowest positions in the hierarchy of evidence. This is because they are vulnerable to many sources of bias. While the cases can be from more than one provider or center, they may often only be reflective of a single set of specific circumstances, including the provider, location, and practices and resources in a specific center, thus significantly limiting the generalizability of the results.[18] They do have a role in describing phenomena, formulating hypotheses, and furthering clinical education.[1,15]

References

1. Howick J. *Introduction to Study Design*; 2002. URL: http://www.cebm.net/wp-content/uploads/2014/06/CEBM-study-design-april-20131.pdf. Accessed December 4, 2017 [WebCite Cache ID 6pfDD1ddH].
2. Wang JJ, Attia J. Study designs in epidemiology and levels of evidence. *Am J Ophthalmol*. 2010;149(3):367−370.
3. Merlin T, Weston A, Tooher R. Extending an evidence hierarchy to include topics other than treatment: revising the Australian'levels of evidence. *BMC Med Res Methodol*. 2009;9(1):1−8.
4. Atkins D, et al. Systems for grading the quality of evidence and the strength of recommendations I: critical appraisal of existing approaches the GRADE Working Group. *BMC Health Serv Res*. 2004;4(1):38.
5. Group, O.L.o.E.W. *The Oxford 2011 Levels of Evidence*. Oxford Centre for Evidence-Based Medicine; 2011. http://www.cebm.net/index. aspx? o= 5653.
6. Higgins JP, et al. *Cochrane Handbook for Systematic Reviews of Interventions*. John Wiley & Sons; 2019.
7. Straus SE, et al. *Evidence-based Medicine E-Book: How to Practice and Teach EBM*. Elsevier Health Sciences; 2018.
8. Noordzij M, et al. Systematic reviews and meta-analyses: when they are useful and when to be careful. *Kidney Int*. 2009;76(11):1130−1136.
9. Stegenga J. Is meta-analysis the platinum standard of evidence? *Stud Hist Philos Sci C Stud Hist Philos Biol Biomed Sci*. 2011;42(4):497−507.
10. Flather MD, et al. Strengths and limitations of meta-analysis: larger studies may be more reliable. *Contr Clin Trials*. 1997;18(6):568−579.
11. Ahmed I, Sutton AJ, Riley RD. Assessment of publication bias, selection bias, and unavailable data in meta-analyses using individual participant data: a database survey. *BMJ*. 2012;344:d7762.
12. Packer M, et al. The effect of carvedilol on morbidity and mortality in patients with chronic heart failure. *N Engl J Med*. 1996;334(21):1349−1355.
13. The cardiac insufficiency bisoprolol study II (CIBIS-II): a randomised trial. *Lancet*. 1999;353(9146):9−13.
14. Effect of metoprolol CR/XL in chronic heart failure: metoprolol CR/XL randomised intervention trial in-congestive heart failure (MERIT-HF). *Lancet*. 1999;353(9169):2001−2007.
15. Noordzij M, et al. Study designs in clinical research. *Nephron Clin Pract*. 2009;113(3):c218−c221.
16. Vogt WP, Gardner DC, Haeffele LM. *When to Use what Research Design*. Guilford Press; 2012.
17. Chido-Amajuoyi OG, et al. Safety concerns or adverse effects as the main reason for human papillomavirus vaccine refusal: national immunization survey−teen, 2008 to 2019. *JAMA Pediatr*. 2021;175(10):1074−1076.
18. Petrisor B, Bhandari M. The hierarchy of evidence: levels and grades of recommendation. *Indian J Orthop*. 2007;41(1):11−15.

Chapter 36

Case series: Design, measures, and classic examples

Suryasnata Bhowmik[1] and Adrija Hajra[2]

[1]*Critical Care Medicine, Nilratan Sircar Medical College, Kolkata, West Bengal, India;* [2]*Department of Internal Medicine, Bringham and Women's Hospital/Harvard Medical School, Boston, MA, United States*

Key points

- Case series is the earliest form of research methodology that has historically led to groundbreaking medical discoveries
- Designing a case series involves reporting collected data from carefully sought cases that demonstrate novel observations in medicine
- Despite multiple limitations, strategic case selection and data collection can offer strong recommendations that may shape medical guidelines

Introduction

Case series are descriptive studies that track multiple cases to "describe their manifestations, clinical courses, and prognoses."[1] Historically, they have been a cornerstone to medical literature, contributing to groundbreaking discoveries: from James Parkinson's 1817 identification of "shaking palsy" (now known as Parkinson's disease) to insights into sickle cell disease, acquired immune deficiency syndrome (AIDS), and Coronavirus disease 2019 (COVID-19).[2–4] While their prominence waned in the late 20th century with the advent of evidence-based medicine, the late 1990s witnessed a renewed interest.[5]

Importance of case series

Building on such rich heritage, case series today serve multiple roles. They facilitate quick reporting of novel presentations, unusual courses, and effects of therapeutic approaches. Fast publication time can alert the community on serious issues like emerging epidemics.[6] The primary objective is to generate leads for plausible hypotheses, setting the stage for analytical studies to conduct further investigations.[7] Insights gained from case series on the natural history, epidemiology, risk factors, and recovery patterns of diseases can help design subsequent studies by guiding inclusion criteria, sample population, follow-up duration, and relevant covariates. Moreover, case series can circumvent the ethical constraints of experimental research, serving as preliminary data sources until comparative evidence is available, especially in treating rare diseases.[8] Refer to Table 36.1 for historical examples that have shaped medical discoveries.

Design of case series

Having established their significance, we now turn to its design. A strategic design can mitigate several inherent limitations. We have deconstructed the design into the following steps.

Case selection

The first step is to define a clear selection criterion for cases to prevent selection bias. For example, describing every patient visiting the health center over a period ensures a diverse and representative sample. Ideally, a case series should have at least four patients, the selection of which depends on the overall study objective:[20]

1. *Intrinsic case series* include unique cases with rare or unusual characteristics to understand presentation, management, or prognosis. For example, the 1976 Ebola virus disease reports depicting unique features helped identify the disease.[21]
2. *Instrumental case series* involve typical cases to understand a broader phenomenon. For example, the case series of patients with severe acute respiratory syndrome

Translational Cardiology. https://doi.org/10.1016/B978-0-323-91790-2.00108-8

TABLE 36.1 Examples of milestones in medical discoveries that began with case reports and series.

Role in medical advancement	Specific examples		
Discovery of new diseases	**Disease/pathogen discovered**	**Year**	**Case series describing**
	Parkinson disease	1817	Patients with a "shaking palsy"[2]
	Sickle cell disease	1910	A woman with severe anemia, multisystem illness, and "freakish" elongated red cells on blood smear[9]
	AIDS	1980s	Kaposi's sarcoma in young homosexuals[3]
	Toxic shock syndrome	1970s	Menstruating women presenting with shock leading to its association with high-absorbent tampons[10]
Discovery of new pathogen	Zika virus	2015	Infant born with microcephaly in Brazil prompting research and identification of Zika as cause of birth defects[11]
	Lyme disease	1970s	Patients in the towns of Lyme and Old Lyme, Connecticut presenting with a cluster of similar manifestations[12]
	SARS-CoV2	2020	Characteristics and outcomes of 41 adults with confirmed COVID-194
Novel treatment for a known condition	*E. coli*-associated hemolytic uremic syndrome	2011	Patients noting significant improvement on eculizumab[13]
	Familial Mediterranean fever	1972	Improvement of disease with colchicine[14]
	Manic-depressive disorder	1949	Effect of lithium on manic episodes[15]
	Infantile hemangioma	2008	Cure of severe hemangiomas with systemic propranolol[16]
Unknown adverse effects of a known drug	Reye syndrome	1963	Children on aspirin dying from fatty degeneration of the viscera[17]
	Phocomelia	1960s	Pregnant women prescribed thalidomide for nausea giving birth to children with limb malformations[18]
Formation of promising hypothesis	Nearly 25% of case series and reports published in The Lancet were followed up with randomized control studies[19]		

in the early 2000s was instrumental in understanding the transmission and pathophysiology of the disease.[22]

3. *Collective case series* include multiple cases that allow cross-comparisons to identify patterns and variances. For example, analysis of pediatric patients during the H1N1 influenza pandemic (2009) enabled comparisons across different parameters like hospitalization and mortality rates.[23]

Literature review

After selecting cases, an extensive literature review using databases like PubMed and Google Scholar with targeted keywords will help determine the cases' value and relevance.

Ethical considerations

Before collecting data, ethical imperatives like anonymity and informed consent are necessary to respect participant privacy and autonomy. Stringent data handling protocols ensure confidentiality and prevent direct or indirect identification.

Data collection

Strategic data collection from selected cases can boost the validity of the study. The efficacy can be enhanced by employing the following methods:

1. Adequate identification and recording of both exposure and outcome. (Medical records provide more accurate identification over self-report.)
2. Selection of diverse cases for a broad view of conditions under investigation. (Case series investigating a new cancer drug includes patients at various stages of the disease.)
3. Use of a combination of qualitative and quantitative techniques (using a mix of medical records, interviews, and questionnaires to collect data).

4. Verification of study conclusions with the participants for alignment between researcher and participants.
5. Collection of data through consistent sources across different cases to enable cross-case analysis and comparison.

Analysis and interpretation

Following data collection, case series can either narratively detail findings, or may use statistical analysis for data pooling. Exploring alternative explanations of observations can help investigate causality. For example, strong inference drawn in cases of adverse drug reactions that resolve with drug cessation and recur with reintroduction.

Writing the report

The final step of design is translating the findings into a report. The 2013 CARE (CAse REport) Guidelines provide a 13-item checklist for crafting detailed and transparent case reports including:[24]

1. *Title:* Include "case series" or "case report" and highlight the significant or unique aspect.
2. *Keywords:* Choose 5–10 words capturing core elements.
3. *Abstract:* Give an overview of the case description and conclusion.
4. *Introduction:* Focus on novelty, background, and summary with the current status of the research topic
5. *Case Presentation*: Detail patient demographics, chief complaints, medical history, clinical findings, timeline, diagnostics, interventions, outcomes, and adverse events (if any).
6. *Discussion and Conclusion:* Discuss how findings contribute to the research question and justify conclusions with medical literature.

The subsequent section provides a step-by-step guide through the entire process from identifying cases to the final publishing of the report.

Get started

Equipped with an understanding of how to design a case series, the following action steps can guide the conduct of a case series. See Table 36.2 for examples of each of these steps in a hypothetical case series in cardiology.

1. Following the words of Powell, "almost all practicing physicians eventually will encounter a case worthy of being reported," keep an eye out for any unusual presentation, rare disease, or novel treatment response.
2. Create specific case selection criteria. Avoid vague questions to prevent ambiguity.
3. Utilize databases like PubMed and Google Scholar to ensure your cases can make a valuable contribution to existing literature.
4. Obtain informed consent from participants by explaining the study and ensuring confidentiality.
5. Collect data on demographics, medical history, clinical findings, diagnostic methods, treatments, and outcomes. Both too little and too large data can decrease validity.
6. Look for any pattern, difference, and unexpected result from the data and be cautious against bias. Employ statistical analysis if applicable.
7. Before moving on to write the report, check multiple journals' guidelines, impact factors, and audience to decide a good fit for your work.
8. Finally, follow the CARE checklist and journal's specifications to draft a comprehensive report.[24]
9. Seek feedback from mentors and colleagues before submission to avoid oversight.
10. Submit for publication and expect revision requests from reviewers.
11. Once published, share the article through professional networks, social media, and conferences to reach a wider audience.

Potential pitfalls

While beginning the research is exciting, being aware of pitfalls at each step can eliminate common issues such as selection bias, information overload, ethical compromise, inept methodology, rigid approach to conclusions, and duplication of existing research. To help navigate through these potential pitfalls, key strategies with explanations are surmised in Table 36.3.[6]

Classic examples

There are multiple historical case series that have shaped medical breakthroughs through the centuries (Table 36.1). Three of them are described here to elucidate their significance.

Case series on early COVID-19 outbreak in China

The case series by Huang et al. (2020)[4] during the early outbreak of COVID-19 is a recent example that illustrates the role of case series in new epidemics. The case series, detailing epidemiological and clinical features of 41 hospitalized patients in Wuhan, provided significant insights, such as common symptoms and prognosis, into the novel nCoV-2019. The swift publication informed global communities of its high transmission rates and mortality, providing an impetus for urgent public health interventions.[4]

TABLE 36.2 Examples of action points for a hypothetical case series in cardiology.

Context: A doctor observes a trend of unexpected arrhythmias after a specific type of mitral valve surgery and wonders whether it is due to postoperative care protocols.

Action	Examples
Identify a research question	"Does the postoperative care protocol influence the incidence of arrhythmias after mitral valve surgery?"
Select cases	All the patients who developed arrhythmias after mitral valve surgery in the past month
Conduct a literature review	Review of existing articles on arrhythmia after mitral valve surgery by combining keywords with Boolean operators. For example: • "mitral valve surgery" AND "postoperative arrhythmias" • "mitral valve repair" OR "mitral valve replacement" AND "arrhythmia incidence" • "postsurgical cardiac care" AND "atrial fibrillation" • "cardiac arrhythmia" NOT "preoperative"
Ethical compliance	Approval from the ethics committee, ensuring anonymization and consent where applicable.
Data collection	Collecting data on all possible confounding factors: • Patient demographics and preoperative data • Details of surgery like type, technique, duration, complications • Postoperative care protocols like medications, rehabilitation • Arrhythmia incidence, type, time of onset postsurgery, duration, severity
Analysis	Descriptive analysis to identify associations between care protocols and arrhythmias
Choose the right journal	A journal focused on cardiothoracic surgery and cardiology like Journal of Cardiology Cases and European Heart Journal
Writing the report	Manuscript detailing the current status of the problem, case description, methodology, findings, and implications, emphasizing the need for further research to find accurate associations for postoperative arrhythmia. If significant associations are found, causality can be explored through subsequent analytic studies.
Seek feedback	Gathering feedback from peers and mentors at a departmental meeting
Submit for publication	Submitting the manuscript to the chosen journal, ready for review
Share your findings	Sharing the research at cardiology conferences and online forums

TABLE 36.3 Key strategies to mitigate potential pitfalls in case series studies.

Key strategy	Explanation
Focused data collection	Collect data that directly answers the research question to avoid information overload
Upstand ethics	Failure can compromise participant trust and confidentiality, and the ethical standing of the study
Maintain rigor in methodology	• Increase external validity by selecting diverse patients offering varied insights • Use both qualitative (questionnaires, already collected healthcare data) and quantitative (interviews, observations) data sources • Validate findings with participants to ensure accuracy and reliability of study conclusions
Stay flexible	Remain open to refining hypotheses and explore alternate explanations based on new or unexpected findings
Maintain transparency	Document the rationale behind case selection, data collection, and chosen methods and mention researcher's degree of involvement
Define the case	Delineate the scope of the study to maintain focus and relevance to the research question An underrepresented sample can skew study findings
Develop in-depth knowledge of the literature	To prevent duplicating existing research without adding new insights

An unusual outbreak of Kaposi's sarcoma leading to discovery of AIDS

In New York City, Hymes et al. reported an unusual outbreak of KS between 1979 and 1981. Generalized lesions and aggressive course reported in young homosexuals contradicted the traditionally observed, rare, indolent, localized disease affecting older individuals. The study hypothesized the possibility of increased KS among homosexuals and association with concurrent STDs. This study drove further research, ultimately leading to the discovery of HIV and AIDS. Case series, thereby, are indispensable for early documentation of atypical disease patterns and generation of hypotheses.[3]

Case series of "fatty degeneration of the viscera, of unknown cause" in children

In 1963, Australian pathologist R. D. K. Reye reported a unique "clinicopathological entity" characterized by rapid-onset, severe encephalopathy, and liver degeneration in children. He named it "fatty degeneration of the viscera, of unknown cause," owing to its hallmark pathological findings. This observation ushered international surveillance efforts and case-control studies that ultimately linked aspirin use to the condition. The disease, now known as Reye syndrome, demonstrates how initial observations through case series can steer research to identify causal relationships. Current guidelines discourage aspirin use in children, markedly reducing its incidence and safeguarding millions of lives annually.[17]

Conclusion

In conclusion, case series serve as a valuable outlet for reporting novel observations and generating hypotheses. However, case series, positioned at the bottom of the evidence pyramid, are susceptible to selection bias, confounding, and lack of external and internal validity. Even then, strong recommendations from case series can impact decision-making and complement analytical studies.[25]

Resources

1. Kirch W, ed. *Case Series. Encyclopedia of Public Health*. Netherlands: Springer; 2008:100–101. https://doi.org/10.1007/978-1-4020-5614-7_327.
2. Huang C, Wang Y, Li X, et al. Clinical features of patients infected with 2019 novel coronavirus in Wuhan, China. *Lancet*. 2020;395(10223):497–506. https://doi.org/10.1016/S0140- 6736(20)30183-5.
3. Hymes KB, Cheung T, Greene JB, et al. Kaposi's sarcoma in homosexual men-a report of eight cases. *Lancet Lond Engl*. 1981;2(8247):598–600. https://doi.org/10.1016/s0140-6736(81)92740-9.
4. Reye RDK, Morgan G, Baral J. Encephalopathy and fatty degeneration of the viscera a disease entity in childhood. *Lancet*. 1963;282(7311):749–752. https://doi.org/10.1016/S0140-6736(63)90554-3.
5. Murad MH, Sultan S, Haffar S, Bazerbachi F. Methodological quality and synthesis of case series and case reports. *BMJ Evid-Based Med*. 2018;23(2):60–63. https://doi.org/10.1136/bmjebm-2017-110853.

References

1. Case series. In: Kirch W, ed. *Encyclopedia of Public Health*. Netherlands: Springer; 2008:100–101. https://doi.org/10.1007/978-1-4020-5614-7_327.
2. Parkinson J. An essay on the shaking palsy. *J Neuropsychiatry Clin Neurosci*. 2002;14(2):223–236. https://doi.org/10.1176/jnp.14.2.223.
3. Hymes KB, Cheung T, Greene JB, et al. Kaposi's sarcoma in homosexual men-a report of eight cases. *Lancet Lond Engl*. 1981;2(8247):598–600. https://doi.org/10.1016/s0140-6736(81)92740-9.
4. Huang C, Wang Y, Li X, et al. Clinical features of patients infected with 2019 novel coronavirus in Wuhan, China. *Lancet*. 2020;395(10223):497–506. https://doi.org/10.1016/S0140-6736(20)30183-5.
5. McDermott MM, Lefevre F, Feinglass J, et al. Changes in study design, gender issues, and other characteristics of clinical research published in three major medical journals from 1971 to 1991. *J Gen Intern Med*. 1995;10(1):13–18. https://doi.org/10.1007/BF02599570.
6. Murad MH, Sultan S, Haffar S, Bazerbachi F. Methodological quality and synthesis of case series and case reports. *BMJ Evid-Based Med*. 2018;23(2):60–63. https://doi.org/10.1136/bmjebm-2017-110853.
7. Misra DP, Gasparyan AY, Zimba O, Yessirkepov M, Agarwal V, Kitas GD. Formulating hypotheses for different study designs. *J Korean Med Sci*. 2021;36(50):e338. https://doi.org/10.3346/jkms.2021.36.e338.
8. Wynn R. Coercion in psychiatric care: clinical, legal, and ethical controversies. *Int J Psychiatr Clin Pract*. 2006;10(4):247–251. https://doi.org/10.1080/13651500600650026.
9. Herrick JB. Peculiar elongated and sickle-shaped red blood corpuscles in a case of severe anemia. 1910. *Yale J Biol Med*. 2001;74(3):179–184.
10. Todd J, Fishaut M, Kapral F, Welch T. Toxic-shock syndrome associated with phage-group-I staphylococci. *Lancet Lond Engl*. 1978;2(8100):1116–1118. https://doi.org/10.1016/s0140-6736(78)92274-2.
11. Kleber de Oliveira W, Cortez-Escalante J, De Oliveira WTGH, et al. Increase in reported prevalence of microcephaly in infants born to women living in areas with confirmed Zika virus transmission during the first trimester of pregnancy - Brazil, 2015. *MMWR Morb Mortal Wkly Rep*. 2016;65(9):242–247. https://doi.org/10.15585/mmwr.mm6509e2.
12. Steere AC, Malawista SE, Snydman DR, et al. Lyme arthritis: an epidemic of oligoarticular arthritis in children and adults in three connecticut communities. *Arthritis Rheum*. 1977;20(1):7–17. https://doi.org/10.1002/art.1780200102.

13. Lapeyraque AL, Malina M, Fremeaux-Bacchi V, et al. Eculizumab in severe Shiga-toxin-associated HUS. *N Engl J Med.* 2011;364(26):2561–2563. https://doi.org/10.1056/NEJMc1100859.
14. Colchicine for familial mediterranean fever. *N Engl J Med.* 1972;287(25). https://doi.org/10.1056/NEJM197212212872514.
15. Cade JFJ. Lithium salts in the treatment of psychotic excitement. *Med J Aust.* 1949;2(10):349–352. https://doi.org/10.1080/j.1440-1614.1999.06241.x.
16. Léauté-Labrèze C, Dumas de la Roque E, Hubiche T, Boralevi F, Thambo JB, Taïeb A. Propranolol for severe hemangiomas of infancy. *N Engl J Med.* 2008;358(24):2649–2651. https://doi.org/10.1056/NEJMc0708819.
17. Reye RDK, Morgan G, Baral J. Encephalopathy and fatty degeneration of the viscera a disease entity in childhood. *Lancet.* 1963;282(7311):749–752. https://doi.org/10.1016/S0140-6736(63)90554-3.
18. Coodin FJ, Uchida IA, Murphy CH. Phocomelia: report of three cases. *Can Med Assoc J.* 1962;87(14):735–739. https://www.ncbi.nlm.nih.gov/pmc/articles/PMC1849647/. Accessed February 8, 2024.
19. Albrecht J, Meves A, Bigby M. Case reports and case series from Lancet had significant impact on medical literature. *J Clin Epidemiol.* 2005;58(12):1227–1232. https://doi.org/10.1016/j.jclinepi.2005.04.003.
20. Abu-Zidan FM, Abbas AK, Hefny AF. Clinical "case series": a concept analysis. *Afr Health Sci.* 2012;12(4):557–562.
21. Emond RT, Evans B, Bowen ET, Lloyd G. A case of Ebola virus infection. *Br Med J.* 1977;2(6086):541–544. https://doi.org/10.1136/bmj.2.6086.541.
22. Zhong NS, Zheng BJ, Li YM, et al. Epidemiology and cause of severe acute respiratory syndrome (SARS) in Guangdong, People's Republic of China, in February, 2003. *Lancet Lond Engl.* 2003;362(9393):1353–1358. https://doi.org/10.1016/s0140-6736(03)14630-2.
23. Libster R, Bugna J, Coviello S, et al. Pediatric hospitalizations associated with 2009 pandemic influenza A (H1N1) in Argentina. *N Engl J Med.* 2010;362(1):45–55. https://doi.org/10.1056/NEJMoa0907673.
24. Gagnier JJ, Kienle G, Altman DG, et al. The CARE guidelines: consensus-based clinical case reporting guideline development. *J Med Case Rep.* 2013;7(1):223. https://doi.org/10.1186/1752-1947-7-223.
25. Guyatt GH, Oxman AD, Vist GE, et al. GRADE: an emerging consensus on rating quality of evidence and strength of recommendations. *BMJ.* 2008;336(7650):924–926. https://doi.org/10.1136/bmj.39489.470347.AD.

Chapter 37

Case control study: Design, measures, and classic examples

Suryasnata Bhowmik[1] and Adrija Hajra[2]

[1]Critical Care Medicine, Nilratan Sircar Medical College, Kolkata, West Bengal, India; [2]Department of Internal Medicine, Bringham and Women's Hospital/Harvard Medical School, Boston, MA, United States

Key points

- Case-control studies retrospectively identify causes or risk factors by comparing individuals with a specific outcome (cases) to those without (controls)
- The odds ratio is the primary measure used to compare the odds of exposure in the case population to the odds of exposure in the control population
- There are multiple possible biases and pitfalls that can be circumvented by cautious selection of cases and controls and judicious analysis of data

Introduction

Case-control studies are observational analytic studies that track individuals with a specified outcome to identify potential causes or risk factors (refer to Fig. 37.1).[1] Ideal for investigating rare or insidious diseases, they offer rapid, efficient, and cost-effective methods for providing hypotheses and insights particularly where cohort studies are unfeasible.[2] Beyond their practical advantages, case-control studies address the ethical challenges of exposing individuals to potential risk factors and facilitate the simultaneous assessment of multiple risk factors, beneficial in diseases without clear hypotheses.[3] Historically, various associations that are now widely accepted were first identified through case-control studies (refer to section "Classical Examples").

Design of case-control studies

Having recognized their significance, we will now explore the design and measures of case-control studies. The first step is the selection of cases (individuals with outcome of interest) and controls (individuals without outcome of interest). At baseline, the exposure status is unknown in both the groups which is compared later during analysis to explore and quantify possible risk factors[3,4] (refer to Fig. 37.2).

Case selection

Outcomes should be defined using objective criteria or through the classification of disease certainty in situations where the diagnosis is uncertain (definite, probable, possible).[4] Selecting cases based on disease prevalence (preexisting cases) or incidence (new cases) affects the study's focus, with the latter preferred for identifying risk factors.[5] Sourcing cases can vary from the general population through mass screening, previous cohort/cross-sectional studies, or hospital records and disease registries.[4]

Selection of control

The control group should represent the relative sizes of the exposed and unexposed segments of the population appropriately and sample independently of the exposure.[6] The number of controls selected can affect study power.[7] Increasing the number of controls relative to cases can enhance statistical power, but benefits diminish beyond a 4:1 ratio.[8]

Selection strategies include the following:[4,9]

1. Base sampling: offers each individual an equal chance of selection, allowing odds ratio to serve as a valid estimate of risk ratio; suitable when the disease is not rare. For example, selecting nondiabetics from a registry of all residents for a diabetes study.
2. Cumulative density sampling: selecting individuals who are free of outcome till study end. Here, the odds ratio

Translational Cardiology. https://doi.org/10.1016/B978-0-323-91790-2.00097-6

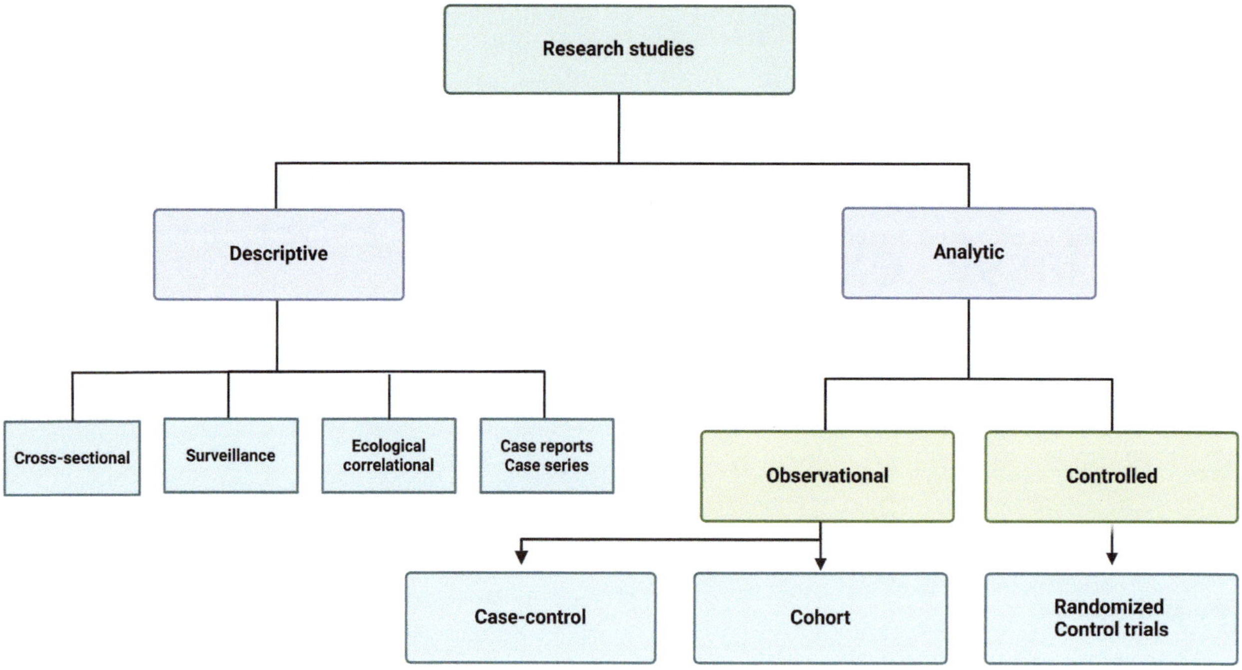

FIGURE 37.1 Classification of research studies. *Created by BioRender.com*

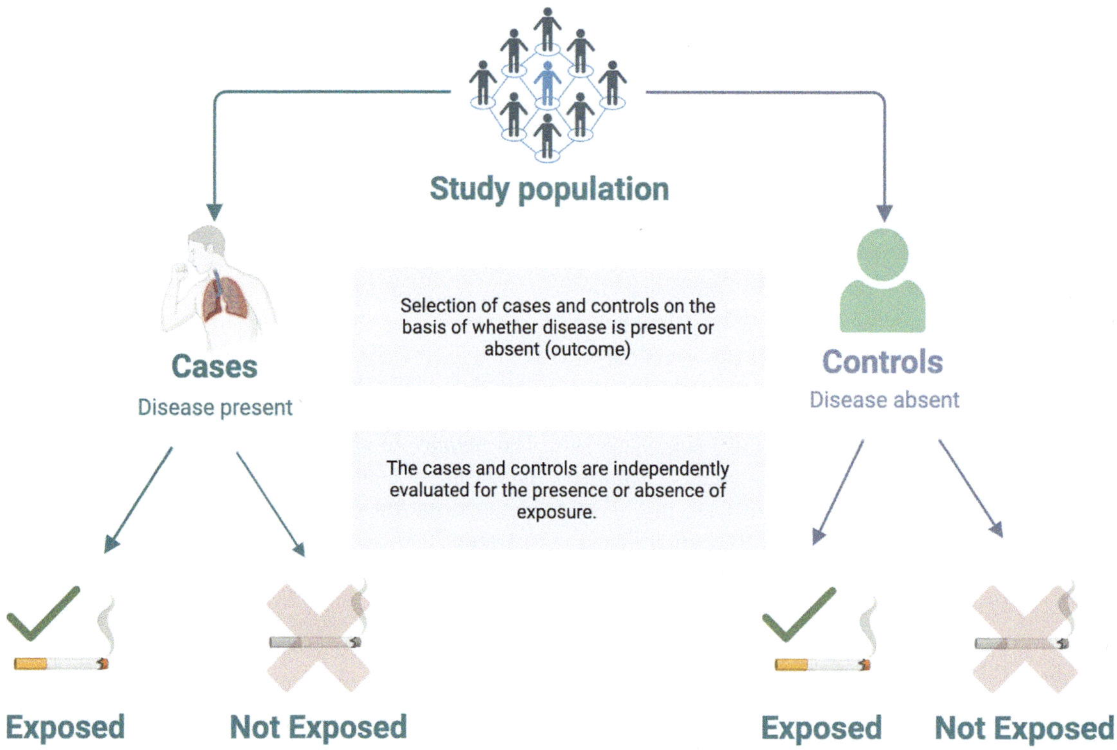

FIGURE 37.2 Schematic overview of design of case-control studies. *Created by BioRender.com*

can be approximated to the risk ratio only under "rare disease assumption."[10]

3. Incidence density sampling: selecting controls and cases concurrently from the same population, allowing a direct estimate of risk ratio without a "rare disease assumption."[10]

The controls can be drawn from various sources including the following:[9]

1. Population: drawn from the same population as the cases, such as a specific geographic area. For example, controls might be randomly chosen from driver's lists.
2. Neighborhood, family, and friends: help control for confounders like socioeconomic status or genotype.
3. Hospitals: selected from the same facility as the cases but should not have diseases related to the exposure under investigation. Choosing controls for a lung cancer study, for example, from a respiratory ward might underestimate the association due to the higher prevalence of smoking in that setting.

Measures

Case-control studies primarily utilize odds ratio (OR) to estimate the relative odds of disease occurrence in exposed versus unexposed individuals. This calculation is straightforward once the exposure status among cases and controls is known (refer to Fig. 37.3).[11]

From the data illustrated in Fig. 37.3, calculating the odds ratio involves the following:

1. Calculating the odds of exposure among cases = a/c = 60/40
2. Calculating the odds of exposure among controls = b/d = 20/80
3. Determining the odds ratio: (a/c)/(b/d) = (an × d)/(c × b) = (60 × 80)/(40 × 20) = 6

The value of the odds ratio can be interpreted as follows:[12]

- OR = 1 indicates no difference in odds between groups.
- OR > 1: suggests increased odds of disease with exposure.
- OR <1: suggests exposure is protective, decreasing odds of disease.

Hence, the above value of odds ratio can be interpreted as: Individuals exposed to the factor (e.g., smoking) have six times the odds of developing the outcome (e.g., lung cancer) compared to unexposed individuals.

Risk ratio versus odds ratio

Risk ratio (RR), calculated in cohort studies, compares the probability of disease in the exposed and unexposed of the *total* population. However, procuring such data of the *entire* population is often unfeasible.[13] Consequently, case-control studies utilize a smaller sample, maintaining a constant number of cases but selecting only limited disease-free controls to calculate OR.

Furthermore, case-control studies cannot measure incidence as controls represent only a fraction of the population, not the total denominator required for calculating incidence rates.[14]

The confidence interval and statistical significance of the OR can be assessed using chi-square tests or Fisher's exact test for smaller sample sizes.[15]

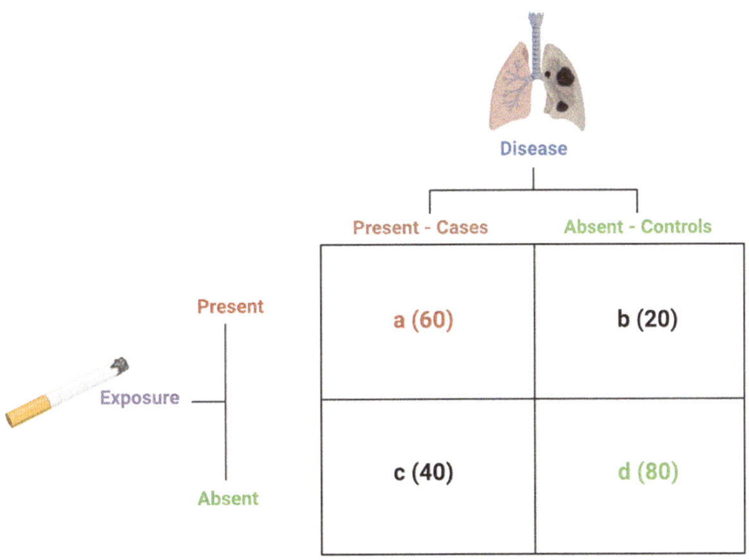

FIGURE 37.3 Distribution of exposure among cases and controls with hypothetical values in a 2 × 2 table. *Created with BioRender.com*

Report writing

After analysis, the reporting of findings to ensure the reproducibility and applicability of research constitutes the final step. The STROBE Statement guides the reporting of case-control studies, ensuring clarity and completeness.[16] Key components include.

1. Title: Includes "case-control study" and a concise summary of the main findings
2. Abstract: An overview on background, objectives, methods, results, and conclusion
3. Introduction: Current understanding of the problem, citing recent references and identifying gaps that the study aims to bridge
4. Methods: Study design, participant selection, definition of variables (e.g., outcome and exposure), and statistical techniques.
5. Results: Present findings including participant demographics and exposure measures.
6. Discussion: Interpret results, discuss limitations, and suggest implications.
7. Funding disclosure: Acknowledge financial support and potential conflicts of interest.

Get started

Equipped with an understanding of the design, the following action steps can guide on how to get started.

1. Start with a literature review to identify knowledge gaps.
2. Formulate an initial broad question ("What could be causing this unprecedented rise of vaginal cancer cases?"), then refine it to your specific area of interest ("Do women with in-utero exposure to DES have greater odds to develop vaginal cancer?").
3. Design the case, justifying the use of a case-control study approach and assessing the feasibility of garnering the requisite number of appropriate cases and controls.
4. Define cases and controls with clear diagnostic and eligibility criteria to ensure a valid comparison.[3]
5. Secure informed consent, ethics review committee approval, and ensure patient confidentiality.
6. Execute the study, choosing appropriate pretested research instruments and data sources, and collect necessary information on desired variables.
7. Analyze the data, defining the exposure of interest while accounting for potential confounders to eliminate bias.
8. Write the report following the STROBE guidelines.[16]

Potential pitfalls

The validity of case-control studies is affected by certain pitfalls that can be attempted to mitigate through awareness and several strategies, including the following:

1. Population-based controls, generally healthier, might lack motivation to recollect details of their exposure accurately. Conversely, cases might overemphasize past exposure, such as overreporting past smoking history in lung cancer patients (recall bias).[17] Using memory aids or adopting a prospective study design can abate this bias. Incident cases, being more motivated, may provide more accurate exposure histories.[18]
2. Ensuring uniform eligibility criteria across cases and controls will prevent "dilution" of any potential association by excluding individuals not at risk of exposure.[4]
3. Substantial differences in baseline characteristics between cases and controls can introduce selection bias and confounding. *Matching* of key characteristics like age, sex, and other significant outcome predictors can diminish these issues.[19] However, cautious limitations of matching parameters to those associated with outcome will prevent *overmatching*.[20]
4. In hospital-based controls, ensuring controls have similar referral patterns as cases will prevent selection bias; for instance, matching patients from a local clinic with those referred from across a state may introduce bias.[21]
5. In convenience-based controls, selecting controls from friends and neighbors might inadvertently include individuals predisposed to the exposure, making their selection inappropriate.[22]
6. Observer or interviewer bias arises when the investigator's knowledge of the participant's case or control status influences the assessment of exposure.[23] This may lead to differential reporting which can be mitigated by *blinding* investigators to the participant status.[24]

Classic examples

Various landmark case-control studies have shaped medical discoveries and public health policies in the past century. In this section, we delve into three such historical examples.

Smoking and lung cancer: Doll and Hill 1950

Amid the stark rise in lung cancer mortality between 1920 and 1940s, Doll and Hill conducted a large-scale study of over 700 lung cancer cases, collecting detailed smoking histories (1948–49). They were compared to an age-and-sex-matched control group without lung cancer. The study reported 0.3% nonsmokers among cases, in contrast to 4.2% nonsmokers in the control group. This significant difference led them to a compelling association between lung cancer and smoking, paving the way for future research and increased awareness.[25]

Modifiable risk factors and coronary heart disease—Yusuf et al. 2004

By the late 20th century, although cardiovascular mortality had declined globally, 80% of its burden remained with developing countries. Prior research, often centered on Caucasians, had limited worldwide generalizability. An international case-control study, the INTERHEART study, sought to address this gap by assessing risk factors for myocardial infarction (MI) across 52 countries, enrolling over 15,000 cases. The study identified nine easily measurable and modifiable factors that accounted for over 90% of cases, with smoking and dyslipidemia being the strongest contributors. The INTERHEART highlighted the preventable nature of MI especially in the young, thus providing a prime example of the potential of case-control studies to uncover global patterns and decrease disease burden.[26]

Diethylstilbestrol and vaginal cancer—Herbst et al

In the late 1960s at Massachusetts General Hospital, physicians identified a rare cluster of vaginal cancer in young women aged 15–22. With no initial hypotheses and considering the disease's rarity, a case-control study was designed with eight cases and 32 age-matched controls. The study enabled analysis of multiple exposures concomitantly that finally linked cancer to in utero exposure of diethylstilbestrol, a drug commonly prescribed at the time to prevent miscarriages. Herbst et al.'s study demonstrated the effectiveness of case-control studies in rare diseases with prolonged latency (in this case, 15–20 years) where no other research methods would be feasible.[27]

Conclusion

The perceived evidence hierarchy, criticized by many, undervalues case-control studies.[6] However, they remain indispensable to public health, offering insights into disease etiology with practical and ethical efficiency.

Resources

1. Schlesselman JJ. *Case-Control Studies: Design, Conduct, Analysis*. Oxford University Press; 1982.
2. Schulz KF, Grimes DA. Case-control studies: research in reverse. *The Lancet*. 2002; 359(9304):431–434.
3. Grimes DA, Schulz KF. Compared to what? Finding controls for case-control studies. *Lancet*. 2005;365(9468):1429–1433.https://doi.org/10.1016/S0140-6736(05)66379-9.
4. Doll R, Hill AB. Smoking and carcinoma of the lung; preliminary report. *Br Med J*. 1950;2(4682):739–748. https://doi.org/10.1136/bmj.2.4682.739.
5. Yusuf S, Hawken S, Ôunpuu S, et al. Effect of potentially modifiable risk factors associated with myocardial infarction in 52 countries (the INTERHEART study): case-control study. *Lancet*. 2004;364(9438):937–952. https://doi.org/10.1016/S0140-6736(04)17018-9.
6. Herbst AL, Ulfelder H, Poskanzer DC. Adenocarcinoma of the vagina. *N Engl J Med*. 1971;284(16):878–881. https://doi.org/10.1056/NEJM197104222841604.

References

1. Setia MS. Methodology series module 2: case-control studies. *Indian J Dermatol*. 2016;61(2):146–151. https://doi.org/10.4103/0019-5154.177773.
2. Vandenbroucke JP, Pearce N. Case—control studies: basic concepts. *Int J Epidemiol*. 2012;41(5):1480–1489.
3. Schulz KF, Grimes DA. Case-control studies: research in reverse. *The Lancet*. 2002;359(9304):431–434.
4. Schlesselman JJ. *Case-Control Studies: Design, Conduct, Analysis*. Oxford University Press; 1982.
5. Miettinen OS. The "case-control" study: valid selection of subjects. *J Chronic Dis*. 1985;38(7):543–548. https://doi.org/10.1016/0021-9681(85)90039-6.
6. Rothman KJ. Six persistent research misconceptions. *J Gen Intern Med*. 2014;29(7):1060–1064. https://doi.org/10.1007/s11606-013-2755-z.
7. Groenwold RHH, van Smeden M. Efficient sampling in unmatched case-control studies when the total number of cases and controls is fixed. *Epidemiol Camb Mass*. 2017;28(6):834–837. https://doi.org/10.1097/EDE.0000000000000710.
8. Grimes DA, Schulz KF. Compared to what? Finding controls for case-control studies. *Lancet*. 2005;365(9468):1429–1433. https://doi.org/10.1016/S0140-6736(05)66379-9.
9. Wacholder S, Silverman DT, McLaughlin JK, Mandel JS. Selection of controls in case-control studies: III. Design options. *Am J Epidemiol*. 1992;135(9):1042–1050.
10. Greenland S, Thomas DC. On the need for the rare disease assumption in case-control studies. *Am J Epidemiol*. 1982;116(3):547–553. https://doi.org/10.1093/oxfordjournals.aje.a113439.
11. Szumilas M. Explaining odds ratios. *J Can Acad Child Adolesc Psychiatry*. 2010;19(3):227–229.
12. Andrade C. Understanding relative risk, odds ratio, and related terms: as simple as it can get. *J Clin Psychiatr*. 2015;76(7):e857–e861. https://doi.org/10.4088/JCP.15f10150.
13. Cummings P. The relative merits of risk ratios and odds ratios. *Arch Pediatr Adolesc Med*. 2009;163(5):438–445. https://doi.org/10.1001/archpediatrics.2009.31.
14. Ranganathan P, Aggarwal R, Pramesh CS. Common pitfalls in statistical analysis: odds versus risk. *Perspect Clin Res*. 2015;6(4):222–224. https://doi.org/10.4103/2229-3485.167092.
15. Kim HY. Statistical notes for clinical researchers: chi-squared test and Fisher's exact test. *Restor Dent Endod*. 2017;42(2):152–155. https://doi.org/10.5395/rde.2017.42.2.152.

16. Vandenbroucke JP, von Elm E, Altman DG, et al. Strengthening the reporting of observational studies in epidemiology (STROBE): explanation and elaboration. *PLoS Med.* 2007;4(10):e297. https://doi.org/10.1371/journal.pmed.0040297.
17. Sedgwick P. Bias in observational study designs: case-control studies. *BMJ.* 2015;350:h560. https://doi.org/10.1136/bmj.h560.
18. Sutton-Tyrrell K. Assessing bias in case-control studies. Proper selection of cases and controls. *Stroke.* 1991;22(7):938–942. https://doi.org/10.1161/01.STR.22.7.938.
19. Horton R. *Encyclopaedic Companion to Medical Statistics.* Google Scholar; 2011. https://scholar.google.com/scholar_lookup?title=Encyclopaedic+Companion+to+Medical+Statistics&author=BS+Everitt&author=CR+Palmer&publication_year=2005&. Accessed February 28, 2024.
20. Elwood M. *Critical Appraisal of Epidemiological Studies.* Google Scholar; 2017. https://scholar.google.com/scholar_lookup?title=Critical+Appraisal+of+Epidemiological+Studies+and+Clinical+Trials&author=M+Elwood&publication_year=2007&. Accessed February 28, 2024.
21. Browner WS. *Designing Clinical Research.* Google Scholar; 2022. https://scholar.google.com/scholar_lookup?title=Designing+Clinical+Research:+An+Epidemiologic+Approach&author=SB+Hulley&author=SR+Cummings&author=WS+Browner&publication_year=2001&. Accessed February 28, 2024.
22. Am G, Se H, Rw H. Selection bias in case-control studies using relatives as the controls. *Int J Epidemiol.* 1989;18(4). https://doi.org/10.1093/ije/18.4.985.
23. Hammer GP, du Prel JB, Blettner M. Avoiding bias in observational studies. *Dtsch Ärztebl Int.* 2009;106(41):664–668. https://doi.org/10.3238/arztebl.2009.0664.
24. Mann CJ. Observational research methods. Research design II: cohort, cross sectional, and case-control studies. *Emerg Med J.* 2003;20(1):54–60.
25. Doll R, Hill AB. Smoking and carcinoma of the lung; preliminary report. *Br Med J.* 1950;2(4682):739–748. https://doi.org/10.1136/bmj.2.4682.739.
26. Yusuf S, Hawken S, Ôunpuu S, et al. Effect of potentially modifiable risk factors associated with myocardial infarction in 52 countries (the INTERHEART study): case-control study. *Lancet.* 2004;364(9438):937–952. https://doi.org/10.1016/S0140-6736(04)17018-9.
27. Herbst AL, Ulfelder H, Poskanzer DC. Adenocarcinoma of the vagina. *N Engl J Med.* 1971;284(16):878–881. https://doi.org/10.1056/NEJM197104222841604.

Chapter 38

Cohort study: design, measures, and classic examples

Binita Ghosh[1], Suryasnata Bhowmik[2] and Dhrubajyoti Bandyopadhyay[3]

[1]*Internal Medicine, SSM Health St. Mary's Hospital, St. Louis, MO, United States;* [2]*Critical Care Medicine, Nilratan Sircar Medical College, Kolkata, West Bengal, India;* [3]*Cardiovascular Disease, Massachusetts General Hospital, Boston, MA, United States*

Key points

- Cohort studies are longitudinal observational studies where subjects are categorized based on their exposure status and followed over time until an outcome develops.
- These studies are utilized to measure incidence rates and relative risk/hazard ratios.
- While cohort studies are generally less prone to biases, issues such as loss to followup and confounders are common challenges that should be carefully addressed during the study design phase.

Introduction

A cohort study is a longitudinal observational study design primarily used to determine the incidence rates of health outcomes. By following subjects over time from exposure to the development of an outcome, cohort studies can establish both the temporal association and the strength of the association (relative risk), providing strong evidence for establishing causality. The cohort study design is popular in translational cardiovascular research for investigating the risk factors of cardiovascular diseases (CVD). The Framingham Heart Study stands as a landmark cohort study in cardiovascular medicine. In this chapter, we will discuss the design, measurements, strengths, limitations, and applications of cohort studies, illustrated with real world examples.

Definition

Cohort studies are a type of observational study in which a "cohort"—a group of subjects with specified characteristics—is divided into two or more groups based on their exposure to a risk factor. These groups are then followed over time to observe the development of a specific outcome.[1]

Clinical questions addressed by cohort studies

1. Measuring cumulative incidence and incidence rates of a health outcome.[1,2]
2. Relative risk and hazard ratio can be calculated. These are tools used to estimate the association strength between a risk factor and a disease. While they do not establish causality, the measured strength of association offers evidence for the establishment of a causal relationship.[1]
3. Study the natural history and prognosis of a health condition.[1]

Subtypes of a cohort study

Cohort studies can be prospective and retrospective (Fig. 38.1; Table 38.1).

Nested case control study

This is a type of case control study nested in a cohort study where the participants of a cohort study who developed an outcome are considered as cases, and the participants who did not develop the outcome are considered as controls. Different exposures can be then analyzed by going back in time.[2]

Get started

The initial step in designing a cohort study is to identify the clinical question that can be addressed by a cohort study. For example, determining the strength of association between a new environemental exposure and the development of cardiovascular diseases. Additionally, the subtype of the

Translational Cardiology. https://doi.org/10.1016/B978-0-323-91790-2.00067-8

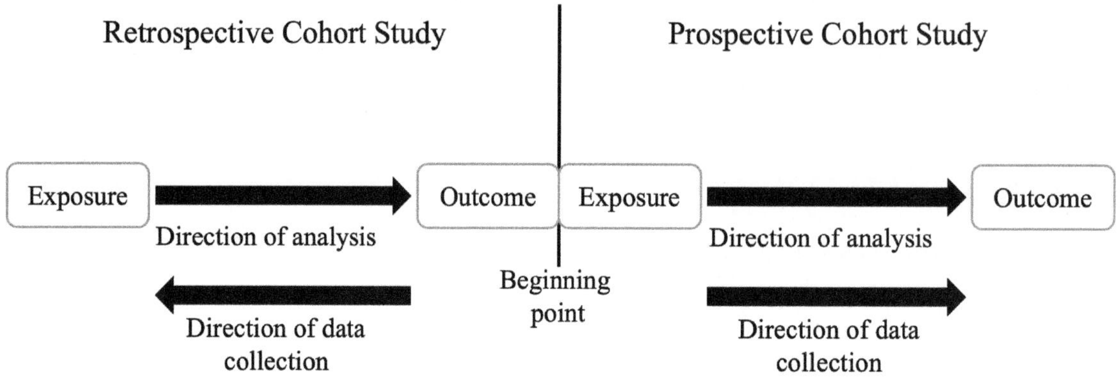

FIGURE 38.1 Prospective and retrospective cohort studies.

TABLE 38.1 Prospective and retrospective cohort studies.

	Prospective cohort study	Retrospective cohort study
Major difference (Fig. 40.1)	None of the participants have developed the outcome of interest yet.[1] Considered relatively superior clinical evidence compared to a retrospective cohort study.[1]	Some of the participants have developed the outcome of interest already.[1]
Design	Select the study sample (individuals without outcome of interest) and divide them into exposure and nonexposure groups based on exposure status.[2] Follow them prospectively over a certain period to monitor for development of the outcome.[1]	Go back in time to a certain point when the subjects had not developed the outcome of interest.[2] Divide the study sample based on outcome and exposure.[2] Analyze the available data based on already developed outcome with respect to exposure status.[1]

cohort study needs to be determined at the beginning of the study for establishing the source of data collection (see Fig. 38.1, Table 38.1). Once the research question has been identified and the subtype of the study has been determined, the following steps, as demonstrated in Fig. 38.2, should be followed to design and execute the cohort study, minimizing confounders and losses to followup.

Measurements

Incidence rate: Incidence is defined as the number of new cases of a health outcome over a specified time in the population at risk.[5]

Incidence = Number of new cases/(Population at risk × time)

For example, if the population of adult males and females in a town is 5000 and over a followup period of 10 years, 50 of them develop cardiovascular diseases (CVD), the incidence rate of CVD in the hypothetical town will be as follows.

Incidence of CVD = Number of new cases of CVD/(Population at risk × followup period)

= 50 new cases / (5000 people × 10 years)

= 50 new cases / (50,000 people year)

= 0.001 cases/people year

= 1 case / 1000 people year

The result signifies that if we follow 1000 people in this town for a year, one new case of CVD will be diagnosed over the period of followup.

Relative risk

Relative Risk = (Risk of outcome in the exposed group)/(Risk of outcome in the nonexposed group)[6]

For example, if 1000 people in the town had hypertension and 4000 people did not have hypertension, and over a followup period of 10 years, 40 of the hypertensive subjects developed CVD, and 10 of the nonhypertensive subjects developed CVD, then the relative risk of CVD among hypertensive subjects in this hypothetical study will be as follows:

Relative risk = (Risk of CVD in hypertension)/(Risk of CVD without hypertension)

= (40/1000) / (10/4000)

= 0.04 / 0.0025

= 16

FIGURE 38.2 A sample research methodology for a cohort study design.[1–4]

The result indicates that hypertensives are 16 times more likely to develop CVD than nonhypertensives.

Other statistical analysis methods, including Kaplan Meier curves, logistic regression, and cox regression, can be used for result generation in cohort studies.[2]

Advantages of a cohort study design

1. Multiple exposures and outcomes can be studied at the same time.[1]
2. Large populations can be studied, yielding higher statistical power and more generalizability of study results compared to clinical trials.[1]
3. Temporal association between exposure and outcome can be assessed.[2]
4. Rare exposures can be studied.[1]
5. The effect of changes in intensity or frequency of an exposure on an outcome can be determined.[1]
6. Cohort studies can analyze health outcomes associated with harmful behavioral or occupational exposures where planning a randomized control trial might not be feasible due to ethical concerns (e.g., alcohol consumption and CVD, occupational chemicals and cancers, etc.).[7]

Disadvantages of a cohort study design

1. The need for prolonged followup can lead to selection bias due to loss of followup.[1]
2. Prolonged periodic followup is time consuming and expensive.[1,2]
3. Although retrospective cohort studies are more time and cost effective in terms of collection of data, recall bias can affect the assignment of exposure. Using data collected for another purpose is associated with the risk of information bias.[1,2]
4. Rare outcomes are difficult to study.[2]

Pitfalls to avoid in a cohort study

Confounding variables

Variables that are associated with both exposure and outcome but do not explain the causality of outcome by exposure are considered confounders. Confounding variables can cause modification of the results of a cohort study. Restriction during study sample selection and stratification, multivariate regression analysis, or propensity score matching during the data analysis phase can reduce the effects of confounders.[1,7]

Loss to followup

Consider loss to followup when analyzing results. In general, less than 5% loss will minimally modify the results, while over 20% loss can significantly affect the results.[4] Measures to minimize loss to followup include enrolling patients with local permanent addresses, frequent reminders for appointments, and active outreach.[2]

Examples of cohort studies in cardiovascular medicine

1. One of the landmark cohort studies in Cardiovascular Medicine is the Framingham Cohort Study (https://www.framinghamheartstudy.org/), which assessed risk factors for CVD. In 1948, the study initially recruited 5209 participants aged 30–62 years old living in Framingham and followed them over time for the development of CVD. Later, they recruited first and second generation participants (offsprings of initial participants), and the study continues to date.
2. The Seven Countries Study (https://www.sevencountriesstudy.com/about-the-study/) is a cohort study that was designed to an find association between diet, lifestyle, physical characteristics, and development of CVD.
3. PERSUADE is an ongoing prospective cohort study in China that examines healthy individuals with noncontrast computed tomography to assess the progression of atherosclerosis. The study expects to develop noninvasive risk stratification strategies for CVD.[8]
4. The Utrecht Cardiovascular Cohort–Second Manifestations of Arterial Disease (UCC-SMART) Study is an ongoing prospective cohort study to analyze risk factors and prognosis of cardiovascular disease. The study has contributed to the 2021 European Society of Cardiology CVD prevention guidelines.[9]
5. LIFECARE prospective cohort study has been designed to understand CVD epidemiology in four Southeast Asian countries.[10]

Conclusion

In conclusion, cohort studies continue to serve multifaceted roles in translational research, from incidence measures to generating evidence for causal links. Despite the requirement for prolonged followups and susceptibility to various biases, proper execution with careful design and implementation can provide significant insights.

References and additional resources

1. Wang X, Kattan MW. Cohort studies: design, analysis, and reporting. *Chest.* 2020;158(1S):S72–S78. https://doi.org/10.1016/j.chest.2020.03.014.
2. Setia MS. Methodology series module 1: cohort studies. *Indian J Dermatol.* 2016;61(1):21–25. https://doi.org/10.4103/0019-5154.174011.
3. Fleiss JL, Tytun A, Ury HK. A simple approximation for calculating sample sizes for comparing independent proportions. *Biometrics.* 1980;36(2):343–346.
4. Dettori JR. Loss to follow-up. *Evid Based Spine Care J.* 2011;2(1):7–10. https://doi.org/10.1055/s-0030-1267080.
5. Tenny S, Boktor SW. Incidence. In: *StatPearls. Treasure Island (FL).* StatPearls Publishing; April 10, 2023.
6. Tenny S, Hoffman MR. Relative risk. In: *StatPearls. Treasure Island (FL).* StatPearls Publishing; March 27, 2023.
7. Mann CJ. Observational research methods. Research design II: cohort, cross sectional, and case-control studies. *Emerg Med J.* 2003;20(1):54–60. https://doi.org/10.1136/emj.20.1.54.
8. Kengne AP, Ntyintyane LM, Mayosi BM. A systematic overview of prospective cohort studies of cardiovascular disease in sub-Saharan Africa. *Cardiovasc J Afr.* 2012;23(2):103–112. https://doi.org/10.5830/CVJA-2011-042.
9. Castelijns MC, Helmink MAG, Hageman SHJ, et al. Cohort profile: the Utrecht Cardiovascular Cohort-Second Manifestations of Arterial Disease (UCC-SMART) study-an ongoing prospective cohort study of patients at high cardiovascular risk in the Netherlands. *BMJ Open.* 2023;13(2):e066952. https://doi.org/10.1136/bmjopen-2022-066952.
10. Shafiq M, Fong AYY, Tai ES, et al. Cohort Profile: LIFE course study in CARdiovascular disease Epidemiology (LIFECARE). *Int J Epidemiol.* 2018;47(5):1399–1400g. https://doi.org/10.1093/ije/dyy168.

Chapter 39

Cross-sectional studies

Adrija Hajra[1], Binita Ghosh[2] and Suryasnata Bhowmik[3]

[1]*Department of Internal Medicine, Bringham and Women's Hospital/Harvard Medical School, Boston, MA, United States;* [2]*Internal Medicine, SSM Health St. Mary's Hospital, St. Louis, MO, United States;* [3]*Critical Care Medicine, Nilratan Sircar Medical College, Kolkata, West Bengal, India*

Key points

- Cross-sectional studies assess the prevalence of both outcome and exposure at the same time within a population.
- While descriptive cross-sectional studies evaluate only prevalence measures, analytical cross-sectional studies extend their scope by analyzing data to identify associations.
- Although these studies cannot establish causal links or temporal associations, they are valuable for gauging the burden and trends of population characteristics.

Introduction

Cross-sectional studies are observational and descriptive epidemiological studies that capture a "snapshot" of the prevalence of diseases, risk factors, or traits within a population at a single point in time. This prevalence data can inform healthcare planning and resource allocation, guide future research, and assist in selecting participants for case-control and cohort studies. Their fast, inexpensive, and straightforward design is ideal for population-based surveys, as well as for conducting validation and reliability studies.[1] Moreover, the predominant attitude of the population evaluated through cross-sectional studies can tailor healthcare recommendations and guidelines. Thus, they are indispensable in medicine with diverse roles, from contributions in larger studies to standalone tools for gauging disease trends and burden.[2]

Design

In cross-sectional studies, researchers simultaneously measure the presence and absence of disease, as well as exposure, within a population at the same time, without follow-up (Fig. 39.1).[3]

Cross-sectional studies are broadly categorized into two main types: descriptive and analytical. Descriptive cross-sectional studies focus solely on determining the prevalence of a health outcome or determinant at a specific time (point prevalence) or over an extended period (period prevalence). Analytical cross-sectional studies go one step further by comparing the disease prevalence between exposed and unexposed groups to explore potential associations. A subtype, the serial cross-sectional study, collects data from different samples of the same population at various times to analyze trends.[3]

Sample size calculation

The size of the sample selected from the population should be large enough to accurately estimate prevalence.[4] It is computed from the following equation[5]:

$$n = \frac{z^2 p(1-p)}{d^2}$$

where n = sample size, z = level of confidence (usually 95% confidence interval), p = expected prevalence obtained from previous studies or a pilot study, d = precision (typically 5%).[6] While sample size calculation in descriptive studies is not contingent upon study power, it is an important factor in analytical studies.

Sampling methods

Selecting participants (sampling) involves either probability sampling, where every individual in the target population has an equal chance of selection, or nonprobability

Translational Cardiology. https://doi.org/10.1016/B978-0-323-91790-2.00099-X

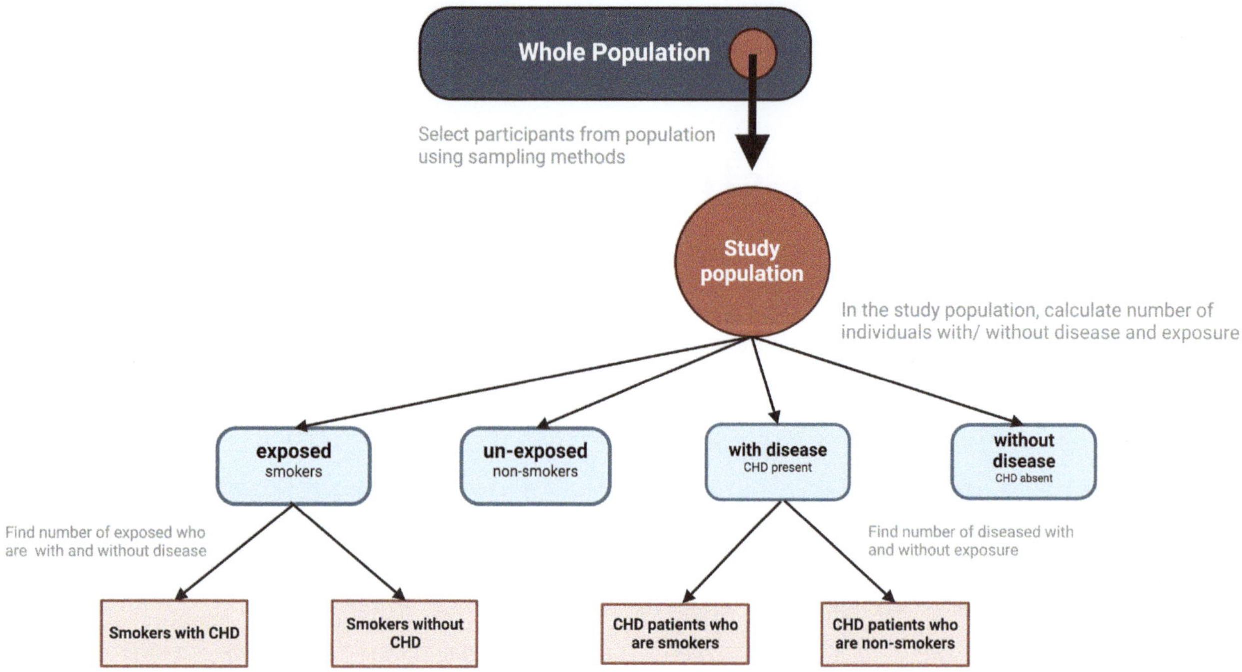

FIGURE 39.1 Flow diagram of cross-sectional study design with action points. *Credit: Created with BioRender.com.*

sampling methods, which may not ensure equal selection chances.[7,8]

Measures

Key measures in cross-sectional studies include calculating the prevalence of disease or exposure and assessing the association between them.

Prevalence calculation

Table 39.1 presents the hypothetical data of the number of individuals with and without the disease (e.g., coronary heart disease, CHD), and with and without exposure (for example, smokers) calculated at the same time from a sample population.

From the data in Table 39.1, we calculate the prevalence of disease (e.g., CHD) and exposure (e.g., smoking) in the population[3]:

Prevalence of outcome (CHD) = (number of individuals with disease/total population) = 100/1000 = 10%

Similarly, prevalence of exposure (e.g., smoking) = (number of smokers/total population) = 200/1000 = 20%

To further analyze the relationship between smoking and CHD, we can find out the number of smokers and nonsmokers with CHD from the population. The revised table with hypothetical values is given in Table 39.2.

TABLE 39.1 Data of the number of individuals with and without the disease.

Total population	Number of individuals with disease (e.g., CHD)	Number of individuals without disease (e.g., CHD)
1000	100	900
	Number of individuals with exposure (e.g., smokers)	Number of individuals without exposure (e.g., nonsmokers)
	200	800

CHD, *coronary heart disease.*

TABLE 39.2 Number of smokers and nonsmokers in patients with and without coronary heart disease.

	CHD present	CHD absent	Total
Smokers	75 a	125 b	200 (a+b) total smokers $= N_1$
Nonsmokers	25 c	775 d	800 (c + d) total nonsmokers $= N_2$
Total	100	900	1000

CHD, *coronary heart disease.*

Prevalence of CHD among smokers = (smokers with CHD/total smokers) = a/(a + b) = 75/200 = 37.5%

Similarly, the prevalence of CHD among nonsmokers = (nonsmokers with CHD/total nonsmokers) = c/(c + d) = 25/800 = 3.12%

These prevalence measures offer insight into the disease burden and impact of risk factors. The variations in disease prevalence across different exposures can highlight high-risk groups, suggesting potential links between disease and exposure.[3]

Prevalence odds ratio

Prevalence odds ratio (POR) is analogous to the odds ratio in case-control studies and is calculated as follows: POR = (a X d)/(b X c), which is used for chronic diseases or persistent risk factors, like heart failure or obesity.[9]

Prevalence ratio

Prevalence ratio (PR) = (Prevalence of CHD among smokers)/(Prevalence of CHD among nonsmokers) = $(a/N_1)/(a/N_2)$.[10] For example, if the prevalence of CHD among smokers is 12 times that among nonsmokers, the PR would be 12, indicating smokers have a 12-fold higher proportion of CHD.

PR is the preferred measure of association for acute diseases, such as an outbreak of viral myocarditis in a cardiac unit. When the disease prevalence is below 10%, POR approximates PR.[11]

Modeling in analytical studies

Modeling involves explanatory regression models to identify variables significantly associated with outcome, or diagnostic prediction models that combine multiple variables to estimate disease likelihood in undiagnosed individuals.[12] For example, an explanatory model examines the relationship between sedentary lifestyle and CHD prevalence. Conversely, a diagnostic model combines several predictors like age, smoking status, dyslipidemia, hypertension, and diabetes to estimate the probability of CHD in an undiagnosed individual.

Get started

With an understanding of the design and measures, follow these steps to conduct your cross-sectional study effectively:

1. Define research question: Identify the health outcome or determinant for investigation, for example, prevalence of hypertension.
2. Set selection criteria: For example, include individuals on antihypertensive medications.
3. Design study: Choose between a descriptive (measuring prevalence) or analytical (examining associations with factors like exercise) approach.
4. Calculate sample size: Use previous studies or pilot data to estimate expected prevalence.[5] For example, for a 5% error margin, 45% expected prevalence, and 95% confidence interval, you would need about 381 ± five participants. Online calculators can aid this process.[13]
5. Sampling strategy: Opt for probability sampling for representativeness or nonprobability sampling for convenience.
6. Collect data: Use questionnaires or conduct interviews to gather data on prevalence.
7. Address biases: Be aware of potential biases (see next section) to improve validity.
8. Analyze data: Calculate prevalence measures for both outcomes and exposures. Use POR for chronic and PR for acute conditions. Employ statistical tools like modeling in analytical studies.
9. Interpret findings: Recognize inherent study limitations and consider suggesting longitudinal studies for further validation.
10. Report findings: Adhere to the Strengthening the Reporting of Observational Studies in Epidemiology (STROBE) guidelines while writing the report.[14] If a diagnostic prediction model is developed, follow the

Transparent Reporting of a multivariable prediction model for Individual Prognosis or Diagnosis (TRIPOD) statement for transparent reporting.[15]

Potential pitfalls

Mitigating potential pitfalls in cross-sectional studies involves strategic planning and methodological adjustments:[12]

1. To prevent selection bias, utilize probability sampling methods to provide every individual with an equal chance of selection. Stratified (selection of a random sample from every nonoverlapping stratum of the population) or cluster sampling (indiscriminate selection of cluster that the population has been randomly divided into) can ensure representation of all population segments.
2. Prevalence-incidence bias, which occurs when certain disease characteristics are inadvertently favored during selection (e.g., selection of only severe heart failure, underrepresenting the actual mix of mild and severe forms in the population), can be mitigated by using appropriate selection criteria to include a spectrum of disease severities and employing stratified sampling.
3. Increase questionnaire response rates by simplifying them, offering incentives, and sending reminders. To counter recall bias, use memory aids and corroborate self-reported data with medical records when feasible.
4. Nonresponse bias—the systematic difference between those who complete a survey and those who do not (e.g., frail patients who are not motivated to complete a survey) can be addressed by examining the characteristics of nonrespondents. Improve survey participation by making them more accessible and engaging.
5. Causal inferences or temporal associations cannot be established in cross-sectional studies since exposure and outcome are ascertained simultaneously.
6. Interpret prevalence with caution, acknowledging that it may be affected by factors like treatment improvements, increased lethality, or selective migration, which affect incidence and disease duration.
7. Survival bias can overrepresent chronic conditions like heart failure due to effective management, potentially overestimating risk factor associations. Adjusting for disease duration or comparative studies can mitigate this bias.
8. Cross-sectional studies cannot measure incidence, as they cannot distinguish between preexisting and new cases at a single time point.
9. Consider multicenter collaborations for rare diseases with low prevalence to increase sample size. Alternatively, focus on known carriers and at-risk populations to improve study power.

Classic examples

To further illustrate the design and importance of cross-sectional studies, this section demonstrates three classical examples of cross-sectional studies.

National Health and Nutrition Examination Survey

The National Health and Nutrition Examination Survey (NHANES) is an exemplary ongoing series of cross-sectional studies assessing the health and nutrition of the US population to provide prevalence data on diseases and risk factors. For example, NHANES data has shown a consistent uptrend in obesity over time, informing policy-making and guiding health initiatives like obesity awareness programs and stringent food labeling laws.[16]

The European Youth Heart Study

Anderson et al.'s analytical cross-sectional study evaluated the link between physical activity and cardiovascular (CVS) risk factors in children. Using accelerometry to measure physical activity, it concluded that increased activity levels were associated with lower cardiovascular risks with children in the lowest quintiles of physical activity having significantly higher CVS risks. Thus, the findings suggested at least 1 h of moderate-intensity physical activity daily in children to prevent risk factor clustering.[17]

ECHOES study: Quality of life among heart failure patients

Hobbs et al. conducted an analytical cross-sectional study, screening over 6000 patients to assess heart failure (HF) prevalence and its effects on quality of life (QoL). The study reported significant QoL impairment among HF patients when compared to the general population. Conversely, asymptomatic or appropriately managed HF patients reported QoL similar to that of the general population. This study highlighted the profound impact of heart conditions and emphasized the value of cross-sectional studies in ascertaining disease burden.[18]

Conclusion

In conclusion, cross-sectional studies continue to serve multifaceted roles in translational research, from prevalence measures to generating hypotheses on causal links. Despite the inability to establish causality and susceptibility to various biases, proper execution with careful design and implementation can provide significant insights.

References

1. Kesmodel US. Cross-sectional studies – what are they good for? *Acta Obstet Gynecol Scand.* 2018;97(4):388–393. https://doi.org/10.1111/aogs.13331.
2. Hennekens CH, Buring JE. Epidemiology in medicine. In: *Epidemiology in Medicine*; 1987, 383-383 https://pesquisa.bvsalud.org/portal/resource/pt./biblio-1078101. Accessed February 11, 2024.
3. Setia MS. Methodology series module 3: cross-sectional studies. *Indian J Dermatol.* 2016;61(3):261–264. https://doi.org/10.4103/0019-5154.182410.
4. Pourhoseingholi MA, Vahedi M, Rahimzadeh M. Sample size calculation in medical studies. *Gastroenterol Hepatol Bed Bench.* 2013;6(1):14–17.
5. Wayne WD, Cross CL. *Biostatistics: A Foundation for Analysis in the Health Sciences.* John Wiley & Sons; 2018.
6. Naing L, Winn T, Rusli BN. Practical issues in calculating the sample size for prevalence studies. *Arch Orofac Sci.* 2006;1:9–14.
7. Shorten A, Moorley C. Selecting the sample. *Evid Base Nurs.* 2014;17(2):32–33. https://doi.org/10.1136/eb-2014-101747.
8. Elfil M, Negida A. Sampling methods in clinical research; an educational review. *Emergency.* 2017;5(1):e52.
9. Tamhane AR, Westfall AO, Burkholder GA, Cutter GR. Prevalence odds ratio versus prevalence ratio: choice comes with consequences. *Stat Med.* 2016;35(30):5730–5735. https://doi.org/10.1002/sim.7059.
10. Coutinho L, Scazufca M, Menezes PR. Methods for estimating prevalence ratios in cross-sectional studies. *Rev Saude Publica.* 2008;42:992–998.
11. Use of the Prevalence Ratio V the Prevalence Odds Ratio as a Measure of Risk in Cross Sectional Studies. PMC; https://www.ncbi.nlm.nih.gov/pmc/articles/PMC1128040/. Accessed 28 February 2024.
12. Wang X, Cheng Z. Cross-sectional studies: strengths, weaknesses, and recommendations. *Chest.* 2020;158(1, Supplement):S65–S71. https://doi.org/10.1016/j.chest.2020.03.012.
13. Sample Size Calculator. https://www.calculator.net/sample-size-calculator.html?type=1&cl=95&ci=5&pp=45&ps=&x=Calculate. Accessed 11 February 2024.
14. Vandenbroucke JP, von Elm E, Altman DG, et al. Strengthening the reporting of observational studies in Epidemiology (STROBE): explanation and elaboration. *PLoS Med.* 2007;4(10):e297. https://doi.org/10.1371/journal.pmed.0040297.
15. Moons KGM, Altman DG, Reitsma JB, et al. Transparent Reporting of a multivariable prediction model for Individual Prognosis or Diagnosis (TRIPOD): explanation and elaboration. *Ann Intern Med.* 2015;162(1):W1–W73. https://doi.org/10.7326/M14-0698.
16. National Center for Health Statistics. National Center for Health Statistics. Division of Health and Nutrition Examination Surveys. In: Ansai N, Lipphardt A, eds. NHANES Interactive Data Visualizations; 2023. https://www.cdc.gov/nchs/nhanes/visualization/index.htm. Accessed February 11, 2024.
17. Andersen LB, Harro M, Sardinha LB, et al. Physical activity and clustered cardiovascular risk in children: a cross-sectional study (The European Youth Heart Study). *Lancet.* 2006;368(9532):299–304. https://doi.org/10.1016/S0140-6736(06)69075-2.
18. Hobbs FDR, Kenkre JE, Roalfe AK, Davis RC, Hare R, Davies MK. Impact of heart failure and left ventricular systolic dysfunction on quality of life. A cross-sectional study comparing common chronic cardiac and medical disorders and a representative adult population. *Eur Heart J.* 2002;23(23):1867–1876. https://doi.org/10.1053/euhj.2002.3255.

Chapter 40

Longitudinal studies: design, measures, and classic examples

Aditi Ujjawal
Saint Vincent Hospital, Worcester, MA, United States

Key points

1. The most common types of longitudinal study designs include cohort study, panel or population-based study, registry study, and experimental study.
2. The most used longitudinal measures are mean, median, standard deviations, percentages, proportions, odds ratios, relative risk, attributable risk, and risk differences.
3. Specialized measures analyzing continuous longitudinal data include mixed-effects models, analysis of response profile, and generalized estimating equations (GEE).
4. Examples of longitudinal studies include the Framingham Heart Study, National Health and Nutrition Examination Survey (NHANES) longitudinal study, and the Multi-Ethnic Study of Atherosclerosis (MESA).

Introduction

A longitudinal study follows a well-defined population pool over time and records biomarker and clinical data from each participant to identify risk factors, characterize disease incidence and phenotype, and study outcomes and prognosis. These studies are expensive and require a long period of time (years to decades) to collect sufficient data to study outcomes of interest. In comparison, cross-sectional study provides data at a single point in time and is used to study the prevalence of a disease. It can only provide information about the correlation of risk factors and disease as compared to longitudinal studies that can establish causation. Cross-sectional studies provide quicker results with the use of fewer resources.

Strengths and limitations

The significant strengths of longitudinal studies are that they are crucial in establishing risk factors, disease progression, and the effect of different interventions. They are most commonly prospective in design, reducing recall bias and establishing temporality and causality. Recording longitudinal data also allows us to measure variation in exposure (for example, current vs. former smoking status) and how it factors into disease causation. Repeated measurements from the same individual eliminate variability or "noise" from the estimation of within-individual change. Longitudinal studies can also provide subjects for a randomized experimental study. The major disadvantages are that conducting a longitudinal study is expensive and requires a large study population and a longer time. It is observational and inferior to an experimental study, which may have a comparison group. Moreover, it is susceptible to sampling, confounding, attrition, and observation bias, all of which can limit external validity.

Types of longitudinal studies

A longitudinal study is classically associated with a prospective cohort study; however, there are other study designs where longitudinal data may be collected. The design of a longitudinal study depends on the research question, disease under study, and availability of resources:

1. Longitudinal cohort study: This design allows researchers to examine how characteristics or exposures at baseline (start of the study) influence outcomes over time.
2. Panel studies: Participants represent a closed population sample who are repeatedly surveyed at intervals, tracking changes in health status, behaviors, and risk factors over time (e.g., National Health and Nutrition Examination Survey). These studies are also referred to as "fixed-effects" or "closed-panel" studies.
3. Twin studies: Monozygotic (identical) and dizygotic (fraternal) twins are compared to disentangle the

Translational Cardiology. https://doi.org/10.1016/B978-0-323-91790-2.00084-8

relative contributions of genetics and environment on health outcomes.

4. Registry-based studies: Utilize existing population-based health registries to track disease incidence, prevalence, and risk factors over time (e.g., cancer registries).
5. Longitudinal case studies: Longitudinal case studies involve an in-depth examination of a small number of individuals or cases over an extended period. These studies provide detailed insights into individual experiences and trajectories of change.
6. Retrospective longitudinal studies: In retrospective longitudinal studies, researchers collect data on past events or exposures from existing records or participants' recall and then follow up with participants over time to collect additional data. This design is helpful in studying outcomes that unfold over a long period.
7. Accelerated longitudinal designs: Accelerated longitudinal designs involve collecting data at multiple time points over a relatively short period, often with frequent assessments. This design allows tracking of developmental processes and short-term changes.
8. Experimental studies: A longitudinal study is classically observational, and data are recorded without manipulating the variables. A randomized controlled trial (RCT), although not classically cited as an example of a longitudinal study, is prospective or forward-looking and offers continuous data over time. An extension study of an RCT, such as the MAVA-LTE extension study, also constitutes a longitudinal study.

Reporting longitudinal data

Analyzing longitudinal data has inherent challenges. These include:

1. *missing data* from subjects who drop out, miss patient visits, or suffer death,
2. *correlated data* collected from the same subject over time, and
3. *individual variability.*

It is essential to describe the sample size at each time point, how subjects were sampled, how frequently exposure and follow-up data was collected, and which statistical methods were used to calculate the results. The American Heart Association has published guidelines on reporting data in clinical studies. Specifically, they recommend using a 33-point checklist for reporting the design and analysis aspects of observational longitudinal studies. Working with a statistician to appropriately characterize the data and minimize false positives and invalid conclusions is important. The following measures are used to report data collected from longitudinal studies.

1. Simple measures analyzing data from one or two time points include mean, median, standard deviations, percentages, proportions, odds ratios, relative risk, attributable risk, and risk differences. The Kaplan−Meier curve often represents survival data.
2. Specialized measures analyzing continuous longitudinal data include mixed-effects models, analysis of response profile, and generalized estimating equations (GEE).
3. Joint models combining longitudinal and survival data analysis, modified Kaplan−Meier analysis, and regression models with time-varying covariates further reduce bias and error.

Potential pitfalls

1. The magnitude of relative risk or survival data alone is insufficient to show causation. Many factors provide evidence for causation, such as statistical significance, odds ratio, relative rest, dose-response relationship, temporality, validity, biological plausibility, and experimental evidence. All these factors must be considered when interpreting data from an observational study.
2. Diagnostic and therapeutic approaches evolve and must be factored into data interpretation. For example, the criteria for myocardial infarction in 1948, when the Framingham Heart Study began, were different than they are now.
3. Reporting longitudinal data using simple summary measures or data from a single time point instead of utilizing all the available data and using specialized models to report the same.

Classic examples

Framingham Heart Study: This is one of the longest-running cohort studies. It began in 1948 and has collected data over three generations from more than 15,000 subjects. The initial cohort comprised 5209 patients aged 30−62 years (2336 men and 2873 women) who underwent regular physical examinations, lifestyle interviews, and laboratory tests to learn about the epidemiology and risk factors for cardiovascular disease. The second generation of 5124 subjects, including the offspring and spouses of the original cohort (2641 women and 2483 men), were enrolled in the year 1971, and finally, in 2002, the third generation of patients were enrolled comprising 4095 subjects (2183 women and 1912 men). The investigators also began collecting data on the Omni cohort—comprising the non-Caucasian population of Framingham, in 1994 and its second cohort in 2003. This study first revealed that elevated blood pressure and cholesterol are important cardiovascular risk factors. Since then, the researchers have collected biomarkers, cardiac imaging, and brain imaging data and developed a genetic database.

Multi-Ethnic Study of Atherosclerosis (MESA) began in 2000 and included a cohort of 6814 participants from six ethnically diverse populations. The baseline examination assessed individuals aged 45−84 years, including 3247 men and 3567 women. Detailed questionnaires, physical examinations, and laboratory tests were recorded to study atherosclerosis progression. MESA notably highlighted the role of inflammation and subclinical atherosclerotic disease and continues to record data from cardiac and brain imaging and genetic profiling to further our understanding of cardiovascular disease progression.

NHANES I Epidemiologic Follow-up Study (NHEFS) was a longitudinal study conducted from 1971 through 1992 that included 14,407 subjects from the NHANES cohort aged 25−74 years. This study collected baseline and longitudinal data on clinical, nutritional, and behavioral factors and their correlation to morbidity, functional limitation mortality, hospital resource utilization, and institutionalization.

The national death index was surveyed to collect data on 3212 patient deaths and causes of death. Follow-up data collection via interviews was undertaken for the 11,195 subjects (90%) alive in 1992.

Women's Ischemia Syndrome Evaluation (WISE) was a multicenter prospective cohort study from 1996 through 2006. It followed 936 women who were referred for coronary angiogram for symptoms of myocardial ischemia. The study aimed to determine the all-cause and cardiac mortality up to 10 years after the index coronary angiogram. Data on patient demographics, risk factors, degree of obstructive coronary artery disease (CAD), hormonal status, and psychosocial factors was recorded. The National Death Index Plus data was used to determine the number of patients who died and their cause of death. Kaplan−Meier estimated mortality rates of 20% all-cause and 12% cardiac mortality were reported at 10 years. The authors concluded that one in five women died from cardiac disease within 9 years of angiography. One in three deaths occurred in women without obstructive CAD. This study highlighted sex differences in clinical presentation and the need for treatment guidance for women who lack obstructive CAD.

Resources and additional reading

1. Althouse AD, Below JE, Claggett BL et al. Recommendations for Statistical Reporting in Cardiovascular Medicine: A Special Report From the American Heart Association. Circulation. 2021; 144(4):e70-e91. https://doi.org/10.1161/CIRCULATIONAHA.121.055393.
2. Diggle PJ, Heagerty P, Liang KY, Zeger SL. Analysis of longitudinal data. Oxford University Press; 2002.
3. Methods and Applications of Longitudinal Data Analysis by Xian Liu, Elsevier, 2016, ISBN: 978-0-12-801,342−7.
4. Fitzmaurice, G., Davidian, M., Verbeke, G., & Molenberghs, G. (Eds.). (2008). Longitudinal Data Analysis (first ed.). Chapman and Hall/CRC. https://doi.org/10.1201/9781420011579
5. Zeger SL, Liang KY, Albert PS. Models for longitudinal data: a generalized estimating equation approach [published correction appears in Biometrics 1989 Mar; 45(1):347]. Biometrics. 1988; 44(4):1049−1060.
6. Gordis Epidemiology (7th ed), Elsevier.

Further reading

1. Mann CJ. Observational research methods. Research design II: cohort, cross sectional, and case-control studies. *Emerg Med J.* 2003;20(1):54−60.
2. Caruana EJ, Roman M, Hernández-Sánchez J, Solli P. Longitudinal studies. *J Thorac Dis.* November 2015;7(11):E537−E540.
3. Schober P, Vetter TR. Repeated measures designs and analysis of longitudinal data: if at first you do not succeed-try, try again. *Anesth Analg.* 2018;127(2):569−575. https://doi.org/10.1213/ANE.0000000000003511.
4. Fitzmaurice GM, Ravichandran C. A primer in longitudinal data analysis. *Circulation.* 2008;118(19):2005−2010.
5. Stevens D, Lane DA, Harrison SL, Lip GYH, Kolamunnage-Dona R. Modelling of longitudinal data to predict cardiovascular disease risk: a methodological review. *BMC Med Res Methodol.* 2021;21(1):283. https://doi.org/10.1186/s12874-021-01472-x.
6. Mahmood SS, Levy D, Vasan RS, Wang TJ. The Framingham Heart Study and the epidemiology of cardiovascular disease: a historical perspective. *Lancet.* 2014;383(9921):999−1008. https://doi.org/10.1016/S0140-6736(13)61752-3.
7. Bild DE, Bluemke DA, Burke GL, et al. Multi-ethnic study of atherosclerosis: objectives and design. *Am J Epidemiol.* 2002;156(9):871−881. https://doi.org/10.1093/aje/kwf113.
8. Gartside PS, Wang P, Glueck CJ. Prospective assessment of coronary heart disease risk factors: the NHANES I epidemiologic follow-up study (NHEFS) 16-year follow-up. *J Am Coll Nutr.* 1998;17(3):263−269. https://doi.org/10.1080/07315724.1998.10718757.
9. Kenkre TS, Malhotra P, Johnson BD, et al. Ten-year mortality in the WISE study (Women's Ischemia Syndrome Evaluation). *Circ Cardiovasc Qual Outcomes.* 2017;10(12):e003863. https://doi.org/10.1161/CIRCOUTCOMES.116.003863.
10. Rader F, Oręziak A, Choudhury L, et al. Mavacamten treatment for symptomatic obstructive hypertrophic cardiomyopathy: Interim results from the MAVA-LTE study, EXPLORER-LTE cohort. *JACC Heart Fail.* 2024;12(1):164−177. https://doi.org/10.1016/j.jchf.2023.09.028.

Chapter 41

Understanding meta-analysis

Vikash Jaiswal[1], Danisha Kumar[2] and Dhrubajyoti Bandyopadhyay[3]

[1]Larkin Community Hospital, South Miami, FL, United States; [2]Dow University of Health Sciences, Karachi, Pakistan; [3]Massachusetts General Hospital, Harvard Medical School, Boston, MA, United States

Key points

- Meta-analysis is a method for quantitatively combining results from multiple studies.
- It increases the precision of estimates by pooling data from various sources.
- Meta-analyses are crucial for resolving conflicting findings and generating new hypotheses.
- They require high-quality primary studies to ensure credible results.
- Common pitfalls include publication bias, variability in study methodologies, and researcher bias.

Introduction to meta-analysis

Gene V. Glass, recognized as a key figure in its development, coined meta-analysis as an "analysis of analyses".[1] This succinct definition encapsulates the essence of meta-analysis. While conventional studies typically focus on analyzing individuals, specimens, nations, or items, meta-analysis shifts the focus to the primary studies themselves.[2]

Meta-analyses are often considered as an advanced iteration of systematic reviews. They are designed to amalgamate findings from prior studies in a quantitative manner. The primary objective of meta-analyses is to synthesize quantitative outcomes reported in selected studies into a singular numerical estimate, thereby encapsulating all individual findings.[3] For instance, meta-analyses can quantify the efficacy of a particular teaching method, the incidence of adverse effects associated with a medical treatment, or the association between lifestyle factors and mental health outcomes across multiple studies. Consequently, meta-analyses are applicable only to studies that provide quantitative data. In comparison to systematic reviews, meta-analyses typically necessitate stricter criteria for selecting the evidence to be summarized. Conducting a meta-analysis generally requires studies to employ consistent methodologies, such as study design, measurement techniques, and interventions.[2]

Meta-analyses offer several potential benefits, including the following.[4]

- Increased precision: Combining data from multiple studies in meta-analysis enhances precision by drawing upon a larger pool of information, offering more robust estimates compared to individual studies.
- Exploration of diverse questions: Meta-analysis broadens the scope by examining effects across diverse populations and interventions not explicitly studied in individual research efforts. This exploration sheds light on the consistency of effects and reveals factors influencing variations in effect estimates.
- Resolution of conflicting findings and hypothesis generation: Meta-analysis evaluates conflicts among studies and quantifies reasons behind discrepancies, resolving controversies and inspiring new research hypotheses.

While meta-analysis offers cost-effective and time-efficient alternatives to primary studies, concerns about the quality and biases of published meta-analyses have emerged. Ioannidis (2016) criticized the proliferation of redundant and potentially misleading meta-analyses, particularly in popular research domains, sometimes influenced by corporate interests or plagued by reproducibility issues.[5] A brief exploration of common pitfalls in meta-analysis is warranted.[6–8]

- "Apples and Oranges" Problem: Meta-analysis combines studies with inherent differences, such as sample characteristics, interventions, and study designs, challenging the creation of consolidated estimates. Without shared properties, such estimates may lose relevance to specific research questions, highlighting the need for alignment with pertinent inquiries.[6]
- "Garbage In, Garbage Out" Problem: The credibility of meta-analysis relies on the quality of included studies.

Translational Cardiology. https://doi.org/10.1016/B978-0-323-91790-2.00109-X

Biased or inaccurate findings in constituent studies can taint the meta-analysis results, emphasizing the importance of assessing study quality and risk of bias.
- "File Drawer" Problem: Publication bias leads to the exclusion of some research findings, skewing meta-analysis conclusions. This bias, favoring positive findings, undermines the integrity of meta-analytic outcomes and complicates estimating its impact.[9]
- "Researcher Agenda" Problem: Meta-analysts' discretion in study selection and analysis, combined with potential personal biases, may influence outcomes. Measures such as preregistration are necessary to mitigate bias arising from researchers' interests.

Methods and results

- **Defining the research question:** When embarking on a study, the initial step is crafting a research question—a pivotal process mirrored in meta-analysis. For a meta-analysis to be effective, it must address a pertinent issue, much like any other research endeavor. To identify such issues, possessing subject-specific knowledge is key. Thus, selecting a research area where one holds background knowledge is advantageous. Reflect on current questions within the field, potential knowledge gaps, and unresolved discussions. Meta-analysis relies on existing research, making a thorough literature review imperative. Explore previous primary studies, their methodologies, and their limitations. Investigate prior reviews and meta-analyses to uncover unresolved issues. Cummings and colleagues (2013) introduced the FINER framework for refining research questions in meta-analysis: Feasible, Interesting, Novel, Ethical, and Relevant. These criteria guide the formulation of focused and effective research questions in meta-analysis.[10] Population, Intervention, Comparison, Outcome (PICO) is another tool used mostly in quantitative evidence synthesis to determine eligibility criteria.[11] For example, in the study by Jaiswal et al., the research question revolves around comparing the long-term cardiovascular outcomes of patients with diabetes mellitus (DM) and multivessel coronary disease who undergo either percutaneous coronary intervention (PCI) or coronary artery bypass graft (CABG).[12] The PICO is P = patients with diabetes mellitus (DM) and multivessel coronary disease, I = percutaneous coronary intervention (PCI), C = coronary artery bypass graft (CABG), and O = long-term cardiovascular outcomes.
- **Literature search:** In line with conducting a thorough literature review, the search process for a meta-analysis demands systematic, reproducible, and transparent methods, ensuring the inclusion of all pertinent studies. Employing a keyword search within electronic databases stands as the most prevalent and crucial strategy.[13] To mitigate biased outcomes stemming from database scope or journal coverage, researchers should utilize a minimum of two different databases.[14–16] Recommended databases include PubMed, Google Scholar, Embase, Scopus, Cochrane, and Web of Science. Crafting a precise research question and employing precise search terms with Boolean operators is advisable.[17] A search string contains different words or terms, which are connected using operators such as AND or OR. For instance, employing terms like (percutaneous coronary intervention OR PCI) AND (coronary artery bypass graft OR CABG) AND (cardiovascular outcomes) AND (patients with diabetes mellitus OR DM) AND (patients with multivessel coronary disease) can enhance search accuracy and relevance while searching for primary studies.
- **Inclusion and exclusion criteria:** In the final stage of defining eligibility criteria, it's crucial to compile a clear list of inclusion and exclusion criteria for application. Eligibility criteria typically stem from the PICO approach, study design, and publication date. Exclusion criteria commonly encompass unrelated, duplicated, unavailable full-text, or abstract-only papers. It is essential to preestablish these exclusions to mitigate researcher bias. Additional reasons for study exclusion may emerge during the search process, often tied to the research question and methodological nuances. Inclusion criteria revolve around articles featuring target patients, examined interventions, or comparisons between studied interventions, among others.
- **Screening process:** In the initial step, remove duplicate references using reference management software such as Zotero or Mendeley.[18,19] Note the total number of references before and after this process. Next, screen references based on titles and abstracts to exclude irrelevant studies. Use caution to avoid discarding potentially useful references prematurely. Retrieve full articles for remaining references and assess their eligibility against predefined criteria. Document reasons for exclusion, especially due to insufficient information or inadequately reported results. Attempt to contact corresponding authors for missing information if necessary. If efforts fail and crucial data is unavailable, consider excluding the study from your analysis. We write down all the details of the inclusion process in a flow diagram. A commonly used template for such a flow chart is the one provided by the PRISMA guidelines.[20]
- **Protocol writing and registration:** Meta-analyses should be registered, particularly for health-related outcomes, ideally through platforms like PROSPERO.[21] Registering protocols early ensures transparency and prevents duplication. It also serves as documented

evidence of the team's plan, research question, eligibility criteria, interventions, quality assessment, and preanalysis strategy.

- **Data extraction and risk of bias assessment**: Once the selection process for meta-analysis studies is complete, data extraction follows suit. As outlined by Cuijpers (2016), three key types of information are typically extracted from chosen articles[22]:
  1. Study characteristics.
  2. Data necessary for effect size calculation.
  3. Assessment of study quality or bias risk.

A common practice in top-tier meta-analyses involves presenting a table that details the characteristics of included studies. It is advisable to extract both adjusted and non-adjusted data during the review process to incorporate the maximum potential confounding factors for later analysis. Two to three independent reviewers should handle the extraction process, categorizing the data into study and patient characteristics, outcomes, and quality assessment (QA) tools. Various QA tools, specific to study designs, are available, such as the ROB-2 Cochrane tool for randomized controlled trials,[23] NIH tool for observational and cross-sectional studies,[24] ROBINS-I tool for nonrandomized trials,[25] QUADAS-2 tool for diagnostic studies, QUIPS tool for prognostic studies, CARE tool for case reports, and ToxRtool for in vivo and in vitro studies. To minimize bias, it is recommended that reviewers independently evaluate the study quality and update the data extraction form before inclusion in the analysis.

- **Statistical analysis:** Qualitative analysis is mostly used to describe data in systematic reviews, while quantitative analysis is divided into two types: MA and network meta-analysis (NMA). Subgroup, sensitivity, cumulative, and meta-regression studies are ideal for determining whether the results are consistent, exploring the influence of certain confounders on the outcome, and identifying the best predictors. Publication bias should be investigated to determine the presence of missing research that might impact the summary. To illustrate basic meta-analysis, we provide an example from Jaiswal et al.'s meta-analysis, where they compare the cardiac death outcome in the PCI group versus the CABG group among DM patients with multivessel disease.[12] An analysis was done to evaluate whether cardiac death or mortality showed an increased risk with PCI or CABG. To do the meta-analysis, we can use free software, such as RevMan,[26] STATA,[27] or R package meta.[28] The authors used Stata. The next step in meta-analysis is the primary analysis, where meta-analysts must decide between two different types of models that are based on different assumptions: fixed effects and random effects.[6] Fixed-effects models assume that all observations share a common mean effect size, which means that differences are only due to sampling error, while random-effects models assume heterogeneity and allow for a variation of the true effect sizes across studies.[6,29,30] In general, the presence of heterogeneity is likely in management meta-analyses because most studies do not have identical empirical settings, which can yield different effect size strengths or directions for the same investigated phenomenon. The authors chose a random effect model. They did an analysis on which group had the higher odds of cardiac mortality: PCI or CABG. There was an increased risk of cardiac death with the PCI procedure. This result is represented through a forest plot. The forest plot of Cardiac death is attached below (Fig. A). We can see the eight studies along with their respective odds ratio (OR 95% CI). We can see a PCI and CABG column which are the two groups that are being compared. The two groups also mention the total sample size and the events that may have taken place in each group. The weights column is the weight of each study. The vertical line is the line of no effect with the value of 0 if continuous measures such as mean difference are being considered, and 1 if dichotomous measures such as odds ratio as in our case are being considered. The square represents the effect size (in this case, OR) of each study. The bigger the box means the study weighed more (i.e., bigger sample size). The diamond shape represents the pooled OR of the six studies.
  Odds ratio (OR): The odds ratio is a measure of the strength of association between two variables. In this context, it indicates the likelihood of an event occurring in one group compared to another. An OR of 1 suggests no difference in the odds of the event between the two groups. An OR greater than one indicates higher odds of the event in the first group compared to the second group, while an OR less than 1 suggests lower odds in the first group. **95% Confidence interval (CI):** The confidence interval provides a range of values within which we are reasonably confident that the true value of the parameter (in this case, the odds ratio) lies. A 95% CI means that if we were to conduct the study multiple times, we would expect the true odds ratio to fall within the specified range of 95 out of 100 times.

Interpreting the pooled OR in the forest plot suggests that the odds of cardiac mortality occurring in PCI are approximately 1.27 times higher than in CABG. The 95% CI (1.02−1.59) indicates that we are 95% confident that the true odds ratio lies between 1.02 and 1.59. This means there is a statistically significant association between the groups, as the interval does not include 1. If the interval included 1, it would imply that there is no significant difference between the groups in terms of the event being studied. We can also confirm statistical significance with *P*-value. *P* values answer the question of how likely it is to observe certain results if a null hypothesis is true. The significance level (α) is a threshold set to determine statistical significance (typically set at 0.05). The *P* value is 0.04 in the forest plot, which means it is less than 0.05, hence there is a statistically significant association. For heterogeneity, we see that $I2 = 0\%$, which means no heterogeneity is detected.

Get started

1. Define a clear research question: Use frameworks like PICO (Population, Intervention, Comparison, Outcome) to ensure clarity and focus.
2. Conduct a thorough literature search: Utilize multiple databases to gather a comprehensive set of studies.
3. Set strict inclusion and exclusion criteria: This helps in maintaining consistency and relevance.

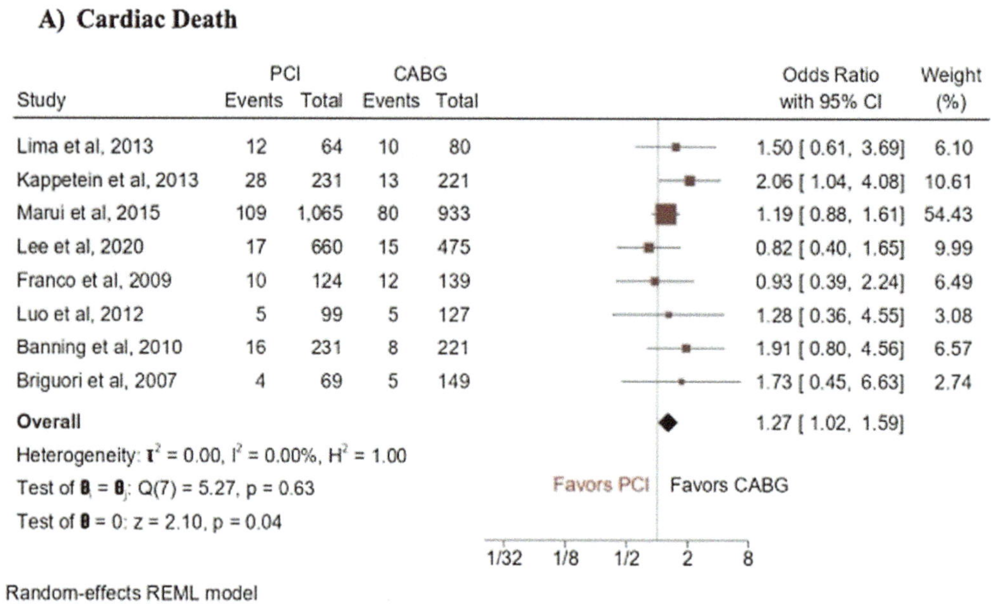

(A) FOREST PLOT OF CARDIAC DEATH

Discussion

Upon completing the fundamental steps of conducting a meta-analysis, the ultimate stage involves presenting its outcomes. Crucially, authors must ensure that all methodological choices are transparent and comprehensible to readers. In quantitative meta-analyses, result tables should encompass vital details and statistical tests, including mean effect sizes, standard errors, confidence intervals, sample sizes, and measures of heterogeneity. These findings, depicted in tables and figures, require verbal clarification within the results and discussion sections. Authors should address the primary research question, discerning relationships between variables or assessing intervention effectiveness. Interpretation should encompass both the significance and practical implications of the results, juxtaposed with prior studies where pertinent. Furthermore, authors should delineate the nuances of the findings, highlighting heterogeneity, moderators, avenues for future research, and theoretical implications.[31]

4. Assess study quality: Use established tools to evaluate the risk of bias in the included studies.
5. Choose appropriate statistical methods: Decide between fixed-effects and random-effects models based on study heterogeneity.

Potential pitfalls

- Combining dissimilar studies ("Apples and Oranges" problem): Ensure included studies are sufficiently similar in terms of design and population.
- Including low-quality studies ("Garbage In, Garbage Out" problem): Perform rigorous quality assessments to exclude biased or poorly conducted studies.
- Ignoring publication bias: Employ methods like funnel plots to detect and adjust for this bias.
- Researcher bias: Pre-register protocols and adhere strictly to them to minimize subjective influences.

Examples

- Quantifying treatment efficacy: For example, comparing the effectiveness of PCI vs. CABG in patients with diabetes and coronary disease.
- Evaluating lifestyle interventions: Assessing the impact of diet and exercise on mental health outcomes across multiple studies.

Resources

- **Books:**
  - Chapter 10: Analysing Data and Undertaking Meta-analyses. https://training.cochrane.org/handbook/current/chapter-10. Accessed on March 7, 2024.
  - Doing Meta-Analysis with R | A Hands-On Guide. Harrer M, Cu P. https://www.taylorfrancis.com/books/mono/10.1201/9781003107347/meta-analysis-mathias-harrer-pim-cuijpers-toshi-furukawa-david-ebert. Accessed on March 7, 2024.
- **Software:**
  - R packages (e.g., "meta"): Open-source tools for conducting meta-analyses.
  - RevMan: Free software by Cochrane for systematic reviews and meta-analyses.
  - Stata: A powerful statistical software for data analysis.
- **Guidelines:**
  - PRISMA (Preferred Reporting Items for Systematic Reviews and Meta-Analyses): A set of guidelines for reporting systematic reviews and meta-analyses.
- **Databases:**
  - Cochrane Library
  - Embase
  - Google Scholar
  - PubMed

References

1. Glass GV. Primary, secondary, and meta-analysis of research. cited 2024 March 7 *Educ Res*. 1976;5(10):3–8. Available from: https://www.jstor.org/stable/1174772.
2. Doing Meta-Analysis with R | A Hands-On Guide. Harrer M, Cu P, cited 2024 March 7. Available from: https://www.taylorfrancis.com/books/mono/10.1201/9781003107347/meta-analysis-mathias-harrer-pim-cuijpers-toshi-furukawa-david-ebert.
3. Lipsey MW, Wilson DB. *Practical Meta-Analysis*. Thousand Oaks, CA, US: Sage Publications, Inc; 2001:247 (Practical meta-analysis).
4. Chapter 10: Analysing Data and Undertaking Meta-analyses. cited 2024 March 7. Available from: https://training.cochrane.org/handbook/current/chapter-10.
5. Ioannidis JPA. The mass production of redundant, misleading, and conflicted systematic reviews and meta-analyses. *Milbank Q*. September 2016;94(3):485–514.
6. Borenstein M, ed. *Introduction to Meta-Analysis*. Chichester, U.K: John Wiley & Sons; 2009:421.
7. Greco T, Zangrillo A, Biondi-Zoccai G, Landoni G. Meta-analysis: pitfalls and hints. *Heart Lung Vessels*. 2013;5(4):219–225.
8. Sharpe D. Of apples and oranges, file drawers and garbage: why validity issues in meta-analysis will not go away. *Clin Psychol Rev*. December 1997;17(8):881–901.
9. Fanelli D. Negative results are disappearing from most disciplines and countries. cited 2024 March 7 *Scientometrics*. March 1, 2012;90(3):891–904. https://doi.org/10.1007/s11192-011-0494-7. Available from:.
10. Cummings SR, Browner WS, Hulley SB. Conceiving the Research Question and Developing the Study Plan.
11. Mattos CT, Ruellas AC de O. Systematic review and meta-analysis: what are the implications in the clinical practice?. cited 2024 March 7 *Dent Press J Orthod*. 2015;20(1):17–19. Available from: https://www.ncbi.nlm.nih.gov/pmc/articles/PMC4373011/.
12. Jaiswal V, Sattar Y, Peng Ang S, et al. Long term outcomes of percutaneous coronary intervention vs coronary artery bypass grafting in patients with diabetes mellitus with multi vessels diseases: a meta-analysis. cited 2024 March 7 *IJC Heart Vasc*. June 1, 2023;46:101185. Available from: https://www.sciencedirect.com/science/article/pii/S2352906723000167.
13. Harari MB, Parola HR, Hartwell CJ, Riegelman A. Literature searches in systematic reviews and meta-analyses: a review, evaluation, and recommendations. cited 2024 March 7 *J Vocat Behav*. April 1, 2020;118:103377. Available from: https://www.sciencedirect.com/science/article/pii/S0001879120300026.
14. Martín-Martín A, Thelwall M, Orduna-Malea E, Delgado López-Cózar E. Google scholar, microsoft academic, Scopus, dimensions, Web of science, and OpenCitations' COCI: a multidisciplinary comparison of coverage via citations. cited 2024 March 7 *Scientometrics*. January 1, 2021;126(1):871–906. https://doi.org/10.1007/s11192-020-03690-4. Available from:.
15. DeSimone JA, Brannick MT, O'Boyle EH, Ryu JW. Recommendations for reviewing meta-analyses in organizational research. cited 2024 March 7 *Organ Res Methods*. October 2021;24(4):694–717. Available from: http://journals.sagepub.com/doi/10.1177/1094428120967089.
16. Mongeon P, Paul-Hus A. The journal coverage of Web of Science and Scopus: a comparative analysis. cited 2024 March 7 *Scientometrics*. January 1, 2016;106(1):213–228. https://doi.org/10.1007/s11192-015-1765-5. Available from:.
17. Gusenbauer M, Haddaway NR. Which academic search systems are suitable for systematic reviews or meta-analyses? Evaluating retrieval qualities of Google Scholar, PubMed, and 26 other resources. cited 2024 March 7 *Res Synth Methods*. 2020;11(2):181–217. Available from: https://onlinelibrary.wiley.com/doi/abs/10.1002/jrsm.1378.
18. Zotero. Your Personal Research Assistant. cited 2024 March 7. Available from: https://www.zotero.org/.
19. Mendeley. Reference Management Software. cited 2024 March 7. Available from: https://www.mendeley.com/?interaction_required=true.
20. The PRISMA 2020 statement: an updated guideline for reporting systematic reviews. Br Med J. cited 2023 November 28. Available from: https://www.bmj.com/content/372/bmj.n71.
21. PROSPERO. cited 2024 March 7. Available from: https://www.crd.york.ac.uk/PROSPERO/.

22. Cuijpers P. *Meta-analyses in Mental Health Research: A Practical Guide*. 2016.
23. Higgins JPT, Altman DG, Gøtzsche PC, et al. The Cochrane Collaboration's tool for assessing risk of bias in randomised trials. cited 2024 March 7 *BMJ*. October 18, 2011;343:d5928. Available from: https://www.bmj.com/content/343/bmj.d5928.
24. Study Quality Assessment Tools. NHLBI, NIH; cited 2024 March 7. Available from: https://www.nhlbi.nih.gov/health-topics/study-quality-assessment-tools.
25. Sterne JA, Hernán MA, Reeves BC, et al. ROBINS-I: a tool for assessing risk of bias in non-randomised studies of interventions. cited 2024 March 7 *BMJ*. October 12, 2016;355. Available from: https://www.bmj.com/content/355/bmj.i4919.
26. RevMan. cited 2024 March 7. Available from: https://training.cochrane.org/online-learning/core-software/revman.
27. Meta-analysis. Stata; cited 2024 March 7. Available from: https://www.stata.com/features/meta-analysis/.
28. Schwarzer G. *Meta: An R Package for Meta-Analysis*. Vol. 7. 2007.
29. Cheung MWL, Vijayakumar R. A guide to conducting a meta-analysis. *Neuropsychol Rev*. June 2016;26(2):121–128.
30. Schmidt FL, Hunter JE. *Methods of Meta-Analysis: Correcting Error and Bias in Research Findings*. cited 2024 March 7. SAGE Publications, Ltd; 2015. Available from: https://methods.sagepub.com/book/methods-of-meta-analysis-3e.
31. DeSimone JA, Köhler Tine, Schoen Jeremy L, eds. *If it Were Only that Easy: The Use of Meta-Analytic Research by Organizational Scholars*; 2019 [Internet]. cited 2024 March 7. Available from: https://journals.sagepub.com/doi/10.1177/1094428118756743.

Chapter 42

Cost-effectiveness study: Design, measures, classic example

Heena Kaushal Asnani
Department of Internal Medicine, Roger Williams Medical Center, Providence, RI, United States

Introduction

CEA stands as a pivotal tool in healthcare policy, providing a comprehensive examination of treatment options to optimize both patient well-being and resource allocation. While clinicians focus on tailoring treatments to individual patients, CEA zooms out to evaluate the broader impact of interventions on entire populations. By scrutinizing the balance between health benefits and economic costs, this approach aims to guide policy decisions, ensuring that effective treatments receive appropriate recognition while preventing unnecessary expenditure on care with limited patient benefit.

Unlike cost-benefit analysis, which quantifies outcomes in monetary terms, CEA typically measures outcomes in natural units (such as life-years gained or quality-adjusted life-years [QALYs]) and compares them against the costs associated with each intervention. The aim of a cost-effective study is to identify which intervention provides the greatest health benefit relative to its cost, assisting decision-makers in allocating resources efficiently to maximize the overall health of a population within given budget constraints.

Benefits

CEA offers several benefits that enhance decision-making in healthcare resource allocation. Firstly, it serves as a crucial tool in identifying overlooked opportunities by pinpointing interventions that, despite being relatively inexpensive, hold significant potential to alleviate the burden of disease. Moreover, it facilitates the reallocation of resources toward more impactful endeavors, highlighting the value of shifting investments from ineffective to effective interventions and from less to more cost-effective options. Raikou et al. investigated the cost-effectiveness of atorvastatin in preventing cardiovascular disease (CVD) among type 2 diabetes patients. Utilizing data from CARDS, they found atorvastatin to be cost-effective, reducing CVD events and long-term costs compared to standard care. This suggests atorvastatin as a viable option for primary CVD prevention in type 2 diabetes, aiding healthcare decisions for managing diabetes and reducing cardiovascular risk.[1]

Furthermore, CEA acknowledges that health interventions aim not only to prevent death or extend life but also to enhance overall well-being. This is particularly evident through measures like the disability-adjusted life year (DALY), which captures both the additional years of life gained and the improved health experienced as a result of an intervention. By assigning disability weights to various health conditions, the DALY provides a comprehensive assessment of the health impact beyond mortality, incorporating factors such as pain, disability, and loss of bodily functions.

DANCAVAS examined the cost-effectiveness of screening men for CVD in European healthcare systems using data from a 5.7-year randomized controlled trial. Screening methods included low-dose computed tomography, limb blood pressure measurement, heart rhythm assessment, and cholesterol/HgbA1c levels. The results showed that younger men without baseline CVD may benefit more. Overall, comprehensive CVD screening proved cost-effective compared to common cancer screening programs, though optimal target groups warrant further clarification.[2]

Additionally, CEA presents advantages over cost-benefit analysis by sidestepping ethical dilemmas and analytical complexities associated with assigning monetary values to health outcomes. By focusing solely on health benefits without monetization, this approach provides a clearer evaluation of interventions' effectiveness and

Translational Cardiology. https://doi.org/10.1016/B978-0-323-91790-2.00112-X

ensures that decisions prioritize improving health outcomes for individuals and populations. Moreover, it allows for a more straightforward comparison of interventions based on their relative cost-effectiveness, streamlining decision-making processes and enabling more efficient resource allocation in healthcare systems.

Steps

Setting clear baseline, appropriate outcomes, and duration

In healthcare or clinical medicine, decide which treatment options or comparators will be analyzed, such as alternative methods for delivering health education, different drugs for specific diseases, alternative medical interventions, or no comparator. Ensure outcomes are consistent across all programs or interventions and clearly defined. Outcome measures can include single or combined health measures and determine the duration over which the analysis is projected and costs are collected. Ensure the time frame is long enough to observe intervention effectiveness and avoid capturing only cyclical or seasonal cost patterns.

EXAMPLE: We have two manuscripts: Manuscript A and Manuscript B. In this example, the outcome measure is the number of citations received by each manuscript over a specific period, say 1 year.

Decision and cost analysis

Quantify programmatic alternatives' impact on health outcomes using decision trees or Markov models. Decision trees depict decision, chance, and end nodes. Markov models are for conditions with recurring events over time. Identify types of costs (provider's or societal viewpoint). List all relevant costs associated with each intervention. Gather comprehensive data on actual expenditures, including donated resources. Categorize costs into development and implementation, and further into resource components.

In our case, these may include research and writing time, editing services, publication fees, promotion costs, etc.

Cost-effectiveness analysis

Calculate cost-effectiveness ratio (CER) by dividing costs by units of effectiveness.

Lower CER indicates higher efficiency.

The difference between average cost-effectiveness ratio (ACER) and incremental cost-effectiveness ratio (ICER) lies in their scope and application within CEA.[3,4]

Average cost-effectiveness ratio

ACER measures the average cost of achieving a unit of effectiveness across all interventions being compared.

It is useful for evaluating the overall efficiency of an intervention or program.

ACER is calculated by dividing the total costs of an intervention by its total effectiveness.

Example with manuscripts: Suppose we are comparing two manuscripts published in different journals. Manuscript A incurred $10,000 in costs and received 100 citations, while Manuscript B incurred $8000 in costs and received 80 citations. The ACER for Manuscript A would be $100 per citation ($10,000/100), and for Manuscript B, it would be $100 per citation ($8000/80). ACER allows us to compare the average cost per citation between the two manuscripts.

Incremental cost-effectiveness ratio

ICER measures the additional cost required to achieve an additional unit of effectiveness when comparing two interventions.

It is useful for assessing the relative efficiency of one intervention over another.

ICER is calculated by dividing the difference in costs between two interventions by the difference in their effectiveness.

Example with manuscripts: Let us continue with Manuscript A and Manuscript B. The ICER would compare the additional cost per citation between the two manuscripts. Suppose Manuscript A costs $2000 more than Manuscript B but received 20 more citations. The ICER would be calculated as ($2000/20) = $100 per additional citation. This indicates that Manuscript A is $100 more expensive for each additional citation compared to Manuscript B.

In summary, ACER provides a general measure of cost-effectiveness across interventions, while ICER allows for a direct comparison of the additional cost per unit of effectiveness between two specific interventions. Both are valuable tools in assessing the cost-effectiveness of interventions, including manuscripts published in academic journals.

Interpreting results

Compare the ACERs of the two manuscripts. In this example, both manuscripts have the same ACER, indicating that they have the same average cost per citation. The systematic review by Li et al. examined the cost-effectiveness of interventions recommended by the American Diabetes Association for preventing and managing diabetes and its complications. Key findings highlighted cost-saving measures like ACE inhibitor therapy for hypertension control and comprehensive foot care, as well as very cost-effective interventions such as lifestyle changes for diabetes prevention and screening for undiagnosed diabetes. The study's implications suggest that prioritizing these interventions could significantly improve healthcare outcomes and cost-effectiveness.[5]

Uncertainty handling

Conduct sensitivity analysis to assess the robustness of the results to variations in key parameters, such as costs or outcomes. This helps to understand the uncertainty associated with the analysis. Address parameter, model, and generalizability uncertainties. Probabilistic sensitivity analysis generates empirical distributions of CERs to explore uncertainty impact.

Decision making

Based on the results of the CEA, make decisions about which intervention is more cost-effective or whether the costs are justified given the outcomes achieved. The quality of the analysis depends on the accuracy and completeness of the data collected, as well as the assumptions made during the analysis. Additionally, interpretation should consider not only the numerical results but also the broader context and implications of the findings. The study by Pignone et al. compared the cost-effectiveness of aspirin, clopidogrel, and combination therapy for secondary prevention of coronary heart disease (CHD). Their analysis showed that aspirin alone was the most cost-effective option, offering similar effectiveness to clopidogrel or combination therapy but at lower costs. Therefore, the study recommended prioritizing aspirin therapy for patients with established CHD. This finding has significant implications for healthcare policy, suggesting that resources can be efficiently allocated by prioritizing aspirin as the first-line antiplatelet therapy for CHD secondary prevention.[6]

Cost-effectiveness thresholds

Cost-effectiveness thresholds are benchmarks used to determine whether a healthcare intervention offers good value for money. They help policymakers, insurers, and other stakeholders decide which interventions to fund based on their costs and the health benefits they provide. These thresholds can vary but typically represent the maximum amount a health system or individual is willing to pay for one unit of health benefit, such as a QALY.[7]

There are two main types of thresholds:

1. **Willingness-to-pay thresholds:** These represent the amount a consumer or society is prepared to pay for a health benefit, balancing this against other competing needs and resource constraints.
2. **Supply-side thresholds:** These consider the opportunity costs of resource allocation. For instance, they estimate the health benefits foregone when funds are diverted from existing interventions to new ones, ensuring that the overall use of resources is optimized.

When choosing a type of cost-effectiveness threshold, the concept of opportunity cost is particularly relevant for providers focused on optimizing health outcomes with available resources. Decision-makers must estimate both the potential health gains from alternative uses of the resources required for a new intervention and the health losses if the new intervention is not implemented.[7]

For decades, U.S. CEAs have often used a $50,000 per QALY benchmark to value health gains, despite its weak theoretical and empirical basis. This study analyzed U.S.-based CEAs from 1990 to 2021 and it found a decline in the use of the $50,000 per QALY threshold (from 19.6% in 1990−99 to 12.6% in 2020−21) and an increase in the $100,000 (from 9.1% to 47.0%) and $150,000 (from 1.5% to 23.6%) per QALY benchmarks. Cancer-related and industry-funded CEAs were more likely to reference the $150,000 threshold. These trends are consistent with the Institute for Clinical and Economic Review's use of higher thresholds.[8,9] The upper limit for prudent healthcare and safety expenditure is set at 140% of Gross domestic product per capita (GDPpc) per QALY. Any program or policy exceeding this cost threshold is likely to statistically shorten more lives than it saves. This value aligns with various healthcare CEA values found in public health literature and serves as a practical guideline for policymakers.[10]

Limitations

Health economic evaluation (CEA) involves assumptions, uncertainties, and data reliance, with sensitivity analysis addressing uncertainties. However, CEA lacks assessment of societal willingness to pay and may overlook external factors. Challenges include health equity neglect and resource misallocation, especially in low- and middle-income countries. GDP per capita often guides cost-effectiveness thresholds. Performing CEA involves aggregating diverse data, with sensitivity analyses crucial for understanding variability.[11] It's vital to identify all intervention costs, including direct, indirect, and future costs. Different perspectives in CEA frame outcomes differently, with societal perspectives crucial for policy decisions. CEA's limitations include oversimplification of health states and difficulties in estimating indirect and societal-level costs. Establishing a threshold for willingness-to-pay per DALY remains ambiguous, with arguments for higher thresholds reflecting modern clinical practice.[12]

References

1. Raikou M, McGuire A, Colhoun HM, et al. Cost-effectiveness of primary prevention of cardiovascular disease with atorvastatin in type 2 diabetes: results from the Collaborative Atorvastatin Diabetes Study (CARDS). *Diabetologia*. April 2007;50(4):733−740.

2. Søgaard R, Diederichsen ACP, Rasmussen LM, et al. Cost effectiveness of population screening vs. no screening for cardiovascular disease: the Danish Cardiovascular Screening trial (DANCAVAS). *Eur Heart J*. November 1, 2022;43(41):4392–4402.
3. Abdul Manaf R, Hanafiah JM. Designing and conducting cost-effectiveness analysis studies in healthcare. *Int J Publ Health Clin Sci*; 2017. https://www.researchgate.net/publication/323412922.
4. Bang H, Zhao H. Cost-effectiveness analysis: a proposal of new reporting standards in statistical analysis. *J Biopharm Stat*. March 4, 2014;24(2):443–460.
5. Li R, Zhang P, Barker LE, Chowdhury FM, Zhang X. Cost-effectiveness of interventions to prevent and control diabetes mellitus: a systematic review. *Diabet Care*. August 2010;33(8):1872–1894.
6. Pignone M, Earnshaw S, Tice JA, Pletcher MJ. Aspirin, statins, or both drugs for the primary prevention of coronary heart disease events in men: a cost–utility analysis. *Ann Intern Med*. March 7, 2006;144(5):326.
7. Bertram MY, Lauer JA, De Joncheere K, et al. Cost–effectiveness thresholds: pros and cons. *Bull World Health Organ*. December 1, 2016;94(12):925–930.
8. Neumann PJ, Cohen JT, Weinstein MC. Updating cost-effectiveness — the curious resilience of the $50,000-per-QALY threshold. *N Engl J Med*. August 28, 2014;371(9):796–797.
9. Neumann PJ, Kim DD. Cost-effectiveness thresholds used by study authors, 1990–2021. *JAMA*. April 18, 2023;329(15):1312.
10. Yanovskiy M, Levy ON, Shaki YY, Zigdon A, Socol Y. Cost-effectiveness threshold for healthcare: justification and quantification. *Inquiry: J Health Care Organ, Provis, Finan*. January 13, 2022;59.
11. Burgers LT, Redekop WK, Severens JL. Challenges in modelling the cost effectiveness of various interventions for cardiovascular disease. *Pharmacoeconomics*. 2014;32(7):627–637.
12. Brooke BS, Kaji AH, Itani KMF. Practical guide to cost-effectiveness analysis. *JAMA Surg*. 2020;155:250–251.

Chapter 43

Diagnostic test evaluation: Design, measures, classic example

Suman Biswas

Department of Imaging, University of Rochester, Rochester, NY, United States

Key points

- Diagnostic tests in cardiology.
- Evaluation process for diagnostic tests.
- Key diagnostic test measures.

By the end of the chapter, the reader should be able to:

- Understand the overall concept of a diagnostic test evaluation process.
- Understand the value of the gold standard, relevant statistical tools, and statistical concepts of diagnostic test evaluation.
- Understand the objective measures of test effectiveness and effect of disease prevalence.
- Grasp the concept with common real-life practical example.

Introduction

Diagnostic tests are pivotal in evidence-based cardiology and crucial in disease diagnosis, health assessment, and guiding treatment strategies. In the United States, approximately 14 billion lab tests drive 70% of clinical decisions annually, highlighting their fundamental role in medical practice. Particularly in cardiology, these tests are vital, providing critical insights into heart health and informing decisions in a field where heart disease remains the leading cause of death, claiming a life every 33 s and imposing an annual economic cost of $239.9 billion. Thus, evaluating diagnostic tests is a crucial aspect of medical research and clinical practice, ensuring that these tests are reliable, accurate, and valuable in the real-world setting.[1–3]

Diagnostic test evaluation: Design[4–11]

The evaluation of diagnostic tests requires careful designing, planning and execution. The evaluation process is often complex, multifaceted, overlapping, and variable according to each case scenario.

Define the objective

a. **Clinical Relevance:** Determine what the test is intended to diagnose or predict. The objective should be clinically relevant, addressing a specific diagnostic question or gap in medical practice.
b. **Target Population:** Identify the population in which the test will be used, including age, race, gender, disease prevalence, and specific risk factors.

Choose the study design

a. **Cross-sectional Studies:** Assess the test's performance by evaluating a specific population at a single point in time, comparing the new test's results with a comparator.
b. **Cohort Studies:** Follow a group of individuals over time to assess how well the test predicts outcomes.
c. **Case-Control Studies:** Compare individuals with a condition to those without, looking backward to see how well the test identified those with the condition.

Select the comparator

a. **Gold-Standard:** The gold standard is the best proven method for diagnosing the condition of interest. The new test's performance is measured against this standard to assess its accuracy and reliability.

Translational Cardiology. https://doi.org/10.1016/B978-0-323-91790-2.00069-1

b. **Without Gold-Standard:** However, a proper gold standard might not be available for ethical, cost, or practical reasons. A surrogate or composite standard, which combines several tests or clinical judgment, may be used in such cases.

Define outcomes

a. **Diagnostic Accuracy Measures:** These include sensitivity (ability to detect the condition when it is present), specificity (ability to exclude the condition when it is not present), positive predictive value (PPV), and negative predictive value (NPV).
b. **Operational Outcomes:** Consider test feasibility, cost, time, ethical aspects, medicolegal aspects, and patient acceptability.

Sample size calculation

Calculate the necessary sample size based on the expected performance of the test, the prevalence of the condition in the target population, and the desired precision of the results.

Ethical considerations

Ensure the study design follows ethical guidelines, including informed consent, confidentiality, and the right to withdraw.

Data collection and analysis

With some rare exceptions, new tests, while developed, often go through several stages of trials. They are summarized in the table in the Diagram 43.1.

a. **Data Collection:** Implement standardized protocols for test administration and data collection to minimize bias.
b. **Statistical Analysis:** Use appropriate statistical methods to analyze the data, including measures of diagnostic accuracy and comparisons with the gold standard or surrogates.

Interpretation and reporting

Interpret the results in the context of the test's intended clinical use, considering statistical significance and clinical relevance. Report the findings transparently, including limitations and the potential impact on clinical practice.

Post-market surveillance

After a test is introduced into clinical practice, ongoing evaluation is necessary to monitor its performance in a broader population and over time.

Test-treatment threshold

Clinicians must consider the threshold at which the likelihood of disease warrants treatment, balancing the risks and benefits based on the available data.

Phase	Description
Preclinical Research	Before testing a new diagnostic test in humans, researchers conduct preclinical studies using in vitro (test tube) methods and in vivo (animal) models to assess its feasibility, accuracy, and potential toxicity or side effects. This stage helps to understand the biological basis of the test and its potential impact on the body.
Phase I	This phase primarily assesses the safety of the diagnostic test and its preliminary effectiveness in detecting the condition it's designed for in a small group of volunteers or patients.
Phase II	Involves a larger group of participants to further assess the test's effectiveness and gather data on how well the test works compared to existing diagnostic methods.
Phase III	Conducted with a larger group of participants to confirm the test's effectiveness, monitor side effects, and compare it to standard or existing diagnostic tests. This phase thoroughly explains the test's accuracy, reliability, and utility in a clinical setting. Once Phase III is completed, the data collected from all stages of the clinical trials are submitted to regulatory bodies (such as the FDA in the United States) for review.
Phase IV	After the test is approved and, on the market, the post-marketing surveillance involves ongoing monitoring of its performance in the general population and any long-term effects or rare adverse reactions not detected in the earlier phases.

DIAGRAM 43.1 Table summarizing the different phases of a clinical trail.

This process ensures that the tests used in clinical practice provide accurate, reliable, and relevant information for patient care, ultimately improving health outcomes.

Diagnostic test evaluation: Measures[12–14]

A. **Sensitivity (true positive rate):** It measures the proportion of true positives (diseased subjects) correctly identified by the test.

B. **Specificity (true negative rate):** It measures the proportion of actual negatives (healthy subjects) correctly identified by the test.
Sensitivity and specificity are independent of disease prevalence and thus measures a test's intrinsic quality.

C. **Positive Predictive Value (PPV):** It indicates the probability that a person has the disease given a positive test result.

D. **Negative Predictive Value (NPV):** It indicates the probability that a person does not have the disease given a negative test result.
Predictive values are influenced by the prevalence of the disease in the population being tested. High disease prevalence increases PPV (more true positives) and decreases NPV (higher chance of false negatives). Low disease prevalence decreases PPV (higher chance of false positives) and increases NPV (more true negatives).

E. **Positive Likelihood Ratio (LR+):** It describes how much the odds of the disease increase when a test is positive.

F. **Negative Likelihood Ratio (LR-):** It describes how much the odds of the disease decrease when a test is negative.
Likelihood ratios (LRs) provide a direct way to use the sensitivity and specificity of a test to estimate how much a test result will change the odds of having a disease. LRs are particularly useful because they are not affected by the prevalence of the disease. Diagram 43.2 summarizes the mathematical aspects.

G. **Receiver Operating Characteristic (ROC) Curves:** The ROC curve is crucial in evaluating a test result represented as binary classifier for diagnostic accuracy, plotting sensitivity versus 1-specificity across various thresholds. Its key feature, the Area Under the Curve (AUC), offers a comprehensive accuracy score.
   AUC = 1: Perfect classifier
   0.5 < AUC <1: Better than random guessing
   AUC = 0.5: No discrimination ability (equivalent to random guessing)
   AUC <0.5: Worse than random guessing (though in practice, one can reverse the model's decisions to get a classifier with AUC >0.5)
This makes the ROC curve indispensable for assessing tests or setting optimal thresholds that balance detecting true positives against false alarms. It's a visual and quantitative guide to understanding a test's precision and comparing its efficacy against others, serving as a benchmark for diagnostic performance evaluation. Diagram 43.3 depicts it visually; it is noteworthy that C curve has larger AUC than A curve.

H. **Advancements in Diagnostic Test Evaluation:** Utilizing multiple similar studies in the metaanalysis test evaluation process achieves higher power and a more robust understanding of the test's effectiveness in its target clinical scenario.[4–7] Advances in imaging technologies and biomarkers continue to evolve, improving diagnostic accuracy and patient outcomes. The rapidly changing Machine Learning and Artificial Intelligence offer promising enhancements in test accuracy and predictive capabilities.[15,16] The field of personalized medicine in cardiology, primarily through genomics and biomarker research, is evolving rapidly for customized treatment.[17]

Diagnostic test evaluation: Classic example

In their meta-analysis, Roberts et al. analyzed data from 37 distinct cohorts across 42 reports, which included 48 evaluations of diagnostic tests. This analysis encompassed a total of 15,263 test outcomes for diagnosing acute heart failure at lower recommended levels:

At lower recommended levels for diagnosing acute heart failure:

A. **B type natriuretic peptide (at 100 ng/L) had:**
A sensitivity of 95% (confidence interval [CI] 93%–96%)
An NPV of 94% (CI 90%–96%)

B. **NTproBNP (at 300 ng/L) had:**
A sensitivity of 99% (CI 97%–100%)
An NPV of 98% (CI 89%–100%)

C. **For MRproANP (at 120 pmol/L), had:**
Sensitivity: 95%–97% (range 90%–98%)
NPV: 90%–97% (range 80%–98%)

Sensitivity decreased and specificity varied at higher threshold values. There was no significant statistical difference in diagnostic accuracy between plasma B type natriuretic peptide and NTproBNP.[18]

- The columns represent the test outcome (positive on the left, negative on the right).
- The rows represent the actual condition of the subjects (disease at the top, no disease at the bottom).

	Test Positive	Test Negative
Disease Present	True Positive (TP)	False Negative (FN)
Disease Absent	False Positive (FP)	True Negative (TN)

- **Sensitivity** measures the proportion of actual positives correctly identified by the test (true positive rate).
$Sensitivity = \frac{True\ Positives}{True\ Positives + False\ Negatives}$
- **Specificity** measures the proportion of actual negatives correctly identified by the test (true negative rate).
$Specificity = \frac{True\ Negatives}{True\ Negatives + False\ Positives}$

- **Positive Predictive Value (PPV)** measures the probability that subjects with a positive screening test truly have the disease.
$PPV = \frac{True\ Positives}{True\ Positives + False\ Positives}$
- **Negative Predictive Value (NPV)** measures the probability that subjects with a negative screening test truly do not have the disease.
$NPV = \frac{True\ Negatives}{True\ Negatives + False\ Negatives}$

- **Positive Likelihood Ratio (LR+)** indicates how much the odds of the disease increase with a positive test result.
$LR+ = \frac{Sensitivity}{1 - Specificity}$
- **Negative Likelihood Ratio (LR-)** indicates how much the odds of the disease decrease with a negative test result.
$LR- = \frac{1 - Sensitivity}{Specificity}$

DIAGRAM 43.2 Mathematical summary of measures of diagnostic tests.

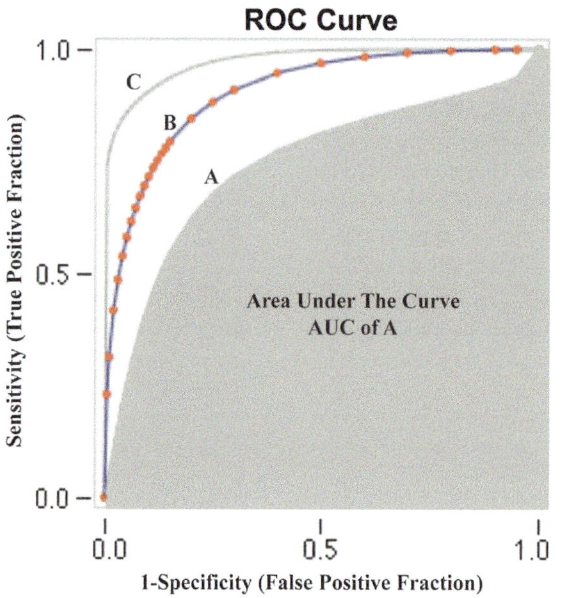

DIAGRAM 43.3 The ROC curve and AUC.

References

1. Centers for Disease Control and Prevention. *Strengthening Clinical Laboratories*. https://www.cdc.gov/csels/dls/strengthening-clinical-labs.html Accessed 20 February 2024.
2. Heart disease and stroke statistics—2023 update: a report from the American Heart Association. *Circulation*. 2023;147. https://www.ahajournals.org/doi/10.1161/CIR.0000000000001123. Accessed February 23, 2024.
3. Centers for Disease Control and Prevention. *Heart Disease Facts*; 2024. https://www.cdc.gov/heartdisease/facts.html. Accessed February 20, 2024.
4. Maw AM, Hassanin A, Ho PM, et al. Diagnostic accuracy of point-of-care lung ultrasonography and chest radiography in adults with symptoms suggestive of acute decompensated heart failure: a systematic review and meta-analysis. *JAMA Netw Open*. 2019;2(3):e190703. https://doi.org/10.1001/jamanetworkopen.2019.0703.
5. Watanabe H, Horita N, Shibata Y, Minegishi S, Ota E, Kaneko T. Diagnostic test accuracy of D-dimer for acute aortic syndrome: systematic review and meta-analysis of 22 studies with 5000 subjects. *Sci Rep*. 2016;6:26893. https://doi.org/10.1038/srep26893.
6. Yang TY, Huang L, Malwade S, Hsu CY, Chen YC. Diagnostic accuracy of ambulatory devices in detecting atrial fibrillation: systematic review and meta-analysis. *JMIR Mhealth Uhealth*. 2021;9(4):e26167. https://doi.org/10.2196/26167.
7. Andrianto A, Mertaniasih NM, Gandi P, et al. Diagnostic test accuracy of Xpert MTB/RIF for tuberculous pericarditis: a systematic review and meta-analysis. *F1000Res*. 2020;9:761. https://doi.org/10.12688/f1000research.22770.1.
8. Driessen RS, Danad I, Stuijfzand WJ, et al. Comparison of coronary computed tomography angiography, fractional flow reserve, and perfusion imaging for ischemia diagnosis. *J Am Coll Cardiol*. 2019;73(2):161–173. https://doi.org/10.1016/j.jacc.2018.10.056.
9. Koplay M, Guneyli S, Akbayrak H, et al. Diagnostic accuracy and effective radiation dose of high pitch dual source multidetector computed tomography in evaluation of coronary artery bypass graft patency. *Wien Klin Wochenschr*. 2016;128(13–14):488–494. https://doi.org/10.1007/s00508-016-1030-9.
10. Veroniki AA, Tsokani S, Agarwal R, et al. Diagnostic test accuracy network meta-analysis methods: a scoping review and empirical assessment. *J Clin Epidemiol*. 2022;146:86–96. https://doi.org/10.1016/j.jclinepi.2022.02.001.
11. Rutjes AW, Reitsma JB, Coomarasamy A, Khan KS, Bossuyt PM. Evaluation of diagnostic tests when there is no gold standard. A review of methods. *Health Technol Assess*. 2007;11(50):iii–51. https://doi.org/10.3310/hta11500.
12. Barrett BJ, Fardy JM. Evaluation of diagnostic tests. In: Parfrey PS, Barrett BJ, eds. *Methods in Molecular Biology*. Clinical Epidemiology; vol. 2249.
13. Obuchowski NA, Bullen JA. Receiver operating characteristic (ROC) curves: review of methods with applications in diagnostic medicine. *Phys Med Biol*. 2018;63(7):07TR01. https://doi.org/10.1088/1361-6560/aab4b1.
14. Molinaro AM. Diagnostic tests: how to estimate the positive predictive value. *Neurooncol Pract*. 2015;2(4):162–166. https://doi.org/10.1093/nop/npv030.
15. Jayakumar S, Sounderajah V, Normahani P, et al. Quality assessment standards in artificial intelligence diagnostic accuracy systematic reviews: a meta-research study. *NPJ Digit Med*. 2022;5(1):11. https://doi.org/10.1038/s41746-021-00544-y.
16. Richens JG, Lee CM, Johri S. Improving the accuracy of medical diagnosis with causal machine learning. *Nat Commun*. 2020;11(1):3923. https://doi.org/10.1038/s41467-020-17419-7.
17. Tada H, Fujino N, Nomura A, et al. Personalized medicine for cardiovascular diseases. *J Hum Genet*. 2021;66(1):67–74. https://doi.org/10.1038/s10038-020-0818-7.
18. Roberts E, Ludman AJ, Dworzynski K, et al. The diagnostic accuracy of the natriuretic peptides in heart failure: systematic review and diagnostic meta-analysis in the acute care setting. *BMJ*. 2015;350:h910. https://doi.org/10.1136/bmj.h910.

Chapter 44

Reliability study

Risheek Kaul[1], Ridhima Kaul[2] and Pradipta Paul[2]

[1]*Department of Cardiology, Westchester Medical Center and New York Medical College, Valhalla, NY, United States;* [2]*Weill Cornell Medicine-Qatar, Education City, Qatar Foundation, Doha, Qatar*

Key points

- Reliability is the reproducibility of a measurement and can be understood mathematically as the expected spectrum of measurements compared to the actual distribution of measurements while accounting for errors.
- Reliability studies are essential to assess new and existing measurement systems or instruments and are useful when the subjects, raters, and test administrators closely resemble the clinic or research practice setting where the instruments will be used.
- Numerous measures of reliability exist with the most common one being internal consistency and interobserver reliability being the most clinically useful.

Introduction and definition

Reliability of a measurement or instrument is an essential part of all translational research. Reliability refers to the relationship between a measurement error and the expected distribution of measurements across multiple time points, observers, and situations.[1,2] Mathematically, it is a ratio of the variability between subjects to the total variability[1] as shown by the equation below:

$$\text{Reliability} = \text{variability between subjects}/\text{total variability}*$$

*where total variability = variability between subjects + measurement error.

Reliability ranges from 0 to 1 with a score of 0 indicating that all variation in a sample is due to error and a score of 1 indicating that all variation is due to differences in the subject.[2,3] Reliability studies are experiments either clinical, industrial, or basic science that are used to assess both new or existing measurement systems[1,4] and historically have been of great importance in cardiology. For example, clinical classification tools such as the New York Heart Association or the Kansas City Cardiomyopathy Questionnaire are instrumental for prognostication of patients with congestive heart failure in both clinical and research settings. Initial data to establish their utility in predicting morbidity and mortality in different groups of patients with heart failure and even in different subsets of populations was based on reliability studies.[5–7] It is pertinent to realize that assessing reliability is one of the initial steps in trying to create a clinical instrument or tool, once reliability has been established efforts need to be made to establish the instrument's validity, feasibility, and acceptability.

Designing a study

Designing a reliability study is based on one essential concept: for the reliability study to be useful, the patients, raters, and test administration in the study must mirror or reflect the real-world context in which the study will be used.[1] With that in mind, the first most important step is to ask an appropriate research question. Ideally, the question must identify the instrument or tool(s) being tested and the manner in which it will impact clinical or research practice.

Once an appropriate question has been presented the next step is to identify a suitable reliability measurement based on the instrument to be tested. A number of different reliability measurements can be utilized, these include the following:

1. Internal consistency reliability: reflective of an individual's responses *within* an instrument and gauges whether or not different items are measuring the same thing.[1] For example, a test measuring shortness of breath at rest, with minimal activity and with strenuous activity is an assessment of internal consistency as one would assume that patients with shortness of breath at rest have a higher chance of shortness of breath with exertion and vice versa. It is measured using the Cronbach alpha test[8] with a result greater than 0.90 indicating a high correlation or reliability of 0.90 for individual comparison and

Translational Cardiology. https://doi.org/10.1016/B978-0-323-91790-2.00040-X

0.70 for a group comparison. Since internal consistency reliability is most commonly used in literature, it is pertinent to understand that the instrument is administered only once and hence cannot account for differences in times of test administration, observers, or situations and is the weakest form of reliability testing.[1]

2. Test-retest reliability: measures the stability of a result over time as participants in a study are requested to respond to a survey at two distinct time points after a prespecified interval period. An example of this would be a quality of life test such as the SF12[9] being used to assess patients' physical and mental well-being at different time points after ICD implantation. A correlation coefficient greater than or equal to 0.70 is considered a strong test–retest reliability.[1,2]
3. Intraobserver reliability: similar to test–retest reliability except that the raters and subjects being rated are fixed. An example of this study design would be if a single physician interprets the maximal diameter of the main pulmonary artery on a cardiac computed tomography scan at multiple time points and then calculates the variation between the measurements as a whole and also between readings giving us an idea of intraobserver reliability. As time is the only variable that changes between measurements, this design usually yields higher reliability estimates.
4. Interobserver reliability: Attempts to measure the stability of results when two or more raters/physicians are scoring the same subject. Continuing with the above example, the physician might request their colleagues to measure the maximal diameter of the main pulmonary artery and subsequently calculate the variation between and within their individual measurements. Reliability studies using inter- and intraobserver reliability have been crucial in establishing the diagnostic validity and reliability of cardiac imaging modalities. The coronary artery disease and data reporting system (CAD-RADS),[10] which now allows for accurate noninvasive classification of CAD is an excellent example of this. Interobserver reliability is considered the broadest and most clinically important reliability test and has been widely used.

When compared to internal consistency reliability, the other three measures of reliability measure the stability of a result across two or more administrations of the instrument. Once the appropriate reliability measure has been selected it is important to select suitable raters. Two key factors play a role when selecting appropriate raters: (1) their level of expertise and (2) their practice settings. Suitable raters are individuals or groups with a high level of training or expertise and belong to a rather limited practice setting for which the instrument being tested has been developed. Although this will improve the reliability assessment, it will hinder the extrapolation of data to clinical or practice settings that are different or have providers with differing levels of expertise.

Once the raters have been picked, usually up to four, however, depending on the type of reliability measure being used, the study subjects are picked. The subjects or patients to be picked should be representative of the clinical practice. Ideally, an increasingly heterogeneous group of subjects with a wide range of clinical conditions should be recruited to increase the robustness of the reliability measure and study. The sample size is determined largely by the degree of precision that the investigators need. Increasing the number of subjects, however, has a greater impact on study precision as opposed to increasing the number of raters. In an effective study design, the raters are selected first based on their generalizability and feasibility and then the number of subjects is selected for their desired precision.[1,11]

It is important to counsel the raters to try and mimic as closely as possible the clinical or research practice environment when assigning their ratings and also provide them with all the necessary information that would be accessed in a real-world setting, including clinical and diagnostic

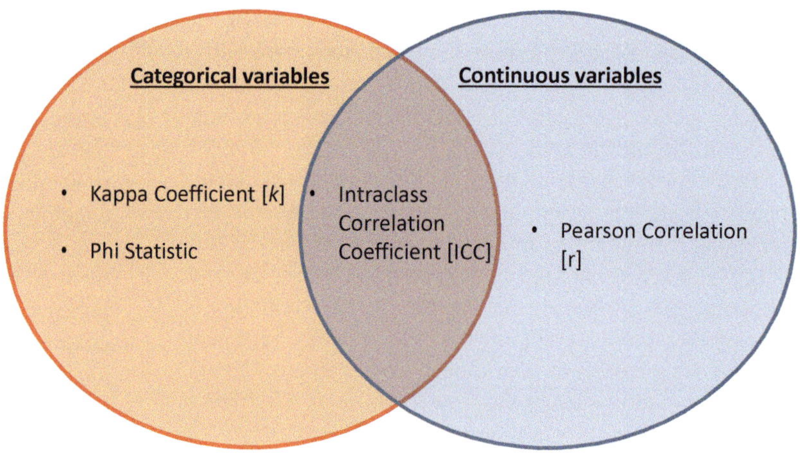

FIGURE 44.1 Statistical measures of reliability for both categorical and continuous variables.

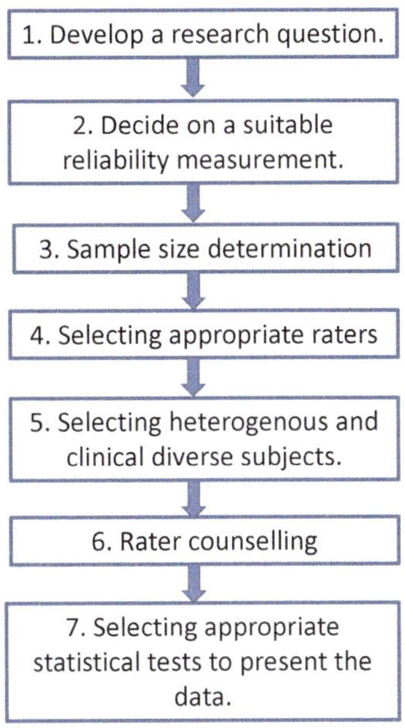

FIGURE 44.2 Flowchart detailing the steps of designing a reliability study.

data. Once the ratings have been assigned, appropriate statistical tests for reliability need to be used to analyze and interpret the data (Fig. 44.1).

Reliability studies can be variable in the cost depending on the measurement or instrument being tested. Funding can be requested through government agencies, industry sponsorship, or drug companies. The process of designing a reliability study has been summarized in Fig. 44.2.

One of the limiting factors in reliability studies is the difficulty of arranging replicated observations because of limited resources, whether it be a small group of clinical specialists who qualify as raters or alternatively, subjects or equipment.[12]

References

1. Karanicolas PJ, Bhandari M, Kreder H, et al. Evaluating agreement: conducting a reliability study. *J Bone Joint Surg Am*. 2009;91(Suppl 3):99–106. https://doi.org/10.2106/JBJS.H.01624.
2. *Basic Methods Handbook for Clinical Orthopaedic Research*. 2019. https://doi.org/10.1007/978-3-662-58254-1. Basic Methods Handb Clin Orthop Res.
3. Dunn G. Design and analysis of reliability studies. *Stat Methods Med Res*. 1992;1(2):123–157. https://doi.org/10.1177/096228029200100202.
4. Donner A, Eliasziw M. Sample size requirements for reliability studies. *Stat Med*. 1987;6(4):441–448. https://doi.org/10.1002/SIM.4780060404.
5. Green CP, Porter CB, Bresnahan DR, Spertus JA. Development and evaluation of the Kansas City Cardiomyopathy Questionnaire: a new health status measure for heart failure. *J Am Coll Cardiol*. 2000;35(5):1245–1255. https://doi.org/10.1016/S0735-1097(00)00531-3.
6. Goldman L, Hashimoto B, Cook EF, Loscalzo A. Comparative reproducibility and validity of systems for assessing cardiovascular functional class: advantages of a new specific activity scale. *Circulation*. 1981;64(6):1227–1234. https://doi.org/10.1161/01.CIR.64.6.1227.
7. Pettersen KI, Reikvam A, Rollag A, Stavem K. Reliability and validity of the Kansas City cardiomyopathy questionnaire in patients with previous myocardial infarction. *Eur J Heart Fail*. 2005;7(2):235–242. https://doi.org/10.1016/J.EJHEART.2004.05.012.
8. Cronbach LJ. Coefficient alpha and the internal structure of tests. *Psychom*. 1951;16(3):297–334. https://doi.org/10.1007/BF02310555.
9. Ware JE, Kosinski M, Keller SD. A 12-Item Short-Form Health Survey: construction of scales and preliminary tests of reliability and validity. *Med Care*. 1996;34(3):220–233. https://doi.org/10.1097/00005650-199603000-00003.
10. Basha MAA, Aly SA, Ismail AAA, et al. The validity and applicability of CAD- RADS in the management of patients with coronary artery disease. *Insights Imag*. 2019;10:117.
11. Shoukri MM, Asyali MH, Donner A. Sample size requirements for the design of reliability study: review and new results. *Statist Meth Med Res*. 2016;13(4):251–271. https://doi.org/10.1191/0962280204SM365RA.
12. Walter SD, Eliasziw M, Donner A. Sample size and optimal designs for reliability studies. *Stat Med*. 1998;17(1):101–110.

Chapter 45

The state of database studies and future research

Juliet Meir[1] and Jayakumar Sreenivasan[2]
[1]*Department of Internal Medicine, Westchester Medical Center, Valhalla, NY, United States;* [2]*Department of Cardiology, Westchester Medical Center, New York Medical College, Valhalla, NY, United States*

Introduction

Databases are currently the main data source on which a majority of retrospective and prospective studies are based. Databases provide a large amount of information about many types of cardiology topics that with the appropriate statistical design can be applied at both the individual and population levels. This chapter discusses the different types of database studies, the concept of big data and its advantages and limitations, and statistical analysis and design. We provide an example of a database registry at the end of this chapter, which can be used as a reference.

Types of database studies

There are five main types of database studies used in cardiology research: population-based, disease-based, procedure/technology-based, government and administration-based (i.e., Medicare and Medicaid or the Veterans Administration), and clinical registries. Population-based databases examine risk factors and the natural history of a particular cardiovascular disease. Disease-based registries look at outcomes for various disease processes, such as heart failure or myocardial infarction. Technology-based databases compile outcomes from wearable and implantable devices. Recently, procedure and technology-based databases have been utilized more frequently given advances in technology and the ability to record information on consumer devices. An example of this is the Apple Heart Study.[1] Administrative databases are compiled by major health insurance and government agencies (i.e., Medicare and Medicaid) and are commonly utilized in the era of big data. Additionally, clinical registry based on electronic medical health records, derived both nationally and internationally, is another database arena that has dominated the big data movement. Large well-maintained registries facilitate the conduct of randomized controlled trials as well. For example, the Thrombus Aspiration in Myocardial Infarction (TASTE) trial was conducted based on the SWEDEHEART Registry[2] and the Study of Access Site for Enhancement for Women trial (SAFE-PCI) that was extracted from the National Cardiovascular Data Registry.[3]

Overview of big data

Databases are extremely valuable and useful because they provide large pools of data that can be analyzed with a large number of patients, leading to more generalizability. Their primary use stems from their ability to generate preliminary evidence, which could be confirmed with subsequent prospective clinical trials.[4,5] Publications utilizing database studies have become increasingly popular in recent years. Of note, 75% of publications looking at database studies were published within the last 4 years.[6]

However, databases are not without limitations. Unlike randomized controlled trials, databases have different reasons for the information collection and differ in how the data is stored and recorded.[7] If not carefully analyzed, the data can be misclassified and be skewed by selection bias. Results of various analyses based on databases are also prone to biases due to unknown confounders because of nonrandomized nature of the database. As a result, extensive statistical analyses and techniques need to be applied to the data before any meaningful conclusions.

Translational Cardiology. https://doi.org/10.1016/B978-0-323-91790-2.00089-7

Advantages of big data

Database studies contribute to research in a number of ways. First, databases reflect the population seen in clinical practice[8] by including all underrepresented patient populations. Second, the number of independent variables collected allows for a large number of measurements and allows researchers to create complex models and predictive algorithms.[9] Third, databases enable one to compare a provider's current clinical care to guideline recommendations. Fourth, databases are particularly useful in studies looking at epidemiology, prevention, and treatment of rare diseases that are otherwise harder to study prospectively. Fifth, databases are useful in studying vulnerable populations such as pregnant women on whom it is difficult to gather information, given the ethical need to avoid and minimize adverse effects on pregnant women. Most of the data looking at pregnant women comes from retrospective and small studies. Sixth, databases can be synchronized to allow for collaboration between multiple departments and disciplines.[10] Lastly, database studies can be valuable tools for prescription drug surveillance and uncovering potential drug repurposing.[11]

Limitations of big data

Although database studies provide a collection of data, there are limitations to their use. Some limitations may include a lack of uniform data entry that leads to missing data and limited generalizability in clinical practice.[12] Another common limitation is difficulty involving data sharing. For instance, in the Cardiology Centers of the Netherlands database, EKG and echocardiogram data were saved in separate databases limiting integrated analysis. Furthermore, there are legal and privacy concerns associated with the analysis of large databases that can lead to privacy breaches.[11] Lastly, there is a potential for selection bias when comparing nonrandomized diagnostic or therapeutic approaches in patients who are dissimilar.

Highlights of big data

Doug Laney was the first to describe big data as the 3 Vs: volume (size of data), variety (structured and unstructured data), and velocity (speed in generating data).[13] Today, big data is defined by additional two characteristics: veracity (trustworthiness of data), and value (the worth extracted from the dataset).[14] An example of maximizing the volume of data extracted from a large database is the Framingham study. Upon establishment in 1948, the study had 5209 patients but has now grown to include >15,000 patients.[15] Variety describes the source of data analyzed, structured, or unstructured. Examples of structured data collected include age, drug dose, ECG, transthoracic echocardiogram values, and genetic data. Examples of unstructured data collected include clinic notes, physical exam findings, differential diagnoses, and medical instructions.[16] A good example highlighting the velocity that can be obtained using big data is seen when looking at stent thrombosis. Given how uncommon stent thrombosis is, randomized controlled trials studying stent thrombosis are often underpowered to detect a significant difference. The Swedish Angiography and Angioplasty Registry changed that by studying 36,000 patients undergoing coronary intervention since 2005 and establishing important findings in the area of drug-eluting stent performance.[17–19]

Registries extracted from big data

Registry, a type of large database, provides big data on everyday practice and is a vital source of information for quality improvement objectives. Some areas of research that benefit from registry analysis include the following: Determination of guideline adherence, performance criteria, compliance, and measuring safety and outcomes.[20] Registries are pivotal in acquiring knowledge and treatment recommendations for rare disease entities. For example, much of what is currently acquired to date on pulmonary arterial hypertension has come from observational studies taken from national or international registries.[21] The most recent pulmonary arterial hypertension treatment algorithm is derived from the 2018 World Symposium on Pulmonary Hypertension, which recommends a risk-based approach to the treatment of pulmonary arterial hypertension.[22–26] Another outcome that came from the registry data is the recommended triple combination treatment with prostacyclin, endothelin receptor antagonist, and phosphodiesterase type 5 inhibitor for patients with high-risk pulmonary arterial hypertension.[27] Furthermore, the benefits of these registries are not only in the large numbers of patients observed but also in their ability to link disease-specific registries with administrative and biobanking data.[21]

Statistical analysis: Linear regression and propensity match

There are two common statistical techniques used in the analysis of database studies: linear regression and propensity score or matching. Linear regression was first defined in 1894 by Sir Francis Galton. Linear regression analysis describes the relationship between two variables and in the process adjusts for confounders.[28] Propensity score is the probability of one of the variables being assigned the treatment or control depending on pretreatment conditions. This statistical analysis aims to define the probability of both treatment groups having a particular exposure as opposed to another. It is usually used when the study looks at two treatment groups.[29] Similar to linear regression techniques, propensity scores cannot account for unknown confounders.[30] Some limitations found when using linear regression, include not being able to use this technique when there are many variables and many potential interactions. In such circumstances, machine-based learning, such as ridge, lasso, and elastic net, allows approaches to restrict the number of variables and interactions in models.[31] Measures such as common statistical stools (i.e., chi-square test, linear regression, analysis of variance, survival analysis, etc.) should be chosen based on the clinical design of the study.[10]

Example of database registry

A recent example of a database registry is the REPORT-HF registry.[32] The REPORT-HF registry is a global, prospective, and observational study that looked at the clinical decisions, outcomes, and management of newly diagnosed HF patients and patients with decompensated HF. This study followed patients for 3 years and looked at postdischarge outcomes and their relation to country-level income in heart failure patients. It enrolled patients from 358 sites in 44 countries and 6 continents. Statistical analysis was performed using Cox univariable and multivariable regression for factors associated with 1-year mortality. For a comparison between groups, analysis of variance or Mann−Whitney *U*-test was used. Novartis Pharmaceutical, which funded the study, was not involved in the design or data collection of the study. Based on the registry it was determined that patients with lower incomes were also generally younger and had their first presentation of heart failure upon hospitalization. Additionally, the highest mortality was seen among patients from the Mediterranean, Africa, and Latin America. This registry allowed for the first global comparison of practice patterns and outcomes for heart failure patients in a large number of countries with various socioeconomic backgrounds.

Future directions

Big data takes advantage of the current information stored in big databases to establish relationships and associations by using common statistical analysis, such as linear regression and propensity matching. There are multiple types of databases that can be assessed nowadays, such as government or administrative, disease-specific, technology-based, registries, and population-based registries. The main limiting factor standing in the way of big data's full integration and discovery stems from the software and linkage of these databases into one infrastructure, which can be comparatively analyzed.

References

1. Turakhia MP, Desai M, Hedlin H, et al. Rationale and design of a large-scale, app-based study to identify cardiac arrhythmias using a smartwatch: the Apple Heart Study. *Am Heart J*. 2019;207:66−75.
2. Lagerqvist B, Frobert O, Olivecrona GK, et al. Outcomes 1 year after thrombus aspiration for myocardial infarction. *N Engl J Med*. 2014;371:1111−1120.
3. Rao SV, Hess CN, Barham B, et al. A registry-based randomized trial comparing radial and femoral approaches in women undergoing percutaneous coronary intervention: the SAFE-PCI for Women (Study of Access Site for Enhancement of PCI for Women) trial. *JACC Cardiovasc Interv*. 2014;7:857−867.
4. Hemingway H, Asselbergs FW, Danesh J, et al. Big data from electronic health records for early and late translational cardiovascular research: challenges and potential. *Eur Heart J*. 2018;39:1481−1495.
5. Attisano T, Silverio A, Stabile E, et al. Safety and feasibility of balloon aortic valvuloplasty in non-TAVI centers: the "BAV for life" experience. *Cathet Cardiovasc Interv*. 2019;93:E63−E70.
6. AlHajji M, Alqahtani F, Alkhouli M. Contemporary trends in the utilization of administrative databases in cardiovascular research. *Mayo Clin Proc*. 2019;94:1120−1121.
7. Denaxas SC, Morley KI. Big biomedical data and cardiovascular disease research: opportunities and challenges. *Eur Heart J Qual Care Clin Outcomes*. 2015;1:9−16.
8. Sardar MR, Badri M, Prince CT, Seltzer J, Kowey PR. Underrepresentation of women, elderly patients, and racial minorities in the randomized trials used for cardiovascular guidelines. *JAMA Intern Med*. 2014;174:1868−1870.
9. Neveol A, Dalianis H, Velupillai S, Savova G, Zweigenbaum P. Clinical Natural Language Processing in languages other than English: opportunities and challenges. *J Biomed Semant*. 2018;9:12.
10. Lee KT, Hour AL, Shia BC, Chu PH. The application and future of big database studies in cardiology: a single-center experience. *Acta Cardiol Sin*. 2017;33:581−587.
11. Silverio A, Cavallo P, De Rosa R, Galasso G. Big health data and cardiovascular diseases: a challenge for research, an opportunity for clinical care. *Front Med*. 2019;6:36.
12. Bots SH, Siegersma KR. Onland-Moret NCet al. Routine clinical care data from thirteen cardiac outpatient clinics: design of the Cardiology Centers of The Netherlands (CCN) database. *BMC Cardiovasc Disord*. 2021;21:287.
13. Laney D. 3D Data Management Controlling Data Volume, Velocity, and Pdf.
14. Raja R, Mukherjee I, Sarkar BK. A systematic review of healthcare big data. *Sci Program*. 2020;2020:1−15.
15. Mahmood SS, Levy D, Vasan RS, Wang TJ. The Framingham Heart Study and the epidemiology of cardiovascular disease: a historical perspective. *Lancet*. 2014;383:999−1008.
16. Weber GM, Mandl KD, Kohane IS. Finding the missing link for big biomedical data. *JAMA*. 2014;311:2479−2480.
17. Sarno G, Lagerqvist B, Frobert O, et al. Lower risk of stent thrombosis and restenosis with unrestricted use of 'new-generation' drug-eluting stents: a report from the nationwide Swedish Coronary Angiography and Angioplasty Registry (SCAAR). *Eur Heart J*. 2012;33:606−613.
18. Grimfjard P, James S, Persson J, et al. Outcome of percutaneous coronary intervention with the Absorb bioresorbable scaffold: data

from the Swedish Coronary Angiography and Angioplasty Registry (SCAAR). *EuroIntervention.* 2017;13:1303–1310.

19. long-term-safety.pdf.
20. Faxon DP, Burgess A. Cardiovascular registries: too much of good thing? *Circ Cardiovasc Interv.* 2016;9:e003866.
21. Weatherald J, Reis A, Sitbon O, Humbert M. Pulmonary arterial hypertension registries: past, present and into the future. *Eur Respir Rev.* 2019;28.
22. Hoeper MM, Kramer T, Pan Z, et al. Mortality in pulmonary arterial hypertension: prediction by the 2015 European pulmonary hypertension guidelines risk stratification model. *Eur Respir J.* 2017;50.
23. Boucly A, Weatherald J, Savale L, et al. Risk assessment, prognosis and guideline implementation in pulmonary arterial hypertension. *Eur Respir J.* 2017;50.
24. Kylhammar D, Kjellstrom B, Hjalmarsson C, et al. A comprehensive risk stratification at early follow-up determines prognosis in pulmonary arterial hypertension. *Eur Heart J.* 2018;39:4175–4181.
25. Weatherald J, Boucly A, Launay D, et al. Haemodynamics and serial risk assessment in systemic sclerosis associated pulmonary arterial hypertension. *Eur Respir J.* 2018;52.
26. Benza RL, Gomberg-Maitland M, Miller DP, et al. The REVEAL Registry risk score calculator in patients newly diagnosed with pulmonary arterial hypertension. *Chest.* 2012;141:354–362.
27. Sitbon O, Jais X, Savale L, et al. Upfront triple combination therapy in pulmonary arterial hypertension: a pilot study. *Eur Respir J.* 2014;43:1691–1697.
28. Kumari K, Yadav S. Linear regression analysis study. *J Pract Cardiovasc Sci.* 2018;4.
29. Salciccioli JD, Crutain Y, Komorowski M, Marshall DC. Sensitivity analysis and model validation. In: *Secondary Analysis of Electronic Health Records.* 2016:263–271.
30. Shah BR, Laupacis A, Hux JE, Austin PC. Propensity score methods gave similar results to traditional regression modeling in observational studies: a systematic review. *J Clin Epidemiol.* 2005;58:550–559.
31. Waldron L, Pintilie M, Tsao MS, Shepherd FA, Huttenhower C, Jurisica I. Optimized application of penalized regression methods to diverse genomic data. *Bioinformatics.* 2011;27:3399–3406.
32. Tromp J, Bamadhaj S, Cleland JGF, et al. Correction to: post-discharge prognosis of patients admitted to hospital for heart failure by world region, and national level of income and income disparity (REPORT-HF): a cohort study. *Lancet Global Health.* 2020;8:e1001.

Chapter 46

Surveys and questionnaires

Adrija Hajra[1] and Heena Kaushal Asnani[2]

[1]*Department of Internal Medicine, Bringham and Women's Hospital/Harvard Medical School, Boston, MA, United States;* [2]*Department of Internal Medicine, Roger Williams Medical Center, Providence, RI, United States*

Introduction

A survey's quality is heavily influenced by the questionnaire design. Insufficiently designed surveys may yield unreliable results that fail to capture the essence of what survey developers are trying to measure, potentially impacting clinical and educational practice. A questionnaire can also gather unquantifiable data such as ideas, beliefs, knowledge, opinions, and behaviors along with quantifiable data. A practitioner should be able to decide whether to implement findings, based on a detailed understanding of the development process when interpreting questionnaire results.

First step

The first step is to figure out if a survey is the best way to answer the research question. The topic of interest should be kept in mind while figuring this out. When it is well understood and involves feedback, opinions, satisfaction, beliefs, or attitudes, a survey might be the most effective way to approach it. If the researcher wants to assess behaviors, observational studies will be a better choice, and when a little is known about the topic of interest, a qualitative analysis should be preferred.

Second step

The second step is a literature review. By reviewing the literature, the researcher will be able to define the construct and determine how it differs from precedent studies. A construct is an idea that researchers are attempting to assess. A similar construct may already exist, and the researcher can look at which survey items or scales were used, and if they fit their construct.[1] If the construct that a researcher is interested in has already been explored with a valid questionnaire, it would be more efficient to use the questionnaire.[2]

Third step

The third step is to find a target audience. It is vital to identify a target audience since the survey will not have any relevance if the inclusion and exclusion criteria are not proper. The construct of interest should be developed in a language that the target audience can easily comprehend.

Fourth step

The fourth step is to conduct interviews/focus groups: These sessions are conducted among a small group of respondents closely similar to the target audience to confirm that respondents understand the construct similarly to the researchers. The perspective of the people is gained through open-ended and closed-ended questions for the purpose of evaluating and defining the construct from multiple perspectives.[2] The questionnaire should be modified according to what is understood and which features are included and excluded by the participants.

Fifth step

The fifth step is to assemble the knowledge gained by literature review and interviews/focus groups: Through focus groups/interviews and literature review, the researcher has shaped the definition of the construct as well as multiple perspectives of it. After merging the ideas from all the previous steps, the list of ideas is compared to see if they match each other, if so, the process is simple. In the event that they do not, it is always a good idea to include the terms and indicators respondents used about the construct in the list.

Translational Cardiology. https://doi.org/10.1016/B978-0-323-91790-2.00063-0

Sixth step

The sixth step is to develop items—This step involves writing survey items that accurately reflect the construct of interest in a language that respondents can understand.[2]

- INTRODUCTION: The survey should begin with a brief introduction that clearly states its objectives and appeals to and attracts respondents.
- STRUCTURE OF THE SURVEY: A logical and structured arrangement of questions, with general questions preceded by specific questions, is recommended. Controversial items should be placed at the end.[3]
- PROBLEMS THAT MAY ARISE WITH QUESTIONNAIRE:[4]
  1. Lexical problems relate to the respondent's understanding of the use of words, meaning, and the context in which the questionnaire is formed. The main cause of this problem is that the researcher overestimates the understanding and vocabulary of the respondents.
  2. Inclusion/Exclusion problems deal with the determination of the scope of the question which can lead to respondents choosing multiple or irrelevant choices.
  3. Temporal problems refer to problems dealing with time periods and time spent.
  4. Logical problems are associated with difficulties respondents face regarding the words that connect two different concepts and the use of presuppositions in the questions.
  5. Computational problems include problems that do not fall into any of the above-mentioned categories. Some examples can be complicated structured questions and questions involving mental calculations and long-term recall.
- LAYOUT OF QUESTIONS:
  1. When designing questions/items, ensure clarity, relevance, and appropriate vocabulary.
  2. While designing questions, the best suitable response format and inclusion of free test options should be kept in mind.[5]
  3. Mixed open- and closed-ended questions with closed-ended ones at the beginning keep respondents interested and encourage high response rates.[3]
  4. Writing questions instead of statements is preferable.
  5. Avoid biased language, negative items, general agreement responses, and double-barreled questions, that is, asking two things in the same question, asking only closed-ended questions, using too few responses (can reduce reliability) and too many responses (do not increase reliability), labeling only the endpoints of response options, labeling the options with both numbers and verbal labels (can cause confusion), and unequally spaced response options.[2,5,6]
  6. It is important to include both positive and negative sides of an issue to reduce bias.[7]
- SCALES: These measure the responses and then the responses are quantified, compared, and analyzed.[8] Examples of different scales are listed below, and they should be used based on what the researcher is trying to assess.
  1. Frequency scales are useful for determining how frequently a target behavior or event occurs.
  2. Thurstone scales allow respondents to gauge their level of opinion about a topic by selecting statements related to the topic.
  3. Likert scale allows the respondent to choose on a continuum from strongly agree to strongly disagree, strongly satisfied to strongly dissatisfied, extremely confident to not at all confident, and so on out of 5–9 choices with neutral point.[2,5,6]
  4. Guttman scales are based on the assumption that individuals who agree with one statement agree with all the previous statements.

Seventh step

The seventh step is to conduct expert validation to ensure that no items or indicators have been omitted, which improves the quality and representativeness of the survey items.[9]

Validity measures the questionnaire's ability to measure what it is supposed to measure, whereas Reliability determines whether it can produce consistent results and is usually tested over time (repeatability), across items (internal consistency), and different researchers (interrater reliability).[3]

Reliability and validity of surveys are not properties of the surveys themselves but of the scores they return and the interpretations they produce.[10]

In a survey, it was suggested that using 6–10 experts could produce a more comprehensive consensus about the construct under assessment.[11]

- Factors calculated are as follows:
  1. Representativeness assesses how well the construct is represented through the items.
  2. Clarity computes how clearly the items can be read.
  3. Relevance measures how every item correlates to the specific aspects of the construct.
  4. Distribution estimates the difficulty of selecting a high score for a given item. If asked, "How confident are you that you can get 100% on the test?" and "How confident are you that you can pass the test?," the respondents will have difficulty in selecting a high score for the former question.[2]

- Components of validity include the following:
  1. Face validity refers to the extent to which the questionnaire seems to measure what it is supposed to measure, in the opinion of respondents, experts, and researchers. It is mainly measured informally, not quantitatively.
  2. Content validity tests to see if the current questionnaire contains all the pertinent items that are necessary to answer the research question. The experts rate the questionnaire items for readability, understanding, precision, simplicity, and comprehensiveness using scales, and the item-rated content validity index (I-CVI) is calculated.
  3. Criterion validity refers to the extent to which the test items correlate with a predetermined standard.
  4. Construct validity of a test is its ability to measure what it claims to measure.[3]
- Components of reliability include the following:
  1. Repeatability evaluates the consistency between scores measured at different times.
  2. Internal consistency is calculated via Cronbach's alpha coefficient (a measure developed using pairwise correlations between scale items with ranges of 0−1 that assesses how scale items relate to one another)
  3. Interrater reliability is measured using kappa statistics that measure the degree of agreement between different raters

Eighth step

The eighth step is to perform cognitive interviews in a small sample size (10−30 participants): This is carried out to determine if respondents interpret the items similarly to the researcher, to have a better understanding of how respondents perceive and interpret questions and to identify potential problems that may arise.

- It involves techniques from psychology to understand human information processing which includes a series of cognitive processes a respondent goes through when responding to a survey such as:
  1. Attention span, word recognition, language processing, memory, and comprehension of an item stem and answer choices
  2. Retrieval of appropriate information from long-term memory
  3. Reasoning and judgment based on comprehension of the item and their memory
  4. Selection of a response.
- Cognitive interviews are carried out through a semi-structured, in-depth interview to identify overall problems with questionnaires. Creating coding criteria that reflect the researcher's intended meaning for each item is a key first step in interpreting the responses gathered during the cognitive interview.
- The three major techniques are think-aloud technique, verbal probing (concurrent and retrospective probing), and observation of the respondent's behavior.[12,13]
  - (i) Think aloud technique asks the respondent to tell the interviewer everything that they are thinking while they go through each item. This helps in gaining insight into the cognitive processes used by the respondent when completing the questionnaire.
  - (ii) Verbal probing asks respondents to paraphrase questions, define meanings of the words used, explain their responses, and identify the areas which they find difficult to understand, interpret, and answer. Concurrent probing asks the respondents about their understanding and interpretation while going through the questionnaire. Retrospective probing involves asking the respondent about their understanding and interpretation after the completion of the questionnaire.
  - (iii) Observation of respondent's behavior during the interview can be important if respondents are flipping the pages, skipping the questions, or changing their appearance while answering the questions. The respondents can then be directly questioned about the difficulties they experienced.

The majority of cognitive interviews actually use a combination of think-aloud, observation, and verbal probing techniques to identify potential errors, and it relies on coding and interpretation of written notes from the interview, not statistical tests.

Ninth step

The ninth step is to analyze the qualitative data obtained in the cognitive interview: There are two ways to analyze results from the cognitive interview.

- Coding: One way to analyze results from cognitive interviews is through coding where predetermined codes are created for common respondent errors, and the frequency of each type of error is tabulated for every item. It is often manageable to interpret the quantitative results through this analysis, but it is also possible to miss errors that cannot be predicted easily, and this technique may not fully explain why an error may be occurring.
- Interaction analysis: It is a qualitative approach to the cognitive interview. In a cognitive interview, interaction analysis typically describes and explains how people interpret and interact during the conversation, so the meanings of responses can be determined by applying this method.

The combination of both is very functional in providing additional information about the "cognitive validity" of the questionnaire.[12]

Tenth step

The 10th step is to conduct pilot testing which increases the validity of the questionnaire even further as it can identify items that may not be functioning the way the researcher intended. In the planned delivery mode, the target population completes the survey, and the data is examined for item ranges and variances, score reliability, and item and composite score correlations.

Survey data collection encompasses various methods: in-person interviews, telephone interviews, mail questionnaires, email surveys, or a combination of these. Response rates are linked to factors like age, disability, and cognitive ability. Nonresponse can stem from survey length and the chosen administration method (e.g., those with writing difficulties might avoid postal surveys).[14]

Email is widely used for web survey invitations due to cost-effectiveness and personalization, favored by the young, high-income, and urban populations. However, challenges include spam, changing email addresses, privacy concerns, and email unavailability for certain groups like low-income individuals, the elderly, and those lacking internet access. Johnston et al.'s study demonstrated that, except for the low-income group (where both phone and email modes got similar responses), email surveys achieved higher response rates (ranging from 57.4% to 66.3%) compared to phone surveys (ranging from 36% to 43%) across all income categories.[15,16]

Pitfalls

Comparing survey methods, mail surveys often have lower response rates than telephone, email, and in-home interviews. This can be attributed to content issues, seasonal trends, language barriers, envelope design, forgetfulness, lack of incentives, long questionnaires, perceived disinterest, and insufficient reminders. Mail surveys, however, offer anonymity for sensitive data and cost less than telephone and in-person interviews. Employing reminder calls can enhance response rates for mailed and emailed surveys. Additionally, personal in-person administration boosts responses but comes at a higher cost.[14,16,17] Thus, understanding the target audience is crucial in selecting the appropriate survey mode.

To further strengthen the survey, a statistical analysis that measures the one-dimensionality and a reliability analysis that measures the internal consistency of the items can be conducted.

References

1. Gehlbach H, Artino AR, Durning S. AM last page: survey development guidance for medical education researchers. *Acad Med.* 2010;85(5):925. https://doi.org/10.1097/ACM.0b013e3181dd3e88.
2. Artino AR, la Rochelle JS, Dezee KJ, Gehlbach H. Developing questionnaires for educational research: AMEE Guide No. 87. *Med Teach.* 2014;36(6):463−474. https://doi.org/10.3109/0142159X.2014.889814.
3. Yaddanapudi S, Yaddanapudi LN. How to design a questionnaire. *Indian J Anaesth.* 2019;63(5):335−337. https://doi.org/10.4103/ija.IJA_334_19.
4. Conrad FBJ. From impressions to data: increasing the objectivity of cognitive interviews. In: *Proceedings of the Survey Research Methods Section of the American Statistical Association.* 1996:1−10. Published online.
5. Artino AR, Gehlbach H, Durning SJ. AM last page: avoiding five common pitfalls of survey design. *Acad Med.* 2011;86(10):1327. https://doi.org/10.1097/ACM.0b013e31822f77cc.
6. Artino AR, Gehlbach H. AM last page. Avoiding four visual-design pitfalls in survey development. *Acad Med.* 2012;87(10):1452. https://doi.org/10.1097/ACM.0b013e31826a06b2.
7. Dillman DA, Smyth JD, Christian LM. *Internet, Mail, and Mixed-Mode Surveys: The Tailored Design Method.* 3rd ed. John Wiley and Sons; 2009.
8. Rattray J, Jones MC. Essential elements of questionnaire design and development. *J Clin Nurs.* 2007;16(2):234−243. https://doi.org/10.1111/j.1365-2702.2006.01573.x.
9. Polit DF, Beck CT. The content validity index: are you sure you know what's being reported? Critique and recommendations. *Res Nurs Health.* 2006;29(5):489−497. https://doi.org/10.1002/nur.20147.
10. American Educational Research Association (AERA. American psychological association (APA), national council on measurement in education (NCME). In: *Standards for Education and Psychological Testing.* American Educational Research Association; 1999.
11. Rubio DM, Berg-Weger M, Tebb SS, Lee ES, Rauch S. Objectifying content validity: conducting a content validity study in social work research. *Soc Work Res.* 2003;27(2):94−104. https://doi.org/10.1093/swr/27.2.94.
12. Nápoles-Springer AM, Santoyo-Olsson J, O'Brien H, Stewart AL. Using cognitive interviews to develop surveys in diverse populations. *Med Care.* 2006;44(11 Suppl 3):S21−S30. https://doi.org/10.1097/01.mlr.0000245425.65905.1d.
13. Drennan J. Cognitive interviewing: verbal data in the design and pretesting of questionnaires. *J Adv Nurs.* 2003;42(1):57−63. https://doi.org/10.1046/j.1365-2648.2003.02579.x.
14. Garcia I, Portugal C, Chu LH, Kawatkar AA. Response rates of three modes of survey administration and survey preferences of rheumatoid arthritis patients. *Arthritis Care Res.* 2014;66(3):364−370. https://doi.org/10.1002/acr.22125.
15. Bandilla W, Couper MP, Kaczmirek L. The mode of invitation for web surveys. *Surv Pract.* 2012;5(3):1−5. https://doi.org/10.29115/SP-2012-0014.
16. Johnston S, Hogg W, Wong ST, Burge F, Peterson S. Differences in mode preferences, response rates, and mode effect between automated email and phone survey systems for patients of primary care practices: cross-sectional study. *J Med Internet Res.* 2021;23(1):e21240. https://doi.org/10.2196/21240.
17. Smyth JD, Olson K, Millar MM. Identifying predictors of survey mode preference. *Soc Sci Res.* 2014;48:135−144. https://doi.org/10.1016/j.ssresearch.2014.06.002.

Chapter 47

Qualitative methods and mixed methods in translational cardiology research

Suman Biswas
Department of Imaging, University of Rochester, Rochester, NY, United States

Key points

- Role of qualitative research in translational research.
- Advantages and limitations of qualitative methods Benefits and challenges of mixed methods.

Introduction

Translational research forms a multidisciplinary bridge from basic science to clinical application, fostering innovation, and improvement in public health. Translational cardiology, traditionally quantitative, faces challenges due to the complex and unpredictable nature of cardiovascular diseases (CVDs). These complexities necessitate qualitative research to capture subjective and nonmeasurable aspects of CVDs. By integrating qualitative and quantitative methods, mixed-method research leverages the strengths of both. This chapter explores the application of these methods in translational cardiology, underscoring their potential and limitations. It highlights how this approach enriches our understanding of patient experiences, health behaviors, socioeconomic factors, and real-life challenges. Furthermore, it illuminates systemic factors that influence the translation of research findings into practice.
By the end of this chapter, the reader should be able to:

- Understand the value and use cases of qualitative and mixed methods in research.
- Understand the overall concept of both methods.
- Understand the pros and cons of these methods.
- Grasp the concept with examples and learn relevant research for further reading.

Qualitative methods

Qualitative and mixed methods have emerged as crucial components in translational research to overcome the limitations of quantitative analysis. While quantitative research is valuable for testing hypotheses and establishing generalizable conclusions through structured and statistical approaches, it may need to fully grasp the intricacies of human behavior and experiences due to its tendency for large sample sizes, higher costs, and a more superficial understanding. Qualitative research, on the other hand, delves into the complex realities of contexts, providing a deeper insight into patient experiences, health behaviors, and socioeconomic challenges. It is beneficial for developing hypotheses, explaining findings in-depth, and assessing intervention acceptability. By complementing quantitative research, qualitative methods enrich clinical research with detailed, context-specific insights, capturing the complexity of real-life situations holistically and inductively. This approach is ideal for exploring new phenomena or revising theories, offering detailed descriptions without proving causality.[1–3]

Example

1. "The Impact of Heart Failure on Patients and Caregivers: A Qualitative Study": McHorney et al. conducted in-person focus groups with 64 patients and 26 caregivers to understand the symptom and symptom impact experiences of heart failure patients and their caregivers. The most commonly reported symptoms were shortness of breath, fatigue/tiredness, swelling of legs and ankles, and trouble sleeping. Patients reported reduced social/family interactions, dietary changes, and difficulty walking and climbing stairs as the most common adverse disease impacts. Note that in this method, the quantitative research methods will not be able to shed light on the topic.[4]
2. Medication adherence and its determinants in patients after myocardial infarction: The study by Pietrzykowski et al. shows how personal and systemic barriers to

Translational Cardiology. https://doi.org/10.1016/B978-0-323-91790-2.00095-2

medication adherence, such as side effects, the complexity of the medication regimen, lack of understanding of the medication's role, and access issues affect patients of myocardial infarction. The study also reveals that facilitators of adherence include solid patient–provider relationships, personalized patient education, and support from family or caregivers. The paper also discusses psychological factors influencing adherence, like motivation, health beliefs, and the impact of patient education on medication understanding and adherence.[5]

Advantages of qualitative methods[6–12]

1. Richness and complexity: Offers detailed insights into complex aspects, emotions, and subtleties that quantitative methods may miss.
2. Adaptability: Allows flexible adjustments and changes in the research direction based on emerging findings.
3. Contextual insights: Excels in revealing the context, underlying reasons, beliefs, and motivations behind phenomena.
4. Exploring new areas: Ideal for investigating less understood or new domains, helping to generate hypotheses for future studies.
5. Focus on perception and experience: Provides a deep understanding of individual experiences and perceptions, offering valuable insights into the subjective aspects of human behavior and social phenomena.

Limitations of qualitative methods[6–12]

1. Subjective nature: The data and outcomes of qualitative research can be highly subjective, potentially impacting the consistency and reliability of results.
2. Resource intensity: Requires significant time and resources for conducting in-depth interviews, focus groups, and observations.
3. Complexity in data analysis: Analyzing qualitative data is complex and demands specialized expertise.
4. Limited generalizability: Small, nonrandom samples may restrict the applicability of findings to broader populations.
5. Risk of bias: The process is prone to biases introduced during data collection and analysis, influenced by the researcher's judgments.
6. Absence of quantitative measures: Lacks statistical data, which limits the ability to measure responses precisely or make broad comparisons and may lead to redundancy in data over time.

Mixed methods

The mixed methods approach integrates both qualitative and quantitative research techniques, leveraging the strengths of each to improve the reliability, validity, and comprehensiveness of research findings. This method is precious in translational research, which aims to apply findings from basic science to enhance human health and well-being. Mixed-methods research supports creating more effective health policies and programs by facilitating a deeper understanding of complex health behaviors and connecting research outcomes with practical applications. Its holistic perspective is crucial for evaluating interventions and programs thoroughly, making it a vital tool in translational research for achieving comprehensive evaluations. Qualitative and mixed methods enhance translational research by providing comprehensive insights that quantitative methods alone might miss, although their findings are typically less generalizable. Mixed methodology is used for needs assessment, intervention development, implementation science, and policy and program evaluation, integrating qualitative insights with quantitative measures to inform, test, and evaluate health interventions and policies comprehensively.[13,14]

Examples

1. Implementing mixed methods in cardiovascular nursing studies: The research utilized a mixed methodology to delve into the intricacies of cardiovascular conditions. It involved gathering both qualitative and quantitative data, merging these forms of data to gain a more comprehensive understanding of the conditions than what could be obtained through solely quantitative or qualitative research.[15]
2. Application of mixed methods in studies on coronary artery disease, diabetes mellitus, and hypertension: This scoping review analyzed the features of mixed methods research publications focusing on coronary artery disease and its primary risk factors. The findings indicate a growing trend in adopting mixed methods research within coronary artery disease and associated risk factors.[16]

Advantages of mixed methodology in research[17–24]

1. Holistic insights: Combines qualitative and quantitative data to explore complex research problems comprehensively.
2. Rich contextualization: Provides a detailed context to research findings, adding depth to numerical data with stories and experiences.
3. Diverse perspectives: Incorporates multiple viewpoints, enhancing the credibility and robustness of findings through validation.
4. Enhancing implementation science: Essential for translating interventions from theory to practice, informed by real-world feedback.

5. Patient-centered approaches: Prioritizes patient experiences and outcomes, fostering stakeholder engagement.
6. Explaining beyond numbers: Offers in-depth narratives that explain the mechanisms behind statistical data.
7. Tracing the journey: Tracks the evolution of programs and interventions, identifying paths of implementation and areas for improvement.
8. Stakeholder engagement: Facilitates understanding of stakeholder perspectives throughout healthcare delivery processes.

Challenges of mixed methodology[17–24]

1. Methodological rigor: Maintaining rigor in qualitative and quantitative aspects is essential, requiring researchers to be proficient and aware of biases.
2. Proficiency: Researchers need expertise in both research paradigms to ensure accurate application and interpretation of methods.
3. Biases: Vigilance against researcher biases and methodological limitations is crucial for credible findings.
4. Integration: Effectively combining qualitative and quantitative data is a crucial challenge but essential for comprehensive insights.
5. Resource intensity: Mixed methods demand significant time, expertise, and financial resources, often requiring interdisciplinary collaboration.
6. Planning: Justifying mixed methods and careful planning is necessary for efficient resource allocation and study success.
7. Sampling issues: Accommodating both data types can complicate sampling strategies, necessitating robust and representative methods.
8. Conflicting results: Researchers must address discrepancies between qualitative and quantitative findings, which can enrich the study's insights.
9. Data integration complexity: Integrating diverse data sources requires sophisticated analysis to draw meaningful, comprehensive conclusions.

Future directions

The future of qualitative research is shaped by two significant trends: meta-analysis and artificial intelligence (AI). Meta-analysis, traditionally used in quantitative research, is being adapted into meta-synthesis for qualitative studies, allowing for the integration and interpretation of findings from multiple studies to generate new insights and enhance the generalizability and credibility of research. On the other hand, AI is revolutionizing qualitative research by improving data collection, analysis, and interpretation efficiency through advanced technologies like machine learning and natural language processing. This speeds up research processes, opens new avenues for exploration, and reduces bias. However, both approaches require careful consideration of ethical issues and data quality to ensure accurate and responsible research outcomes.[25–30]

References

1. Rahman MS. The advantages and disadvantages of using qualitative and quantitative approaches and methods in language "testing and assessment" research: a literature review. *JEL*. 2016;6(1):102. https://doi.org/10.5539/jel.v6n1p102.
2. Bakken S. Quantitative and qualitative methods advance the science of clinical workflow research. *J Am Med Inf Assoc*. 2023;30(5):795–796. https://doi.org/10.1093/jamia/ocad056.
3. Lefèvre H, Moro MR, Lachal J. Research in adolescent healthcare: the value of qualitative methods. *Arch Pediatr*. 2019;26(7):426–430. https://doi.org/10.1016/j.arcped.2019.09.012.
4. McHorney CA, Mansukhani SG, Anatchkova M, et al. The impact of heart failure on patients and caregivers: a qualitative study. *PLoS One*. 2021;16(3):e0248240. https://doi.org/10.1371/journal.pone.0248240.
5. Pietrzykowski Ł, Michalski P, Kosobucka A, et al. Medication adherence and its determinants in patients after myocardial infarction. *Sci Rep*. 2020;10(1):12028. https://doi.org/10.1038/s41598-020-68915-1.
6. Lakshman M, Sinha L, Biswas M, Charles M, Arora NK. Quantitative vs qualitative research methods. *Indian J Pediatr*. 2000;67(5):369–377. https://doi.org/10.1007/BF02820690.
7. Flemming K, Briggs M. Electronic searching to locate qualitative research: evaluation of three strategies. *J Adv Nurs*. 2007;57(1):95–100. https://doi.org/10.1111/j.1365-2648.2006.04083.x.
8. McIlvennan CK, Morris MA, Guetterman TC, Matlock DD, Curry L. Qualitative methodology in cardiovascular outcomes research: a contemporary look. *Circ Cardiovasc Qual Outcomes*. 2019;12(9):e005828. https://doi.org/10.1161/CIRCOUTCOMES.119.005828.
9. Nordfonn OK, Morken IM, Lunde Husebø AM. A qualitative study of living with the burden from heart failure treatment: exploring the patient capacity for self-care. *Nurs Open*. February 8, 2020;7(3):804–813. https://doi.org/10.1002/nop2.455.
10. Tenny S, Brannan JM, Brannan GD. Qualitative study. In: *StatPearls. Treasure Island (FL)*. StatPearls Publishing; September 18, 2022.
11. Al Hamid A, Ghaleb M, Aljadhey H, Aslanpour Z. A systematic review of qualitative research on the contributory factors leading to medicine-related problems from the perspectives of adult patients with cardiovascular diseases and diabetes mellitus. *BMJ Open*. 2014;4(9). https://doi.org/10.1136/bmjopen-2014-005992. e005992-e005992.
12. Manja V, Guyatt G, You J, Monteiro S, Jack S. Qualitative study of cardiologists' perceptions of factors influencing clinical practice decisions. *Heart*. 2019;105(10):749–754. https://doi.org/10.1136/heartjnl-2018-314339.
13. Liamputtong P. *Handbook of Research Methods in Health Social Sciences*. Berlin Heidelberg: Springer; 2019.
14. Halcomb E, Hickman L. Mixed methods research. *Nurs Stand*. 2015;29(32):41–47. https://doi.org/10.7748/ns.29.32.41.e8858.
15. Dickson VV, Page SD. Using mixed methods in cardiovascular nursing research: answering the why, the how, and the what is next. *Eur J Cardiovasc Nurs*. 2021;20(1):82–89. https://doi.org/10.1093/eurjcn/zvaa024.

16. Campbell DJT, Tam-Tham H, Dhaliwal KK, et al. Use of mixed methods research in research on coronary artery disease, diabetes mellitus, and hypertension: a scoping review. *Circ Cardiovasc Quality Outcomes*. 2017;10(1):e003310. https://doi.org/10.1161/CIRCOUTCOMES.116.003310.
17. Wasti SP, Simkhada P, van Teijlingen ER, Sathian B, Banerjee I. The growing importance of mixed-methods research in health. *Nepal J Epidemiol*. 2022;12(1):1175–1178. https://doi.org/10.3126/nje.v12i1.43633.
18. Shorten A, Smith J. Mixed methods research: expanding the evidence base. *Evid Base Nurs*. 2017;20(3):74–75. https://doi.org/10.1136/eb-2017-102699.
19. Smajic E, Avdic D, Pasic A, Prcic A, Stancic M. Mixed methodology of scientific research in healthcare. *Acta Inf Med*. 2022;30(1):57–60. https://doi.org/10.5455/aim.2022.30.57-60.
20. Yardley L, Bishop FL. Using mixed methods in health research: benefits and challenges. *Br J Health Psychol*. 2015;20(1):1–4. https://doi.org/10.1111/bjhp.12126.
21. Skamagki G, King A, Carpenter C, Wåhlin C. The concept of integration in mixed methods research: a step-by-step guide using an example study in physiotherapy. *Physiother Theory Pract*. 2024;40(2):197–204. https://doi.org/10.1080/09593985.2022.2120375.
22. McKenna L, Copnell B, Smith G. Getting the methods right: challenges and appropriateness of mixed methods research in health-related doctoral studies. *J Clin Nurs*. 2021;30(3–4):581–587. https://doi.org/10.1111/jocn.15534.
23. Bishop FL. Using mixed methods research designs in health psychology: an illustrated discussion from a pragmatist perspective. *Br J Health Psychol*. 2015;20(1):5–20. https://doi.org/10.1111/bjhp.12122.
24. Tariq S, Woodman J. Using mixed methods in health research. *JRSM Short Rep*. 2013;4(6). https://doi.org/10.1177/2042533313479197.
25. Levitt HM. How to conduct a qualitative meta-analysis: tailoring methods to enhance methodological integrity. *Psychother Res*. 2018;28(3):367–378. https://doi.org/10.1080/10503307.2018.1447708.
26. Lachal J, Revah-Levy A, Orri M, Moro MR. Metasynthesis: an original method to synthesize qualitative literature in psychiatry. *Front Psychiatr*. 2017;8:269. https://doi.org/10.3389/fpsyt.2017.00269.
27. Beam AL, Drazen JM, Kohane IS, Leong TY, Manrai AK, Rubin EJ. Artificial intelligence in medicine. *N Engl J Med*. 2023;388(13):1220–1221. https://doi.org/10.1056/NEJMe2206291.
28. Chubb J, Cowling P, Reed D. Speeding up to keep up: exploring the use of AI in the research process. *AI Soc*. 2022;37(4):1439–1457. https://doi.org/10.1007/s00146-021-01259-0.
29. Xu Y, Liu X, Cao X, et al. Artificial intelligence: a powerful paradigm for scientific research. *Innovation*. 2021;2(4):100179. https://doi.org/10.1016/j.xinn.2021.100179.
30. Anis S, French JA. Efficient, explicatory, and equitable: why qualitative researchers should embrace AI, but cautiously. *Bus Soc*. 2023;62(6):1139–1144. https://doi.org/10.1177/00076503231163286.

Part VI

Clinical trials

Amr Salama

Chapter 48

Clinical trials

Turki Al Garni[1,2] and Amr A. Arafat[3]

[1]*Interventional Cardiology, Division of Cardiology, Hamilton Health Sciences, McMaster University, Hamilton, ON, Canada;* [2]*Interventional Department, Prince Sultan Cardiac Center, Riyadh, Saudi Arabia;* [3]*Cardiac Surgery Department, Prince Sultan Cardiac Center, Riyadh, Saudi Arabia*

Definitions

A clinical trial is an experimental study in which a treatment or intervention is assigned to patients and compared to the standard treatment or placebo.

A controlled trial means the inclusion of another group for comparison; these groups are studied in parallel with the treatment group. Randomized means assigning treatment by chance. Therefore, all randomized trials are controlled, but not necessarily all controlled trials are randomized.

A clustering clinical trial is a clinical trial where the included are not individual participants. It is used commonly in research related to education or public health. The randomization unit could be a school or a district.

Motivation to perform clinical trials

In actual practice, patients who had specific treatment are different from those who did not. This is called confounding by indication, and this is the essence of the standard of care. Confounders are factors that affect the outcome and are distributed unequally between groups. For example, suppose you want to evaluate the effect of smoking on myocardial infarction in an observational study. In that case, you may have two unequal groups where smokers are younger and nonsmokers are older. In this case, you have a higher incidence of myocardial infarction in nonsmokers, which is attributed to the unequal age distribution between both groups. In observation studies, you can adjust for measured confounders in the analysis. However, we still have a problem with the unmeasured confounders. Clinical trials adjust for both measured and unmeasured confounders by randomizing the treatment; consequently, treatment assignment is no longer dependent on patients' characteristics. Therefore, you create two identical groups that are different only in the treatment assigned. For this reason, a clinical trial is best to estimate the causal relationship between the treatment and the outcome.

Types of randomization

Simple randomization

In this randomization method, every new patient is randomized to the treatment group individually. It can be done using coin flips or random number generators. The disadvantage of this method is the possibility of creating two unequal groups, especially in small studies.

Blocked randomization

In this method, randomization is done within blocks to have equal numbers of patients in each treatment group. The block size should vary to avoid unmasking of the treatment.

Response-adaptive randomization

Response-adaptive randomization is done by changing the randomization ratio from 1:1 over time to randomly include more patients in the arm that is performing better.

Blinding

Blinding is masking treatment assignment to ensure fair conduction of the study postrandomization. So, the physicians, patients, and investigators are unaware of the treatment assignment. Blinding can be done by using a placebo. Placebo is a pharmacologically inactive substance that resembles the active treatment. Placebo is used because people behave differently when they are studied. This is called the placebo or Hawthorn effect. In a study, compliers on a placebo had better survival compared to noncompliants.

Translational Cardiology. https://doi.org/10.1016/B978-0-323-91790-2.00091-5

Equipoise

Equipoise is the uncertainty about the benefits and risks of treatment. There is sufficient expected benefit to test the treatment on some patients; still, there is a doubt about withholding the treatment from other patients. Equipoise may present for a short period, which is called the window of opportunity.

Advantages

- Clinical trials have the highest hierarchical order among other studies
- The causative relationship between the treatment and the outcome is examined in clinical trials. The two treatment arms are almost identical in all baseline characteristics because of randomization. The only difference between the groups is the treatment given. So the causation of the outcome can be attributed to the treatment.

Limitations

- Clinical trials are expensive compared to other study designs
- Clinical trials are time-consuming since the investigators need time to recruit the patients and follow them longitudinally.
- Not all clinical issues can be randomized, such as randomizing the effect of pollution on cardiovascular outcomes.
- Randomization could be unethical, such as randomizing smoking
- In cluster trials, contamination can occur as participants can talk to each other.

Designing phase III clinical trial

- **Choose the research question.** It should be valid and ethical and can be answered from the design. For example, in the Partner I trial, the research question was to compare transcatheter versus surgical aortic valve replacement in high-risk patients who are still candidates for surgical intervention.
- **Choose the treatment and control groups.** In the Partner I trial, the treatment group is the transcatheter aortic valve replacement, and the control is the surgical aortic valve replacement.
- **Choose the study participants.** Define the inclusion criteria; the participants could benefit from the treatment, and the benefit can be detected. Do not include participants where treatment is harmful (for example, an allegory of the treatment). Restrictive enrollment can lead to precise comparisons but less generalizable results.
  In the Partner I trial, the investigators randomized 699 high-risk patients with severe aortic stenosis to transcatheter or surgical aortic valve replacement. Transcatheter was performed with a balloon-expandable bovine pericardial valve using either a transfemoral or a transapical approach.
- **Choose the primary outcome.** It should be objective, measurable, and of clinical importance.
  The primary outcome in the Partner trial was all-cause death within 1 year.
- Secondary outcomes are less objective but clinically important. Secondary outcomes usually require a larger sample size.
- Other endpoints could be surrogate endpoints that are used to substitute the clinical outcomes; they benefit from reducing the study duration. Composite events are considered to have occurred if any one of several different outcomes occurred. For example, the composite endpoint of stroke, myocardial infarction, and death is commonly used in cardiovascular trials.
- Write the protocol, get the ethical committee approval, and register the clinical trial in one of the international clinical trial registries such as clinicaltrials.gov.
- Calculate the sample size. An adequate sample size is essential to perform the clinical trial. Too small a sample is insufficient to answer the research question, and too large a sample size will put more participants at risk of intervention. The sample size is calculated for the primary outcome of the study. Components of sample size calculation are the *P*-value (alpha level), which is usually set at .05, power (type II error), which is usually set at a level of $\geq$80%, and the effect size, which is the minimal difference to detect between both study groups. Several free software is available to calculate the sample size, such as PS for power and sample size calculation.
- **Choose the randomization method;** it could be simple or blocked randomization. Randomization can be stratified based on specific patients' characteristics or participation centers.
- In some trials, investigators can conduct a run-in phase to check the patient's compliance. Noncompliant patients are excluded from the study.
- Blinding: patients or investigators can be blinded to the assigned treatment.
- Follow-up: ensure comparable follow-up between groups. Minimize withdrawals and postrandomization exclusion.
- The data safety monitoring board is a group of scientists that are given full unblinded access to the data and can perform interim analysis and terminate the study if required.

- Analysis of the results: either use intention-to-treat analysis where patients are analyzed according to the original randomization group or as treated where they are analyzed according to the treatment received.

Further reading

1. Smith CR, Leon MB, Mack MJ, et al. Transcatheter versus surgical aortic-valve replacement in high-risk patients. *N Engl J Med.* 2011;364(23):2187–2198. https://doi.org/10.1056/NEJMoa1103510.

Chapter 49

Nonrandomized controlled clinical trials

Turki Al Garni[1,2,3], Amr A. Arafat[4], Bader Alqarni[5] and Abdulrahman Mubarak Alqahtani[6]

[1]Department of Medicine, McMaster University, Hamilton, ON, Canada; [2]Interventional Cardiology, Division of Cardiology, Hamilton Health Sciences, McMaster University, Hamilton, ON, Canada; [3]Interventional Cardiology Department, Prince Sultan Cardiac Center, Riyadh, Saudi Arabia; [4]Cardiac Surgery Department, Prince Sultan Cardiac Center, Riyadh, Saudi Arabia; [5]Jeddah First Health Cluster, Alsolaimaniah, Jeddah, Saudi Arabia; [6]Interventional Cardiology Department, King Salman Cardiac Center, King Fahd Medical City, Riyadh, Saudi Arabia

Definition

In nonrandomized clinical trials, patients are not assigned to the treatment groups by chance. However, patients can choose the treatment group, or they can be assigned by the researchers.

Advantages

- Sometimes randomization is not ethical or not feasible. In this case, a nonrandomized trial with a proper control group is an option.
- In real-world practice, many patients could have several comorbidities that could affect the outcome of the treatment. In controlled randomized trials, these patients are usually excluded from randomization. Nonrandomized trials have more relaxed inclusion criteria; therefore, the lower restrictions in the nonrandomized trials relatively increase their external validity compared to randomized trials.

Limitations

- Because of the nonrandomized nature of this design and allocation to a treatment group is judged by patients or researchers, the baseline characteristics of the treatment and control groups could vary widely. The study could be also confounded by indication since patients who will receive the treatment are different from those who did not. Patients who receive the treatment could be sicker with more comorbidities. Therefore, proper adjustment of the baseline characters is challenging and might not be possible in all cases if patients who received the treatment are completely different from the control group.

Designing a nonrandomized trial

The following points should be considered when designing a nonrandomized trial. The investigator should do a comprehensive collection of baseline variables and all possible confounders that can influence the outcomes for both groups. It is recommended to observe both treatment groups at the same period of time to adjust for the temporal effect, since the disease characteristics, diagnostic criteria, and concomitant treatment change over time and could affect the study outcomes if both groups were included from different time periods. The investigators should collect the primary outcomes at multiple times for both groups, including one before the start of the intervention. This is beneficial to separate the effect of treatment from other environmental factors. Additionally, the data collection procedure should be the same for both groups.[1]

An example of a nonrandomized trial is the nonrandomized trial comparing nonischemic heart preservation and static cold preservation for heart transplantation.

Estimating the causal effect

There are several methods to estimate the causal effect in this design. Examples of these methods are multivariable analysis, propensity score matching, instrumental variable analysis, and comparing changes over time between the intervention and control groups (sometimes called "difference in differences"). A mixture of these methods can be used.

There are several methods of multivariable regression analysis depending on the type of study outcomes. Linear regression can be used for continuous outcomes, logistic regression for binary outcomes, and Cox regression for time-to-event outcomes. In propensity score matching,

Translational Cardiology. https://doi.org/10.1016/B978-0-323-91790-2.00071-X

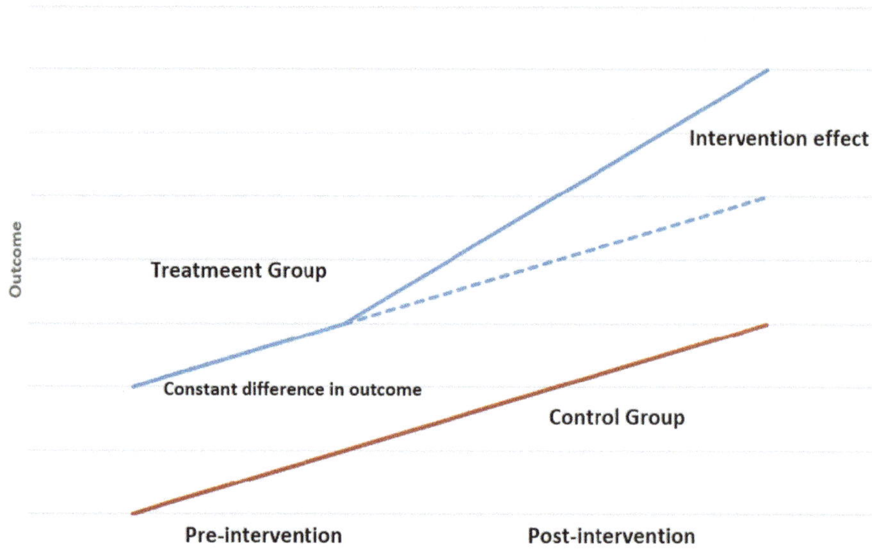

FIGURE 49.1 Graphical explanation to difference in difference.

every patient is given a score based on the concomitant comorbidities. Then, matching is performed based on this score. The main limitation of propensity score matching and multivariable regression is that they match on measured variables only. Several unmeasured variables could have affected the outcomes and are not included in the analysis.

Instrumental variable analysis matches measured and unmeasured variables if properly selected. The instrumental variable is a variable that has a strong association with the treatment but not with the outcome. Measuring the association between the instrumental variable and the outcome can assess the unbiased relation between the treatment and the outcome. For example, to assess the relation between the use of aprotinin (serine protease inhibitor) and mortality after coronary artery bypass grafting, an instrumental variable of surgeons' preference to use aprotinin is selected. Assessing the relation between surgeons' preference and mortality measures the unbiased relation between aprotinin use and mortality.[2]

Another example of choosing the proper instrumental variable is studying the relationship between alcohol consumption and ischemic heart disease. The direct relation between alcohol and ischemic heart disease might be confounded by positive and negative health-related behaviors. The alcohol dehydrogenase gene makes alcohol intake unpleasant. Assessing the relationship between the gene and ischemic heart disease indirectly measures the relation between alcohol and ischemic heart disease.[3]

The difference in difference design is suitable for longitudinal data. In difference in designs, the trends in associated outcomes between both groups are compared over time before and after treatment. Any significant difference in trends can be interpreted as an intervention effect (Fig. 49.1). Similar to the propensity score, a difference in difference increases the risk of bias in the presence of unmeasured confounds.

References

1. Bonell CP, Hargreaves J, Cousens S, et al. Alternatives to randomisation in the evaluation of public health interventions: design challenges and solutions. *J Epidemiol Community Health.* 2011;65(7):582–587. https://doi.org/10.1136/jech.2008.082602.
2. Schneeweiss S, Seeger JD, Landon J, Walker AM. Aprotinin during coronary-artery bypass grafting and risk of death. *N Engl J Med.* 2008;358(8):771–783. https://doi.org/10.1056/NEJMoa0707571.
3. Nilsson J, Jernryd V, Qin G, et al. A nonrandomized open-label phase 2 trial of nonischemic heart preservation for human heart transplantation. *Nat Commun.* 2020;11(1):2976. https://doi.org/10.1038/s41467-020-16782-9.

Chapter 50

Historical controlled studies

Turki Al Garni[1,2,3], Amr A. Arafat[4], Bader Alqarni[5] and Monirah Albabtain[6]

[1]Department of Medicine, McMaster University, Hamilton, ON, Canada; [2]Interventional Cardiology, Division of Cardiology, Hamilton Health Sciences, McMaster University, Hamilton, ON, Canada; [3]Interventional Cardiology Department, Prince Sultan Cardiac Center, Riyadh, Saudi Arabia; [4]Cardiac Surgery Department, Prince Sultan Cardiac Center, Riyadh, Saudi Arabia; [5]Jeddah First Health Cluster, Alsolaimaniah, Jeddah, Saudi Arabia; [6]Pharmacy Department, Prince Sultan Cardiac Center, Riyadh, Saudi Arabia

Definition

The historically controlled study is a study design in which a new treatment or intervention is compared to historically adequately documented active treatment or intervention in comparable patients.

Advantages

- A historically controlled trial is useful for a life-threatening disease where the control arm is not ethical
- A historically controlled trial requires a smaller sample size
- As the result of the smaller sample size and historically chosen control group, this study design has a lower cost
- In historically controlled studies, recruitment for rare diseases is feasible

Limitations

- The control group is retrospectively studied with its inherent selection and referral biases.
- A historically controlled trial is an unblinded study
- This study design is prone to selection bias
- It is difficult to adjust the baseline data between groups
- The characters of the patients change over time.
- Disease diagnostics and definitions could have changed
- The treatment of disease changes over time, and concurrent medications may differ between groups.
- Missing data may affect the results significantly.

Design of a historically controlled trial

Several factors should be considered when choosing a historical control. These factors include baseline characteristics, healthcare environment, standard of care, and psychological effects.

Data can come from different sources, including medical charts, published observational studies or clinical trials, natural history studies, and registries.

Patients' selection is prone to bias even if selected from clinical trials, as these trials don't present real-world data. The inclusion of pragmatic clinical trials is an option as these patients present a real-world clinical experience. It is also possible to use patients as their own control if the complete history of the patients is well documented.

Methods to improve the study design

The control groups should have recently received the defined treatment as previously defined. The inclusion criteria and the outcome measures should be the same for both groups. It is also essential to unify the prognostic criteria in both groups. The investigators can also consider matching during analysis and using special statistical methods such as propensity score matching and sensitivity analysis. The treatment group can be matched with patients who received different treatments during the same period to decrease the temporal effect.[1]

An example of a historically controlled study is the study by El-Chami and associates, who compared the performance of the Micra transcatheter pacemaker to a reference data set of patients with transvenous pacemakers. The study included 2667 patients in the transvenous control group versus 1817 patients with an attempted Micra implantation procedure. The intervention group included patients intended to be implanted with a market-approved Micra device at 179 centers in 23 countries. Those patients were compared with patients with de novo conventional transvenous pacemaker systems who were recruited from

Translational Cardiology. https://doi.org/10.1016/B978-0-323-91790-2.00079-4

six recent Medtronic trials of dual-chamber pacing. Micra is a single-chamber pacemaker; therefore, events related to the right atrium were excluded from the conventional dual-chamber pacemaker group to approximate a single-chamber dataset. The study endpoints were device and procedure-related complications that occurred early (before 30 days) or late (after 30 days). The main study objective was to assess the safety of the new device (Micra); therefore, the study endpoints were compared between groups after excluding events related to the right atrium only from the conventional group. Because the study was not randomized and both groups were not thoroughly matched (except for right atrium-related complications), the authors recommended performing a randomized clinical trial to compare the performance of the new device compared to the conventional transvenous pacing after they had established the potential safety of the new device in this study.[2]

References

1. Ghadessi M, Tang R, Zhou J, et al. A roadmap to using historical controls in clinical trials – by Drug Information Association Adaptive Design Scientific Working Group (DIA-ADSWG). *Orphanet J Rare Dis*. 2020;15:69.
2. El-Chami MF, Al-Samadi F, Clementy N, et al. Updated performance of the Micra transcatheter pacemaker in the real-world setting: a comparison to the investigational study and a transvenous historical control. *Heart Rhythm*. 2018;15(12):1800–1807. https://doi.org/10.1016/j.hrthm.2018.08.005.

Chapter 51

Crossover studies

Turki Al Garni[1,2], Amr A. Arafat[3,4] and Monirah Albabtain[5]

[1]Interventional Cardiology, Division of Cardiology, Hamilton Health Sciences, McMaster University, Hamilton, ON, Canada; [2]Interventional Cardiology Department, Prince Sultan Cardiac Center, Riyadh, Saudi Arabia; [3]Cardiac Surgery Department, Prince Sultan Cardiac Center, Riyadh, Saudi Arabia; [4]Cardiothoracic Surgery Department, Tanta University, Tanta, Egypt; [5]Pharmacy Department, Prince Sultan Cardiac Center, Riyadh, Saudi Arabia

Definition

Crossover studies are longitudinal studies where each participant is assigned to a sequence of two or more interventions or treatments rather than one intervention or treatment (Fig. 51.1).

Advantages

- Each participant will present his/her own control. Therefore, the confounding effect of other measured or unmeasured variables is reduced. Confounders are factors that affect the outcome and are distributed unequally between groups. The baseline data are almost identical in this study design since the patients serve as their own control.
- A crossover study design is useful for chronic or stable conditions. These conditions are suitable for long-term treatment and for using more than one treatment regimen, and the treatment should not lead to a complete cure.[1]
- The design is also suitable for bioequivalence trials. Bioequivalence is a study comparing the bioavailability, efficacy, and safety of two substances with the same dose and administration route. In these trials, the blood concentration is determined for a new drug and a reference medication.
- The sample size for the crossover design is much smaller than the parallel study designs since each participant is allocated to the treatment and control groups. Additionally, the treatment response variability from the mean response in the same patient is lower than interpatient variability.

Limitations

- The carryover effect limits this study design. The effect of one treatment may persist during the second treatment. This limitation requires a washout period. The washout period depends on the bioavailability of the drug and needs expert opinion to determine the optimal washout period.
- The order of treatment may affect the outcome. This limitation can be overcome by assigning treatment to participants in random order. If we compare treatment A to B, patients should be assigned randomly to start on A or B. Therefore, some patients will start on A followed by B, and others will start on B followed by A.
- The order of treatment can also affect reporting the side effects. If participants receive the treatment with more side effects first, this could lead to overreporting with the subsequent treatment.
- Crossover studies are not suitable for medications that have long-term or irreversible effects.
- This study design is unsuitable for acute diseases or diseases of short duration.

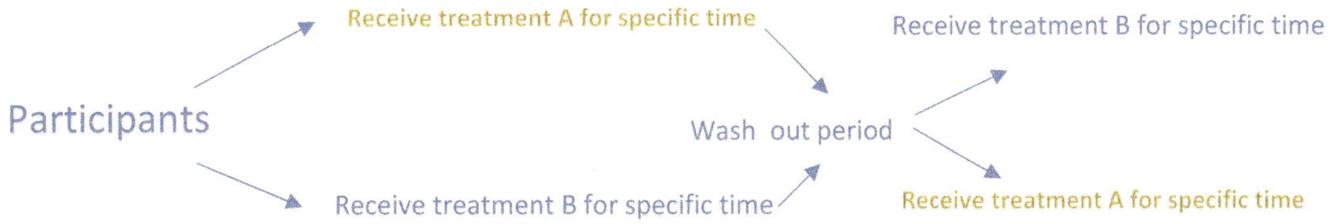

FIGURE 51.1 Schematic representation of a crossover study design. Each participant receives both treatments (A and B) with an intervening washout period to minimize carryover effects.

Translational Cardiology. https://doi.org/10.1016/B978-0-323-91790-2.00081-2

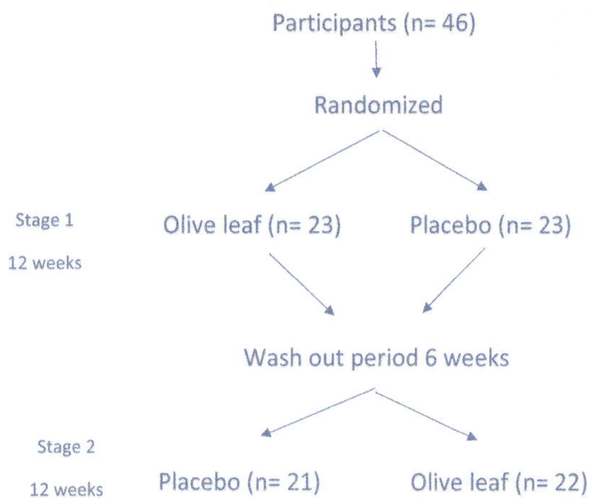

FIGURE 51.2 Flowchart of a crossover trial examining the effects of Olive leaf versus placebo on insulin sensitivity in patients with diabetes. After a 12-week treatment period and a 6-washout period, patients switched treatments.

Designing crossover studies and examples

In crossover studies, all participants receive the same treatment for the same period.

de Bock and associates (2013) randomized 46 patients to receive either Olive leaf or a placebo to manage diabetes. The patients received both treatments (n = 23 in each group) for 12 weeks, then it was followed by 6 weeks of washout. The placebo patients received olive leaf for 12 weeks after the washout period and vice versa (Fig. 51.2). The primary outcome was insulin sensitivity, assessed via a 75 g oral glucose tolerance test evaluated at 0, 30, 60, 90, and 120 min. In this study, the authors presented the baseline data for all patients as one group; since all patients received the treatment and placebo at a specific time. The statistical analysis of this study used tests for correlated outcomes (see below). The study outcomes were repeated measure variables; therefore, linear mixed models were used to assess the main treatment effect. The authors included the baseline data in the model as covariates in addition to the possible confounders that could have affected the outcome, such as concurrent medications. The authors also adjusted for the baseline level of the study outcomes before starting the patients on either placebo or the treatment.[2]

Reporting and analysis

Crossover trials require prestudy registration in a clinical trial registry. An extension of the CONSORT checklist and diagram was developed to include crossover studies. For analysis of the crossover studies, tests for matched data should be considered. For continuous variables, a paired *t*-test or Wilcoxon matched-pair signed-rank tests are appropriate for two measures comparison. Categorical data can be compared with McNemar (binary variable) or Friedman test (three or more levels of variables). Other tests for repeated measures can be considered, such as repeated measures analysis of variance (-ANOVA) and mixed, fixed, and random effect regressions.[3]

References

1. Sibbald B, Roberts C. Understanding controlled trials. Crossover trials. *BMJ*. 1998;316(7146):1719. https://doi.org/10.1136/bmj.316.7146.1719.
2. de Bock M, Derraik JGB, Brennan CM, Biggs JB, Morgan PE, et al. Olive (*Olea europaea* L.) leaf polyphenols improve insulin sensitivity in middle-aged overweight men: a randomized, placebo-controlled, crossover trial. *PLoS One*. 2013;8(3):e57622. https://doi.org/10.1371/journal.pone.0057622.
3. Dwan K, Li T, Altman DG, Elbourne D. CONSORT 2010 statement: extension to randomised crossover trials. *BMJ*. 2019;366:l4378. https://doi.org/10.1136/bmj.l4378.

Chapter 52

Withdrawal trials (randomized discontinuation trials)

Turki Al Garni[1,2], Amr A. Arafat[3,4] and Maha M. AlRasheed[5]

[1]Interventional Cardiology, Division of Cardiology, Hamilton Health Sciences, McMaster University, Hamilton, ON, Canada; [2]Interventional Cardiology Department, Prince Sultan Cardiac Center, Riyadh, Saudi Arabia; [3]Cardiac Surgery Department, Prince Sultan Cardiac Center, Riyadh, Saudi Arabia; [4]Cardiothoracic Surgery Department, Tanta University, Tanta, Egypt; [5]Department of Clinical Pharmacy, College of Pharmacy, King Saud University, Riyadh, Saudi Arabia

Introduction

Withdrawal trials, also known as randomized discontinuation trials, are essential components of clinical research aimed at evaluating treatment efficacy and informing medical decisions. These trials involve randomly assigning patients to either continue or discontinue a specific treatment, allowing researchers to observe the effects of treatment withdrawal on patient outcomes. Understanding the significance of withdrawal trials is crucial for their role in shaping evidence-based medicine and guiding clinical practice across various medical specialties.

Definitions

Withdrawal trials are clinical trials where all participants receive active treatment for a specific time (open-label phase). Then, responders are randomized to receive either active treatment or placebo (withdrawal of active treatment).

Predictive enrichment techniques are used to select participants who have the highest chance to benefit from the treatment. Previous studies that showed a higher response in specific subgroups could be beneficial. Noncompliant patients, or those reporting adverse events, are not considered for study enrollment.[1]

Advantages

- Withdrawal trials include responders only.
- This design assesses the recurrence of symptoms after stopping the treatment.
- Withdrawal design also assesses the compliance of participants before enrollment. The open-label phase is similar to the run-in phase.
- The inclusion of responders only increases the power of the study for a given sample size.
- The withdrawal study design has an ethical advantage by reducing the time on placebo since it includes responders only.

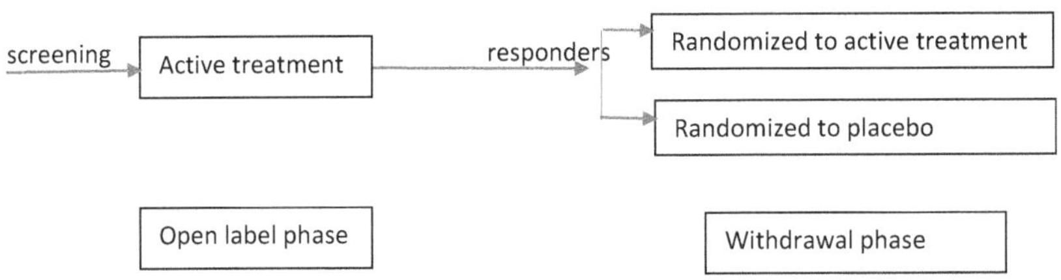

Translational Cardiology. https://doi.org/10.1016/B978-0-323-91790-2.00023-X

- Withdrawal trials assess if the treatment needs to be continued or can be stopped.
- Subjects who have disease relapse can be started on another therapy.
- Withdrawal trial is one of the methods to reduce dropout from clinical trials.

Limitations

- Withdrawal design is suitable for chronic diseases but not limited to them
- Withdrawal design is not suitable for curable or potentially curable during the run-in phase.
- Withdrawal of treatment from responders and assigning them to a placebo could lead to deterioration
- In patients who deteriorated after withdrawal of the active treatment may not return to the prerandomization state after reinstitution of the active treatment.

Designing a withdrawal clinical trial

The withdrawal study design follows the same steps of designing a randomized clinical trial, including the choice of the research question, inclusion and exclusion criteria, and study registration. An example of a withdrawal clinical trial is (randomized withdrawal study of patients with symptomatic neurogenic orthostatic hypotension responsive to droxidopa)[2] This study included 181 participants in the open-labeled phase; all of them had the active treatment (droxidopa). There were 101 responders defined as one point or more improvement on the symptoms questionnaire and an improvement in systolic blood pressure of 10 mmHg or more at 3 min poststanding. The responders were randomized to placebo versus active treatment. The study analysis was conducted on participants of the withdrawal phase-only who were randomized to placebo versus active treatment.

Another design was developed to avoid the withdrawal of active treatment from responders and put them on a placebo. REPLACE is a randomized open-label study in which patients were randomized to continue phosphodiesterase type 5 inhibitors (PDE5i) for pulmonary hypertension or switch to riociguat.[3] Data from a RESPITE study indicated that the transition from PDE5i to riociguat may be feasible and beneficial in patients not adequately responding to PDE5i. Therefore, patients were randomly assigned to continue PDE5-i therapy or switch to riociguat.[4]

Pitfalls

The main concern for withdrawal trials is choosing the proper condition and treatment to be investigated. The investigators should choose the withdrawal trials for chronic disease, and the treatment should be for long-term use and noncurative. During the open-label, the investigators should pay attention to patients' monitoring and assessing the study endpoints so that nonresponders can be identified rapidly.

References

1. Li J, Zhao L, Tian L, et al. A predictive enrichment procedure to identify potential responders to a new therapy for randomized, comparative controlled clinical studies. *Biom.* 2016;72:877–887. https://doi.org/10.1111/biom.12461.
2. Biaggioni I, Freeman R, Mathias CJ, et al. Randomized withdrawal study of patients with symptomatic neurogenic orthostatic hypotension responsive to droxidopa. *Hypertension.* 2015;65(1):101–107. https://doi.org/10.1161/HYPERTENSIONAHA.114.04035.
3. Hoeper MM, Al-Hiti H, Benza RL, et al. Switching to riociguat versus maintenance therapy with phosphodiesterase-5 inhibitors in patients with pulmonary arterial hypertension (REPLACE): a multicentre, open-label, randomised controlled trial. *Lancet Respir Med.* 2021;9(6):573–584. https://doi.org/10.1016/S2213-2600(20)30532-4.
4. Hoeper MM, Simonneau G, Corris PA, et al. RESPITE: switching to riociguat in pulmonary arterial hypertension patients with inadequate response to phosphodiesterase-5 inhibitors. *Eur Respir J.* 2017;50(3):1602425. https://doi.org/10.1183/13993003.02425-2016.

Chapter 53

Factorial study design

Turki Al Garni[1,2], Amr A. Arafat[3] and Abeer Bakhsh[4]

[1]Interventional Cardiology, Division of Cardiology, Hamilton Health Sciences, McMaster University, Hamilton, ON, Canada; [2]Interventional Cardiology Department, Prince Sultan Cardiac Center, Riyadh, Saudi Arabia; [3]Cardiac Surgery Department, Prince Sultan Cardiac Center, Riyadh, Saudi Arabia; [4]Heart Function and Transplant, Prince Sultan Cardiac Center, Riyadh, Saudi Arabia

Introduction

Factorial study design is a powerful research approach used to simultaneously investigate multiple interventions or variables within a single study. By examining the effects of two or more factors and their interactions, factorial designs allow researchers to efficiently explore the independent and combined effects of different treatments or interventions. This flexible and efficient design offers several advantages, including the ability to assess multiple interventions within the same study, control for confounding variables, and maximize the use of resources. Understanding the principles and applications of factorial study design is essential for researchers seeking to address complex research questions and optimize study efficiency.

Definition

In clinical studies, treatment or intervention can be called a factor. Factorial design is a study design that allows testing multiple treatments or interventions on the same subjects, which permits an assessment of potential interactions among the treatments. The simplest form of factorial design is 2 × 2 design. In this design, two treatments are evaluated. If we study drugs A and B, we will have four groups. The groups will receive the following treatment: (Active drug A and B), (active drug A and placebo B), (placebo A and active B), and (placebo A and B).

In a factorial design, increasing the levels or numbers of factors will require a higher sample size to have a high probability of detecting the specified effect size (power). The power of the factorial design depends on the level of each factor, the number of factors, the number of possible combinations, and the significance level (Type I error).

Advantages

- The factorial design allows studying the interaction between several treatments and whether the effect of one treatment depends on the other (effect modification).
- The factorial design can evaluate multiple interventions on the same individuals.
- The factorial design may achieve results more quickly by evaluating more than one factor in a single study. This advantage hastens the approval of new medical therapies.
- Factorial design is more efficient than one-factor design if there is an interaction. Evaluating the effect of multiple factors on the outcomes yields accurate results similar to real-life experience and increases the study's external validity.
- The study of more than one factor in a single study lowers the cost of conducting research compared to the conduction of multiple studies with one treatment at a time.

Limitations

- If there is an interaction between treatments, the power is decreased compared to a single-factor design.
- If the number of factors increases, the required sample size will increase and, consequently, the study's cost.

Designing factorial design and examples

The simplest factorial design is a 2 × 2 design with two levels of treatment A crossed by two levels of treatment B. This design yields four treatment regimens.

Placebo A + Placebo B
Placebo A + Treatment B
Treatment A + Placebo B
Treatment A + Treatment B

Translational Cardiology. https://doi.org/10.1016/B978-0-323-91790-2.00059-9

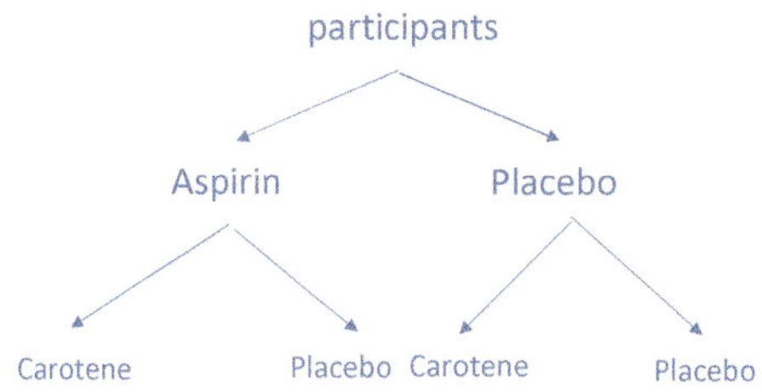

FIGURE 53.1 The physician health study flowchart.

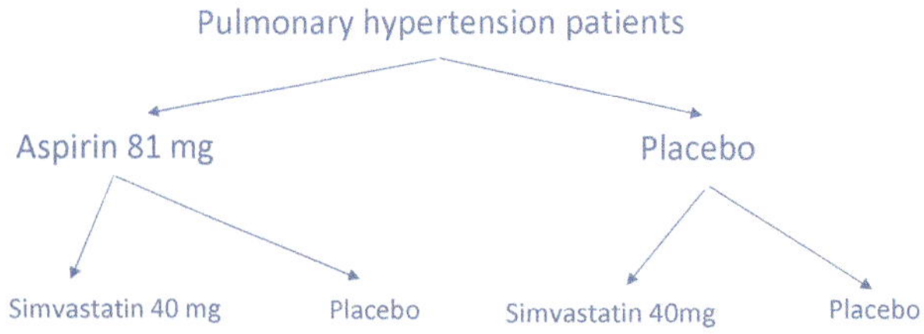

FIGURE 53.2 The flowchart of Kawut et al.'s study of aspirin and simvastatin in patients with pulmonary hypertension.

In the physician health study, physicians were randomized to receive aspirin versus placebo. Then, each arm was assigned to active carotene versus placebo carotene (Fig. 53.1).

Another factorial design example is the randomized clinical trial of aspirin and simvastatin for pulmonary arterial hypertension.[1] The authors of this study randomized 92 patients with pulmonary hypertension to aspirin 81 mg or its placebo and simvastatin 40 mg or its placebo (Fig. 53.2).

The primary endpoint was a 6-minute walking distance after 6 months of treatment in this study. The authors performed an intention-to-treat analysis for the study endpoints. They did not find a significant interaction between aspirin and simvastatin. Therefore, they analyzed aspirin versus placebo and simvastatin versus placebo separately.

Pitfalls of factorial design

The investigators should be aware that increasing the number of factors will increase the number of possible treatment combinations. This will make the execution of the trial and the statistical analysis challenging. As the number of factors increases, the sample size will increase, and interpretation of higher-order interactions will be difficult. A higher treatment combination also requires an increase in the block size, which will affect the homogeneity of the treatment combination and consequently will increase the errors and decrease the precision. Therefore, it is recommended to use the simplest 2×2 factorial design.[2]

Analysis of factorial trials

In analyzing factorial studies, tests such as analysis of variance (ANOVA) and regression analysis with testing the interaction between treatment arms should be considered. ANOVA that tests the relationship between two or more independent variables and a response variable is called factorial ANOVA. In addition to testing the relationship between the independent and response variables, factorial ANOVA assesses if two or more factors affect a single response variable, that is, it tests the interaction between these independent variables.[3]

Reference

1. Kawut SM, Bagiella E, Lederer DJ, et al. Randomized clinical trial of aspirin and simvastatin for pulmonary arterial hypertension: ASA-STAT. *Circulation.* 2011;123(25):2985–2993. https://doi.org/10.1161/CIRCULATIONAHA.110.015693.
2. Stampfer MJ, Buring JE, Willett W, Rosner B, Eberlein K, Hennekens CH. The 2 × 2 factorial design: its application to a randomized trial of aspirin and U.S. physicians. *Stat Med.* 1985;4:111–116. https://doi.org/10.1002/sim.4780040202.
3. Montgomery AA, Peters TJ, Little P. Design, analysis and presentation of factorial randomised controlled trials. *BMC Med Res Methodol.* 2003;3:26. https://doi.org/10.1186/1471-2288-3-26.

Chapter 54

Group allocation

Marcela de Almeida Lopes[1], Pedro Henrique M. Craveiro de Melo[2] and André Luiz Nunes Gobatto[3,4]

[1]HCor, Research Institute, São Paulo, Brazil; [2]Cardiovascular Research Foundation, New York, NY, United States; [3]Hospital da Cidade, Salvador, Bahia, Brazil; [4]Hospital São Rafael, Salvador, Bahia, Brazil

Key points

1. Group allocation is an effective and straightforward approach for comparing treatment outcomes among groups or individuals.
2. It's crucial to prevent confounding bias, selection bias, and observer bias in this type of study.
3. Various measures such as randomization, blinding, and allocation concealment are best employed to mitigate bias in this type of study.
4. Randomization helps ensure that groups are comparable in all relevant characteristics at the outset of the study, thereby reducing the potential for bias.
5. Blinding, both of participants and researchers, can minimize the influence of preconceived expectations on study outcomes.

Introduction

Random allocation is used to assign individuals to different research groups, treatment groups, and control groups. This allows researchers to control all known and unknown factors that may affect results in treatment groups and control groups.

Allocation concealment prevents researchers from influencing which participants are assigned to a given intervention group (Herbison et al., 2011). This process must be included in the experiment for the success of any RCT.

Simple randomization is easy to implement and inexpensive and may be a good option for large trials. Group randomization is a good option when researchers want to reach the number of participants in each group or provide a balance of significant covariates. These methods will be discussed in this chapter, along with their advantages and disadvantages. It is essential to select a method that will produce valid results for the study.

Choosing the best randomization method

There are several randomization methods, each with its advantages and disadvantages: Simple randomization, block randomization; stratified randomization; and adaptative randomization.

Simple randomization is the traditional method of group allocation in clinical trials. In simple randomization, a random list is generated, usually by computer, determining the sequence of participant group allocation. Every participant will be allocated based on this list. The randomization list must be blinded from the investigators, the assistant team, and the participants to keep allocation concealment. For example, allocation may be available only for the pharmacist delivering the investigational drug or placebo in a trial studying a new drug. The advantages of simple randomization are that it is inexpensive and straightforward to use—several computer programs (e.g., Stata) or internet websites (e.g., randomizer.org) can generate the randomization list.

Additionally, it keeps allocation concealment since every participant has an equal chance of being randomized to the treatment or control group. The major disadvantage of simple randomization is its use with small sample sizes, especially less than 100 participants. In this scenario, there is a big chance of imbalance. For example, in a trial including 20 patients, one group could end up with eight patients and the other with 12.

In small samples, simple randomization can imbalance significant baseline covariates between groups since it does not consider any baseline characteristic when randomizing participants. For example, in a trial of a new drug to treat hypertension, planning to include 60 participants, simple randomization might result in more participants with severe hypertension in the treatment group than in the control

Translational Cardiology. https://doi.org/10.1016/B978-0-323-91790-2.00029-0

(A)

(B)

FIGURE 54.1 (A) Simple randomization of eight participants (*black* for control, *blue* for intervention). This random sequence resulted in an unbalance between groups (5 × 3). (B) Randomization into blocks of four, resulting in four participants in each group.

group, just by chance. As a result, outcomes might be worse for the treatment group than for the control group, even if the intervention was effective, because participants in the treatment group were sicker than those in the control group. Therefore, simple randomization is not a good option for trials with small sample sizes or when a balance for key covariates is desired.

In blocked randomization, participants are randomized within blocks instead of randomizing each participant individually. The blocks have a predetermined size, and within each block, the number of participants is balanced in each group. For example, in a trial including 16 participants to test a new drug (D) versus placebo (P), the block sequence generated might be DDPP, DPDP, PPDD, or PDPD. Blocked randomization has a disadvantage: if the trial is not blinded, and researchers know the assignment of previous participants and the block size, they might be able to predict the allocation of the next participant. This can be corrected by using different block sizes in the same trial (e.g., 4 and 6). Blocked randomization will lead to a balance in sample size between groups; however, the groups may still be unbalanced for important covariates since baseline characteristics are not considered in the randomization process (Fig. 54.1).

When baseline covariates strongly impact study outcomes, stratified randomization might be used to balance these covariates between groups. In stratified randomization, eligible participants are first separated into strata according to a baseline characteristic, for example, gender or severity of the disease. Each stratum has a separate randomization list, and after being categorized into one of the strata, the participant is randomized to the control or treatment arm. In this method, each participant in each stratum has an equal chance of being allocated to the control or treatment group. If the sample size is big enough, there will be a balance in the number of subjects allocated to control or treatment within each stratum. Variables typically used for stratified randomization include the site center in multicentric studies and disease severity. Therefore, the major advantage of stratified randomization is the balance of important known variables. However, it should be done with a few strata, preferably using dichotomous variables, or it might jeopardize allocation concealment since the allocation starts to be predictable, especially in small sample sizes.[2] A general recommendation for stratification is to use a limited number of strata in randomization. This will balance other factors; consider using regression adjustments later if there happen to be

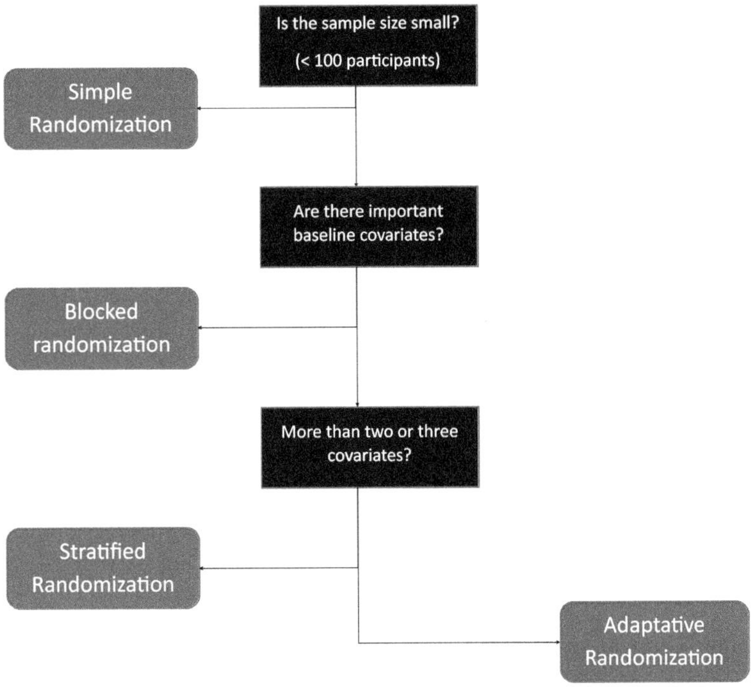

FIGURE 54.2 Flow chart of the options of randomization method based on sample size and number of important baseline covariates to be balanced among the treatment arms. *Adapted from Fregni F, Illigens BM, eds.* Critical Thinking in Clinical Research: Applied Theory and Practice Using Case Studies. *Oxford, UK: Oxford University Press; 2018.*

imbalances in some baseline factors; use a permuted blocks design with random block sizes for each stratum.

Adaptive randomization uses algorithms including baseline covariates and the allocation of previous participants to allocate the following participants.[3] The minimization method is the most used method of adaptative randomization and uses a computerized algorithm that lists the baseline covariates of interest for the participant, evaluates the balance (or imbalance) of these covariates among the participants already included in each of the study groups, and then allocates the participant to the control or the treatment arm to minimize the imbalances.[2] The advantage of adaptative randomization is that it balances important known variables between groups. However, it must be computer-generated, which means it must be available at randomization time. Also, stratified randomization might jeopardize allocation concealment in small trials once the allocation group starts to be predictable, depending on the number of variables and participants.[2]

Therefore, what is the best randomization method? For large trials, any method would fit well; in small trials, any method might be inaccurate. Simple randomization is easy and inexpensive but can lead to considerable group imbalances in trials with small sample sizes. Blocked randomization balances the number of participants in each group but not the baseline covariates. Stratification is usually combined with blocked randomization to provide a balance between groups in the number of participants and significant covariates. However, depending on the number of covariates used for stratification and the sample size, this method could lead to imbalances, going back to the problem of the simple randomization method. Adaptive randomization methods, especially minimization, allow for more baseline covariates and produce balanced groups of both the number of participants and covariates, but are more complex and expensive. Fig. 54.2 shows a flowchart for helping to select the most appropriate randomization technique.

Research team

A scientific research team is a group of qualified and trained people, aware of their roles and responsibilities, working together to complete a research project successfully. Typically, the research team is composed of personnel directly engaged in the conduction of the research (e.g., principal investigator, coinvestigator, research assistants, and data managers) and associated staff (clinicians, pharmacists, veterinarians, technicians, and officers). Facilities experienced with FDA's enforcement policies and Good Clinical Practices studies may count on a wider range of contributors, including the quality assurance team and regulatory team.[4]

The well-desired introduction of randomization and allocation can impact the facility's staff, demanding more personnel engaged with a single study or changing well-established routines. Part of this impact can be mitigated by securing rationalization and awareness regarding the positive impact of practices on the study's internal validity.

Funding

Funding has long been considered a barrier to implementing randomization and blinding in translational research. Conversely, randomized confirmatory studies have a higher potential to be prioritized by public and private funders in assessing applications for new projects. Also, not implementing bias-reducing methods can delay the development of new drugs and upcoming technologies, increasing the overall costs.

Preclinical randomized controlled trials have been proposed to overcome some limitations of independent animal studies and reduce the gap between translational research and clinical trials.[5]

Since simple or even sophisticated randomization methods are freely available online, most of the attention on planning a study's budget should be focused on concealment allocation. Truly secure methods for concealment will impact the overall costs, increasing the staff needed to execute independently different stages of the study (i.e., one scientist cannot be part of randomization/coding, experimental interventions, and data recording/analysis). Also more likely is the need for external consultants, especially statisticians.

References

1. Herbison P, Hay-Smith J, Gillespie WJ. Different methods of allocation to groups in randomized trials are associated with different levels of bias. A meta-epidemiological study. *J Clin Epidemiol.* 2011;64(10):1070–1075. https://doi.org/10.1016/j.jclinepi.2010.12.018.
2. Kang M, Ragan BG, Park JH. Issues in outcomes research: an overview of randomization techniques for clinical trials. *J Athl Train.* 2008;43(2):215–221. https://doi.org/10.4085/1062-6050-43.2.215.
3. Fregni F, Illigens BM, eds. *Critical Thinking in Clinical Research: Applied Theory and Practice Using Case Studies.* Oxford, UK: Oxford University Press; 2018.
4. Center for Clinical and Translational Science (CCTS). *Responsibilities of the Research Team*: CCTS at the Ohio State University. Retrieved May 3, 2022, from: https://ccts.osu.edu/content/responsibilities-research-team.
5. Wouters OJ, McKee M, Luyten J. Estimated research and development investment needed to bring a new medicine to market, 2009–2018. *JAMA.* 2020;323(9):844–853. https://doi.org/10.1001/jama.2020.1166.

Chapter 55

Hybrid design

Zachariah B. Nealy
Riverside Regional Medical Center, Newport News, VA, United States

Key points

- Hybrid design allows for two different studies to be performed simultaneously often with complementary goals that can be joined for the sake of achieving a larger powered goal
- Hybrid design can be more complex than traditional study design and often involve more stringent subgroup assignment with multiple phases to accomplish different goals while also utilizing limited resources more meticulously
- The most common and delineated types are effectiveness–implementation studies that can be used early or late in the intervention development pipeline to accomplish both goals simultaneously

Hybrid study design refers to developing a novel study design as a result of combining two previously known study designs. The goal of these studies is to overcome any of a number of problems. These can include cost, patient recruitment, reducing time to completion, including complex patients earlier in the drug development process, or even reducing the limitations of one of the two combined trial designs. Combining two different designs often is performed with a specific purpose in mind related to the intervention being tested. Examples of hybrid models may include superiority–noninferiority combinations, effectiveness–implementation, and retrospective-prospective trial designs. In designing these combinations, there is the opportunity to conduct these studies in parallel or in sequential phases. Underpowering can be a risk with running arms in parallel due to too few patients in each arm, but also this can occur with running studies in sequential phases if the rate of patient dropout is dependent on the length of time receiving the intervention. This should be kept in mind when designing the study. Ultimately, though, the benefits of hybrid design are its flexibility in the initial design and, to an extent, its ability to adjust the design as the study goes on and needs adaptations as well. Hybrid studies provide the ability to compartmentalize and improve the efficiency of the trial by utilizing complementary study designs. As such, this will be study-dependent as its benefits are a function of the individual parts as well. The limitations will be similar in this manner. A further benefit is that hybrid studies can yield evidence for efficacy in a single trial by assigning patients to appropriate subgroups after a broad patient recruitment. In comparison to conventional design, which has narrow definitions for recruitment for the study and does not have subgroups in mind during early design, this provides a more pragmatic approach that can improve study outcomes and statistical significance without further analytic techniques.[1] Overall, there are some typical design models utilized to explain these as well as other combinations.

There are several common design features to help allocate resources effectively. The simplest design for a hybrid design model would be, for parallel design, to assign patients to a subgroup based on the perceived intervention protocol. An example of this might be where one arm was a superiority study of adding the intervention to standard of care (SOC) compared to SOC alone, while the other arm was a primarily noninferiority design where the intervention in addition to the SOC is compared to an alternative accepted intervention along with the SOC. The goal of this hybrid study could be to attempt to show noninferiority with the aggregate of the two studies, noninferiority with reduced adverse effects in one subgroup, and superiority in the other subgroup. This would potentially account for possible failure to prove the hypothesis or underpowering of one of the two arms while still having the statistical significance of the aggregate hypothesis. Another simple design may be performed in sequential phases by studying a disease- or disease resolution–exposure relationship by combining retrospective cohort and prospective case-control designs. This could be done by performing a series of retrospective cohort studies of the disease–exposure relationship to inform a prospective case-control study of those willingly already exposed.[2] Such studies could

Translational Cardiology. https://doi.org/10.1016/B978-0-323-91790-2.00041-1

answer long-term monitoring questions that prior studies might not be able to answer in a shorter timeframe.

Hybrid studies often utilize several more complex designs, including randomized block design, Solomon four-group design, and switched replication design. Randomized block design assigns subjects to subgroups based on similarities so that each subgroup is more different than any two members of a subgroup would be. This can be useful for monitoring interventions if it is estimated that this difference will affect the outcome. The Solomon four-group design takes this a step further by dividing patients into groups based on the concern that the pretest may sensitize the posttest measurement. So, a Solomon four-group design would include two groups that receive a pretest and posttest (with and without the intervention) and then two groups with only the posttest (with and without the intervention). An example may include examining the effect that physical rehabilitation has on 6 min walk tests where the researchers are concerned that the pretest may motivate some patients to increase their exercise outside of physical rehabilitation if they know their pretest scores which would influence the posttest results. Another example would be if knowledge of the pretest itself impacts the posttest results. Switched replication design is useful for studies with limited numbers of patients because it allowa patients to act as their own controls. In this design, the traditional design would enroll two groups with two phases with three measurement periods (before phase 1, after phase 1, and after phase 2). In the first phase, one group would receive the intervention and the other is a control, and then, during the next phase, the two groups switch so that the one receiving the intervention stops and the previous control starts receiving the intervention. This allows for monitoring after the intervention for adverse events and withdrawal effects and, most importantly, allows an individual to act as their own control.[3] As such, incorporating these designs can improve a hybrid study greatly (Fig. 55.1).

Effectiveness–implementation hybrid design methodology has grown in popularity. Currently, there are three hybrid design types based on which study is the primary concern: (type 1) studies evaluating intervention strategies primarily while also observing implementation information secondarily, (type 2) studies aiming to evaluate both, and (type 3) studies evaluating implementation strategies primarily and monitoring intervention effects secondarily. Moving from type 1 to type 3 sees a shift in focus from patient and clinical unit to system, provider, and facility.[4] An example of type 1 hybrid design would be the Coordinated Anxiety Learning and Management (CALM) study. This studied the effect of a combination intervention by anxiety clinical specialists (ACS) including flexibility so they would treat anxiety disorders, coexisting depression, and/or coexisting alcohol abuse, utilizing on-location ACS for the first assessment, utilizing computer-based psychotherapy programs, and an online status tracker related to progress in treating anxiety.[5] The CALM study was designed so it could be reproduced and implemented at additional sites as well. In addition, they also ran a parallel study where they interviewed practitioners and staff about possible barriers to care and factors that improved care for future implementation. The effectiveness study was performed on 1004 patients with controls for comparison and demonstrated a small to medium effect size, while the implementation study showed how it could have been performed more optimally in the future. Effectiveness in this study was heavily studied, while implementation evaluation was done afterward. Combinations like this allow for early implementation information to be derived while effectiveness is being tested. Studies like this can be performed once efficaciousness has been proven so that studies can move forward earlier (Fig. 55.2).

Type 2 effectiveness–implementation hybrid studies will place equal importance on both aspects. The Enhancing

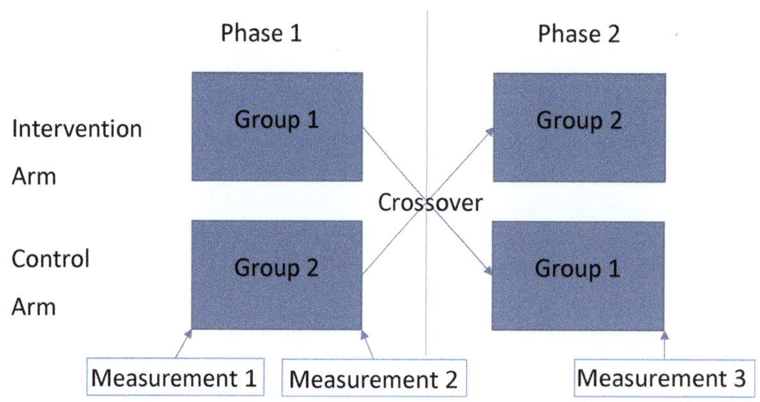

FIGURE 55.1 Visual schematic of switched replication design to demonstrate measurement period and the utilization of patients as their own controls; useful for certain hybrid studies.

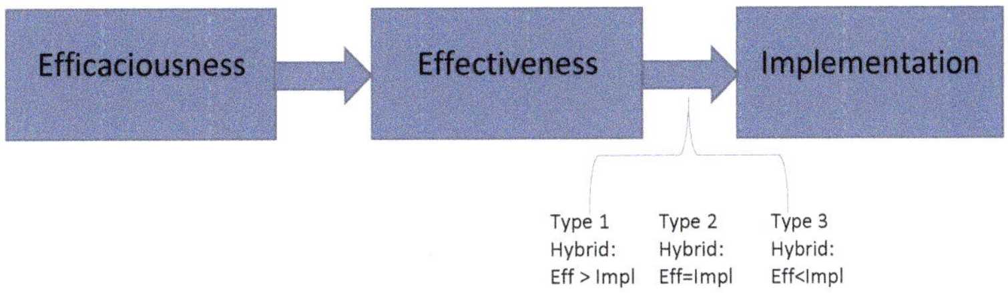

FIGURE 55.2 The intervention development process and where each type of effectiveness–implementation hybrid studies fit in; Eff = effectiveness, Impl = implementation.

Quality-of-care in Psychosis study developed a care model for schizophrenia patients while evaluating the implementation of said model in a small controlled trial. The effectiveness aspect evaluated the model while the implementation aspect examined how to facilitate its use in the clinic. In this case, there was attention to patient-level factors such as symptoms, adverse effects, and other clinical factors compared to usual care of schizophrenia to evaluate effectiveness. Implementation was continually studied and reimplemented using formative evaluations to continually find barriers and solutions for adaptation of the intervention. Because of this, the study could evaluate effectiveness under ideal circumstances at the end of the trial as well as pragmatically at the beginning of the trial. The focus on both aspects is what would make this a type 2 trial.[6]

Type 3 hybrid studies focus more on implementation compared to effectiveness. The Adaptive Implementation of Effective Programs Trial (ADEPT) is an example of type 3 hybrid studies. ADEPT evaluated the effect of sequential multiple assignment randomized trial design to determine which clinical locations adapt to the replicating effectiveness programs for implementing evidence-based medicine practices. The implementation aim was to determine how adding internal facilitators to ensure the implementation of evidence-based medicine practices affected patient's mental health-related quality of life and mood symptoms. The effectiveness aim was whether an internal facilitator affected the same patient care outcomes compared to sites without them. In this sense, this trial is not evaluating effectiveness as these are evidence-based practices. It is evaluating the effectiveness of the implementation measures though so this causes the hybrid nature of the design. The implementation aims are the main point of this trial, as a result. In this sense, this leans more toward implementation but does provide a small aspect of evaluating the effectiveness of the interventions.[7]

Hybrid studies in terms of personnel roles and budgetary concerns will be the same as the individual parts. As such, budgeting will be constrained and must be planned ahead of time as the budget may be split between the two arms.

Action items

1. Find a topic of interest for study and develop a potential hypothesis (such as if an intervention works or if a new measure of a disease is representative of disease activity).
2. Develop a primary hypothesis related to this activity (such as one related to effectiveness, superiority, noninferiority, exposure–disease development, etc.).
3. When examining the primary hypothesis, look into possible limitations or ways the original study could fail (such as in terms of failure to reject the null hypothesis or additional factors that could either affect the exposure–disease relationship or implementation factors that could be barriers to the effectiveness or time constraints to implementation or cost or recruitment difficulties).
4. Pick another type of study design that could compliment the primary goal or could help to account for shortcomings and limitations of the original study (such as performing a retrospective study to compliment the prospective exposure–disease development/improvement, examining implementation barriers, adding noninferiority arm, even adding efficaciousness arm to an effectiveness arm if necessary, etc.).
5. Determine how to properly allocate resources (patients' assignment, cost, intervention, time, need for controls) as well as potential factors that can affect the outcome (pretest knowledge, intervention exposure, etc.) between the two arms of the study to determine the need for additional complexity to the study design (such as Solomon four-group design, randomized block design, switched replication design, etc.).
6. Review the trial design up to this point to determine if the hypotheses are compatible, achievable, complimentary, and feasible with current resources. If so, attempt to write an IRB protocol for submission and a grant proposal for submission. If not, review and make adjustments.
7. Once outcomes have been collected, consider using the Fisher's combination test to achieve significance when aggregating data, if appropriate.

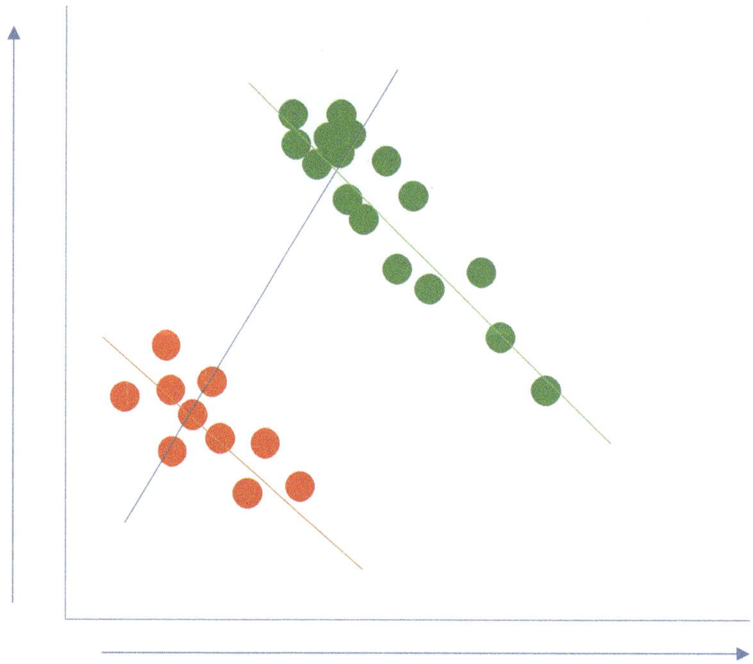

FIGURE 55.3 Simpson's paradox: When graphing each of the subgroups individually, their combined trends are the opposite of the aggregate trend.

Pitfalls

- The more different the individual aspects of the trial are, the more difficult it will be to conduct and collect data on; when developing the individual studies for the hybrid, make sure they provide complimentary goals, like running a superiority and noninferiority study together or effectiveness and implementation studies together.
- Effectiveness–implementation hybrid studies should have an appropriate understanding of the efficaciousness and some information about effectiveness and barriers to implementation to improve chances of success.
- Hybrid studies can run the risk of being underpowered in cases where subgroups are used to test various hypotheses related to running the intervention in various manners; one way of avoiding this is utilizing a switched replication design to allow patients to act as their own controls.
- When analyzing outcomes from the study, an important aspect to keep in mind during analysis is Simpson's paradox, which can directly hurt the conclusions of the study unless they are explicitly addressed; this is especially a problem for superiority–noninferiority studies; see Fig. 55.3 for visual depiction.

References

1. Soon GG, Nie L, Hammerstrom T, Zeng W, Chu H. Meeting the demand for more sophisticated study designs. A proposal for a new type of clinical trial: the hybrid design. *BMJ Open*; January 1, 2011. https://www.ncbi.nlm.nih.gov/pmc/articles/PMC3191591/.
2. Page M, Walker HL. Design and operational considerations for hybrid retrospective-prospective studies. *Value Health*; 2016. Retrieved March 6, 2022, from https://www.medpace.com/wp-content/uploads/2016/12/Design-and-Operational-Considerations-for-Hybrid-Retrospective-Prospective-Studies.pdf.
3. Bill Pelz HCCC. *Research Methods for the Social Sciences*: Lumen. Retrieved March 6, 2022, from: https://courses.lumenlearning.com/atd-herkimer-researchmethodsforsocialscience/chapter/chapter-10-experimental-research/.
4. Curran GM, Bauer M, Mittman B, Pyne JM, Stetler C. Effectiveness-implementation hybrid designs: combining elements of clinical effectiveness and implementation research to enhance public health impact. *Med Care*; March 2012. https://www.ncbi.nlm.nih.gov/pmc/articles/PMC3731143/.
5. Curran GM, Sullivan G, Mendel P, et al. Implementation of the calm intervention for anxiety disorders: a qualitative study. *Implement Sci*; March 9, 2012. https://www.ncbi.nlm.nih.gov/pmc/articles/PMC3319426/.
6. Brown AH, Cohen AN, Chinman MJ, Kessler C, Young AS. Equip: implementing chronic care principles and applying formative evaluation methods to improve care for schizophrenia: queri Series. *Implement Sci*; February 15, 2008. https://www.ncbi.nlm.nih.gov/pmc/articles/PMC2278162/.
7. Kilbourne AM, Almirall D, Eisenberg D, et al. Protocol: adaptive implementation of effective programs trial (ADEPT): cluster randomized smart trial comparing a standard versus enhanced implementation strategy to improve outcomes of a mood disorders program. *Implement Sci*; September 30, 2014. https://www.ncbi.nlm.nih.gov/pmc/articles/PMC4189548/.

Chapter 56

Large, pragmatic trials

Zachariah B. Nealy
Cardiovascular Division, Department of Medicine, University of Virginia Health, Charlottesville, VA, United States

Key points

- Large pragmatic trials utilize flexible implementation in usual care settings to mirror real-world scenarios with the goal of assessing effectiveness in noncontrolled settings
- The PRECIS-2 tool is an effective way of assessing the necessary features to determine the pragmatism of a study (i.e., how much the study mirrors real-world conditions)
- Because of the limitations, larger size studies in multiple areas and repeat trials in various locations may be necessary to truly answer the question of intervention effectiveness

Benefits of large pragmatic trials include maximizing a therapy's effectiveness and establishing causality assuming efficaciousness was proven already. Other benefits include simple structuring and real-world conditions that include uneven compliance, various comorbidities included, and variable practitioner experience levels and preferences.[3] Less stringent inclusion and exclusion criteria allow opportunities for older and more complex patients that otherwise would be excluded. As such, pragmatic trials fall near the later phases of translational studies (Fig. 56.1).

Pragmatic trials are inherently limited by lower internal validity in exchange for higher external validity.[4] Smaller single-location trials can be confounded by heterogeneity from the trial setting (i.e., treatments for city A may not be applicable to city B's population). Increasing sample size and implementing cluster randomization help reduce this problem.[1] As pragmatic trials are less controlled, the placebo effect and Hawthorne effect can occur with less concrete outcomes. Additionally, establishing direct causality is difficult without prior explanatory trials.[5] Fewer follow-ups can limit measuring effect for certain interventions so easily obtainable outcomes or short study length may be necessary. Thus, careful study design is important for success with pragmatic trials.

PRECIS-2 is a validated scoring system for evaluating trial pragmatism.[6] Its implementation is subjective but has good interrater reliability. The criteria and their description are as follows:

- Eligibility criteria: As few inclusion and exclusion criteria as necessary to study a realistic setting regardless of risk, perceived responsiveness, expected compliance or comorbidity.
- Intervention flexibility: Addresses freedom in implementation by provider and patient.
- Practitioner expertise: Minimal for ease of implementation.
- Comparison interventions: Usual practice or best alternative.
- Comparison intervention practitioner expertise: Minimal to allow a wide range of practitioners.
- Follow-up intensity: Minimally different from usual care.
- Primary trial outcomes: Concrete and clinically meaningful.
- Patient compliance and practitioner adherence:No or few additional measures for compliance.
- Analysis: Includes all patients regardless of compliance or eligibility; ideally, concrete metrics to reduce the need for adjustment to the prior criteria and can be done with extraction from electronic medical records or surveys if easily obtainable.[7]

Examples of well-designed pragmatic trials include Post-MI Free Rx Event and Economic Evaluation (MI FREEE), Randomized Controlled Trial of Supervised and Directly Observed Treatment of Tuberculosis (DOT), and Thrombus Aspiration during ST-Segment Elevation Myocardial Infarction (TASTE) trial. MI FREEE examines the effect that full insurance coverage of beta-blockers, ACE inhibitors, ARBs, and statins had compared to standard coverage on the rate of repeat vascular events and hospitalizations. They included patients with recent myocardial infarction and excluded those with health savings accounts and Medicare coverage. Patients were

Translational Cardiology. https://doi.org/10.1016/B978-0-323-91790-2.00007-1

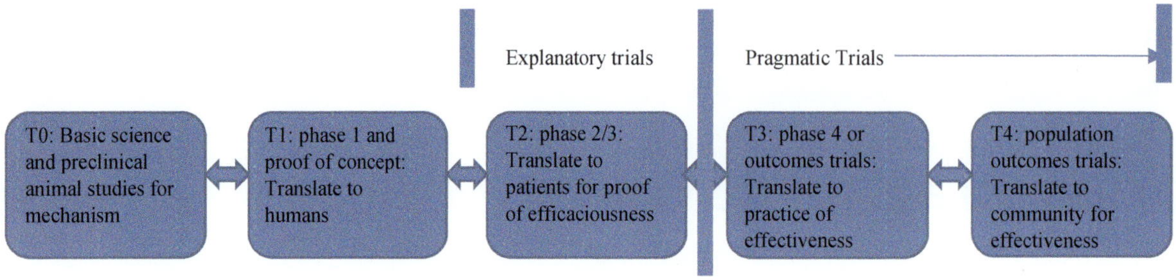

FIGURE 56.1 Classification of phases of translational research and where pragmatic trials fit in.

randomized in clusters at the insurance provider level, which eliminates the need for informed consent. Prescribed medications (including those covered), follow-up, provider expertise, and compliance were provider and patients determined. Primary outcomes were concrete. The analysis included all patients and data was extracted from the insurance company claims database minimizing provider and patient burden. MI FREE is an ideal pragmatic design (Table 56.1).[8]

DOT evaluated a clinic-level intervention with direct patient observation for taking prescriptions compared to home supervision to determine the effect on treatment completion and infection resolution. They included newly diagnosed tuberculosis cases receiving a new treatment and excluded drug-resistant tuberculosis cases, severe tuberculosis cases, and anyone who was unable to be supervised. This trial's eligibility criteria were less pragmatic by necessity due to the inability to perform the intervention on some patients. Allowing patients' choice in observation days and for surrogate medication collection allowed strong intervention flexibility. Furthermore, they included patients nonadherent to the study protocol in the analysis under a different group. Practitioner expertise, intervention expertise, follow-up intensity, and treatment comparison were unchanged from usual care. Adjusting adherence was the goal of the study, but no additional measures, like contacting the patient outside of these times, were included. Overall, the study is highly pragmatic.[9]

The TASTE trial was a hospital-level study that evaluated if thrombectomy affected 30-day mortality in STEMI patients compared to usual care. Eligible patients had recent STEMIs. Patients were recruited from a Swedish national coronary angiography and angioplasty registry and were randomized at enrollment. The study excluded patients requiring bypass grafting for their STEMI treatment, which limited pragmatism mildly. The study design minimized the need for follow-up. Practitioner expertise was high given it utilized interventional cardiologists, but this was standard care for STEMIs so no additional training was required. The comparison intervention was standard care. The remainder of hospital, pharmaceutical, and clinic care was delivered flexibly by additional providers. Patient compliance was untampered and practitioners were discouraged from crossing patients over but were still included in the final analysis. The national registry as a part of their nationalized healthcare system allowed easy monitoring of the primary outcome and data extraction. Though some aspects were less pragmatic, overall, it was more pragmatic than explanatory.[10]

Personnel preparation and training for pragmatic research is simpler as the conditions mirror real-world scenarios. The usual care team and the setting are sufficient, and the optimal team would be that found in a resource-limited family medicine clinic. Stakeholders include patients, practitioners, the community, and the health system. Outcomes should be easily analyzed by the real-world practitioner, though statisticians and information specialists may aid in analysis and data extraction, respectively.[2] NIH provides opportunities for funding larger-scale pragmatic trials with a preference to cluster randomized trials, as does PCORI. Grant conditions are stringent and limited to questions of major public health importance. Preference is for well-characterized interventions in randomized controlled designs delivered within the existing healthcare framework. They prefer the collection of outcome data using electronic medical records and favor high-volume centers.[11] Overall, the landscape for pragmatic trials provides an opportunity for translational researchers.

Getting started

1. Look for clinical questions based on either previously proven, efficacious interventions or existing ideas with robust designs.
2. Utilize PRECIS-2 when coming up with methods and designs as this will help later on during analysis and publication. This may involve liberalizing eligibility and intervention flexibility. The goal of the trial is that it may be performed in real-world settings, so when attempting to determine a measure of outcomes, make sure your metric assumes minimal follow-up, expertise, and adherence for chances of success.

TABLE 56.1 Brief summary of pragmatic trials in reference to PRECIS-2 criteria.

Trial	Eligibility criteria	Intervention flexibility	Practitioner expertise	Comparison intervention	Intervention expertise	Follow-up	Outcomes	Compliance/ adherence	Analysis
MI FREEE	Recent MI and participating insurance plan; no exclusion	Entirely up to provider	No additional training. Comparable to standard of care	Full insurance coverage of Rx versus standard coverage	No additional expertise or training	No changes to existing practices	Hospitalization and vascular events filed to insurance; concrete and easily obtainable from electronic health records	No additional measures	Included all patients; partitioned by original randomization
DOT	New Tb; excluded multidrug resistant, severe disease, and those unable to be observed	Timing of when to attain medicine is determined by the patient with the option for family to attain medicine as long as they're within the boundaries defined by the trial	No additional training. Comparable to standard of care	Direct observation compared to family observation at home (clinic standard)	No additional expertise or training	Follow-up mildly increased for trial patients but could opt out and be included in the analysis	Treatment completion and infection resolution determined by smear; concrete and based on lab work	Increased significantly but this was the intervention. Otherwise no additional measures	All patients included and partitioned into a different group
TASTE	Recent STEMI recruited from Swedish registry; excluded patients receiving CABG	Up to provider but they were discouraged from crossing patients over between intervention and control group	Interventional cardiologists but otherwise standard care team for STEMIs	Aspiration thrombectomy versus standard of care	No additional training or expertise	Only followed for 30 days and no additional follow-up	30-day mortality; concrete and obtainable from registry	No additional measures	All patients included; crossed over if necessary

3. Search through the electronic medical record or regularly collected clinic metrics already in existence (such as labs, vital signs, or other process flow measurements) to find concrete outcomes that will be easy to obtain.
4. If, in designing the trial, the measurements may bias a subgroup later on, make a note of this for analysis later. Selection bias is important to consider so all possible patients to represent the population should be included, but subgroup analysis may identify a clinically significant affected population as well.
5. Review the trial design up to this point to see if it still answers the question posed forth. If so, attempt putting together an IRB submission and begin trial once a determination is made. If not, then review and make adjustments as needed.

Pitfalls

- Pragmatic trial design is not permission for the Laissez-faire method development as it is an appropriate balance between permissive attitudes toward the study and inability to make conclusions about the results.
- Consent is still necessary when able in these types of studies and will be needed for publication in certain cases.
- Outcomes should ideally be concrete and measurable in a manner that can be repeated in alternative settings for their replication of the trial.
- The applicability of the results may be on a smaller and more local scale than larger explanatory trials but may be still externally valid in that area. As such, the larger the sample size, the more likely the study is to be applicable to other populations or even inspire trials in other locations.
- Cluster randomization can improve patient recruitment but is also prone to selection bias and covariate imbalance if sites drop.
- Causality can be established without prior explanatory trials but requires robust study design and usually subgroup analysis to account for confounders.

Additional resources

For cluster randomization, including some of its limitations and ways to overcome these: Nancy Haff MD. Improving Health Care Quality. JAMA Network Open. https://jamanetwork.com/journals/jamanetworkopen/fullarticle/2709710. Published October 26, 2018. Accessed January 23, 2022.

For access of PRECIS-2's online tool: https://www.precis-2.org.

For NIH funding opportunities applications for pragmatic trials: https://www.nia.nih.gov/research/blog/2021/09/pragmatic-clinical-trials-update-nia-funded-real-world-research-improve.

For PCORI funding opportunities applications for pragmatic trials: https://www.pcori.org/funding-opportunities/announcement/pragmatic-clinical-studies-cycle-2-2021.

References

1. Patsopoulos NA. A pragmatic view on pragmatic trials. *Dialogues Clin Neurosci*; 2011. https://www.ncbi.nlm.nih.gov/pmc/articles/PMC3181997/. Accessed January 23, 2022.
2. Holtrop JS, Glasgow RE. *Pragmatic Research: An Introduction for Clinical Practitioners*. OUP Academic; 2020. https://academic.oup.com/fampra/article/37/3/424/5875565. Accessed January 23, 2022. Published July 23, 2020.
3. Pawson R. Pragmatic trials and implementation science: grounds for divorce? *BMC Med Res Methodol*; 2019. https://bmcmedresmethodol.biomedcentral.com/articles/10.1186/s12874-019-0814-9. Accessed January 23, 2022.
4. Ford I, Drain PK, et al. Pragmatic trials. *N Engl J Med*; 2016. https://www.nejm.org/doi/full/10.1056/NEJMra1510059. Accessed January 23, 2022.
5. Ernst E, Canter PH. Limitations of "pragmatic" trials. *Postgrad Med*; 2005. https://pmj.bmj.com/content/81/954/203. Accessed January 23, 2022.
6. Loudon K, Zwarenstein M, Sullivan FM, Donnan PT, Gágyor I, Hobbelen HJSM, Althabe F, Krishnan JA, Treweek S. The precis-2 tool has good interrater reliability and modest discriminant validity. *J Clin Epidemiol*; 2017. https://pubmed.ncbi.nlm.nih.gov/28603007/#:~:text=Conclusion%3A%20We%20have%20assessed%20the,%2C%20systematic%20reviewers%2C%20and%20funders. Accessed January 23, 2022.
7. Thorpe KE, Zwarenstein M, Oxman AD, et al. A pragmatic–explanatory continuum indicator summary (PRECIS): a tool to help trial designers. *J Clin Epidemiol*; 2009. https://www.sciencedirect.com/science/article/abs/pii/S0895435609000481. Accessed January 23, 2022.
8. Choudhry NK, Al E, et al. Full coverage for preventive medications after myocardial infarction. *N Engl J Med*; 2011. https://www.nejm.org/doi/full/10.1056/nejmsa1107913. Accessed January 24, 2022.
9. Zwarenstein M, Schoeman JH, Vundule C, Lombard CJ, Tatley M. Randomised controlled trial of self-supervised and directly observed treatment of tuberculosis. *Lancet*; 1998. https://www.sciencedirect.com/science/article/pii/S0140673698040227?casa_token=DqL6sa__RsQAAAAA%3A2MaWDIFljuXCONW8kDxDSvrD289KiT-l0RmRMhwP7JJ6q04HawP69KuUixwg_iJdMx4CvC9. Accessed January 24, 2022.
10. Fröbert O, Al E, et al. Thrombus aspiration during st-segment elevation myocardial infarction. *N Engl J Med*; 2013. https://www.nejm.org/doi/full/10.1056/nejmoa1308789. Accessed January 24, 2022.
11. Pragmatic clinical trials. *NCCIH funding opportunities for RM-16-019*. National Center for Complementary and Integrative Health; 2017. https://www.nccih.nih.gov/news/events/pragmatic-clinical-trials-nccih-funding-opportunitiesfor-rm16-019. Accessed January 25, 2022.

Chapter 57

Equivalence and noninferiority trials: Design, measures, and classic example

Naira Link Woite[1], Anderson C. Woite Silva[2] and Bruno B. Lima[3]

[1]Harvard School of Public Health, Boston, MA, United States; [2]University of Sao Paulo, Sao Paulo, Brazil; [3]Vanderbilt University Medical Center, Nashville, TN, United States

Key points

- Randomized clinical trials (RCTs), the gold standard design used to implement new therapies into clinical practice, frequently need to compare the new intervention to the standard-of-care (SoC) treatment instead of to placebo due to ethical considerations.
- Equivalence trials aim to demonstrate that a new intervention is equivalent to another. They often use the SoC as a baseline comparison. To do so, an equivalence margin should be defined beforehand, if the treatment effect of the new treatment falls above or below this margin, it is considered not equivalent. The equivalence margin is always computed above and below the null value.
- Noninferiority trials are used to demonstrate that a new intervention is not inferior to another. Just like equivalence trials, this type of study tends to use the SoC as the comparison intervention. Here, however, a noninferiority margin is defined. This margin is only computed above or below the null value, depending on how efficacy is measured. A one-sided test can be used to demonstrate noninferiority.

Introductory section/why it matters

Randomized clinical trials (RCTs) are well accepted as the gold standard to compare the effect of therapies and therefore implement novel drugs or interventions into clinical practice. For many researchers, a placebo-controlled trial is the best model to estimate the effect of a new treatment. However, in many cases, using a placebo is not the most ethical choice, given that well-established standard-of-care therapies (SoC) already exist. As a result, nowadays the goal of RCTs, most often, is to demonstrate that the new treatment outperforms the SoC. However, superiority in the treatment effect may not be the sole purpose of a new treatment. Lower costs, easier adherence (due to a better route of administration or simpler dosing regimens), or less side effects are also important goals to be achieved, as long as the treatment effect can be maintained at a reasonable level. Equivalence and noninferiority trials are specially designed to address these issues. They seek to demonstrate that a new intervention is "as good as" or "not inferior" in effect to the SoC, which has been previously established.

In summary, equivalence and noninferiority studies, unlike conventional comparative studies which focus on the efficacy difference between two interventions, adapt traditional hypothesis testing methods, allowing a new treatment to be compared with a reference treatment without the need to demonstrate superiority. The design and statistical aspects of equivalence and noninferiority trials, illustrated with pertinent examples, will be explored in this chapter (Fig. 57.1).

Equivalence trials

Equivalence trials aim to demonstrate that the efficacy of two treatments is equivalent. It is a useful design to compare a generic drug to an original one, for example. To achieve this, the trial aims to establish that the difference in efficacy between both treatments is zero. However, this is virtually impossible given the variance of estimations. Therefore, this type of trial requires establishing an equivalence margin (M) both below and above zero. After defining the margin, the hypothesis testing is performed with the null and alternative hypotheses inverted when compared to the conventional superiority trials. In other words, the null hypothesis is that the effect of both

Translational Cardiology. https://doi.org/10.1016/B978-0-323-91790-2.00017-4

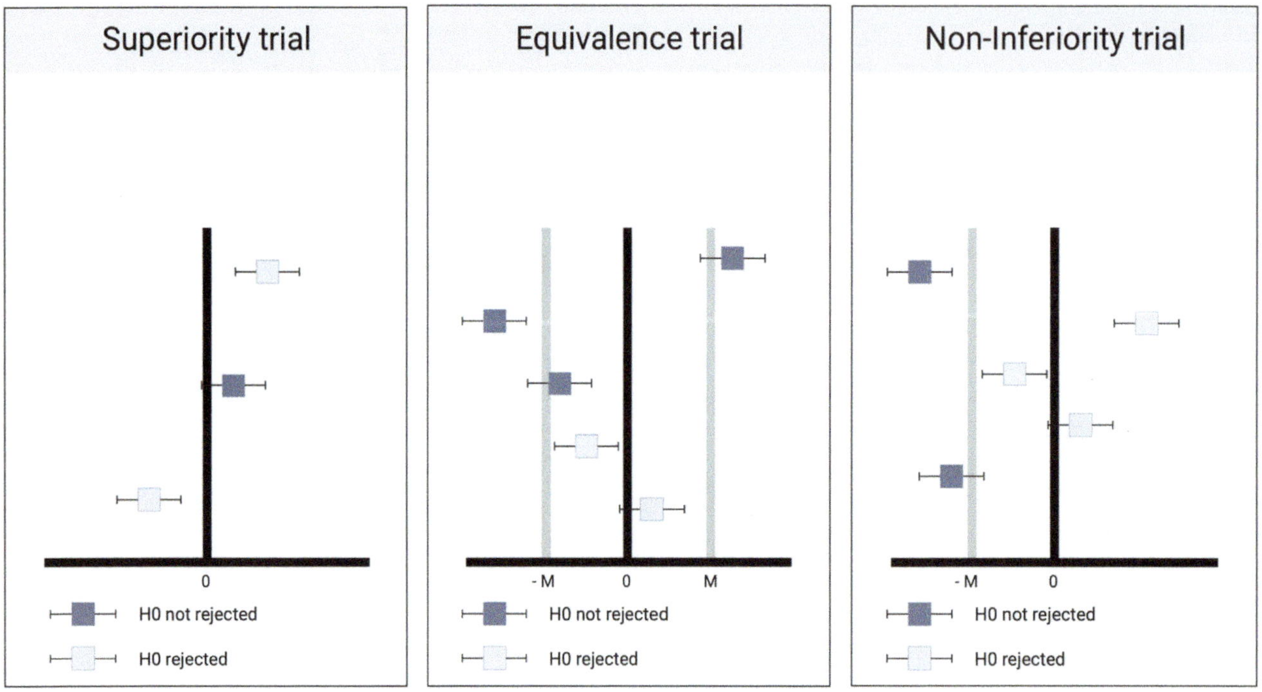

FIGURE 57.1 Comparison between superiority, equivalence, and noninferiority trials. M = equivalence or noninferiority margin. H0 = null hypothesis. *Adapted from Fregni, Felipe, Illigens Ben MW, Critical Thinking in Clinical Research: Applied Theory and Practice Using Case Studies. New York, NY: Oxford University Press; 2018.*

treatments is different (H0: A≠B), and the alternative hypothesis is that the effect of both treatments is the same (Ha: A = B).[1]

One of the greatest challenges is defining the M, which can be described as the maximum clinically acceptable difference in return for the secondary benefits of the new therapy. This margin should ideally be smaller than a clinically meaningful difference. On the other hand, an M that is too small may result in a sample size that is too large.[2]

Significance, benefits, and limitations

Equivalence trials, while perhaps less immediately appealing than traditional superiority trials, hold substantial significance in clinical research. These trials play an important role in establishing that a new therapy is as effective as the existing standard while offering other benefits beyond efficacy, like lower costs and fewer side effects. They ensure patients receive optimal care without compromising on treatment quality. Moreover, guarantee that a new treatment is also not superior, which may result in increased toxicity of a compound. Overall, equivalence trials are crucial for validating new treatments that are not only innovative but also reliably effective.

However, these trials also face limitations, including potential changes in the standard of care that could make equivalence demonstrations irrelevant, and "biocreep" (see below), a phenomenon where a series of trials gradually diminish efficacy standards. Ensuring statistical rigor in this context is also challenging. The lack of superiority requirements can lead to less strict analysis. For example, if the margin is too large, researchers could incorrectly conclude equivalence, when the new treatment is actually inferior.

Statistical considerations

In equivalence trials, both an "intention to treat" (ITT) and a "per protocol" (PP) analysis are crucial to ensure study validity. In these studies, the ITT approach is not considered a conservative approach, such as in superiority trials, because it tends to increase the chance of a type I error by making the treatments appear more similar. On the other hand, per-protocol analysis can lead to broader confidence intervals due to the reduced number of participants.

Another important consideration is the effect of confounders in these trials. As confounding variables tend to make the novel therapy and the standard therapy to seem more similar, they may cause the researcher to incorrectly conclude equivalence.[3]

Noninferiority

Noninferiority trials are designed to show that a new treatment is at least as good as and not worse than another one. It allows for a small, nonclinically significant difference—the noninferiority margin (M). Just like equivalence trials, other advantages may justify the implementation of the new drug or intervention, even if it is necessary to sacrifice a fraction of the effect of the SoC. In this study design, we perform a one-sided hypothesis under the null hypothesis that the treatment difference exceeds M. The alternative hypothesis, therefore, states that the treatment difference is less than M, potentially indicating the new treatment's noninferiority or even superiority, though superiority is not the primary aim.

In this design, M must be carefully defined and can be interpreted as how much the control can outperform the new treatment without the latter being considered inferior. Interpreting results of noninferiority trials can be complex, as the same treatment effect difference could be deemed noninferior or inferior based on the study's design. The essence of understanding it is to focus on the distinction between statistical significance in superiority trials and clinical relevance in noninferiority trials (e.g., a statistically significant difference may not be clinically relevant) (Fig. 57.2).[4]

Significance, benefits, and limitations

Noninferiority trials are significant in clinical research for similar reasons as equivalence trials: they allow the introduction of treatments that are not inferior to standard care but with other potential benefits like reduced side effects, cost, or better dosing schedule. However, in contrast to equivalence trials, this design also leaves room for drug effect improvement, allowing for superiority demonstration. Moreover, they balance the need for innovation with the assurance of treatment effectiveness. The primary benefit of such trials is the ethical and practical feasibility of comparing a new treatment with an active control rather than a placebo, especially when withholding standard therapy would be unethical.

However, they also have limitations like the risk of "biocreep" (see below), assay sensitivity (see below), and the necessity to establish clinical equipoise, which means that a true ambivalence about the efficacy of the new therapy exists in the research community.[5] Additionally, defining an appropriate noninferiority margin is complex and crucial, as too large a margin could accept a significantly less effective treatment, whereas too small could make the trial unfeasible.

Statistical considerations

Noninferiority trials are usually believed to require larger sample sizes than superiority trials. However, it will strongly depend on the inferiority margin (M)'s magnitude. A smaller M necessitates a larger sample size to maintain the power of the study. Factors like data stratification, which can reduce variance, and the choice between ITT and PP analyses also influence sample size requirements. Just like equivalence trials, these trials must be powered

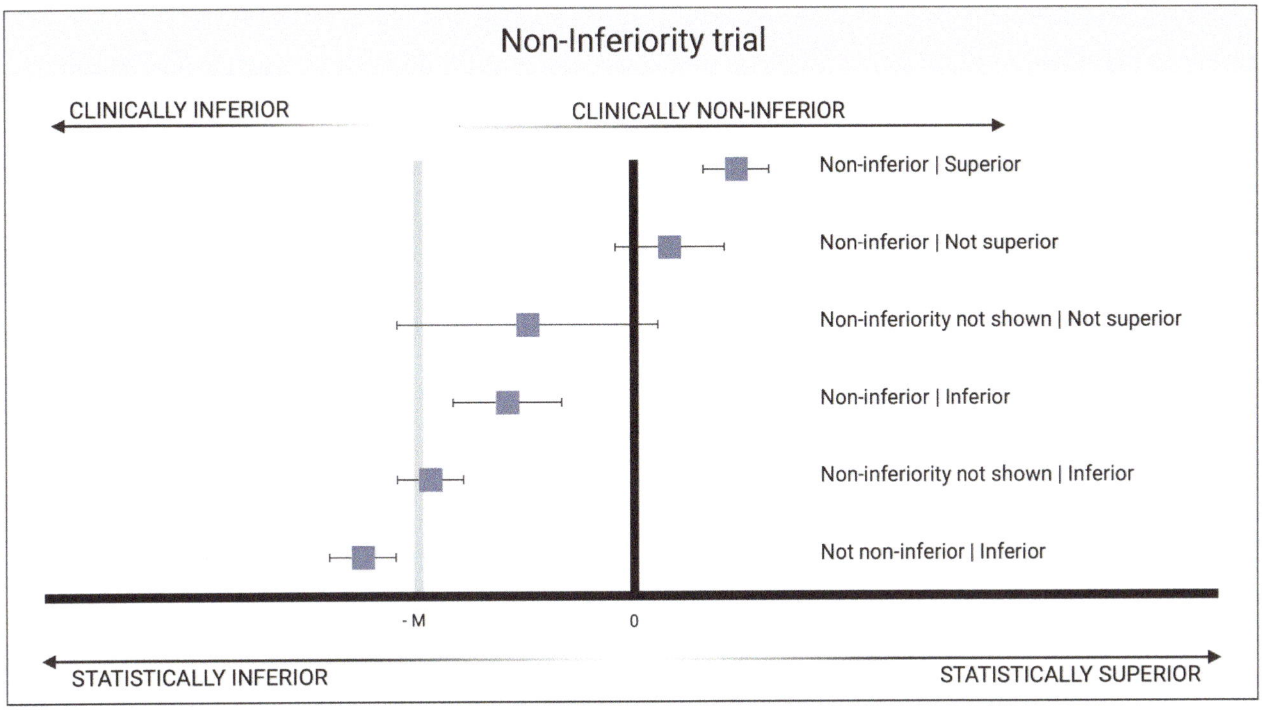

FIGURE 57.2 Difference between clinical relevance and statistical significance. *Adapted from Fregni, Felipe, Illigens Ben MW, Critical Thinking in Clinical Research: Applied Theory and Practice Using Case Studies. New York, NY: Oxford University Press; 2018.*

adequately for both ITT and PP to ensure robust and reliable results.[4]

Here another important consideration is the importance of a careful evaluation of the study design when facing studies described as noninferiority trials. Some trials described as noninferiority may, in fact, be reinterpreted as superiority trials that failed to demonstrate an advantage for the new therapy.[6] Proper definition of the noninferiority margin, sample size calculations, and detailed reporting of the confidence intervals are essential for the validity of noninferiority trials.

Differences between equivalence and noninferiority

Equivalence and noninferiority study designs are similar in many aspects, as both compare a new treatment to an existing standard without aiming to demonstrate superiority. However, they have different objectives and implications. Equivalence trials aim to show that the new treatment's efficacy is statistically similar to the standard, which is crucial when the new treatment offers other benefits, such as improved safety or convenience. Therefore, a therapy that is rendered "equivalent" should be clinically indistinguishable from the standard therapy, even though they are not exactly the same. In contrast, noninferiority trials seek to demonstrate that the new treatment is not significantly worse than the standard, allowing for some degree of reduced efficacy in exchange for other advantages, while also leaving room for demonstrating that a treatment is better. These distinctions affect the choice of the noninferiority or equivalence margin, study power, and the interpretation of results (Table 57.1).[7]

Biocreep

Biocreep is defined as the cumulative decline in effectiveness that can occur when a series of new drugs, each proven noninferior to its predecessor rather than a placebo. This may result in a new standard of care that is actually not superior to a placebo. For instance, if drug A is established as effective against a placebo and subsequently, drug B is shown to be noninferior to drug A based on fewer side effects, drug B may become the new standard. Later, if drug C, with an easier dosing regimen, is tested and found noninferior to drug B, it might become the standard treatment. This sequence can lead to a new standard (drug C) that is potentially not superior to the original placebo because it was never directly compared to it.[8] This phenomenon, biocreep, highlights a potential gradual reduction in treatment efficacy over time through successive noninferiority trials.

Assay sensitivity

Assay sensitivity is defined as the trial's ability to distinguish between an effective therapy and one that is not effective. Its absence can lead to incorrect conclusions. In superiority trials, the lack of assay sensitivity will probably lead to a negative result, as the trial will not be able to show a statistically significant benefit over control. Therefore, they inherently confirm assay sensitivity. On the other

TABLE 57.1 Equivalence and noninferiority studies: null and alternative hypotheses, error types, and error.

	Superiority study	Equivalence study	Noninferiority study
Desired outcome	New treatment is better than old treatment or placebo	Both treatments are similar	New treatment is not unacceptably worse than old treatment
Null hypothesis (H_0)	Difference between treatments = 0	Difference between treatments ≠ 0	Difference between treatments ≤ −M
Alternate hypothesis (Ha)	Difference between treatments ≠ 0	Difference between treatments = 0	Difference between treatments > −M
Type 1 error (alpha)	Affirming both treatments are different when no difference exists	Affirming both treatments are equivalent when a difference exists	Affirming the new treatment is noninferior when it is inferior
Type 2 error (beta = 1-power)	Affirming both treatments are equivalent when a difference exists	Affirming both treatments are different when no difference exists	Affirming the new treatment is inferior when it is noninferior
95% CI	Must not include 0	Must lie between −M and +M	Lower limit must lie above −M
Type of analysis	ITT preferable	Both ITT and PP should be performed	Both ITT and PP should be performed

Adapted from Walker J. Non-inferiority statistics and equivalence studies. *BJA Educ.* 2019;19(8):267–71. Fregni, Felipe, Illigens Ben MW, Critical Thinking in Clinical Research: Applied Theory and Practice Using Case Studies, New York, NY: Oxford University Press; 2018.

hand, a noninferiority trial that lacks assay sensitivity may wrongly generate a positive result. This can happen in two ways (1): if it fails to detect real differences between the new treatment and the active control, or (2) if it wrongly claims a treatment is noninferior when it is ineffective compared to a placebo. This concern can be addressed by including a placebo arm when ethically permissible, allowing for the assessment of both noninferiority to the active control and superiority over placebo.[9]

Constancy assumption

The constancy assumption is an important premise in noninferiority trials. It relies on the idea that the active control's effectiveness, previously proven superior to placebo, remains unchanged over time. However, in real-world scenarios, this premise can be challenged by factors like resistance to antibiotics or advancements in medical practice.

A possible way to uphold this assumption is to demonstrate that the active control would still outperform a placebo in the current trial context. Again, a three-arm trial with a placebo could directly verify this if ethical considerations allow it. If this approach is not possible, closely mirroring the methodology of earlier trials that established the active control's superiority, such as selection criteria, blinding, randomization, treatment regimens, and measurement techniques, can help mitigate potential discrepancies.[10]

Budget considerations

The budget for equivalence and noninferiority trials is similar to that of standard superiority studies. Governmental agencies or intramural grants are common funding sources for investigator-initiated or academically interesting trials, while pharmaceutical or biotechnological companies often finance sponsor-initiated trials.

Getting started with an equivalence or noninferiority study

Recognizing a knowledge gap, asking the research question, and deciding on the study design

Recognizing a knowledge gap and formulating the research question are the first and one of the most important steps in all research. In this case, it is important to carefully rule out a superiority design and establish why a noninferiority or equivalence design may be adequate. What are the advantages provided by the new intervention? Is there a necessity to aim specifically for equivalence (e.g., the new intervention is neither better nor worse than the SOC)? It is important to point out here that these trials are versatile and not limited to drugs or procedures; they can address a wide range of research topics, such as duration of treatment.

Choosing the active control

In these designs, the active control should be carefully chosen. It must be a SoC with strong evidence of efficacy, supported by robust data. If these assumptions do not hold, a trial design may be infeasible. Challenges arise if the standard of care shows only a small effect or if the efficacy estimate lacks precision. It is crucial for researchers to ensure that the chosen active control would be proven superior to placebo in the current study. Therefore, investigators need to identify the most suitable active control, keeping in mind that regulatory approval alone does not mean that it will be appropriate for the trial.

Choosing the noninferiority/equivalence margin

Selecting the noninferiority/equivalence margin (M) is one of the most important aspects of these trial designs. This step relies heavily on the historical performance of the chosen active control against a placebo. The M will directly impact study design, sample size, statistical analysis, and interpretation. The first consideration is that the margin must not exceed the entire effect of the active control over placebo (M1), considering historical data and variability in estimates. Ensuring that the constancy assumption holds is essential; otherwise, a new drug might be wrongly deemed noninferior.

Establishing that the new treatment is not inferior to M1 assures that its effect is greater than zero but does not allow us to say that the drug has a clinically meaningful effect. Therefore, the second consideration involves choosing a margin smaller than M1 (M2), to preserve some efficacy of the active control. M2 represents the maximum acceptable loss of effect. Noninferiority is established if the new drug's effect is above −M2, and equivalence is established if the drug's effect is between −M2 and +M2. Strategies for determining M2 include clinical judgment and fixed margin approaches, each with its own advantages. Check the Resources section for more information.

Sample size calculation, statistical analysis, and missing data

In these trials, sample size determination is crucial and is directly influenced by the equivalence margin. Smaller margins require larger sample sizes. It is also important to consider that ITT and PP analyses have distinct implications in these trials. ITT is generally more conservative in superiority studies but can be considered less conservative in equivalence/noninferiority trials. Conversely, PP analysis includes only those who followed the protocol and does not consider patients who dropped out of the analysis, which might be related to the treatment. Equivalence/noninferiority is most clearly established when both ITT and PP analyses concur. Discrepancies suggest exclusion bias, weakening study conclusions unless adequately explained.[11]

Finally, while in superiority trials, missing data is associated with type II errors due to decreased sample size and increased variance, in equivalence or noninferiority trials, this effect may make treatments look more similar and increase type I error.

Potential pitfalls

- Design the study as an equivalence or noninferiority trial from the start, rather than converting a failed superiority trial.
- Ensure adequate sample size, accounting for potential dropouts, using available sample size calculators.
- Perform both intent-to-treat and per-protocol analyses for more credible results.
- Remove confounding variables to prevent incorrect conclusions about equivalence or noninferiority.

Real-world examples

Example 1—Equivalence trial

The Facilitated Angioplasty with Tirofiban or Abciximab study[12] was a randomized controlled multicenter open-label equivalence trial. It aimed to test the equivalence of high-dose bolus tirofiban versus abciximab during primary percutaneous coronary intervention (PPCI) in terms of ST-segment resolution (STR). The trial enrolled 692 patients with ST-segment elevation myocardial infarction, randomized 1:1 to receive either abciximab or high-dose bolus tirofiban. The primary endpoint was the rate of complete ($\geq$70%) STR 90 min after the first balloon inflation.

The sample size was calculated based on achieving a power of approximately 0.80 to establish equivalence for a prespecified equivalence margin. The methodology included both intention-to-treat and per-protocol analyses to strengthen the validity of the findings. However, the study failed in showing the equivalence of high-dose bolus tirofiban and abciximab as adjunctive therapy to PPCI. It demonstrated the challenges and complexities involved in designing and conducting equivalence trials, particularly in terms of selecting appropriate margins and managing statistical analysis to accurately interpret the data. The findings highlight the need for careful consideration in trial design, especially in choosing the noninferiority/equivalence margins and ensuring adequate statistical power.

Example 2—Noninferiority trial

The "Rivaroxaban versus Warfarin in Nonvalvular Atrial Fibrillation" study[13] was a randomized, double-blind noninferiority trial comparing rivaroxaban to warfarin for stroke prevention in atrial fibrillation patients. Participants were at moderate-to-high risk for stroke, and the primary efficacy endpoint was a composite of stroke and systemic embolism. Sample size calculations aimed for adequate power to establish noninferiority with a prespecified margin. Analyses included per-protocol, safety-as-treated, and intention-to-treat populations. The methodology incorporated both noninferiority and superiority testing, with careful attention to patient adherence and treatment discontinuation, illustrating the complexities of designing and interpreting noninferiority trials.

Example 3—Noninferiority trial

The "Cryoballoon or Radiofrequency Ablation for Paroxysmal Atrial Fibrillation" study[14] was a multicenter, randomized, noninferiority trial comparing cryoballoon ablation to radiofrequency ablation. It aimed to determine if cryoballoon ablation was noninferior to radiofrequency ablation in treating drug-refractory paroxysmal atrial fibrillation. Key methodology included a randomized assignment, stratification by center and age, and a mix of efficacy and safety endpoints. Statistical analysis focused on time-to-event data, employing log-rank tests and Cox proportional-hazards models. The trial also incorporated both modified intention-to-treat and per-protocol cohorts to validate its findings.

Resources

- Fregni, Felipe, and Illigens Ben M W. Critical Thinking in Clinical Research: Applied Theory and Practice Using Case Studies, Oxford University Press, New York, NY, 2018. Chapter 15
- US Department of Health and Human Services; Food and Drug Administration; Center for Drug Evaluation and Research (CDER); Center for Biologics Evaluation and Research (CBER). Guidance for industry non-inferiority clinical trials 2010.
- Althunian TA, de Boer A, Groenwold RHH, Klungel OH. Defining the non-inferiority margin and analyzing non-inferiority: An overview. Br J Clin Pharmacol. 2017 Aug; 83(8):1636–42.

References

1. Walker E, Nowacki AS. Understanding equivalence and noninferiority testing. *J Gen Intern Med*. February 2011;26(2):192–196.

2. Walker J. Non-inferiority statistics and equivalence studies. *BJA Educ*. August 2019;19(8):267−271.
3. US Department of Health and Human Services; Food and Drug Administration; Center for Drug Evaluation and Research (CDER); Center for Biologics Evaluation and Research (CBER). *Guidance for Industry Non-inferiority Clinical Trials*. 2010.
4. Schumi J, Wittes JT. Through the looking glass: understanding non-inferiority. *Trials*. May 3, 2011;12:106.
5. Freedman B. Equipoise and the ethics of clinical research. *N Engl J Med*. July 16, 1987;317(3):141−145.
6. Le Henanff A, Giraudeau B, Baron G, Ravaud P. Quality of reporting of non-inferiority and equivalence randomized trials. *JAMA*. March 8, 2006;295(10):1147−1151.
7. Christensen E. Methodology of superiority vs. equivalence trials and non-inferiority trials. *J Hepatol*. May 2007;46(5):947−954.
8. Everson-Stewart S, Emerson SS. Bio-creep in non-inferiority clinical trials. *Stat Med*. November 30, 2010;29(27):2769−2780.
9. Hey SP, Weijer C. Assay sensitivity and the epistemic contexts of clinical trials. *Perspect Biol Med*. 2013 Winter;56(1):1−17.
10. Koopmeiners JS, Hobbs BP. Detecting and accounting for violations of the constancy assumption in non-inferiority clinical trials. *Stat Methods Med Res*. May 2018;27(5):1547−1558.
11. Althunian TA, de Boer A, Groenwold RHH, Klungel OH. Defining the non-inferiority margin and analysing non-inferiority: an overview. *Br J Clin Pharmacol*. August 2017;83(8):1636−1642.
12. Marzocchi A, Manari A, Piovaccari G, et al. Randomized comparison between tirofiban and abciximab to promote complete ST-resolution in primary angioplasty: results of the facilitated angioplasty with tirofiban or abciximab (FATA) in ST-elevation myocardial infarction trial. *Eur Heart J*. December 2008;29(24):2972−2980.
13. Patel MR, Mahaffey KW, Garg J, et al. Rivaroxaban versus warfarin in nonvalvular atrial fibrillation. *N Engl J Med*. September 8, 2011;365(10):883−891.
14. Kuck KH, Brugada J, Fürnkranz A, et al. Cryoballoon or radiofrequency ablation for paroxysmal atrial fibrillation. *N Engl J Med*. June 9, 2016;374(23):2235−2245.

Chapter 58

Adaptive designs

Guilherme Leite B. Gonçalves[1], Júlia Lemos Lima Verde[1] and Bruno B. Lima[2]

[1]*Federal University of Ceará School of Medicine, Fortaleza, Ceará, Brazil;* [2]*Vanderbilt University School of Medicine, Nashville, TN, United States*

Key points

- Adaptive trials are more efficient because they allow personalization of the treatment, answer faster, and change whenever it is necessary (prespecified and guided by a predefined protocol).
- Adaptive clinical trial designs allow the study to be continued.
- It is ethical, as it enables, for example, inopportune treatments to be stopped or more effective treatments to be inserted.

Introduction

To encourage the development of new drugs, the FDA introduced a set of incentives in 2006, among which adaptive design was included as a clinical research format. This approach enables the creation of research in a more flexible format, where therapies and tests, for instance, can be modified throughout the course, allowing for a faster pace of research progress. For example, upon observing that a particular medication is more effective for a certain age group or gender, or even in combination with another drug, the initial study protocol can be customized and edited accordingly. Moreover, it facilitates addressing multiple research questions simultaneously, unlike traditional methods.

Despite traditional clinical trials boasting a robust history, they adhere to rigid protocols that often hinder adjustments encountered during the project's progression, thereby diminishing their efficacy.

Get started

Given all the possibilities offered by adaptive studies, it represents a necessary innovation in the conduct of clinical research. The significance of these clinical trials encompasses efficiency in resource utilization, real-time adaptability, patient-centric focus, accelerated drug development, employment of robust statistical methods, and ethical considerations.

Efficiency in resource utilization

The ability to modify aspects such as sample size and treatments based on stored and observed data allows for better management of time, resources, and involved participants.

Real-time responsiveness

It is known that new discoveries emerge daily. In light of this, adaptive studies become even more useful as their capacity to adapt in real-time allows for the adjustment of parameters in response to scientific advancements. This enables the refinement of protocols and optimization, even in more invasive processes.

Enhanced patient-centric approach

The adaptive nature of these trials allows for a more patient-centric approach. Tailoring treatments based on interim results, such as identifying specific patient subgroups that benefit most from therapy, contributes to personalized medicine and the improvement of patient outcomes.

Accelerated drug development

The traditional development of medications is often slowed down. In the methodology of adaptive design, due to the ease of altering protocols, this development can be accelerated. This is of utmost importance in health crises, such as the COVID-19 pandemic.

Optimized statistical rigor

Despite being characterized by greater flexibility compared to traditional studies, this methodology is anchored in robust statistical models to maintain the validity and integrity of the research.

Translational Cardiology. https://doi.org/10.1016/B978-0-323-91790-2.00038-1

Ethical considerations

This model enables a greater enhancement of ethical practices, as the ability to modify protocols in response to discoveries ensures increased protection of patients, thus prioritizing the well-being and safety of participants in the respective study.

In summary, adaptive clinical trials represent a significant improvement in clinical research as they provide a format that allows for the incorporation of new data from other studies. The increasing popularity of this approach is a clear indicator of the importance of having a flexible, patient-centric model with ethical robustness. The gradual adoption of this model signifies greater accessibility to scientific advances for the general population.

Subtypes of adaptive design trials

There are several models of adaptive trial designs; some of the most common ones are as follows:

- **Adaptive randomization:**
  It allows the adjustment of the randomization ratio of participants to different treatment arms based on current results. This allows for the allocation of more participants to the treatment showing greater efficacy. Usually useful when there is a need to allocate more participants to the more promising treatment arm as the trial progresses.
- **Group sequential design:**
  This model incorporates interim analyses at predefined points during the trial to allow for early stopping, sample size adjustment, or modification of the study parameters based on emerging data. It can be used for long-term trials, making it possible to flexibilize and increase efficiency in decision-making.
- **Adaptive dose-finding design:**
  This model is based on the possibility of adjustment of the dosage of the investigational drug based on interim safety and efficacy data. It can be useful to identify the optimal therapeutic dose. Usually found in early-phase clinical trials, aiming to determine the appropriate dosage is a primary objective.
- **Biomarker-adaptive design:**
  This method incorporates biomarker information to guide treatment assignments or modifications based on patient characteristics. If there is prior knowledge suggesting certain patient subgroups may respond differently to treatments, it is really helpful.
- **Sample size re-estimation:**
  It is a style that allows for a reassessment of the sample size based on interim results to ensure adequate statistical power. Often used when there is uncertainty about the anticipated effect size or event rate.
- **Adaptive seamless design:**
  It enables the integration of different phases of a clinical trial into a single adaptive study, streamlining the transition to confirmatory trials. Employed in studies that aim to optimize the efficiency of drug development, for example.
- **Umbrella and basket trials:**
  Umbrella trials test multiple treatments within a single disease type, while basket trials assess a single treatment across multiple disease types. Those models facilitate the evaluation of targeted therapies.
- **Adaptive enrichment design:**
  Modifies the patient population inclusion criteria to enrich the trial for certain subgroups more likely to benefit from the treatment. Useful for identifying and focusing on patient populations most likely to respond to a specific intervention.

It is worth emphasizing that these models can be combined and modified according to the needs of the study project. The selection of the format to be used will depend on the questions to be answered, the specific characteristics of the study, and the nature of the required interventions (Table 58.1).

TABLE 58.1 Subtypes of adaptive designs and the occasions in which they are mainly used.

Design	Mainly chosen ...
Adaptive randomization	... when there is a need to allocate more participants to the more promising treatment arm as the trial progresses.
Group sequential design	... for long-term trials, making it possible to flexibilize and increase efficiency in decision-making.
Adaptive dose-finding design	... in early-phase clinical trials aiming to determine the appropriate dosage is a primary objective.
Biomarker-adaptive design	... if there is prior knowledge suggesting certain patient subgroups may respond differently to treatments.
Sample size reestimation	... when there is uncertainty about the anticipated effect size or event rate.
Adaptive seamless design	... in studies that aim to optimize the efficiency of drug development.
Umbrella and basket trials	... when there is a need to facilitate evaluation of targeted therapies.
Adaptive enrichment design	... for identifying and focusing on patient populations most likely to respond to a specific intervention.

Potential pitfalls

Limited experience

Adaptive designs are newer concepts, and there might be limited experience available compared to traditional trial designs. This lack of long-term data can make it challenging to evaluate potential issues.

Inflated type I error

The conduct of multiple interim analyses increases the risk of false positives, necessitating stringent statistical methods to mitigate such occurrences.

Operational challenges

Changes to the study protocol can modify various elements, ranging from the data collection method to the training of operators. This underscores the complexity involved in the utilization of adaptive designs.

Communication challenges

The use of adaptive design imposes some operational challenges, particularly regarding communication with the various stakeholders of the study, as every modification needs to be clearly communicated to all involved in the project.

Despite the challenges, this methodology has been gaining increasing popularity due to its potential to combine efficiency and flexibility. Nevertheless, it is crucial that all collaborators are well trained and aligned with the purpose and direction of the study to ensure its success and steer clear of potential pitfalls.

Examples of ...

1. I-SPY 2 TRIAL (Investigation of Serial Studies to Predict Your Therapeutic Response with Imaging and Molecular Analysis 2):
   - I-SPY 2 is a phase II, multicenter, adaptive clinical trial focused on evaluating the efficacy of novel agents in the neoadjuvant treatment of high-risk breast cancer. It aims to identify biomarker signatures associated with response to different treatments. The trial uses a Bayesian adaptive randomization design, allowing real-time adjustment of treatment assignment probabilities based on accumulating patient outcomes and biomarker data.
2. BATTLE Trial (Biomarker-integrated Approaches of Targeted Therapy for Lung Cancer Elimination):
   - BATTLE is a phase II/III clinical trial designed to identify effective targeted therapies for patients with advanced nonsmall cell lung cancer (NSCLC). It focuses on integrating biomarker analysis into the treatment decision-making process. BATTLE employed an adaptive randomization design, allowing for real-time adjustments to treatment arms based on interim analyses of biomarker data. The trial aimed to identify which targeted therapies were most effective for specific molecular subtypes of NSCLC.
3. FIRE-3 Trial (AIO KRK-0306):
   - The FIRE-3 trial is a phase III study comparing two different chemotherapy regimens in combination with cetuximab (an anti-EGFR monoclonal antibody) as first-line treatment for metastatic colorectal cancer. The trial aimed to determine which chemotherapy backbone provided superior efficacy in combination with cetuximab. FIRE-3 employed an adaptive design that allowed for the modification of the primary endpoint during the course of the trial. The primary endpoint was initially the overall response rate but was later changed to overall survival based on accumulating data.

Resources

Adaptive design in clinical trials offers a flexible approach that allows modifications to the trial procedures based on interim data analyses. Here are some key resources and considerations for implementing adaptive designs in clinical trials.

(1) FDA Guidance: The FDA provides comprehensive guidance on using adaptive designs for clinical trials. This document includes principles for designing, conducting, and reporting adaptive trials, as well as the types of information to submit for FDA evaluation. This guidance covers Bayesian adaptive designs and complex trials that rely on computer simulations (https://www.fda.gov/regulatory-information/search-fda-guidance-documents/adaptive-design-clinical-trials-drugs-and-biologics-guidance-industry).

(2) Design Considerations: When planning adaptive trials, it is crucial to consider the complexity involved. Key aspects include obtaining funding, setting up trial protocols, regulatory and ethical approvals, development of randomization systems, and conducting interim analyses. These steps often require more detailed planning and resources compared to traditional fixed designs (https://www.bmj.com/content/360/bmj.k698).

Further reading

1. Barker AD, Sigman CC, Kelloff GJ, et al. I-SPY 2: an adaptive breast cancer trial design in the setting of neoadjuvant chemotherapy. *Clin Pharmacol Ther.* 2009;86(1):97–100. https://doi.org/10.1038/clpt.2009.68.

2. Kim ES, Herbst RS, Wistuba II, et al. The BATTLE trial: personalizing therapy for lung cancer. *Cancer Discov.* 2011;1(1):44–53. https://doi.org/10.1158/2159-8274.CD-10-0010.
3. Heinemann V, von Weikersthal LF, Decker T, et al. FOLFIRI plus cetuximab versus FOLFIRI plus bevacizumab as first-line treatment for patients with metastatic colorectal cancer (FIRE-3): a randomized, open-label, phase 3 trial. *Lancet Oncol.* 2014;15(10):1065–1075. https://doi.org/10.1016/S1470-2045(14)70330-4.
4. Mahajan R, Gupta K. Adaptive design clinical trials: methodology, challenges and prospect. *Indian J Pharmacol.* 2010;42(4):201–207. https://doi.org/10.4103/0253-7613.68417.

Chapter 59

Blinding: Who? and how?

Zaid Al Jebaje

Department of Medicine, Division of Cardiology, University of Rochester Medical Center, Rochester, NY, United States

Introduction

Blinding was first introduced in the 1950s, and it is estimated that 60% of clinical trials are labeled as "blinded."[3,4] The effect of blinding has been well studied; it is estimated that the odds ratio of having a significant intervention effect between double-blinded trials and open-label ones reaches 17%.[5] Therefore researchers are encouraged to explore blinding in their projects as much as possible. Also, readers may consider looking for that particular aspect in trial design when evaluating published literature.

To understand blinding, we need to touch on another concept: randomization; the two are often applied together, which might create some confusion regarding their purpose; randomization is also considered essential for blinding, so it is rare to find a double-blinded trial without randomization.[6]

While randomization tries to combat selection bias in the early recruitment phases of a study, blinding aims to overcome ascertainment and performance bias, which tend to occur at later stages of a trial and can affect the analysis of a research outcome drastically.

Another concept to mention is allocation concealment, which indicates that the person or process involved in randomization is not aware or not involved in the following stages of the study.

Who should be blinded in a research study?

Optimally, a researcher would want to blind as many individuals in their study as possible to overcome differences in intervention or assessment of outcomes. We can divide specific groups to target in your research: the participants or study subjects, the intervention group (usually clinicians), the data abstraction or collection group (also known as assessors or observers), and finally, the analysts or statisticians. If a study subject was aware of whether they were receiving an intervention or, for example, a placebo, their adherence and reporting of outcomes or side effects may differ; the same applies to clinicians who might direct their attention to the treatment group creating ascertainment bias. Data collectors or analysts might treat the treatment group differently, subjecting the study to performance bias.

Most large multicenter RCTs are reviewed periodically by an independent group called the "Data and Safety Monitoring Board" or DSMB; their role is to provide recommendations regarding the study design and the safety of the study subjects. Whether DSMB should be blinded or not is still a controversy, but most trialists prefer unblinding this establishment as their input may determine if the trial is to be continued or terminated.

When only one group is subjected to blinding, the study is labeled "single-blinded," when two groups (usually the study subjects and the providers) are included in the blinding process. The analysis is termed "double-blinded." The latter model is considered superior because it eliminates conscious and unconscious observer bias. To ensure a higher level of accuracy and objectivity of clinical data, a research team might choose to triple-blind their study, extending the blinding process to data analysts. This model is less often adopted as it is complex and usually not applicable in large population studies.

How do we blind?

Both the feasibility and success of a blinding technique should be assessed by the research team throughout the study. The method depends on the studied population and the type of intervention, and it ranges from using a placebo in medical research to sham operations in surgical trials. The research team must be careful that the blinding does not confuse data analysts when assessing outcomes. In some situations, blinding becomes difficult or ethically unfeasible, as in cases where psychotherapy is the studied intervention. Hence, trialists should look for creative

Translational Cardiology. https://doi.org/10.1016/B978-0-323-91790-2.00018-6

methods to overcome bias or standardize their research design to avoid the challenges of no blinding.

In conclusion, blinding is an essential methodological feature in any trial design. Researchers should consider blinding as many groups as possible to improve the study's validity and overcome bias. If blinding is not feasible or unethical, other techniques should be followed to solve issues resulting from the lack of blinding.

References

1. Karanicolas PJ, Farrokhyar F, Bhandari M. Practical tips for surgical research: blinding: who, what, when, why, how? *Can J Surg*. 2010;53(5):345–348.
2. Moustgaard H, et al. Impact of blinding on estimated treatment effects in randomized clinical trials: meta-epidemiological study. *BMJ*. 2020;368:l6802.
3. Ortiz V, Ellenberg SS, Weinberg EM. Blinding in clinical trials for chronic liver diseases. *Semin Liver Dis*. 2021;41(2):163–171.
4. Devereaux PJ, et al. An observational study found that authors of randomized controlled trials frequently use concealment of randomization and blinding, despite the failure to report these methods. *J Clin Epidemiol*. 2004;57(12):1232–1236.
5. Schulz KF, et al. Empirical evidence of bias. Dimensions of methodological quality associated with estimates of treatment effects in controlled trials. *JAMA*. 1995;273(5):408–412.
6. Schulz KF, Grimes DA. Blinding in randomized trials: hiding who got what. *Lancet*. 2002;359(9307):696–700.

Chapter 60

Phases of clinical trials

Pratik Agrawal
Division of Cardiology, WVU Heart & Vascular Institute, Morgantown, WV, United States

Phase 0

This is a preclinical phase in which animal testing or in vitro studies are conducted following new drug discovery. The primary purpose of Phase 0 is to evaluate the safety of new treatments before use in human subjects. Drug pharmacodynamics and pharmacokinetic characteristics are recorded in this phase. Upon completion of preclinical testing, a drug sponsor can file for an Investigational New Drug application to propose human testing in a clinical trial.

Phase I

After a new drug is approved for human testing in a clinical trial, Phase I of clinical trials can be commenced to establish the safety and dosages of the new drug or treatment by studying a small number of human subjects. This phase primarily examines drug tolerance and pharmacodynamic and pharmacokinetic properties such as maximally tolerated dose, bioavailability, and body compartment distribution in addition to allowing a preliminary assessment of drug activity. There are several approaches to estimate the maximally tolerated dose, namely the simple dose escalation method, rule-based designs with "step-up/step-down" approach (e.g., traditional $3+3$ design), and model-based designs with the Bayesian approach (e.g., continual reassessment method).[1–3] Furthermore, numerous variations have been proposed to these methods to study pharmaceuticals depending on the field of research.

Phase I typically involves 20–50 healthy human volunteers and is completed over a duration of a few months. Sometimes patient populations can also be studied in Phase I of clinical trials, for example, to study the safety of new cytotoxic drugs (chemotherapy agents) in patients with cancer. The main limitation of Phase I is the small number and homogenous nature of study participants. Results of the Phase I study are extrapolated to a heterogenous population to evaluate drug safety and advance the clinical trial to Phase II.

Phase II

This phase is a systematic investigation to assess the efficacy, or biological effect, of a new drug or treatment. A cohort of human participants is selected after careful enrollment based on strict inclusion and exclusion criteria. Since there is commonly an uncertainty regarding exact dose–response curves following Phase I studies, there can be several intervention arms studied in Phase II each consisting of different doses of the pharmaceutical agent within the maximally tolerated dose range. Drug efficacy can be compared by studying the pretreatment status and post-treatment status, or alternatively, through the comparison of treated humans with a concurrent or historical control group. There are various approaches to designing a Phase II trial, which can include a traditional two-staged design (where additional participants are enrolled only if a prespecified minimum level of biological activity is achieved in early participants), multiple-staged, or sequential staged design.[1]

Phase II typically involves around 50–250 human volunteers and is completed over a duration of a few months to ~2 years. Naturally, Phase II trials are not adequately powered to detect a significant difference in major clinical outcomes related to the use of new pharmaceutical agents. Hence, appraisal of the treatment effect in Phase II has to rely on many surrogate endpoints such as relevant lab tests and biomarker levels, changes in blood pressure, improvement in related symptom assessment scales, and so on. One of the main purposes of the Phase II trial is to understand whether further research and development of the new drug is suitable or feasible. If the new drug or treatment satisfies the efficacy measures in Phase II,

Translational Cardiology. https://doi.org/10.1016/B978-0-323-91790-2.00046-0

the study moves forward to a Phase III trial in which a larger cohort of human participants is enrolled.

Phase III

As outlined previously, the Phase III trial is designed to overall assess the clinical value of a new drug or treatment by studying the efficacy and adverse events in a larger human population. Generally, the exclusion criteria are not as strict as in Phase II trials to allow for a larger and more heterogeneous study population, which can simulate real-world clinical outcomes data to a considerable degree.

Phase III typically involves a few hundred to a few thousand human participants and is completed over a period of a few years or more. The most common type of Phase III trial is a comparative efficacy trial where a study drug or treatment is compared to standard therapy or placebo. Examples of randomized controlled trial designs include parallel-group studies, crossover studies, factorial designs, etc. Randomized controlled trials can also be classified according to study hypothesis, for example, superiority, equivalency, or noninferiority trials. Proper randomization techniques, blinding, stratification, control of confounders, and/or other biases between treatment groups are crucial measures to enhance the quality of a Phase III trial. Statistical analysis is most commonly and preferably performed via "intention-to-treat" analysis, while "per-protocol" analysis can be complementary to the findings. The *Consolidated Standards of Reporting Trials guideline* was established to improve trial reporting and is widely adopted by medical journals and relevant societal institutions.[4]

If the studied drug is proven to be safe and efficacious in the intended population, the drug sponsor can file a New Drug Application, following which the FDA conducts an extensive review of study data and supportive evidence to decide if the drug can be approved and marketed for specific indications. Although data obtained from Phase I and Phase II studies provide the foundation, the essence of a high-quality clinical trial ultimately leading to potential drug approval depends on the robustness of Phase III design and study protocol. Guidance regarding developing clinical trials protocols can be found in the document *SPRINT 2013 statement: Defining standard protocol items for clinical trials*, which serves as a practice guideline to aim for the completeness of trial protocols.[5]

Phase IV

Even though new drugs or treatments can be approved based on short- or intermediate-term outcomes, it is imperative to study the long-term adverse events, especially for drugs intended for chronic conditions, which is done in Phase IV and is often widely known as postmarketing surveillance. Phase III trials are statistically powered to detect an adverse event rate of approximately 1:100 to 1:1000 subjects (based on Hanley's Rule of 3),[6] while pharmaceutical agents once approved by the FDA can be potentially used by millions in the target population. This also underlines the importance of Phase IV trials to monitor for less common and rare adverse events and to establish the long-term safety profile of drugs following approval and marketing.

References

1. Friedman LM, Furberg CD, DeMets DL, Reboussin DM, Granger CB. *Fundamentals of Clinical Trials*. 5th ed. Springer International Publishing; 2015. https://doi.org/10.1007/978-3-319-18539-2.
2. Le Tourneau C, Lee JJ, Siu LL. Dose escalation methods in phase I cancer clinical trials. *J Natl Cancer Inst*. 2009;101(10):708–720. https://doi.org/10.1093/jnci/djp079.
3. Yin G, Yuan Y. Bayesian model averaging continual reassessment method in phase I clinical trials. *J Am Stat Assoc*. 2009;104(487):954–968. https://doi.org/10.1198/jasa.2009.ap08425.
4. Schulz KF, Altman DG, Moher D, CONSORT Group. CONSORT 2010 statement: updated guidelines for reporting parallel group randomised trials. *BMJ*. March 23, 2010;340:c332. https://doi.org/10.1136/bmj.c332.
5. Chan A-W, Tetzlaff JM, Altman DG, et al. SPIRIT 2013 Statement: defining standard protocol items for clinical trials. *Ann Intern Med*. 2013;158:200–207. https://doi.org/10.7326/0003-4819-158-3-201302050-00583.
6. Eypasch E, Lefering R, Kum CK, Troidl H. Probability of adverse events that have not yet occurred: a statistical reminder. *BMJ*. 1995;311(7005):619–620. https://doi.org/10.1136/bmj.311.7005.619.

Chapter 61

The IDEAL framework

William Plath[1,2], Mudathir Ibrahim[1,2], Joel Horovitz[2] and Peter McCulloch[1]

[1]*Department of General Surgery, Maimonides Medical Center, Brooklyn, NY, United States;* [2]*Nuffield Department of Surgical Sciences, University of Oxford, Oxford, United Kingdom*

Key points

- The IDEAL Framework offers a structured pathway for evaluating surgical operations, invasive medical devices, and complex therapeutic interventions, addressing unique challenges in surgical innovation.
- The IDEAL Framework takes the innovation from concept to long-term assessment through the following sequential stages: Pre-IDEAL, Idea, Development, Exploration, Assessment, and Long-term study.
- Recommendations specific to each stage of innovation are provided to guide researchers in conducting high-quality studies, thereby improving the evidence base for surgical innovation and ultimately enhancing patient care.

Introduction

The development and assessment of surgical and interventional techniques have historically followed a less regulated and more variable path when compared to the structured approach typically seen in drug development[1]—largely because it has been recognized that surgical innovation is more difficult to evaluate. While pharmaceutical innovation has long adhered to formalized stages and regulatory guidelines, surgical innovation is characterized by a different set of challenges and, until recently, no tailored evaluation methods[2] to address these existed. The journey from conceptualization to the widespread acceptance of a new surgical procedure or medical device represents a complex process that has called for innovative methodologies to match the intricacies of the surgical field.

One of the impetuses for a new methodology was the low esteem for surgical investigators which culminated in Horton's Lancet editorial in April of 2009—"Surgical research or comic opera; questions but few answers."[3] It was in recognition of these unique challenges that the Idea, Development, Exploration, Assessment, Long-term Study (IDEAL) Framework emerged.[4] Designed to address the specific complexities of surgical innovation, this framework offers a structured pathway for the evaluation of surgical operations, invasive medical devices, and other complex therapeutic interventions. Unlike pharmaceuticals, where the path to evaluation is relatively well established, surgical innovations have often grappled with suboptimal study designs, inadequate reporting, and a lack of consensus on standards for high-quality trials.[5]

The IDEAL Framework and Recommendations are the product of extensive collaboration and research. Originating from a series of conferences held at Balliol College, Oxford from 2007 to 2009, the IDEAL Collaboration brought together surgeons and methodologists to delve into the challenges of conducting high-quality trials in surgery. Unlike the abundant randomized controlled trials (RCTs) in pharmaceutical research, the scarcity of high-quality trials supporting surgical innovations highlighted the need for a tailored approach to evaluation. This unique framework acknowledges that surgical innovations vary significantly from their pharmaceutical counterpart and therefore could not conform to the existing evaluation pathway and guidelines. The Balliol participants agreed that a structured evaluation pathway for surgical innovation had to be based on an understanding of the life cycle of new procedures and devices. Modeling after the phases of clinical trials for pharmaceuticals, the IDEAL framework delineates the unique stages of surgical innovation, from the initial preclinical idea to the long-term surveillance studies (Table 61.1), identifies the key question that needs to be resolved at each stage, and provides recommendations on how to properly and safely evaluate the innovation before advancing to follow-up stages. The

Translational Cardiology. https://doi.org/10.1016/B978-0-323-91790-2.00006-X

TABLE 61.1 Summary of the IDEAL framework phases.

	Stage 0	Stage 1	Stage 2a	Stage 2b	Stage 3	Stage 4
	Pre-IDEAL	**Idea**	**Development**	**Exploration**	**Assessment**	**Long-term follow-up**
Design	Nonhuman studies	First-in-human	Small selected group	Multicenter cohort	Large multicenter study	Surveillance
Objective	Concept investigation	Proof of concept	Stabilization of innovation	Feasibility, safety, and clinical outcomes	Comparative evaluation	Late and rare events
Patients	None	Single to few	10s	100s	100s+	100s+
Surgeons and sites	Single	Single	Few	Multiple	Multiple	Multiple
Study type	Case report	Case report	Case series	Prospective and collaborative cohort studies	Pivotal trial (e.g., Randomized controlled trial)	Database and registries

principles on which evaluation recommendations are made are simple: evaluation must be consistent with the four pillars of medical ethics (beneficence, nonmaleficence, autonomy, and justice) and transparency on what was done—why and what the outcomes were should be maximized. The most rigorous and appropriate methods should be used to answer the key question at each stage.

As a result, the IDEAL Framework and Recommendations represent a pivotal shift in the evaluation of surgical operations, invasive medical devices, and complex therapeutic interventions. In a field characterized by unique challenges, this framework has the potential to enhance the evidence base, facilitate consensus among stakeholders, and ultimately improve the quality of care provided to patients.

The IDEAL Framework and Recommendations

(Table 61.2)

Pre-IDEAL (stage 0)—Exploration of early concepts

The pre-IDEAL stage involves essential research conducted before the first trial of the innovation in a human. The original version of IDEAL did not include this stage, which is why it is now termed stage 0. The nature, aims, and outcomes of these studies may vary greatly depending on the intervention type, and recommendations are therefore couched in very general terms based on the fundamental principles of IDEAL. Studies that may be appropriate at this stage include cadaver and animal studies, modeling, material testing, and simulations. Pre-IDEAL studies should focus on proving that the concept accomplishes the intended purpose and provide some understanding of the safety and reliability of the design. Any studies that can predict avoidable risks of harm to the first patient should be conducted. By the end of the pre-IDEAL stage, researchers should be able to identify predictable risks.

Recommendations for the pre-IDEAL stage[6]

- All predictable risks to patients should be investigated before human studies begin, using a precautionary approach based on a prior risk assessment.
- Guidelines on best scientific practices and ethics specific to the types of study should be followed where available.
- A minimum dataset describing technical consistency should be made public before first-in-human testing.

Real-world example of the pre-IDEAL stage

Design of a device for transanal endoscopic surgery without Pneumorectum using 3D printing. Pre-Ideal Phase[7]

- Aim: Analyze and validate the design evolution of a device for transanal endoscopic surgery without pneumorectum
- Design: Iterative design improvements of prototype performed on simulators with viscera and cadavers
- Findings: The device improved with changes in design, materials and mechanics

Idea (stage 1)—First-in-human use

The Idea stage marks the first use of a new procedure or device in a patient and serves as a proof of concept. The innovation at this stage may be planned following a pre-IDEAL stage, or as an emergency where preplanning is not feasible. The study includes a highly selective small patient pool (single to few) and is performed by very few or, in most cases, a single surgical innovator. Patient selection criteria must be explained, and outcomes typically include technical success, adverse events, and short-term clinical measures. Reporting in this stage is often in the form of case reports. The innovator should provide a detailed technical description of the innovation and clinical outcome, which ensures transparency regardless of clinical success. Once the idea stage confirms proof of concept, technical achievement, safety, and potential efficacy, the innovation is considered ready to progress to the next stage of evaluation.

Recommendations for stage 1 (idea)[6]

- Provide full details of patient selection, technique and outcomes, and patients not selected during the time frame, and why
- Use standard well-defined measures for reporting outcome and patient characteristics
- Use a structured reporting system, for example, SCARE checklist
- Make the above information available to peers regardless of outcome—*this is ethically important to avoid others from repeating mistaken interventions which harmed patients, but regulatory or legislative frameworks may be needed to protect those who conduct research ethically and report such interventions in good faith.*

Real-world example of stage 1 study

First clinical experience of intraoperative high-intensity focused ultrasound (HIFU) in patients with colorectal liver metastases: a phase I–IIa study[8]

- Aim: The value in treating colorectal liver metastases with HIFU has not been studied
- Design: Single center prospective study on six and nine patients (phase I and II respectively)
- Findings: HIFU was feasible, safe, and effective in ablating areas of the liver scheduled for resection

TABLE 61.2 IDEAL Framework Recommendations at each stage of the lifecycle.

	IDEAL Framework Recommendations
Stage 0: Pre-IDEAL	• All predictable risks to patients should be investigated before human studies begin • Guidelines on best scientific practice and ethics specific to the types of study should be followed where available • A minimum dataset describing technical consistency should be made public before first-in-human testing
Stage 1: Idea	• Provide full details of patient selection, technique and outcomes, and patients not selected during the time frame, and why • Use standard, well-defined measures for reporting outcomes and patient characteristics • Use a structured reporting system, e.g., SCARE checklist • Make the above information available to peers regardless of outcome
Stage 2a: Development	• Make protocol for study available • Use standard well-defined measures for reporting outcomes and patient characteristics • Report and explain all exclusions • Report all cases sequentially with annotation and explanation of when and why changes to indication or procedure took place • Display main outcomes graphically to illustrate the above

Stage 2b: Exploration	• Make protocol for study available
	• Use standard well-defined measures for reporting outcomes and patient characteristics
	• Participate in collaborative multicenter cooperative data collection, incorporating feasibility issues such as:
	  ◦ Estimating effect size,
	  ◦ Defining intervention quality standards,
	  ◦ Evaluating learning curves,
	  ◦ Exploring subgroup differences,
	  ◦ Eliciting key stakeholder values and preferences,
	  ◦ Analysis of adverse events
	• Preplanned consensus meeting before progressing to an RCT to identify feasibility and ability to recruit, intervention and comparator definitions, appropriate patient selection criteria, and primary endpoint.

Stage 3: Assessment	• Register on an appropriate international register
	• Use standard well-defined measures for reporting outcomes and patient characteristics
	• Incorporate information about patient and clinician values and preferences in consent information and outcome measure design
	• reporting guidelines:
	  ◦ CONSORT update of 2010 with extension for nonpharmacological treatments
	  ◦ COMET
	  ◦ TIDieR
	  ◦ SPIRIT (for RCT protocol design)

Stage 4: Long-term follow up	• Registries may begin from the earliest stages of human use • Registry datasets should be defined by the clinical community with patient input • Datasets should be simple, cheap, and easy to collect • Curation of registries by the clinical community is desirable • Funding of registries should be agreed between government and commercial interests but kept separate from curation • Consent for the use of registry data in research should be broad and where possible automatic

Development (stage 2a)—Stabilization of the technique

In the development stage, procedures undergo iterative modifications to reach a stable version. At this stage, the intervention is evolving, and any modifications should be clearly detailed with the rationale for the changes. The development stage typically involves a small number of patients (typically less than 30) and is performed by a selected group of surgical innovators, usually within a single institution. Study designs in this stage are usually single-center prospective case series. Outcomes at this stage should focus on safety considerations, technical success, and procedural consistency. Transparency is essential, especially when explaining why specific cases were rejected or when modifications to the technique occurred. This stage aims to develop a stable procedure for the subsequent stage.

Recommendations for stage 2a[6]

- Make protocol for study available
- Use standard, well-defined measures for reporting outcomes and patient characteristics
- Report and explain all exclusions
- Report all cases sequentially with annotation and explanation of when and why changes to indication or procedure took place
- Display main outcomes graphically to illustrate the above

Real-world example of stage 2a study

Perioperative and functional results for robot-assisted radical cystectomy (RASC) with totally intracorporeal neobladder in male patients via the Vesica Patavina (Ves.Pa.) Technique: IDEAL Stage 2a Report[9]

- Aim: Refine the novel Ves.Pa. technique of intracorporeal neobladder construction in a RARC
- Design: Single center prospective study of consecutive patients ($n = 20$) performed by a single surgeon
- Findings: RARC with the refined Ves.Pa. technique was found to be safe, feasible, and replicable with a reduced risk of complications and improved short-term functional outcomes

Exploration (stage 2b)—Bridge to a pivotal trial

The exploration stage focuses on gaining more experience with the innovation, in a broader group of surgeons and patients, with the ultimate goal of preparing for a future RCT or comparative study. The innovation should be more stable at this stage, with only minor variations. This stage involves more patients (typically in the hundreds) with less selective criteria and is performed by many surgical operators throughout multiple institutions. Multicenter prospective collaborative studies and feasibility RCTs are typical study designs that are aimed at achieving consensus between the operators. Outcome measures should focus on safety, clinical outcomes, and the feasibility of a future RCT. By the end of this stage, the study should demonstrate the technique is stable and all the prerequisites needed for a pivotal RCT should be complete.

Recommendations for stage 2b[6]

- Make protocol for study available
- Use standard well-defined measures for reporting outcome and patient characteristics
- Participate in collaborative multicenter cooperative data collection, incorporating feasibility issues such as:
  - estimating effect size
  - defining intervention quality standards
  - evaluating learning curves
  - exploring subgroup differences
  - eliciting key stakeholder values and preferences
  - analysis of adverse events
- Preplanned consensus meeting before progressing to an RCT to identify feasibility and ability to recruit, intervention and comparator definitions, appropriate patient selection criteria, and primary endpoint

Real-world example of stage 2b study

Endoscopic pilonidal sinus treatment (EPSiT): A prospective multicenter trial[10]

- Aim: Evaluation of (EPSiT for its effectiveness in treating pilonidal disease
- Design: Multicenter prospective study of 250 patients conducted at secondary and tertiary colorectal surgery centers
- Findings: The EPSiT procedure was found to be safe and effective, with improved short and long-term outcomes compared to more invasive procedures.

Assessment (stage 3)—Pivotal study/RCT

The assessment stage involves a definitive comparative evaluation, typically against the current standard treatment. The study usually includes a substantial number of patients (hundreds to thousands) with numerous surgical operators at multiple centers. The preferred study design is a large multicenter RCT; however, in some instances, appropriate alternatives may be considered. The intervention is compared with the current standard therapy, aiming to identify specific clinical outcomes. The endpoints of this stage should show evidence of the intervention's relative effectiveness and identification of aspects that will require long-term monitoring.

Recommendations for stage 3[6]

- Register on an appropriate international register (e.g., clinicaltrials.gov)
- Use standard well-defined measures for reporting outcome and patient characteristics
- Avoid the risk of investigator bias by excluding operators from data collection and from the informed consent process, using neutral blinded observers, trained nurses, and/or computer decision support programs—or having investigators undergo special training
- Incorporate information about patient and clinician values and preferences in consent information and outcome measure design
- Reporting guidelines:
  - CONSORT update of 2010 with extension for non-pharmacological treatments
  - COMET
  - TIDieR
  - SPIRIT (for RCT protocol design)

Real-world example of stage 3 study

Minimally invasive versus open distal pancreatectomy (LEOPARD): A multicenter patient-blinded RCT[11]

- Aim: Compare the time to functional recovery of minimally invasive versus open distal pancreatectomy (OPD) in patients with left-sided pancreatic tumors
- Design: Multicenter patient-blinded randomized controlled superiority trial of 111 patients at 14 centers
- Findings: Minimally invasive distal pancreatectomy reduces time to functional recovery compared with ODP

Long-term study (stage 4)—Identifying rare and later outcomes

The long-term study stage involves the use of continuous surveillance through registries and databases to identify late or rare safety outcomes, trends, and changes in procedure use. Registries are essential for device surveillance and life cycle assessments, and their introduction can begin as early as stage 1. Registries aim to ensure "real-world" outcomes are evaluated, promoting transparency, safety, and the equitable uptake of innovation. Outcome measures focus on long-term effects, rare events, and quality assurance.

Recommendations for stage 4[6]

- Registries may begin from the earliest stages of human use
- Registry datasets should be defined by the clinical community with patient input
- Datasets should be simple, cheap, and easy to collect
- Curation of registries by the clinical community is desirable
- Funding of registries should be agreed between government and commercial interests but kept separate from curation
- Consent for the use of registry data in research should be broad and where possible automatic

Real-world example of stage 4 study

Transesophageal echocardiography (TEE) and risk of respiratory failure in patients who had ischemic stroke or transient ischemic attack (TIA): An IDEAL phase 4 study[12]

- Aim: Determine if TEE was associated with an increased risk of respiratory failure in patients who had ischemic stroke or TIA
- Design: Multicenter retrospective cohort study of 99,081 patients, age >65 years old who were hospitalized for out-of-hospital stroke or TIA
- Findings: TEE was not associated with an increased risk of subsequent respiratory failure in the cohort of patients

Get started

- **Understand the stages:** Familiarize yourself with the IDEAL Framework's sequential stages—Pre-IDEAL (Stage 0), Idea (Stage 1), Development (Stage 2a), Exploration (Stage 2b), Assessment (Stage 3), and Long-term Study (Stage 4)—to recognize where your innovation or study fits.
- **Identify your stage:** Determine the current stage of your surgical innovation by assessing the following key factors:
  - Study type
  - Design structure
  - Numbers of patients
  - Objective of study

Potential pitfalls—How IDEAL may help avoid common mistakes in surgical evaluation

- Failure to conduct appropriate preclinical research before the first-in-human trial of an innovation. This exposes patients to preventable risk and unnecessary harm and may be considered unethical. It also defeats the purpose of meaningful innovation aimed to protect and extend lives.

- Failure to report an unsuccessful first-in-human trial of an innovation. This contributes to publication bias and may permit unintentional future trials of potentially harmful interventions by other proceduralists or preclude the modification and improvement of a potentially lifesaving intervention.
- Skipping IDEAL stages 2a and b and progressing directly to stage 3 because of successful outcomes at stage 1 may lead to inadequate evaluation and suboptimal evidence. Direct evaluation of surgical innovation in an RCT without initial stabilization, wider acceptance and agreement on critical definition and outcomes, and evaluation of proceduralist learning curve may result in failed or poor-quality and suboptimal RCT.
- Surgical devices are characteristically different from procedures. In that, they are subject to numerous iterations and modifications in the pre-IDEAL phase (stage 0) and are almost at maturity and well stabilized by the first-in-human IDEA phase (stage 1). Therefore, IDEAL stage 2a may be omitted or combined with stage 2b in the evaluation of surgical devices. However, the same is not true for surgical procedures.
- Forced evaluation of surgical innovation using established pharmaceutical evaluation pathways. There are characteristics inherent to surgical procedures and devices that pose challenges for evidence-based evaluation with the existing evaluation pathways. IDEAL remains the most complex framework available for evaluating surgical innovations.

Additional reading

1. McCulloch P, Altman DG, Campbell WB et al. No surgical innovation without evaluation: the IDEAL recommendations. *The Lancet.* 2009; 374(9695):1105−1112. doi:https://doi.org/10.1016/S0140-6736(09)61116-8
2. McCulloch P, Cook JA, Altman DG, Heneghan C, Diener MK. IDEAL framework for surgical innovation 1: the idea and development stages. *BMJ.* 2013; 346(jun18 3):f3012-f3012. doi:https://doi.org/10.1136/bmj.f3012
3. Ergina PL, Barkun JS, McCulloch P, Cook JA, Altman DG. IDEAL framework for surgical innovation 2: observational studies in the exploration and assessment stages. *BMJ.* 2013; 346(jun18 3):f3011-f3011. doi:https://doi.org/10.1136/bmj.f3011
4. Cook JA, McCulloch P, Blazeby JM, Beard DJ, Marinac-Dabic D, Sedrakyan A. IDEAL framework for surgical innovation 3: randomized controlled trials in the assessment stage and evaluations in the long term study stage. *BMJ.* 2013; 346(jun18 3):f2820-f2820. doi:https://doi.org/10.1136/bmj.f2820
5. Pennell CP, Hirst AD, Campbell WB et al. Practical guide to the Idea, Development and Exploration stages of the IDEAL Framework and Recommendations. *British Journal of Surgery.* 2016; 103(5):607−615. doi:https://doi.org/10.1002/bjs.10115
6. Hirst A, Philippou Y, Blazeby J et al. No Surgical Innovation Without Evaluation: Evolution and Further Development of the IDEAL Framework and Recommendations. *Annals of Surgery.* 2019; 269(2):211−220. doi:https://doi.org/10.1097/SLA.0000000000002794

References

1. McCulloch P. Randomised trials in surgery: problems and possible solutions. *BMJ.* 2002;324(7351):1448−1451. https://doi.org/10.1136/bmj.324.7351.1448.
2. Cook JA. The challenges faced in the design, conduct and analysis of surgical randomised controlled trials. *Trials.* 2009;10(1). https://doi.org/10.1186/1745-6215-10-9.
3. Horton R. Surgical research or comic opera: questions, but few answers. *Lancet.* 1996;347(9007):984−985.
4. McCulloch P, Altman DG, Campbell WB, et al. No surgical innovation without evaluation: the IDEAL recommendations. *Lancet.* 2009;374(9695):1105−1112.
5. Robinson NE, Fremes SE, Hameed I, et al. Characteristics of randomized clinical trials in surgery from 2008 to 2020. *JAMA Network.* 2021;4(6). https://doi.org/10.1001/jamanetworkopen.2021.14494. e2114494-e2114494.
6. Hirst A, Philippou Y, Blazeby J, et al. No surgical innovation without evaluation: evolution and further development of the IDEAL framework and recommendations. *Ann Surg.* 2019;269(2):211−220.
7. Rodríguez-García JI, Sierra-Velasco JM, García-Santos G, Villazon MM, Flórez L, Fernández-Rodrıguez M. Transanal endoscopic surgery with a 3D printed device. *Tech Coloproctol.* 2021;25(8):965−969.
8. Dupré A, Melodelima D, Pérol D, et al. First clinical experience of intra-operative high intensity focused ultrasound in patients with colorectal liver metastases: a phase I-IIa study. In: Lau WYJ, ed. *PloS One.* Vol. 10. 2015:e0118212. https://doi.org/10.1371/journal.pone.0118212.
9. Moro F, Fabio Z, Tonet E, Alessandro M, Betto G, Novara G. Perioperative and functional results for robot-assisted radical cystectomy with totally intracorporeal neobladder in Male patients via the vesica Patavina (Ves.Pa.) technique: IDEAL stage 2a report. *Eur Urol Open Sci.* 2023;57:8−15. https://doi.org/10.1016/j.euros.2023.09.001.
10. Meinero P, Stazi A, Carbone A, Fasolini F, Regusci L, La Torre M. Endoscopic pilonidal sinus treatment: a prospective multicentre trial. *Colorectal Dis.* 2016;18(5):O164−O170.
11. de Rooij T, van Hilst J, van Santvoort H, et al. Minimally invasive versus open distal pancreatectomy (LEOPARD). *Ann Surg.* 2019;269(1):2−9.
12. Bruce SS, Navi BB, Zhang C, et al. Transesophageal echocardiography and risk of respiratory failure in patients who had ischemic stroke or transient ischemic attack: an IDEAL phase 4 study. *BMJ Surg Interv Health Technol.* 2022;4(1). https://doi.org/10.1136/bmjsit-2021-000116. e000116-e000116.

Part VII

Clinical: Preparation

Zakaria Almuwaqet, Bruno Lima

Chapter 62

Patient perspectives

Luiz Campedelli[1] and Filipe A. Moura[2,3]

[1]*Division of Pulmonary, Allergy, Critical Care, and Sleep Medicine at the University of Pittsburgh, Pittsburgh, PA, United States;* [2]*Division of Cardiovascular Medicine, Brigham and Women's Hospital, Harvard Medical School, Boston, MA, United States;* [3]*TIMI Study Group, Division of Cardiovascular Medicine, Brigham and Women's Hospital, Harvard Medical School, Boston, MA, United States*

Key points

1. Patients provide unique insights that can lead to improvement in healthcare delivery and higher satisfaction with treatment outcomes.
2. It is crucial to acknowledge and incorporate participants' viewpoints in the development of clinical research, an effort that has gained increasing encouragement from policymakers and healthcare institutions.
3. Patient-reported outcomes (PROs) are a valuable method in clinical research to capture participants' outcomes related to symptoms, functional health, and well-being.
4. PROs can help inform clinical practice, health policy, and are used to support drug approval and pricing decisions.

Introduction

Historically, clinical trials have predominantly operated with minimal patient involvement in their development and execution. While research conducted without patient engagement may at first pass appear more convenient, it can lead to costs in the form of reduced recruitment and retention, heightened dissatisfaction, oversight of potential adverse drug events, diminished trust in healthcare providers, and the formulation of burdensome research queries that lack relevance for patients. Hence, as the ultimate beneficiaries of clinical research, patients provide unique insights that can lead to more effective interventions, improved healthcare delivery, and higher satisfaction with treatment outcomes.

Over the past few decades, policymakers and healthcare institutions, such as the US Food and Drug Administration (FDA) and the National Cancer Institute (NCI), have intensified efforts to encourage patient engagement in clinical research.[1,2] It is crucial to acknowledge their viewpoint and incorporate their voices throughout the entire clinical research process. In support of this, the FDA has also conducted disease-specific patient-focused drug development meetings, particularly in diseases with important unmet clinical therapy needs.

In this chapter, we will briefly explore key tools and strategies that have proven invaluable in amplifying patient input in research. We will also delve into successful case studies illustrating how to enhance patient engagement and challenges in its implementation in clinical research.

Understanding the patient's perspective

There are multiple reasons for a patient's participation in a study. They often navigate a multifaceted web of considerations that include disease burden, potential health benefits, safety concerns, financial cost or compensation, and overall study logistics. In the context of a clinical trial, a patient's perspective refers to the insights, experiences, and priorities they have for the research study. Capturing a patient perspective will require consulting with them, often by patient-reported outcomes (PROs), to ensure that the trial design, conduct, and outcome measures are relevant and meaningful to the patient population.

There are other elements of clinical trials that can be particularly confusing and appear to put patients at a disadvantage, such as the fact that a proportion of patients (i.e., those randomized to receive a placebo) will not receive the desired therapy they are seeking. In a questionnaire with cancer survivors, 57% of patients related that they would fear receiving a placebo if enrolled in a clinical trial and 61% reported feeling uncomfortable with the concept that their treatment assignment would be decided by chance.[3]

The financial aspect is another factor that might bring a burden on the patient and discourage retention.[4] Hence, it is important to carefully explain how costs related to the trial will be handled, to prevent loss of follow up.

Translational Cardiology. https://doi.org/10.1016/B978-0-323-91790-2.00026-5

Another challenge lies in the realm of medical terminology. A mere 12% of the American population possesses proficient health literacy.[5] This underscores the importance of the investigator's role in presenting crucial documents, such as informed consent, in a manner that is comprehensible to patients. It is essential for investigators to "translate" complex elements, particularly for those with limited educational backgrounds.

Patient-reported outcomes

PROs are measurable aspects of a patient's health status that are directly self-measured, without the influence of a clinician. It is comprised of different components that assess outcomes related to symptoms, functional health, psychological issues, and well-being. It has proven to be a valuable tool, capable of including the patient's perspective in clinical trials.

Its significance is amplified when considering that some studies suggest that up to 50% of symptomatic adverse effects go undetected by physicians in clinical trials.[6] This occurs because the threshold for a doctor to report a drug's adverse effects is often higher than that for a patient, potentially leading to an underestimation of the frequency and severity of adverse effects if the drug label relies solely on clinician reports.[6]

Considering this, the NCI has developed the Patient-Reported Outcome-Common Terminology Criteria for Adverse Events (PRO-CTCAE). It consists of an item library of 78 symptomatic toxicities derived from the CTCAE reporting system, which allows the investigator to select the symptomatic adverse events to assess the most important symptomatic toxicities for the therapy being studied. The patient, however, will have the ability to report toxicities not selected in the routine assessment through the write-in section of the PRO-CTCAE.[7] Of note, there are several ways in which the use of PROs can have shortcomings in trial design and reporting as well as extrapolation (see Table 62.1 for details).

Authors have previously identified several pathways in which PRO results may be impactful. These include informing clinical practice and guidelines, informing health policy, supporting drug approval and pricing decisions, supporting reimbursement decisions, informing clinical and shared decision-making, and informing consent for treatment.[8]

Newer PROs have proven to be very helpful and facilitate adhesion, such as the electronic PROs, which allow patients to answer the questionnaire via the computer, instead of via the traditional paper-based model. This not only saves healthcare teams time by eliminating the need to transfer data from paper to computer but also aligns with patients' preferences.[9]

Patient advocacy groups

The landscape of patient advocacy has undergone significant transformation over the years. Initially, in the 1930s, it primarily served as a source for patient education and fundraising for research. In the 1980s, notably during the HIV/AIDS crisis, patients began taking a more proactive role in research, self-educating about their conditions and challenging the system, particularly in response to concerns about the slow pace of drug approval.[10]

Today, advocacy plays a much bigger and more active role in all aspects of clinical research. This role can be synthesized into five main attributes of patient advocacy: (1)safeguarding, which aims to protect patients from healthcare error; (2)apprising, which aims to educate the patient about his disease and rights; (3)value, which secures the patient's free will and protects its privacy; (4)mediating, which makes the voice of the patient to be heard by the healthcare team; and(5) championing social justice in the provision of healthcare, which confronts policies, inequalities and other problems of the healthcare system.[11]

Patient advocates have become an indispensable part of the patient's collective voice, particularly in the realm of cancer research and cardiometabolic disease. Their contributions not only protect from potential research-related harm but also drive the healthcare industry to enhance the quality of research, benefiting both current and future patients.

Examples of successful patient engagement efforts

One of the biggest success examples regarding active patient participation in clinical research is the international group Outcome Measures in Rheumatology (OMERACT). OMERACT initially started as a rheumatology conference that was created with the goal of improving "the accuracy and responsiveness to change of clinically relevant (to patient and clinician) endpoints."[12]

Over time, OMERACT progressively increased patient participation in all areas. What began as contributions primarily centered around workshop discussions in its early meeting evolved into patients delivering plenary presentations, crafting consensus statements, and becoming coauthors of peer-reviewed publications a decade later.[13] This dynamic partnership between patients and physicians has produced remarkable outcomes, including the incorporation of fatigue as a core outcome measure in clinical trials for Rheumatoid arthritis, which was previously overlooked by healthcare professionals until patients actively contributed to the research process.[13] Besides the already mentioned benefits in the outcome core set area, OMERACT participants also listed other significant areas

TABLE 62.1 Patient-reported outcome (PRO) trials—barriers to impact across different trial stages.

Trial design and planning	Trial conduct and analysis	Trial reporting
Inadequate patient involvement in PRO selection	Low PRO compliance rates	Authors not using/citing guidelines to report PRO trials
Lack of clear PRO objectives	Lack of personnel training on administration of PRO instruments	Failure to report the a priori PRO hypothesis
Inappropriate choice of PRO instrument	Lack of communication between researchers and administrators	Failure to report baseline PRO compliance
Inadequate PRO instrument validation	Lack of standardization of the PRO questionnaire administration process	Failure to report rationale for the chosen PRO instrument
Insufficient consideration of the PRO timing	Lack of patient adherence to the PRO component of the study	Failure to report mode of administration of the PRO instrument
Poor integration of PROs into the clinical trial design		Failure to report timing of PRO assessment
Lack of PRO-specific hypotheses		Failure to report methods of PRO data collection
Failure to consider PRO data collection feasibility		Failure to report clinical significance of PRO findings
Lack of consideration for the PRO data analysis plan		Reporting levels of missing PRO data
Insufficient planning for PRO data management		Failure to report statistical methods dealing with missing PRO data
Inadequate budgeting for PRO measures		Failure to report generalizability of PRO trial results
Lack of training for study personnel on PRO measures		Selective reporting of PRO results
Insufficient consideration of cultural and language differences in PRO measures		Discrepancies between PRO protocol and PRO trial report
		Failure to report PRO data in the main trial publication
		Late publication of PRO trial results and in a different journal
		Journal word restrictions

PRO, patient-reported outcome.
This table provides a comprehensive overview of the various barriers that can occur at each stage of a patient-reported outcome trial, from design and planning through conduct and analysis, to reporting. The aim is to enable stakeholders to identify and address these barriers to improve the quality and impact of PRO research.
Adapted from Rivera SC, Kyte DG, Aiyegbusi OL, Slade AL, McMullan C, Calvert MJ. The impact of patient-reported outcome (PRO) data from clinical trials: a systematic review and critical analysis. *Health Qual Life Outcome.* 2019;17(1):156.

of improvement due to the patient-physician partnership, such as improvement in the research agenda, improvement in PROs, and improvement in communication and motivation.

In similar fashion, in the setting of a major unmet need in clinical trial science for nonalcoholic steatohepatitis, the Metabolic Multi Organ Science Accelerating Innovation in Clinical Trials is a global forum that is only in its second year (as of 2023) of bringing together cardiology, endocrinology, and hepatology clinical trialists, sponsors, regulatory experts, payers, and patient representatives to foster collaboration.[14] Patient representatives have provided valuable perspectives for the generation of future consensus statements and helping guide future changes in clinical trial design and regulatory guidance on approval for future NASH therapies (of which there currently are none approved).

Barriers to patient involvement in research

Despite past success in patient involvement with research and drug development, their role in clinical research is still quite limited and often fragmented. Patient participation in study design and conduct remains challenging to implement and is not widespread. In fact, a recent investigation found that only 7 of 150 clinical trial studies had patient representation, which occurred mostly only during the execution phase.[15]

Another concern with patient involvement in research is tokenism, which is when patient engagement with clinical research is merely symbolic and only done to give the appearance of involvement without actually valuing their input in a meaningful way.[16] This is a practice that can undermine the authenticity of the participatory process, as it fails to truly integrate patients' unique perspectives and experiential knowledge.[16,17] Agencies like the Patient-Centered Outcomes Research Institute have tried to circumvent this by only funding research that leads to knowledge translation to patients or meaningful patient engagement in research design and implementation.

References

1. U.S. Department of Health and Human Services FDA Center for Drug Evaluation and Research, U.S. Department of Health and Human Services FDA Center for Biologics Evaluation and Research, U.S. Department of Health and Human Services FDA Center for Devices and Radiological Health. Guidance for industry: patient-reported outcome measures: use in medical product development to support labeling claims: draft guidance. *Health Qual Life Outcome*. December 2006;4(1):79.
2. Basch E, Reeve BB, Mitchell SA, et al. Development of the national cancer institute's patient-reported outcomes version of the common terminology Criteria for adverse events (PRO-CTCAE). *J Nat Cancer Inst*. September 29, 2014;106(9). dju244–dju244.
3. Cancer Support Community. Negative Perceptions of Cancer Clinical Trials Persist. Available from: www.CancerSupportCommunity.org.
4. Unger JM, Hershman DL, Albain KS, et al. Patient income level and cancer clinical trial participation. *J Clin Orthod*. February 10, 2013;31(5):536–542.
5. Cutilli CC, Bennett IM. Understanding the health literacy of America: results of the national assessment of adult literacy. *Orthop Nurs*. January 2009;28(1):27–32.
6. Basch E, Jia X, Heller G, et al. Adverse symptom event reporting by patients vs clinicians: relationships with clinical outcomes. *J Natl Cancer Inst*. December 2, 2009;101(23):1624–1632.
7. Kluetz PG, Slagle A, Papadopoulos EJ, et al. Focusing on core patient-reported outcomes in cancer clinical trials: symptomatic adverse events, physical function, and disease-related symptoms. *Clin Cancer Res*. April 1, 2016;22(7):1553–1558.
8. Rivera SC, Kyte DG, Aiyegbusi OL, Slade AL, McMullan C, Calvert MJ. The impact of patient-reported outcome (PRO) data from clinical trials: a systematic review and critical analysis. *Health Qual Life Outcome*. December 2019;17(1):156.
9. Richter JG, Becker A, Koch T, et al. Self-assessments of patients via Tablet PC in routine patient care: comparison with standardised paper questionnaires. *Ann Rheum Dis*. December 1, 2008;67(12):1739–1741.
10. Perlmutter J, Bell SK, Darien G. Cancer research advocacy: past, present, and future. *Cancer Res*. August 1, 2013;73(15):4611–4615.
11. Abbasinia M, Ahmadi F, Kazemnejad A. Patient advocacy in nursing: a concept analysis. *Nurs Ethics*. February 2020;27(1):141–151.
12. Tugwell P, Boers M. Developing consensus on preliminary core efficacy endpoints for rheumatoid arthritis clinical trials. OMERACT Committee. *J Rheumatol*. March 1993;20(3):555–556.
13. De Wit M, Abma T, Koelewijn-van Loon M, Collins S, Kirwan J. Involving patient research partners has a significant impact on outcomes research: a responsive evaluation of the international OMERACT conferences. *BMJ Open*. 2013;3(5):e002241.
14. https://mosaic-nash.tmacademy.org.
15. Benizri N, Hallot S, Burns K, Goldfarb M. Patient and family representation in randomized clinical trials published in 3 medical and surgical journals: a systematic review. *JAMA Netw Open*. September 9, 2022;5(9):e2230858.
16. Majid U. The dimensions of tokenism in patient and family engagement: a concept analysis of the literature. *J Pat Exp*. December 2020;7(6):1610–1620.
17. Murad MH, Shah ND, Van Houten HK, et al. Individuals with diabetes preferred that future trials use patient-important outcomes and provide pragmatic inferences. *J Clin Epidemiol*. July 2011;64(7):743–748.

Chapter 63

Budgeting

Leticia Maria Tedesco Silva and Kamil Detyniecki
Department of Neurology, University of Miami Miller School of Medicine, Miami, FL, United States

Key points

- Creating a reasonable budget can increase the chances of receiving funding and the overall success of your project.
- Accounting for your institution's rates and policies on sponsored research is a crucial part of the budget development.
- A reasonable budget contains an adequate description of expenses and the justification for those expenses in each grant project period.

Introduction

A crucial step in preparing a research project is creating a budget that incorporates all the costs required to achieve the study objectives. A well-thought, carefully drafted budget could increase your chances of funding, as it shows the reviewer you understand the full scope of your work. Underestimating costs could compromise project execution. Therefore, it is imperative to work collaboratively with your institution to ensure your budget includes all the institutional fees and study-specific costs.

Developing a reasonable budget

The first thing to consider when developing a budget is identifying the limits for each type of expense, the overall funding limits, and if there are any funding restrictions. For federal grants, this information can be found on the Funding Opportunity Announcement. Reviewing and following the governing cost principles applicable to the recipient and the type of award requested will increase the chances of being funded. The National Institutes of Health (NIH) applies five tests on each cost (allowability, allocability, reasonableness, necessity, and consistency) before approving the item. In this phase, it is important to list all your expected relevant costs and check if they would pass the tests. The funding agency may disallow the cost if it is not compliant with the principles or surpasses the established limits.

Writing a budget justification provides more in-depth detail and reason for each cost and is often considered by reviewers as a good indicator of the feasibility of the research. It is also an opportunity for you to convince the reviewers of the necessity and reasonability of your requests.

The next step is to discuss with the grants or sponsored programs office within your own institution the facilities and administrative (F&A) costs rate.

Facilities and administrative costs

The F&A costs (previously known as indirect costs or IDC) are costs that are incurred for common or joint objectives and, therefore, cannot be identified readily and specifically with a particular sponsored project, an instructional activity, or any other institutional activity. Facilities costs include plant operation and maintenance such as janitorial services, depreciation or use allowance of buildings and equipment, and library expenses. Administrative costs include departmental administration expenses, general administration (accounting and payroll), and sponsored project administration (personnel and other costs of offices whose responsibility is the administration of sponsored projects).

Most institutions establish a yearly negotiated F&A rate for federal-sponsored projects. The F&A rate may vary depending on the type of research project (observational or interventional) and the type of funding agency (federal or private; for-profit or not-for-profit).

The F&A rate is usually applied to a modified total direct cost base that excludes from your direct costs the items not subject to F&A (such as costs allocated to buying new equipment). It is necessary to verify with your institution what type of costs are subject to F&A.

Translational Cardiology. https://doi.org/10.1016/B978-0-323-91790-2.00057-5

Direct costs

Direct costs are attributable specifically to the execution of a particular sponsored project. Examples of direct costs include personnel, materials and supplies, equipment, and transportation.

Personnel

Costs with personnel are usually the most significant expense on a budget. For NIH studies, all personnel from the applicant's organization dedicating effort to the project should be listed on the personnel budget with their base salary and effort, even if they are not requesting salary support. Include name (or "to be disclosed"), position, and percentage of time on the project. Remember to comprise the pay raise estimates depending on the duration of your project. Some funding agencies may establish salary caps.

One metric commonly used for the estimation of effort is person months. To calculate, for example, the effort of a principal investigator in person months, multiply the percentage of the effort associated with the project times the number of months of the principal investigator's appointment at the institution. If the principal investigator's appointment with the institution consists of a 9-month academic year and the principal investigator will devote 9 months at 30% time/effort, then the effort would be 2.7 person-months ($9 \times 0.3 = 2.7$).

A full-time equivalent (FTE) is the budgetary equivalent of one position, continuously filled full-time for the entire fiscal year, and may comprise any combination of part-time and full-time positions. The workload that qualifies as "full-time" varies by position. A full-time employee's FTE is represented as 1, and a half-time employee's is represented as 0.5. An employee's FTE is not used to estimate effort. Calculating FTEs can help plan resources and workloads across full and part-time personnel.

Fringe benefits must be charged to the grant if the fund is used to pay salaries and wages. The fringe benefit rate generally includes social security, unemployment/worker's compensation, retirement, and health insurance. The sponsored programs office at your institution establishes the rate, and the amount can vary depending on the personnel's type of contract (full or part-time).

It is essential to include the personnel responsible for all aspects of data management (e.g., creating a data management plan, choosing and setting up the appropriate data capture system, designing the Case Report Forms, performing the statistical analysis, generating data reports, and coordinating data acquisition and management with other centers). Do not forget to incorporate the personnel involved in the data collection and entry. Some studies may also require a research coordinator.

Examples of personnel that can be involved in the research project:

- Key/senior personnel: individuals who participate in the scientific development or execution of a project in a significant, measurable way, regardless of whether they are requesting salaries or compensation.
- Postdoctoral associates
- Graduate students: tuition remission is usually an allowable cost and may be included in the graduate student compensation limits.

Equipment

Articles with more than 1 year of useful life and a cost greater than $5000 per unit are considered equipment. This section is not subject to F&A in most cases. Before including an equipment in your budget, ensure it is not already available in your institution's facilities. If your institution has the equipment you are requesting, you will need to justify why the existing equipment is not sufficient for the project execution and how the new equipment use will be allocated specifically to your project. It is recommended to list a reasonable cost for the equipment, preferably after a quotation with two different companies. Remember that funding for equipment such as computers and printers is typically derived from the F&A costs since they can be used for multiple projects and personal use.

Travel

Anticipated travel related to the project should have its details (destination, number of people traveling, dates, and duration) stated in the budget justification. Explaining how the travel is directly related to the project is important. For example, if requesting funds to present the results of the project at a conference, the budget justification should cite the name and location of the conference, the reason for the travel, and which members of the study team will be presenting the results.

Materials and services

Items not categorized as equipment are considered supplies. Examples include chemicals, animals, drugs, assays, lab glassware, and syringes. Estimated costs can be based on the monthly expenditures of the laboratory you will be using.

Consultants

Someone who is an expert in the field and not an employee of your institution can provide consulting services for your project. The budget justification should include the services to be performed by the consultant(s), the number of days of anticipated consultation, the expected compensation rate, travel reimbursement, and other related costs. It may be required to obtain a letter from each consultant signaling their willingness to act as a consultant to the project.

Consortiums/subawards

When another institution is supposed to perform activities related to your research project, you will need to establish a subaward or consortium agreement between your organization and the outside organization. This agreement should explicit the willingness to comply with the grant's scientific, administrative, financial, and reporting requirements. Those agreements must be included in the application to support the applicant's readiness to conduct the proposed research.

Remember that the grantees are responsible for their subawardees project-related actions. Therefore, if the subawardee requires more money to carry out its part of the project, it is the grantee's duty to find a budget solution. The grantee should also answer questions from the subawardee about human subjects or animal research, report possible financial conflicts of interest of the subawardees, and ensure the subawardee complies with all required human subjects and animal research certifications and licensing.

Timeline

Correctly estimating the duration of the project can be challenging. It is advised to be realistic about how much can be accomplished each year of the grant. For some cases, accounting for unforeseen circumstances and increasing the maximum number of years expected to the project's conclusion could be a solution.

Considering that the budget is based on an estimate of the expenses incurred in performing the proposed project, including inflation or cost escalation, is a fair part of the estimate. Some institutions apply an inflation rate to some budget categories.

Detailed versus modular budgets

A detailed budget is the most common budget format requested by funding agencies. When using a detailed budget format, you must specify and justify all your direct costs.

The modular budget uses specific modules (or increments) in which direct costs are requested. The NIH requires a modular format budget for applications with budgets of up to $250,000 direct costs per budget period. When using this format, instead of submitting a detailed line-item budget, funds are requested in modules of $25,000 to a maximum of $250,000 a year, excluding F&A and consortium costs. A modular budget must justify only personnel and consortium costs. Additional justification can be necessary in special cases, such as when variations in the number of modules requested annually occur.

Final tips

While writing your application for funding, it is expected and advisable to place a lot of effort into the scientific aspects of your project. However, you should not forget that the budget is fundamental for securing the execution of your proposed investigations, being an important part of your application. Not allocating enough time to plan your budget, especially if your project includes subawards to other institutions, could severely delay your application. Some institutions have dedicated research offices to review grant budgets, a resource that can help investigators clarify and improve their proposed budgets. Asking for feedback from project collaborators on the proposed budget can also decrease the chances of missing concealed research needs. Finally, do not lower your budget, expecting that doing this could increase your chances of funding. Ask for the budget you need to execute the project and nothing less or more than that.

Further reading

1. Higdon J, Topp R. How to develop a budget for a research proposal. *West J Nurs Res*. 2004;26(8):922–929. https://doi.org/10.1177/0193945904269291.
2. Bhatti JA, Akhtar U, Butt BA, Janjua NZ. Presenting budget in a research grant application. *J Pak Med Assoc*. 2013;63(6):793–796.
3. Fahrenholtz M, Salvati DR, Scott LB, Goldstein AM, Keswani SG. Writing an effective National Institutes of Health (NIH) budget: how to get the money for your science. *Surgery*. 2022;171(2):342–347. https://doi.org/10.1016/j.surg.2021.06.002.
4. National Institutes of Health (NIH). Grants & Funding Develop Your Budget. Retrieved from https://grants.nih.gov/grants/how-to-apply-application-guide/format-and-write/develop-your-budget.htm. Accessed 3 January 2023.
5. National Institute of Allergy and Infectious Diseases (NIAID). Create a Budget. Retrieved from https://www.niaid.nih.gov/grants-contracts/create-budget. Accessed 3 January 2023.
6. Fahrenholz M, Cheng LS, Olutoye II O, et al. Ancillary documents for NIH grant applications: the pages beyond the science. *Surgery*. 2023;173(2):479–484. https://doi.org/10.1016/j.surg.2022.09.038.

Chapter 64

Ethics and review boards

Diego Laurentino Lima[1], Victor Perim[2] and Raquel Nogueira[1]

[1]*Department of General Surgery, Montefiore Medical Center and the Albert Einstein College of Medicine, Bronx, NY, United States;* [2]*Medical School, the Federal University of the State of Rio de Janeiro (UNIRIO), Rio de Janeiro, Brazil*

Key points

- Distinguishes between clinical practice and clinical research.
- Provides ethical frameworks for participant protection.
- Discusses the role of Institutional Review Boards (IRBs)

Introduction

Clinical practice involves enhancing patient's well-being through evidence-based practice or experience, while also having a reasonable expectation of success. Clinical research, on the other hand, focuses on gathering knowledge by testing different hypotheses and generating conclusions that improve medical care and public health. In general, the risks involved with interventions of clinical practice are justified by the expectation of benefit to the individual. Conversely, although the participant might get quality patient care during clinical research, it is designed not on the individual goal, but instead, on testing the project premises in the most precise and ethical way (Gallin et al., 2018; Brody and Miller, 2003).

Clinical research versus clinical practice

Clinical practice involves enhancing patient's well-being through evidence-based practice or experience, while also having a reasonable expectation of success. Clinical research, on the other hand, focuses on gathering knowledge by testing different hypotheses and generating conclusions that improve medical care and public health. In general, the risks involved with interventions of clinical practice are justified by the expectation of benefit to the individual. Conversely, although the participant might get quality patient care during clinical research, it is designed not on the individual goal, but instead, on testing the project premises in the most precise and ethical way.[1,2]

Ethics and clinical research

The improvement of patient care and public health heavily depends on clinical research conducted with human beings. Providing healthcare providers with the understanding of different disease processes as well as what is the best treatment option for a certain condition helps to advance medical care. With that in mind, and bypassing the philosophical debate on whether clinical research should be an option or an imperative, it is consensual that ethical limits need to be set.[3–5]

Considering that, the main preoccupation lies in the fact that few individuals need to accept the research burden of inconvenience and risk to benefit others. In that sense, by respecting the rights, welfare, dignity, and freedom of choice of subjects, we minimize the risk of exploitation while ensuring that their willingness to participate contributes to scientific knowledge. That way, the questions of "why" and "how" to do clinical research can be ethically responded.[1]

History of ethics in clinical research

Research has historical roots as an unregulated and somewhat cruel practice. Experimentation and clinical practice had little to no distinction, mostly because no systematic evidence supported most interventions at the time. Despite being performed in the hope of achieving benefit for their patients, research had little ethical consideration, and frequently caused more harm than good.

In addition to this already unregulated scenario, around World War II it became clear that medical advances could

Translational Cardiology. https://doi.org/10.1016/B978-0-323-91790-2.00044-7

benefit society when performed at a larger scale and with a systematic approach. This gave birth to an era of the "unashamedly utilitarian" when the individuals enrolled in research were those that were considerably available and could provide some value to society. This led to orphanages, prisons, and homes for the emotionally vulnerable to become places suitable for conducting research. [1]Influenced by the experiments done by Nazis during this time, public and professional concerns started to raise around ethics on clinical research.[6,7]

In the United States, during the 1960s and 1970s, a great debate around ethics in research sparked after a series of scandals involving research projects became public, such as the syphilis studies at the US Public Health Service in Tuskegee and the hepatitis B studies at Willowbrook.[7] As a consequence, after several congressional hearings, the 1974 Research Act and the creation of the US National Commission for protection of Human Subjects of Biomedical and Behavioral Research lead to a shift in how clinical research was done in the country.[5]

The commission produced several reports and recommendations around clinical research, with the Belmont Report being the most notable one. This report became the basis for highlighting the need for protection of individuals from exploitation and harmful activities, serving as a base for future regulations like the common rule, which governs the management of most research projects today.

During the 1990s, studies regarding the human deficiency virus (HIV) have showed that even participating as a subject in research could be an advantage. That is because it became clear that for certain populations, like oncological and HIV patients, participating in research meant a light of hope in their prognosis, and selection criteria could mean life or death. This meant that everyone needed to be represented equally, especially underrepresented minorities.[8–10]

Research ethics codes and regulations

Mostly influenced by the landmark crisis of the 20th-century research scene, several reports and documents were drafted to address the most pressing matters involving ethics in clinical research. The overarching theme of all of these was the assumption that clinical research was pivotal for advancing medical care. The first one to address such matters was the Nuremberg Code, written after all the Nazi cruelties were exposed during the Nuremberg trial. It focused on 10 points regarding human experimentation that stated its value while also highlighting the importance of things such as informed consent and the prioritization of the welfare and rights of individuals.[11]

Under Nuremberg's influence, several other documents were written that addressed different aspects of this matter such as the Declaration of Helsinki, emphasizing that research participation should not mean a disadvantage in medical care, as well as addressed issues involving individual's incapacity of giving informed consent, such as children and the cognitively impaired. Similarly, the Belmont report describes that all research should follow the principles of beneficence, respect for persons, and justice. Beneficence meant maximizing benefits while minimizing harm, respect implied compliance to the autonomous decision-making of individuals, and justice demanded an equal distribution of benefits and burdens.

Following this trend, the Council of International Organizations of Medical Sciences (CIOMS), the World Health Organization (WHO), and later federal agencies issued different reports and created regulations that guide how most clinical research should be ethically done. The most notable is the Common Rule, which applies to 17 different agencies to this day.[12]

Ethical framework for clinical research

A systematic framework was created to help satisfy, in a practical approach, the different aspects of research that may be considered ethical: social and scientific value, scientific validity, fair subject selection, favorable risk/benefit ratio, independent view, informed consent, and respect for enrolled individuals.[13]

- *Value and validity*: The study should propose a clinically or socially valuable question that contributes to improving health (value) while also having a valid methodology, ensuring that the participation of the subject is valuable.
- *Fair subject selection*: Choosing individuals for participation in research should be based on the scientific question and considerations of risk, benefit, and vulnerability. Additionally, there must be attention to the exploitation of vulnerable populations and the equitable distribution of burdens and benefits.
- *Favorable risk/benefit ratio*: The potential benefits should outweigh the risks imposed on participants by the nature of the study design. Moreover, the principle of beneficence states that people should be protected from deliberate harm while also maximizing the potential benefits.
- *Independent review*: Research projects should be reviewed by independent individuals with varied expertise and no personal interest in the research under scrutiny, such as Institutional Review Boards (IRBs) and Research Ethics Committees.
- *Informed consent*: A tool for demonstrating respect for the person's autonomy and respecting their choice of participating. The process should involve the disclosure of the study information, comprehension of the information given, voluntariness with the decision, and authorization.[14,15]

- *Respect for enrolled subjects*: Appropriate clinical monitoring and attention to the participant's well-being, taking note of adverse effects, and potential injuries, as well as the confidentiality of enrolled information. Also, one should be able to withdraw from the project at any time without penalty.

Ethics in randomized controlled trials

Randomized controlled trials (RCTs) are considered the "gold standard" for testing drugs and interventions. The goal of an RCT is to compare one intervention A with one or more alternatives B or C in a controlled, randomized, and usually blinded scenario.[1] The main ethical justification lies in an "honest null hypothesis" considering that, if proven that none of the alternative interventions have any significant difference, the subjects are not submitted to any treatment known in advance to be inferior. There is controversy regarding the meaning of said "equipoise"[16] and about what should count as convincing evidence in relation to "*P* values," nevertheless, it still remains the method for testing different approaches in a way that reduces confounding.[16–19]

To achieve the level of confidence and ethics produced by RCTs, different parameters need to be taken into consideration, such as equipoise, choice of controls, randomization, blinding, and sharing of preliminary information.

- *Equipoise*: Guarantees that no subject will be submitted to any previously known inferior intervention, comparing only truly equivalent therapies with unknown superiority.
- *Choice of controls*: The outcome variables used in RCTs influence directly the conclusions drawn from the study, as well as validity and generalizability. Special ethical considerations need to be made when using placebos.
- *Randomization*: Random assignment of subjects helps to eliminate confounding and attempt to equalize both intervention arms to uncontrollable variables of each subject.
- *Blinding*: the random assignment and blinding methods are used to reduce bias and elevate the validity of the study. In some projects, however, subjects and investigators may be able to guess which arm they are in only based on aspects like adverse effects. In rare cases, there may be a need to break blinding due to possible clinical complications related to the intervention.
- *Sharing preliminary information*: RCTs may take a long time and may achieve statistical significance before the end of the study. With that in mind, study monitors are tasked to monitor preliminary results that may pose a clear threat to subjects or just prove that there is a clear advantage of one intervention over the other.

The use of a placebo may impose different ethical dilemmas in the investigator's framework. That is because placebo use may deprive the individual of the appropriate treatment and its absence in RCTs can hinder the capacity to prove the safety and efficacy of a certain drug. The workaround usually consists of the comparison between the proposed intervention against an already proven to be an effective treatment in the market, preserving the equipoise principle. This way, the use of placebo is usually revered for (1) there is no proof to be an effective treatment for the studied condition; (2) withholding treatment does not impose major risks or burdens; (3) there are methodological benefits of using placebo paired with negligible downsides to it is consensual, however, that if the outcome of not using treatment or using placebo means death or serious morbidity, placebo controls should not be used.[20]

Institutional review boards

The local body of oversight in clinical research is the IRB. These organizations ensure that ethical guidelines and regulatory procedures are followed at the institutional level. It serves the purpose to hold investigators accountable and provide reassurance that all approved projects take into account the values of beneficence, justice, and respect.

Historically, as said previously, research and clinical practice had loose definitions as well as minimal oversight. This freedom of practice leads to the advancement of science in disregard for ethical principles, especially during the 20th century. In 1946 the Nuremberg trial served as a base for the Nuremberg code, which sparked the discussion around what is ethically accepted in clinical research.[21] Despite its importance, many physicians, especially in the United States, judged that it imposed limitations to the way research as well as clinical practice was performed, arguing it was influenced by European standards. Despite the controversy and the pressure of different scandals of the 1960s and 1970s, the NIH created the Clinical Center (CC) in Bethesda, Maryland, which developed the policy for research with human subjects. The CC took into account the protection of participants by requiring the projects to undergo prospective review from independent subjects. This was the starting point to what would then become the IRBs.[22]

The subsequent forms of prospective review had increasing levels of protection for research subjects, overseeing not only healthy participants and the requirement of informed consent, culminating in what would then be the reference for distinguishing therapeutic medicine from research: The Belmont Report. This document served as a base for all ethical considerations required by the DHHS and NIH as well as the IRBs, varying from the federal to the institutional level.

All clinical research should undergo independent prospective review by an IRB as a mechanism for promoting ethical research. By that, research is considered any systematic investigation designed to develop or contribute to a generalized knowledge involving a human subject (a living individual through which the investigator obtains information). There are six groups of exceptions that share a common rationale of excluding the need for approval by IRBs to projects that do not expose individuals to physical, social, or psychological harm other than those encountered in everyday life.

When a primary investigator proposes a research project, he or she must submit a research protocol, which contains a written description on the methodological approach used in said research, as well as discuss issues related to the protection of subjects. The criteria of approval by IRBs are in accordance with seven minimal regulatory criteria:

1. The proposed research design is scientifically sound and will not unnecessarily expose subjects to risk.
2. Risks are reasonable in relation to anticipated benefit.
3. Risk to subjects is minimized.
4. Subject selection is equitable.
5. Informed consent is provided by research subjects or their legally authorized representative.
6. Additional safeguards are required for subjects likely to be vulnerable to coercion or undue influence.
7. Subject privacy and confidentiality are maximized.

Additionally, IRBs are required to continue monitoring already approved projects on an annual basis or sooner if a risk of harm is presented to the subjects. The IRB member may judge the project as no longer safe or ethically reasonable and choose to suspend, modify, or even terminate it.

References

1. Gallin J, Ognibene E, Johnson L. Principles and practice of clinical research. In: *Principles and Practice of Clinical Research*. Elsevier; 2018:19–200.
2. Brody H, Miller FG. The clinician-investigator: unavoidable but manageable tension. *Kennedy Inst Ethics J*. 2003;13(4):329–346. https://pubmed.ncbi.nlm.nih.gov/15049297/.
3. Engel GL. The need for a new medical model: a challenge for biomedicine. *Science*. 1977;196(4286):129–136. https://pubmed.ncbi.nlm.nih.gov/847460/.
4. Jonas H. Philosophical reflections on experimenting with human subjects. *Biomed Ethics Law*; 1979:239–264. https://link.springer.com/chapter/10.1007/978-1-4615-6561-1_16.
5. Beauchamp TL, Childress JF, Vaughn L, Pierce J. *Principles of Biomedical Ethics*. 8th ed. 2019:65.
6. Paul W. *Nazi Medicine and the Nuremberg Trials: From Medical Warcrimes to Informed Consent*. 2006.
7. Rothman DJ. Ethics and human experimentation. Henry Beecher revisited. *N Engl J Med*. 1987;317(19):1195–1199. https://pubmed.ncbi.nlm.nih.gov/3309660/.
8. Bleyer A, Montello M, Budd T, Saxman S. National survival trends of young adults with sarcoma: lack of progress is associated with lack of clinical trial participation. *Cancer*. 2005;103(9):1891–1897. https://pubmed.ncbi.nlm.nih.gov/15795902/.
9. Hébert-Croteau N, Brisson J, Lemaire J, Latreille J. The benefit of participating to clinical research. *Breast Cancer Res Treat*. 2005;91(3):279–281. https://pubmed.ncbi.nlm.nih.gov/15952061/.
10. Albert J, Jeff S. *The Social Impact of AIDS in the United States*; 1993. https://pubmed.ncbi.nlm.nih.gov/25121219/.
11. *The Nuremberg Code*. 1946.
12. Council for International Organizations of Medical Sciences. *International Ethical Guidelines for Biomedical Research Involving Human Subjects*. 2002.
13. Emanuel EJ, Wendler D, Grady C. What makes clinical research ethical? *JAMA*. 2000;283(20). Available from: http://jama.jamanetwork.com/.
14. Faden RR, Beauchamp TL, King NMP. *A History and Theory of Informed Consent* [cited 2022 Sep 13]; 1986:392. Available from: https://global.oup.com/academic/product/a-history-and-theory-of-informed-consent-9780195036862.
15. Mandava A, Pace C, Campbell B, Emanuel E, Grady C. The quality of informed consent: mapping the landscape. A review of empirical data from developing and developed countries. *J Med Ethics*. 2012;38(6):356–365. https://pubmed.ncbi.nlm.nih.gov/22313664/.
16. Freedman B. Equipoise and the ethics of clinical research. *N Engl J Med*. 1987;317(3):141–145. https://pubmed.ncbi.nlm.nih.gov/3600702/.
17. Levine RJ. *Ethics and Regulation of Clinical Research* [cited 2022 Sep 13]; 1986. Available from: https://philpapers.org/rec/LEVEAR-2.
18. Miller FG, Brody H. Clinical equipoise and the incoherence of research ethics. *J Med Philos*. 2007;32(2):151–165.
19. Wasserstein RL, Lazar NA. The ASA's statement on p-values: context, process, and purpose. *Am Stat*. 2016;70(2):129–133. https://www.tandfonline.com/doi/abs/10.1080/00031305.2016.1154108.
20. Levine R. *The Use of Placebos in Randomized Clinical Trials*. IRB Rev Hum Subjects; 1985.
21. Sass HM. The nuremberg code, German law, and prominent physician-thinkers. *JAMA*. 1997;277(9). http://www.ncbi.nlm.nih.gov/pubmed/9042833.
22. Emanuel EJ. *The Oxford Textbook of Clinical Research Ethics* [cited 2022 Sep 13]; 2008:827. Available from: https://global.oup.com/academic/product/the-oxford-textbook-of-clinical-research-ethics-9780199768639.

Chapter 65

Regulatory considerations for new drugs and devices

Anant Jain and Jayakumar Sreenivasan
Department of Medicine, Westchester Medical Center, New York Medical College, Valhalla, NY, United States

Key points

- FDA drug and device approval process.
- Challenges in FDA approval.
- Postmarket surveillance and recalls.

Introduction

Any drug can have an adverse event associated with its use. Most governments around the world have developed regulatory agencies to oversee the development and marketing of drugs and medical devices. The scope and the regulations imposed by different agencies vary; however, they require similar premarketing processes to be completed before any drug or device approval. In the United States, the Food and Drug Administration (FDA) is a federal agency that is responsible for protecting and promoting public health. It fulfills its role through control and supervision of food safety, tobacco products, dietary supplements, medications, medical devices, vaccines, biopharmaceuticals, cosmetics, blood transfusions, animal foods and feed, and veterinary products. In the year 2021, the FDA approved 50 novel medications and 41 novel medical devices.[1,2]

FDA drug approval

A drug is defined by the FDA as "articles (other than food) intended to affect the structure or any function of the body of man or other animals" and "intended for use in the diagnosis, cure, mitigation, treatment, or prevention of disease."[3]

FDA's Center for Drug Evaluation and Research (CDER) is responsible for drug approval and usually follows a five-step process: discovery/concept, preclinical research, clinical research, FDA review, and FDA post-market safety monitoring.

1. **Discovery/concept**: Research for a new drug begins at this stage in the laboratory. Once a potential drug is recognized, information regarding the mechanism of action, pharmacokinetics, and its potential benefits is collected through experiments.[4]
2. **Preclinical research**: The drug of interest then undergoes laboratory and animal testing to answer basic questions about safety, especially on dosing and toxicity levels. These findings are then reviewed and the decision on whether the drug should be tested on people is made at the end of this stage.[5]
3. **Clinical research**: At this stage studies or trials are conducted on people. The various substages are summarized in Table 65.1. The company must submit an Investigational New Drug (IND) application to the FDA, which includes data regarding animal studies, drug manufacturing, and study protocol, before proceeding with clinical research.[6]
4. **FDA review:** If the drug's effectiveness for its intended use and safety is demonstrated in the abovementioned process, the company can file a new drug application (NDA). The review team at the FDA has 6–10 months to decide whether to approve the drug.[7]
5. **Postmarket surveillance**: It is the practice of monitoring the safety of a pharmaceutical drug or medical device after it has been released on the market. The FDA reviews reports of problems with drugs and can decide to add warnings regarding dosage or other major issues.[8]

Translational Cardiology. https://doi.org/10.1016/B978-0-323-91790-2.00019-8

TABLE 65.1 Sub-stages of clinical research.

Phase	Objective	Study population	Duration
I	Evaluate safety	10s of participants (usually healthy volunteer)	Weeks—months
II	Test efficacy and further evaluate safety	100s of participants (including those with condition)	Months—year
III	Confirm efficacy and monitor adverse effects	1000s of participants (including those with condition)	Years

Expedited programs

The FDA utilizes several approaches to promote the development of certain drugs, especially drugs that may represent the first available treatment for an illness or ones that have a significant benefit over existing drugs. To make these therapies promptly available, these approaches involve swift drug development and review. These include fast track designation, breakthrough therapy designation, accelerated approval pathway, and priority review designation, which are summarized in Table 65.2.[9]

FDA device approval

The medical devices sold in the United States are regulated by the FDA's Center of Device and Radiological Health (CDRH). The FDA uses a risk-based classification system for medical devices based on the intended use of the device and the risk it entails to the patient. Medical devices are classified into Classes I (Adhesive Bandages, wheelchairs, tongue depressors), II (Catheters, contact lenses, needle), and III (pacemakers, coronary stents). The level of risk increases from Class I to Class III and the class to which the device is assigned determines the type of premarket proposal for approval.

TABLE 65.2 Expedited drug development and review designations.

Type of pathway	Fast track designation	Breakthrough therapy designation	Accelerated approval pathway	Priority review designation
Qualifying criteria (Data should demonstrate)	A drug that is intended to treat a serious condition AND *demonstrate the potential to address unmet medical need*	A drug that is intended to treat a serious condition AND evidence indicates that the drug may demonstrate *substantial improvement on a clinically significant endpoint(s) over available therapies*	A drug that treats a serious condition AND generally provides a meaningful advantage over available therapies AND demonstrates an *effect on a surrogate endpoint that is reasonably likely to predict clinical benefit or on a clinical endpoint that can be measured earlier than irreversible morbidity or mortality (IMM) that is reasonably likely to predict an effect on IMM or other clinical benefit* (i.e., an intermediate clinical endpoint)	An application (original or efficacy supplement) for a drug that treats a serious condition AND, if approved, would provide a *significant improvement in safety or effectiveness, prevention, or diagnosis of a serious condition*
Type of data required	Preliminary nonclinical, mechanistic, or clinical data	Preliminary clinical data	Not specified; sponsor should make justification of alternate endpoint-based scientific support	Data contained in the final NDA submission

General controls (prohibitions against adulteration, misbranding, mislabeling) are the baseline requirements of the Food, Drug, and Cosmetic (FD&C) Act, which apply to all three medical device classes. When there is insufficient evidence to establish general controls for Class II devices, "special controls" provide an acceptable guarantee of safety and effectiveness. Class III devices undergo the premarket approval process (PMA) to establish safety and efficacy as they are the riskiest and can cause substantial harm.

Common pathways for device approval

PMA

PMA is the most stringent regulatory category for medical devices which applies to Class III devices. Through the PMA, the FDA determines if there is sufficient valid scientific evidence to assure the device's effectiveness for its intended use and safety. The evidence can vary from objective trials without matched controls to well-controlled investigations.[10] The PMA includes a section on clinical investigations involving human subjects which contains information about safety and effectiveness, adverse reactions and complications, device failures and replacements, and so on.[11] "Supplementary applications" must be submitted for any change in the device affecting its safety and effectiveness by the manufacturer.

Premarket notification (501k)

Class I and II devices that have the same intended use with similar safety and efficacy compared to another legally marketed device and are not exempt require a premarket notification (501k) before clearance. The sponsors must provide documentation to support their substantial equivalence claims to a legally marketed device.[12]

Exempt

If the FDA determines that a 510(k) is not required to provide assurance of safety and effectiveness for the device, a device may be exempt from 510(k). These include most Class I devices, some Class II devices, and preamendment devices.

Humanitarian device exemption

A medical device can receive humanitarian device exemption (HDE) if it benefits in the treatment or diagnosis of a condition that affects less than 8000 individuals in the United States per year. HDE is subject to certain restrictions, however, is exempt from some effectiveness requirements of the FD&C Act.[13]

DeNovo pathway

This pathway classifies novel medical devices as Class I or Class II for which there is no legally marketed predicate device and general controls alone, or general and special controls, provide reasonable assurance of safety and effectiveness. Devices that are classified into class I or class II through a de novo pathway can be used as the basis for future premarket notification [510(k)] submissions, when relevant.[14]

Expanded access

As with investigational drugs, the FDA has pathways for permitting the use of an investigational (unapproved) device for a condition for which there is no alternative therapy available. These include emergency use, compassionate use, and treatment use and are summarized in Table 65.3[15]

Issues with drug and device approval process

The FDA approval is based on evidence that the benefits of the drug/device for indicated use outweigh the risks for most patients. The quality of clinical trial evidence used by the FDA seems inconsistent, variable across indications, and can be heavily biased. FDA does not conduct its own independent studies on drug safety and effectiveness rather relies on data and studies provided by the manufacturers to make the judgment when approving drugs.

TABLE 65.3 Investigational devices.

Type of pathway	Emergency use	Compassionate use	Treatment use
Criteria	Life-threatening or serious condition No time to get FDA approval	Patient with a serious condition or disease Patient does not otherwise qualify for inclusion in the clinical study, but the physician believes they may benefit	Life-threatening or serious disease Controlled clinical trial Sponsor actively pursuing market approval

Pivotal trials are the most critical of clinical trials and are routinely used as the basis for the approval of drugs and devices. As per a study, the approval of therapeutics for cardiovascular disease, diabetes mellitus, and hyperlipidemia often relied upon at least three trials while cancer indications were based on a single trial.[16] A study of high-risk therapeutic medical devices found that pivotal studies usually enrolled less than 300 patients and were often designed without blinding, comparators, or primary endpoint follow-up exceeding 1 year[17]

The evidence requirements for device approval tend to be flexible; hence, many approved devices may lack clinical trial data supporting their safety and effectiveness. A systematic review of 123 studies for high-risk cardiovascular devices that received PMA between 2000 and 07 showed that only 33 of 123 studies (27%) were randomized and 17 of 123 (14%) were blinded. Also, for 77 original PMAs for cardiac implantable electronic devices from 1979 through 2012, there have been more than 5800 PMA supplements filed.[18] Accumulating PMA supplements over subsequent models of high-risk devices may result in a device that differs substantially from the originally approved device.[19]

An overwhelming majority (>95%) of medical devices currently in the US market were cleared by the FDA through the 510(k) process; however, the 501k program remains controversial. The FDA reported that it requests clinical data for less than 10% of 510(k) submissions.[20] Because of its lower scrutiny of the 510(k)-program compared to PMA, it provides an opportunity for companies to forego intensive testing and clinical trials.

While the expedited programs facilitate earlier access to drugs/devices that are urgently needed by society, they may also undermine efficacy and safety. A study of all FDA approvals granted for Breakthrough Therapy designation from 2012 through 2017 suggests that pivotal trials supporting these approvals commonly lacked randomization, double-blinding, and control groups, used surrogate markers as primary endpoints, and enrolled small numbers of patients.[21]

Postmarket surveillance helps recognize the safety events of novel pharmaceuticals and biologics that occur once they are in the market. Moreover, 32% of 222 novel therapeutics approved by the FDA from 2001 through 2010 were affected by a postmarket safety event.[22] In the case of medical devices, it is plausible that by the time evidence is generated and appropriate disclosures are made, the market may have moved on to future product generations.

The FDA may recall a device when it finds a reasonable prospect that it would cause serious adverse health consequences or death.[23] In a recent study, 3012 devices (10.7%) with 510(k) clearance and 84 devices (27.1%) with PMA were recalled.[24] While the FDA may suggest a recall, it has no legal authority to enforce a drug recall and can only suggest it. The responsibility for pulling dangerous products off the market ultimately falls on the manufacturer.

References

1. Center for Drug Evaluation and Research. *Novel Drug Approvals for 2021*: U.S. Food and Drug Administration. https://www.fda.gov/drugs/new-drugs-fda-cders-new-molecular-entities-and-new-therapeutic-biological-products/novel-drug-approvals-2021. Accessed 31 December 2021.
2. Center for Devices and Radiological Health. *Device Approvals.* U.S. Food and Drug Administration; 2021. https://www.fda.gov/medical-devices/recently-approved-devices/2021-device-approvals. Accessed December 31, 2021.
3. *Human Drugs Overview for Imported Products*: U.S. Food and Drug Administration. https://www.fda.gov/industry/regulated-products/human-drugs. Accessed 31 December 2021.
4. *Step 1: Discovery and Development*: U.S. Food and Drug Administration. https://www.fda.gov/patients/drug-development-process/step-1-discovery-and-development. Accessed 31 December 2021.
5. *Step 2: Preclinical Research*: U.S. Food and Drug Administration. https://www.fda.gov/patients/drug-development-process/step-2-preclinical-research. Accessed 31 December 2021.
6. *Step 3: Clinical Research*: U.S. Food and Drug Administration. https://www.fda.gov/patients/drug-development-process/step-3-clinical-research. Accessed 31 December 2021.
7. *Step 4: FDA Drug Review*: U.S. Food and Drug Administration. https://www.fda.gov/patients/drug-development-process/step-4-fda-drug-review. Accessed 31 December 2021.
8. *Step 5: FDA Post-Market Drug Safety Monitoring*: U.S. Food and Drug Administration. https://www.fda.gov/patients/drug-development-process/step-5-fda-post-market-drug-safety-monitoring. Accessed 31 December 2021.
9. *Expedited Programs for Serious Conditions—–Drugs and Biologics.* U.S. Food and Drug Administration; 2014. https://www.fda.gov/regulatory-information/search-fda-guidance-documents/expedited-programs-serious-conditions-drugs-and-biologics. Accessed December 31, 2021.
10. *Medical Device Classification Procedures. 21CFR §860.7.* 2018.
11. *Premarket Approval (PMA)*: U.S. Food and Drug Administration. https://www.fda.gov/medical-devices/premarket-submissions-selecting-and-preparing-correct-submission/premarket-approval-pma#data. Accessed 31 December 2021.
12. *Premarket Notification 510(K)*: U.S. Food and Drug Administration. https://www.fda.gov/medical-devices/premarket-submissions-selecting-and-preparing-correct-submission/premarket-notification-510k. Accessed 31 December 2021.
13. Humanitarian Device Exemption (HDE). *Program Guidance.* U.S. Food and Drug Administration; 2019. https://www.fda.gov/regulatory-information/search-fda-guidance-documents/humanitarian-device-exemption-hde-program. Accessed December 31, 2021.
14. *De Novo Classification Request.* U.S. Food and Drug Administration; 2021. https://www.fda.gov/medical-devices/premarket-submissions/de-novo-classification-request. Accessed December 31, 2021.
15. *Expanded Access for Medical Devices*: U.S. Food and Drug Administration. https://www.fda.gov/medical-devices/investigational-device-exemption-ide/expanded-access-medical-devices#treatment. Accessed 31 December 2021.

16. Downing NS, Aminawung JA, Shah ND, Krumholz HM, Ross JS. Clinical trial evidence supporting FDA approval of novel therapeutic agents, 2005–2012. *JAMA*. 2014;311(4):368–377. https://doi.org/10.1001/jama.2013.282034.
17. Rathi VK, Krumholz HM, Masoudi FA, Ross JS. Characteristics of clinical studies conducted over the total product life cycle of high-risk therapeutic medical devices receiving FDA premarket approval in 2010 and 2011. *JAMA*. 2015;314(6):604–612. https://doi.org/10.1001/jama.2015.8761.
18. Zheng SY, Dhruva SS, Redberg RF. Characteristics of clinical studies used for US Food and Drug Administration approval of high-risk medical device supplements. *JAMA*. 2017;318(7):619–625. https://doi.org/10.1001/jama.2017.9414.
19. Darrow JJ, Avorn J, Kesselheim AS. FDA regulation and approval of medical devices: 1976–2020. *JAMA*. August 3, 2021;326(5):420–432. https://doi.org/10.1001/jama.2021.11171.
20. *Evaluating Substantial Equivalence in Premarket Notifications [510(k)]*. U.S. Food and Drug Administration; 2014. https://www.fda.gov/regulatory-information/search-fda-guidance-documents/510k-program-evaluating-substantial-equivalence-premarket-notifications-510k. Accessed December 31, 2021.
21. Puthumana J, Wallach JD, Ross JS. Clinical trial evidence supporting FDA approval of drugs granted Breakthrough therapy designation. *JAMA*. 2018;320(3):301–303. https://doi.org/10.1001/jama.2018.7619.
22. Downing NS, Shah ND, Aminawung JA, et al. Postmarket safety events among novel therapeutics approved by the US food and drug administration between 2001 and 2010. *JAMA*. 2017;317(18):1854–1863. https://doi.org/10.1001/jama.2017.5150.
23. *Recall Authority, 21 USC §360h(e)(1)*. 2019.
24. Dubin JR, Simon SD, Norrell K, Perera J, Gowen J, Cil A. Risk of recall among medical devices undergoing US food and drug administration 510(k) clearance and premarket approval, 2008–2017. *JAMA Netw Open*. May 3, 2021;4(5):e217274. https://doi.org/10.1001/jamanetworkopen.2021.7274.

Chapter 66

Funding approaches

Levy Jo D. Manuntag[1], Samantha Eugene D. Manuntag[2] and Jayakumar Sreenivasan[3]

[1]*Department of Internal Medicine, Westchester Medical Center/New York Medical College, Valhalla, NY, United States;* [2]*Department of Internal Medicine, University of Reno School of Medicine, Reno, NV, United States;* [3]*Department of Cardiology, Westchester Medical Center, New York Medical College, Valhalla, NY, United States*

Key points

1. To understand the options a researcher has to obtain funding, including the agencies that provide them.
2. To discuss the kinds of grants available and how to select the appropriate kind of grant for the intended research.
3. To be able to write a strong research proposal and avoid the common pitfalls that lead to rejection.

Introduction

Obtaining proper funding for research is an important step in pursuing one's academic goals. Before a researcher can submit a proposal, he or she must have an idea that is compatible with a sponsor's objectives. Once the proper aims have been met, funding may come from either a public or private agency, with different types of grants available depending on the researcher and the kind of work he or she is looking to do. When writing a grant proposal, it is important to state both the objectives of the researcher and how those align with the organization, as well as avoiding ill-prepared applications that may reflect poorly on the applicant.

How to obtain funding

There are a multitude of ways to obtain funding and the research idea will need to be proposed to the right agencies to increase its chances of being financed. Once the plan has been established, finding the appropriate sponsorship will greatly influence how the grant proposal will be written and presented.

The two main avenues by which a researcher can acquire financial backing are public funding and private funding. The former subsidizes grants through government (national, state, county, and city) or other public sectors taken from taxpayers to support and advance growing needs within the country, such as health, education, and social programs. The latter provides monetary assistance through companies and foundations within the private sector and will fund projects that support their visions.[1] Through this chapter, we will explore the different means to look for funding, as well as the characteristics of a strong grant proposal.

Public agencies

In the United States, the US government has a number of branches and individual agencies that have funded research for military advantage since World War II. While each organization has differing goals and objectives for their research, the US government remains the largest contributor of research funding.[1,2] Under the federal government, the National Science Foundation (NSF) is the agency interested in progressing science without regard for utility or political influence and instead focuses on the pursuit of expanding knowledge. On the other hand, the development of biomedical science has become the responsibility of the National Institutes of Health (NIH), where 27 individual but cooperative centers unify under the goal to extend and improve the nation's health. Another major agency is the Department of Defense who support both its own laboratories and outside institutions with the objective of strengthening the country's military. While these three agencies are the mainstays of research funding, there are several others with their own intentions. The full list of these organizations as well as different types of grants offered ways to apply can be found at Grants.gov.[2]

Translational Cardiology. https://doi.org/10.1016/B978-0-323-91790-2.00003-4

Specifically, under the NIH, there are a variety of grants available depending on the type of investigator. Individual research grants fund one large project depending on the kind of grant offered. This includes a fully funded project, smaller grants, developmental grants, and those related to small business advancements needing government assistance. Another type is a multi-project grant that includes many interrelated project proposals with different investigators focusing around a central idea. As for beginning researchers, career development and training awards are offered to those within the early years of their profession to provide supplementation to their salaries and funds and to encourage the long-term pursuit of research. These grants are typically reserved for students, fellows, and new faculty and are limited to US citizens or resident aliens.[3]

Private agencies

Corporations and foundations are the main sources of private grants, contributing approximately $8 billion dollars yearly in the United States. Given that there is less regulation, grants from these sources are often less formal when applied and can be used to fund research that is more experimental in nature. Also, due to the more casual nature of the submission, funding can be obtained through networking and personal contacts, as well as pitching ideas directly to companies.[4]

Corporations provide financial support in a for-profit manner that promotes their business and furthers their monetary gains.[4] There are increasing need for university-industry partnerships, encouraged by the government and universities, so that universities can grow fiscally and industries are able to increase profits. While there are increasing formal application and criteria needs, informal means of procuring funding can be done through interacting with personal contacts, such as a former employer, colleague, or former coworkers who have moved into the industry.[5]

On the other hand, trusts and foundations with philanthropic goals contribute money to policymaking, social reform, and research that is in line with their mission.[4,5] Of note, financial aid can be awarded through community-based participatory research, in which organizations can work with investigators within the community. This partnership is a means to promote and translate clinical research to the public and can serve as a stepping stone for early-career scholars.[6] By obtaining smaller grants, a researcher can build a foundation for his or her career by demonstrating the ability to carry out a planned proposal, thereby providing a foundation for future applications.

Before sending out proposals, the investigation into which sources would be the most fitting to fund the research should be done. Within the United States, the *Foundation Center Online* (https://fconline.foundationcenter.org/) is a database that can be accessed to search for key topics, such as subject matter or geographical preference. Researchers should also consider looking into international organizations, such as The Asia Foundation (https://asiafoundation.org/), for more opportunities. In the end, it is best to check the websites of foundations that fall within the research's scope and, if available, contact the appropriate personnel before submitting a proposal.

How to write a grant proposal

Writing a grant application can be difficult, oftentimes standing in the way of a strong research study if done poorly. The first step is to start early when writing the initial draft to give enough time to begin writing and looking into multiple grants and their assessment criteria.[7] If possible, finding a grants administrator before and during the drafting process can provide valuable guidance through the process, as well as a biostatistician for their input on methodology.[7] When in the early stages, it is necessary to find the right grant and have the proposal align with the funder's central objectives.[2,7] Once this has been done, the researcher should read through the application instructions and draft an individualized proposal based on their guidelines. While many of these criteria are similar across multiple organizations, it is important to personalize each proposal to answer specific queries posed by the funding body. It may also help to reach out directly to the program officer in charge of the grant with questions or to a colleague who has obtained a similar grant with the organization in which the researcher is applying.[2,8]

When writing a good proposal, there are important questions that one must answer, also known as the 3Ps, as listed below:[7,9]

- Person—Who is the researcher and what are his/her qualifications? Does he/she have the skills to conduct a successful study at the current stage of his or her career?
- Project—What is the importance of this research and how will it address an existing problem or challenge it in a new, groundbreaking way? What are the potential outcomes that could change science or society as a whole?
- Place—What institution is this researcher coming from? What academic background does his or her institution have and what manpower and facilities are available that would allow them to conduct a successful study?

The content of the proposal should be clear and concise, translating easily to both expert researchers and laymen alike. An informal review of the NIH grant review process showed that the most important sections of the application were the hypothesis/aims of the study, methodology, and preliminary work; thus, these areas should receive ample time and review before submission.[10] Alongside the above criteria, there is a growing demand for equity, diversity, and inclusion in research, as the majority of valued research is based in high-income countries with the majority of editors of academic journals being men. Since the majority of global health burdens affect low- and middle-income countries, research reflecting these populations is more important than ever, as is the need for increasing female leadership in scientific exploration.[11]

Common pitfalls

No matter how interesting or innovative an idea is, if it is presented in a poorly written or haphazard way then it is unlikely to be funded. Common mistakes include grammatical or spelling errors, which can be easily fixed with proofreading software or a third-party editor. A poorly structured proposal with vague descriptions, unnecessary wordiness, or inconsistencies within the text can also account for rejection.[12–14] These mistakes show carelessness that may insinuate the researcher as disorganized or careless, reflecting negatively on his or her ability to handle a complex project. It is also best to avoid the use of acronyms or jargon as reviewers of the proposal may include those outside of the field of research. It is best to simplify ideas for all types of readers.[8,10,13]

Within the content of the proposal, other issues include that the aims of the study lack focus, are not fully developed, or may be overpromising.[10] Or perhaps the study is not innovative or novel in its goals and does not fall within the funder's research goals, or the budget does not accurately reflect the activities covered within the narrative.[12,13] These problems can be remedied by revisions and feedback from coinvestigators and outside personnel to clarify the direction of the proposal. By carefully mapping out the application to efficiently meet the reviewer's assessment criteria in a succinct manner, the researcher will be able to maximize his or her chances to obtain funding.

References

1. Bauer DG. *The "How to" Grants Manual: Successful Grantseeking Techniques for Obtaining Public and Private Grants*. Lanham, MD: Rowman & Littlefield; 2021.
2. Berry D. *Gaining Funding for Research a Guide for Academics and Institutions*. Berkshire, England: Open University Press; 2010.
3. American College of Cardiology. https://www.acc.org/education-and-meetings/meetings/meeting-items/2020/09/22/19/18/webinar-how-can-ec-cardiologists-build-a-successful-research-career-grant-research; 2020. Accessed December 28, 2021.
4. Devine EB. The art of obtaining grants. *Am J Health Syst Pharm*. 2009;66(6):580–587. https://doi.org/10.2146/ajhp070320.
5. Kenway J, Al E. *Winning & Managing Research Funding*. Gardners Books; 2007.
6. Tendulkar SA, Chu J, Opp J, et al. A funding initiative for community-based participatory research: lessons from the harvard catalyst seed grants. *Prog Commun Health Partnersh*. 2011;5(1):35–44.
7. Nelson AW, Lamb AD, Gnanapragasam VJ. Applying for research funding. part 2 – writing a grant application. *J Clin Urol*. 2016;9(4):230–233. https://doi.org/10.1177/2051415816630212.
8. Lindgreen A, Di Benedetto CA, Verdich C, et al. How to write really good research funding applications. *Ind Market Manag*. 2019;77: 232–239. https://doi.org/10.1016/j.indmarman.2019.02.015.
9. Arthurs OJ. Think it through first: questions to consider in writing a successful grant application. *Pediatr Radiol*. 2014;44(12): 1507–1511. https://doi.org/10.1007/s00247-014-3053-6.
10. Inouye SK, Fiellin DA. An evidence-based guide to writing grant proposals for clinical research. *Ann Intern Med*. 2005;142(4):274. https://doi.org/10.7326/0003-4819-142-4-200502150-00009.
11. Nafade V, Sen P, Pai M. Global Health Journals need to address equity, diversity and inclusion. *BMJ Glob Health*. 2019;4(5). https://doi.org/10.1136/bmjgh-2019-002018.
12. Mikal JP, Rumore G. *10 Common Grant-Writing Mistakes*. The Chronicle of Higher Education; 2018. https://www.chronicle.com/article/10-common-grant-writing-mistakes/. Accessed December 28, 2021. Published July 22, 2020.
13. Ward D. *The Top 10 Grant-Writing Mistakes*. Principal; 2002. https://eric.ed.gov/?id=EJ642992. Accessed December 28, 2021. Published November 30, 2001.
14. Giladi AM, Hume KM, Chung KC. Factors impacting successfully competing for research funding. *Plast Reconstr Surg*. 2014;134:59. https://doi.org/10.1097/01.prs.0000455401.44917.98.

Chapter 67

Conflict of interest

Diego Laurentino Lima[2], Carlos Andre Balthazar da Silveira[1], Ana Caroline Dias Rasador[3] and Raquel Nogueira[1]

[1]Bahiana School of Medicine and Public Health, Brotas, BA, Brazil; [2]Department of General Surgery, Montefiore Medical Center and the Albert Einstein College of Medicine, Bronx, NY, United States; [3]Department of Surgery, Montefiore Medical Center, Bronx, NY, United States

Key points

- Impact of Conflict of Interest (COI) on research and medical practice.
- Increasing focus on COI and transparency.
- Legislative and policy responses.

Introduction

Conflict of interest (COI) is defined as a "compromise of a person's objectivity when that person has a vested interest," not restricted to the research setting or medical profession.[1] The COI may occur when a person benefits financially or nonfinancially from the results of a professional activity by receiving financial support or gaining social prestige. Regardless of the type, when applied to research and biomedical sciences, the presence of COI may generate biases during the interpretation of data and application of interventions, possibly implicating patients' health and outcomes.[2,3]

The 80s was an emblematic decade regarding COI, in which several studies and commentaries were published to assess the impacts of this problem in the research field.[4,5] Since then, there has been an increasing need for transparency in most fields, especially biomedical research and medical practice, where this is an essential subject of debate. Compared to 2006, where the number of publications assessing the COI of the authors was 600, in 2017, 30,000 publications were assessing the same matter, compared to up to 400,000 in the last year (2022).[6,7] This urgent need for transparency resulted from the perception that few policies existed to assess the COI of physicians and researchers and few policies to manage the possible relationship between these professionals and the industries involved.[8,9]

Examples of conflicts of interest

There are multiple examples of COI. However, one of the most important and discussed is the financial COI. The financial COI is a well-recognized definition in which the researchers, especially those involved in clinical research, receive financial benefits from affiliated industries to perform various activities.[10,11] In these cases, physicians may be asked to give teaching lessons, advice, and provide research expertise in exchange for reimbursement. Later, these researchers may also be asked to report favorable outcomes to either the intervention or control being applied during the research process. Sometimes, the researcher can work directly as a consultant for the industries or be engaged in the company's membership panel, commonly sharing the same interests as the other members.[11,12] Thereby, the researcher may act in favor of the company and may distance himself from the patient's health care and the validity of the research project.[12] Dr. Arnold Relman, former editor of The New England Journal of Medicine, published an article in 1984 that already reported the problem of the financial connections made by the pharmaceutical and medical industries to gather academic researchers to perform studies that favored their products.[4,13] This way, the physicians and researchers keep being enrolled in a capitalist setting that, by subsidizing the whole research process, only acknowledges the presentation of favorable outcomes and has great disinterest in the unfavorable ones since the last ones can negatively impact the industry profits.[4]

Previous evidence has already shown that when the researcher is financially involved with industries to perform a study, this study has a higher chance of favoring the company that is sponsoring the research[14,15]; also, this study is more likely to be inferior in terms of quality[16,17]

Translational Cardiology. https://doi.org/10.1016/B978-0-323-91790-2.00055-1

and has lower chance to be published.[18] In 1998, Stelfox et al.[19] published an article assessing the adherence of physicians to calcium channel antagonists' prescription and its relationship with the financial connections between these doctors and industries. This study showed that doctors who adhered to calcium channel antagonist prescriptions presented a higher chance of having a financial relationship with companies in the field and pharmaceutical manufacturers. Also, this study showed that only 2 of the 70 studies assessed reported the COI of the authors on the published manuscript. Another study from Barnes and Bero,[16,20,21] published in The New England Journal of Medicine, made a regression analysis on the studies about passive tobacco smoking. This study showed that three-quarters of the studies at the time, which concluded that passive smoking did not impair health quality, were from authors affiliated with tobacco industries.

The funding necessary to perform research studies, especially the ones regarding important updated topics, will probably derive from companies with secondary interests in the favorable outcomes of the products being studied, enlightening some authors about the possibly inherent association between funding and COI.[22] Conversely, Korn[23] reported, in a published commentary, that the company's research funding is, most times, a vital aspect for the academic centers to keep working and performing updated high-quality research, and therefore, some level of COI is inevitable and must be managed by the academic centers. In these cases, the researchers frequently experience difficulties maintaining the trust and confidence of the readers and the government while constructing the research study and receiving fundamental funding, which allows constant productivity.

Another kind of COI is the nonfinancial COI, which is much more subtle and challenging to identify than the financial one and is not necessarily illegitimate or associated with negative secondary interests.[23] In this type of COI, due to interest in benefits from the publication, such as validity, recognition, prizes, or even due to the desire to generate impactful evidence to improve literature and medical knowledge, the researcher gets intellectually attached to a point of view, which compromises the judgment and impacts negatively in generating trustful evidence.[3,23] Similar to the financial COI, it is inherent to medical practice and biomedical research and can equally impair the construction of the research study. However, it does not receive the same reprisal as the financial COI. The author reports that, due to public awareness and aversion to financial COI, the responsibility for validating the research interests and biases most times is delegated to the government, while the management of nonfinancial COI receives much less attention and is performed by academic centers and researchers.[23]

The British Medical Journal showed that many authors report, during the submission of a paper, that they do not have COI despite having them. According to the journal, this probably relates to the fact that most researchers fear that the public will not trust the study and its results or even make negative assumptions about the author. Also, the journal reports that in some cases, the author may feel confident about not impairing judgment with COI, so they prefer not to declare it, avoiding the same negative assumptions. The presence of COI between a researcher and the industries does not have to be proof that this person's judgment is compromised since most of the advances in medical technologies, vaccines, and medications used worldwide were only possible because of the support of research funding.[13]

Dr. Fineberg, in a viewpoint published in The Journal of the American Medical Association (JAMA), reported that "the preservation of trust is the essential purpose of policies about conflict of interest." The physicians and researchers, therefore, should acknowledge the presence of COI during the construction of research studies, making visible to the public, in general, the possible impacts of this financial conflict in the report of data and conduction of the study. This, however, is an important point of subjectivity in the topic since the editors and public readers may interpret the impact of this COI differently, considering the amount of money involved or the grade of involvement between the industries and the researcher.[22]

Since the abolition of COI in medical practice and biomedical sciences research production is unlikely to happen, the academic centers, research groups, industries, and government must develop new policies to assess this COI during the production and report of data from research studies. Several journals report that the best way to improve the reduction of bias during the interpretation and report of data is to acknowledge the COI but also develop new policies to assess these COI during the production of research evidence.[13,24,25]

Establishing guidelines and regulations to manage conflicts of interest is suggested in a field where trust and objectivity are paramount. The Physician Payment Sunshine Act, established in the United States in 2010, is an example of legislation that mandates transparency in the relationships between physicians and teaching hospitals with pharmaceutical and medical device companies, being an enforcement of COI report. Under this act, these entities must report financial transactions and gifts they receive, making this information publicly accessible. It serves as a cornerstone, reinforcing the importance of transparency and ethical conduct in medical research and healthcare, ultimately benefiting the scientific community and the patients.[24,26] Honesty in the relations between the researchers, industries, and the patient must be the final aim of those efforts.

References

1. Faerstein E. A dictionary of epidemiology. *Cad Saúde Pública*. 2016;32. https://api.semanticscholar.org/CorpusID:9682827.
2. Clark J. Does conflict of interest disclosure worsen bias? *PLoS Med*. April 24, 2012;9(4):e1001210.
3. Norris SL, Holmer HK, Ogden LA, Burda BU. Conflict of interest in clinical practice guideline development: a systematic review. *PLoS One*. October 19, 2011;6(10):e25153.
4. Relman AS. The new medical-industrial complex. *N Engl J Med*. October 23, 1980;303(17):963–970.
5. Relman AS. Dealing with conflicts of interest. *N Engl J Med*. September 19, 1985;313(12):749–751.
6. https://www.ncbi.nlm.nih.gov/pubmed/?term=conflict+of+interest. Accessed 18 October 2023.
7. Annane D, Charpentier B. Do I have a conflict of interest? Yes. *Intensive Care Med*. October 20, 2018;44(10):1741–1743.
8. Vogeli C, Koski G, Campbell EG. Policies and management of conflicts of interest within medical research institutional review boards: results of a national study. *Acad Med*. April 2009;84(4):488–494.
9. Weissman JS, Koski G, Vogeli C, Thiessen C, Campbell EG. Opinions of IRB members and chairs regarding investigators' relationships with industry. *J Empir Res Hum Res Ethics*. March 1, 2008;3(1):3–13.
10. Thompson DF. Understanding financial conflicts of interest. *N Engl J Med*. August 19, 1993;329(8):573–576.
11. Laupacis A. On bias and transparency in the development of influential recommendations. *Can Med Assoc J*. January 31, 2006;174(3):335–336.
12. DeAngelis CD. Conflict of interest and the public trust. *JAMA*. November 1, 2000;284(17):2237.
13. Smith R. Beyond conflict of interest. *BMJ*. August 1, 1998;317(7154):291–292.
14. Rochon PA. A study of manufacturer-supported trials of nonsteroidal anti-inflammatory drugs in the treatment of arthritis. *Arch Intern Med*. January 24, 1994;154(2):157.
15. Bero LA, Galbraith A, Rennie D. The publication of sponsored symposiums in medical journals. *N Engl J Med*. October 15, 1992;327(16):1135–1140.
16. Rochon PA. Evaluating the quality of articles published in journal supplements compared with the quality of those published in the parent journal. *JAMA, J Am Med Assoc*. July 13, 1994;272(2):108.
17. Bero LA, Rennie D. Influences on the quality of published drug studies. *Int J Technol Assess Health Care*. March 10, 1996;12(2):209–237.
18. Blumenthal D. Withholding research results in academic life science. *JAMA*. April 16, 1997;277(15):1224.
19. Stelfox HT, Chua G, O'Rourke K, Detsky AS. Conflict of interest in the debate over calcium-channel antagonists. *N Engl J Med*. January 8, 1998;338(2):101–106.
20. Cho MK. The quality of drug studies published in symposium proceedings. *Ann Intern Med*. March 1, 1996;124(5):485.
21. Davidson RA. Source of funding and outcome of clinical trials. *J Gen Intern Med*. May 9, 1986;1(3):155–158.
22. Fineberg HV. Conflict of interest. *JAMA*. May 2, 2017;317(17):1717.
23. Korn D. Conflicts of interest in biomedical research. *JAMA*. November 1, 2000;284(17):2234.
24. Wilson M. The Sunshine Act: commercial conflicts of interest and the limits of transparency. *Open Med*. 2014;8(1):e10–e13.
25. Brems JH, Davis AE, Clayton EW. Analysis of conflict of interest policies among organizations producing clinical practice guidelines. *PLoS One*. April 30, 2021;16(4):e0249267.
26. O'Connor CM. Conflict of interest and transparency. *JACC Heart Fail*. January 2015;3(1):94.

Chapter 68

How to do subject recruitment?

Maham A. Waheed[1] and Anam Akbar Waheed[2,3]

[1]*Maimonides Medical Center, Brooklyn, NY, United States;* [2]*Medstar Washington Hospital Center, Washington, DC, United States;* [3]*Medstar Georgetown University Hospital, Washington, DC, United States*

Importance of appropriate subject recruitment

The Department of Health and Human Services (DHHS) as well as the US Food and Drug Administration (FDA) require that clinical trials have equitable and representative recruitment of subjects. This is crucial to have a trial result that is valid, representative of the population and therefore applicable clinically to the population. Efficient patient recruitment is an important key factor in determining the success of a clinical research trial. Low rates of recruitment and retention as well as unbalanced recruitment can result in negative implications, such as decreased statistical power and validity of scientific results as well as longer clinical trial durations, and increased financial constraints.

Delays in participant recruitment in turn then contribute significantly to a delay in trial completion. It has been reported that more than 80% of clinical trials are unable to reach their recruitment targets by specified time periods and that study durations have almost doubled beyond their planned enrollment durations due to low recruitment rates. This not only extends the duration of the study period, but adds considerably to the financial, ethical, and scientific constraints associated with extending study trial periods.

Identifying recruitment targets

For successful trial recruitment, the investigator needs to have access to an adequate number of persons who fit the study inclusion criteria and are representative of the population. He or she then needs to obtain informed consent and follow up on those recruited. The persons recruited need to be willing to enroll and participate in the study.

The Clinical Trials Transformation Initiative (CITI) has proposed an upstream approach to recruitment planning to help mitigate barriers to recruitment. The CITI framework mainly focuses on three target areas: trial design and protocol development, trial feasibility and site selection, and the recruitment and planning stages, respectively.

In the trial design and protocol development phase, it is important to clearly define recruitment targets and timelines. Specifically, the inclusion and exclusion criteria need to be defined, as well as the factors such as race, gender, and socioeconomic status across which we need adequate representation for valid results. Inclusion of special populations such as pregnant women and their recruitment should also be carefully considered when designing the trial. Recruitment of special populations and vulnerable populations is covered separately in another chapter of this book.

Patient or disease based registries are great resources that aid clinicians in retrospective identification of potential subjects for enrollment in prospective clinical trials. Registries such as population based cancer registries, disease specific registries, national health databases, and web-based disease-neutral registries, provide large databases for the identification of eligible participants for clinical trials. Their utilization for the purposes of enrollment allows for efficient, timely recruitment. A study conducted in 2013, explored the use of patient and disease registries over 10 years, as a means of recruitment for prospective clinical trials and discovered that registries provide large databases for efficient recruitment of potential subjects.[1] The ability for retrospective identification and eligibility for enrollment in various trials from a large sample population, using patient registries, allows study investigators to research different diseases such as cancer, diabetes, asthma, and stroke.[1,2]

Site selection should be done to optimize recruitment targets. The CITI framework recommends an evidence based trial feasibility analysis to assess disease prevalence, site infrastructure, site capabilities, and target population

Translational Cardiology. https://doi.org/10.1016/B978-0-323-91790-2.00106-4

access. There should also be a plan for suitable site trial surveillance and follow-up. At the sites, primary care physicians or subspecialists who can help refer patients to the trial should be engaged and educated on the trial which would increase patient engagement and resources.

Subject recruitment phase and reducing barriers

Researchers then need to optimize and plan out effective recruitment strategies early during study planning to widely reach out to potential participants who meet inclusion criteria. Anticipated barriers to recruitment should be addressed in the study design. Engagement of healthcare providers, patient advocacy groups, disease-specific advocacy groups, and community leaders can all help bolster recruitment. Advertising via flyers, phone calls, and television can also be done to increase the target population's awareness of the trial. For instance, when studying dietary effects on high blood pressure, the investigators of the DASH study employed multiple recruitment strategies such as mass mailing, mass media advertising, and worksite screenings. Of these, the mass mailing strategy provided the highest yield of participants.[3]

The process of recruitment should be closely overseen by a supervising party who can monitor and evaluate the process of recruitment at regular intervals and assist in navigating potential barriers to increase timely enrollment and achieve early recruitment goals. For this purpose, the FDA requires an Institutional Review Board (IRB) to review clinical studies and ensure they comply with the IRB regulations and thus help maintain equity in the subject recruitment process. For instance, the study investigators conducting the Survival and Ventricular Enlargement Trial (SAVE) studied the effect of captopril, an angiotensin-converting enzyme inhibitor, on morbidity and mortality in patients with left ventricular dysfunction after myocardial infarction. They conducted a randomized, double blinded, placebo controlled trial with a large sample size of 2231 patients. These patients were recruited over a course of 3 years from 112 different participating hospital sites in the United States and Canada, based on strict eligibility criteria as enlisted in the study protocol. The principal investigator served as the supervising party for all the co-investigators and oversaw the overall execution of the clinical trial.[4]

The IRB ensures that the confidentiality of all research subjects is maintained and that their rights are protected. Additionally, the IRB also overseas materials and methods, any direct recruitment advertisements, as well as the informed consent form used to aid in subject recruitment to ensure that there is voluntary participation and no coercion. The tools used for recruitment must provide accurate information in regard to the study with a clear delineation of inclusion criteria to allow prospective participants to determine their eligibility for enrollment. It must be devoid of any coercive material and explain the risks and benefits of participation in simple terms as described by the study protocol.

Protocol complexity is recommended to be kept to what is necessary to reduce the burden of participation. Willingness to participate is often affected by complicated study questionnaires (complex trials), lengthy study durations, invasiveness of the study intervention, and potential risks associated with the intervention or fears regarding blinded study drugs and outcome measures. Transparent and clear explanations of risks and benefits, and the availability of personnel to answer questions and concerns can significantly allay fears.

Additionally, allocating a recruitment budget and resources to aid in the patient enrollment process can also help during the recruitment process. Financial incentives for trial participants can also help augment early recruitment goals. Incentives such as transport to and from a trial site for a trial visit, meal coupons, and discounts on other medications for the disease being investigated are other incentives that can be offered to augment recruitment. The investigators of the DASH study also offered financial reimbursement to study participants and 6% of the individuals enrolled did list financial incentives as a major reason for participation.[3]

Key points

- Subject recruitment is essential for the validity, generalizability, and success of clinical trials.
- It is crucial for study investigators to plan out sample recruitment before trial initiation.
- Subject recruitment should be planned clearly and intentionally in the protocol development phase.
- A clear inclusion and exclusion criteria should be developed.
- It is important to recruit subjects representing different races, genders, and socioeconomic statuses to have valid trial results.
- Early identification and navigation of barriers to enrollment is vital.

References

1. Tan MH, Thomas M, MacEachern MP. Using registries to recruit subjects for clinical trials. *Contemp Clin Trials*. March 1, 2015;41:31–38.
2. Trotter JP. Patient registries: a new gold standard for "real world" research. *Ochsner J*. September 21, 2002;4(4):211–214.

3. Vollmer WM, Svetkey LP, Appel LJ, et al. Recruitment and retention of minority participants in the DASH controlled feeding trial. *Ethn Dis.* April 1, 1998;8(2):198–208.
4. Pfeffer MA, Braunwald E, Moyé LA, et al. Effect of captopril on mortality and morbidity in patients with left ventricular dysfunction after myocardial infarction: results of the survival and ventricular enlargement trial. *N Engl J Med.* September 3, 1992;327(10):669–677.

Further reading

1. Huang GD, Bull J, McKee KJ, Mahon E, Harper B, Roberts JN, CTTI Recruitment Project Team. Clinical trials recruitment planning: a proposed framework from the clinical trials transformation initiative. *Contemp Clin Trials.* March 1, 2018;66:74–79.
2. Anderson DL. *A Guide to Patient Recruitment and Retention.* Centerwatch; 2004.
3. Sullivan J. *Subject Recruitment and Retention: Barriers to Success.*
4. Lamberti MJ, Mathias A, Myles JE, Howe D, Getz K. Evaluating the impact of patient recruitment and retention practices. *Drug Inf J.* September 2012;46(5):573–580.
5. Bielski RJ, Lydiard RB. Therapeutic trial participants: where do we find them and what does it cost? *Psychopharmacol Bull.* 1997;33(1):75.
6. Ross S, Grant A, Counsell C, Gillespie W, Russell I, Prescott R. Barriers to participation in randomised controlled trials: a systematic review. *J Clin Epidemiol.* December 1, 1999;52(12):1143–1156.
7. Hunninghake DB, Darby CA, Probstfield JL. Recruitment experience in clinical trials: literature summary and annotated bibliography. *Contr Clin Trials.* December 1, 1987;8(4):6–30.
8. Sussman AL, Cordova C, Burge MR. A comprehensive approach to community recruitment for clinical and translational research. *J Clin Transl Sci.* August 2018;2(4):249–252.
9. Kost RG, Corregano LM, Rainer TL, Melendez C, Coller BS. A data-rich recruitment core to support translational clinical research. *Clin Transl Sci.* April 2015;8(2):91–99.
10. Recruiting Study Subjects. *Guidance for Institutional Review Boards and Clinical Investigators.* FDA.gov; January 1998.
11. Newcomb PA, Love RR, Phillips JL, Buckmaster BJ. Using a population-based cancer registry for recruitment in a pilot cancer control study. *Prev Med.* January 1, 1990;19(1):61–65.
12. Bolen S, Tilburt J, Baffi C, et al. Defining "success" in recruitment of underrepresented populations to cancer clinical trials: moving toward a more consistent approach. *Cancer: Interdiscipl Int J Am Cancer Soc.* March 15, 2006;106(6):1197–1204.

Chapter 69

Data management

Ana Carolina do A.H. de Souza[1] and Jade Iwasaka-Neder[2]

[1]Department of Radiology and Medicine (Cardiology), Brigham & Women's Hospital/Harvard Medical School, Boston, MA, United States;
[2]Department of Radiology, Boston Children's Hospital/Harvard Medical School, Boston, MA, United States

What is research data, and why should we manage it?

Research data can be defined as "any information or evidence that supports research findings."[1] In clinical and translational research, data refers to any information collected and generated during a research project, such as patient demographics, laboratory results, imaging studies, or treatment outcomes. Data can exist in different formats, including datasets, notes, research notebooks, codebooks, video recordings, audio, data pipelines, contextual data (metadata), and more.[2] Due to the complexity of handling an increasing amount of data, the FAIR Guiding Principles were developed to provide a framework to ensure that research-derived information is findable, accessible, interoperable, and reusable, thus making scientific results more transparent and reproducible.[3,4] Proper documentation and management of data and other information are critical to ensuring the research project is conducted ethically, accurately, and efficiently.[3,4]

Data management plans

Data management plans (DMPs) outline how research teams handle data during and after a project.[5] These formal documents should align with the FAIR Guiding Principles and provide a clear and comprehensive framework for collecting, organizing, and storing data, as well as ensuring its security.[6] DMPs include a detailed description of the rationale and context for data collection, methods used for acquiring, structuring, and organizing data and metadata files, strategies for validation and quality assurance, confidentiality, and use conditions. DMPs should also cover how data will be shared and made accessible for future research, promoting transparency, reproducibility, and reusability.[6] Many funding agencies require researchers to submit a DMP as part of their grant applications or renewals, including the National Science Foundation (NSF) and the National Institutes of Health (NIH). The specific aspects covered by the DMP vary depending on the funding agency and the nature of the research project. Fig. 69.1 displays the main sections commonly included in a DMP. As the number of funding entities and research institutions requiring DMPs continues to increase,[7] DMPs have become a vital tool for researchers to avoid errors in data handling and ensure that their work meets the standards of scientific integrity.

Online data management plan tools

Online tools and templates are available to help researchers create a DMP both for personal use and for grant applications. Examples include the following:

1. *DMPTool*: The DMPTool is a free online tool that provides guidance on the components of a DMP and includes fillable templates for various American and international funding agencies. Users can also share customized plans with their organizations or publish them for open access. This tool has templates for over 46 agencies and is recommended by several academic centers. Available at dmptool.org.
2. *DMPOnline*: The European equivalent to the DMPtool, developed by the Digital Curation Center in partnership with the Universities of Edinburgh and Glasgow. Has similar features to the DMPtool.[8] Available at dmp.su.se.
3. *DART Rubric*: Guides researchers on writing DMPs specifically for the NSF with detailed explanations for different required sections and types of projects. Available at osf.io/mgjpp.

Principles of data collection, cleaning, and formatting

Once data collection has started, it should be organized, formatted, and stored securely for future analyses,

Translational Cardiology. https://doi.org/10.1016/B978-0-323-91790-2.00092-7

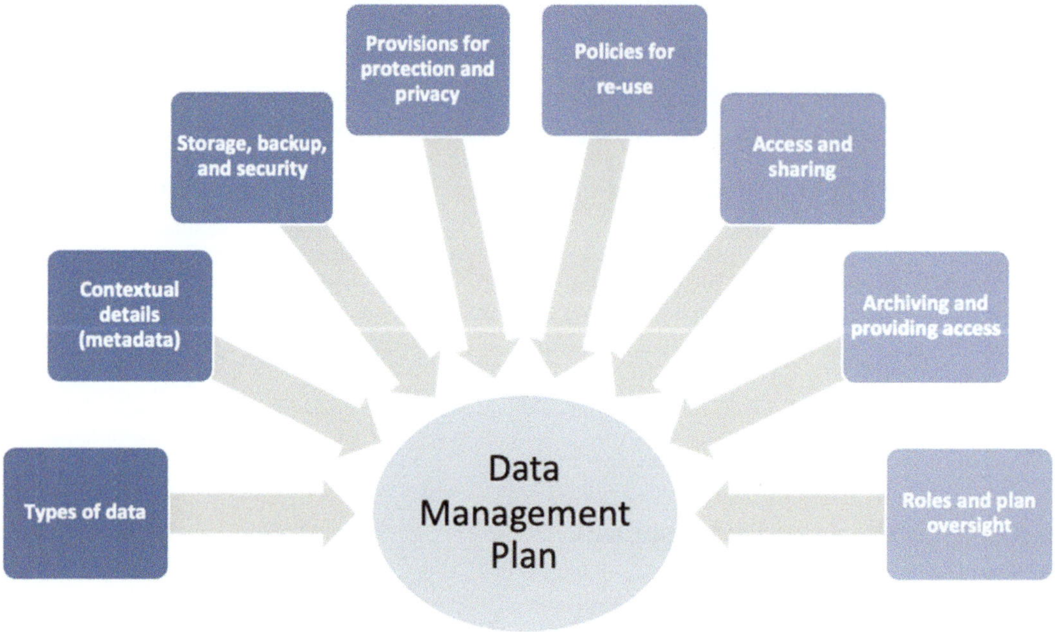

FIGURE 69.1 Main elements of a data management plan (DMP).

transmission, and preservation.[9] Data cleaning and formatting is the first step to creating structured datasets ready for analysis and interpretation. Datasets are collections of quantitative or qualitative values frequently organized under columns and rows. The tidy data principles suggest a standardized way to organize the collected data to facilitate analyses.[10] According to these principles, each variable is represented in a column (e.g., age, ethnicity, body mass index), each observation in a row (e.g., individual study participants), and each cell contains an individual value. Every column and row should be labeled with a distinct title, and data should be entered in a consistent format across the dataset, including categories, times, and dates.[9] A data dictionary with variable meanings should always be available for reference. Finally, reviewing the dataset for potential errors, missing values, or outliers is important.[10,11]

Special considerations should be made about research involving clinical data, which can be obtained prospectively or retrospectively through medical records consultation. According to the Health Insurance Portability and Accountability Act (HIPAA) Privacy Rule, any protected health information (PHI) should be deidentified before secondary use, to protect individuals' privacy.[12] PHI includes any information related to a person's physical or mental conditions and their provision of care that can be used to identify the individual. HIPAA classifies 18 direct identifiers as PHI, including names, birth dates, social security numbers, and others. An anonymization plan should be made before data collection, and the level of deidentification will be based on the plans for access and sharing of the data.[13,14] Secure, HIPAA-compliant storage options should also be defined ahead of time, as detailed in the next sessions.

Data storage: Where and for how long?

The best data storage strategy depends on multiple factors that include data type and size (e.g., text files, DICOM images, codes), frequency of access and modification, and storage needs. The first step is to determine the best formats for data storage depending on the specific project. Whenever possible, files should be saved in interoperable formats that allow for open, unencrypted access (e.g., text file vs. Microsoft Word) and use a file naming convention. Naming conventions should be established and documented in advance and provide a structured way to organize and find files.[15] Since those are usually arranged by name, the most important identifying information should come first, and additional details such as date, type of experiment, and researcher name/initials can be sequentially added. Naming conventions should also include a system for version control, which can be as simple as numbering the files as significant changes are done (e.g., v01, v02).[15]

Potential storage locations include, but are not limited to (1) electronic notebooks (e.g., LabArchives); (2) online platforms (e.g., REDcap, Flywheel); (3) code repositories; and (4) shared cloud storage tools (e.g., Dropbox, Google Drive). To prevent data loss, having backups is essential. The 3-2-1 rule suggests that three copies of the complete data are made, in two different media being at least one located off-site.

The second step is determining for how long the data should be stored. Academic institutions and funding agencies have distinct policies on the duration of data retention, and these regulations should be considered when building the DMP. Funding agencies and research centers usually determine the minimum period that data must be stored after the completion of the research project. The NIH, for example, requires retaining the data for a minimum of 3 years after the end of the grant award.[16] Other institutions may have different policies regarding data storage and sharing, so researchers need to be aware of them.

Data security and privacy

Data security maintenance aims to avoid alteration and unauthorized access to the data, to maintain intellectual property, and to preserve the participant's right to privacy in the case of research involving human subjects. Researchers must comply with relevant ethical and regulatory guidelines, such as obtaining approval from an Institutional Review Board (IRB). Every research team member is responsible for guaranteeing data security, and access to the data should be given only to the personnel directly involved in the work. While all devices containing research data should be encrypted and protected by strong passwords and antimalware programs, paper records and other physical materials should be placed in a locked space with limited access by external individuals. Additionally, institutions should implement regular vulnerability testing to protect devices and systems from attacks and data theft.

Data sharing and publication

Publishing scientific papers has become the mainstem for measuring academic success and achieving professional promotions. Despite being a well-established way to disseminate one's work, peer-reviewed manuscripts are just one of the options for sharing a project's results. Similar to data retention guidelines, funding agencies, institutions, and scientific journals have different requirements on which data should be shared and where. These materials can be published in trusted data repositories, such as Dataverse (available at dataverse.org), Dryad (available at datadryad.org), Figshare (available at figshare.com), and Mendeley Data (available at data.mendeley.com). The re3data (available at re3data.org) is an open registry of research data repositories across disciplines that can be used to consult and compare the best repositories for one's work.

Institutions and researchers on multicenter partnerships can also create their own registries to collect data on specific outcomes of interest.[17,18] These registries work as centralized systems to organize and collect information about patient populations with a particular disease, condition, or exposure. The process of creating a data registry typically involves defining the scope of the registry, what information will be collected, structuring the database, and planning data security strategies, as well as funding sources.[19]

Data registries in cardiology and other fields offer valuable resources for clinical research, allowing investigators to evaluate outcomes, track trends, develop policies, and inform clinical decision-making given the heterogeneity and dimension of the databases.[20] A robust example is the American College of Cardiology National Cardiovascular Data Registry (NCDR), which was idealized in 1987 and currently aggregates multiple registries in different areas, such as the Chest Pain—MI Registry and the CathPCI Registry.[21,22]

Conclusion

Data management plays a pivotal role at all stages of clinical and translational research. It should be planned at the beginning of every new project by creating a DMP to guide the collection, organization, storage, and sharing of research data. There are a variety of free management resources readily available online, some of which were briefly introduced in this chapter. Adopting a structured and comprehensive data management approach can ensure that the results are ethical, accurate, and reproducible by other researchers. Ultimately, appropriate data management practices lead to a more reliable and sustainable system for advancing medical research and improving human health.

References

1. Borgman CL. *Big Data, Little Data, No Data: Scholarship in the Networked World*. MIT Press; 2017.
2. Sarker IH. Machine learning: algorithms, real-world applications and research directions. *SN Comput Sci*. 2021;2(3):160.
3. Wilkinson MD, Dumontier M, Aalbersberg IJJ, et al. The FAIR guiding principles for scientific data management and stewardship. *Sci Data*. 2016;3:160018.
4. *FAIR principles. GO FAIR*. https://www.go-fair.org/fair-principles/. Accessed 1 March 2023.
5. Writing a data management & sharing plan. NIH Sci Data Shar. https://sharing.nih.gov/data-management-and-sharing-policy/planning-and-budgeting-for-data-management-and-sharing/writing-a-data-management-and-sharing-plan. Accessed 14 February 2023.
6. Office of The Director, National Institutes of Health. Final NIH policy for data management and sharing. *NOT-OD-21-013*; 2020. https://grants.nih.gov/grants/guide/notice-files/not-od-21-013.html.
7. Williams M, Bagwell J, Nahm Zozus M. Data management plans: the missing perspective. *J Biomed Inform*. 2017;71:130−142.
8. Sallans A, Donnelly M. DMP online and DMPTool: different strategies towards a shared goal. 2012;7(2):123−129.
9. Longwood Data Management. *Collect & Analyze*. https://datamanagement.hms.harvard.edu/collect-analyze. Accessed 21 May 2024.
10. Wickham H. Tidy data. *J Stat Softw*. 2014;59:1−23.

11. Broman KW, Woo KH. Data organization in spreadsheets. *Am Stat.* 2018;72(1):2–10.
12. Act A. Health insurance portability and accountability act of 1996. *Publ Law.* 1996;104:191.
13. Office for Civil Rights (OCR). *Guidance Regarding Methods for De-identification of Protected Health Information in Accordance With the Health Insurance Portability and Accountability Act (HIPAA) Privacy Rule. Health Information Privacy*: U.S. Department of Health & Human Services. https://www.hhs.gov/hipaa/for-professionals/privacy/special-topics/de-identification/index.html#rationale. Accessed 21 May 2024.
14. Fetzer DT, West OC. The HIPAA privacy rule and protected health information: implications in research involving DICOM image databases. *Acad Radiol.* 2008;15(3):390–395.
15. Goldman J. Research data management. File names. https://datamanagement.hms.harvard.edu. Accessed 21 May 2024.
16. National Institute of Health Office of Extramural Research. Record retention and access. *NIH Grants Pol Statement.* https://grants.nih.gov/grants/policy/nihgps/html5/section_8/8.4.2_record_retention_and_access.htm. Accessed 21 May 2024.
17. Crooijmans C, Jansen TPJ, Konst RE, et al. Design and rationale of the Netherlands registry of invasive coronary vasomotor function testing (NL-CFT). *Int J Cardiol.* 2023;379:1–8.
18. Cossío F, Schurz H, Engström M, et al. VAI-B: a multicenter platform for the external validation of artificial intelligence algorithms in breast imaging. *J Med Imag.* 2023;10(6):061404.
19. Agency for Healthcare Research and Quality/AHRQ. *Registries for Evaluating Patient Outcomes: A User's Guide.* Government Printing Office; 2014.
20. McNamara RL. Cardiovascular registry research comes of age. *Heart.* 2010;96(12):908–910.
21. Brindis RG, Fitzgerald S, Anderson HV, Shaw RE, Weintraub WS, Williams JF. The American college of cardiology-national cardiovascular data registry (ACC-NCDR): building a national clinical data repository. *J Am Coll Cardiol.* 2001;37(8):2240–2245.
22. Weintraub WS, McKay CR, Riner RN, et al. The American College of Cardiology national database: progress and challenges. *J Am Coll Cardiol.* 1997;29(2):459–465.

Chapter 70

Quality control in translational research

Matheus José Barbosa Moreira[1] and Carlos Alberto Campello Jorge[2]
[1]*Federal University of Rio Grande do Norte, Rio Grande do Norte, Brazil;* [2]*Department of Medicine, Federal University of Mato Grosso, Mato Grosso, Brazil*

Key points

The following topics will be discussed in this chapter:

- Definition and cornerstones for quality control systems;
- Components and structure of a quality management plan;
- How to prepare your trial documents and workflow for a quality control visit;

Introductory section

Quality control systems are based on the principles of good clinical practice (GCP). GCP is an international standard of quality defined by the International Conference on Harmonization (ICH). It includes the ethical and scientific requirements to register any pharmaceutical product or device undergoing human subject testing. Since 1970, the Food and Drug Administration has developed GCP, which includes a variety of steps during each and every phase of the trial:

- Protocol development by the study team according to standard quality principles;
- Institutional Review Board (IRB) review of the protocol with particular focus on participant privacy and health protection;
- Attainment of informed consent from each participant before participation in any activities within the study;
- Incorporation of standard methods for the collection of data as well as their submission to a central system, usually in the form of a case report form (CRF);
- Auditing and monitoring of the generated and submitted data;
- Making sure the data is complete, accurate, and consistent.

Get started

The primary objective of the quality control system is to protect the rights of human volunteers at all stages of the study. To ensure this, the GCP outlined the following rules:

- Each trial protocol that includes the testing of drugs or devices on humans must be evaluated by an ethical committee (EC). This evaluation is based on preclinical and clinical data regarding the proposed trial;
- The safety and well-being of the volunteer should always take precedence over the interests of science and society;
- Ethical principles need to be followed in accordance with the Declaration of Helsinki;
- A qualified physician or, when appropriate, a qualified dentist should be responsible for the care and health decisions made during a clinical trial;
- It is important that the team conducting the trial is qualified, trained, and has experience in handling the task assigned to them;
- Trial information should be recorded, handled, and stored in a manner that makes accurate verification possible.

In a trial, the primary responsibility falls on the sponsor, who monitors recruitment sites. Another vigilance measure undertaken by regulatory agencies, such as the FDA, is auditing. Clinical trials are audited to ensure data quality and compliance with treatment and follow-up rules established by the GCP, good laboratory practices, and the protection of human subjects—described in the 21st Code of Federal Regulations, Chapter I, subchapter A, part 50.

Translational Cardiology. https://doi.org/10.1016/B978-0-323-91790-2.00064-2

Auditing and monitoring

Auditing and monitoring are mechanisms that take part in the compliance program of vigilance. Cooperatively, these mechanisms serve the primary purpose of identifying operational weaknesses and fragilities in the program early in a trial. To maintain system improvements, internal and external reviewers should be utilized to detect and prevent irregularities within an organization and to ensure that existing errors are identified and corrected in a timely manner.

Maintaining this cycle of improvement requires nurturing a nonpunishment culture among the trial's team members and their relationship with any Contract Research Organization or sponsor. The ideal goal of 100% compliance is desirable, but it is also unrealistic. Compliance systems are meant to be a continuous process of supervision, leading to continuous improvement of activities and workflows.

Checklist for assessing quality practices in a quality management plan

- Guarantee that the source document is secure, up-to-date, and readable.
- Ensure that the delegation form includes all the staff members, including their duties according to their abilities, training, and licenses according to the protocol's procedures.
- Ensure protocols are written clearly and know which procedures are to be implemented and when these procedures will occur.
- Make sure the organization's monitoring plan is being followed and that data collection is being applied.
- If there are clinical research standard operating procedures (SOPs), including human subject protection and data management, make sure they are being followed.
- Understand how case report forms are developed, and medical and research records secured.
- Ensure compliance with regulations and guidelines.
- Make sure staff members are checking their own work and not relying on others, as guaranteeing data quality is everyone's responsibility. However, always provide a corrective environment and not a punitive one, focusing on task efficiency and on finding out the reason the problem occurred to fortify a preventive plan, documenting the whole process.
- Know about and correctly prepare for the procedures to be performed when receiving a quality inspection at the study site.

Types of visits/auditing

Sponsor may perform a qualification prestudy visit, an initiation visit, monitoring visits, and a closure visit. The **qualification visit** aims to inspect the potential site facilities and the staff's ability to conduct the research, to meet with the study team, and to evaluate the site's readiness for the initiation of the research. The **initiation visit** is performed to review the study's protocol and to resolve any queries that the research staff might have. It occurs before trial activation and the first subject's enrollment, and it might take a whole day to ensure the team's readiness.

The **monitoring visits** can begin right after the first subject's enrollment, at specific times or at different study phases, or continuously by remote monitoring. It aims to review protocol application and data accuracy and to identify areas for improvement. The **closure visit** occurs when the follow-up of all subjects has been completed with the objective of reviewing records, regulatory documents, and data retention guidelines.

Sponsor audits may occur to ensure documentation completeness and site organization for inspections and to suggest actions for performance improvement. They occur due to the result of monitoring findings, such as high patient enrollment, high staff turnover, abnormal number of adverse effects (high or low), and subject enrollment rates not compatible with the research site demographics, and even to assure monitoring practices and quality. The main goals of regulatory audits are to ensure data integrity, verify submitted information, and protect human subjects.

Potential pitfalls: Preparing for a quality control visit

A thorough review of all protocol versions and approvals, including Institutional Review Boards, is essential. Be sure that all original participant consents have been entered in the medical record and are signed and dated according to the institution's policies. In addition to the licenses and Curriculum Vitae of the investigators, review the laboratory certifications and other necessary documents, such as the investigative brochure versions, the statement of the investigator, and the delegation of the responsibility log.

Assure that all laboratory and medical image reports as well as practitioner's notes are included correctly in the medical records. Check the informed consent documentation and the parameters of the documented procedures. All missing data should be registered and obtained whenever possible. Ascertain the PI's attestation of both laboratory and procedure reports, necessary signatures, and their availability.

Inform the pharmacy of the auditing schedule by reviewing other forms and records, such as CRFs and pharmacy records. Ensure the integrity and up-to-dateness of CRFs, and verify the reliability of adverse events reports, including medication use records (with start and end dates), study-related medications, and concomitant medications. The pharmacy record should include dispensing information, along with a drug count and registration if there are any discrepancies. If applicable, verify the Investigational Agent Accountability Record.

How to conduct a successful quality visit

- Welcome the monitor/auditor and guide them to the appropriate rooms, orienting them to important areas such as bathrooms and telephones, and providing them with information as needed;
- Whenever a reviewer first contacts a research site, discuss the formats of medical and research records;
- Ensure that all electronic systems, forms, and regulatory files needed are available in the monitor room;
- Schedule appointments with PI, pharmacy, and other requested sectors, accompanying the reviewer to the designated places at the scheduled times;
- Verify that medical and research records are stored securely if the activity will take more than 1 day;
- Previously review and constantly improve the main shortcomings, such as issues with the informed consent form, as well as failure to follow the protocol, to keep standardized and accurate records, to report adverse events, or to account for the study's drug disposition, even if it is not an investigational new drug research.

Further reading

1. Macefield RC, Beswick AD, Blazeby JM, Lane JA. A systematic review of on-site monitoring methods for health-care randomised controlled trials. *Clin Trial.* February 2013;10(1):104–124. https://doi.org/10.1177/1740774512467405.
2. Fukushima M. Gan to Kagaku Ryoho. 1996;23(2):172–182.
3. Center for Clinical and Translational Science (CCTS). *Responsibilities of the Research Team.* CCTS at the Ohio State University. Retrieved Sep 10, 2022, from: https://ccts.osu.edu/content/responsibilities-research-team.
4. Ottevanger PB, Therasse P, van de Velde C, et al. Quality assurance in clinical trials. *Crit Rev Oncol Hematol.* 2003;47(3):213–235. https://doi.org/10.1016/s1040-8428(03)00028-3.
5. Valania M. *Quality Control and Assurance in Clinical Research.* Applied Clinical Trials Online; 2006. Retrieved Sep 10, 2022, from https://www.appliedclinicaltrialsonline.com/view/quality-control-and-assurance-clinical-research.
6. Johnson J, Gupta NV. Recent advances in quality management of clinical trials. *Int J Pharm Pharmaceut Sci.* 2013;5(3):34–38.
7. International Conference on Harmonisation of Technical Requirements for Registrations of Pharmaceuticals for Human Use. *Guidelines for Good Clinical Practice E6(R1)*; June 10, 1996. Retrieved from http://www.ich.org/fileadmin/Public_Web_Site/ICH_Products/Guidelines/Efficacy/E6_R1/Step4/E6_R1__Guideline.pdf Johnson.

Chapter 71

Research in acute care settings

Matheus Simonato[1], Beatriz Marinho[2], Marcela de Almeida Lopes[3] and Michael G. Nanna[1]

[1]*Department of Internal Medicine, Yale School of Medicine, New Haven, CT, United States;* [2]*School of Medicine, Universidade Estacio de Sa, Rio de Janeiro, Brazil;* [3]*HCor Research Institute, São Paulo, Brazil*

Introduction

Research in acute care settings is defined as research conducted in patient populations seeking care for events that were originally unexpected or unplanned. One such example would be a randomized clinical trial of an intervention for the treatment of acute myocardial infarction leading to cardiogenic shock. Many of the treatments for patients presenting acutely with cardiovascular disease are supported by only limited, if any, evidence.[1] This issue, which is not exclusive to cardiovascular disease, represents a ripe opportunity for research, given the many unanswered questions that still exist in clinical practice.

However, there are a number of challenges that surround the development of prospective research in acute care settings that may serve as barriers to carrying out meaningful investigations.[2] First, given the acute nature of the diagnoses requiring further study, there are significant time constraints due to a limited window to identify the appropriate patients for enrollment, activate the research team, obtain informed consent, and implement the intervention. Second, the ability of patients in the acute care setting to consent may be limited by their diagnosis, requiring specific attention to how best to conduct the informed consent process. Third, many of the patients that present in this specific research setting have significant social vulnerabilities. Understanding these challenges and ways to appropriately address them is essential to the successful and ethical implementation of acute care research.

Get started

Many of the aspects of acute care research are similar to research in other settings. Acute care investigators will naturally need a study question, hypothesis, statistical planning, and protocol. Therefore, we will focus on the unique characteristics and challenges of acute care research in this brief chapter.

Timing

As mentioned above, one of the key characteristics of acute care research is its time-sensitive nature. Researchers planning interventional (including pharmacological and procedural interventions) acute care research must develop a well-designed plan that details how to deliver study interventions in an agile and efficient way, requiring the concerted collaboration of the involved stakeholders: investigators, emergency department, critical care physicians, nursing, ancillary staff, patients and their families. Similarly, investigators must have the support of highly trained research staff, potentially available 24 h a day, with specific skills to successfully implement research processes in acute care settings.[3] One recent example of a successful acute care research effort in an interventional trial is the RESILIENT study in Brazil. Investigators planned to randomize 600 stroke patients to thrombectomy within 8 h of symptom onset.[4] This specific trial was conducted in a developing country where the government was "not convinced"[4] of the feasibility of implementation of the strategy in its public healthcare system. Even in this complex environment, investigators were able to implement a protocol that required brain imaging within 30 min and arterial access within 60–90 min of randomization in 12 reference hospitals.[4] Coordination of efforts for early identification of potentially eligible subjects, effective recruitment, mobilization of the research team to complete informed consent and clinical stakeholders to implement the study protocol, and follow-up of study participants to ensure fidelity of study endpoints are all necessary steps to the successful implementation of an acute care research project.

Translational Cardiology. https://doi.org/10.1016/B978-0-323-91790-2.00104-0

Informed consent

As is the case with all interventional research, the process of informed consent is essential and provides an opportunity to clearly explain to the patient their diagnosis and the rationale for planned interventions, including expected risks and benefits of participating in the study versus undergoing standard care. However, in acute care situations, patients may be unconscious or unable to fully rationalize the situation at hand given the multitude of potential influences and stressors. In many situations, surrogate decision-makers become necessary. Research on this topic has shown surrogate decision-makers are motivated by the potential for research to help others, the importance of research for medical progress, and trust in the medical team.[5] Nevertheless, surrogate decision-makers may also refuse consent due to being too anxious to consider research, fears that the patient would receive the experimental treatment, and concerns about the risks of the study.[5] In the fast-paced environment of acute care research, care must be taken to use the time allotted for informed consent in the most efficient way possible while ensuring that the appropriate ethical standards are met. One must note that patients in acute care settings are most interested in recovering their health, returning home, and having a diagnosis and that the majority of patients feel that physicians do not understand what matters most to them.[6] According to a study specific to acute care settings, consent forms need to be short (around three to four pages), constructed in a logical sequence fit for review with the investigator, and have simplified language to avoid information overload in a stressful situation.[7]

There are however situations when informed consent may be waived. In the United States (US) for conscious subjects research that has a waiver of consent needs to pass the "minimal risk" standard.[8] The US Food and Drug Administration (FDA) has released a guidance document detailing waiver of consent for patients unable to consent, which would include life-threatening situations requiring urgent intervention with unproven or unsatisfactory available treatments and need for further data, with unfeasible informed consent due to medical condition, inability to reach a representative in a timely manner, and expected benefit to the participant.[9] In some cases, where there is an expectation that complete informed consent may not be feasible, consultation with the FDA regarding exemption from informed consent (EFIC) should be carried out. A recent systematic review has shown that in the 41 trials approved by the FDA with EFIC, there were high rates of racial disparities and adverse effects, and only 8% of trials led to a positive primary endpoint.[10]

In the German ISAR-REACT 5 trial, where over 4000 patients with acute coronary syndrome were randomized to ticagrelor or prasugrel, investigators note in the protocol that for patients unable to consent, two physicians not involved with the study or one another would assess the "assumed will" of the patients and also contact relatives.[11] These patients would have additional protections and closer oversight and, if the patient recovers, the trialists would obtain informed consent retrospectively.[11] If informed consent was denied, the patient would be excluded from the study.[11] Interestingly, in the Dutch COACT trial, which randomized 552 patients with cardiac arrest to immediate coronary angiography versus delayed angiography, investigators detailed in the protocol the use of deferred consent after either neurological recovery at 3 days or reaching the legal representative.[12] This strategy was selected given the clinical equipoise of the studied treatments and also avoiding selection bias.[12] Investigators must study their local laws and regulations before determining what is the best approach regarding informed consent for their study.

Ethical concerns

Finally, investigators engaging in acute care research must consider the care of special populations when designing their studies. For instance, patients who do not have access to medical insurance will have less access to medical care, which has important repercussions for diagnosis, treatment, and clinical outcomes.[13] In addition, physicians in acute care settings will encounter patients who have different forms of social vulnerability,[2] including minorities, non-English speaking populations, undocumented patients, patients with substance dependence, and incarcerated patients, among others. Important ethical principles for research, which are not specific to acute care settings, have been established in the 1978 Belmont Report.[14] Additionally, there are legal regulations in the United States regarding the protection of human subjects in research in the Title 45 Code of Federal Regulations (CFR) Chapter 46,[15] also known as the Common Rule. There are also specific regulations for research in pregnant women and fetuses (subpart B), prisoners (subpart C), and children (subpart D).[15] It is recommended that researchers become familiarized with these regulations and work closely with their institutional review boards to ensure that their planned research is ethical and aligned to these principles.

Potential pitfalls

As highlighted above, acute care research has multiple challenges that need to be addressed when designing a scientific investigation. Therefore, it is not surprising how challenging it can be to meet enrollment goals and satisfy statistical power requirements. Trialists have attempted to find alternatives for designing studies in acute care settings aiming to facilitate this process. One example is the design and implementation of pragmatic trials. Pragmatic trials are those that focus on evaluating the effectiveness of a therapy to change outcomes in more typical, "real-world," practice settings, also known as clinical effectiveness.[16] These trials have broad and simple inclusion criteria, aiming to obtain results that are more generalizable, and studying outcomes that are routinely collected in the healthcare record and do not require complex adjudication (such as all-cause mortality).[16] Pragmatic trial designs tend to better facilitate enrollment and the implementation of the study protocol in the acute care setting. Examples of pragmatic trials include the ADAPTABLE trial evaluating low-dose versus full-dose aspirin[17] and multiple trials during the COVID-19 pandemic,[16] especially the British RECOVERY trials.[18] A classic example of a pragmatic trial in the acute care setting is the Swedish TASTE trial, which randomized over 7000 patients with ST-segment elevation myocardial infarction to thrombus aspiration versus usual care.[19]

Funding

Several different organizations may offer funding for projects in the acute care setting. Investigators interested in such an endeavor should seek institutions that support the specific disease that they are targeting in their study. For cardiovascular disease-related trials, US investigators may consider seeking funding from the National Institutes of Health's National Heart, Lung, and Blood Institute, the American Heart Association, and the American College of Cardiology, among others. Other funders that can be considered include the American Academy of Neurology and the Society of Critical Care Medicine. Ultimately, identifying the right partnership with a study sponsor is among the most critical steps in carrying out a successful acute care research project.

References

1. Fanaroff AC, Califf RM, Windecker S, Smith SC, Lopes RD. Levels of evidence supporting American College of Cardiology/American Heart association and European society of cardiology guidelines, 2008–2018. *JAMA*. 2019;321(11):1069–1080. https://doi.org/10.1001/JAMA.2019.1122.
2. Wilson MP, Guluma KZ, Hayden SR, eds. *Doing Research in Emergency and Acute Care: Making Order Out of Chaos*. Wiley-Blackwell; 2015.
3. Schuckman S, Babcock L, Spinner C, et al. Acute care research competencies for clinical research professionals. *J Clin Transl Sci*. 2020;4(6):485. https://doi.org/10.1017/CTS.2020.38.
4. Martins SO, Mont'Alverne F, Rebello LC, et al. Thrombectomy for stroke in the public health care system of Brazil. *N Engl J Med*. 2020;382(24):2316–2326. https://doi.org/10.1056/NEJMOA2000120/SUPPL_FILE/NEJMOA2000120_DATA-SHARING.PDF.
5. Mehta S, Pelletier FQ, Brown M, et al. Why substitute decision makers provide or decline consent for ICU research studies: a questionnaire study. *Intensive Care Med*. 2012;38(1):47–54. https://doi.org/10.1007/S00134-011-2411-1.
6. van den Ende ES, Schouten B, Kremers MNT, et al. Understanding what matters most to patients in acute care in seven countries, using the flash mob study design. *BMC Health Serv Res*. 2021;21(1):1–11. https://doi.org/10.1186/S12913-021-06459-4/TABLES/2.
7. Dickert NW, Bernard AM, Brabson JAM, et al. Partnering with patients to bridge gaps in consent for acute care research. 2020;20(5):7–17. https://doi.org/10.1080/15265161.2020.1745931.
8. Beskow LM, Lindsell CJ, Rice TW. Consent for acute care research and the regulatory "gray zone.". 2020;20(5):26–28. https://doi.org/10.1080/15265161.2020.1745950.
9. United States Food and Drug Administration. *Guidance for Institutional Review Boards, Clinical Investigators, and Sponsors Exception From Informed Consent Requirements for Emergency Research*; 2011. http://www.fda.gov/cder/guidance/index.htm. Accessed August 12, 2022.
10. Feldman WB, Hey SP, Kesselheim AS. A systematic review of the food and drug administration's 'exception from informed consent. *Pathway*. 2018;37(10):1605–1614. https://doi.org/10.1377/HLTHAFF.2018.0501.
11. Schüpke S, Neumann F-J, Menichelli M, et al. Ticagrelor or prasugrel in patients with acute coronary syndromes. *N Engl J Med*. 2019;381(16):1524–1534. https://doi.org/10.1056/NEJMOA1908973/SUPPL_FILE/NEJMOA1908973_DATA-SHARING.PDF.
12. Lemkes JS, Janssens GN, van der Hoeven NW, et al. Coronary angiography after cardiac arrest without ST-segment elevation. *N Engl J Med*. 2019;380(15):1397–1407. https://doi.org/10.1056/NEJMOA1816897/SUPPL_FILE/NEJMOA1816897_DATA-SHARING.PDF.
13. Zhou RA, Baicker K, Taubman S, Finkelstein AN. The uninsured do not use the emergency department more—they use other care less. *Health Aff*. 2017;36(12):2115. https://doi.org/10.1377/HLTHAFF.2017.0218.
14. United States Department of Health and Human Services. *Ethical Principles and Guidelines for the Protection of Human Subjects of Research - the National Commission for the Protection of Human Subjects of Biomedical and Behavioral Research: Notice of Report for Public Comment*. 1979.
15. United States Department of Health and Human Services. *45 Code of Federal Regulations 46*. https://www.hhs.gov/ohrp/regulations-and-policy/regulations/45-cfr-46/index.html. Accessed 22 August 2022.
16. Casey JD, Beskow LM, Brown J, et al. Use of pragmatic and explanatory trial designs in acute care research: lessons from COVID-19. *Lancet Respir Med*. 2022;10(7):700–714. https://doi.org/10.1016/S2213-2600(22)00044-3.
17. Jones WS, Mulder H, Wruck LM, et al. Comparative effectiveness of aspirin dosing in cardiovascular disease. *N Engl J Med*. 2021;384(21):1981–1990. https://doi.org/10.1056/NEJMOA2102137/SUPPL_FILE/NEJMOA2102137_DATA-SHARING.PDF.
18. Dexamethasone in hospitalized patients with covid-19. *N Engl J Med*. 2021;384(8):693–704. https://doi.org/10.1056/NEJMOA2021436/SUPPL_FILE/NEJMOA2021436_DATA-SHARING.PDF.
19. Fröbert O, Lagerqvist B, Olivecrona GK, et al. Thrombus aspiration during ST-segment elevation myocardial infarction. *N Engl J Med*. 2013;369(17):1587–1597. https://doi.org/10.1056/NEJMoa1308789.

Chapter 72

Special populations

Bruno B. Lima[1], Marcela de Almeida Lopes[2] and Zakaria Almuwaqqat[3]

[1]*Division of Cardiology, Department of Medicine, Vanderbilt University Medical Center, Nashville, TN, United States;* [2]*HCor, Research Institute, São Paulo, Brazil;* [3]*Department of Medicine (Cardiology), Emory University School of Medicine, Atlanta, GA, United States*

Introductory section/Why it matters

"Vulnerable subjects," "special populations," and "minorities" are terms that have been used interchangeably in clinical and translational research. However, inconsistent terminology could negatively impact the design implementation in studies that target underrepresented groups. Therefore, it is essential to define and understand those terms before discussing any practical considerations.

"Vulnerable subjects," as defined by the Code of Federal Regulations for the Protection of Human Subjects (also known as 45 CFR 46), include pregnant women (Subpart B), prisoners (Subpart C), and children (Subpart D). Furthermore, those mentioned in 45 CFR 46.111(b) also include mentally disabled or economically or educationally disadvantaged persons because they are at risk for undue influences in a research environment.

The terms "minorities" and "diverse populations" have been used to describe historically underrepresented ethnic or racial groups, women, LGBTQ+ community members, or other populations often overlooked in clinical and translational research studies.

"Special populations" is a broader term that, according to Winter et al., can be defined by age (minors younger than 18 years of age or elderly adults older than 65), historically underrepresented ethnic or racial groups, and people who live in rural areas.

Despite continuing efforts to stimulate adequate representation of special populations in clinical and translational research, current participation rates do not accurately reflect the US demographic characteristics. Moreover, increased participation warrants a sufficient sample size for the results to apply to the diverse groups that researchers aim to serve. Nevertheless, studying special populations poses significant challenges for the recruitment and retention of participants.

Potential pitfalls

The inclusion of special populations helps translational researchers succeed in improving health outcomes for everyone. Some groups in the general population may require a dedicated study because some risks or benefits need specific attention for certain people, or they may need a different dose or treatment schedule. For these reasons, the inclusion of special populations provides opportunities for health improvements that are not easy to predict but are sure to occur.

For instance, the use of medicines in the elderly requires special consideration due to the frequent occurrence of underlying diseases, the use of other drugs simultaneously, and the consequent risk of medicines interaction.

Alternatively, studies involving children are necessary to improve their treatments. Children represent a vulnerable population—they have developmental, physiological, and psychological differences from adults.

Pregnant women, in general, should be excluded from studies where the treatment is not for use in pregnancy. For studies that include pregnant women because the treatment is intended for use during pregnancy, exceptional attention to follow up on the pregnancy, the fetus, and the child is needed.

Although ethnic differences among populations may cause differences in a treatment's safety, efficacy, dosage, or schedule, many therapeutical interventions have comparable characteristics and effects across different groups. Therefore, the regulatory authority or the study sponsor may be concerned that differences in ethnic factors could alter the efficacy or safety of the treatment in the population in a specific group.

Translational Cardiology. https://doi.org/10.1016/B978-0-323-91790-2.00037-X

Get started

One of the most critical stages in a clinical or translational research project is recruiting and retaining the study participants. Additionally, losses to follow-up could ultimately undermine the results of a study.

Potential obstacles to including special populations (Fig. 72.1) are lack of trust in researchers and government agencies, time and scheduling constraints, concerns about becoming a "guinea-pig," inability to take time off work, and transportation to and from the research site.

In addition, when including underage participants in clinical research, federal regulations mandate the consenting process to be written in clear language so the children and their parents can understand that.

Yancey et al. recommend some targeted approaches to decrease participant loss, specifically in underrepresented and minority groups:

- Intensive follow-up and contact with subjects;
- Retain interviewers, field staff, and study staff over time;
- Involve staff from the targeted community;
- Provide social support by offering accessible locations for study visits and data collection;
- Ensure timely incentive payments and accessibility of project staff;
- Encourage study staff's knowledge of community dynamics and project leadership/staff visibility and involvement in the community.

Examples of ...

Studies involving special populations sometimes require unique adaptations to accommodate the participants' cultural context. Developing culturally tailored materials, making the recruitment material available in the participant's language, and having research team members fluent in that language could significantly improve recruitment and retention.

Culturally targeted research strategies often need to go beyond language or recruitment materials to accommodate the participants' reality appropriately. For example, Larkey et al. demonstrated that trained Hispanic lay advocates (called Embajadoras) brought more women at referral and enrollment than untrained Hispanic women and were more successful at enrollment than untrained Anglo controls. Embajadoras also distributed more brochures than control groups for a large-scale national prevention study, the Women's Health Initiative (WHI).

It is also critical to address the barriers that disproportionately affect participation in clinical research by the socioeconomically disadvantaged. By adopting community-engaged research practices, researchers can also improve the lack of trust, fear of exploitation, and low levels of familiarity with medical research. Ejiogu et al. developed door-to-door outreach efforts, neighborhood meetings, and mobile exam centers to increase enrollment for African Americans who participated in a longitudinal aging study.

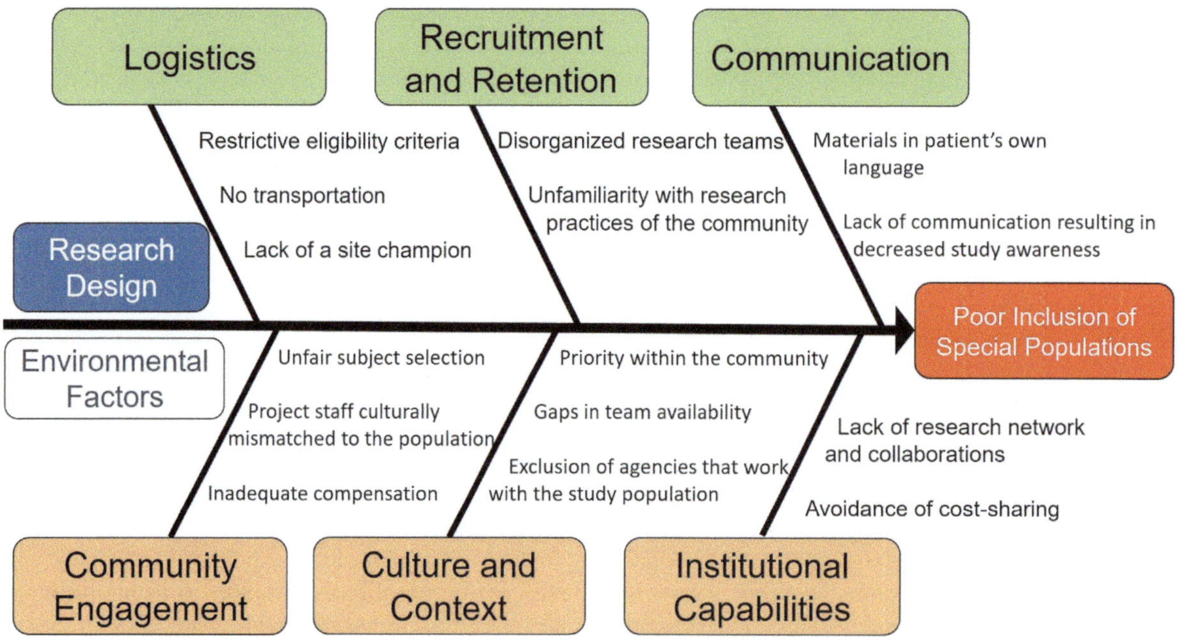

FIGURE 72.1 Causative features for poor inclusion of special populations in translational research studies.

Payment for research participants is considered one of the essential strategies to engage and retain special populations in research. However, there is evidence of differences in perceptions of fair payment by racial/ethnic groups. For example, Walter et al. found that requested fees differed significantly, with Hispanics requesting more payment than non-Hispanic Whites.

Aside from payment for participation, other essential factors associated with improved retention based on a systematic review by Treweek et al. including 45 clinical trials were telephone reminders, opt-out procedures for contacting potential participants, and open trial designs.

Standard research team personnel

A typical research team comprises the principal investigator (PI), who oversees all aspects of a clinical trial from protocol design, recruitment, data collection, analysis, and interpretation of results. Still, he can delegate tasks to other research team members. The PI ensures that all research team members have appropriate education, training, and qualifications to assume delegated study tests. In addition, all study members are responsible for ensuring that the conduct of the study is compliant with institutional, state, federal, and industry guidance and regulations.

The coinvestigators may perform all or some PI functions, but they do not accept primary responsibility for the research study. Instead, the coinvestigator is under the supervision of the PI and is responsible for performing study-related procedures and making critical study-related decisions in compliance with the ethical conduct of the study.

The data coordinator is responsible for the overall data management of a research study. Data points for analysis must be extracted from multiple source documents and entered into specific databases. In addition, they work closely with the research team to identify key data points for collection and analysis in the study development process.

The research coordinator/nurse oversees and coordinates the daily activities of clinical research studies. They work closely with the clinical teams and investigators to ensure that all protocol-required procedures and visits occur according to protocol-specified guidelines.

In addition, research coordinators usually manage participant enrollment and ensure compliance with the protocol and other applicable regulations. Some of the tasks performed by research coordinators include participant recruitment, obtaining informed consent, and educating participants on the details of the research study. Other activities might also include assessing participant eligibility, facilitating participant care and follow-up per protocol, creating source documentation, assessing toxicities/adverse events, and reporting serious adverse events per IRB and sponsor requirements.

Resources

In the past few years, securing research funding has been a significant challenge for investigators in the translational research field at all career stages. First, consider how your research falls within the funding agency's mission as you contemplate applying for financing from different federal, state, or private agencies (Fig. 72.2). If so, determine which grant mechanism to pursue since each agency has many awards for scientists at different career levels with various research needs.

Patient-Centered Outcomes Research Institute (PCORI) is an independent nonprofit, nongovernmental research organization with a mandate to improve the quality and relevance of evidence available to help patients, caregivers, clinicians, employers, insurers, and policymakers make informed health decisions. Specifically, PCORI funds comparative clinical effectiveness research, or CER, and supports work that will improve the methods used to conduct such studies.

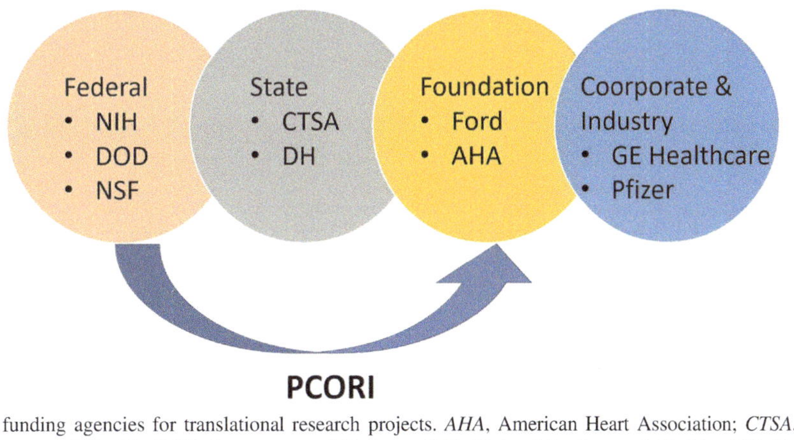

FIGURE 72.2 Grants and funding agencies for translational research projects. *AHA*, American Heart Association; *CTSA*, Clinical and Translational Science Awards; *DH*, Department of Health; *DOD*, Department of Defense; *NIH*, National Institutes of Health; *NSF*, National Science Foundation; *PCORI*, Patient-Centered Outcomes Research Institute.

Further reading

1. Center for Clinical and Translational Science (CCTS). Responsibilities of the Research Team; n.d.. CCTS at the Ohio State University. Retrieved May 3, 2022, from: https://ccts.osu.edu/content/responsibilities-research-team.
2. Edwards E. Funding your research: going beyond the standard NIH R01. *Nat Center Complement Integr Health*; May 13, 2016. Retrieved May 3, 2022, from https://www.nccih.nih.gov/research/blog/funding-your-research-going-beyond-the-standard-nih-r01.
3. Ejiogu N, Norbeck JH, Mason MA, Cromwell BC, Zonderman AB, Evans MK. Recruitment and retention strategies for minority or poor clinical research participants: lessons from the healthy aging in neighborhoods of diversity across the life span study. *Gerontologist.* 2011;1(Suppl 1):S33−S45.
4. Larkey LK, Staten LK, Ritenbaugh C, et al. Recruitment of hispanic women to the women's health initiative. The case of embajadoras in Arizona. *Contr Clin Trials*. 2002;2:289−298.
5. Treweek S, Lockhart P, Pitkethly M, et al. Methods to improve recruitment to randomized controlled trials: cochrane systematic review and meta-analysis. *BMJ Open*. 2013;3.
6. Walter JK, Burke JF, Davis MM. Research participation by low-income and racial/ethnic minority groups: how payment may change the balance. *Clin Transl Sci.* 2013;3:363−371.
7. Winter SS, Page-reeves JM, Page KA, Haozous E, Solares A, Cordova N. Inclusion of special populations in clinical research: important considerations and guidelines. *J Clin Transl Res*. 2018;4(1):56−69.
8. Yancey AK, Ortega AN, Kumanyika SK. Effective recruitment and retention of minority research participants. *Annu Rev Publ Health*. 2006;27:1−28.

Chapter 73

Report forms: Harm and quality of life

Gabriel Cavalcante Lima Chagas[1], Amanda Ribeiro Rangel[1], Débora Moreira Lima[2] and Bruno B. Lima[3]

[1]*Federal University of Ceará, Fortaleza, Ceará, Brazil;* [2]*Walter Cantídio University Hospital, Fortaleza, Ceará, Brazil;* [3]*Division of Cardiology, Department of Medicine, Vanderbilt University Medical Center, Nashville, TN, United States*

Key points

- Understand harm, error, adverse effects, and complications within cardiology.
- Acknowledge the correlation between harm and quality of life.
- Comprehend how to prevent, identify, and mitigate harm and evaluate the quality of life in cardiology.

Introduction

Harm and quality of life (QoL) are intrinsically connected concepts. Quality of life is defined by the World Health Organization (WHO) as the individuals' perceptions of their position in life in the context of the culture and value systems in which they live and in relation to their goals, expectations, standards, and concerns.[1] Harm consists of any negative effect on a patient's health or well-being, whether or not it is evident to the patient. It is imperative to develop activities in attempting to lessen the severity and impact of harm in the quality of care, which requires the understanding of its common causes and burdens.

The burdens of harm

The extent of harm experienced by a patient can vary depending on its severity and duration. Medical errors were estimated to be the third leading cause of death in the United States, resulting in over 250,000 deaths annually.[2]

Low-value health care, defined as healthcare services that provide no, or minimal, benefit to the patient if the potential harms, costs, alternatives, or patient preferences are taken into consideration, is an important component of waste, accounting for $76 to $101 billion annually.[3,4]

The underlying reasons for harm

Harm in the context of healthcare can arise from multifactorial etiologies. Errors are defined as the failure of a planned action to be completed as intended or the use of a wrong, inappropriate, or incorrect plan to achieve an aim. Adverse events (AEs) can be characterized as an injury caused by medical management or complications instead of the underlying disease. AEs can be preventable or nonpreventable.[5]

The increased complexity in health care systems requires the development of error-proof environments to improve patient safety. Several strategies can be implemented to prevent errors, such as automation of procedures to remove human effort, designed workspace to increase medication safety measures, standardization and simplification of processes, continuous education, and training of the health care team. The WHO created The Global Patient Safety Action Plan 2021–2030, which aims to eliminate avoidable harm in health care. This proposal outlines different approaches for four key groups or categories of partners: governments, healthcare facilities and services, stakeholders, and the WHO Secretariat.[6]

Adverse event classification systems in cardiology

An ideal classification system for AEs should provide an easily and widely applicable tool for accurate and precise grading and reporting for clinical, research, and educational purposes.[7] In clinical research, systematic and reproducible reporting of AEs allows different studies and healthcare settings to be comparable with one another and facilitates the generation of secondary analyses.[8]

Translational Cardiology. https://doi.org/10.1016/B978-0-323-91790-2.00098-8

The National Cancer Institute (NCI) of the National Institutes of Health (NIH) first published standardized definitions, known as the Common Terminology Criteria for Adverse Events (CTCAE), for the assessment and reporting of AEs in research in oncology in 2003,[9] with several updates published thereafter. However, the CTCAE is employed in fields other than oncology as well, including cardiology. The CTCAE categorizes AEs based on their severity, with grades ranging from 1 to 5 with unique definitions of clinical impact and intervention requirement (Table 73.1).[10]

In cardiology, a universally recognized and standardized system for classifying AEs and patient harm in general is currently lacking. Instead, there are several commonly used systems for classifying different procedures and interventions in research studies, clinical trials, and regulatory reporting in cardiology (e.g., Academic Research Consortium [ARC] classification for percutaneous coronary interventions [PCIs],[11] Bleeding Academic Research Consortium [BARC] classification for bleeding complications in PCI or coronary artery bypass grafting [CABG],[12] Valve Academic Research Consortium [VARC] criteria for transcatheter aortic, mitral, and tricuspid therapies[13]).

Strategies for mitigating harm in cardiology

In the United States, the burden of low-value healthcare is significant in terms of costs and patient-centered outcomes. For example, nearly half of the patients receive at least one low-value test or procedure annually, which increases the risk of preventable complications from subsequent cascades of care and the costs to patients and society.[3]

Several factors contribute to the genesis and persistence of low-value care in healthcare systems, including patient expectations and demands, limited evidence-based practice, time constraints, workflow pressures, defensive medicine, and misaligned financial incentives, among others.[3]

The social-ecological model of health promotion proposes that interventions to foster high-value, patient-centered care require a multifaceted approach involving patient-, clinician-, health system-, payer-, and policy-level solutions.[3]

Within the context of medical practice in cardiology, appropriate use criteria (AUC) are developed by key specialties (e.g., American Heart Association [AHA], American College of Cardiology [ACC]) and subspecialty societies to help guide the appropriate use of cardiovascular tests and procedures, including coronary revascularization imaging and peripheral artery intervention.[3,14,15] In research, reducing low-value care also requires complex strategies that emphasize evidence-based practice and efficient resource utilization, including adherence to society guidelines, consideration of pragmatic trial designs, definition of research objectives aligned with clinical relevance and patient-centered care, regular review and evaluation of ongoing and completed studies, and transparent reporting for data sharing and collaboration.[16–18]

The importance of quality of life

Quality of life is a multidimensional concept that includes subjective evaluations. Currently, it is considered an important indicator for assessing the efficacy of both diagnostic and therapeutic interventions.[19]

Health-related quality of life (HRQoL) is defined by the Centers for Disease Control and Prevention (CDC) as an individual's or a group's perceived physical and mental health over time.[20] Some definitions fail to distinguish between HRQoL and QoL. A potential solution is to characterize HRQoL as the way health is empirically estimated to affect the QoL.[21]

Understanding QoL is important for improving patient care. Patient-reported QoL assessments have the potential to uncover areas necessitating care modifications and improvements; additionally, they can provide insight into therapies that yield minimal benefits.[19] Engaging patients in their healthcare can facilitate the identification of potential errors and contribute to the prevention of AEs since it increases the comprehension of the therapeutic proposal.[22]

The analysis of QoL is also important for economic reasons. It not only allows one to evaluate whether the employed treatment and the resources spent correlate positively with patient satisfaction but may also be useful in optimizing resource allocation.[23]

Therefore, healthcare providers endeavor to mitigate harm and optimize patient outcomes. This involves carefully weighing the potential benefits and risks of medical interventions, implementing safety measures, providing patient education, and closely monitoring patients for any signs of harm or AEs.[6]

Evaluating the quality of life in cardiology

Quality of life can be evaluated through many validated measurement tools, which commonly encompass both subjective and objective measures.[19] Questionnaires are the most commonly used method of evaluation.[23]

There is no best tool in an absolute sense for QoL assessment. The following criteria are considered when selecting and evaluating the suitability of QoL instruments[1]: appropriateness: a measure that aligns with the particular purpose and research[2]; reliability: the degree to which an instrument is free from random error[3]; validity:

TABLE 73.1 Grading of Common Terminology Criteria for Adverse Events (CTCAE)[10] according to clinical impact and intervention requirement.

	Common Terminology Criteria for Adverse Events (CTCAE)				
	Grade 1	**Grade 2**	**Grade 3**	**Grade 4**	**Grade 5**
Clinical impact	Mild	Moderate	Severe but not immediately life-threatening	Life threatening	Death
Intervention required	Clinical or diagnostic observations only	Minimal, local, or noninvasive intervention indicated	Hospitalization or prolongation of hospitalization indicated	Urgent intervention indicated	Not applicable

the degree to which an instrument measures what it claims to measure[4]; precision: the accuracy of the distinction made by the measure[5]; interpretability: the possibility of assigning meaning to scores[6]; responsiveness: the ability to detect changes over time[7]; acceptability: how acceptable is the completion of a measure for respondents; and[8] feasibility: the extension of effort, burden, and disruption to staff arising from using a measure.[24]

Generic HRQoL measures usually contain the dimensions of physical, emotional, and social functioning, as well as global perceptions of health and well-being.[25] The generic questionnaires most commonly used in cardiovascular diseases include the WHO Quality Of Life Assessment Instrument (WHOQOL), Medical Outcomes Study 36-item Short-Form Health Survey (SF-36), Nottingham Health Profile (NHP), and Euro-Quality of Life Questionnaire (EuroQoL, EQ-5D).[23]

Specific HRQoL measures focus on an area of primary interest, where the instrument may be specific to a population of patients (e.g., adults or elderly), a disease (e.g., heart failure), a problem (e.g., symptoms), or a certain function (e.g., activity level).[24] The most commonly used specific questionnaires for cardiovascular diseases include the MacNew Heart Disease Health-Related Quality of Life Questionnaire (MacNew), Seattle Angina Questionnaire (SAQ), Minnesota Living with Heart Failure (MLHF) questionnaire, Atrial Fibrillation Severity Scale (AFSS).[23]

Different questionnaires can be used for the same disease. The utilization of HRQoL measures in the field of cardiology is limited by the paucity of available information to determine the most appropriate instrument and the uncertain reliability of the published data.[26]

Conclusions

A thorough understanding of how to prevent, identify, and mitigate harm in both clinical practice and research is paramount for the clinician-researcher to promote high-value, patient-centered care. By incorporating these practices in cardiology, clinician-researchers play a crucial role in fostering a healthcare culture that prioritizes patient safety and QoL and achieves optimal outcomes.

Get started

(1) Summarize potential contributing factors to patient harm in cardiology.
(2) Elaborate on key strategies to identify and mitigate harm in cardiology, especially in preventable causes.
(3) Determine how patient-reported QoL assessment can improve clinical practice and research in cardiology.

Potential pitfalls

(1) Misjudging the multifaceted and complex causes of harm.
(2) Not involving the patients, clinicians, health system, payers, and policymakers in strategies for health promotion and harm reduction.
(3) Underestimating the importance of patient-reported outcomes and metrics of patient QoL in clinical practice and research.

Real-world examples

(1) An example of the impact of baseline risk assessment in the prevention of harm in treatment—Phillips, R. et al. Impact of Cardiovascular Risk on the Relative Benefit and Harm of Intensive Treatment of Hypertension. J Am Coll Cardiol. 2018; 71(15):1601–1610.[24]
(2) An example of harm-benefit balance in health-related QoL following a cardiovascular intervention—Croog, S. et al. The effects of antihypertensive therapy on the quality of life. N Engl J Med. 1986; 314(26):1657–1664.[27]
(3) An example of cost-effectiveness assessment of cardiac therapies—Eze-Nliam C. et al. Cost-effectiveness Assessment of Cardiac Interventions: Determining a Socially Acceptable Cost Threshold. Interv Cardiol. 2014; 6(1):45–55.[28]

Resources

(1) Kini, V. et al. Strategies to Reduce Low-Value Cardiovascular Care: A Scientific Statement From the American Heart Association. *Circ Cardiovasc Qual Outcomes*. 2022; 15(3):e000105.[3]
(2) Bowman, L. et al. Randomized Trials Fit for the 21st Century: A Joint Opinion From the European Society of Cardiology, American Heart Association, American College of Cardiology, and the World Heart Federation. *Circulation*. 2023; 147(12):925–929.[18]
(3) Haraldstad, K et al. A systematic review of quality of life research in medicine and health sciences. Quality of Life Research. 2019 Oct 1; 28(10):2641–50.[19]

References

1. [cited 2023 Jun 25]. *The World Health Organization Quality of Life (WHOQOL)*; 2012. Available from: https://www.who.int/publications/i/item/WHO-HIS-HSI-Rev.2012.03.
2. Makary MA, Daniel M. Medical error—the third leading cause of death in the US. *BMJ*. May 3, 2016;353:i2139.

3. Kini V, Breathett K, Groeneveld PW, et al. Strategies to reduce low-value cardiovascular care: a scientific statement from the American Heart Association. *Circ Cardiovasc Qual Outcome.* March 2022;15(3):e000105.
4. Verkerk EW, van Dulmen SA, Westert GP, Hooft L, Heus P, Kool RB. Reducing low-value care: what can we learn from eight de-implementation studies in the Netherlands? *BMJ Open Qual.* September 2022;11(3):e001710. http://bmjopenquality.bmj.com/content/11/3/e001710.abstract.
5. [cited 2023 Jun 25]. *Conceptual Framework for the International Classification for Patient Safety*; 2009. Available from: http://www.who.int/patientsafety/taxonomy/ICPS_Statement_of_Purpose.pdf.
6. World Health Organization. *Global Patient Safety Action Plan 2021−2030:Towards Eliminating Avoidable Harm e Health Care* [cited 2023 Jun 25]. World Health Organization; 2021:1−96. Available from: https://www.who.int/teams/integrated-health-services/patient-safety/policy/global-patient-safety-action-plan.
7. Khalilzadeh O, Baerlocher MO, Shyn PB, et al. Proposal of a new adverse event classification by the society of interventional radiology standards of practice committee. *J Vasc Interv Radiol.* October 2017;28(10):1432−1437.e3.
8. Zhang S, Liang F, Tannock I. Use and misuse of common terminology criteria for adverse events in cancer clinical trials. *BMC Cancer.* 2016;16(1):392.
9. Trotti A, Colevas AD, Setser A, et al. CTCAE v3.0: development of a comprehensive grading system for the adverse effects of cancer treatment. *Semin Radiat Oncol.* July 2003;13(3):176−181.
10. U.S. Department of Health and Human Services, National Institutes of Health, National Cancer Institute. *Common Terminology Criteria for Adverse Events (CTCAE) Version 5.0*; 2017:1−147. Available from: https://ctep.cancer.gov/protocoldevelopment/electronic_applications/docs/CTCAE_v5_Quick_Reference_8.5x11.pdf.
11. Cutlip DE, Windecker S, Mehran R, et al. Clinical end points in coronary stent trials: a case for standardized definitions. *Circulation.* May 2007;115(17):2344−2351.
12. Mehran R, Rao SV, Bhatt DL, et al. Standardized bleeding definitions for cardiovascular clinical trials: a consensus report from the Bleeding Academic Research Consortium. *Circulation.* June 2011;123(23):2736−2747.
13. Committee W, Généreux P, Piazza N, Alu MC, Nazif T, Hahn RT, et al. Valve academic research consortium 3: updated endpoint definitions for aortic valve clinical research. *J Am Coll Cardiol.* June 2021;77(21):2717−2746. https://doi.org/10.1016/j.jacc.2021.02.038.
14. Bailey SR, Beckman JA, Dao TD, et al. ACC/AHA/SCAI/SIR/SVM 2018 appropriate use criteria for peripheral artery intervention. *J Am Coll Cardiol.* January 22, 2019;73(2):214−237.
15. Winchester DE, Maron DJ, Blankstein R, et al. ACC/AHA/ASE/ASNC/ASPC/HFSA/HRS/SCAI/SCCT/SCMR/STS 2023 multimodality appropriate use criteria for the detection and risk assessment of chronic coronary disease. *J Am Coll Cardiol.* June 27, 2023;81(25):2445−2467.
16. Calvert M, Kyte D, Mercieca-Bebber R, et al. Guidelines for inclusion of patient-reported outcomes in clinical trial protocols: the SPIRIT-PRO extension. *JAMA.* February 2018;319(5):483−494.
17. Usman MS, Van Spall HGC, Greene SJ, et al. The need for increased pragmatism in cardiovascular clinical trials. *Nat Rev Cardiol.* 2022;19(11):737−750.
18. Bowman L, Weidinger F, Albert MA, Fry ETA, Pinto FJ. Randomized trials fit for the 21st century: a joint opinion from the European Society of Cardiology, American Heart Association, American College of Cardiology, and the World Heart Federation. *Circulation.* March 2023;147(12):925−929.
19. Haraldstad K, Wahl A, Andenæs R, et al. A systematic review of quality of life research in medicine and health sciences. *Qual Life Res.* October 1, 2019;28(10):2641−2650.
20. *HRQOL Concepts*: CDC. Available from: https://www.cdc.gov/hrqol/concept.htm.
21. Karimi M, Brazier J. Health, health-related quality of life, and quality of life: what is the difference? *Pharmacoeconomics.* July 1, 2016;34(7):645−649. https://pubmed.ncbi.nlm.nih.gov/26892973/.
22. Naik AD, Dyer CB, Kunik ME, McCullough LB. Patient autonomy for the management of chronic conditions: a two-component reconceptualization. *Am J Bioeth.* February 2009;9(2):23−30.
23. Gierlaszyńska K, Pudlo R, Jaworska I, Byrczek-Godula K, Gąsior M. Tools for assessing quality of life in cardiology and cardiac surgery. *Kardiochir Torakochirurgia Pol.* 2016;13(1):78−82.
24. Phillips RA, Xu J, Peterson LE, Arnold RM, Diamond JA, Schussheim AE. Impact of cardiovascular risk on the relative benefit and harm of intensive treatment of hypertension. *J Am Coll Cardiol.* April 2018;71(15):1601−1610.
25. Lin XJ, Lin IM, Fan SY. Methodological issues in measuring health-related quality of life. *Tzu Chi Med J.* March 2013;25(1):8−12.
26. Garin O, Herdman M, Vilagut G, et al. Assessing health-related quality of life in patients with heart failure: a systematic, standardized comparison of available measures. *Heart Fail Rev.* May 1, 2014;19(3):359−367.
27. Croog SH, Levine S, Testa MA, et al. The effects of antihypertensive therapy on the quality of life. *N Engl J Med.* June 1986;314(26):1657−1664.
28. Eze-Nliam CM, Zhang Z, Weiss SA, Weintraub WS. Cost-effectiveness assessment of cardiac interventions: determining a socially acceptable cost threshold. *Interv Cardiol (Lond).* 2014;6(1):45−55.

Chapter 74

Subject adherence

André De Souza[1,2] and Alice Deberaldini Marinho[3]
[1]*The Warren Alpert Medical School of Brown University, Providence, RI, United States;* [2]*The Miriam Hospital, Providence, RI, United States;*
[3]*Federal University of the State of Rio de Janeiro, Rio de Janeiro, Brazil*

Introduction

Subject adherence or compliance in translational research describes the degree to which patients and research subjects act by the advice or instructions of their healthcare provider or researchers, correctly following the protocol. Participant adherence to regimens is an essential part of every trial and is a fundamental issue to be considered when designing and conducting clinical research. However, it remains a major challenge regardless of the simplicity or complexity of the study design, especially when the follow-up period lasts for months or years.

There are two primary types of adherence in clinical research: follow-up adherence and regimen or protocol adherence. The first refers to following the schedule of assessment measures within planned time windows until the endpoint is reached. The second refers to pursuing the assigned regimen consistently. Participants must get the treatment they are assigned to receive as described in the protocol. Otherwise, the intervention being analyzed is no longer the one intended, and the results no longer reflect the primary objective of the trial. Therefore, adequate participant adherence is critical to the evaluation of the efficacy of the medication.

Get started

Nonadherence and its consequences

Nonadherence or poor adherence is a major problem in clinical research and affects the outcomes of trials by reducing their statistical power. Hence, it may threaten researchers' ability to complete investigations and require that more participants be enrolled to achieve adequate power for the results to be valid. As a consequence, the trial's financial cost and duration will be increased, delaying important results. Participants who are enrolled but do not complete a trial (study attrition) can also cause biased results and affect survival analyses such as Kaplan–Meier.

Although the literature is contradictory regarding the effects of demographic factors on adherence, Driscoll et al.[1] found some characteristics to be predictive of attrition, such as older age, male gender, lower education, functional impairment, poorer cognitive performance, lower verbal intelligence, comorbid conditions, and worse physical health.

Regimen adherence problems can involve both errors of omission and commission. Adherence problems can be random or nonrandom, transitory or long-term, or attributable to other factors. The failure to follow treatment instructions may be attributed to several reasons. Some common motives for nonadherence are poor communication, forgetfulness, unintentional underdosing or overmedicating, and unpleasant side effects.

Adherence can modify individual outcomes in research. Better outcomes have been reported for patients with good adherence to treatment regimens when compared to those with poor adherence in studies of cardiac disease, cancer, and psychiatric disorders. Therefore, nonadherence affects treatment benefits and the health of subjects. There are also statistical challenges associated with problematic adherence since inadequate adherence contributes to statistical variance and increases the risk of type II errors, possibly leading to erroneous conclusions.

Assessing adherence

The assessment of participant compliance in clinical trials can be done in many different ways. One method is tablet counting, with patients receiving more tablets than necessary and returning unused tablets at the next clinic visit. Adherence is measured by the number of tablets returned. The second strategy is measuring the plasma or urine concentration of the drug. However, this method depends

Translational Cardiology. https://doi.org/10.1016/B978-0-323-91790-2.00002-2

on the pharmacokinetics and pharmacodynamics of the drug. In pharmacological studies, markers (e.g., inert molecules, radioactive substances, and stable isotopes) may be used to quantify exposure to agents. Furthermore, patients can utilize diaries to write when they took the drug and their symptoms during the day.

Another approach is direct observation, trying to ensure complete compliance. Nonetheless, it is a time-consuming process and requires a high level of effort from both the participant and the study staff. One more way to monitor adherence is by calculating the number of medication pills expected to have been taken by all participants combined and the actual number of medication pills taken. The "actual" divided by the "expected" gives a general sense of treatment adherence, but it neglects the daily variation pattern.

Lastly, smart packaging and smart pills using microchips with monitoring systems can be used to track when the medicine is removed from the packaging and when the pill is swallowed by the participant. The integration of these systems within the pill has economic implications with an increase in cost, complicating the implementation and not enabling greater usage of this technology.

Increasing adherence

Researchers should, when feasible and when it does not introduce bias, include strategies to enhance adherence through all research phases in clinical trials. Adherence-enhancing efforts can be multifactorial, addressing health, social, research, and logistical matters. These strategies must be incorporated equally between the control and treatment groups so that they do not impact the outcomes.

One strategy for enhancing medication adherence might be the introduction of an automated phone call, text message, or email that alerts participants when it is time to take the medication. The same method of reminders can be used for impending study visits. Up-front scheduling can also be included in this approach, with a written document detailing the dates and times of every visit.

Another preventative approach for promoting adherence is prerandomization screening, trying to identify potential subjects at risk for study attrition so that those likely to cooperate and who understand study burdens enroll while those at risk are screened out. This approach can be "run-in," which places potential subjects on a prescribed regimen to assess their fidelity, or "test-dosing," identifying those who may have difficulty tolerating adverse drug effects. The evaluation of other types of behavioral and psychosocial data based on variables that are related to adherence can also be used (e.g., anxiety, depression, motivation, and barriers to participation).

In the Oncology setting, we can use predictive biomarkers to evaluate the response to some therapy regimens, such as immune checkpoint inhibitors in lung cancer. There is a biomarker value of PD-L1 expression on tumor cells by immunohistochemistry (IHC) to predict response to PD-1 axis blockade. Furthermore, it is also possible to predict severe fluoropyrimidine-associated toxicity with DPD deficiency. DPYD variants c.1679T>G and c.1236G>A/HapB3 were clinically relevant predictors to evaluate toxicity in a systematic review and meta-analysis published by Meulendijks et al.[2]

Participants in clinical trials expose themselves to risks and have significant responsibilities in following protocols, deserving recognition for their altruistic roles. Types of study in which subjects are relegated to a control group that receives no perceived benefit can lead to study attrition. In that sense, some type of compensation or reward can be provided to subjects to make participation and adherence more attractive, including monetary payments, gifts, or free health or child care. The government can also offer incentives to subjects who adhere closely to regimens in government-sponsored research.

Other efforts to improve compliance include intervening early when adherence problems emerge, simplifying medication packaging and regimens (e.g., limiting the number of medications prescribed simultaneously), improving patient education, transparency of the informed consent document, and a strong relationship among the study coordinator, care providers, and participants.

Regarding the adherence of minority research participants, intensive follow-up and contact with subjects seem to improve retention. It is also important to provide social support, have the same interviewers over time, staff from the targeted community, accessible locations for intervention implementation, regular telephone reminders, and timely incentive payments.

Intention to treat versus per-protocol

One potential solution to the problem of nonadherence in trials is a statistical concept called intention-to-treat (ITT) analysis. The analytic strategy of ITT is a comparison of the treatment groups that includes all patients as originally allocated after randomization, ignoring poor or nonadherence, protocol deviations, patient withdrawal, and anything that happens after randomization. ITT analysis is generally conservative, avoiding overoptimistic estimates of the efficacy of an intervention, and maintains the prognostic balance generated from the original randomization. Due to this cautious approach, ITT also minimizes type I errors and is recommended to avoid bias in superiority trials.

Some arguments against ITT are valid. Because the analysis is too cautious and decreases the likelihood of type I errors, the risk of type II errors is increased. Also, heterogeneity can be introduced if nonadherence, dropouts, and compliant subjects are mixed in the final analysis. Additionally, the interpretation might become difficult if participants cross over to opposite treatment arms.

One of the alternatives to ITT analysis is the per-protocol (PP) analysis. PP is a comparison of treatment groups that include only patients who completed the treatment originally allocated. The PP population is a subset of the ITT population who completed the study without any major protocol violations, excluding anyone who did not adhere to treatment, switched groups, or missed measurements. ITT tends to make the two treatments look similar, whereas PP removes patients who do not complete treatment, being able to reflect differences between both arms of the trial. For superiority trials, the PP analysis is considered supportive since it can lead to bias if done alone. However, it has been suggested that in non-inferiority trials both ITT and PP analyses need to be used.

There is also the modified ITT concept (mITT), which is a subset of the ITT population and allows the exclusion of some randomized subjects. However, because there is a lack of consistent guidelines for its application, the definition of mITT is irregular and arbitrary. The mITT approach differs from the ITT analysis by eliminating patients or reassigning them to a different study group than the group to which they were randomized. Studies using mITT are more likely to be positive than those following an ITT approach.

In situations in which adherence is incomplete, ITT analyses may make unsafe interventions appear to be safe. Additionally, if the poor adherence was due to toxicity, ITT can result in an effective intervention appearing to be ineffective. Therefore, in these situations, the best analytic strategy is PP. Alternately, ITT gives unbiased estimates for the causal effects of treatment assignment, being used as a tool for clinical outcomes such as survival analysis.

Conclusions

Nonadherence is a major problem in translational research and affects the outcomes of trials. Some tools can be used to assess and increase adherence, aiming to reduce bias in the results of clinical trials. Decisions should be guided by the principles of the ITT and PP strategies. When the ITT and PP analyses come to the same conclusions, confidence in the study results is increased.

References

1. Driscoll K, Killian M, Johnson S, Silverstein J, Deeb L. Predictors of study completion and withdrawal in a randomized clinical trial of a pediatric diabetes adherence intervention. *Contemp Clin Trial.* 2009;30(3):212–220.
2. Meulendijks D, Henricks LM, Sonke GS, Deenen MJ, Froehlich TK, et al. Clinical relevance of DPYD variants c.1679T>G, c.1236G>A/HapB3, and c.1601G>A as predictors of severe fluoropyrimidine-associated toxicity: a systematic review and meta-analysis of individual patient data. *Lancet Oncol.* 2015;16(16):1639–1650.

Further reading

1. Gallin J, Ognibene F, Johnson L. *Principles and Practice of Clinical Research.* 4th ed. USA: Academic Press; 2017.
2. Robiner W. Enhancing adherence in clinical research. *Contemp Clin Trial.* 2005;26(1):59–77.
3. Pullar T, Kumar S, Feely M. Compliance in clinical trials. *Ann Rheum Dis.* 1989;48(10):871–875.
4. Cummins G. Smart pills for gastrointestinal diagnostics and therapy. *Adv Drug Deliv Rev.* 2021;177:113931.
5. Yancey A, Ortega A, Kumanyika S. Effective recruitment and retention of minority research participants. *Annu Rev Publ Health.* 2006;27(1):1–28.
6. Page S, Persch A. Recruitment, retention, and blinding in clinical trials. *Am J Occup Ther.* 2013;67(2):154–161.
7. Uruga H, Mino-Kenudson M. Predictive biomarkers for response to immune checkpoint inhibitors in lung cancer: PD-L1 and beyond. *Virchows Arch.* 2021;478(1):31–44.
8. Detry M, Lewis R. The intention-to-treat principle. *JAMA.* 2014;312(1):85.
9. Gupta S. Intention-to-treat concept: a review. *Perspect Clin Res.* 2011;2(3):109.
10. Shah P. Intention-to-treat and per-protocol analysis. *Can Med Assoc J.* 2011;183(6).
11. Hernán MA, Robins JM. Per-protocol analyses of pragmatic trials. *N Engl J Med.* 2017;5;377(14):1391–1398.

Chapter 75

Intuitive survival analysis

Ahmed Alnajar[1,2], Isildinha M. Reis[2] and J. Sunil Rao[2]
[1]The Division of Cardiothoracic Surgery, The Department of Surgery, The University of Miami Miller School of Medicine, Miami, FL, United States;
[2]Division of Biostatistics, The Department of Public Health Sciences, The University of Miami Miller School of Medicine, Miami, FL, United States

Key points

- Researchers studying time-to-event outcomes should be familiar with the survival analysis framework in clinical and translational research.
- Multiple methods could be utilized in different research settings and no one size fits all.
- Understanding the steps involved in creating Kaplan–Meier curves will help researchers effectively analyze and plot their data.

Why it matters?

Research in cardiology frequently looks at the effect of a medication or procedure on progression-free survival or overall survival. However, researchers often observe incomplete information about some individuals included in their studies due to administrative censoring or losses to follow-up. Ignoring these individuals will result in discarding valuable incomplete information, which will lead to inadequate estimation of the probability of survival over time. Therefore, it is essential when carrying out a survival analysis to include the censored observations.

Introduction

Survival analysis is also called time-to-event analysis. It uses data in which "the time to an event" is the outcome of interest. The outcome of survival data is not necessarily linked to mortality but time until a specific event occurs; thus, this type of analysis can be applied to a wide variety of situations where the time-to-event outcome is present, such as:

- Time until recurrence of a disease
- Time to recover from a disease
- Time to discharge from hospital
- Time to onset of treatment response
- Time until the relief of a symptom (e.g., palpitation)
- Time to recovery of platelet count after bone marrow transplantation
- Time until device failure (e.g., left ventricular assist devices)
- Time until organ failure (e.g., heart graft transplantation[1])

Since these clinical outcomes will occur in the future, when researchers analyze time-to-event outcomes, they are unlikely to be able to wait until the event has occurred to all the participants (e.g., until all study patients provide full follow-up data). In addition, while participants are likely to enter the study at different time points, some may not adhere to their follow-up schedules, or some may even withdraw from the study. This will lead to participants "censoring" at the date of last documented event-free status (e.g., date last known to be alive, date of last documented progression-free status, or date of consent withdrawal) since incomplete information is more valuable than no information at all. In survival analysis, however, by censoring, we do not discard the information obtained from individuals who are still event-free and the time to occurrence of the "no-event" (i.e., censored time) becomes as important as the time to an event. In clinical studies, the four main analytic techniques in time-to-event data are to:

- Estimate of survival curves -> Kaplan–Meier estimate.

Translational Cardiology. https://doi.org/10.1016/B978-0-323-91790-2.00075-7

- Estimate of survival rates and median survival time - > Kaplan–Meier estimate.
- Compare two or more treatment groups - > Log-rank test.
- Evaluate the influence of risk (or protective) factors on the likelihood of the event of interest over time - > Cox regression.

To plan a study, interpret survival curves from a paper, or critically appraise literature, clinicians do not need to be mathematically savvy, but should be at least familiar with the above survival techniques. In one of our previous studies,[2] couple of anonymous peer reviewers revealed that they do not understand the application of survival analysis in our study, one mistakenly confused "actual" with "actuarial" survival estimates and the other requested underestimated survival proportion by ignoring censored observations. In this chapter, we will not delve into the mathematics behind different survival analysis techniques; instead, we will briefly introduce the theoretical framework and practical approaches in different scenarios. In addition, we will point out many available resources that provide a thorough explanation of the principles behind the statistical tests involved in survival (cited in this chapter, but formulas were deliberately left out for simplification).

Censoring

In a study, censoring occurs when the event of interest is not observed for some individuals. For the censored individuals, we don't know the exact survival time (or event time), but when the individuals themselves are not observed for having survival times (or event times) smaller or larger than certain values, truncation occurs instead of censoring.

Types of censoring

Right censoring

Most survival data include or assume right-censored observations, or observations censored in the future (Fig. 75.1A). In which, a participant's survival time is said to be right-censored if all that is known about the time of the event is that it did not occur before or at a certain time.[3] Common reasons for censoring are:

- The participant is lost to follow-up or dropped from the study (e.g., participant refuses further participation in the study, disappears [loss of contact], or moves away).
- Administrative censoring is said to occur when the study is closed, or statistical analysis is conducted before all participants experience the event.

Left censoring

It is less common to have left-censored data, or observations censored from the past, but that could happen if the true survival time is less than or equal to the observed survival time.[4] For example, when studying breast cancer recurrence among women who had surgical removal of the primary tumor. If some patients die from breast cancer at 12 months following mastectomy, for whom we do not have the date of recurrence, all we know is that recurrence occurred sometime between the mastectomy and before the date of death. Thus, the time to recurrence is—at most—12 months.

Interval censoring

Survival time is said to be interval-censored if it is only known that it occurred during a time interval. For example, an individual may have had SARS-CoV-2 tests at regular times (e.g., at 3, 6, and 9 weeks following the local pandemic announcement). At first, the individual tested negative at the 3-week time point but tested positive at the 6-week test. In such a case, the individual's true survival time occurred after the 3-week and the before 6-week tests. Interval censoring incorporates the information of both right-censoring and left-censoring.

Truncation

Truncation occurs, as indicated above, due to sampling bias; in such cases, only those individuals whose data are available can be observed and their data are analyzed.[5]

Types of truncation

Similar to censoring, left truncation occurs when participants enter a study at a particular age (not necessarily the origin for the event of interest) and are followed from this delayed entry time until the event occurs or until the participant is censored, while right truncation occurs when only individuals who have experienced the event of interest are observable. It is easy to understand why truncation causes bias since there will be underestimated the risk of events. For example, when studying congenital heart diseases (CHDs) in adults, the standard analysis does not account for the fact that many patients were excluded before they survived into adulthood. An example of adjusting for truncation in survival analysis was performed by Oliver et al. to find independent predictors of excess mortality in adults with CHD.[6] However, the methods for adjusting for truncation and left censoring are less widely known and infrequently applied in general longitudinal studies.

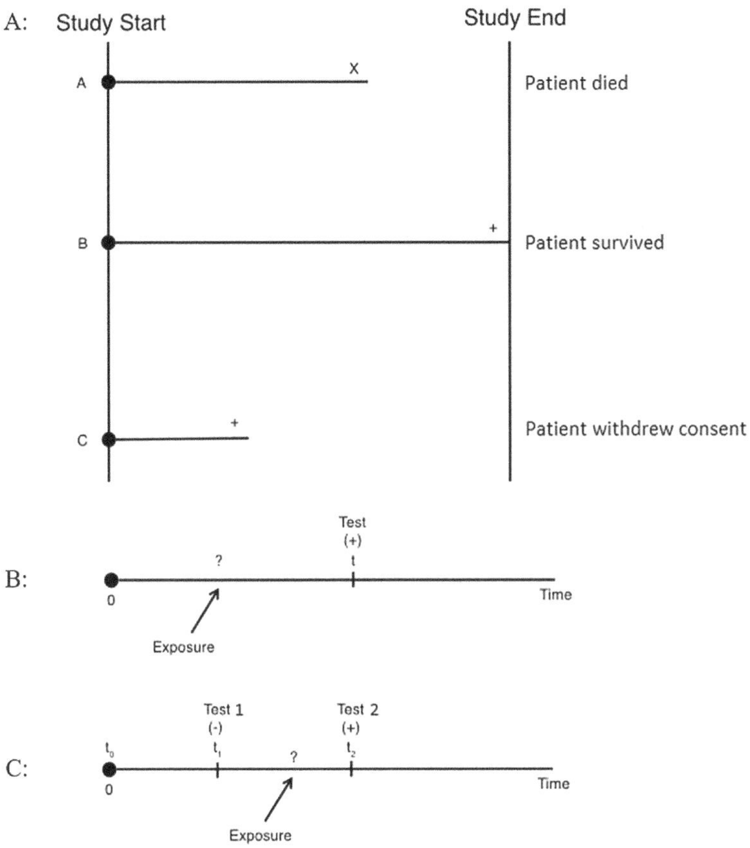

FIGURE 75.1 Censored date types. (A) *Right* censored data. (B) *Left*-censored data. (C) Interval censoring (t_0: Study start, t_1 and t_2: Different time intervals in which the patient was tested).

Methods used to estimate survival

Kaplan–Meier method

The Kaplan–Meier Method (KM) is also called the Product-Limit method, which computes the **Actual** estimates of survival rates at each distinct survival time (ungrouped/participant survival data). Survival data consist of two pieces of information for time-to-event analysis (Table 75.1):

1. The time for each participant.
2. An indicator of whether the time belongs to a censored observation or an event observation.

This information will allow for an estimation of the survival probabilities for the population at each time an event occurs. The survivor function produced using the KM method gives the probability that a person survives longer than some specified time, such as survival rates at a 2-year interval (e.g., 20% will likely survive 2 years following an intervention). If estimable, the KM method also can provide an estimate of the median survival time, which correspond to 50% survival. From a KM survival plot, an estimate of the median survival time is obtained by locating 50% on the Y-axis and reading over and down to the X-axis for the time component. In other words, the median survival is the smallest time at which the survival probability drops to 0.5 (50%) or below (e.g., by median survival of 6 years, half the patient population who received that intervention survive, half die). In another more specific example, researchers studied survival time for 12 patients, with 1-year enrollment during 2019, and 2 years of additional follow-up (up to 2022). Each patient was enrolled in the study at a different time between 2019 and 2020 (Fig. 75.2A). However, during our analysis, we looked at the absolute time for each patient (Fig. 75.2B), reordered the data (Fig. 75.2C), then used the KM method to draw a survival curve (Fig. 75.2D). We then estimated survival probability at a certain time point (e.g., 64.9% at 2 years) and median survival time (e.g., 30 months).

Life-table method

The Life-Table method, also called the **Actuarial** method, computes time interval survival estimates when grouped survival data are only available; that is, the actual survival times are unknown. The life-table summarizes the experiences of participants over the predefined follow-up intervals in a cohort study or a clinical trial. For example, in a study, data available for analysis could be several events and the number of censored observations in equally spaced intervals (e.g., 0 to 6, >6 to 12, >12–18 months).

TABLE 75.1 Time-to-event data table that are collected and analyzed. The most important columns that represent time-to-event are time_month (an indicator of time) and status (an indicator of events).

Table A

Characteristics			Operative data			Last follow up	
mrn	Age	Sex	Date of Surgery	Surgery	XC time	Follow up date	Mortality status
x8910xx	60	Male	7/4/2017	AVR+TVR	214	11/7/2021	Alive
x0838xx	62	Female	8/4/2017	MVR	44	12/10/2022	Alive
x2063xx	54	Female	3/9/2018	AVR	83	7/22/2022	Died
x5651xx	56	Male	8/25/2017	MVR	77	5/2/2022	Alive
x3705xx	58	Male	5/5/2017	AVR+MVR	225	4/30/2019	Lost to follow up
x3786xx	75	Male	6/19/2017	AVR	117	7/18/2022	Died
x3833xx	69	Male	1/9/2018	MVR	42	5/10/2022	Alive
x4394xx	64	Male	3/26/2018	TVR	34	5/10/2022	Alive
x4556xx	66	Male	5/22/2018	TVR	88	1/31/2022	Alive
x4992xx	77	Female	6/26/2018	AVR	65	12/11/2021	Alive

Table B

Id	Age	Is_Male	AVR	MVR	TVR	XC_Time	Time_Month	Status
1	60	1	1	0	1	214	52.9	0
2	62	0	0	1	0	44	65.13333	0
3	54	0	1	0	0	83	53.2	1
4	56	1	0	1	0	77	57.03333	0
5	58	1	1	1	0	225	24.16667	0
6	75	1	1	0	0	117	61.83333	1
7	69	1	0	1	0	42	52.73333	0
8	64	1	0	0	1	34	50.2	0
9	66	1	0	0	1	88	45	0
10	77	0	1	0	0	65	42.13333	0

AVR, aortic valve replacement; *MVR*, mitral valve replacement; *TVR*, tricuspid valve replacement; Time_Month, calculated from "Date of Surgery" and "Follow-up date"; *XC*, cross clamp.
(A) Data collected by the research group in raw format. It's not uncommon that researchers forget identifiers (mrn/dates data) and use colors to help them visualize and organize their tables, but they are meaningless during the analysis. (B) Clean table data ready for analysis. Clean up process included replacing identifiers with linking variables, replacing column names with indicators consisting of one word or words separated by under scroll sign (_), unfold compact columns into indicator columns to restrict each column to 1 useful information. Disclosure: identifying columns (mrn/dates) in the Table contain made-up numbers for the purpose of illustration.

Hazard function

The hazard function gives the instantaneous potential per unit time for the event to occur, given that the individual has survived up to a certain time point. While the survivor function focuses on surviving, the hazard function focuses on the event occurrence (e.g., death, pacemaker failure, allograft rejection, etc.), given survival up to a certain time point. A time-specific hazard rate is the risk of failure, given that the participant has survived up to the specific time.

Comparing survival curves

In clinical studies, we are often interested in assessing whether there are differences in survival among different

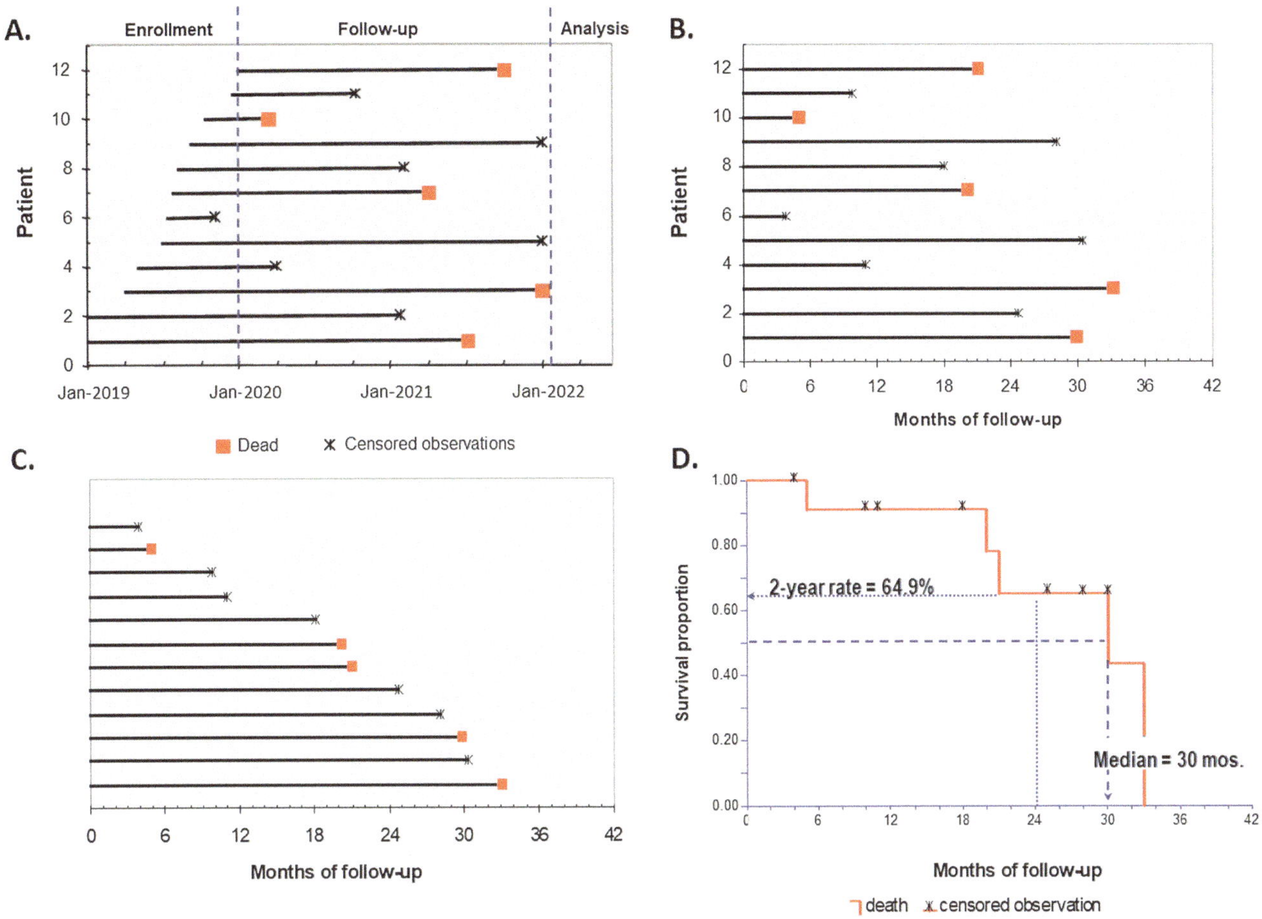

FIGURE 75.2 To construct a Kaplan–Meier curve in a study of 12 patients with one enrollment year and two follow-up years. (A) At the time of analysis, there were 5 deaths and seven censored observations, censoring occurred for two **on-study** patients who survived passed the follow-up time (id: 5 and 9) and **five off-study** due to refusal or loss to follow-up (id: 2, 4, 6, 8, and 11). (B) All patients assumed to start the study at the same time (time 0). (C) Sorted times give a structure to the KM curve. (D) The final form of a Kaplan–Meier survival plot.

groups of participants. For example, in a study to compare survival outcomes of heart transplantation between males and females, Weiss et al. found that female recipients, irrespective of donor sex, had lower overall survival than men (3.6% lower cumulative survival at 5 years). They estimated survival curves for each group, considered separately, using the KM method, and they compared the groups statistically using the log-rank test, which found a significant difference (P-value < .003).[7]

The log-rank test is popular for testing the null hypothesis of no difference in survival between two or more independent groups. The test compares the entire survival experience between groups and can be thought of as a test of whether the survival curves are identical (overlapping) or not. While the log-rank test does not necessarily hold proportional-hazards assumption, it does have little power when the survival or hazard functions cross,[8] especially if crossing occurs after 80% survival as suggested by Hsieh and Chen's simulation.[9] Thus, other tests may be used to evaluate the survival rate differences,[10] such as Fleming–Harrington (if crossing occurs at 80% ~ 60% survival and censored rate ≤50%), two-stage (if crossing at 80% ~ 60% survival and censored rate>50%, or crossing at 60 ~ 20 survival), or Gehan–Wilcoxon (if crossing at 0% ~ 20% survival[9]) methods.

Regression analysis

In survival analysis, we are often interested in assessing the effect of several risk factors individually or simultaneously to relate these factors to survival time, thereby producing hazard ratios (HRs). HRs are produced when comparing 2 hazard rates. A hazard rate has defined the risk of failure, given that the participant has survived up to a specific time. A probability must lie in the range of 0–1.

Cox (proportional hazards) regression

Sir David Cox, who died on January 18, 2022 (during the time of writing this chapter), was arguably the most influential statistician of the latter half of the 20th century, thanks to the Cox regression model,[11] a semiparametric regression framework for identifying factors that influence the time to an event occurring.[12] A semiparametric model has no assumptions about the shape of the baseline hazard function. However, it has other assumptions (e.g., independence, changes in predictors produce proportional changes in the hazard regardless of time). We assume that the hazards are proportional over time using Cox regression, which implies that the effect of a risk factor is constant over time. There are several approaches to assess the proportionality assumption; some are based on statistical tests and others involve graphical assessments. If either a statistical test or a graphical analysis suggests that the hazards are not proportional over time, then the Cox regression model is not appropriate,[13] and adjustments must be made to account for nonproportionality, such as stratifying the data into groups (i.e., within groups the hazards are proportional, and different baseline hazards are estimated in each stratum [as opposed to a single baseline hazard as was the case for the model presented earlier]). If adjustments are made to account for disproportionality, then the Cox model can still be used to assess one predictor's hazards ratio in univariable analysis and many independent predictors to adjust the hazards ratio in a multivariable model. Conversely, another approach would be to use a parametric regression to deal with time-varying survival.

Parametric regression

There are parametric regression models used in survival analysis that assume specific distributions for the survival times: exponential, Weibull, normal, and lognormal.

The exponential distribution is the only distribution that has a constant hazard/failure rate. Thus, a constant hazard rate over time (constant hazard function) implies exponential distribution of survival times. The Weibull distribution is the generalization of the exponential distribution, which allows the hazard/failure rate to vary over time. There are well-recognized statistical techniques for exploring the effect of several explanatory variables on survival assuming parametric distribution. Some may even allow us to estimate the parameters of the distribution and time-varying hazards.[14] Hsich et al. identified 3 phases of mortality risk following heart transplantation: early post-transplantation, constant, and late.[15] They found the recipient's sex not to be a significant risk factor predicting mortality. However, they also found different significant risk factors for mortality for each hazard phase such as CHD, renal, and hepatic function in the early phase that were not significant in other phases. Age, however, had a varying relationship over different hazard phases: older age affects the early and late phase, while younger age affects the constant phase.

Competing risks

The competing risks issue occurs when there are several possible outcome events of interest.[16,17] For example, a prospective study may be conducted to assess risk factors for time to incidents of cardiovascular disease events. For cardiovascular disease, one outcome may preclude witnessing another outcome (e.g., we cannot witness 30-day MI in patients die on the operating table). Researchers measure whether each of the component outcomes occurs during the study observation period as well as the time to each distinct event. The goal of the analysis is to determine the risk factors for each specific outcome and whether the outcomes are correlated. During regression, the influence of a covariate (risk factor) can be evaluated in relation to a cause-specific hazard for the specific outcomes, which was demonstrated using Fine–Gray regression model by multiple studies.[18–21]

Random survival forests

The random forest is an algorithm attributed to machine learning (a form of artificial intelligence) that aggregates numerous, independently trained decision trees built from different samples and takes their majority vote to improve the predictive accuracy of the module used with the given data. There are several benefits to the random forest algorithm, such as allowing for the assessment of a covariate relationship with the primary endpoint without model assumptions of distribution or linearity using a full nonparametric model. The random forest has the disadvantage of interpretation difficulty when compared to the standard Cox regression models. Developing techniques, such as partial plots and interaction analyses, are being employed to make machine learning interpretable. Hsich et al.[15] utilized a random survival forest to generate "variables of importance" assessing the strength of the predictability signal of each independent variable. The authors also determined that heart transplantation recipients' sex was not as important as other variables in predicting mortality following heart transplantation. However, they found an important relationship between hepatic or renal dysfunction's effects on mortality, which reflects the severity of illness with end-organ dysfunction at the time of transplantation.

Get started

- In order to assess survival outcomes, as researchers, you should develop protocols that allow you to retain incomplete information and analytical strategies that utilize it. As we emphasized in previous chapters (FUNDAMENTALS), your first steps should be to develop an appropriate research question and to calculate adequate power in collaboration with a statistician.

- When developing a protocol and drafting the consent, let your Institutional Review Board know that you will remove all identifying information of patients who withdraw from the study and communicate that clearly in the consent.
- Know what type and how many predictors of survival you want to include in your study. Some predictors will violate the proportional hazard assumption and some fit better with parametric modeling. Aim to use one predictor for 10 events (meaning if a study has 100 patients and the outcome of interest occurred in 30 overtime, you could use a maximum of three predictors simultaneously in a multivariable model). Predictor selection is another topic outside of the scope of this chapter, but what is necessary to understand is that while there are models that can help with selection—such as the Least Absolute Shrinkage And Selection Operator; also called LASSO—what matters is what is clinically meaningful and important.
- When presenting study results to the readers, a quantitative description of the extent of follow-up within compared groups is required. This includes the number of observations and percentages within each of the main groups, as well as the total time spent at risk (from study entry to the event or censoring) below the X-axis.
- When presenting study results, consider plotting the median survival time: the estimated time point at which 50% of the study population has experienced the event. However, if the median survival time cannot be estimated (e.g., if >50% of the study population has not experienced the event), in the report of result, indicate the estimates of a lower percentile of the survival times, such as the 25th percentile or the average survival time.

Pitfalls to avoid

- Failure to use the survival model in time-to-event data will result in unreliable and biased estimates and model interpretations could be dangerously misleading.
- When describing a study's result, do not only report survival time and its estimates at certain time points, but also provide a summary of follow-up times using median survival time. Incomplete and diverse follow-up time between the primary comparison groups can lead to informative censoring and bias, and this bias may go undetected.[22]
- When analyzing small sample-sized data, unreliable estimations will be problematic on their own. In survival analysis, when the number of events is not sufficient or in other words, the effective sample size is small, the results of these models may not be accurate. Thus, a Bayesian analysis—which includes informative priors from previous experience or studies—should be considered to include models that have certain conditions for the data.[23]

References

1. Alnajar A. *Identifying Factors Associated With Long-Term Graft Survival Following Heart Transplantation Using Statistics and Machine Learning Methods*. University of Miami; 2021.
2. Lamelas J, Alnajar A. Early outcomes for surgical minimally invasive SAPIEN 3 transcatheter mitral valve replacement. *Ann Thorac Surg*. 2021;112(2):494–500.
3. Leung KM, Elashoff RM, Afifi AA. Censoring issues in survival analysis. *Annu Rev Public Health*. 1997;18:83–104.
4. Rai S, Mishra P, Ghoshal UC. Survival analysis: a primer for the clinician scientists. *Indian J Gastroenterol*. 2021;40(5):541–549.
5. Cain KC, Harlow SD, Little RJ, et al. Bias due to left truncation and left censoring in longitudinal studies of developmental and disease processes. *Am J Epidemiol*. 2011;173(9):1078–1084.
6. Oliver JM, Gallego P, Gonzalez AE, et al. Risk factors for excess mortality in adults with congenital heart diseases. *Eur Heart J*. 2017;38(16):1233–1241.
7. Weiss ES, Allen JG, Patel ND, et al. The impact of donor-recipient sex matching on survival after orthotopic heart transplantation: analysis of 18 000 transplants in the modern era. *Circ Heart Fail*. 2009;2(5):401–408.
8. Liu Y, Yin G. Partitioned log-rank tests for the overall homogeneity of hazard rate functions. *Lifetime Data Anal*. 2017;23(3):400–425.
9. Hsieh J-J, Chen H-Y. A testing strategy for two crossing survival curves. *Commun Stat Simulat Comput*. 2017;46(8):6685–6696.
10. Li H, Han D, Hou Y, Chen H, Chen Z. Statistical inference methods for two crossing survival curves: a comparison of methods. *PLoS One*. 2015;10(1):e0116774.
11. Cox DR. Some statistical methods connected with series of events. *J Roy Stat Soc B*. 1955;17(2):129–157.
12. In J, Lee DK. Survival analysis: Part I - analysis of time-to-event. *Korean J Anesthesiol*. 2018;71(3):182–191.
13. Stensrud MJ, Hernán MA. Why test for proportional hazards? *JAMA*. 2020;323(14):1401–1402.
14. Blackstone EH, Naftel DC, Turner ME. The decomposition of time-varying hazard into phases, each incorporating a separate stream of concomitant information. *J Am Stat Assoc*. 1986;81(395):615–624.
15. Hsich EM, Blackstone EH, Thuita LW, et al. Heart transplantation: an in-depth survival analysis. *JACC Heart Fail*. 2020;8(7):557–568.
16. Austin PC, Lee DS, Fine JP. Introduction to the analysis of survival data in the presence of competing risks. *Circulation*. 2016;133(6):601–609.
17. McCaw ZR, Claggett BL, Tian L, et al. Practical recommendations on quantifying and interpreting treatment effects in the presence of terminal competing risks: a review. *JAMA Cardiol*. 2022;7(4):450–456.
18. Bakir NH, Finnan MJ, Itoh A, et al. Competing risks to transplant in bridging with continuous-flow left ventricular assist devices. *Ann Thorac Surg*. 2021;114.
19. Alvarez JA, Orav EJ, Wilkinson JD, et al. Competing risks for death and cardiac transplantation in children with dilated cardiomyopathy. *Circulation*. 2011;124(7):814–823.

20. Kilic A, Mathier MA, Hickey GW, et al. Evolving trends in adult heart transplant with the 2018 heart allocation policy change. *JAMA Cardiol.* 2021;6(2):159–167.
21. Wang Y, Leifheit EC, Krumholz HM. Trends in 10-year outcomes among medicare beneficiaries who survived an acute myocardial infarction. *JAMA Cardiol.* 2022;7.
22. May S, McKnight B. Graphics and statistics for cardiology: survival analysis. *Heart.* 2017;103(5):335–340.
23. Taheri Soodejani M, Tabatabaei SM, Mahmoudimanesh M. Bayesian statistics versus classical statistics in survival analysis: an applicable example. *Am J Cardiovasc Dis.* 2021;11(4): 484–488.

Chapter 76

Monitoring committee in clinical trials

Bruno B. Lima[1], Marianna Daibes Rachid de Andrade[2], Julia Tireli Rocha[3], Lara Rodrigues Da Matta Calegari[4] and Vinicius Vieira Neves[5]

[1]*Noninvasive Cardiovascular Imaging Program, Department of Radiology and Medicine (Cardiology), Brigham & Women's Hospital/Harvard Medical School, Boston, MA, United States;* [2]*Spaulding Neuromodulation Center, Spaulding Rehabilitation Hospital/Harvard Medical School, Boston, MA, United States;* [3]*Faculdade de Medicina de Jundiai, Jundiai, Brazil;* [4]*Faculdade Evangélica Mackenzie do Paraná, Curitiba, Brazil;* [5]*Department of Behavioral Science and Health, University College London (UCL), London, United Kingdom*

Key points

- Role of Monitoring Committees: These committees ensure participant safety, evaluate intervention effectiveness, and maintain the integrity and accuracy of research findings.
- Independence and Confidentiality: Monitoring committees must operate independently from study investigators and protect trial data and participant information.
- Ethical Considerations: Balancing benefits and risks, maintaining confidentiality, and upholding informed consent and human subject protection are critical.
- Challenges: Ensuring independence, managing conflicts of interest, and making decisions in complex situations are significant challenges for monitoring committees.

Introduction

Clinical trials, especially in cardiology, are vital for validating the safety and efficacy of new treatments. Central to these trials are monitoring committees—groups of independent experts who supervise the trial's execution, guaranteeing adherence to the protocol and participant safety, with the power to halt the trial for severe safety issues.

Monitoring committees uphold trial quality and integrity. Their rigorous oversight ensures accurate, reliable results, contributing to optimal patient care. Key responsibilities of these committees in cardiology trials include the following:

- Ensuring participant safety and rights
- Evaluating intervention effectiveness
- Guaranteeing research findings' accuracy and reliability
- Providing recommendations to investigators

By effectively fulfilling these roles, monitoring committees ensure the safe and effective deployment of new treatments, thereby advancing patient care in cardiovascular disease.

Definition and purpose

The monitoring committee, also known as a data monitoring committee (DMC) or data and safety monitoring board (DSMB), is an autonomous team of experts tasked with scrutinizing clinical trial data to safeguard participants' safety and well-being.

Comprising clinicians, statisticians, and other experienced professionals in clinical trials, this committee is appointed by the study sponsor. Crucially, it remains independent from the study investigators, distancing itself from the trial's day-to-day operations and thereby avoiding influence from investigators when making recommendations.

The committee conducts regular meetings to assess trial data, focusing on intervention safety, effectiveness, and trial progress. It may also review data quality and the investigators' adherence to the trial protocol.

Postreview, the committee can make recommendations to the study sponsor. These could involve continuing the trial as planned, suggesting modifications, or recommending termination. Additionally, the committee may propose the collection of further data or the execution of additional analyses.

The committee's role is instrumental in promoting participant safety and well-being in clinical trials. Its recommendations can contribute to the safe and ethical conduct of the trial, ensuring the generation of accurate and reliable results.

Translational Cardiology. https://doi.org/10.1016/B978-0-323-91790-2.00014-9

Roles and responsibilities

The monitoring committee carries out several pivotal functions in a clinical trial:

- Safety monitoring: It reviews data to identify potential safety risks to participants, including adverse events and fatalities. If risks are detected, the trial may be recommended for suspension or modification.
- Efficacy monitoring: The committee evaluates the effectiveness of the intervention based on primary and secondary endpoints. If insufficient evidence of effectiveness is found, the trial may be advised for suspension or alteration.
- Reviewing trial conduct: The committee ensures the trial is ethically run and protocol deviations are properly handled, reviewing data related to participant recruitment, intervention administration, and data collection. If conduct issues arise, the trial could be recommended for suspension or modification.
- Making recommendations: Based on their review, the committee advises whether the trial should continue, be altered, or be terminated, aiming to protect participants' interests.

Ethical considerations

Monitoring committees face crucial ethical considerations in clinical trials, primarily balancing the benefits and risks to participants, ensuring data and participant confidentiality, and upholding informed consent, human subject protection, and scientific integrity.

- Balancing benefits and risks: The committee must carefully assess potential trial benefits, such as improved health or quality of life, against potential risks like side effects, complications, or death. It's paramount that the trial's benefits outweigh its risks before proceeding.
- Maintaining confidentiality: The committee must protect trial data and participant information from unauthorized access, thereby preserving participant privacy.
- Upholding informed consent: The committee is tasked with ensuring all participants have been granted informed consent, meaning they have a comprehensive understanding of the trial's risks and benefits before agreeing to participate.
- Protecting human subjects: Ensuring the trial subjects are protected from physical, psychological, and social harm is another ethical imperative of the committee.
- Safeguarding scientific integrity: The committee must ensure the trial maintains scientific soundness, necessitating the collection and analysis of data in an impartial, unbiased manner.

By meticulously attending to these ethical considerations, monitoring committees can ensure that clinical trials are conducted safely, ethically, and with scientific rigor.

Challenges and solutions

Monitoring committees are instrumental to clinical trials, yet they encounter numerous challenges, including:

- Ensuring independence: These committees must operate independently from study investigators, a challenge when investigators appoint committee members. To ensure independence, it is crucial that committee members don't hold any financial or personal links to the investigators or study sponsor.
- Controlling potential conflicts of interest: Even without financial or personal connections to the study investigators or sponsor, committee members may have biases toward the intervention under review. These potential conflicts of interest should be disclosed to the sponsor and other committee members.
- Deciding in complex, uncertain situations: Committees often encounter convoluted scenarios and uncertainty. For instance, they might need to decide whether a trial should proceed if there's evidence of participant harm. These situations require the committee to balance the risks and benefits to protect participants' interests.

To mitigate these challenges, there should be clear guidelines and standard operating procedures detailing the committee's roles and responsibilities, how to handle potential conflicts of interest, and decision-making processes in uncertain situations. Further, ongoing training for committee members can provide them with the necessary skills and knowledge to fulfill their roles effectively.

Early termination of clinical trials for futility

There are instances where clinical trials are terminated early for various concerns. Two notable examples include the Randomized Ablation-based atrial Fibrillation rhythm control versus rate control Trial in patients with heart failure and high burden Atrial Fibrillation (RAFT-AF) and the Functional Testing Underlying Coronary Revascularization (FUTURE) trial. RAFT-AF focused on managing patients with atrial fibrillation, a condition characterized by irregular heartbeats, and heart failure. At the same time, FUTURE aimed to evaluate the efficacy of fractional flow reserve in guiding coronary revascularization.

In the RAFT-AF trial, the DMC intervened when only 33% of patients were enrolled, requesting a futility analysis. Upon reviewing this analysis, the committee recommended halting enrollment but continuing follow-up for a minimum of 2 years. The early termination was due to lower-than-expected enrollment and perceived futility. While the rhythm-control strategy did not prove superior in cardiovascular outcomes, functional and serum biomarkers did show significant improvements with this approach, despite some concerns over high periprocedural complications.

Conversely, the FUTURE trial was stopped by its DSMB due to a significant increase in early mortality among patients undergoing coronary angiography with revascularization guided by fractional flow reserve readings. An interim data check revealed more than double the risk of death in the group following the fractional flow reserve-guided strategy compared to the control group. The trial was terminated after enrolling 933 out of the targeted 1700 patients. This highlights the critical role of monitoring in clinical trials to ensure the safety and well-being of the participants.

Conclusions

Monitoring committees are an integral part of clinical trials, safeguarding participant safety and upholding trial integrity. They vigilantly review data to promptly identify safety concerns, potentially leading to trial modifications or cessation. They also maintain trial integrity by enforcing protocol adherence and ensuring unbiased data collection and analysis.

The committee's crucial role extends to translational cardiology, where they validate the accuracy and reliability of trial results, aiding the development of novel treatments for cardiovascular disease. As clinical trials evolve, particularly with emerging technologies like wearable devices and artificial intelligence, monitoring committees' roles will continue to adapt, maintaining safety, ethics, and scientific soundness in these trials. Their steadfast commitment contributes significantly to accurate, reliable trial results that ultimately improve patient health outcomes in cardiovascular disease.

Get started

1. Review all necessary components of the CMC application.
2. Collect all documentation and data supporting the identity, safety, and stability of the drug product.
3. Ensure all required documentation is in place, including a copy of the FDA's GxP inspection (GMP inspection documents for all manufacturing or packaging sites used).

Potential pitfalls

- Insufficient Information: Not providing enough information to delineate the drug's safety and stability.
- Failure to Submit Changes: Not submitting or providing adequate information regarding changes in the manufacturing process and sites as an amendment to the IND in a timely manner.
- Incomplete Records: Not providing a complete batch record and stability data after using an alternate manufacturing process.
- Outdated Inspections: Not ensuring that GxP inspections are up-to-date for all sites used in the drug's manufacturing process.

Examples of ...

Early Termination for Futility.

- RAFT-AF Trial: Terminated early due to lower-than-expected enrollment and perceived futility, although functional and serum biomarkers showed improvements.
- FUTURE Trial: Stopped due to a significant increase in early mortality among patients undergoing coronary angiography with fractional flow reserve-guided strategy.

Resources

1. FDA Guidance Documents: https://www.fda.gov/regulatory-information/search-fda-guidance-documents
2. Further Reading: https://onlinelibrary.wiley.com/doi/10.1002/9781118445112.stat04932

Further reading

1. Armstrong PW, Califf RM. Data and safety monitoring boards: academic credit where credit is due? *JAMA*. 2013;310(15):1563–1564.
2. Ball G, Piller LB, Silverman MH. Continuous safety monitoring for randomized controlled clinical trials with blinded treatment information: Part 1: ethical considerations. *Contemp Clin Trials*. 2011;32:S2–S4.
3. Vaz M, Srinivasan K. Ethical challenges & dilemmas for medical health professionals doing psychiatric research. *Indian J Med Res*. 2014;139(2):191.
4. Lad PM, Dahl R. The data and safety monitoring board in sponsored pediatric clinical trials: practical applications. *J Hosp Adm*. 2015;4(1).
5. Demets DL. *Data Monitoring Committees*. Wiley StatsRef: Statistics Reference Online; 2014.
6. Parkash R, Wells GA, Rouleau J, et al. Randomized ablation-based rhythm-control versus rate-control trial in patients with heart failure and atrial fibrillation: results from the RAFT-AF trial. *Circulation*. 2022;145(23):1693–1704.
7. Rioufol G, Dérimay F, Roubille F, et al. Fractional flow reserve to guide treatment of patients with multivessel coronary artery disease. *J Am Coll Cardiol*. 2021;78(19):1875–1885.

Part VIII

Regulatory basics

Aditi Nayak

Chapter 77

FDA overview

Alysa Vereen[1] and Guileonne Ngnitewe[2]
[1]Priovant Therapeutics, Durham, NC, United States; [2]Regulatory Promotional Review ProPharma Group, Madison, WI, United States

Key points

- **FDA Regulation:** The FDA oversees drugs, biologics, medical devices, food, cosmetics, and tobacco to ensure their safety, efficacy, and security, which is crucial for public health protection.
- **Drug Approval Process:** This process includes stages like IND application, Phase 1–3 studies, and the submission of NDA or BLA, with the FDA reviewing these under PDUFA and issuing EUAs during emergencies, such as the COVID-19 pandemic, showcasing its adaptability.
- **Role of Regulatory Affairs Professionals:** They ensure compliance with evolving regulations, advise on scientific and regulatory requirements, and facilitate communication with agencies like the FDA, which is essential for the safety and success of products.

Purpose and overview

Introduction

Translational research is largely positioned around developing innovative scientific products into meaningful technologies. Comparably, this scientific discipline ties directly into regulatory affairs and engaging regulatory agencies, such as the US Food and Drug Administration (FDA). The FDA is responsible for protecting public health by ensuring the safety, efficacy, and security of human and veterinary drugs, biological products, and medical devices, as well as US food supply, cosmetics, and products that emit radiation. While similar regulatory bodies exist in other countries and parts of the world, this chapter will focus on providing an overview of the role of the FDA.

The big picture

The FDA has an important role in improving both human and animal health. They are an agency within the US Department of Health and Human Services and function under the Office of the Commissioner and four directorates that oversee the core functions of the agency: Medical Products and Tobacco, Foods and Veterinary Medicine, Global Regulatory Operations and Policy, and Operations. The FDA represents an essential role in supporting the development of new and novel products and technologies. In executing its mission, the FDA is tasked with advancing the public health by facilitating efforts to increase public accessibility to information that is accurate, timely, and evidence-based. Overall, these efforts allow for a multidisciplinary approach across all phases of product development, which results in the highest quality for patients and the protection of the public.

Why it matters

Although the general public understands FDA's fundamental mission in protecting public health, the various mechanisms by which they guide the development of technology and medical products to the market are less familiar. The FDA's regulatory authority is quite broad and can be confusing for consumers, clinicians, and the industry to navigate. As such, the strategic planning and coordination across the various phases of the translational environment are also often overlooked.

Introduction to the FDA

History

The first regulation established to protect the public in the United States dates back to the early 1900s and led to several laws that still impact regulatory affairs. Those laws include the following:

- **The 1906 Food and Drugs Act**

This act was the first comprehensive federal food and drug regulation in the United States that resulted in the

Translational Cardiology. https://doi.org/10.1016/B978-0-323-91790-2.00043-5

federal government prohibiting **adulterated** or **misbranded** drugs and food.

- **The Federal Food, Drug, and Cosmetic Act of 1938**

This act was passed after a legally marketed toxic elixir killed 107 people, including many children. This tragedy led in the FDA to require mandatory safety evidence for all new drugs and standards for food, as well as mandating factory inspections.

- **The Kefauver-Harris Amendments of 1962**

Within a few years of the widespread use of thalidomide in Europe, Australia, and Japan, approximately 10,000 children were born with **phocomelia**, a rare birth defect that leads to missing or underdeveloped upper and lower limbs. In addition to limb reduction anomalies, other effects later attributed to thalidomide included congenital heart disease, malformations of the inner and outer ear, and ocular abnormalities. Thalidomide was banned in most countries after the tragedy, but the FDA's **pharmacovigilance** at the time prevented the drug's marketing and use in the United States. Still, these events led to even stricter rules for manufacturers to show proof of drug safety and effectiveness.

- **The Medical Device Amendments of 1976**

These rulings resulted from the US Senate finding that faulty medical devices had caused 10,000 injuries, including 731 deaths. This law set the standard for safety and effectiveness for new devices moving forward.

Modern day drug/device approval process

For a drug, biological, or device to be approved today, the following processes are required.

- **Investigational New Drug process**

An Investigational New Drug (IND) application is a formal request submitted to the FDA seeking permission to administer an **investigational drug** or biologic to human subjects. Federal regulations mandate that the initial IND submission undergo review within 30 days of receipt by the FDA. Following this period, the IND is either approved for further clinical investigation or placed on a clinical hold. The primary objective of the IND process is to evaluate the safety profile of the investigational product.

- **NDA/BLA process**

After successful completion of Phase 1−3 studies, a New Drug Application (NDA) or a Biologic License Application (BLA) is filed. The Prescription Drug User Fee Act (PDUFA) process, which is a law that allows the FDA to collect fees from drug manufacturers to fund the new drug approval process, requires standard review of the drug or biologic to be completed in 10−12 months. After that, the FDA either approves the therapeutic or provides a "Complete Response" letter to the Sponsor. A Complete Response means that the therapeutic is not approvable in its current form and that the FDA has identified deficiencies. Many applications require additional submissions, resulting in additional FDA review time.

- **Emergency Use Authorization**

In exceptional situations, the FDA can authorize the emergency use of an unapproved medical product or an unapproved use of an approved product following a declaration of emergency by the US Department of Health and Human Services Secretary. The Emergency Use Authorization (EUA) ends once the product is approved or the emergency declaration is lifted. An example of an EUA in action was the early administration of the COVID-19 vaccines.

FDA branches and divisions

Over the years, the FDA has made momentous efforts to modernize its structure. As recently as March 2019, the FDA implemented a reorganization plan to advance the mission of protecting and promoting public health. From realigning existing offices to consolidating others, the new and improved structure's overall aim is to improve efficiency throughout the Agency.

FDA currently functions as an agency within the US Department of Health and Human Services under the Office of the Commissioner and four directorates that provide oversight of the core functions. A current overview of the most accurate representation of the FDA's structure can be found at https://www.fda.gov/about-fda/fda-organization/fda-organization-charts.

Office of the Commissioner

The Office of the Commissioner (OC) provides centralized direction across the agency to ensure efforts are managed effectively. This office also directs programs, facilitates coordination, and provides strategy and advice to agency leadership over scientific, legislative, and policy-related activities.

The four directorates: FDA's field offices

There are four directorates: (1) Office of Medical Products and Tobacco, (2) Office of Foods and Veterinary Medicine, (3) Office of Global Regulatory Operations and Policy, and (4) Office of Operations. Each of these offices has distinct roles and responsibilities that contribute to FDA's mission.

Office of Medical Products and Tobacco

- This office provides advice to the Commissioner on programs and issues related to medical and tobacco products.

Office of Foods and Veterinary Medicine

- This office encompasses the Center for Food Safety and Applied Nutrition (CFSAN) and the Center for Veterinary Medicine (CVM).

Office of Global Regulatory Operations and Policy

- This office is integral in enforcing federal laws that govern biologics, cosmetics, dietary supplements, drugs, food, medical devices, radiation-emitting products, tobacco products, and veterinary medicine products with potentially harmful effects to consumers.

Office of Operations

- This office works to coordinate emergency preparedness and response activities for events that involve FDA-regulated products. The Office of Operations also offers support services across the FDA and its centers.

Navigating FDA in translational medicine

Navigating the FDA in translational medicine can be daunting due to the numerous divisions, offices, and centers within its structure. Bridging the gap between laboratory research and clinical application poses challenges for clinicians, who may find it difficult to locate the right resources for translating scientific discoveries into practical use with human subjects and patients. The following steps outline strategies for identifying regulations, guidance documents, and the relevant FDA office to support your product, project, or study.

- **Define Your Product, Project, or Study**: Clearly define the nature and purpose of your translational medicine endeavor. Identify the specific product, project, or study you are working on, including its intended use and target population.
- **Research Applicable Regulations and Guidance Documents**: Conduct thorough research to identify relevant regulations and guidance documents issued by the FDA. Focus on regulations and guidance specific to your product category (e.g., drugs, biologics, medical devices) and intended use.
- **Utilize FDA Resources and Databases**: Explore FDA websites and databases to access comprehensive regulatory information. Utilize resources such as the FDA's Guidance Documents Database, Regulatory Procedures Manual, and Code of Federal Regulations (CFR) to find relevant information.
- **Engage with FDA Contacts and Support Services**: Reach out to FDA contacts, such as regulatory affairs specialists or project managers, for guidance and assistance. Take advantage of FDA support services, such as the Office of Regulatory Affairs (ORA) or the Center for Biologics Evaluation and Research (CBER), to obtain tailored support for your specific needs.
- **Establish Communication Channels with FDA Offices**: Establish communication channels with relevant FDA offices early in your project or study. Attend FDA workshops, webinars, or conferences related to your area of interest to network and gather information. Consider scheduling meetings or consultations with FDA officials to discuss your project and seek guidance on regulatory requirements.

Regulatory principles

With the constant changes happening within healthcare, a regulatory affairs professional is essential to mitigate the risks that come with the drug, biologic, and medical device approval process.

Regulatory professionals are responsible for the following:

- Tracking the continuously evolving regulations in all regions where the company plans to distribute its products
- Advising on legal and scientific restraints and requirements
- Collecting, gathering, and evaluating scientific data
- Presenting registration documents to regulatory agencies and carrying out any subsequent negotiations necessary to obtain or maintain marketing authorization for the products concerned
- Providing strategic and technical advice at the highest level within their companies, significantly contributing to both the commercial and scientific success of the organization
- Assisting the company in avoiding issues resulting from poorly maintained records, flawed scientific reasoning, or inadequate data presentation

Additionally, the regulatory affairs department will often take part in the development of product marketing concepts and is commonly required to approve packaging and advertising before it is used commercially, ensuring compliance with all regulations.

- Here are a few examples:
  - A regulatory affairs professional updating someone in another department on the recent changes to a disease state FDA Guidance
  - A regulatory affairs strategy professional having conversations with the FDA to get a new indication for a device
  - A regulatory advertising and promotion professional saving the company millions of dollars by catching a false claim about an indication, such as saying a medication can be used to treat other diseases it is not approved for

What is an FDA guidance?

- Guidance documents represent the FDA's current thinking on a topic, but they do not create or confer any rights for or on any person and do not operate to bind FDA or the public. An alternative approach may be used if it satisfies the requirements of the applicable statutes and regulations.
- FDA guidances explain the agency's interpretation of, or policy on, a regulatory issue. The FDA prepares guidances primarily for industry, but also for other stakeholders, and uses them to address such matters including the design, manufacturing, and testing of regulated products, scientific issues, content and evaluation of applications for product approvals, as well as inspection and enforcement of policies.
- Although guidances are not legally binding, they show stakeholders one sure way to reach their regulatory goal, however, stakeholders are free to use other approaches that satisfy the relevant law and regulations

When it goes wrong

Do's and Don'ts: Tips for Efficiently Bringing Your Product to Market

DO	DON'T
Do Conduct Thorough Research: Understand the regulatory requirements specific to your product category (e.g., drug, biologic, medical device). Stay updated on FDA guidelines and industry standards.	**Don't Underestimate the Importance of Quality**: Do not compromise on product quality to cut costs or save time. Poor quality control can lead to recalls and damage to your reputation.
Do Develop a Clear Regulatory Strategy: Plan your regulatory pathway early in the development process. Include timelines for key milestones such as IND, NDA, or BLA submissions.	**Don't Rush the Development Process**: Avoid rushing clinical trials or data collection to meet deadlines. Ensure all data is accurate and thoroughly vetted.
Do Engage with the FDA Early and Often: Schedule pre-submission meetings to discuss your development plan and regulatory strategy. Respond promptly to FDA inquiries and requests for additional information.	**Don't Disregard International Regulations**: If planning to market your product internationally, do not ignore the regulatory requirements of other countries. Ensure your product meets global regulatory standards.
Do Invest in Regulatory Affairs Expertise: Hire or consult with experienced regulatory affairs professionals. Keep your team informed about regulatory changes and updates.	**Don't Neglect Regulatory Requirements**: Avoid skipping any regulatory steps or requirements, even if they seem minor. Ignoring regulations can lead to delays, fines, or rejection of your product.
Do Conduct Comprehensive Clinical Trials: Design trials that meet FDA requirements for safety and efficacy. Collect and analyze data rigorously.	**Don't Misbrand Your Product**: Ensure all labeling and marketing materials are accurate and comply with FDA regulations. Avoid making unapproved claims about your product.
Do Prepare Detailed and Accurate Submissions: Ensure all regulatory submissions are complete and accurate. Provide all required data and documentation.	**Don't Forget to Train Your Team**: Do not assume your team is fully aware of all regulatory requirements. Regularly train and update your staff on compliance and regulatory changes.

Get started

In your biotechnology program:

⇒ What programs are in place to mitigate risk?
⇒ Have you sought out a regulatory affairs professional?
⇒ What stage are you in your clinical trial?
⇒ During which steps will you need to implement a regulatory affairs professional?

Potential pitfalls

- **Understanding FDA Regulations:** A lack of understanding can lead to non-compliance and delays in product development or marketing approval.
- **FDA Approval Process Complexity:** The multifaceted process, including IND application, clinical trials, and NDA submission, is daunting and time-consuming, requiring careful planning and resources.
- **Navigating FDA's Organizational Structure:** The complexity of the FDA's multiple offices and directorates may pose challenges in identifying appropriate communication and regulatory guidance channels.

- **Consequences of Non-Compliance:** Failing to adhere to FDA regulations and guidelines can result in product recalls, fines, or legal actions, endangering public health and business interests.
- **Accessing FDA Resources:** Clinicians and researchers might struggle to find FDA resources and guidance documents tailored to their specific needs, hindering the translation of scientific findings into products and therapies.

Examples

- **Marketing Compliance:** A pharmaceutical company preparing a promotional campaign discovers that some materials make unapproved claims about a new drug's effectiveness. A regulatory advertising and promotion professional is crucial to ensure compliance, preventing enforcement action and protecting the company's reputation.
- **Navigating the IND Process:** Without a clear understanding of the FDA's regulatory pathways, a pharmaceutical company might face delays in initiating clinical trials or have their IND application rejected, slowing down drug development.
- **Global Market Expansion:** A biotech startup aiming to expand globally may struggle to obtain international market approvals without a dedicated regulatory affairs team to monitor and interpret changing regulations, limiting their growth potential and competitive advantage.

Resources

https://www.fda.gov/about-fda/fda-basics/fda-fundamentals

https://www.usp.org/sites/default/files/fda-exhibit/legislation/1906.html

https://academic.oup.com/toxsci/article/122/1/1/1672454

https://rarediseases.info.nih.gov/diseases/12123/phocomelia#:~:text=Phocomelia%20is%20a%20rare%20birth,attached%20directly%20to%20the%20trunk

https://ncats.nih.gov/translation/spectrum

https://www.topra.org/topra/topra_member/what_is_regulatory_affairs.aspx

https://www.fda.gov/regulatory-information/search-fda-guidance-documents

https://www.fda.gov/files/drugs/published/Best-Practices-for-Communication-Between-IND-Sponsors-and-FDA-During-Drug-Development.pdf

https://www.ncbi.nlm.nih.gov/books/NBK195047/

Chapter 78

IND/New Drug Application

Sergio Rodriguez Labra

Department of Chemistry, The Scripps Research Institute, San Diego, CA, United States

Key points

- Learning about the regulatory requirements to conduct clinical trials and commercialize a product early on will help guide your drug development process most efficiently.
- Whenever possible, take advantage of the FDA's expedited and assistance programs to accelerate commercialization.
- Engage with the FDA ahead of time to confirm the specific requirements your particular IND/NDA application will require, help bolster them, and maximize their chances of approval.

Introduction

The use and testing of therapeutics in the United States are regulated by the FDA, which requires their safety, efficacy, and quality controls to be demonstrated by new drug applicants through the Investigational New Drug (IND) application before they are allowed to be used in human trials.[1]

The path of bringing a preclinical drug candidate to market involves significant health risks, costs, and prolonged duration, which has pushed the regulatory agencies and the markets to emphasize safety and solid preclinical development before moving to clinical trials.[2] IND application submissions are sent to the Center for Drugs Evaluation and Research (CDER) to each be critically reviewed by a team of scientific experts to ensure strict margins of safety are maintained during the clinical trial.[3] Thus, successfully obtaining IND authorization represents a milestone of great significance; propelling the investigational drug product and its sponsor company or organization into a significantly more valuable asset able to attract new investment, partnership, licensing, and acquisition opportunities during the clinical trials.

If and when the data collected from the trials demonstrate the safety and efficacy of the new drug in its intended use, a New Drug Application (NDA) is filed with the FDA for them to review and weigh the risks and benefits of allowing it to treat the relevant rest of the public. Once the NDA is approved, the drug can be immediately publicly marketed and sold within the United States.

Becoming well acquainted with the intricacies and resources available for IND and NDA submissions allows one to plan and build a compelling application time- and cost-efficiently to maximize the chances for drug candidate clinical authorization and public marketing approval.

Is an IND required?

In general, IND authorization is required if any investigation intends to eventually support any change of prescription, indication, or advertisement of a new or currently marketed drug in the United States. The FDA's code of federal regulations "21 CFR 312.2" contains the precise requirements and should further guidance be sought to determine IND requirement/exemption, an academic institution's IRB Regulatory Guidance Office and/or the FDA can provide it.[4]

The IND filing process

The process formally begins when the sponsor files an IND application with the FDA. The IND sponsor is the individual, pharmaceutical company, academic institution, governmental agency, or organization that takes responsibility for the clinical study requiring the IND. The sponsor selects the qualified principal investigator(s)—who is in charge of initiating and conducting the study on each of the centers involved in the study—and ensures the investigation is properly monitored in line with FDA requirements. Sponsor–Investigator designations are allowed if there is a single investigator and they hold the IND.

There are two IND categories: research and commercial. Research INDs are often submitted by physician

Translational Cardiology. https://doi.org/10.1016/B978-0-323-91790-2.00058-7

researchers while Commercial INDs are generally submitted by companies or nonprofit entities like the NIH with the intent of bringing a new pharmaceutical product to market.[5]

There are two other subcategories of INDs relevant to academic research: Exploratory IND and Expanded Access (Treatment) IND. Exploratory INDs are first-in-human studies, or single and low-dose studies conducted on small healthy volunteer cohorts; given their lower risk profile, these studies may be approved with more flexible preclinical data benchmarks. Expanded access is required when it is determined that there is no comparable or satisfactory therapy available to diagnose or treat a patient's condition, and the patient is unable to participate in a clinical trial or obtain the investigational drug under another IND.[5]

All IND (and NDA) applications are reviewed by a committee or a division specializing in the therapeutic area. The IND main sections and relevant advice are summarized in Fig. 78.1. Important to note is that an IND is dynamic after FDA submission and should be modified and expanded upon throughout the clinical investigation. A complete IND application may also involve product labeling and previous human experience with the investigational product sections if the product has been previously clinically investigated or marketed outside the United States, in which case all the detailed relevant clinical information regarding safety and efficacy from anywhere in the world should be provided. Submission via the Electronic Submission Gateway is the encouraged faster processing method. An IND becomes effective 30 days after the submission is received by the FDA unless the FDA requests modifications and places the study on hold until they are addressed. Thus, ample time should be planned to allow to build a strong complete application.

The NDA filing process

The final stage of the traditional drug development pipeline is the submission of an NDA application to the FDA to formally propose the product's sale and marketing in the United States. The application components are listed in Fig. 78.2,[6] which compiles all the drug's data gathered during the preclinical and IND-authorized clinical trials (e.g., Phase I through II/III) for the FDA to evaluate:[7]

- The drug's safety and efficacy in its proposed use(s), and whether the benefits outweigh the risks.
- The appropriate labeling or package insert the drug should have if approved.
- The drug manufacturing methods and controls' ability to preserve the drug's identity, strength, quality, and purity

FDA resources

It is essential to engage early on with the FDA as the lead drug product nears the maturity to consider applying to conduct clinical trials. The FDA reviewing officials offer three different kinds of meetings with drug product sponsors: Type A, B, and C. Type B, which includes all milestone meetings like the pre-IND meeting is granted within 60 days of request. Once one is confident about the data supporting a potential IND application, the pre-IND meeting provides the best opportunity to determine if the collected nonclinical data up to that point is enough to support a Phase I or II study in humans and to discuss the scope and design of the study. Note that the complete background package of the necessary data and information disclosed (the pre-IND Briefing packet) is submitted to the FDA at least 4 weeks before the meeting date.[7]

The FDA also offer different programs to expedite the application review; reduce its cost; and/or reduce its minimum clinical evidence requirements for approval (Table 78.1).[8,9] These are meant to help address unmet medical needs in treating serious conditions, though as such, some designations may involve additional requirements such as a proposed viable commercial development plan earlier in the development process. The FDA reviews these applications on a case-by-case basis and eligibility is encouraged to be discussed with the FDA before officially soliciting—at the time of the IND submission for most.

Financing

In addition to possible lawyer and consultant fees, the sponsor pays a fee to the FDA at the time the IND is submitted. They can be paid online through pay.gov. The fee rate changes every year so it is encouraged to check in advance the applicable fees at the FDA website and ensure that the IND submission will have a high chance of being approved without delays through a pre-IND meeting with the FDA.

In addition to the fee reductions/waivers in some of its expedited programs, the FDA provides support for small businesses through the CDER Small Business and Industry Assistance (SBIA), which helps small businesses (<500 full-time employees) navigate the regulation of human drug products and assists in finding the resources to submit a successful IND. The FDA can also grant a fee waiver if it determines along the Small Business Administration (SBA) that it qualifies for one, after which it would have 1 year to submit the IND.[10] For its part, the NDA carries a significantly higher fee cost associated with its submission. Per the Prescription Drug User Fee Act (PDUFA) of 2017, the base fee is on the order of \$1.6–3.1 million USD[11]; though it also allows for similar fee waivers and reductions. More details on the FDA's guidance resources.[12]

Get started

- Find out if your lead drug studies will require an IND, and which IND type you will need.

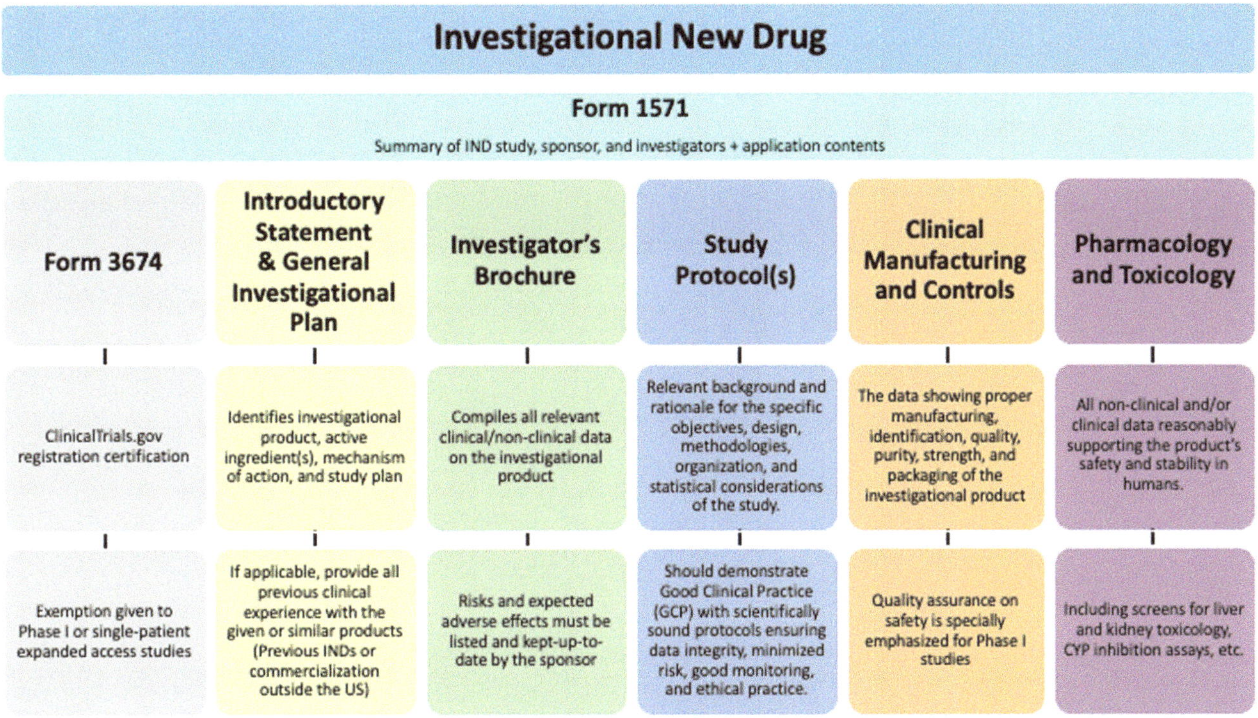

FIGURE 78.1 IND main sections and relevant advice summary.

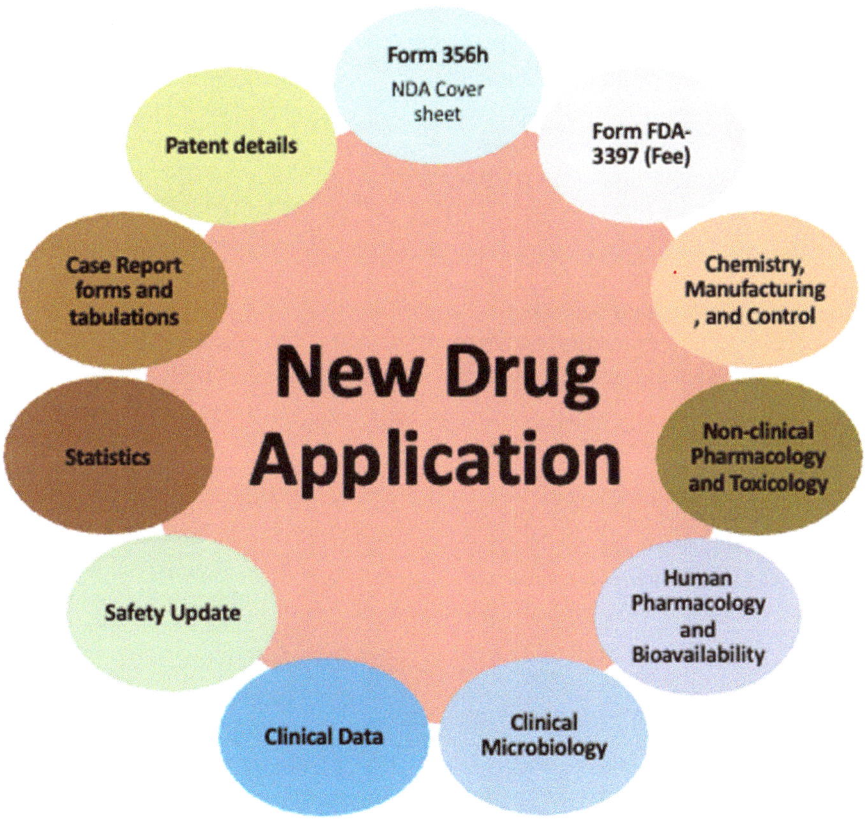

FIGURE 78.2 NDA application components.

TABLE 78.1 FDA expedited programs available for early clinical investigations.

FDA program	Qualifying criteria	Features	References for precise details
Fast track designation	Drug intended to treat a serious condition AND data (including nonclinical) demonstrates the potential to address an unmet medical need OR Drug has been designated as a qualified infectious disease product	Development and review are expedited, including the use of rolling review (FDA begins reviewing portions of the application before completion)	Section 506(b) of the FD&C Act
Breakthrough therapy designation	Drug intended to treat a serious condition AND preliminary clinical evidence indicates that the drug may demonstrate substantial improvement on a clinically significant endpoint(s) over available therapies	Intensive guidance on efficient drug development, proactive involvement of senior management, rolling review, and consideration for priority review	Section 506(a) of the FD&C Act
Orphan drug designation	Drug intended to treat a condition affecting ≤200,000 people in the United States	Marketing application fee is waived and other financial development incentives	21 CFR parts 316.20—316.30 of the FD&C Act
Accelerated approval pathway	Drug that treats a serious condition AND provides a meaningful advantage over available therapies AND demonstrates an effect on a surrogate endpoint likely to predict clinical benefit	Approval based on an effect on the surrogate endpoint or intermediate clinical endpoint.	21 CFR parts 314, 601, and Section 506(c) of the FD&C Act
Priority review	Drug that treats a serious condition AND	Expedites the NDA (or similar filings) evaluation	21 CFR 312 of the FD&C Act

- Get well-acquainted with the IND components so you can ensure the appropriate preclinical studies (e.g., animal studies, toxicology, etc.) are designed and performed to build up the lead drug's application.
- Explore available funding and guidance support opportunities to prepare your application filings.

Additional resources

1. The US Food and Drug Administration Website
   a. The US FDA maintains all its up-to-date resources and guides on its website. Particular sections of interest are:
      i. Contact information: https://www.fda.gov/about-fda/contact-fda
      ii. Educational resources, including presentations by their officials: https://www.fda.gov/Training/learningportal/ucm417363.htm#humandrug
      iii. Information and resources for Industry (Drugs): https://www.fda.gov/Drugs/ResourcesForYou/Industry/default.htm
      iv. Guidances (Drugs): https://www.fda.gov/Drugs/GuidanceComplianceRegulatoryInformation/Guidances/default.htm
      v. Investigational New Drug (IND) Application: https://www.fda.gov/Drugs/DevelopmentApprovalProcess/HowDrugsareDevelopedandApproved/ApprovalApplications/InvestigationalNewDrugINDApplication/default.htm
      vi. New Drug Application: https://www.fda.gov/drugs/types-applications/new-drug-application-nda
      vii. Prescription Drug User Fees (for NDAs): https://www.fda.gov/forindustry/userfees/prescriptiondruguserfee/default.htm
2. The Electronic Code for Federal Regulations (e-CFR)
   a. The general database with the current rules published in the Federal Register with all the specific definitions and details. Relevant sections involve:
      i. The IND Application regulations found on Part 312: https://www.ecfr.gov/cgi-bin/text-idx?SID=e09c55087458aadb6c203352c7ebd4f7&mc=-true&node=pt.21.5.312&rgn=div5
      ii. The current Title 21 outline. Below the relevant subchapters: https://www.ecfr.gov/cgi-bin/text-idx?gp=&SID=e09c55087458aadb6c203352c

7ebd4f7&mc=true&tpl=/ecfrbrowse/Title21/21chapterI.tpl

1. Subchapter A: General (Parts 1−99)
2. Subchapter C: Drugs: General (Parts 200−299)
3. Subchapter D: Drugs for Human use (Parts 300−399)

3. The FDA's Pre-IND Consultation Program
   a. This is a special official resource to request advice regarding the evidence support and protocol design for studies specifically treating infections: https://www.fda.gov/Drugs/DevelopmentApprovalProcess/HowDrugsareDevelopedandApproved/ApprovalApplications/InvestigationalNewDrugINDApplication/Overview/default.htm
4. ClinicalTrials.gov
   a. The US National Library of Medicine's searchable database of private and public clinical studies conducted around the world. Containing over 200 thousand trials, past, and present, it is a great resource where to find example studies before and at the Investigational Plan development stage: https://clinicaltrials.gov/
5. The Electronic Submissions Gateway
   a. This is the FDA central electronic regulatory transmission portal where the IND would be submitted: https://www.fda.gov/ForIndustry/ElectronicSubmissionsGateway/default.htm

References

1. U.S. Food and Drug Administration. *FDA and Clinical Drug Trials: A Short History*; 2018. https://www.fda.gov/AboutFDA/WhatWeDo/History/ucm304485.htm. Accessed June 7, 2019.
2. Kesselheilm AS, Wang B, Franklin JM, Darrow JJ. Trends in utilization of FDA expedited drug development and approval programs, 1987-2014: cohort study. *Bmj*. 2015;351.
3. Meadows M. *Promoting Safe & Effective Drugs for 100 Years*; 2006. https://www.fda.gov/AboutFDA/WhatWeDo/History/ProductRegulation/ucm2017809.htm. Accessed May 7, 2019.
4. U.S. Food and Drug Administration. *Application User Fees for Combination Products*; 2005. https://www.fda.gov/downloads/RegulatoryInformation/Guidances/UCM147118.pdf. Accessed January 9, 2022.
5. Oner BS, Labra SR, Fehr S. FDA drug regulation: investigational new drug applications. *Acad Entrepr Medical Health Sci*. 2019;1(3):7.
6. Center for Drug Evaluation and Research. U.S. Food and Drug Administration. *Providing Regulatory Submissions in Electronic Format — NDAs*; 1999. https://www.fda.gov/files/drugs/published/Guidance-for-Industry–Providing-Regulatory-Submissions-in-Electronic-Format—New-Drug-Applications.pdf. Accessed January 8, 2022.
7. U.S. Food and Drug Administration. *Formal Meetings between the FDA and Sponsors or Applicants of PDUFA Products*; 2017. https://www.fda.gov/drugs/types-applications/new-drug-application-nda. Accessed January 7, 2022.
8. U.S. Food and Drug Administration. *Designating an Orphan Product: Drugs and Biological Products*; 2018. https://www.fda.gov/ForIndustry/DevelopingProductsforRareDiseasesConditions/HowtoapplyforOrphanProductDesignation/default.htm.
9. Center for Drug Evaluation and Research. U.S. Food and Drug Administration. *Guidance for Industry: Expedited Programs for Serious Conditions — Drugs and Biologicvs*; 2014>. https://www.fda.gov/media/86377/download.
10. U.S. Food and Drug Administration. *User Fee Waivers, Reductions, and Refunds for Drug and Biological Products*; 2011. https://www.fda.gov/downloads/Drugs/GuidanceComplianceRegulatoryInformation/Guidances/ucm079298.pdf. Accessed January 9, 2022.
11. U.S. Food and Drug Administration. *Prescription Drug User Fee Amendments*; 2022. https://www.fda.gov/industry/fda-user-fee-programs/prescription-drug-user-fee-amendments. Accessed January 9, 2022.
12. Center for Drug Evaluation and Research. U.S. Food and Drug Administration. *Prescription Drug User Fee Act Waivers, Reductions, and Refunds for Drug and Biological Products*; 2019. https://www.fda.gov/media/131797/download. Accessed January 8, 2022.

Chapter 79

Regulatory basics of medical device regulation

Fernando Aguel[1] **and Michael John**[2]

[1]*Heart Failure and Circulatory Support Regulatory Affairs, MCRA, Washington, DC, United States;* [2]*Head of Cardiovascular Regulatory Affairs, MCRA, Washington, DC, United States*

Key points

- Importance of regulatory planning.
- Device classification and regulatory pathways.
- Investigational Device Exemption (IDE) and pre-submissions.

Why it matters

Medical device developers tend to be creative problem solvers and leverage strong engineering or clinical (or both) skills and know how to develop medical device therapies for patients. While a strong engineering and clinical background in medical device development is certainly key to launching a successful medical device line, one must not overlook the importance of developing a regulatory plan to achieve marketing authority for the device. A well-planned regulatory approach may help streamline marketing authorization, while a poorly planned regulatory approach can lead to added costs and delays in getting to market. This chapter presents some of the basics of medical device regulation as a foundation to develop a regulatory approach.

Regulatory basics of medical device regulation

Products considered medical devices

When developing a medical product, it is important to start at the basics of whether the device is considered a medical device. FDA guidance document "Classification of Products as Drugs and Devices & Additional Product; Classification Issues: Guidance for Industry and FDA Staff" provides an interpretation of the definition of a medical device.[1]

Section 201(h) of the FD&C Act (21 USC 321(h)) provides that the term "device" means:

an instrument, apparatus, implement, machine, contrivance, implant, in vitro

reagent, or other similar or related article, including any component, part, or

accessory, which is–

(1) recognized in the official National Formulary, or the United States

Pharmacopeia, or any supplement to them,

(2) intended for use in the diagnosis of disease or other conditions, or in the cure,

mitigation, treatment, or prevention of disease, in man or other animals, or

(3) intended to affect the structure or any function of the body of man or other

animals, and which does not achieve its primary intended purposes through chemical action

within or on the body of man or other animals and which is not dependent upon

being metabolized for the achievement of its primary intended purposes.[1]

It is additionally important to note that "the term 'device' does not include software functions excluded pursuant to section 520(o)."[2] As the digital health industry expands, it is important to note that the FDA has expanded on the regulation of software in and as medical devices in

Translational Cardiology. https://doi.org/10.1016/B978-0-323-91790-2.00024-1

the FDA Guidance: Policy for Device Software Functions and Mobile Medical Applications.[3]

It is important to note that some products may be considered a medical device not only based on the device's physical properties and function but also on how that device is marketed. For example, it is possible for an electric blanket to simply be a consumer product, even when used in a hospital setting. At the same time, when a blanket is marketed with language implicitly or explicitly implying a medical claim, that may change whether the product is considered a medical device by the FDA. In the previous example, an electric blanket that is marketed as a blanket to maintain normothermia triggers the blanket to be considered a regulated medical device.

Regulation of a medical devices

After one has determined if a product is a medical device, it is equally important to know whether the FDA actively regulates the medical device. Whether a device is actively regulated will depend on the risk level a device carries, among other considerations including its intended use.

While both FDA centers will provide input into the regulation of the devices, the lead center will be determined by the product's primary mode of action.[4] In the above example of a drug-eluting coronary stent, the primary mode of action has historically been determined to be the stent/device, so CDRH is assigned as the lead center.

Device classification

Once one has determined that a product is a regulated medical device under the CDRH jurisdiction, it is important to determine how CDRH will regulate the device (i.e., what type, if any, submission needs to be sent to FDA and how CDRH will review that submission). How CDRH regulates the device will depend on the device's classification. Devices will fall into either class 1 devices, class 2 devices, or class 3 devices, generally depending on their level of risk. For example, class 1 devices are the least risky devices and are subject only to general controls such as labeling requirements and the vast majority are exempt from an FDA determination of substantial equivalence or approval before being marketed.

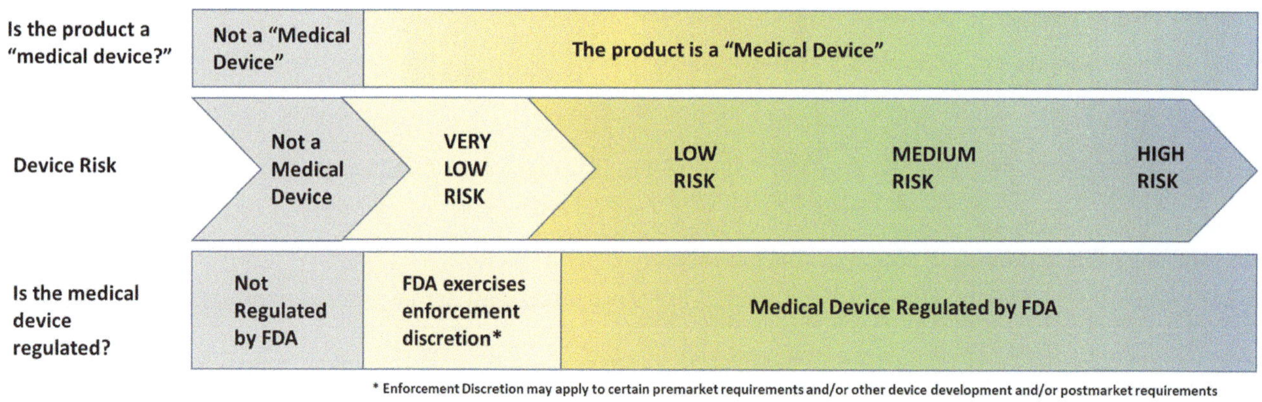

Where medical devices are regulated

When considering how a medical device is regulated, it is also important to note that a medical device/product may be a combination product, meeting the definition of both a device and another medical product, such as a drug or biologic. Most combination devices are reviewed both by FDA's Center for Devices and Radiological Health (CDRH) and the center that regulates the other aspects of the device. For example, in the case of a drug-eluting coronary stent, the device is reviewed both by CDRH and FDA's Center for Drug Evaluation and Research (CDER).

Generally, class I devices are exempt from premarket notification (510(k) submission), while class II devices require either a 510(k) determination of substantial equivalent or a De Novo application to be legally marketed in the US FDA has, however, exempted some class II devices from premarket submissions to FDA.[5] FDA has also identified "Reserved Medical Devices," which are class I devices, but that require 510(k) submission.[6] It is important to note that most class I devices and all class II devices are NOT exempt from GMP requirements.[5]

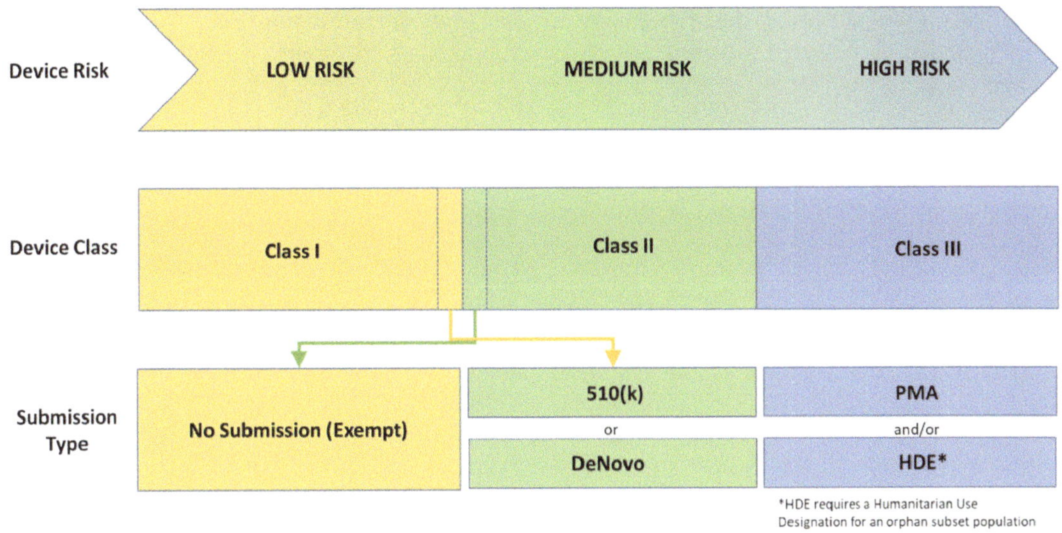

It should be noted, however, that regardless of the device classification (class 1, 2, or 3) or exemption status, the vast majority of regulated medical devices (with the exception of class I devices that are exempt from certain requirements) are required to be manufactured in facilities with a quality management system compliant with the quality system regulations (QSR) currently found in 21 CFR 820.[7] It should be noted that the FDA issued a final rule harmonizing the QSR with the International Organization for Standardization (ISO), ISO 13485:2016 Medical devices—Quality management systems—Requirements for regulatory purposes.[8] In addition, all devices are required to adhere to medical device reporting requirements,[9] and other requirements such as for medical device reports of corrections and removals,[10] unless FDA has specifically noted otherwise.

Clinical data collection

Investigational devices are devices that are not legally marketed devices that are intended to be used in clinical studies, or devices that are legally marketed that are intended to be studied for an intended use other than what they are authorized for. For regulatory purposes, clinical studies for investigational devices are separated into significant risk device studies and nonsignificant risk device studies.

Investigational Device Exemption (IDE).

The guidance document titled "Information Sheet Guidance For IRBs, Clinical Investigators, and Sponsors Significant Risk and Nonsignificant Risk Medical Device Studies" lists the following criteria for defining a significant risk (SR) device study and a nonsignificant risk (NSR) device study.

"Under 21 CFR 812.3(m), an SR device means an investigational device that:

- Is intended as an implant and presents a potential for serious risk to the health, safety, or welfare of a subject;
- Is purported or represented to be for use supporting or sustaining human life and presents a potential for serious risk to the health, safety, or welfare of a subject;
- Is for a use of substantial importance in diagnosing, curing, mitigating, or treating disease, or otherwise preventing impairment of human health and presents a potential for serious risk to the health, safety, or welfare of a subject; or
- Otherwise presents a potential for serious risk to the health, safety, or welfare of a subject."[11]

An NSR device study is one that does not meet the definition of an SR device study.

These criteria include somewhat subjective criteria, so understanding who makes the decision for whether a device study is SR or NSR is important to understand. A study sponsor should make an initial assessment for whether a device study is SR or NSR. An institutional review board (IRB) must then review and concur or modify the SR or NSR determination. At any point during the determination process, the sponsor or IRB may contact the FDA for help in making this determination. If FDA has already made a determination of SR or NSR for a device study, that determination is final. While it is important to consider the device description when deciding whether a device study is SR or NSR, there are other factors that should be considered, including the marketing status of the device (whether the device is being used in an approved manner) and other study components such as the procedures included as part of the clinical protocol used for screening, device use or data gathering such as for endpoint data.

Investigational Device Exemptions (IDE) are submissions to the FDA in which the agency will evaluate whether the sponsor has demonstrated that the investigational device is reasonably safe for subjects and that the welfare and rights are adequately protected to start the clinical study. Although both SR and NSR device studies require IRB approval to conduct a device study, an IDE is required for device studies determined to be SR device studies, whereas an IDE is not required for an NSR device study.

It is common for the FDA to require clinical data from a device study to evaluate certain marketing applications. Some of the clinical data may come from SR device studies and some may come from NSR device studies. If device study data is intended to be used for supporting a marketing application, a sponsor may obtain FDA feedback on a study plan or protocol to ensure that the type of data being captured in a study could support a marketing application before commencing a device study. The most common method for obtaining device study feedback is through the FDA's presubmission program. This presubmission program can also be used to align with the FDA on the preclinical data plan to support safety to commence an IDE or a marketing submission. The FDA guidance document titled "Requests for Feedback and Meetings for Medical Device Submissions: The Q-Submission Program; Guidance for Industry and Food and Drug Administration Staff" outlines the types of presubmissions and scopes of each.[12]

Medical devices classification

As mentioned earlier in this chapter, the FDA classifies medical devices into one of three regulatory classes based on risk but also includes the level of control needed to ensure that those devices are safe and effective. Class I devices and class II devices are subject to general controls and both general and special controls, respectively. Class I and class II devices generally require premarket notification (510(k) submission) to the FDA to gain clearance, unless that product type of exempt from premarket notification (most but not all class I devices are 510(k) exempt and the large majority of class II devices require 510(k) clearance). Class III devices are subject to general controls and premarket approval.

Premarket notification—510(k)

Premarket notification or a 510(k) submission is broadly considered the marketing mechanism for class II devices, even though we have already described in more detail the times when a 510(k) might be required for class I devices, and times when class II devices may not require 510(k) submission. The 510(k) paradigm is one which relies on demonstrating "substantial equivalence" to a legally marketed class I or class II device. In terms of demonstrating safety and effectiveness, it relies on what can be described as transitive property (in mathematics). That is, if device D is substantially equivalent to device C and Device C is substantially equivalent to device B and Device B is substantially equivalent to device A, then Device D must be *as* safe and effective as device A. The question that this brings up is how does FDA evaluate "substantial equivalence."

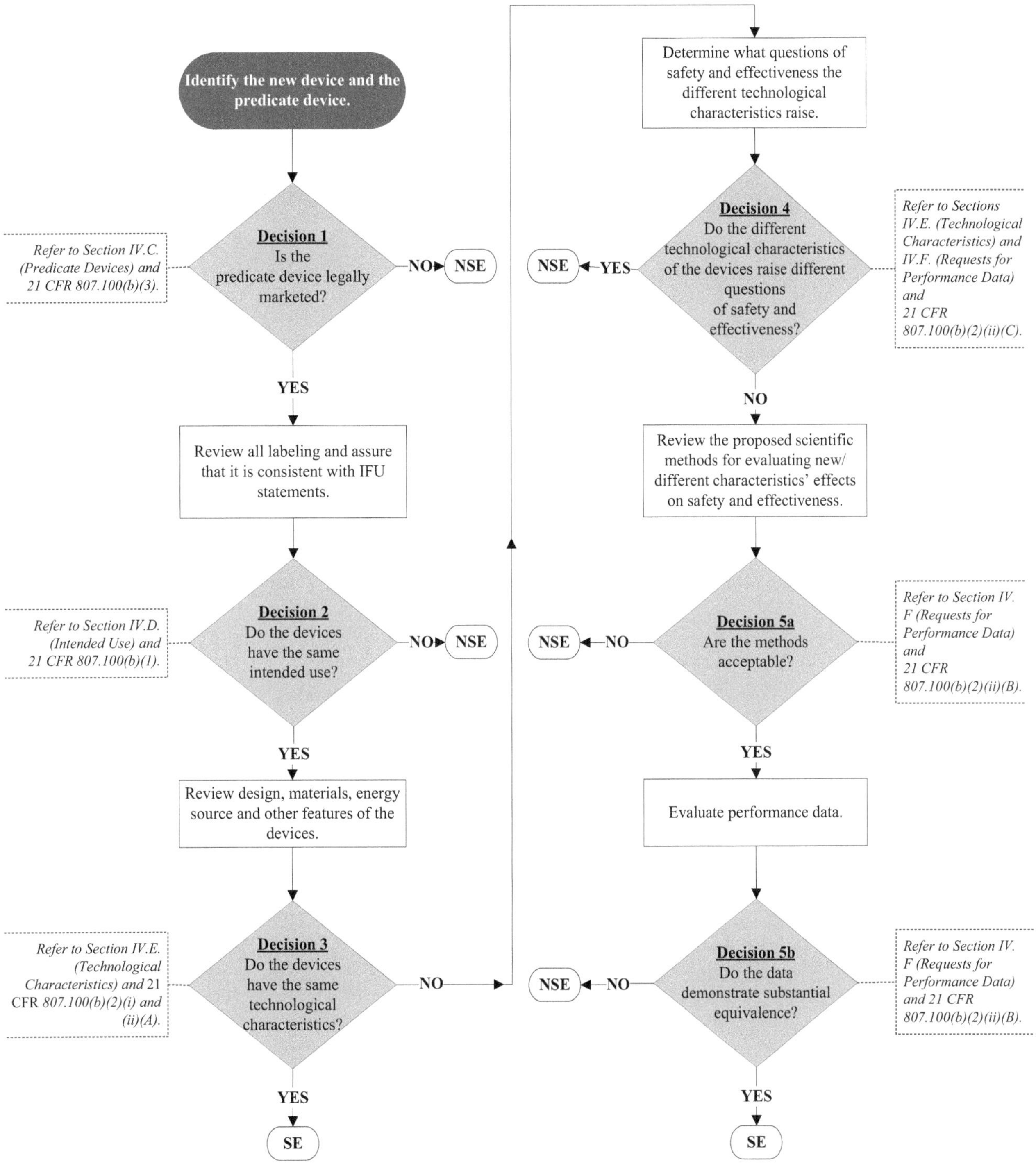

SE = "Substantially Equivalent"
NSE = "Not Substantially Equivalent"
IFU = "Indications For Use"

This Flowchart is not intended to be used as a 'stand-alone' document and should only be considered in conjunction with the accompanying text in this guidance.

FDA has developed a flow chart (below) to help with how to demonstrate substantial equivalence.[13]

This FDA flow chart appropriately references regulatory definitions, but that can make the flow chart difficult to understand and interpret. A simplified way to interpret the decision-making process is to divide the review process into three main topics. Topic 1 is whether there is a valid predicate device, that is, whether a predicate device has been identified to compare the new device to. Topic 2 is whether the predicate is appropriate. An appropriate predicate device will have the same intended use as the new device and the differences in technology do not introduce new questions of safety or effectiveness. This is a commonly misunderstood topic. Intended use is not the same as an indication for use, but it is related to it. A new device need not have identical indications for use, it just needs to have the same intended use to demonstrate substantial equivalence.

De Novo submission

A De Novo submission is a way to gain marketing authorization for a class I or II device when there is no predicate device to compare to for establishing substantial equivalence. A De Novo submission is the FDA's mechanism for classifying new device types as class I or class II devices. For a device that has no legally marketed predicate device, the FDA recommends that a manufacturer discuss with the review division whether a De Novo is the appropriate pathway (class I or class II) or whether a premarket approval (class III) is the appropriate pathway. FDA considers the level of risk of a device when making decisions on new device types as well as what level of control is needed for that device type. Letters granting a De Novo application include the identified risks to health and the special controls (for class II devices) that align with those risks to health that the device type must meet to be legally marketed. Once a De Novo application is granted (i.e., the new device type is classified and a new regulation is created), De Novo can be used as a predicate device for future 510(k) submissions.

A De Novo submission is not appropriate for a class III device. Class III devices require premarket approval to be legally marketed.

Just as with all class I and II devices (whether exempt from 510(k) or cleared via 510(k), devices authorized for marketing via the De Novo pathway need to be manufactured in accordance with good manufacturing practices, and follow other relevant reporting regulations. However, similar to 510(k) submissions, De Novo submissions do not generally require manufacturing information to be included in the submissions for FDA review.

FDA has issued guidance on the process for the submission and review of a De Novo classification request.[14]

Premarket approval

Premarket approval (PMA) is the most common marketing pathway for class III devices. These can include implanted or durable ventricular assist devices. Every PMA must have an independent data set to support approval. The regulatory bar for approval for a PMA is the demonstration of "reasonable assurance of safety and effectiveness" using "valid scientific evidence."[15] Valid scientific evidence can take multiple forms, including data from a randomized controlled trial, a single-arm clinical investigation, well-documented case histories, and even real-world evidence supporting the safety and effectiveness of the device. In many cases, PMA approval will include a condition of approval that will require some form of postmarket data collection.

As part of a PMA application review, the FDA will also review the quality management system in place to manufacture the subject device and final labeling of the subject device, among many other parts. FDA has a series of guidance documents relevant to the content and review of PMA applications.[16]

Humanitarian device exemption

Humanitarian device exemptions (HDEs) are marketing authorizations that are generally reserved for class III devices that have received humanitarian use device (HUD) designation. A HUD designation is reserved for "medical device intended to benefit patients in the treatment or diagnosis of a disease or condition that affects or is manifested in not more than 8000 individuals in the United States per year."[17] The regulatory bar for an HDE approval is demonstration of reasonable assurance of safety and probable benefit."[18] Similarly to PMA submissions, in many cases, HDE approval will include a condition of approval that will require some form of postmarket data collection. For HDEs, this postmarket data may be able to serve as the basis for PMA approval, such as in the case of the Berlin Heart EXCOR Pediatric Ventricular Assist Device.[19]

Just as with class I and II devices, devices approved for marketing via the PMA pathway or HDE pathway need to be manufactured in accordance with good manufacturing practices and follow other relevant reporting regulations. However, unlike for class I and II devise, GMP procedures must be submitted to FDA for approval as part of the PMA or HDE submissions, respectively.

Breakthrough Device Designation

Manufacturers can consider requesting Breakthrough Device Designation (BDD), generally for novel devices or novel indications for use. Generally, devices that are good candidates for a BDD will likely require a De Novo application or a PMA application, although there are exceptions and some devices granted BDD have come to market via the 510(k) pathway.

The FDA has issued a guidance document on the BDD program, including the requirements a device must demonstrate to be granted BDD.[20]

Devices granted BDD may be able to negotiate flexibility on pre–post market balance; that is, the amount or type of data needed for premarket approval or clearance and what data may be collected in the postmarket setting. In some cases, depending on the specific device, the FDA may evaluate the following strategies for pre–post market balance, each strategy with its own trade-offs.

1. Same endpoints and follow-up, but less patients; continue enrollment and follow-up in postmarket
2. Same endpoints but less follow-up—event accumulation can be a challenge and may result in more patients but less follow-up. May be useful when follow-up time and not patient recruitment is the bottleneck; continue follow-up in postmarket
3. Different endpoints (e.g., QOL or functional assessments) and sample size; continue to enroll and follow patients in postmarket to assess for objective clinical outcomes (CVRX)

It is important to note however, that the breakthrough devices program does not change the statutory requirement for approval; that is, a device must still demonstrate reasonable assurance of safety and effectiveness (PMA, De Novo) and substantial equivalence (510(k)—premarket notification).

The CVRX device trial, BeAT-HF, was designed using the strategy outlined in point 3,[21] and the CVRX device received PMA approval in 2019.[22]

Q-submissions/presubmissions

Although presubmissions are a specific type of Q-submissions that can be submitted to the FDA to request written feedback or a meeting with the FDA, the two terms are often used interchangeably, although one is a subset of the other.[12] Presubmissions submitted to the FDA can be narrow in scope (e.g., requesting FDA feedback on a large animal study protocol) or broad in scope (e.g., requesting general feedback on the entire in-vitro testing suite for a transcatheter heart valve). These should generally contain a detailed device description, background information relevant to the questions posed to the FDA and a set of questions that a submitter is asking for feedback on.

Three common presubmission missteps

FDA encourages early interaction on regulatory strategy and review of testing protocols before starting testing. Manufacturers will sometimes not seek alignment with the FDA early on. Failure to align with the FDA on testing approach can lead to the need to perform repeat testing, which can really affect manufacturers' milestones. It is generally a good idea to align with FDA on test procedures for in-vitro testing that may be tricky or that may have long lead times. This is especially true for areas such as biocompatibility testing and animal studies.

Another common pitfall is to not include sufficient information for the FDA to review and understand the device or test procedure. Overall, the amount of detail included in feedback from the FDA is commensurate with the amount of detail you provide to the FDA, including a detailed device description.

Lastly, it is not reasonable to expect the FDA to design studies for you. Rather, the FDA will generally review and provide feedback on your proposal. FDA does not design clinical/nonclinical studies.

Getting started

Predictability

Important to know the regulatory pathway since this may impact the time to market for devices, from less than 1 year to 5–7 years. The device development and regulatory plan should be considered early. The earlier a manufacturer engages with the FDA on a regulatory plan and the specifics of testing required, the more predictable the regulatory pathway becomes.

Precedent

Precedent is important. This is true for identifying a predicate device for a 510(k) submission as well as for determining the testing necessary to include in an IDE or marketing submission. Precedent can help identify regulatory strategies that have worked in the past for your device type.

It is important however to note that even with precedent, FDA is using a total product lifecycle approach to regulating devices. That is to say, when presented with new information, whether from postmarket data or newly identified failure modes identified via recall for a product type, the FDA may incorporate this information to inform what the premarket requirements need to be moving forward, accounting for the new information. One example of this type of TPLC approach is the example of reusable water-based heater–cooler devices used in operating rooms. Before a signal of device contamination leading to patient infections, the FDA had not previously asked for cleaning and maintenance validation in the premarket submissions because the devices are nonsterile and the water inside the system was not in apparent contact with the patient. However, since the occurrences of infections were identified, the FDA has adapted to require cleaning and disinfection validation, adding water-based heater—cooler devices to the list of devices identified in the Reprocessing Medical Devices in Health Care Settings: Validation Methods and Labeling guidance document.[23]

Additional resources

If one has questions about the regulation of a device and has not yet contacted the FDA, one can contact The CDRH Division of Industry and Consumer Education (DICE) for general questions about medical device regulation.[24]

References

1. *Classification of Products as Drugs and Devices & Additional Product; Classification Issues: Guidance for Industry and FDA Staff*; September 2017. https://www.fda.gov/media/80384/download.
2. *How to Determine if Your Product is a Medical Device*; 2022. https://www.fda.gov/medical-devices/classify-your-medical-device/how-determine-if-your-product-medical-device.
3. *Policy for Device Software Functions and Mobile Medical Applications Guidance for Industry and Food and Drug Administration Staff*; September 28, 2022. https://www.fda.gov/media/80958/download.
4. Frequently Asked Questions About Combination Products. https://www.fda.gov/media/106799/download.
5. *Medical Device Exemptions 510(k) and GMP Requirements*; 2024. https://www.accessdata.fda.gov/scripts/cdrh/cfdocs/cfpcd/315.cfm.
6. *Reserved Medical Devices*; 2024. https://www.accessdata.fda.gov/scripts/cdrh/cfdocs/cfpcd/3151.cfm.
7. CFR. *Code of Federal Regulations Title 21 PART 820 Quality System Regulation*; 2023. https://www.accessdata.fda.gov/scripts/cdrh/cfdocs/cfcfr/CFRSearch.cfm?CFRPart=820.
8. *Medical Devices; Quality System Regulation Amendments*; 2024. https://www.federalregister.gov/documents/2024/02/02/2024-01709/medical-devices-quality-system-regulation-amendments.
9. CFR. *Code of Federal Regulations Title 21 Part 803 Medical Device Reporting*; 2023. https://www.accessdata.fda.gov/scripts/cdrh/cfdocs/cfcfr/CFRSearch.cfm?CFRPart=803.
10. CFR. *Code of Federal Regulations Title 21 Part 806 Medical Device Reports of Corrections and Removals*; 2023. https://www.accessdata.fda.gov/scripts/cdrh/cfdocs/cfcfr/CFRSearch.cfm?CFRPart=806.
11. *Information Sheet Guidance for IRBs, Clinical Investigators, and Sponsors Significant Risk and Nonsignificant Risk Medical Device Studies*; January 2006. https://www.fda.gov/media/75459/download.
12. *Requests for Feedback and Meetings for Medical Device Submissions: The Q-Submission Program; Guidance for Industry and Food and Drug Administration Staff*; 2023. https://www.fda.gov/media/114034/download.
13. *The 510(k) Program: Evaluating Substantial Equivalence in Premarket Notifications [510(k)] Guidance for Industry and Food and Drug Administration Staff*; 2014. https://www.fda.gov/media/82395/download.
14. *De Novo Classification Process (Evaluation of Automatic Class III Designation) Guidance for Industry and Food and Drug Administration Staff*; 2021. https://www.fda.gov/media/72674/download.
15. *PMA Clinical Studies*; 2023. https://www.fda.gov/medical-devices/premarket-approval-pma/pma-clinical-studies#:~:text=Although%20the%20manufacturer%20may%20submit,device%20is%20safe%20and%20effective.
16. *PMA Guidance Documents*; 2023. https://www.fda.gov/medical-devices/premarket-approval-pma/pma-guidance-documents.
17. *Guidance for Industry and Food and Drug Administration Staff Humanitarian Use Device (HUD) Designations*; 2019. https://www.fda.gov/media/130442/download.
18. Humanitarian Device Exemption (HDE). *Program Guidance for Industry and Food and Drug Administration Staff*; 2019. https://www.fda.gov/media/74307/download.
19. Summary of Safety and Effectiveness Data: EXCOR Pediatric Ventricular Assist Device. https://www.accessdata.fda.gov/cdrh_docs/pdf16/P160035B.pdf.
20. *Breakthrough Devices Program Guidance for Industry and Food and Drug Administration Staff*; 2023. https://www.fda.gov/media/162413/download.
21. Zile MR, Abraham WT, Lindenfeld J. First granted example of novel FDA trial design under Expedited Access Pathway for premarket approval: BeAT-HF. *Am Heart J*. 2018;204:139—150.
22. FDA Approval Order: BAROSTIM NEO System. https://www.accessdata.fda.gov/cdrh_docs/pdf18/P180050A.pdf.
23. *Reprocessing Medical Devices in Health Care Settings: Validation Methods and Labeling Guidance for Industry and Food and Drug Administration Staff*; 2017. Issued March 17, 2015 https://www.fda.gov/media/80265/download.
24. *Division of Industry and Consumer Education*; 2019. https://www.fda.gov/about-fda/cdrh-offices/division-industry-and-consumer-education.

Chapter 80

Digital health technologies in clinical trials—Regulatory basics

Sanjeev P. Bhavnani
Division of Cardiology, Scripps Clinic, San Diego, CA, United States

Key points

- Clinical trial designs are rapidly evolving and incorporate a broad spectrum of new methodologies, including those related to device development and early validation as well as large, decentralized, and electronic health record-embedded trials to evaluate the effectiveness of digital health technologies (DHTs) on the process of care and healthcare outcomes.
- The US Food and Drug Administration (FDA) provides oversight of DHTs in several areas, including the manufacture and marketing of DHTs that are medical devices and the conduct and review of proposed clinical studies that utilize DHTs. Consistent with this role, the FDA provides guidance on the regulatory status of DHTs and offers guidance for the requirements and conduct of clinical trials that use DHT for remote data acquisition, the considerations for using DHTs in real-world data and real-world evidence generation, and outlines the diversity and equity requirements for broad enrollment in clinical trials of new products.
- Important factors to consider when designing clinical trials that incorporate DHTs include patient-generated health data and how patient engagement and clinical workflows are incorporated into trial methodologies.

Introduction—Why it matters?

The development of digital health technologies (DHTs) has increased rapidly over the past decade. This growth, particularly in cardiovascular (CV) medicine, has included a variety of DHTs—such as wearable sensors, smartphone-connected products, microfluidic lab-on-a-chip technologies, nanosensors, and handheld-ultrasound—and is poised to improve patient care and the efficiency of healthcare delivery.[1] The applications of DHTs and the data generated from patients as part of a remote-monitoring strategy have emerged as important facilitators of clinical research.[2] In the context of trial designs, DHTs can be used to shorten clinical study timeframes[3] and enhance enrollment through decentralized designs that recruit large groups of study subjects virtually on smartphone-applications.[4] Interactive patient engagement through study subject use of DHTs, real-time physiologic measurements, and in the collection of patient-reported outcomes (PROs) are novel methods that can better inform our understanding of how a disease phenotype evolves over time[5] and how such diseases are represented through new digital approaches.[6]

The objective of this chapter is to highlight various types of trial methodologies and the regulatory and clinical considerations for conducting research incorporating DHTs. It is not the objective of this chapter to recommend any specific digital trial methodologies or specific technologies that can be incorporated within the design of clinical trials. Understanding the diversity of trials leveraging DHTs and the various nuances in performing digital health-enabled research is essential for future evidence generation and those factors that can ultimately impact patient care.[7]

Regulatory advances for DHT-enabled clinical trials

The use of DHTs within clinical trials encompasses a regulatory framework from planning and selecting DHT technologies, managing and analyzing data acquired from DHTs, and developing novel digital endpoints and outcomes that are consistent with the study outcomes of interest. While this framework serves as a foundation for the design and execution of clinical trials in general, when considering the use of DHTs (i.e., in trials for the development or evaluation of novel drugs or devices), it is

Translational Cardiology. https://doi.org/10.1016/B978-0-323-91790-2.00105-2

important to understand that the selection of a DHT must be suitable to address a clinical outcome and that the DHT is "fit-for-purpose."[8] Fit-for-purpose, a regulatory requirement, is essential and means that the DHT has a sufficient level of validation to support its use and interpretability in the clinical investigation. To ensure a DHT is fit-for-purpose encompasses the following key requirements:

- Understanding the disease or condition in question and aligning how DHTs capture the measurements important to clinical trials and to study participants (e.g., blood pressure, glucose, ECG).
- The DHT used in a clinical investigation must meet acceptable feasibility and performance characteristics such as accuracy, precision, and consistency of measurements over time, and the uniformity of measurements across similar DHTs (e.g., different brands of blood pressure devices used in different study subjects enrolled in the same clinical trial).
- Identifying the clinical benefits and risks including the context of use in a clinical trial (i.e., how a study subject interacts with a DHT, how a study subject feels and functions using a DHT).
- Selecting or developing the clinical outcome assessment (e.g., patient-reported outcome, observer-reported outcome, clinical-reported outcome, or performance outcome) that can be adequately measured and captured using DHTs.

In the aggregate, clearly describing the concepts of interest, context of use, a clear rationale, and sufficient evidence to justify the use of DHTs in clinical trials ensures, in part, that regulatory pathways for DHTs in clinical trials are broadly met.[9,10]

Leveraging these aforementioned priorities, three recent regulatory developments for the considerations on how DHTs are used within clinical trials are discussed herein and include the following: when using DHTs for remote data acquisition in clinical trials; real-world data using DHTs for remote patient monitoring (RPM); and equitable participation to ensure diversity of inclusion in trial participation with studies using DHTs.

DHT trials—From study design to clinical translation

Many trials are designed with the primary purpose of evaluating DHTs for marketing as medical devices. Others utilize DHTs as tools to conduct clinical research whose primary purpose is to evaluate other medical products. While these may seem like separate uses of DHTs, there are overlaps with how DHTs are used to execute either type of investigation.[11] Such considerations include, but are not limited to, those related to the digital platform's usability; data characteristics necessary for use in clinical and regulatory decision-making; cybersecurity; data privacy, data sharing and governance; management of large volumes of diverse data; and interoperability of data sources.[12] For example, trials that evaluate a DHT to understand its measurement correlation with a particular cardiovascular disease (e.g., atrial fibrillation, heart failure, hypertension, or ischemic heart disease) or those trials that aim to validate a digital biomarker may employ similar methodologies for how DHTs are used for remote data acquisition such as app-based enrollment and how data is collected and analyzed on digital platforms. Similarly, clinical trials that aim to assess the effectiveness of care delivery such as medical treatments in acute myocardial infarction[13] or a new clinical intervention after vascular surgery[14] may incorporate DHTs to remotely monitor study subjects and to collect patient-generated health data (PGHD) as part of the trial's primary or secondary outcomes. Taken together, there are several domains for how DHTs are applied in clinical trials. These may be further stratified into those that leverage established trial methodologies, such as randomized and pragmatic designs, and others that have developed new methodologies such as decentralized trials, point-of-care, and electronic health record (EHR) embedded studies.

These domains are listed in Table 80.1 with representative DHT trial examples and specific trial methodologies. The purpose of this list is to be illustrative of the types of trial designs and how DHTs are being used in different investigations with different patient populations and outcomes. Having an understanding of the spectrum of digital health trial methodologies, as shown in Table 80.1, is especially important for the field of cardiology whose practice is highly varied with no two patient populations or two clinics having the same exact requirements for how to incorporate DHTs within clinical workflows.[1,2] This heterogeneity, itself, may suggest that the translation of DHT trial methods to patient care should be variable and that pragmatic approaches that incorporate specific practice-based patient and clinical requirements with those DHTs that meet such requirements may be best poised to inform future clinical and translational research.[3,4,7]

Remote patient monitoring and real-world data with DHTs

Real-world data (RWD) is defined as data relating to patient health status and/or the delivery of health care routinely collected from a variety of sources. Real-world evidence (RWE) is the clinical evidence about the usage and potential benefits or risks of a medical product derived from analysis of RWD.[12,30] In the context of RWD, DHTs used within remote patient monitoring (RPM) supplemented by existing clinical information such as those within EHRs can provide a more comprehensive assessment of a patient's health and the trajectory of a chronic disease. For example,

TABLE 80.1 Examples of DHTs in clinical trials, trial designs, and results with clinical implications.

Note that the included list should not be taken as comprehensive but rather illustrative of concepts on DHTs being used in trials. The lexicon used in this table that describes the types of trials was not intended to represent definitive descriptions of or labels for such trial designs.

DHT evaluated and its function	Trial aim, trial design, and sample size	Results	Conclusion and clinical implications
Examples of trials related to preclinical studies and technology development			
Title: *Continuous monitoring of deep-tissue hemodynamics with stretchable ultrasonic phased arrays* Trial focused on technology development **Citation**: Wang et al.[15]			
Prototype skin-conformal ultrasonic phased array for the monitoring of hemodynamic signals from tissues up to 14 cm beneath the skin. Doppler spectra from cardiac tissues, central blood flow waveforms, and estimates of cerebral blood supply in real time	Prototype development and testing, including testing with healthy volunteers. Sample size: not provided	"The cardiac phases acquired by the stretchable ultrasonic phased array match well with the cardiac cycle traced by the ECG. The measured tissue Doppler provides a direct quantitative assessment of specific myocardium, which complements conventional ECG and stethoscope measurements."	"The continuous direct insights on anatomies and functions of cardiac tissues offered by the stretchable ultrasonic phased array may be valuable for acute disease prognosis and postsurgery monitoring. Doppler ultrasonography is generally limited to point-of-care applications. The stretchable ultrasonic phased array allows monitoring of blood flow in major arteries and veins quantitatively and continuously. This study suggests that quantitative blood flow measures, typically limited to collection in point-of-care settings, may be feasible to collect continuously utilizing a stretchable ultrasonic phased array"
Title: *Hearing the heart validation of a novel digital health earbud technology to measure cardiac time intervals through infrasonic hemodynography* Trial focused on technology development **Citation**: Wheeler et al.[16]			
Prototype earbud device for measuring infrasound signals of arterial blood flow Time-series and continuous infrasound waveforms	Prototype development and validation among patients with cardiovascular disease. Validation was compared to echocardiography and hemodynamic waveforms with concordance overlaid data visualizations to CV time intervals. Sample size: 5	>450,000 time-series data points across 257 CV cycles demonstrated high correlation ($R = 0.87$, mean absolute error 14.7 ms, and bias of 7.2 ms) between infrasound and cardiac catheterization for the measurement of left ventricular ejection time.	"In a first-in-person study, high accuracy was demonstrated between infrasound, echocardiography, and hemodynamic waveforms to capture CV time intervals including CV performance measures. Further studies are underway to validate IH and the earbud sensor towards noninvasive hemodynamic monitoring"
Title: *Individualized studies of triggers of paroxysmal atrial fibrillation: the I-STOP-AFib randomized clinical trial* Trial incorporating n-of-1 design **Citation**: Marcus et al.[17]			
Smartphone-ECG device for patient-triggered, 30-s, single-lead ECG.	Randomized, n-of-1 design to evaluate self-selected atrial fibrillation triggers and correlation to atrial fibrillation (AF) related quality of life. Meta-analysis of n-of-1 overall study subjects. Sample size: 446	No difference in AF-related quality of life (AFEQT) however, those randomized to n-of-1 testing self-reported 40% fewer AF-events at 4 weeks (adjusted hazard ratio 0.60 [95% CI 0.43–0.83, $P < .001$] with meta-analysis demonstrating that exposure to alcohol was associated with AF events.	"n-of-1 Testing of AF triggers did not improve AF-associated quality of life but was associated with a reduction in AF events. Acute exposure to alcohol increased AF risk, with no evidence that other exposures, including caffeine, more commonly triggered AF"

Continued

TABLE 80.1 Examples of DHTs in clinical trials, trial designs, and results with clinical implications.—cont'd

DHT evaluated and its function	Trial aim, trial design, and sample size	Results	Conclusion and clinical implications
Examples of trials related to clinical utilization			
Title: *Postoperative remote automated monitoring and virtual hospital-to-home care system following cardiac and major vascular surgery: user testing study* Trial focused on user testing **Citation**: McGillion et al.[18]			
Vital sign devices such as blood pressure, oximetry, weight and temperature, and wearable telemetry devices for heart rate and heart rhythm.	Postoperative remote automated monitoring and virtual hospital-to-home care with digital health remote monitoring after major cardiovascular surgery with patients monitored both during the incident hospitalization and transition to home. User testing included nurses and patients with qualitative and quantitative measures of user performance and user acceptance of digital remote monitoring. Sample size: 37 (26 nurses and 11 patients)	Majority of nurses (18/26) and patients (8/11) were able to independently complete device-related tasks for activation and continued utilization for remote monitoring. Identified limitations and areas for improved patient education including troubleshooting device connections for continuous transmission of data and adherence to specific remote monitoring protocols.	"User testing indicated a high degree of user acceptance to vital sign systems among nurses and patients. Key insights were provided that informed refinement of clinical workflow training and systems implementation"
Title: *SMART Markers: collecting patient-generated health data as a standardized property of health information technology* Trial focused on SMART-on-FHIR integration **Citation**: Sayeed et al.[19]			
Mobile device software framework for the collection of patient-generated health data from DHTs.	Development of a patient-reported outcomes measurement information system and collection of patient-generated data connected to electronic health record (EHR) information via SMART-on-FHIR* integration methods. *Fast Healthcare Interoperability Resources. Sample size: N/A	FHIR-based repository sorting of various patient-generated health data devices linked with EHR metadata (demographic and clinical information) was captured and integrated at the point of care.	An open-sourced programming platform for app developers to import digital health and patient-generated health data into existing or new smartphone-based health applications. Such methods have implications for the inclusion of patient-generated health data into EHRs.
Title: *Association of wearable device use with pulse rate and healthcare use in adults with atrial fibrillation* Trial incorporating electronic health record DHT measurements **Citation**: Wang et al.[20]			
Commercially available smartwatches and wearable devices for heart rate and heart rhythm monitoring.	Healthcare outcomes and rates of healthcare use among patients with atrial fibrillation who use digital devices for heart rate and rhythm monitoring compared to those who do not use devices followed in an integrated healthcare system. Propensity matching of monitored and nonmonitored AF patients through natural language processing of electronic health	Despite similar baseline heart rate (75 bpm), atrial fibrillation patients who used wearable devices for monitoring pulse rate and detecting arrhythmia had greater healthcare utilization and a composite of healthcare visitations, cardioversion, and atrial fibrillation ablation compared to those who did not use wearable devices (17% vs. 7%, $P = .001$).	"This study found that follow-up healthcare use among individuals with AF was increased among those who used wearables compared with those with similar pulse rates who did not use wearables. Given the increasing use of wearables by patients with AF, prospective, randomized, long-term evaluation of the

	record data and relevant diagnostic codes. Sample size: 16,320 (348 used wearable devices were matched to 15,972 who did not use wearable devices)		associations of wearable technology with health outcomes and health care use is needed."
Title: *Continuous wearable monitoring analytics predict heart failure hospitalization: The LINK-HF Multicenter Study* DHT trial incorporating multiple modalities **Citation**: Stehlik et al.[21]			
Wearable, multimodality sensor-based patch that simultaneously collects: ECG, accelerometry, skin impedance and temperature to derive heart rate and heart rate variability, arrhythmia burden, respiratory rate, and others.	Biometric data was continuously measured and transferred over a smartphone cellular connection to a cloud-based analytical platform that used machine learning and similarity networking to generate differences in biometric parameters in over time. Machine-learned results were then correlated to clinical endpoints including heart failure hospitalization and emergency room visitations. Sample size: 100	Machine learning on multimodal biometric data streams generated an algorithm at the per-patient level that was predictive of worsening heart failure and heart failure hospitalization (AUC 0.89) with a corresponding sensitivity between 76% and 88% and a specificity of 85%. Changes in biometric parameters were detected at a median of 6.5 days before hospitalization.	"Multivariate physiological telemetry from a wearable sensor can provide accurate early detection of impending rehospitalization with a predictive accuracy comparable to implanted devices. The clinical efficacy and generalizability of this low-cost noninvasive approach to rehospitalization mitigation should be further tested"
Examples of trials related to clinical workflow assessment			
Title: *smartADHERE Investigators. Efficacy of a centralized, blended electronic, and human intervention to improve direct oral anticoagulant adherence: Smartphones to improve rivaroxaban ADHEREnce in atrial fibrillation (SmartADHERE) a randomized clinical trial* Electronic health record-embedded prospective digital health trial **Citation**: Turakhia et al.[22]			
Medication management smartphone application enabled with adaptive text messaging	Data collection of medication and prescription refill tracking coupled with semiautomated text messaging to determine medication compliance. A phone and coach-based system was used to monitor severe nonadherence. A common data dashboard was developed as the medium for healthcare personnel to monitor anticoagulation adherence, nonadherence, and side effects among patients with atrial fibrillation. This combination of centralized data collection and clinical evaluation constituted a "blended adherence intervention." Sample size: 139	There was no difference in anticoagulation adherence over 6 months between patients randomized to medication management versus standard care (proportion of days covered by anticoagulation 0.86 vs. 0.88, $P = .62$).	"No differences were observed in the primary or secondary adherence outcomes, with adherence high in both groups, despite prescreening for at-risk individuals. SmartADHERE illustrates the challenges of open-label behavioral and digital health trials, where enrollment itself may lead to selection bias or treatment effect"

Continued

TABLE 80.1 Examples of DHTs in clinical trials, trial designs, and results with clinical implications.—cont'd

DHT evaluated and its function	Trial aim, trial design, and sample size	Results	Conclusion and clinical implications
Title: *Patient-reported outcomes from patients with heart failure participating in the future patient telerehabilitation program: data from the intervention arm of a randomized controlled trial* Trial focused on the use of digital patient-reported outcomes **Citation**: Skov Schacksen et al.[23]			
Technology toolbox and Heart Portal consisting of: - Patient educational content on heart failure - Communication platform for patients to design their own telerehabilitation program - Patient-reported outcome questionnaires, and - DHTs including blood pressure, heart rate, oximetry, weight, and sleep monitoring.	Patient-reported outcome questionnaires were collected using a research-based electronic data capture that was digitally aggregated on the Heart Portal. Patient-reported outcomes included sleep quality (Spiegel), measures of physical limitations, symptoms, self-efficacy, social interaction, and quality-of-life (Kansas City Cardiomyopathy Questionnaire), and psychological well-being. Sample size: 140	Over a duration of 1-year, digital monitoring was associated with increases in clinical and social well-being and measures of quality-of-life among heart failure patients randomized to telerehabilitation. A significant increase in clinical and social well-being and quality of life were found in those participating in the telerehabilitation program ($P < .05$).	"There was a significant increase in clinical and social well-being and quality of life during the 1-year period of participating in a telerehabilitation program. These results suggest that patient-reported outcome questionnaires may be used as a tool for patients in a telerehabilitation program that can both monitor and guide patients in mastering their own symptoms."
Title: *Point-of-care screening for heart failure with reduced ejection fraction using artificial intelligence during ECG-enabled stethoscope examination in London, UK: a prospective, observational, multicenter study.* Trial focused on point-of-care screening **Citation**: Bachtiger et al.[24]			
Artificial intelligence-based ECG stethoscope for point-of-care screening of heart failure with reduced ejection fraction.	Single lead ECG via 4 auscultatory points on the chest. The ECG is embedded with a deep learning, convolutional neural network algorithm to predict low ejection fraction. Data was acquired at the point-of-care immediately before a transthoracic echocardiogram. Sample size: 1050	After pairing the ECG stethoscope with echocardiographic measurement of ejection fraction, the deep learning algorithm acquired at an optimal combination of two positions at the point-of-care detected low ejection fraction (<40%) with high accuracy compared to ejection fraction obtained at echocardiography (AUC of 0.85).	"A deep learning system applied to single-lead ECGs acquired during a routine examination with an ECG-enabled stethoscope can detect LVEF of 40% or lower. These findings highlight the potential for inexpensive, noninvasive, workflow-adapted, point-of-care screening, for earlier diagnosis and prognostically beneficial treatment."
Examples of trials related to clinical outcomes and population health			
Title: *Better Effectiveness After Transition—Heart Failure (BEAT-HF) Research Group. Effectiveness of remote patient monitoring after discharge of hospitalized patients with heart failure: The Better Effectiveness After Transition—Heart Failure (BEAT-HF) Randomized Clinical Trial* Cardiovascular outcome trial **Citation:** Ong et al.[25]			
Telemonitoring with multiple DHTs including: ECG, heart rate, oximetry, weight, and blood pressure. Post-discharge patient-reported outcome measures were collected via telephone calls by registered nurse coaches	Telemonitoring with DHTs used in a structured fashion (once daily patient measurements). Patient-generated health data was analyzed by clinical teams for real-time decisions such as diuretic dosing and guideline-directed medical therapy titration among heart failure patients. Sample size: 1437	Compared to standard care, telemonitoring was not associated with a difference in a 30-day, 90-day, or 180-day rate of heart failure readmission or mortality (adjusted hazard ratio 1.03 [95% CI 0.88 −1.20, $P = .74$].	"Among patients hospitalized for HF, combined health coaching telephone calls and telemonitoring did not reduce 180-day readmissions"

Title: *A randomized trial of pocket-echocardiography-integrated mobile health device assessments in modern structural heart disease clinics.* Trial focused on clinical decision support using DHTs in global health settings **Citation**: Bhavnani et al.[26]			
Point-of-care use of multiple DHTs including smartphone ECG, handheld ultrasound, smartphone-connected blood pressure, oximetry, and actigraphy devices and portable brain natriuretic peptide laboratory testing.	Creation of 'smartphone clinics' in a resource-limited area of a developing country with endemic rates of rheumatic heart disease. These smartphone clinics were equipped with point-of-care DHTs that were used by local healthcare workers to assess the severity of structural heart disease. All DHT findings were uploaded to the local EHR of a healthcare system for clinical decisions. Sample size: 253	Patients with rheumatic and structural heart disease randomized to smartphone clinics, compared to usual care, an initial assessment of the severity of structural heart disease with DHTs was associated with a shorter time to the referral for therapy including valvuloplasty and/or surgical valve replacement leading to a 60% relative risk reduction of a hospitalization and/or death at 1 year.	"An initial DHT diagnostic strategy was associated with a shorter time to definitive therapy among patients with rheumatic heart disease in a resource-limited area was associated with improved outcomes."
Title: *Impact of practice-based management of pulmonary artery pressures in 2000 patients implanted with the CardioMEMS sensor* Trial focused on monitoring by an implantable sensor **Citation:** Heywood et al.[27]			
Implantable sensor for continuous, ambulatory measurements of pulmonary artery pressure.	Comparison of real-world practice of pulmonary artery (PA) pressure changes to those pressure changes observed from the CHAMPION randomized clinical trial.	Compared to randomized trial patients, real-world practice patients demonstrated a greater cumulative mean reduction in mean pulmonary artery pressures [434 versus 150.1 mmHg, $P < .001$] over an average duration of follow-up of 6 months after implantation.	"The first 2000 general-use patients managed with hemodynamic-guided heart failure care had higher PA pressures at baseline and experienced greater reduction in PA pressure over time compared with the CHAMPION clinical trial. These data demonstrate that general use of implantable hemodynamic technology in a nontrial setting leads to significant lowering of PA pressures."
Title: *Promoting cardiovascular health and wellness among African Americans: community participatory approach to design an innovative mobile-health intervention* Community-based participatory research trial **Citation**: Brewer et al.[28]			
Culturally and linguistically appropriate smartphone application for lifestyle interventions in CVD.	Mixed methods, community-based participatory research (CBPR) for user-centric design of a culturally relevant CVD application in an African American church community. Phased approach for user feedback, content development, iterative testing, and final development. Sample size: 23	Content development factors included those leading to healthy behaviors, goal setting, and self-assessments in risk factors such as diabetes, hypertension, obesity, and stress management. Dietary information and physical activity emerged as important engagement factors for app utilization.	"This study illustrates integration of formative and CBPR approaches to design a culturally relevant, mobile health lifestyle intervention to address CV health disparities among African Americans. Given the positive app perceptions, our study supports the use of an iterative development process by others interested in implementing a mobile health lifestyle intervention for racial/ethnic minority communities."
Title: *Apple heart study investigators. Large-scale assessment of a smartwatch to identify atrial fibrillation* Decentralized digital health trial **Citation:** Perez et al.[29]			

Continued

TABLE 80.1 Examples of DHTs in clinical trials, trial designs, and results with clinical implications.—cont'd

DHT evaluated and its function	Trial aim, trial design, and sample size	Results	Conclusion and clinical implications
Smartwatch-photoplethysmography-based heart rate measurements and wearable ECG patch for arrhythmia detection.	Decentralized, nationwide, pragmatic recruitment with devices sent to study subjects. All trial requirements were completed on a smartphone app configured for trial information, consent, and data collection. Sample size: 419,297 (overall). Analyzed sample size: 2161 (study subjects who received a notification)	Among study subjects who received smartwatch-based notifications of an irregular heart and were sent ECG patches, 450 returned patches with analyzable data. Of those, 34% were observed to have atrial fibrillation on subsequent ECG-patch monitoring.	"The probability of receiving an irregular pulse notification was low. Among participants who received notification of an irregular pulse, 34% had atrial fibrillation on subsequent ECG patch readings and 84% of notifications were concordant with atrial fibrillation. This siteless (no on-site visits were required for the participants), pragmatic study design provides a foundation for large-scale pragmatic studies in which outcomes or adherence can be reliably assessed with user-owned devices."
Title: *Digital care transformation: interim report from the first 5000 patients enrolled in a remote algorithm-based cardiovascular risk management program to improve lipid and hypertension control* Trial focused on using DHTs for remote patient monitoring **Citation**: Scirica[30]			
Home-based blood pressure monitor linked to a software program integrated within a single institution's electronic health record.	Pre–post design to compare blood pressure and cholesterol control with patient-generated health data (PGHD) acquired with patients at home. Review of PGHD by care navigators and medication adjustment by pharmacists in a remote algorithm-based cardiovascular (CV) risk management program. Sample size: 5000	Among patients with established cardiovascular disease (CVD) enrolled in the remote algorithm-based program, a reduction in LDL of 24 mmHg (pre and post of 133 vs. 109, respectively, $P < .001$) and reduction in systolic blood pressure of 30 mmHg (pre and post of 165 vs. 135, respectively, $P < .001$) was observed.	"Our program provides an end-to-end solution including patient identification, improved data collection, education, and defined treatment pathways with the ability to prescribe and titrate therapy. When implemented at scale, and with optimized data collection, this model provides a robust framework that can be deployed in learning health systems."
Title: Validation of an algorithm for continuous monitoring of atrial fibrillation using a consumer smartwatch Trial using bring your own DHT device **Citation**: Avram et al.[31]			
Repository of a variety of consumer wearables (such as actigraphy and sleep sensors) and DHTs (such as ECG and blood pressure devices).	"Bring your own device" methodology with participants enrolling with the wearable or DHT device of their choice. Aggregation of patient-generated health data into a centralized database that is linked to a participants' electronic health record. Sample size: 204	A nested analysis of 204 participants with over 80,000 h of a smartwatch-photoplethysmography (PPG)-based ECG measurements and patch-based ECG monitoring (28-day duration telemetry), smartwatch ECG demonstrated a high correlation of smartwatch-ECG to patch-based ECG ($r = 0.98$) for quantifying atrial fibrillation burden with a sensitivity of 97% and specificity of 99%.	"Our findings demonstrate that a novel algorithm using a commercially available smartwatch can continuously detect AF with excellent performance and that confirmation with smartwatch-PPG ECG further enhances specificity. In addition, our smartwatch with the PPG algorithm can estimate the AF burden. Further research is needed to determine whether this algorithm is useful in screening for AF in select at-risk patients."
Title: A digital biomarker of diabetes from smartphone-based vascular signals Trials focused on the development of novel biomarkers **Citation**: Avram et al.[32]			

Smartphone-based photoplethysmography	Development of a deep neural network (DNN) to detect vascular signals in diabetes. PPG waveforms were obtained via study participants placing their index fingers on the camera of an iPhone, and these waveforms were used as the sole input for the DNN. A 39-layer convolutional deep neural network was developed and was trained and validated on a random sample of individuals who self-reported a diagnosis of diabetes. Sample size: 61,857. Divided into A primary cohort (sample size 53,870), validation cohort (sample size 7806), and clinical cohort (prospectively enrolled sample size of 181)	The DNN achieved an area under the curve (AUC) for prevalent diabetes of 0.766 in the primary cohort (95% CI: 0.750–0.782; sensitivity 75%, specificity 65%) and 0.740 in the validation cohort (95% CI: 0.723–0.758; sensitivity 81%, specificity 54%). Within a regression analysis that included baseline demographics (age, gender, race/ethnicity, and body mass index) the AUC was 0.830 (95% CI: 0.787–0.873) and the DNN score remained independently predictive of diabetes. The performance of the DNN in the clinic cohort demonstrated a similar AUC of 0.740 (95% CI: 0.723–0.758).	"In this large-scale study and validation across three distinct cohorts, we show that smartphone-measured PPG, analyzed with deep learning, can serve as an independent, noninvasive digital biomarker of prevalent diabetes. Importantly, the ability of this PPG biomarker to predict diabetes was independent of standard risk factors and comorbidities, and discrimination further improved when adding easily obtainable covariates such as age, gender, race/ethnicity and BMI"
Title: Remote speech analysis in the evaluation of hospitalized patients with acute decompensated heart failure Trials focused on the development of novel biomarkers **Citation**: Amir et al.[33]			
Voice capture on a smartphone with data collected on a speech app. This data was transmitted to a cloud-based server for analysis. Smartphone-app-based speech analysis.	The speech application was configured to record patterns of voice among acute decompensated heart failure (ADHF) patients who were prompted to record standardized sentences at the time of hospitalization. Five speech measures were selected as the ones considered to be most promising. Each of these had some combination of the following characteristics: high temporal and/or spectral resolution, autoregressive model for the spectrum, nonlinear amplitude and frequency mapping, symmetric version of a nonlinear spectra ratio, and Euclidian distance. Similarity between speech measures was used to calculate differences in voice between different states of heart failure and the change from admission day to discharge day. Recordings were obtained from heart failure patients who repeated five sentences 3–4 times and at the time of admission for acute decompensated heart failure and upon discharge after receiving treatment. Sample size: 40	1484 recordings were analyzed in unsupervised clustering and demonstrated a difference in speech measures among 94% of recordings before and after heart failure treatment.	"In this study using a novel speech recognition system in patients demonstrated its ability to identify voice alterations reflective of changes in ADHF clinical status and showed large changes in speech measures between the congested (admission) and decongested (discharge) states. Specifically, distinct speech vocalization features were identified in hospital admission versus discharge recordings of 40 patients with ADHF, with 95% of patients showing significant shifts in at least 4 of the 5 tested speech measures"

AHR, adjusted hazard ratio; *AUC*, area under the receiver operator curve; *CI*, confidence interval; *DHT*, digital health technologies; *PPG*, photoplethysmography; *RPM*, remote patient monitoring; *SMART-on-FHIR*, Substitutable Medical Applications and Reusable Technologies-Fast Health Interoperability Resource.

DHTs including implantable sensors such as pulmonary artery pressure measurements[32,34] and wearables such as smart device-based EKG technologies[35] for use among heart failure patients and patients with atrial arrhythmias; respectively, through patient-generated health data (PGHD) and RPM can enable a frequent assessment of physiologic parameters. It can also shorten the time to detection of disease recurrence and potentially to determine the effectiveness of treatments or interventions in a shorter duration when compared to standard care or scheduled study visitations. By providing the opportunity for continuous data collection and analysis, RPM utilizing DHTs offers opportunities to design new clinical workflows; to develop clinical decision support (CDS) systems relevant to clinical trials; and to determine the outcomes resulting from lifestyle modification, pharmacotherapy, and/or medical interventions.[36] DHTs used within such RPM programs (examples provided in Fig. 80.1) are emerging and are being designed in a fashion that may have the potential to generate RWE with hybrid clinical trials such as pragmatic clinical trials,[37] the analysis of RPM data as part of EHRs,[38] and to determine the effectiveness of DHTs coupled with clinical decisions on chronic disease management.[36] Trials that incorporate factors important to patients and their use of DHTs from the time of study initiation and throughout trial execution may lead to a better understanding of those measures that lead to successful DHT use over time.[2,3]

Scirica and colleagues[39] demonstrated risk reduction among 5000 elderly patients with CVD who were enrolled in nonrandomized, prospective, observational RPM-based CV-risk management trial. The study paired DHTs (BP, weight, and activity monitoring) and RPM with care navigators, pharmacists, and physicians for when to initiate and titrate medical therapy, and it was designed to be virtual with no office-based encounters. Initial results at 6-month demonstrated that nearly 50% of all enrolled patients achieved lipid profile reduction to target CVD-guideline recommendations with some patients achieving a reduction in systolic blood pressure of 30 mmHg (overall reduction from a baseline of 165 to 135 mmHg) on follow-up. Upon secondary analyses,[40] workflow automations for RPM data collection and analysis, as well as multichannel communications including text messaging and communications through patient-health records with data linked directly to the institution's EHR, emerged as key factors to improve patient-provider interactions and information sharing. While this study was designed at a large, urban, integrated healthcare system, similar disease management programs and trials are underway including programs that harness designs that include study participants across different institutions, multiple geographies, and varied socio-economic classes and enroll patients within the course of care delivery that use clinical workflows in contrast to study workflows. Harnessing clinical workflows and real-world patient-provider interactions present unique opportunities to use DHTs at the point of care and to capture DHT measurements in parallel to clinical trial outcomes.[31,41,42]

Equitable participation in DHT-enabled clinical trials—Regulatory perspectives

Developing and promoting policies to ensure adequate enrollment of historically underrepresented and vulnerable subgroups in clinical trials is among the highest priorities for clinical, research, and regulatory agencies who are collectively tasked with ensuring diverse, equitable, and inclusive practices drive broad eligibility for participation in DHT-enabled clinical trials.[36] Rural communities, women, the elderly, and people from most racial and ethnic minorities have historically been underrepresented in clinical trials of either investigational DHTs or those studies using approved or cleared DHTs in the evaluation of new products or for real-world evidence. For example, in a large nationwide, decentralized study to determine the utility of a commercially available smartwatch to detect atrial fibrillation, less than 6% of the enrolled cohort of over 400,000 participants (mean age 41 ± 13 years) were above the age of 65 and less than 30% were nonwhite ethnicities.[29] While smartwatch users and those who enroll in decentralized studies using such devices may have different demographic characteristics than those who are at a higher risk of atrial fibrillation, the absence of multimorbidity risk and the lack of inclusion of a diverse group with a different profile of social determinants of health should be taken as an opportunity to design new frameworks and policy for the design and execution of decentralized and DHT-enabled studies that characterize common diseases where disparities exist. Toward these objectives, the FDA has proposed requirements to include diversity plans that broaden the eligibility criteria and provide an outline to augment enrollment practice and trial designs (e.g., by using telehealth and/or RPM) to reach diverse communities and to improve the rates of participation from underrepresented racial and ethnic populations.[36,11] Merging technology-based approaches with innovative methods to improve data collection for underrepresented cohorts and to upskill clinical researchers on culturally and linguistically appropriate services can come together to enhance equity in the next generation of DHT-enabled clinical trials.

Concluding remarks

New technologies and innovative trial approaches and processes that leverage DHTs and their associated multidisciplinary data sources are modifying clinical trial designs by increasing the quantity and quality of data that can be acquired from study participants and are enhancing

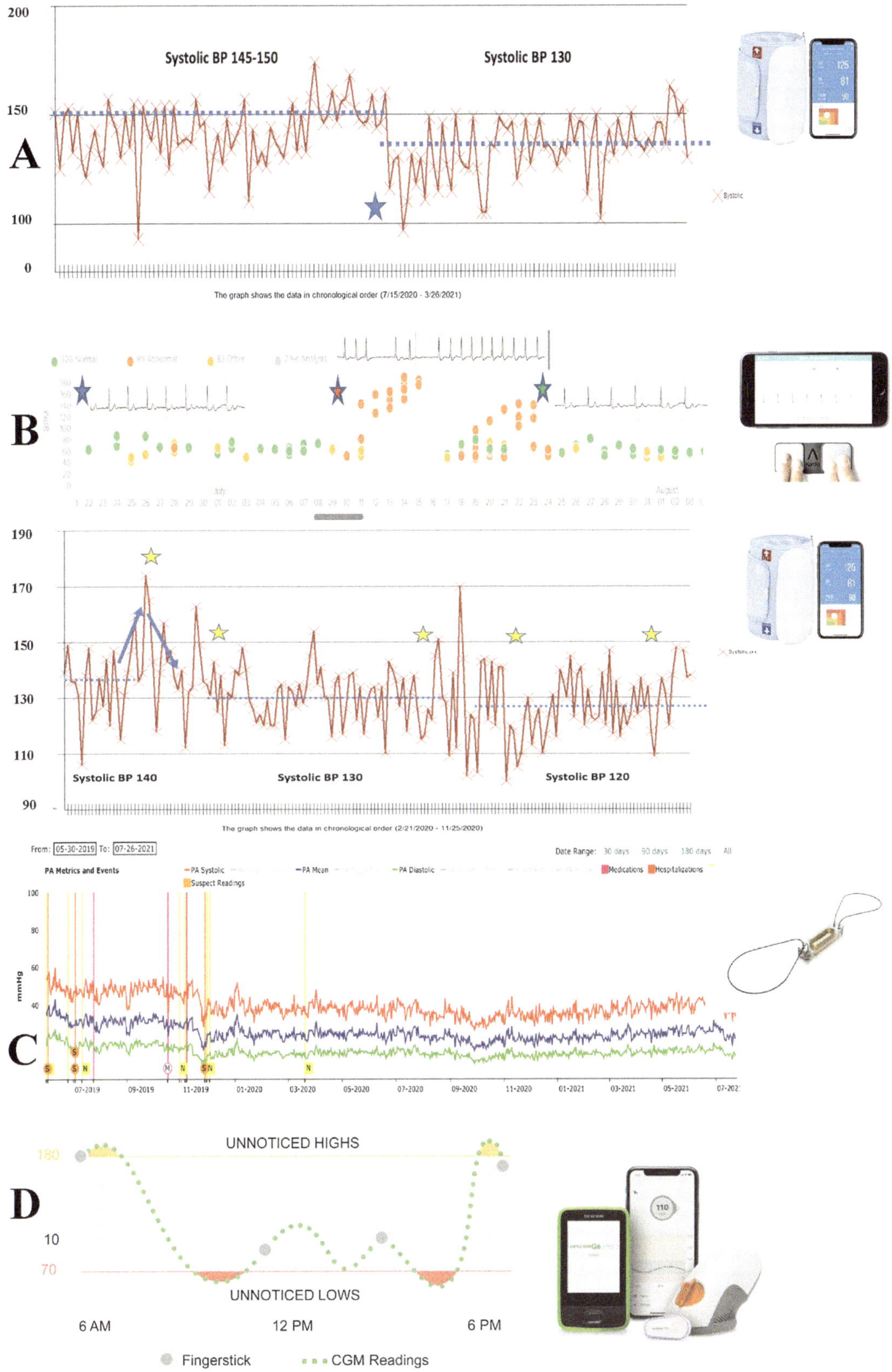

FIGURE 80.1 Real-world data and remote patient monitoring with DHTs (examples provided from reference[2]) (A) Remote hypertension DHT trial—68-year-old man with ischemic heart disease and obesity (BMI 45) on multiple antihypertensive therapies. Began intermittent fasting (*blue* star) resulting in an improvement in blood pressure control. (B) Multimodality DHT trial with Smartphone-ECG and home-based BP monitoring—74-year old with persistent atrial fibrillation (AF) and heart failure (ejection fraction 25%). Underwent AF ablation (blue star), recurrence of rapid AF and initiation of antiarrhythmic therapy with amiodarone (*red* star). Repeat AF ablation (*green* star) to sinus rhythm. Titration of heart failure therapy (*yellow* stars) with blood pressure reduction. (C) Pulmonary artery sensor guided DHT trial—70-year old with heart failure (ejection fraction 20%) due to ischemic cardiomyopathy. Sensor implantation after a hospitalization (2019) with medication titration based on pulmonary artery pressures (*orange* and *yellow* lines) resulting in reduction in pulmonary pressure from 60 mmHg to less than 40 mmg and sustained throughout 2020–21 and improvements in symptoms. (D) Continuous glucose monitoring DHT trial—64-year old with type-2 diabetes and ischemic heart disease. Observations of AM and PM hypoglycemia with continuous monitoring that required adjustments in medical therapies to prevent further hypoglycemic episodes.

efficiency and patient participation. Such innovations are collectively expanding the tools and resources available to help answer new research questions that can be answered. As DHTs become increasingly ubiquitous and data platforms more sophisticated, it is important to employ robust clinical, research, and regulatory science approaches to understand how DHTs can be used to advance trial methodologies, and how continuously streaming digital health data will be integrated to inform new patterns of health and disease. In doing so, a new ontology of clinical trials can emerge to better inform the quality of care across the cardiovascular care continuum.

Get started

- Understand the diversity of clinical trials that incorporate DHTs and how digital methodologies such as remote patient monitoring are advancing trial designs and how patient-reported outcomes are collected during trial follow-up.
- It is important that a DHT be "fit for purpose" and that the verification and validation of general-purpose consumer DHTs is achieved within a given trial and when used in specific clinical and regulatory applications.
- New decentralized and remote monitoring clinical trials that incorporate DHTs are emerging as potentially important developments to recruit large and diverse populations of study participants.

Pitfalls to avoid

1. Avoid the assumption that digital methodologies can apply to all trial designs.
2. Patient participation in clinical trials that incorporate DHTs, while anticipated, may be variable given that practice patterns and the use of DHTs across different patients and clinical settings may be different.
3. While DHTs represent an opportunity to promote health equity through improving patient outcomes and patient recruitment and retention, it is important to take into consideration issues related to diversity, usability, and access, and that the results generated from trials that use DHTs are applicable and have a direct translation to the patient population with the condition being studied.

References

1. Bhavnani SP, Narula J, Sengupta PP. Mobile technology and the digitization of healthcare. *Eur Heart J*. May 7, 2016;37(18):1428–1438. https://doi.org/10.1093/eurheartj/ehv770.
2. Bhavnani SP. Digital health: opportunities and challenges to develop the next-generation technology-enabled models of cardiovascular care. *Methodist Debakey Cardiovasc J*. 2020 October-December;16(4):296–303. https://doi.org/10.14797/mdcj-16-4-296.
3. Inan OT, Tenaerts P, Prindiville SA, et al. Digitizing clinical trials. *NPJ Digit Med*. July 31, 2020;3:101. https://doi.org/10.1038/s41746-020-0302-y.
4. Marquis-Gravel G, Roe MT, Turakhia MP, et al. Technology-enabled clinical trials: transforming medical evidence generation. *Circulation*. October 22, 2019;140(17):1426–1436. https://doi.org/10.1161/CIRCULATIONAHA.119.040798.
5. Shameer K, Badgeley MA, Miotto R, Glicksberg BS, Morgan JW, Dudley JT. Translational bioinformatics in the era of real-time biomedical, health care and wellness data streams. *Brief Bioinform*. January 2017;18(1):105–124. https://doi.org/10.1093/bib/bbv118.
6. Weissler EH, Naumann T, Andersson T, et al. The role of machine learning in clinical research: transforming the future of evidence generation. *Trials*. August 16, 2021;22(1):537. https://doi.org/10.1186/s13063-021-05489-x.
7. Bhavnani SP, Parakh K, Atreja A, et al. 2017 roadmap for innovation-ACC health policy statement on healthcare transformation in the era of digital health, big data, and precision health: a report of the American College of Cardiology task force on health policy statements and systems of care. *J Am Coll Cardiol*. November 28, 2017;70(21):2696–2718. https://doi.org/10.1016/j.jacc.2017.10.018.
8. https://www.fda.gov/about-fda/center-devices-and-radiological-health/cdrh-mission-vision-and-shared-values.
9. https://www.fda.gov/medical-devices/premarket-submissions-selecting-and-preparing-correct-submission/investigational-device-exemption-ide.
10. https://www.fda.gov/regulatory-information/search-fda-guidance-documents/fda-decisions-investigational-device-exemption-clinical-investigations.
11. https://www.fda.gov/regulatory-information/search-fda-guidance-documents/digital-health-technologies-remote-data-acquisition-clinical-investigations.
12. *Use of Real-World Evidence to Support Regulatory Decision-Making for Medical Devices*: FDA.
13. Shan R, Ding J, Weng D, et al. Early blood pressure assessment after acute myocardial infarction: insights using digital health technology. *Am J Prev Cardiol*. September 2020;3:100089. https://doi.org/10.1016/j.ajpc.2020.100089.
14. McGillion MH, Parlow J, Borges FK, et al, PVC-RAM-1 Investigators. Post-discharge after surgery virtual care with remote automated monitoring-1 (PVC-RAM-1) technology versus standard care: randomised controlled trial. *BMJ*. September 30, 2021;374:n2209. https://doi.org/10.1136/bmj.n2209.
15. Wang C, Qi B, Lin M, et al. Continuous monitoring of deep tissue hemodynamics with stretchable ultrasonic phased arrays. *Nat Biomed Eng*. 2021;5(7):749–775.
16. Wheeler C, Patel S, Waldman CE, et al. Hearing the heart- validation of a novel digital health earbud technology to measure cardiac time intervals through infrasonic hemodynography. *Circulation*. 2021;144:A11669.
17. Marcus GM, Modrow MF, Schmid CH, et al. Individualized studies of triggers of paroxysmal atrial fibrillation: the I-STOP-AFib randomized clinical trial. *JAMA Cardiol*. November 2021;14:e215010. https://doi.org/10.1001/jamacardio.2021.5010.
18. McGillion M, Ouellette C, Good A, et al. Postoperative remote automated monitoring and virtual hospital-to-home care system

following cardiac and major vascular surgery: user testing study. *J Med Internet Res*. 2020;22:e15548.

19. Sayeed R, Gottlieb D, Mandl KD. SMART Markers: collecting patient-generated health data as a standardized property of health information technology. *NPJ Digit Med*. January 23, 2020;3:9. https://doi.org/10.1038/s41746-020-0218-6.
20. Wang L, Nielsen K, Goldberg J, et al. Association of wearable device use with pulse rate and health care use in adults with atrial fibrillation. *JAMA Netw Open*. May 3, 2021;4(5):e215821. https://doi.org/10.1001/jamanetworkopen.2021.5821.
21. Stehlik J, Schmalfuss C, Bozkurt B, et al. Continuous wearable monitoring analytics predict heart failure hospitalization: the LINK-HF multicenter study. *Circ Heart Fail*. March 2020;13(3):e006513. https://doi.org/10.1161/CIRCHEARTFAILURE.119.006513.
22. Turakhia M, Sundaram V, Smith SN, et al, SmartADHERE Investigators. Efficacy of a centralized, blended electronic, and human intervention to improve direct oral anticoagulant adherence: smartphones to improve rivaroxaban ADHEREnce in atrial fibrillation (SmartADHERE) a randomized clinical trial. *Am Heart J*. July 2021;237:68–78. https://doi.org/10.1016/j.ahj.2021.02.023.
23. Skov SC, Dyrvig AK, Henneberg NC, et al. Patient-reported outcomes from patients with heart failure participating in the future patient telerehabilitation program: data from the intervention arm of a randomized controlled trial. *JMIR Cardio*. July 2, 2021;5(2):e26544. https://doi.org/10.2196/26544.
24. Bachtiger P, Petri CF, Scott FE, et al. Point-of-care screening for heart failure with reduced ejection fraction using artificial intelligence during ECG-enabled stethoscope examination in London, UK: a prospective, observational, multicentre study. *Lancet Digit Health*. January 5, 2022;(21):00256–00259. https://doi.org/10.1016/S2589-7500(21)00256-9.
25. Ong MK, Romano PS, Edgington S, et al. Better effectiveness after transition–heart failure (BEAT-HF) research group. Effectiveness of remote patient monitoring after discharge of hospitalized patients with heart failure: the better effectiveness after transition – heart failure (BEAT-HF) randomized clinical trial. *JAMA Intern Med*. March 2016;176(3):310–318. https://doi.org/10.1001/jamainternmed.2015.7712.
26. Bhavnani SP, Sola S, Adams D, Venkateshvaran A, Dash PK, Sengupta PP. ASEF-VALUES investigators. A randomized trial of pocket-echocardiography integrated mobile health device assessments in modern structural heart disease clinics. *JACC Cardiovasc Imag*. April 2018;11(4):546–557. https://doi.org/10.1016/j.jcmg.2017.06.019.
27. Heywood JT, Jermyn R, Shavelle D, et al. Impact of practice-based management of pulmonary artery pressures in 2000 patients implanted with the CardioMEMS sensor. *Circulation*. April 18, 2017;135(16):1509–1517. https://doi.org/10.1161/CIRCULATIONAHA.116.026184.
28. Brewer LC, Hayes SN, Caron AR, et al. Promoting cardiovascular health and wellness among African-Americans: community participatory approach to design an innovative mobile-health intervention. *PLoS One*. August 20, 2019;14(8):e0218724. https://doi.org/10.1371/journal.pone.0218724.
29. Perez MV, Mahaffey KW, Hedlin H, et al, Apple Heart Study Investigators. Large-scale assessment of a smartwatch to identify atrial fibrillation. *N Engl J Med*. November 14, 2019;381(20):1909–1917. https://doi.org/10.1056/NEJMoa1901183.
30. *Submitting Documents Using Real-World Data and Real-World Evidence to FDA for Drug and Biological Products*: FDA.
31. Avram R, Ramsis M, Cristal AD, et al. Validation of an algorithm for continuous monitoring of atrial fibrillation using a consumer smartwatch. *Heart Rhythm*. September 2021;18(9):1482–1490. https://doi.org/10.1016/j.hrthm.2021.03.044.
32. Avram R, Olgin JE, Kuhar P, et al. A digital biomarker of diabetes from smartphone-based vascular signals. *Nat Med*. October 2020;26(10):1576–1582. https://doi.org/10.1038/s41591-020-1010-5.
33. Amir O, Abraham WT, Azzam ZS, et al. Remote speech analysis in the evaluation of hospitalized patients with acute decompensated heart failure. *JACC Heart Fail*. January 2022;10(1):41–49. https://doi.org/10.1016/j.jchf.2021.08.008.
34. Desai AS, Bhimaraj A, Bharmi R, et al. Ambulatory hemodynamic monitoring reduces heart failure hospitalizations in "real-world" clinical practice. *J Am Coll Cardiol*. May 16, 2017;69(19):2357–2365. https://doi.org/10.1016/j.jacc.2017.03.009.
35. Goldenthal IL, Sciacca RR, Riga T, et al. Recurrent atrial fibrillation/flutter detection after ablation or cardioversion using the AliveCor KardiaMobile device: iHEART results. *J Cardiovasc Electrophysiol*. November 2019;30(11):2220–2228. https://doi.org/10.1111/jce.14160.
36. https://www.fda.gov/regulatory-information/search-fda-guidance-documents/enhancing-diversity-clinical-trial-populations-eligibility-criteria-enrollment-practices-and-trial.
37. Asch DA, Troxel AB, Goldberg LR, et al. Remote monitoring and behavioral economics in managing heart failure in patients discharged from the hospital: a randomized clinical trial. *JAMA Intern Med*. June 1, 2022;182(6):643–649. https://doi.org/10.1001/jamainternmed.2022.1383.
38. Lee NS, Anastos-Wallen R, Chaiyachati KH, Reitz C, Asch DA, Mehta SJ. Clinician decisions after notification of elevated blood pressure measurements from patients in a remote monitoring program. *JAMA Netw Open*. January 4, 2022;5(1):e2143590. https://doi.org/10.1001/jamanetworkopen.2021.43590.
39. Scirica BM, Cannon CP, Fisher NDL, et al. Digital care transformation: interim report from the first 5000 patients enrolled in a remote algorithm-based cardiovascular risk management program to improve lipid and hypertension control. *Circulation*. 2021;143(5):507–509. https://doi.org/10.1161/CIRCULATIONAHA.120.051913.
40. Gordon WJ, Blood AJ, Chaney K, et al. Workflow automation for a virtual hypertension management program. *Appl Clin Inform*. October 2021;12(5):1041–1048. https://doi.org/10.1055/s-0041-1739195.
41. https://clinicaltrials.gov/ct2/show/NCT03713333.
42. *ASE 30th Annual Scientific Sessions*. 2019. Page 111 section P2–071.

Chapter 81

A short overview of the orphan drugs development

Yugandhar Kothapalli, Christina M. Mapp and Uma S. Singh

Department of Pharmaceutical and Biomedical Sciences, College of Pharmacy, The University of Georgia, Athens, GA, United States

Introduction

Globally, rare diseases collectively make a significant impact, and it is estimated that more than 350 million people worldwide suffer from rare diseases.[1] However, there is no single justification and definition of rare diseases. The United States Food and Drug Administration (FDA) defines it as any disease or condition that affects fewer than 200,000 people or about 1 in 1500 people, while in Europe, it is defined as those affecting fewer than 5 in 10,000 population pool.[2] There are currently over 8000 rare diseases that affect more than 30 million people in the United States alone, and more are being discovered each day. Out of which, only 5% of patients are receiving specific authorized or licensed treatments.[3,4] Many rare diseases still have very limited or no therapeutic options, especially conditions caused by genetic mutations.[5] To reduce the morbidity and mortality caused by rare diseases, accessibility of drugs to patients is critical. This short report is focused on orphan drugs. Orphan drugs are medicines or vaccines invented to treat rare diseases. In recent years, many countries have employed regulations, a combination of legislation, and policies for orphan drugs.[6] The implemented regulations and policies may enable easy accessibility of orphan drugs to patients, though they are significantly different in various countries. Unfortunately, the two countries with the largest populations, China and India, lack orphan drug policy, which has a substantial impact on their populations.

Process/pathway

Recent advances in DNA sequencing, particularly whole genome, and exome sequencing, have improved the identification of genes that cause rare diseases. Advanced tools like proteomics, epigenetics, system biology, traditional gene mapping, genomic sequencing, and other multiomics approaches have accelerated the identification of specific gene mutations responsible for the generation of rare diseases.[7] After identifying biological/drug targets, the next step is to develop potential therapeutic agents by inventing new chemical entities or via drug repurposing. The drug discovery process for orphan drugs also involves multiple-step procedures, including the synthesis of large compound libraries, high throughput screening, and an accurate assay method to understand the interaction mechanisms between the biological/drug target and newly discovered therapeutic agents. After evaluation of initial interactions, structure-activity relationship (SAR) studies are used to optimize the drug candidate.[8–10] The overall drug discovery process for identifying a new orphan drug is complicated, expensive, involves longer timelines, and has low success rates. On the other hand, drug repurposing or repositioning practices are cost-effective methods to develop specific orphan drugs.[11] It facilitates the use of approved drugs with known biological pathways and safety profiles. However, there are some limitations and hurdles that certainly need structured guidelines and better incentives.[12,13] To overcome these challenges encountered by all stakeholders, the United States created a flexible regulatory approach called the accelerated approach (AA) pathway.[14] The AA pathway is exponentially essential for the development of orphan drugs.[7] Due to the challenges of the discovery of orphan drugs, it makes a clear sense that it is difficult to address these serious problems by a single institution, program, or country. In effort to improve the process of drug development for rare diseases, the International Rare Diseases Research Consortium (IRDiRC) was established in 2011.[15] The IRDiRC aims to unite various funding bodies, companies, patient advocacy organizations, and research institutions to reach its goal of 1000 new therapies for rare diseases by 2027.

Translational Cardiology. https://doi.org/10.1016/B978-0-323-91790-2.00107-6

Common challenges

Understanding the intricacies of rare diseases, such as their diverse manifestations and distribution across both small and large geographic areas, presents a significant challenge. In the discovery of orphan drugs for rare diseases, a poor understanding of the disease and its origins are significant challenges for the scientific communities. Because these conditions affect fewer people than more common disorders, there is significantly less information available regarding the causes and mechanisms of rare diseases. Without a fundamental understanding of the mechanism for these diseases, it is difficult for researchers to determine a therapeutic endpoint. Small and geographically dispersed populations also make it challenging to conduct clinical trials for the drug candidates that have been determined to be effective. Additionally, orphan drug development is an expensive process, but due to the relatively low populations affected, it is unlikely that these costs will be replenished by drug profits alone. However, orphan drugs are sometimes the first in a new drug class that may provide treatments for other conditions, or the drug itself may be effective in treating other conditions, as two valuable examples are mentioned in Table 81.1. Therefore, it is concluded that the efforts of orphan drug repurposing may also lead to additional opportunities for profit.

A growing understanding of many rare diseases at the molecular and genetic levels and incentives through the Orphan Drug Act (ODA) have allowed for new opportunities to innovate new therapeutic agents for rare diseases. ODA was passed in 1983; it provides a 50% tax credit for costs associated with the research and design of the drug and 7 years of exclusive marketing (compared to the general 5-year period for undesignated new medicines).[19] After the endorsement of this act, there has been a noticeable increase in the effort to develop orphan drugs (Fig. 81.1).

ODA also established the FDA Office of Orphan Products Development (OOPD). Its mission is to accelerate the evaluation and development of products such as drugs, biologics, and devices that demonstrate promise for diagnosing and treating rare diseases and conditions. For example, in 2020, the FDA approved 32 novel drugs and biologics with orphan drug designation, allowing orphan-designated products to account for 58% of all new drug approvals in the United States.[21] Despite tremendous support from public/nonprofit research funders and the pharmaceutical industry, 95% of rare diseases still lack an approved treatment.[3] Therefore, it is crucial to continuously foster innovation for orphan drugs and address the many challenges involved in rare disease research. To date, many patients are still seeking treatments, and the variety of therapeutic categories in which drugs are approved is also increasing (Fig. 81.2).

TABLE 81.1 First-in-class orphan drugs[16]: Examples of two drugs that were first in the orphan drug classes, later these drugs were effective against other conditions.

Rare disease	Orphan drug	Drug class	Other uses for drug class
Chronic myeloid leukemia	*Imatinib*[17]	Tyrosine kinase inhibitor	To treat a wide range of cancer, including breast, pancreatic, and colorectal cancer
Cushing's disease	*Pasireotide diaspartate*[18]	Somatostatin analog	To treat acromegaly, neuroendocrine tumors, certain diabetic complications

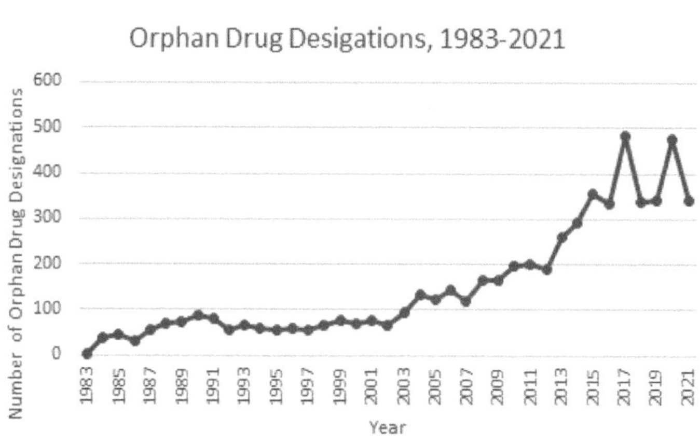

Note: The indicated number of orphan drug designations in 2021 is current as of December 3, 2021.

FIGURE 81.1 Orphan drug designations.[20] This line graph demonstrates the upward trend in orphan drug designations after passing the ODA.

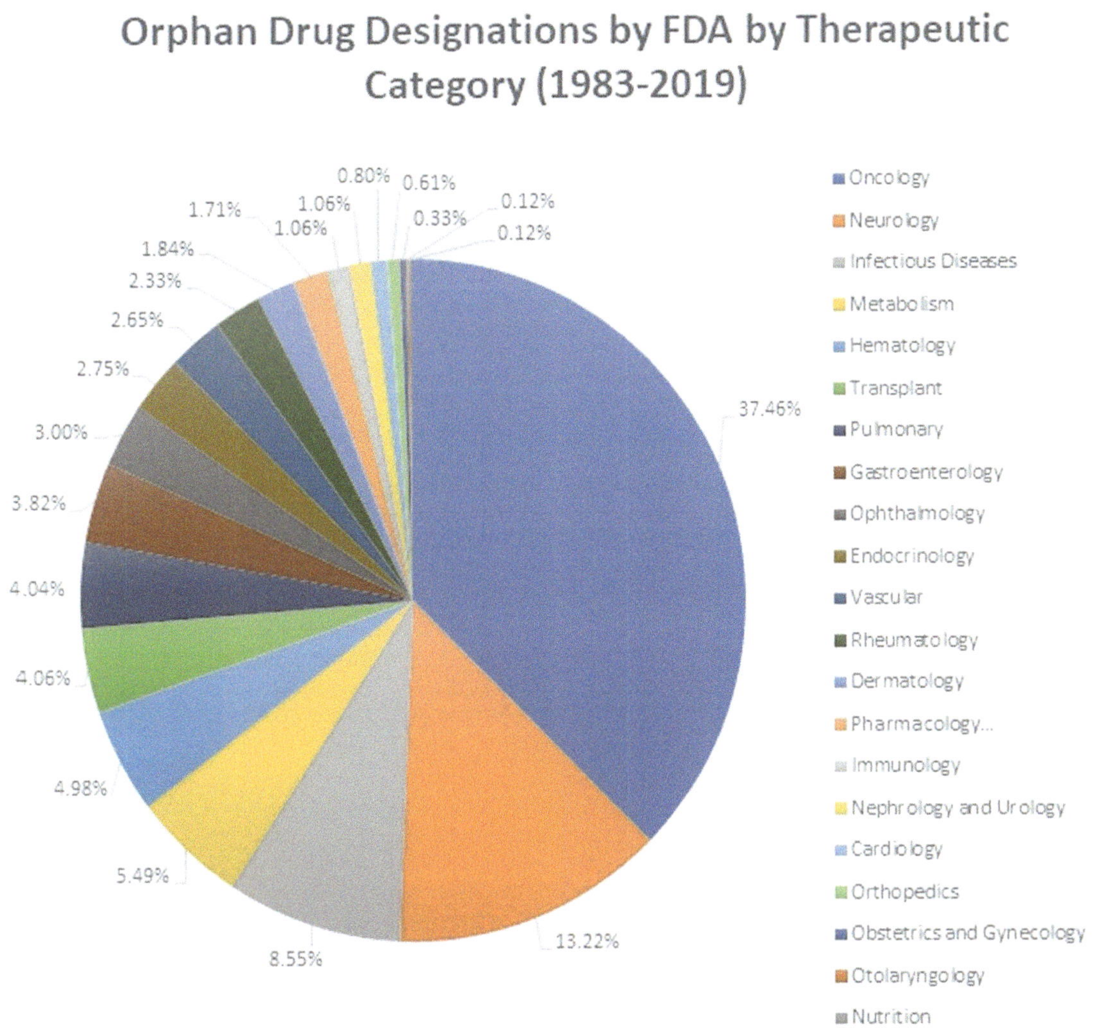

FIGURE 81.2 **Orphan drug designations by therapeutic category.**[20] Many orphan drug designations from 1983 to 2019 have been in oncology, while designations for cardiology, orthopedics, obstetrics, otolaryngology, and nutrition each made up less than 1% of the total designations during this period.

Budgetary considerations and funding options

Due to smaller patient populations, recovery of research and development costs for orphan drugs through market sales is often unattainable. Financial incentives such as research grants, tax credits, marketing exclusivity, user fee waivers, and corporate tax reductions help to improve profit outcomes. Additionally, nonfinancial incentives such as fast-track approval and prelicensing access assist in recovering the costs of development.

Conclusions

Rare diseases collectively affect a large patient population globally. Orphan drugs are the only options for the treatment of these diseases. However, significant challenges in development and low potential for profits are associated with these classes of drugs. Implementation of ODA in the United States and Orphan Drug regulations in Europe and other countries have accelerated orphan drug discovery. In recent years, a collaborative effort to incentivize research and drug development for these diseases has been encouraged. Nevertheless, much advancement in financial support and government participation and regulation is needed in the discovery and development of orphan drugs. A continuation of efforts and the creation of incentives in many countries are required that may benefit thousands of patients around the world with the standing rare diseases.

Additional reading

1. Blin. O; Lefebvre, M-N.; Rascol, O. et al. "Orphan drug clinical development." https://www.sciencedirect.com/science/article/abs/pii/S0040595720300093

2. Gammie, T. et al. "Access to Orphan Drugs: A Comprehensive Review of Legislations, Regulations and Policies in 35 Countries." https://pubmed.ncbi.nlm.nih.gov/26451948/
3. Bouwman M. L. et al. "Regulatory issues for orphan medicines: A review." https://www.sciencedirect.com/science/article/pii/S2211883719300796

Acknowledgment

We thank Ransom A. Jones for his inputs and corrections of this short report.

References

1. Klimova B, Storek M, Valis M, Kuca K. Global view on rare diseases: a mini review. *Curr Med Chem*. 2017;24(29):3153–3158.
2. Sharma A, Jacob A, Tandon M, Kumar D. Orphan drug: development trends and strategies. *J Pharm BioAllied Sci*. 2010;2(4):290–299.
3. Fetro C, Scherman D. Drug repurposing in rare diseases: myths and reality. *Therapie*. 2020;75(2):157–160.
4. Ramírez W. *Rare Diseases : Prevalence, Treatment Options and Research Insights*. Hauppauge, NY: Nova Science Publishers, Incorporated; 2017.
5. Boycott KM, Ardigo D. Addressing challenges in the diagnosis and treatment of rare genetic diseases. *Nat Rev Drug Discov*. 2018;17(3):151–152.
6. Gammie T, Lu CY, Babar ZUD. Access to orphan drugs: a comprehensive review of legislations, regulations and policies in 35 countries. *PLoS One*. 2015;10(10):e014002.
7. Kerr K, McAneney H, Smyth LJ, Bailie C, McKee S, McKnight AJ. A scoping review and proposed workflow for multi-omic rare disease research. *Orphanet J Rare Dis*. 2020;15(1):107.
8. Rana P, Chawla S. Orphan drugs: trends and issues in drug development. *J Basic Clin Physiol Pharmacol*. 2018;29(5):437–446.
9. Burton A, Castano A, Bruno M, et al. Drug discovery and development in rare diseases: taking a closer look at the tafamidis story. *Drug Des Dev Ther*. 2021;15:1225–1243.
10. Tambuyzer E, Vandendriessche B, Austin CP, et al. Therapies for rare diseases: therapeutic modalities, progress and challenges ahead. *Nat Rev Drug Discov*. 2020;19(2):93–111.
11. Sardana D, Zhu C, Zhang ML, Gudivada RC, Yang L, Jegga AG. Drug repositioning for orphan diseases. *Brief Bioinf*. 2011;12(4):346–356.
12. van den Berg S, de Visser S, Leufkens HGM, Hollak CEM. Drug repurposing for rare diseases: a role for academia. *Front Pharmacol*. 2021;12:746987.
13. Pushpakom S, Iorio F, Eyers PA, et al. Drug repurposing: progress, challenges and recommendations. *Nat Rev Drug Discov*. 2019;18(1):41–58.
14. Kakkis ED, O'Donovan M, Cox G, et al. Recommendations for the development of rare disease drugs using the accelerated approval pathway and for qualifying biomarkers as primary endpoints. *Orphanet J Rare Dis*. 2015;10. https://doi.org/10.1186/s13023-014-0195-4.
15. Hechtelt Jonker A, Hivert V, Gabaldo M, et al. Boosting delivery of rare disease therapies: the IRDiRC orphan drug development guidebook. *Nat Rev Drug Discov*. 2020;19(8):495–496.
16. Phillips MI, Lee JR. Recent orphan drugs that are first-in-class. *Expert Opin Orphan Drug*. 2014;2(8):759–763.
17. Cohen MH, Moses ML, Pazdur R. Gleevec (TM) for the treatment of chronic myelogenous leukemia: US food and drug administration regulatory mechanisms, accelerated approval, and orphan drug status. *Oncol*. 2002;7(5):390–392.
18. Arnaldi G, Boscaro M. Pasireotide for the treatment of Cushing's disease. *Expet Opin Invest Drugs*. 2010;19(7):889–898.
19. Herder M. What is the purpose of the orphan drug act? *PLoS Med*. 2017;14(1):e1002191.
20. Miller KL, Fermaglich LJ, Maynard J. Using four decades of FDA orphan drug designations to describe trends in rare disease drug development: substantial growth seen in development of drugs for rare oncologic, neurologic, and pediatric-onset diseases. *Orphanet J Rare Dis*. 2021;16:265.
21. Woodcock J. *Rare Disease Day 2021: FDA Shows Sustained Support of Rare Disease Product Development during the Public Health Emergency*. U.S. Food and Drug Administration; 2021. Published https://www.fda.gov/news-events/fda-voices/rare-disease-day-2021-fda-shows-sustained-support-rare-disease-product-development-during-public. Accessed December 10, 2021. Updated 01 March 2021.

Chapter 82

Regulatory, product development, resources, and financial challenges associated with creating and commercializing combination products

Kurt Dasse and Priscilla Petit

Inspired Therapeutics LLC, Merritt Island, FL, United States

Key points

The purpose of this chapter is to describe what constitutes a combination product and the process for obtaining regulatory approvals. The chapter presents three real-world examples of contemporary combination products and the challenges related to obtaining funding for combination product development.

Why it matters

The strategy used to develop a combination product is complex from the product development, regulatory, and financial perspectives. Unlike a clear regulatory path for a drug, device, or biologic, it may be unclear which center within the FDA should serve as the lead for premarket approval and the strategy to obtain regulatory approval. Companies often struggle to attract a strategic partner from a drug or medical device corporation since each of these organizations is mainly interested in their area of expertise.

Introduction

The Federal Food, Drug and Cosmetic Act was amended by Congress in 1990 to introduce the concept of combination products. A combination product was viewed as a product comprised of two or more FDA-regulated components (i.e., drug/device, biologic/device, drug/biologic, or drug/device/biologic) combined for use. The Office of Combination Products (OCP) was ultimately established within the Commissioner's Office of the FDA in 2002 in accordance with Section 204 of the Medical Device User Fee and Modernization Act of 2002 (MDUFMA).[1]

Combination products generally proceed through a single approval pathway, either as a drug, biologic, or device. The classification is based upon the primary mode of action of the product which determines which FDA center will have primary jurisdiction over the product (CDER for drugs, CBER for biologics, CDRH for devices). The "primary mode of action" (PMOA) of the combination product is based on section 503(g) (1) of the Act.[2] The PMOA of a combination product may not be apparent for all products. In such cases, the company is encouraged to submit a "request for designation" (RFD) to the FDA Office of Combination Products. The RFD process is outlined in 21 CFR Part 3. After submitting the RFD, the company will obtain an agency determination on which FDA center they should work with to develop the new combination product.

Definition of combination product: Combination products are defined in 21 CFR 3.2(e).[2] The term combination product includes the following:

1. A product comprising two or more regulated components, that is, drug/device, biologic/device, drug/biologic, or drug/device/biologic, that are physically, chemically, or otherwise combined or mixed and produced as a single entity.
2. Two or more separate products packaged together in a single package or as a unit and comprising drug and device products, device and biological products, or biological and drug products.
3. A drug, device, or biological product packaged separately that according to its investigational plan or proposed labeling is intended for use only with an approved individually specified drug, device, or biological product where

Translational Cardiology. https://doi.org/10.1016/B978-0-323-91790-2.00094-0

both are required to achieve the intended use, indication, or effect and where upon approval of the proposed product the labeling of the approved product would need to be changed, for example, to reflect a change in intended use, dosage form, strength, route of administration, or significant change in dose; or

4. Any investigational drug, device, or biological product packaged separately that according to its proposed labeling is for use only with another individually specified investigational drug, device, or biological product where both are required to achieve the intended use, indication, or effect.

Different types of combination product: The following table identifies nine different types of combination products.[1–11] Each product described below has a device combined with a drug or a biologic, a drug combined with a biologic, or all three (drug/device/biologic) combined.

Type	Description	Common example(s)
1	Convenience kit or copackage: drug and device are provided as individual constituent parts within the same package	Drug or biological product vials packaged with device(s) or accessory kits (empty syringes, auto-injectors, transfer sets), first aid or surgical kits containing devices and drugs
2	Prefilled drug delivery device/system: drug is filled into or otherwise combined with the device AND the sole purpose of the device is to deliver drug	Prefilled drug syringes, auto-injectors, metered-dose inhalers, dry powder inhalers, nasal spray, pumps, transdermal systems, prefilled iontophoresis systems, or microneedle "patch"
3	Prefilled biologic delivery device/system: biological product is filled into or otherwise combined with the device AND the sole purpose of the device is to deliver biological product	Vaccine or other biological product in a prefilled syringe, auto-injector, nasal spray, transdermal systems, or microneedle patch preloaded with biological product
4	Device coated/impregnated/otherwise combined with drug: device has an additional function in addition to delivering the drug	Drug pills embedded with sensors, contact lenses coated with a drug, drug-eluting stents, drug-eluting leads, condoms with spermicide, dental floss with fluoride, antimicrobial coated catheters/sutures, bone cement with antibiotics
5	Device coated or otherwise combined with biologic: device has an additional function in addition to delivering the drug	Live cells seeded on or in a device scaffold, extracorporeal column with column-bound protein
6	Drug/biologic combination	Antibody–drug conjugates, progenitor cells combined with a drug to promote homing
7	Separate products requiring cross-labeling	Light-activated drugs or biological products not copackaged but labeled for use with a specific light source device
8	Possible combination based on cross-labeling of separate products	Drug/biological product under development utilizes a device, but unclear whether the final product will require that the two be cross-labeled
9	Other type of part 3 combination product (e.g., drug/device/biological product) Combination product not otherwise described	All 3 articles are combined in a single product (e.g., a prefilled syringe containing an antibody–drug conjugate), a device to manufacture a biologic also includes a drug or biologic in the kit, or the product contains two different combination product types (e.g., type 1 and type 2 are provided together

Examples of combination products: Examples of combination products where the components are physically, chemically, or otherwise combined (see 21 CFR 3.2(e) (1)) include the following:

- A monoclonal antibody combined with a therapeutic drug
- A device coated or impregnated with a drug or biologic
  - Drug-eluting stent, pacing lead with steroid-coated tip, catheter with antimicrobial coating, condom with spermicide, transdermal patch
- Prefilled drug delivery systems (syringes, insulin injector pen, metered dose inhaler)

Examples of copackaged combination products where the components are packaged together (see 21 CFR 3.2(e) (2)) include the following:

- A drug or vaccine vial packaged with a delivery device
- A surgical tray with surgical instruments, drapes, and anesthetic or antimicrobial swabs

- First-aid kits containing devices (bandages, gauze), and drugs (antibiotic ointments, pain relievers)

Finally, an example of product that may be cross-labeled as a combination product with components that are separately provided but specifically labeled for use together includes a photosensitizing drug with an activating laser/light source (see 21 CFR 3.2(e) (3) or (e) (4)).

FDA guidance for early development of innovative combination products

FDA recommends that developers of combination products recognize that additional data and information may be necessary to address the scientific and technical issues raised by the new use of the constituent in the combination product. Issues may be raised by combining the constituent parts or by new uses for the constituent in the combination product, such as a new indication for use, a different target population, a new route of administration, or by different local or systemic exposure profiles once the products are combined. Developers are advised to consider the following questions early on[4]:

- Are the constituent parts already approved for an indication?
- Is the indication for a given constituent part similar to that proposed for the combination product?
- Does the combination product broaden the indication or intended target population beyond that of the approved constituent part?
- Does the combination product expose the patient to a new route of administration or a new local or systemic exposure profile for an existing indication?
- Is the drug formulation different than that used in the already approved drug?
- Does the device design need to be modified for the new use?
- Is the device constituent used in an area of the body that is different than its existing approval?
- Are the device and drug constituents chemically, physically, or otherwise combined into a single entity?
- Does the device function as a delivery system, a method to prepare a final dosage form, and/or does it provide active therapeutic benefit?
- Is there any other change in design or formulation that may affect the safety/effectiveness of any existing constituent part or the combination product as a whole?
- Is a marketed device being proposed for use with a drug constituent that is a new molecular entity?
- Is a marketed drug being proposed for use with a complex new device?

Market size: The market for combination products has been forecasted to exceed $177 billion by 2024.[5]

Real-world examples of the path to regulatory approval for combination products

Case study 1: Inhaled nitric oxide is a classic example of a combination product. There are a variety of ways to generate nitric oxide all of which require a delivery system (device) and the drug. Two approaches have been undertaken to develop a commercially viable inhaled nitric oxide (iNO) system. The first approach involved creating nitric oxide gas in tanks.[6,7] The tanks contained a high concentration (100−800 ppm) of nitric oxide (NO).[8] They were connected to a delivery system that controlled the dose provided to the patient. The initial delivery system delivered 20−80 ppm of gas to a pediatric patient by adding oxygen (O_2) to the nitric oxide to dilute the mixture for the treatment of primary pulmonary hypertension of the newborn (PPHN).

The second approach was quite different and creative. Rather than using tanks of nitric oxide, the system relied on taking nitrogen dioxide (NO_2) and flowing it over a cartridge coated with an antioxidant.[9] The antioxidant stripped one atom of oxygen from the NO_2 creating pure NO.

A concern with the first approach using tanks of NO was the risk of creating the formation of nitrogen dioxide as a biproduct that could be inhaled by the patient. The pathophysiology of inhaled NO_2 was reviewed by Petit et al. and was shown to have the potential for significant negative sequelae when inhaled.[10] In contrast, the second approach was specifically designed to minimize the risk of delivering NO_2 to the patient by only delivering NO with trace or no NO_2 in the mixture.

An interesting dilemma for the second approach was that the drug was manufactured at the bedside. This was a challenge for the FDA because there was no precedent for how to inspect the manufacturing process when the drug was produced in this manner. This is an example of how the creation of a combination product is fraught with challenges despite its potential advantages.

The regulatory path for inhaled NO was determined to be an NDA with the CDER serving as the lead center along with the advice of CDRH.

Obtaining funding for a combination product during the preclinical stage of the development was difficult. Neither pharmaceutical nor device companies were willing to fund the second project because it was a combination product comprised of both a device and a drug. Likewise, investors were uniformly weary of investing mainly due to the regulatory path.

Case 2: Drug-eluting stents—A key consideration in the development of drug-eluting stents was to identify the primary mode of action. A drug-eluting stent is designed to dilate diseased arteries, and therefore, the PMOA is the device's ability to open the artery.[11] The drug provides a secondary PMOA to reduce inflammation and restenosis of an artery.[11] The product was submitted through CDRH as a medical device. Because the secondary PMOA is linked to the drug component, CDER also participated as an FDA center.

Case 3: Drug-eluting disc—The PMOA for a drug-eluting disc is to provide chemotherapy for a brain tumor.[11] A secondary action is to provide localized drug delivery. In this case, the lead center was CDER with CDEH as an advisory center.

Get started

Start with establishing the core intellectual property for the product and the freedom to operate. Draft a product development plan. Identify the regulatory path before embarking on the research and development. Creating a quality system is essential. Have a manufacturing plan and involve manufacturing engineering early in the process. Define a clear exit strategy.

Pitfalls: Classic examples of failure modes include the following

- Device companies trying to be drug companies and vice versa.
- Failure to involve quality and regulatory experts who know both spaces.
- Inability to obtain adequate funding due to pharma companies that are reluctant to invest in device companies and vice versa

Conclusions

The path to commercial approval of combination products may be considered the most challenging among the various types of medical products. The combination of challenges with product development, determining the PMOA and the appropriate regulatory center, and identifying true experts with drug, biologic, and device expertise raises the complexity of getting to market. However, for those who succeed, the market opportunity can be significant and absolutely rewarding.

References

1. https://www.sciencedirect.com/topics/engineering/combination-product.
2. https://www.fda.gov/combination-products/about-combination-products/frequently-asked-questions-about-combination-products#examples.
3. https://www.fda.gov/combination-products/about-combination-products/combination-product-definition-combination-product-types.
4. https://www.fda.gov/media/75273/download.
5. https://www.globenewswire.com/news-release/2019/04/30/1812682/0/en/177-7-Billion-Drug-Device-Combination-Products-Market-by-Product-Region-Global-Forecast-to-2024.html.
6. Roberts JD, Polaner DM, Lang P, Zapol WM. Inhaled nitric oxide in persistent pulmonary hypertension of the newborn. *Lancet*. October 3, 1992;340(8823):818–819. https://doi.org/10.1016/0140-6736(92)92686-a.
7. https://pubs.asahq.org/anesthesiology/article/130/3/435/19201/Nitric-Oxide-Story.
8. https://www.accessdata.fda.gov/drugsatfda_docs/nda/99/20845_INOmax_Approv.pdf.
9. https://patents.google.com/patent/US8226916B2/en.
10. Petit PC, Fine DH, Vásquez GB, Gamero L, Slaughter MS, Dasse KA. The pathophysiology of nitrogen dioxide during inhaled nitric oxide therapy. *ASAIO J*. 2017 Jan/Feb;63(1):7–13. https://doi.org/10.1097/MAT.0000000000000425. .
11. https://cdn.ymaws.com/www.casss.org/resource/resmgr/CMC_Euro_Speaker_Slides/2016_CMCE_ChesworthTim.pdf.

Chapter 83

CMC and GxP—Cardiology

Melissa Lou Silva[1,2], Luiza G. Schmitt[3,4], Thomaz R. Mostardeiro[4,5] and Bruno B. Lima[6]

[1]Department of Radiology, University of California San Diego, San Diego, CA, United States; [2]Department of Medicine, Universidade Estadual do Ceará, Fortaleza, Ceará, Brazil; [3]Department of Radiation Oncology, University of Texas Southwestern, Dallas, TX, United States; [4]Department of Medicine, Universidade Federal de Santa Maria, Santa Maria, Rio Grande do Sul, Brazil; [5]Department of Radiology, University of Texas Southwestern, Dallas, TX, United States; [6]Division of Cardiovascular Medicine, Department of Internal Medicine, Vanderbilt University Medical Center, Nashville, TN, United States

Key points

- Following the discovery phase, processing and developing a new drug necessitate the submission of an Investigational New Drug (IND) application to the US Food and Drug Administration (FDA).
- The Chemistry, Manufacturing, and Control (CMC) section of the IND application holds paramount significance in the process of developing new drugs and securing regular authorization and approval.
- The CMC information is essential to maintain the stability and reliability of pharmaceutical products, ensuring that they are biologically equivalent to the same products used in clinical trials.
- Good practices policies and guidelines (GxP) ensure compliance of drug products with FDA guidelines.

Introductory section/why it matters

The journey to drug discovery, from preclinical testing, clinical trials to regulatory submissions, comprises several critical phases in the evolution, approval, and authorization of new drugs. The CMC section offers comprehensive details regarding the drug's chemistry, manufacturing processes, and the controls implemented to ensure its quality and consistency.

This chapter provides a broad overview of the CMC aspect within regulatory submissions, encompassing the typical inclusion of information, compliance with regulatory requirements, and the pivotal role of CMC in conjunction with the minimum standard practices mandated for pharmaceutical manufacturers throughout the drug development process.

CMC session

In the United States, securing government authorization for the use of a previously unmarketed human drug product necessitates the submission of an IND application to the FDA.[1] The CMC component within an IND application plays a decisive role, serving as a critical foundation to ensure the successful development and regulatory approval of pharmaceutical products.[2]

CMC information is an integral part of submissions like New Drug Application (NDA), Biologics License Applications, Annual Reports, and Supplements presented to the US FDA. The CMC section and its content format must strictly adhere to the FDA regulations and principles for drug approval. Pharmaceutical and biotechnology companies navigating the regulatory landscape must adhere to specific sections outlined in the Code of Federal Regulations (CFR) Title 21 CFR, including the following:

- Title 21 CFR Part 314: Applications for FDA Approval to Market a New Drug.[3]
- Title 21 CFR Part 210: Current Good Manufacturing Practice in Manufacturing, Processing, Packing, or Holding of Drugs: General.[3]
- Title 21 CFR Part 211: Current Good Manufacturing Practice for Finished Pharmaceuticals.[3]
- Title 21 CFR Part 600: Biological Products: General.[3]

Within an IND application, the CMC section contains detailed information about different aspects. It is imperative to submit enough information to guarantee the proper identification, quality, purity, and strength of the investigational drug throughout the life cycle of an IND application and across each phase of the clinical investigation (Phase 1 for safety, proof of concept, and exploration; Phase 2 for safety, efficacy, and dose-ranging; and Phase 3 for efficacy confirmation, dose-ranging, and safety).[4] By the CFR Title 21 CFR Part 312.23(a) (7) (i), it is acknowledged that the specific information for assuring the proper attributes of the

Translational Cardiology. https://doi.org/10.1016/B978-0-323-91790-2.00093-9

investigational drug will vary with factors such as the investigative phase, proposed duration, dosage form, and available information.[3]

As the clinical development progresses, sponsors are strongly urged to engage with the CMC review team, fostering discussions on any issues that might impact the acceptability of the NDA. Ensuring the quality of drugs is essential for many reasons. Firstly, it is crucial to guarantee that the marketed drug possesses features similar to those demonstrated as safe and effective. Secondly, maintaining stringent quality standards throughout the production process is essential to meet regulatory requirements. Third, confirming that the drug used aligns with its labeled description prevents potential safety issues or adverse effects. Remarkable changes, such as alterations in sterilization methods, container closure systems impacting product quality, or shifts from synthetic to biological sources of a drug substance, necessitate prompt amendments under the same IND without a 30-day waiting period. Conversely, non–safety-related CMC changes or updates can be reported in annual reports, ensuring that the FDA stays informed of significant developments promptly. This dedication to quality ensures the protection of public health and safety, preserving the integrity of the pharmaceutical industry.

Conforming to Title 21 CFR Part 312, the components of an IND application include the CMC for:[5]

- Drug Description: includes identification, quality, purity, and strength of the investigational drug.
- Drug Substance: Details about the composition, characteristics, and properties of the active pharmaceutical ingredient.
- Drug Product: Information encompassing the formulation, composition, and manufacturing processes of the final pharmaceutical product.
- Placebo Formulation (if applicable): Details regarding any placebo used in the study, including its composition and manufacturing.
- Labeling Information: Pertinent details about product labeling, ensuring clarity and accuracy in conveying information to end-users.
- Environmental Analysis: Assessment of the environmental impact of the investigational drug or biological product, highlighting potential effects on the surroundings.

CMC section components of an IND application are illustrated in Fig. 83.1. Compliance to such regulatory standards is paramount for the successful approval of drug applications and the subsequent marketing of products in the United States pharmaceutical market. The CMC timeline corresponding to the IND is depicted in Fig. 83.2, outlining the strategic integration of CMC policies throughout the investigational process.

IND and NDA[6]

When preparing a drug for regulatory approval, there are several differences between an IND application and an NDA. A few of them are succinctly summarized below:

- Not all phases of an IND submission require "application" of the International Council for Harmonisation Quality Guidelines. For the characterization and Pharmaceutical Development Information of the Drug Substance (DS), some data to support the proposed structure in early IND phases may be sufficient, whereas full characterization is required for an NDA.
- The specifications for the Drug Substance and Drug Product in IND submissions may be based on tentative acceptance criteria, such as safety levels of solvents from a few small IND arrays, whereas multiple pilot

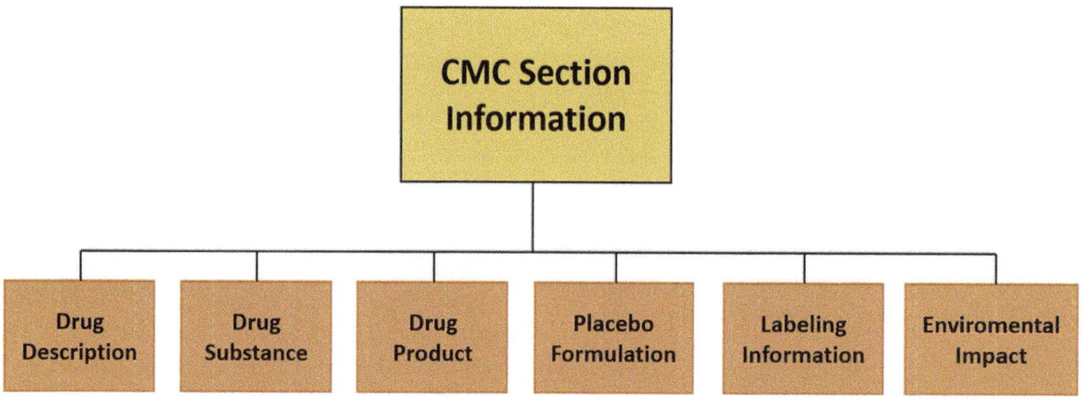

FIGURE 83.1 A illustrative diagram demonstrating the CMC section components of an IND application.

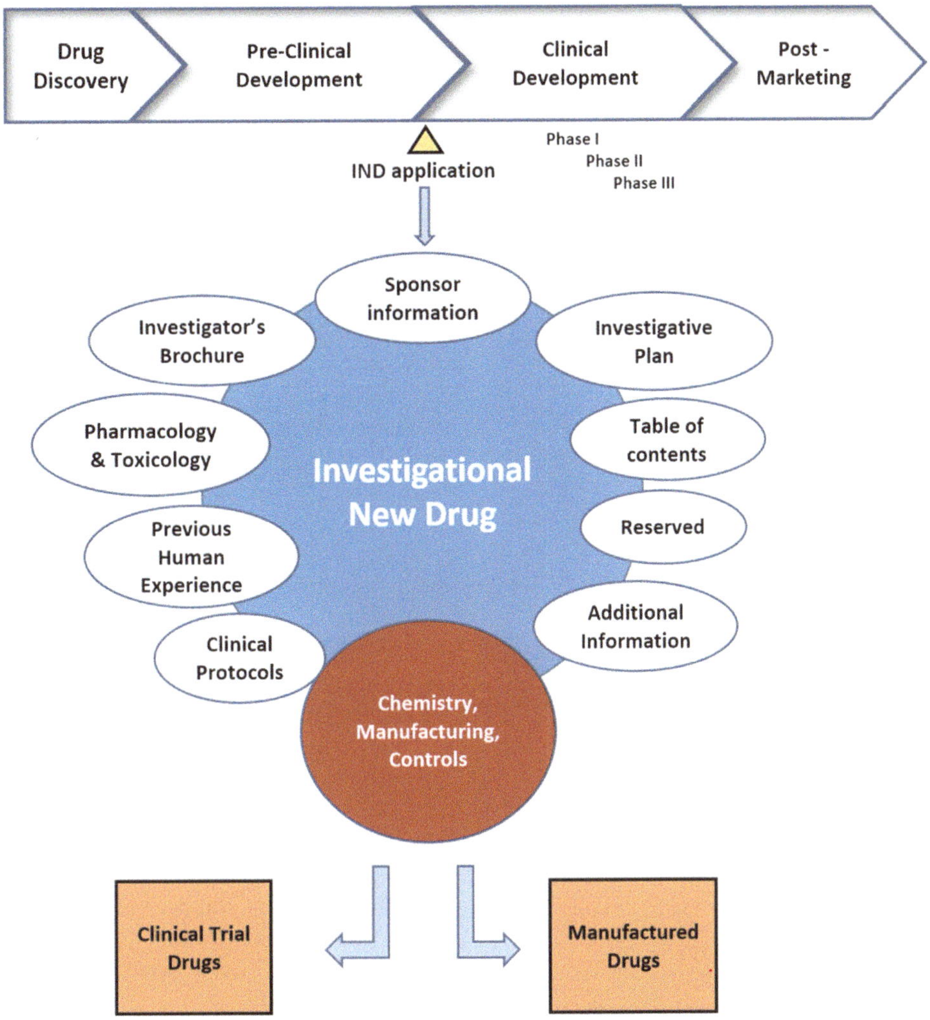

FIGURE 83.2 A schematic figure demonstrating the role of CMC practices in maintaining the fidelity of manufactured drugs to clinical trial drugs, ensuring the effectiveness and safety of the drug products.

or full-scale arrays and statistical analysis are required for an NDA.

- Analytical procedures in IND submissions may only require scientifically sound methods without full validation, whereas full validation is necessary for an NDA.
- Regarding impurities, identification is necessary for both IND and NDA submissions, while identification plus qualification are required for NDA submissions.
- Requirements vary in the process validation. An IND does not necessitate process validation, while it can be completed after NDA approval. Stability protocols for an IND's Phase 1 do not need a detailed protocol, unlike for an NDA. Stability data and shelf-life requirements differ as well.
- Throughout the IND phase, the data are gathered to support the clinical studies' duration, whereas in an NDA, it is to support the marketed drug.

CMC section challenges[5]

- Insufficient information and assessment-related subject risk.
- The potential risks associated with the transition from clinical trials to the larger market production may lead to changes in manufacturing, impacting drug quality, its physical and chemical properties, as well as its purity.[7]
- CMC lacks data related to maintaining drug stability throughout the developmental process.
- Challenges in providing comprehensive data supporting the "comparability of toxicology and proposed clinical lots".
- Insufficient information regarding raw materials of animal origin.
- Impurity may lead to harm and safety concerns

- Insufficient information on product transfer policies between manufacturing or packaging sites which may result in approval delays.

Good Practices

Good Practice Guidelines and Regulations, often abbreviated as GxP, constitute a set of established guidelines and regulations. Their purpose is to guarantee that biopharmaceutical products maintain consistent high quality during manufacturing, control, storage, and distribution, are suitable for their intended use, and comply with the requirements of marketing authorization or clinical trial authorization.[8,9]

The FDA has outlined GxP practices, encompassing various regulations. The "G" stands for "Good" and "P" stands for "Practice". The "x" varies depending on the application. It can be "M" for "Manufacturing", as in "Good Manufacturing Practices", "D" for "Distribution", "R" for "Review", "C" for "Clinical", "S" for "Storage", "L" for "Laboratory", etc. Sometimes, a "c" is found in front of the abbreviation. This simply signifies that the present version of the guidelines is the latest one. For example, "cGCP" is an abbreviation for "Current Good Clinical Practices".[10]

GxP regulations include those requirements outlined in the US FDA CFR Title 21 Part 11 and EudraLex Volume 4dGMP Guidelines.[11] In brief, critical components are listed below:[12]

- Strict procedures to guarantee quality control of pharmaceuticals.
- Precise and current documentation of manufacturing processes in the pharmaceutical industry.
- Authentication of the purity, potency, and quality of every component involved in the pharmaceutical manufacturing process.
- Dependable storage practices for pharmaceutical products.

Good Manufacturing Practice (GMP) regulations, which describe the minimum standards that a pharmaceutical manufacturer must meet in its production processes, are critical components of GxP practices, particularly when manufacturing processes shift from small to large scale.[7] Strict adherence to GMP regulations set by the FDA ensures that the drug series made in the large-scale clinical setting are the same drugs used in smaller series during preclinical settings. FDA also intended to help ensure that Active Pharmaceutical Ingredients meet the quality and purity characteristics that they purport or are represented to possess.[13]

When a manufacturer applies for approval to market a new or generic drug, the FDA reviews the application and assesses the manufacturer's compliance with the cGMP regulations.[14] Before approving the CMC application, the manufacturer's compliance with GMP regulations must be reviewed. Drug manufacturing facilities should take this into account to ensure that every aspect of production, including starting materials, premises, equipment, staff qualifications, and personal hygiene, adheres to a set of manufacturing standards.[14]

Crucially, it is recommended to establish specific recording systems to illustrate compliance with these procedures.[15] These regulations ensure that a product is safe for use and contains the ingredients and strength it claims to have.

Furthermore, it is important to note that there is no single regulatory entity or administration overseeing drug manufacturing guidelines and regulations. Each country has its own guidelines and regulators, although the requirements are similar across countries.

Although the goals of CMC and GxP may overlap, CMC is specifically concerned with the development and production of pharmaceutical drug products, while GxP encompasses the manufacturer's practices, drug distribution, and storage (GMP, GDP, and GSP, respectively). Fig. 83.3 illustrates the interface between CMC and GxP.

Summary

CMC and the GxP processes are critical elements in a successful IND application. The GxP has several components (GCP, GDP, GLP, GMP, and GSP) to secure approval for the IND application. Prompt submission of any changes to the drug manufacturing process is essential to anticipate additional delays in the approval of the CMC application. Comprehensive documentation and data related to the drug product's identity, safety, stability, and manufacturing compliance guarantee the security and consistent production of pharmaceuticals.

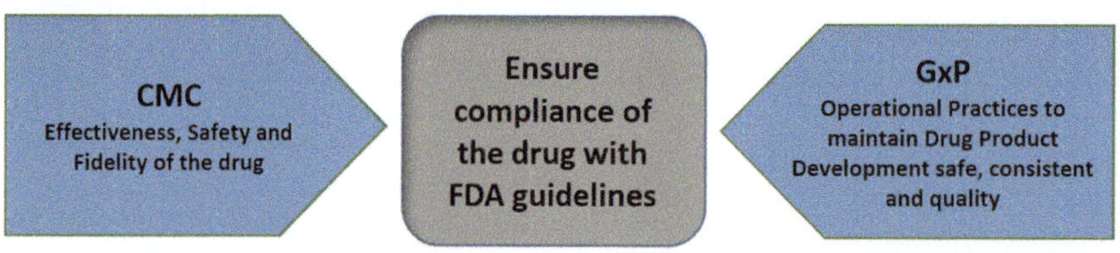

FIGURE 83.3 A schematic diagram demonstrating the interface between CMC and GxP with both promoting safe, consistent and quality drug production.

Get started

1. Review all the necessary components of the CMC application.
2. Collect all documentation and data supporting the identity, safety, and stability of the drug product.
3. Ensure all required documentation is in place, including a copy of the FDA's GxP inspection (GMP inspection documents for all manufacturing or packaging sites used).
4. Submit the CMC and GxP information as part of the IND application.
5. Respond promptly to amendment requests to enhance the likelihood of application approval.

Potential pitfalls

- Insufficient information provided to delineate the drug's safety and stability.
- Failure to submit timely or adequate information regarding changes in the manufacturing process and sites as an amendment to the IND.
- Incomplete batch records and stability data after utilizing an alternate manufacturing process.
- Outdated GxP inspections for all sites involved in the drug's manufacturing process.

Resources

- US Food and Drug Administration: CFR, Code of Federal Regulations, Title 21—Food and Drugs; Part 211—Current Good Manufacturing Practice for Finished Pharmaceuticals.
- US Food and Drug Administration: CFR, Code of Federal Regulations, Title 21—Food and Drugs; Part 312.

References

1. Chemistry, Manufacturing and Controls (CMC). *Revive*. https://revive.gardp.org/resource/chemistry-manufacturing-and-controls-cmc/?cf1%E2%81%844encyclopaedia.
2. Understanding Chemistry, Manufacturing and Controls (CMC). *Element*; 2022. Retrieved November 23, 2023, from https://www.element.com/nucleus/2022/understanding-chemistry-manufacturing-and-controls.
3. CFR. *Code of Federal Regulations Title 21*. FDA; 2023. Retrieved November 23, 2023, from https://www.accessdata.fda.gov/scripts/cdrh/cfdocs/cfcfr/CFRSearch.cfm?CFRPart=314&showFR=1.
4. *IND Applications for Clinical Investigations: Chemistry, Manufacturing, and Control (CMC) Information*: FDA. https://www.fda.gov/drugs/investigational-new-drug-ind-application/ind-applications-clinical-investigations-chemistry-manufacturing-and-control-cmc-information.
5. FDA. *Code of Federal Regulations Title 21*. FDA US Food & Drug Administration CFR; 2020.
6. *Types of Applications*. https://www.fda.gov/drugs/how-drugs-are-developed-and-approved/types-applications.
7. DiFeo TJ. Safety and efficacy: the role of chemistry, manufacturing, and controls in pharmaceutical drug development. *Drug Dev Ind Pharm*. 2004;30(3):247–257. https://doi.org/10.1081/DDC-120030428.
8. *Guidance for Industry, Q7A Good Manufacturing Practice Guidance for Active Pharmaceutical Ingredients*: FDA. https://www.fda.gov/regulatory-information/search-fda-guidance-documents/guidance-industry-q7a-good-manufacturing-practice-guidance-active-pharmaceutical-ingredients.
9. GxP Regulations for Healthcare. *ClearData*. https://www.cleardata.com/platform-services/gxp-regulations/.
10. Lauwers R. *What is GxP?*; 2018. Retrieved November 23, 2023, from https://www.aodocs.com/blog/what-is-gxp.
11. *Current Good Manufacturing Practice (CGMP) Regulations*: FDA. https://www.fda.gov/drugs/pharmaceutical-quality-resources/current-good-manufacturing-practice-cgmp-regulations.
12. *Good Clinical, Laboratory, and Manufacturing Practices (GxP)*; 2024. Retrieved November 23, 2023, from https://docs.microsoft.com/en-us/compliance/regulatory/offering-gxp.
13. *Good Manufacturing Practice*: European Medicine Agency. https://www.ema.europa.eu/en/human-regulatory/research-development/compliance/good-manufacturing-practice.
14. Dutton G. *cGMP Issues Are Increasing in Pharma*; 2014. Retrieved November 23, 2023, from https://www.lifescienceleader.com/doc/cgmp-issues-are-increasing-in-pharma-0001.
15. DiFeo TJ. Drug product development: a technical review of chemistry, manufacturing, and controls information for the support of pharmaceutical compound licensing activities. *Drug Dev Ind Pharm*. 2003;29(9):939–958. https://doi.org/10.1081/DDC-120025452.

Chapter 84

Non-US regulatory

Guileonne Ngnitewe[1] and Alysa Vereen[2]

[1]*Regulatory Promotional Review ProPharma Group, Madison, WI, United States;* [2]*Priovant Therapeutics, Durham, NC, United States*

Key points

- To provide access to patients worldwide and also to yield market revenue, understanding non-US regulatory affairs is essential.
- Navigating the regulatory requirements in different regions is a must for patients worldwide to have access to novel therapeutics.
- Harmonization of the global regulatory authorities can ensure that safe, effective, and high-quality medicines are developed and registered in the most resource-efficient manner

Introduction/why it matters

Translational research is largely positioned around developing innovative scientific products into meaningful technologies to benefit patients worldwide. In order to market these therapies and make them accessible globally, getting regulatory approval with a country's regulatory agency is the stepping stone to providing that access. When it comes to the global pharmaceutical market revenue from 2010 to 2022, US sales have increased year-over-year and have reached roughly 50% in recent years, making the rest of the world's market revenue (China, Europe, Japan, Latin America, Africa/Asia/Australia) totaling about 50% combined. In order to provide access to patients worldwide and also to yield market revenue, understanding non-US regulatory affairs is essential.

North America (Canada)

In Canada, the regulatory authority that provides oversight on drug and health products (including natural products), medical devices, food, and other health products is known as Health Canada. Health Canada sets rules that are set out in laws and regulations for a broad range of products sold in Canada, including their product use, import, labeling, packaging, advertising, manufacturing, and environmental safety. To assist with implementing their laws and regulations, Health Canada maintains regular communication with businesses and has guidance documents available to help companies clearly understand rules and responsibilities.

China

China's regulatory body is known as the National Medical Products Administration (NMPA), and it is responsible for regulating drugs (including traditional Chinese medicines and ethno-centric medicines), medical devices, and cosmetics in China. Of note, the NMPA was previous known as the China State Food and Drug Administration until it was renamed in 2018. renamed in 2018 to NMPA. The NMPA develops the registration system and conducts strict review and approval for the marketing of these products in China.

Europe

In Europe, the regulatory body that governs countries in the European Union (EU) is known as the European Medicines Agency (EMA). The EMA's mission is to foster scientific excellence in the evaluation and supervision of medicines, for the benefit of public and animal health in the EU.

It's important to note that despite the EMA's involvement in regulatory affairs in the EU, decisions such as evaluating clinical trials, along with the evaluation of medical devices, food, and cosmetics, are decided by each individual country. Also, the EMA does not evaluate the initial marketing authorization of all medicines in the EU, as the majority of medicines available in the EU are authorized by each country.

Like the EU, the European Free Trade Association (EFTA) is an intergovernmental organization in Europe, and it consists of Iceland, Liechtenstein, Norway, and Switzerland. EFTA's mission is to promote free trade and

Translational Cardiology. https://doi.org/10.1016/B978-0-323-91790-2.00028-9

economic integration to the benefit of its four countries, and to benefit their trading partners around the globe. Each country has its own regulatory agency.

The regulatory authorities responsible for overseeing drugs and medical devices vary by country within the EU and EFTA. In the EU, countries like Austria, Belgium, and Bulgaria are overseen by their respective Federal Offices for Safety in Health Care and Drug Agencies for drugs, while the regulation of medical devices falls under similar agencies. Similarly, countries in the EFTA, such as Iceland and Norway, have their own Medicines Agencies for drug regulation and Medical Products Agencies for medical devices. For countries outside these unions, like Albania and Serbia, regulatory bodies such as the Agency for Medicinal Product and Medical Devices oversee both drugs and medical devices.

Among all the countries in the Europe region, the top country for pharma production in 2021 was Switzerland, followed by France, Italy, Germany, and the United Kingdom.

Japan

The Japanese regulatory agency is the Pharmaceuticals and Medical Devices Agency (PMDA), which works alongside its Ministry of Health, Labour and Welfare. The PMDA's mission is to protect the public health by assuring safety, efficacy, and quality of pharmaceuticals and medical devices in Japan. In addition, they conduct scientific reviews of marketing authorization application of pharmaceuticals and medical devices, along with monitoring their post-marketing safety. Of note, the PMDA is responsible for compensating people who have had adverse drug reactions and infections by pharmaceuticals or biological products.

Africa

Every country in Africa, except for the Sahrawi Republic, has a National Medicines Regulatory Authority, but only 7% have a moderately developed capacity, while over 90% have minimal to no capacity. The African Medicines Agency (AMA) is a proposed specialized regulatory agency to bridge this gap. Similar to the EMA, the AMA is intended to facilitate the harmonization of medical regulation throughout the AU. The proposed AMA is intended to be the regulatory body for medicines, traditional medicines, and medical devices in the AU.

Asia and Australia

In East Asia, regulatory oversight for drugs and medical devices is conducted by various agencies. China is regulated by the NMPA for both drugs and medical devices, while Japan operates under the PMDA. South Korea is overseen by the Ministry of Food and Drug Safety for both categories, while Taiwan's Food and Drug Administration (FDA) manages regulations. In South Asia, India's Central Drugs Standard Control Organization handles both drugs and medical devices, as does the General Directorate of Pharmaceutical Affairs in Afghanistan. In Southeast Asia, countries like Malaysia have the National Pharmaceutical Regulatory Agency for drugs and the Medical Device Authority for medical devices. Western Asia and the Middle East see a similar trend, with countries like Israel and Turkey regulating both drugs and medical devices through their respective health ministries. North and Central Asia, as well as Oceania, follow similar patterns with specialized agencies overseeing both drugs and medical devices in most cases. Each regulatory authority ensures the safety, efficacy, and quality of pharmaceuticals and medical devices within its jurisdiction.

In Oceania, Australia's Therapeutic Goods Administration oversees the regulation of both drugs and medical devices, ensuring they meet stringent safety and efficacy standards before being marketed. Similarly, New Zealand's Medicines and Medical Devices Safety Authority plays a vital role in regulating pharmaceuticals and medical devices to safeguard public health. These agencies conduct thorough evaluations of products, assess their risks and benefits, and monitor their safety throughout their lifecycle. Additionally, Papua New Guinea's Pharmaceutical Services Standards Branch, under the National Department of Health, is responsible for ensuring the quality and safety of pharmaceuticals and medical devices within its jurisdiction.

Latin America

In the Latin America and Caribbean (LAC) region, regulatory reliance, the act whereby a national regulatory authority in one jurisdiction may take into account and give significant weight to assessments performed by another national regulatory authority or trusted institution, often takes precedence. Regulatory documents regarding marketing authorization were found in 20 LAC regulatory's websites, covering 34 countries, and of those, 7 countries do not accept regulatory reliance, 13 authorities (Argentina, Colombia, Costa Rica, Dominican Republic, Ecuador, El Salvador, Guatemala, Mexico, Panama, Paraguay, Peru, Uruguay, and the unique Caribbean Regulatory System for 15 Caribbean States) rely on marketing authorizations issued by the EMA, FDA, and Health Canada. 10 countries in the LAC region rely on marketing authorizations from Australia, Japan, and Switzerland. Some key countries' regulatory authorities—Argentina's Argentinian Health Authority (ANMAT), Brazil's Brazilian Health Authority (Anvisa), Chile's Chilean Public Health Institute (ISP), and Mexico's Mexico Ministry of Health (COFEPRIS)—are reference authorities for 8 LAC regulators.

Similar to Africa, the Latin American and Caribbean Medicines Agency (AMLAC) is a proposed regional regulatory organization. The goal of AMLAC is to improve the regulation, quality, safety and efficacy of medicines in Latin America and the Caribbean, in order to strengthen the public health capacities of the region.

International Council for Harmonisation of technical requirements for pharmaceuticals for human use

The International Council for Harmonisation of Technical Requirements for Pharmaceuticals for Human Use (*ICH*) is an organization that exists to bring together global regulatory authorities and the pharmaceutical industry to discuss the scientific and technical aspects of drug registration. Its mission is to achieve great harmonization worldwide to ensure that safe, effective, and high-quality medicines are developed and registered in the most resource-efficient manner. ICH achieves its mission through the creation of ICH guidelines, developed by scientific consensus with regulatory and industry experts. Europe, the United States and Japan are the founding regulatory members of ICH.

Get started

- In your biotechnology program:
  - What programs are in place to mitigate risk when seeking access to your therapeutic worldwide?
  - Have you sought out a regulatory affairs professional?
  - Will you prioritize regions for having access to your therapeutic? If so, why?

Potential pitfalls

- Different regions have varying *regulatory requirements*, making global approval a complex task. Regulatory professionals must navigate these differences effectively.
- Maintaining accurate and complete data is vital for regulatory compliance in any region. Be sure to stay organized from the beginning.
- Regulatory teams often work with pricing and market access teams to strategize drug pricing and reimbursement strategies; although thinking of finance is important, the importance of getting access to patients should always come first.
- Continuous training of a regulatory affairs profession is essential to stay updated with evolving regulations globally.
- Do not assume that just because one region has approved a product that other region(s) will follow.

Examples

- An example of the importance of harmonization in non-US regulatory affairs is how a recent report shows that over 85% of *medical technology companies* lack consistent methods to share product information across global regulatory systems. This inability to share documents globally means companies cannot ensure the reliability or accuracy of product information across markets, which can hinder patient access to these therapies. Harmonization could be a potential solution to resolve this.
- Another example that applies non-US regulatory and the importance of companies seeking approval in multiple markets is that it can lead to global innovation. For instance, *Hong Kong* has plans to boost its research and development capabilities and is considering setting up its own drug regulatory approval system, in order to attract pharmaceutical companies.
- Just because a company is able to get regulatory approval in one region, it does not mean that other regions will follow. One example of this is a *rare disease* drug being approved in the United States, but not being approved in Europe.

Resources (additional reading, consultants, contractors)

References

1. Doerr P, Valentin M, Nakashima N, et al. Reliance: a smarter way of regulating medical products - the IPRP survey. *Expet Rev Clin Pharmacol.* 2020;14(2):173–177. https://doi.org/10.1080/17512433.2021.1865798.
2. Definition of Harmonization. Simpla. www.simpla-project.eu. http://www.simpla-project.eu/en/guidelines/introduction-and-problem-setting/definition-of-harmonization/#:~:text=Harmonization%20means%20working%20on%20those.
3. *Registration and Marketing Authorization (MA): Indicators and Fact Sheets.* www.who.int; 2018. https://www.who.int/publications/m/item/02-gbt-ma-rev-vi-ver-1nov2018-final-adjusted.
4. What is Regulatory Affairs? TOPRA. www.topra.org. https://www.topra.org/topra/topra_member/what_is_regulatory_affairs.aspx.
5. What is Public Health Authority FAQ: Act for Public Health. Accessed 30 November 2023. https://actforpublichealth.org/public-health-authority/#:~:text=Public%20health%20authority%20refers%20to.
6. *Global Pharma Market Revenue by Regional Share 2010–2022.* Statista; 2024. https://www.statista.com/statistics/275535/distribution-of-global-pharmaceutical-market-revenue/#:~:text=Global%20pharmaceutical%20market%20%2D%20revenue%20distribution%202010%2D2022%2C%20by%20region&text=In%202022%2C%20the%20North%20American. Accessed November 30, 2023.
7. Canada H. What health Canada does as a regulator. Aem. https://www.canada.ca/en/health-canada/corporate/mandate/regulatory-role/what-health-canada-does-as-regulator.html; April 4, 2016.
8. *Main Responsibilities of the National Medical Products Administration.* english.nmpa.gov.cn; 2019. http://english.nmpa.gov.cn/2019-07/18/c_377587.htm.

9. *The Essential List of Regulatory Authorities in Europe*; 2020. www.raps.org. https://www.raps.org/news-and-articles/news-articles/2020/1/the-essential-list-of-regulatory-authorities-in-eu.
10. European Medicines Agency. *What We Do - European Medicines Agency*. European Medicines Agency; August 2019. Published https://www.ema.europa.eu/en/about-us/what-we-do.
11. EFTA. *The European Free Trade Association | European Free Trade Association*; 2016. Efta.int. Published https://www.efta.int/about-efta/european-free-trade-association.
12. Top 5 Countries for Pharma Production in Europe: PharmaBoardroom. Accessed November 30, 2023. https://pharmaboardroom.com/articles/top-5-pharma-producers-in-europe/.
13. *The Pharmaceutical Industry in Figures 2021*; 2021. www.efpia.eu. https://www.efpia.eu/publications/downloads/efpia/the-pharmaceutical-industry-in-figures-2021/.
14. Outline of PMDA. *Pharmaceuticals and Medical Devices Agency*; 2004. www.pmda.go.jp. https://www.pmda.go.jp/english/about-pmda/outline/0005.html.
15. Ncube BM, Dube A, Ward K. Establishment of the African Medicines Agency: progress, challenges and regulatory readiness. *Journal of Pharmaceutical Policy and Practice*. 2021;14(1). https://doi.org/10.1186/s40545-020-00281-9.
16. *The Essential List of Regulatory Authorities in Asia*. Raps.org; 2023. Published https://www.raps.org/News-and-Articles/News-Articles/2015/4/The-Essential-List-of-Regulatory-Authorities-in-As. Accessed November 30, 2023.
17. Durán CE, Cañás M, Urtasun MA, et al. Regulatory reliance to approve new medicinal products in Latin American and Caribbean countries. *Rev Panam Salud Públic*. 2021:1–10. https://doi.org/10.26633/rpsp.2021.10.
18. Exteriores S de R. *Colombia, Cuba and Mexico advance the Latin American and Caribbean Medicines Agency initiative*. gob.mx; 2023. https://www.gob.mx/sre/prensa/colombia-cuba-and-mexico-advance-the-latin-american-and-caribbean-medicines-agency-initiative?idiom=en. Accessed November 30, 2023.
19. ICH. *ICH Official Web Site: ICH*. Ich.org; 2019. Published https://www.ich.org/page/mission.
20. *Overview of ICH International Council for Harmonisation of Technical Requirements for Pharmaceuticals for Human Use*; 2023. https://admin.ich.org/sites/default/files/2023-11/OverviewOfICH_2023_1115.pdf. Accessed November 30, 2023.
21. Elopre JL. Navigating the Complex World of Global Regulatory Affairs in Oncology. *Proventa International*; September 1, 2023. https://proventainternational.com/navigating-the-complex-world-of-global-regulatory-affairs-in-oncology/. Accessed November 30, 2023.
22. Exclusive. Hong Kong can attract pharmaceutical giants by "having own drug regulatory system." *South China Morning Post*; June 30, 2023. https://www.scmp.com/news/hong-kong/health-environment/article/3225997/hong-kongs-pandemic-fight-shows-need-boost-rd-health-minister-says-outlining-plans-turn-city. Accessed November 30, 2023.
23. New Research finds 87% of Medtechs Lack Consistent Methods to Share Product Information Across Global Regulatory Teams. Yahoo! Finance. Accessed November 30, 2023. https://finance.yahoo.com/news/research-finds-87-medtechs-lack-120300507.html?guce_referrer=aHR0cHM6Ly93d3cuZ29vZ2xlLmNvbS8&guce_referrer_sig=AQAAAAxU-bquYktACqKm2s_bLbpG6yWQBPIepq0fyTMAjaoX6-GujJRLNI9Vxxjccy_4EI232E1R6-Ek0h4u103gURpV8A9BCiMAxvlHcGzmbhEhBNp3x1E06DcMsZ6q3zjeC1OUer1xAUpHD9tTa8seHCjsdy1HsbtMgSUlbnQWNtgV&guccounter=1.
24. Waldron J. *Ipsen still out of regulatory luck as European Commission Rejects Rare Disease Drug*. Fierce Biotech; 2023. Available at: https://www.fiercebiotech.com/biotech/ipsen-still-out-regulatory-luck-european-commission-rejects-rare-disease-drug-approval. Accessed November 30, 2023.

Chapter 85

Postmarket drug safety monitoring

Precious Idorenyin Anthony
All Saints University School of Medicine, Roseau, Dominica

Key points

- Postmarket drug safety surveillance is necessary to detect any adverse effect associated with drug use that was not evident in premarket studies and testing.
- In the United States, postmarket surveillance of drugs is carried out by the FDA through the use of databases that collect information on drug safety.
- The Sentinel Initiative is a database launched by the FDA to collect information from consumers, healthcare workers, and manufacturers on drug and product safety.

Introductory section/why it matters

Postmarket drug safety surveillance is a practice that is of great necessity due to the limited information available on drug safety attained from premarket testing and trials. Some side effects of drugs as well as their adverse cross reactivity with other drugs and substances are discovered after the drug has been released to the market. In addition to this, there is the possibility of contamination and mislabeling of drugs by manufacturers and distributors alike. All these factors confirm the need for continual monitoring of drugs after their release. Drug safety monitoring is carried out in the United States by the Food and Drug Administration, FDA. The FDA collects information from consumers, healthcare workers, and companies involved in the manufacturing, processing, and distribution of medications. They investigate, collate, and publish data for access to the public, ensuring drug safety. The system used to collect data is called The FDA Adverse Event Reporting System (FAERS). The information that is gathered and proven can then be published to an FDA-established database called Sentinel. The FDA also publishes information on public recalls of drugs, voluntarily made by manufacturers. This system allows the FDA to ensure that the drugs that remain in circulation and are accessible to the public are safe for use.

Introduction

Postmarket safety monitoring of drugs involves the continual review of medications already released on the market to ensure the safety of the medications for consumer use. The necessity for this process stems from the fact that at initial production and approval for the usage of new medications, the available knowledge of drug safety is limited to drug testing and clinical trials.[1] Side effects that were not discovered during these phases of drug testing could still develop over time as consumers begin to use these medications. There could also be issues with labeling and contamination that occur during the processing and distribution of these drugs that could potentially increase the risk of adverse effects. There are various regulatory bodies in different countries responsible for the postmarket safety monitoring of drugs. These bodies are responsible for communicating arising drug safety issues to the professionals dispensing these medications and to the consumers. The regulatory bodies can also respond to new evidence received on the dangers of distributed medicines in a variety of ways including updating product information, issuing prescription guidelines and black box warnings, and in some cases, taking the drug off the market completely.[1]

The need for the continual monitoring of the safety of drugs after release to the markets is emphasized by cases of the discovery of adverse effects of medications after the release of drugs into the market.

Data collection

The United States Food and Drug Administration, FDA is tasked with the responsibility of the regulation of postmarket drug safety in the country. To do this, the FDA has established systems to support the ease of reports of adverse events associated with drug use. The FDA collects reports on the adverse effects of drugs directly from the consumers, healthcare workers, and manufacturers of these products.

Translational Cardiology. https://doi.org/10.1016/B978-0-323-91790-2.00035-6

Manufacturers are mandated to report any observed adverse effect of their products and to forward any reports that have been made to them by the consumers and distributors of their products. To organize the reports, the FDA created a database called the FAERS. The database collects data on the adverse effects of drugs that come from healthcare workers, manufacturers, and consumers.[2] In addition to the reports made on the FAERS database, the FDA also collects data from the Vaccine Adverse Event Reporting System (VAERS), Manufacturer submitted safety reports, Medical literature, and even drug utilization databases.[3] The information submitted is then investigated by the FDA with the help of clinical studies to verify the data. If an adverse effect is deemed to be existent, the FDA may then implement directives to ensure public safety.[4] To ensure the proper surveillance of drugs released to the market, the FDA mandates certain bodies to submit postmarketing safety information to the FDA. These bodies can be categorized as those associated with drugs submitted for approval and those associated with unapproved drugs. To this effect, companies responsible for the production of a drug that has been submitted for FDA approval along with the manufacturing, and distribution teams of these drugs are mandated to submit postmarketing safety information. This is also required of marketers and label mentioned distributors of any unapproved prescription or over the counter drugs.[5]

Active surveillance

In 2022, the FDA implemented the Postmarketing Adverse Drug Experience (PADE) Inspections Compliance Program For Human Drug And Therapeutic Biological Products. Under this, the FDA requires continuous surveillance and documentation of every Approved New Drug Application, Abbreviated New Drug Application, Biologies License Application, and Combination Product. This documentation is submitted in the form of reports. One of these reports includes Individual Case Safety Reports that document an individual or patient specific adverse event that occurs due to the use of medication.[6] These reports should include details such as the patient's name, the medication that induced the adverse effect, the specific adverse effect or report of death, and an identifiable reporter.[5] The reporting entities are required to evaluate the submitted complaint of adverse effects, categorizing them as either: spontaneous or solicited (not spontaneously reported), serious or nonserious based on the outcome of the adverse effect, and expected or unexpected based on the warnings on the current US label. The FDA requires the timely submission of periodic and annual medication reports for review. Reports like this would contain any complaints of adverse effects submitted during the stipulated period. These entities are tasked with the responsibility of evaluating any safety concerns raised against their products These measures ensure the continual postmarket surveillance by the FDA for all medications. Reports of adverse events are expected to be made by the entities to the FDA no later than 15 days after receiving the reports. If reports are received later than the stipulated time, the entities are required to provide justifications.[6]

FDA's Sentinel Initiative

The FDA launched a database that could access multiple data sources to retrieve and accumulate information on drug safety data in 2008, called the Sentinel Initiative. This initiative birthed the launch of the full Sentinel System which has been up and running since 2016. It is currently the largest multisite database for sharing medical product safety.[7] The FDA is committed to the continuous expansion of the system and put forth plans in September 2019 to expand the system on three different fronts. These fronts would manifest as 3 separate operating centers. These are outlined in Table 85.1.

The goal is for the FDA to create an efficient source of data concerning drug and medical product safety, allowing this information to be used to ensure public safety and for further medical research.

TABLE 85.1 Sentinel district coordinating centers.

Sentinel centers	Purpose
The Sentinel Operations Center	- This would be responsible for developing partnerships with organizations involved in health, pharmacy, medicine, and statistics to leverage these connections to support the goal of ensuring product safety.
The Innovation Center	- Tasked with the responsibility of creating and implementing better and more efficient ways to gather and protect pertinent information from electronic medical records.
The Community Building and Outreach Center	- This would be responsible for disseminating information to relevant parties such as stakeholders and would provide information on the proper use of the Sentinel System and data.[7]

Manufacturer inspections

Under the Compliance Program, the FDA also conducts inspections of the manufacturing entities of products that would be accessed on the markets. The inspections are conducted to ensure that these corporations are compliant with FDA stipulated rules of conduct. Inspections could be prescheduled or unannounced. Inspections can be carried out to determine if FDA approval applicants and non-applicants have documentation stipulating their process of the surveillance, receipt, evaluation, and reporting of postmarketing safety information. This is required by the FDA, with laid out guidelines to evaluate any raised safety concern.[6] The different types of FDA inspections are summarized in Table 85.2.

Supplemental applications

These are required from manufacturers who wish to alter any information previously included in their New Drug Application. As postmarket drug safety surveillance yields new information about a medication, manufacturers may wish to alter certain characteristics of the drug such as formulation, dosage, or labeling to reflect the information gathered. These changes would need to be approved by the FDA and would be submitted in the supplemental applications.[9]

Drug advertisements

Another measure that the FDA undertakes to ensure post-market drug safety is the regulation of drug advertisements. While over the counter drug ads are regulated by the Federal Trade Commission, the FDA is responsible for the regulation of prescription drug ads. The FDA requirements for prescription drug ads depend on the type of advertisements. Types of drug ads that are approved by the FDA are summarized in Table 85.3.

Generic drugs

The regulation of generic drugs is another protective measure undertaken by the FDA to ensure postmarket drug safety. The FDA issues a patent to approved drugs allowing the manufacturer the sole right to produce and market this drug. Once the patent time period is up, other manufacturers can produce the generic form of the drug. These manufacturers are required to file an Abbreviated New Drug Application and to conduct bioequivalence studies to prove that their drug is comparable to the original in dosage, strength, quality, performance, and intended use. This way, the FDA ensures that the generic forms of a drug match the quality and safety of the original.[9]

Potential pitfalls in the postmarket surveillance of drug safety

- Clinicians and patients may be unaware of potential confounding factors that are involved in adverse events and may wrongly attribute these adverse events to the drug.
- Incidences of underreporting can limit available data on the frequency of adverse drug reactions.
- Noncoerced/spontaneous reports that are made directly to the systems such as the FAERS and Vaccine Adverse Event Reporting System (VAERS), may not contain all the necessary information required for the proper investigation of these adverse events.

Real world examples of product recalls published by the FDA

- In March 2023, Camber Pharmaceuticals recalled Atovaquone Oral Suspension due to possible contamination with Bacillus Cereus. The recall was reported on the FDA site in the same month[11]

TABLE 85.2 Types of FDA inspections.

Type of inspection	Purpose
Surveillance inspections	- Carried out by the FDA to assess the quality and standards of production of FDA approved products. - These inspections are carried out on products already on the market and ensure postmarket safety.
For cause inspections	- A quality control process that is usually carried out following a previous violation or complaint.
Application -based inspections	- Carried out as part of the process of approving the release of a new drug. - It is carried out to ensure that the manufacturing process of the product is in compliance with FDA standards and that data submitted on the product is accurate.
Others	- The FDA also carries out inspections to ensure that any clinical or nonclinical research carried out on new products is up to standard in terms of reliability, and integrity and is compliant with FDA procedures.[8]

TABLE 85.3 Types of drug advertisements.

Type of advertisement	Details
Product claim ads	- These give detailed information about a drug, including its risk and benefits. - Depending on the media type, details on drug adverse effects must either be listed or contact details where this information could be obtained must be provided to the public
Reminder ads	- These are recognized by the FDA as ads that provide the name of a drug without detailing its risks or adverse effects. - These ads cannot be used for drugs with known serious risks as the details of these drugs must be provided to the public in all ad forms.
Help seeking ads	- These ads encourage individuals suffering from a specific condition to contact their healthcare providers for more information on their conditions. While they may mention a drug company's name, they do not tell viewers about a specific drug[10]

- In May 2023, the FDA published a recall made by Novis PR LLC of G Suppress GX, an expectorant, nasal decongestant, and cough suppressant for pediatric patients. The product was recalled because it contained an undeclared anesthetic/analgesic product inside which was composed of 60% Ethyl Alcohol and % Benzocaine, both products which could be harmful to the pediatric populace.[12]
- On August 16th, 2021, The FDA published a Pfizer recall of Chantix, a medication that aids with smoking cessation. It was recalled due to levels of a nitrosamine, N nitroso varenicline, that were above the Pfizer established acceptable intake level. Nitrosamine is a potential carcinogen making it especially dangerous at increased levels.[13]

Additional resources

- **Recalls, Market Withdrawals, & Safety Alerts | FDA** https://www.fda.gov/safety/recalls-market-withdrawals-safety-alerts
- The Sentinel Initiative homepage https://www.sentinelinitiative.org/
- FDA Adverse Event Reporting System (FDAERS) Dashboard

Get started

- The reader should understand the importance of post-market drug safety surveillance.
- The reader should understand the role of the FDA in drug safety surveillance.

References

1. Hooimeyer A, Bhasale A, Perry L, et al. Regulatory post-market drug safety advisories on cardiac harm: a comparison of four national regulatory agencies. *Pharmacol Res Perspect*. December 9, 2020;8(6). https://doi.org/10.1002/prp2.680.
2. Center for Drug Evaluation and Research. *FDA Adverse Event Reporting System (FAERS) Public Dashboard*. FDA; 2021. Available from: https://www.fda.gov/drugs/questions-and-answers-fdas-adverse-event-reporting-system-faers/fda-adverse-event-reporting-system-faers-public-dashboard.
3. Food and Drug Administration. *Drug and Biologics Safety Surveillance Best Practice Statement Center for Drug Evaluation and Research (CDER) Center for Biologics Evaluation and Research (CBER) US Food and Drug Administration*. U.S. Food and Drug Administration; January 2024. https://www.fda.gov/media/130216/download. Accessed July 2, 2023.
4. Center for Drug Evaluation and Research. *Questions and Answers on FDA's Adverse Event Reporting System (FAERS)*. FDA; 2018. Available from: https://www.fda.gov/drugs/surveillance/questions-and-answers-fdas-adverse-event-reporting-system-faers.
5. Center for Drug Evaluation and Research. *Postmarketing Adverse Event Reporting Compliance Program*. U.S. Food and Drug Administration; May 5, 2020. www.fda.gov/drugs/surveillance/postmarketing-adverse-event-reporting-compliance-program.
6. FDA. *Postmarketing Surveillance and Epidemiology: Human Drug and Therapeutic*. U.S Food and Drug Administration; 2022. Available from: https://www.fda.gov/media/84969/download.
7. Center for Drug Evaluation and Research. *FDA's Sentinel Initiative*. FDA; 2023. Available from: https://www.fda.gov/safety/fdas-sentinel-initiative.
8. FDA. *Types of FDA Inspections*. FDA; 2023. Available from: https://www.fda.gov/inspections-compliance-enforcement-and-criminal-investigations/inspection-basics/types-fda-inspections.
9. FDA. *Step 5: FDA Post-Market Drug Safety Monitoring*. FDA; 2018. Available from: https://www.fda.gov/patients/drug-development-process/step-5-fda-post-market-drug-safety-monitoring.
10. Center for Drug Evaluation and Research. *Basics of Drug Ads*. FDA; 2015. Available from: https://www.fda.gov/drugs/prescription-drug-advertising/basics-drug-ads.
11. FDA. *Camber Pharmaceuticals, Inc. Issues Voluntary Nationwide Recall of Atovaquone Oral Suspension, USP 750mg/5ml Due to Potential bacillus Cereus Contamination in the Product*. FDA; 2023. Available from: https://www.fda.gov/safety/recalls-market-withdrawals-safety-alerts/camber-pharmaceuticals-inc-issues-voluntary-nationwide-recall-atovaquone-oral-suspension-usp.

12. Center for Drug Evaluation and Research. *Pfizer Expands Voluntary Nationwide Recall to Include Four Additional Lots of CHANTIX® (Varenicline) Tablets Due to N- Nitroso Varenicline Content.* FDA; 2021. Available from: https://www.fda.gov/safety/recalls-market-withdrawals-safety-alerts/pfizer-expands-voluntary-nationwide-recall-include-four-additional-lots-chantixr-varenicline-tablets.
13. FDA. *Novis PR LLC Issues Voluntary Recall of G-Supress Dx Pediatric Drops Due to Incorrect Packaging.* FDA; 2023. Available from: https://www.fda.gov/safety/recalls-market-withdrawals-safety-alerts/novis-pr-llc-issues-voluntary-recall-g-supress-dx-pediatric-drops-due-incorrect-packaging.

Chapter 86

Postmarket medical device safety monitoring

Adam J. Prince[1], Aaron V. Kaplan[1] and Emily P. Zeitler[1,2]
[1]*Dartmouth Health, Lebanon, NH, United States;* [2]*The Dartmouth Institute, Lebanon, NH, United States*

Key points

- Effective postmarket surveillance incorporates many tools, including the MAUDE database, FDA-mandated postapproval studies, comprehensive registries, and clinician case reports
- FDA has the authority to require postmarket surveillance as a condition of approval, order postmarket surveillance after market authorization, determine surveillance plan adequacy and progress, and inspect device manufacturer facilities
- Postmarket monitoring can be a useful tool to decrease premarket FDA requirements, optimize the balance between public safety and device innovation, and identify late adverse events related to medical devices
- Once commercially available, adverse events related to medical devices are required to be reported by manufacturers via medical device reports (MDRs) to the Manufacturer and User Facility Device Experience (MAUDE) database, where they are available to the general public

Introduction

Postmarket device surveillance (PMDS) is the process of obtaining and analyzing safety and effectiveness data about a marketed medical device following market authorization. PMDS further characterizes the device performance outside the strict confines of a premarket clinical study and monitors for device failures. A clear understanding of a device's performance following market authorization is important for device manufacturers, those using medical devices (e.g., device operators), patients, payers, and regulators. PMDS requirements differ by country; this chapter focuses on postmarket surveillance mechanisms, including requirements for medical devices marketed in the United States that are within the jurisdiction of the US Food and Drug Administration (FDA). Regulation of medical devices in other regions such as Europe and Asia is beyond the scope of this chapter.

Postmarket device surveillance reflects an acknowledgment by the FDA and the medical device community that the entirety of device safety concerns, failure mechanisms, and adverse events cannot be known at the time of FDA market authorization ("approval"). A comprehensive approach to PMDS can provide a better understanding of device performance. This in turn facilitates FDA's public health mission to bring innovative medical devices to market to meet public needs in a timely way. This paradigm reflects an implicit contract with the public that good faith efforts are made to ensure device safety and effectiveness when used in a real-world setting. The type and seriousness of safety questions that remain at the time of FDA market authorization range widely, and postmarket surveillance must be "fit for purpose" depending on the specifics of the device and condition being treated.[1] Importantly, the FDA does not regulate medical practice, so once a device has been market authorized and is commercially available, it may be used in ways or for populations not previously evaluated. Through this process of device diffusion into clinical practice, there may be added data about device safety and effectiveness, which may have important regulatory consequences. These consequences may include label expansions, requirements for further formal study, or other outcomes.

Just as premarket device evaluation can take on a variety of forms, postmarket device surveillance contains various mechanisms for surveillance, which serve different purposes. Surveillance mechanisms vary from formal and compulsory postmarket clinical trials to individual reports of device events in the medical literature, and various methods in between. Each of these plays a role in

Translational Cardiology. https://doi.org/10.1016/B978-0-323-91790-2.00012-5

understanding how a device performs after approval. This process informs future device development and identifies areas that require further study. Some of the major methods and tools related to postmarket device surveillance are discussed in this chapter along with some illustrative examples that have contributed to the evolution of FDA medical device surveillance (Table 86.1).

Manufacturer and User Facility Device Experience

FDA maintains a registry known as the Manufacturer and User Facility Device Experience (MAUDE) database for medical device reports (MDRs) with events of serious injury, death, or device malfunction.[2] MDRs are provided by mandatory reporters, who are required to submit to MAUDE if they receive information of device-related events, and voluntary reporters, who are not required to (but may) submit events to the database. The database is updated monthly and serves primarily as a repository of complications during and immediately following device-based procedures. The public-facing MAUDE interface can be searched via the Internet based on a variety of characteristics (Fig. 86.1).

Voluntary MDR reporters include patients, physicians, device users, and other individuals or groups who wish to report device events. Mandatory reporters include device manufacturers, distributors, and device user facilities. If a device manufacturer receives the report of a device-associated adverse event, the manufacturer is required to determine whether the report meets the criteria for an MDR, and if so, to generate and submit an MDR. The Code of Federal Regulations (CFR) contains information for manufacturers looking to determine whether an adverse event requires an MDR, as well as the expected timeline for MDR generation.[3] MDRs must be submitted within 30 days of adverse report awareness, though in cases of "unreasonable risk of substantial harm to public health" or specific FDA request, this is shortened to 5 days. Since manufacturers may not be directly involved in all cases involving their devices, this process is biased toward events involving high risk.[4]

MAUDE is a vital source for adverse event reports since it is publicly searchable. It is particularly powerful for the identification of novel failure mechanisms, which may occur so infrequently that it would not be reasonably measurable in time- and size-limited premarket trials. However, MAUDE is imperfect as a source of device malfunction information for various reasons. First, it relies mostly on the voluntary reporting of information at the "front lines" of adverse events by patients, physicians, and other healthcare providers, many of whom are unfamiliar with MAUDE. The motivation for and perceived burden for creating medical device reports differs across these reporting groups, as does the ability to transmit technically nuanced medical events; thus, MAUDE is known to be incomplete, at best, in its capture of adverse events.[5] Second, the MAUDE database does not capture data on device usage and therefore is unable to accurately provide a denominator from which rates of events can be determined. Third, events and associated details reported to MAUDE are not obtained in a standardized fashion, verified, or independently adjudicated. Protected information, such as trade secrets or medical information, is withheld from public reporting in MAUDE. MAUDE is best for identifying safety events occurring at the time of implantation. Failures that occur after implantation or in general use especially outside of the device user facility are less likely to be captured by MAUDE.

Product performance reports

Product performance reports (PPRs) are publicly available manufacturer-compiled comprehensive analyses of medical device events. These reports are a complement to the MAUDE database. While event reporting in this context is isolated to the manufacturer, there is the benefit of a known "denominator." Like other sources of postmarket surveillance, findings from PPRs allow manufacturers, innovators, operators, and regulators to identify opportunities for device improvement. PPRs are generally available online directly through the manufacturer and organized by the device (Fig. 86.2).[6–8]

Postapproval studies

Postapproval studies (PAS) are clinical studies that are developed by the FDA in collaboration with a device manufacturer at the time of device approval. PAS can take on many forms, as outlined in postmarket device surveillance guidance published by the FDA, with the most recent update finalized in 2022.[9] Devices marketed in the United States are classified by the FDA based on risk, ranging from least risk in Class I (e.g., toothbrushes) to most risk in Class III (e.g., surgically implanted heart valves). Since 1990, manufacturers of class II and class III devices have been required to conduct postmarket device surveillance. This requirement is governed by Section 522 of the Federal Food, Drug, and Cosmetic Act, and includes any class II or class III device that is implanted for at least 1 year, utilized by pediatric patients, or a device with significant health risks upon malfunction.[11] Postmarket monitoring can take on a variety of forms, as determined by the FDA with agreement by the device manufacturer. PAS have typically been single-arm studies focused on confirming results observed in the preapproval pivotal trials. A PAS may also be designed to incorporate data collection requirements imposed by other bodies, such as the Centers for Medicare

TABLE 86.1 A selection of postmarket surveillance mechanisms.

Resource	Description/Definition	Advantages	Disadvantages
Manufacturer and User Facility Device Experience (MAUDE)	FDA-maintained database of medical device adverse event reports since the 1990s. MAUDE is publicly available, online, and searchable for the prior 10 years. Device reports are a combination of voluntary and compulsory event reports from device manufacturers, facilities, and device users.	- Centralized, publicly accessible. - Reporting available to wide variety of users including patients and physicians - Timely awareness of submitted events. - Free text included to allow for reporting of novel device failures or complications - Best suited to compiling complications or events occurring during or immediately after device-based procedures	- Unknown denominator of devices in use, so cannot calculate event rates - Primarily relies on voluntary reporting, so the database is incomplete - Manufacturers required to submit event reports of which they are aware, so the database is skewed toward the most serious events - May include redundant events - Lack of identifiers in the submission process, so characteristics of the event, patient, and operator are incompletely known and include only those elements included in the report - Events are not independently verifiable - Late events less likely to be included
Medical Product Safety Network (MedSun)	An adverse event reporting program launched in 2002 by the FDA to identify, understand, and solve problems with medical devices based on formal partnerships with a set of facilities. Participating facilities use an Internet-based system to submit adverse medical device events. These partnerships are also leveraged by the FDA to survey specific device users regarding specific concerns.	- Public access to findings - Reporting is by trained staff to provide a high level of relevant detail - Includes nonhospital facilities which allows for reporting of noncritical adverse events - Formal partnerships include the opportunity to perform a "deep dive" into a specific device-related issue with those facilities/users with the most relevant knowledge and experience	- Reporting only by specific clinical sites - Does not address all approved devices - More labor and cost-intensive than other reporting systems such as MAUDE - Frequently a "reactive" form of surveillance as opposed to "proactive"
Product Performance Report (PPR)	Publicly available manufacturer-compiled and curated comprehensive report of medical device events typically with a set group of predefined potential adverse events specific to a device type or model	- Publicly available summary reports - Includes the "denominator" of devices in service so rates can be calculated	- Novel or unexpected device failures may not be included - Data contributing to the reports are not publicly available, so manufacturer-reported findings cannot be independently verified - Patient and facility information is not included
Device registries	An organized system that uses observational study methods to collect uniform data (clinical and other) to evaluate specified outcomes for a population defined by a particular disease, condition, or exposure	- Ability to perform long-term follow-up for rare or late events - Improves understanding of device performance in clinical practice	- In most cases, there is a significant burden of data collection on healthcare providers and facilities that must be managed and considered - Data quality varies

Continued

TABLE 86.1 A selection of postmarket surveillance mechanisms.—cont'd

Resource	Description/Definition	Advantages	Disadvantages
	and that serves one or more predetermined scientific, clinical, or policy purposes.[a]	- Lower burden of oversight and human subjects' protections compared with traditional post approval study. - Centralized collection and reporting of events which may be available to investigators with postmarket clinical questions - Can serve as a platform for FDA label expansion - Can be combined with other data sources (e.g., claims) to assess longitudinal outcomes	- Issues of data ownership and availability must be resolved - Managing and maintaining a clinical registry can require substantial resources
Traditional postapproval studies (PAS)	A study conducted after FDA market authorization to improve understanding of device safety and effectiveness in clinical practice. PASs can take a variety of forms including but not limited to registry, clinical trial, or nonclinical studies and are designed in collaboration with the FDA to ensure post-market surveillance questions will be addressable.	- Opportunity to build rigor into study design to improve ability to make safety and effectiveness determinations - Ability to build in opportunities for planned analyses for future label expansion - Can be combined with or adapted to postmarket data collection requirements for payers as in the coverage with evidence development (CED) program	- As in premarket clinical trials, a traditional trial-based PAS may not reflect "real world" use of the medical device due to a controlled environment with exclusion and inclusion criteria. - Often significant cost and time investment required - Lag time in awareness of adverse events. - May miss late or rare events.
Operator and investigator observations	Observations related to device performance in clinical practice outside of formalized data collection mechanisms reported in the medical literature or other publicly accessible forums (e.g., social media).	- Entirely egalitarian reporting opportunity reflecting experience in "real world" practice - Source of information related to late device performance - Source of information related to novel or rare device events not otherwise conforming to formalized data collection mechanisms - Publicly available	- No oversight and limited ability to adjudicate findings or reports - Events may fall well outside of intended device use and intended device population - Depends on voluntary operator reporting - Important observations may not come to the attention of manufacturers or regulators in a timely fashion or ever

[a] *R.E. Gliklich and N.A. Dreyer, eds., Registries for Evaluating Patient Outcomes: A User's Guide, second ed. (Rockville, MD: Agency for Healthcare Research and Quality, 2010).*

FIGURE 86.1 Screenshot of the Manufacturer and User Facility Device Experience (MAUDE) database search. The MAUDE database serves as a publicly available repository of medical device reports, submitted by voluntary and mandatory reporters. MAUDE is maintained by the FDA and provides valuable insights into device-related adverse events, with medical device reports searchable by defect, device, and date.

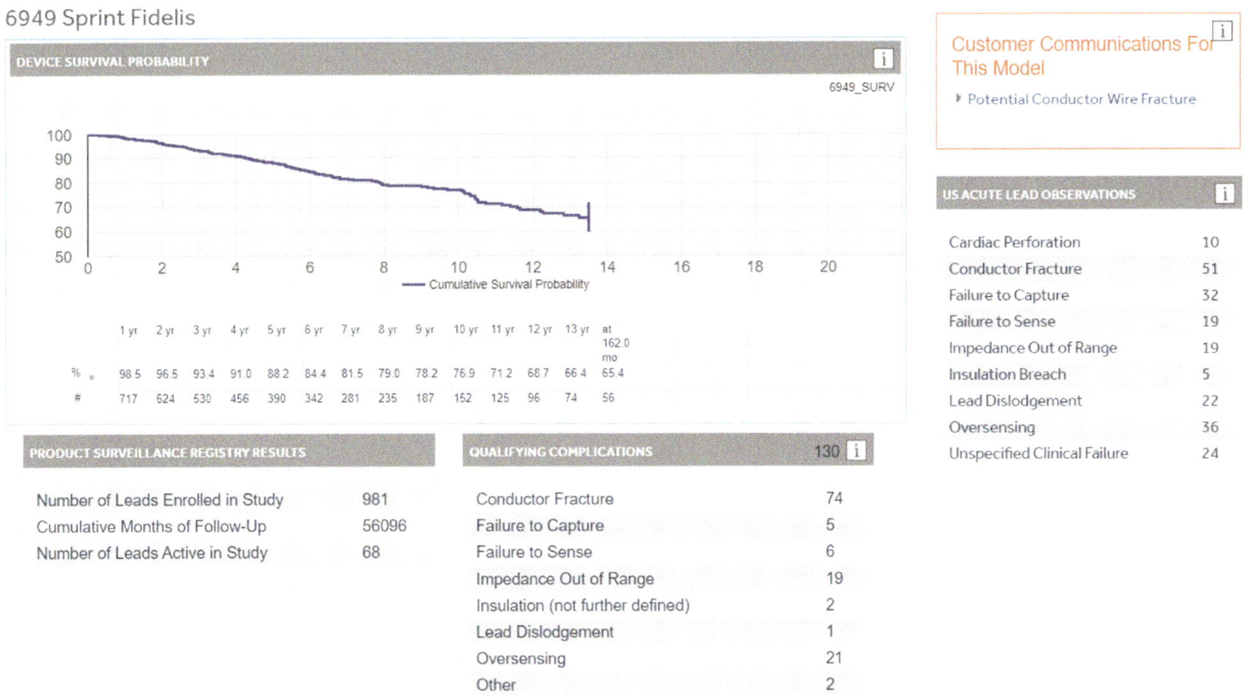

FIGURE 86.2 Excerpt from the product performance report for the Sprint Fidelis defibrillator Lead. Product performance reports are curated and reported by the device manufacturer and are intended to include all distributed devices organized by model. PPRs report device performance in clinical practice based on predetermined events.

and Medicaid Services (CMS). This is discussed further below.

The paradigm of postmarket surveillance reflects an acknowledgment that risks and uncertainties remain at the time of market authorization, and there is a benefit to approving devices to address unmet clinical needs. Development of a formal PMDS plan improves the ability of FDA and manufacturers to gain an understanding of the frequency and severity of adverse events associated with device use, device effectiveness in clinical practice, and rare events or late complications, all in hopes of decreasing harm through early identification of device-associated complications. A premarket agreement can be leveraged to balance the burden of data collection between the pre- and postmarket periods by building in an extension to a premarket study in the event the device becomes market authorized.[12] Furthermore, data collected through postmarket surveillance can better characterize device performance and identify design foci for device iteration. If thoughtfully designed, a postapproval study can serve as a platform to study use in expanded populations.[13]

As noted above, the FDA may provide premarket approval contingent upon a postapproval study,[14] and has the authority to take enforcement action if agreed-upon postapproval monitoring requirements are not met. In addition, when data become available that identify new or unanticipated device failures, the PMDS process can be used to implement ad hoc studies to further characterize these newly identified safety issues.

Developments that may lead to unplanned postmarket surveillance, with or without an associated recall, include unexpected adverse events related to unexpected frequency or mechanism of device complications. Often, these events are first recognized via reports in the medical literature, and are not part of the MAUDE database, as was the case with the Riata lead fractures (discussed below). Typically, if the FDA determines that postmarket surveillance is required to address a public health question related to a device, the FDA can issue a "522 Order" mandating the development of a surveillance plan within 30 days. This typically results in cooperation between the FDA and the device manufacturer to design a plan to adequately address outstanding questions. Specific guidelines for postmarket surveillance are provided in the Code of Federal Regulations and FDA guidance.[15,16] Manufacturers have up to 15 months from the date of order issuance to begin surveillance. If 15 months have elapsed from 522 Order issuance with no implemented surveillance plan (including manufacturers who have no accepted plan), this is considered a failure to comply.

The postapproval study design is guided by the information required to address the public health question. Surveillance mechanisms can range from randomized clinical trials to observational studies, national registries, bench studies, or animal studies. Basic details of postapproval studies under FDA jurisdiction are entered into a publicly available database that includes information about study status.[17] More recently, some manufacturers have designed postapproval studies as a continuation of a premarket study design to reduce the logistical and financial costs of conducting two separate studies. For example, the Barostim Neo System (CVRx, Minneapolis, MN) was approved with Breakthrough Device designation[18] (previously the expedited access pathway) based on surrogate endpoints in the premarket portion of the trial and "hard" clinical endpoints in the postmarket phase with an adaptive statistical design.[19] Regardless of the design, postmarket surveillance studies are subject to the same requirements as premarket studies including human subjects' protections via institutional review board, and FDA oversight as in the investigational device exemption program.

Mandated postapproval medical device studies historically required significant financial investment. An analysis of 277 studies from 2005 to 2013 estimated a median cost of nearly \$2.2 million per study; 138 cardiovascular specialty-specific studies were estimated to cost a median of nearly \$2.4 million.[20] As noted above, postapproval studies can be a catalyst for label change and simultaneously serve as a platform for data collection to support label expansion. A review of FDA postapproval studies from 2005 to 2011 found 223 studies across 158 medical devices, and for completed studies, a device label change was requested in 53% of cases.[21]

Once a postapproval study is approved and initiated, manufacturers are responsible for interim report submission every 6 months for the first 2 years, then annually beginning in year 2. The status of postapproval studies is publicly available.[22] In parallel to requirements for premarket studies, the final postmarket surveillance report is submitted within 3 months of study completion, or the end of postmarket surveillance. In the event that the interim or final report is found inadequate by FDA, the agency may either provide a letter of deficiency or request for a review process. Findings in the interim or final reports that lead to new safety or effectiveness concerns may lead to labeling changes, safety warnings or communications, additional postmarket surveillance requirements, or, rarely, device recall.

The 2022 FDA guidance on postmarket surveillance includes some important updates to the earlier version, including an emphasis on tracking enrollment milestones to help with the timely completion of the surveillance plan. There is also a stated goal for collaboration between the device manufacturer and FDA for surveillance plan submission and review within 60 days of the 522 order. The process for finalizing an FDA guidance is purposefully long and involves input from many stakeholders.

Postmarket surveillance for coverage and payment

Sometimes, data collection is required as a condition of coverage in the Centers for Medicare and Medicaid Services (CMS) Coverage with Evidence Development (CED) program, and this mechanism has emerged as a vehicle for performing postmarket surveillance. This mechanism is used by CMS when there are remaining questions about the device after FDA market authorization, which may be based on studies that do not well represent Medicare beneficiaries.[23] For example, as a condition of transcatheter mitral valve replacement (TMVR) coverage by CMS, follow-up information must be entered into an approved registry designed, in part, to address CMS outstanding questions for Medicare beneficiaries. These questions include, for example, "When performed outside a controlled clinical study, how do outcomes and adverse events compare to the pivotal clinical studies?" and "What are the long term (≥5 year) outcomes and adverse events?"[24] Studies and registries conducted to meet CED requirements are subject to FDA oversight with regard to matters such as human subjects' protections and event reporting. Theoretically, data collected through a CED study can lead to complete withdrawal of coverage by CMS or even revocation or alteration of FDA market authorization.[24a]

Manufacturer inspections

In addition to postmarket surveillance, the FDA is also tasked with the inspection of device manufacturer facilities.[25] In routine inspections, manufacturers will be notified ahead of time (at least 5 days prior in domestic locations, and frequently longer notification time for international inspections). The initial notification includes a detailed reason(s) for inspection, and records may be requested from the manufacturer at that time. FDA inspections "for cause" may occur with no prior notification. Throughout the inspection, the FDA will communicate regularly to ensure that the manufacturer is aware of the inspection findings.

Manufacturers may request feedback from the FDA regarding a device inspection if the resultant FDA response involves a public health priority, systemic or major action, or an emerging safety issue; these concerns are expected to cause risk of morbidity or mortality if unaddressed. After receipt of the proposed mitigation by the manufacturer, the FDA can provide nonbinding feedback; however, regulatory action may still result.[11]

In the event that a device is found to be defective or dangerous to public health, there are several actions that may be taken. Either a device manufacturer or FDA may initiate a device recall to remove a device from circulation. In many cases, device manufacturer recalls are voluntary and may be preceded or accompanied by a device "advisory." In the event that there is no voluntary recall for a device that is deemed a health risk, FDA may order the manufacturer to perform a mandatory recall. Manufacturers may conduct a voluntary recall at any time to mitigate a health risk, though they must report the recall to FDA. For FDA-ordered recalls, there is a tiered system. Class I recall is the most serious and implies that the use of a device "will cause serious adverse health consequences or death." A class II recall implies that device use may cause reversible adverse health consequences or nonserious health consequences. Finally, a class III recall is issued for products that are not likely to lead to adverse health consequences. In the event of a recall, FDA provides guidance for communication from the manufacturer to the public.[26]

The future of postmarket surveillance

Postmarket surveillance will remain an integral component of the medical device life cycle, due to the need for balance between healthcare advancement and identification of adverse events that result from device use. As seen with the creation of MAUDE and the MedSun network (discussed in Table 86.1), centralized registries remain vital for global access to device use information. These sources will continue to identify areas for improvement and drive device innovation and iteration.[27] Given the inherent flaws of the MAUDE database (Table 86.1), it is likely that there will be continued improvements to the reporting process for medical device reports, to verify and more efficiently extract relevant clinical information and ultimately increase device safety and effectiveness. As more emphasis is placed on cost-effective and informative postmarket surveillance methods, trial designs will continue to be optimized to best position devices for accurate assessment, and even for expanded indications. Case Study 1 reviews the role of postmarket surveillance in the growth of transcatheter aortic valve replacement.

Considerable efforts have been made to redesign the postmarket surveillance process to reduce the burdens associated with data collection and curation. By seamlessly incorporating surveillance into clinical practice, manufacturers and regulators can harness the power of real-world evidence coupled with rapid advances in computing power. Such a system could help inform decision-making by patients, physicians, regulators, and payers.

The impact of post market device monitoring: Three real world examples

Case study 1—Transcatheter aortic valve replacement: A postmarket success story

Postmarket surveillance studies can be utilized to generate data that support the expansion of indications for a medical device. One notable example of label expansion success occurred with transcatheter aortic valve replacement (TAVR), which was developed as an alternative to surgical aortic valve replacement. The first TAVR device was FDA approved in 2011 for patients with severe aortic stenosis at prohibitive risk for a surgical valve replacement, contingent on the rigorous postmarket surveillance requirements that were designed to simultaneously meet the coverage with evidence development requirements for CMS. This gave rise to the Society of Thoracic Surgeons/American College of Cardiology (STS/ACC) Transcatheter Valve Therapy (TVT) Registry.[28] Initial 2011 approval was followed 1 year later for patients at high surgical risk,[29] and in 2016 for patients at intermediate surgical risk.[30] In 2019, landmark clinical trials reported at least noninferior outcomes for patients with severe aortic stenosis and low surgical risk receiving TAVR in comparison to surgical valve replacement.[31,32] Based on these clinical trial data, as well as clinical outcomes evaluated from the postmarket clinical registry, the FDA expanded the indication for the Sapien and CoreValve TAVR procedures to include patients at low surgical risk. The approval was linked to the manufacturer agreement to expand monitoring in these trials to 10 years through the centralized registry for 10 years.[33] Expanding indications have been a factor in the explosion of TAVR volume over the past decade, which has increased from approximately 4600 procedures in 2012 to nearly 73,000 procedures in 2019.[34] Thus, this is an example of combining dedicated randomized premarket studies with single arm studies embedded within the manufacturer-agnostic STS/ACC TVT Registry to ultimately support label expansion. Moreover, in addition to patient population expansion, these studies evaluated alternative access methods for introducing TAVR valves into the vasculature.[34]

Fig. 86.3 demonstrates the impact of expanded indications on TAVR volume in the United States over the past decade, particularly following the expansion to moderate risk patients in 2016. The increase in TAVR volume highlights the tremendous growth and improved access that well-designed postmarket surveillance studies can facilitate.

Case study 2—Fidelis and Riata implantable cardioverter-defibrillator leads

Medtronic's Sprint Fidelis implantable cardioverter-defibrillator (ICD) lead was FDA approved in 2004 as a supplement to a lead originally approved in 1993 with multiple intervening device iterations. As a supplement, no significant clinical testing was required or performed. Subsequently, it was discovered that some Sprint Fidelis models had a higher-than-expected failure rate due to lead fracture. Leads subject to this fracture-mediated failure were at risk for increased impedance, oversensing, inappropriate shocks, and inability to defibrillate or pace when needed.[35,36] Recognition of the elevated risk of adverse events with these leads led to a software release by the manufacturer to improve detection of lead failure,[37] along with an FDA Class I recall in 2007, 3 years after FDA market authorization.[38] Following this Class I recall, Medtronic voluntarily stopped sales of these leads, though 268,000 leads had been implanted globally. Medtronic maintains active surveillance of these leads through a product surveillance registry that initially enrolled 981 patients.[39] An estimated 25,000 Fidelis leads are still in service globally as of 2024 and are involved in ongoing monitoring.

Contemporaneously, in 2008, 1 year after the Fidelis recall, many models of the St. Jude Riata ICD lead were noted to have another novel failure mechanism occurring more frequently than anticipated, based on reports trickling in from the scientific literature. These lead models were implanted in nearly 80,000 patients in the United States, and like the Sprint Fidelis lead, were market authorized based on a supplement to a lead approved many years and generations prior without significant interim clinical testing. Based on the novel failure mechanism, the uncertainty related to the relationship between mechanical and electrical failure of the lead, and the significant risks associated with invasive lead revision, legitimate questions were raised by regulators and practicing physicians as to the correct approach to managing these leads in situ. There were no mandated surveillance studies at the time, which led to criticism of the surveillance process and calls for more active vigilance in postmarket surveillance.

The events related to the Medtronic Sprint Fidelis and St. Jude Riata ICD lead recalls have led to changes in postmarket surveillance for these devices, including dramatic increases in the number of leads and the duration of follow-up required as a condition of approval.[40,41]

Case study 3—NC Trek RX and NC Traveler RX coronary dilation catheters

Abbott Vascular distributed nearly 14,000 4–5 mm NC Trek RX and NC Traveler RX Coronary Dilation Catheters throughout the United States from August 2019 to January 2020. Due to increased heat exposure during device manufacture, balloons produced during these dates had a higher likelihood of deflation failure. Following receipt of 13 complaints of this issue, as well as 1 report of a related death, Abbott sent a Field Safety Notice at the end of January 2020 to all recipients of affected catheter lots, which contained a list of all affected catheters by part and lot number with instructions to return all unused devices. Abbott stopped shipping affected catheters and designated field representatives to facilitate catheter identification and appropriate returns. Each Field Safety Notice also contained a check box for return to Abbott or

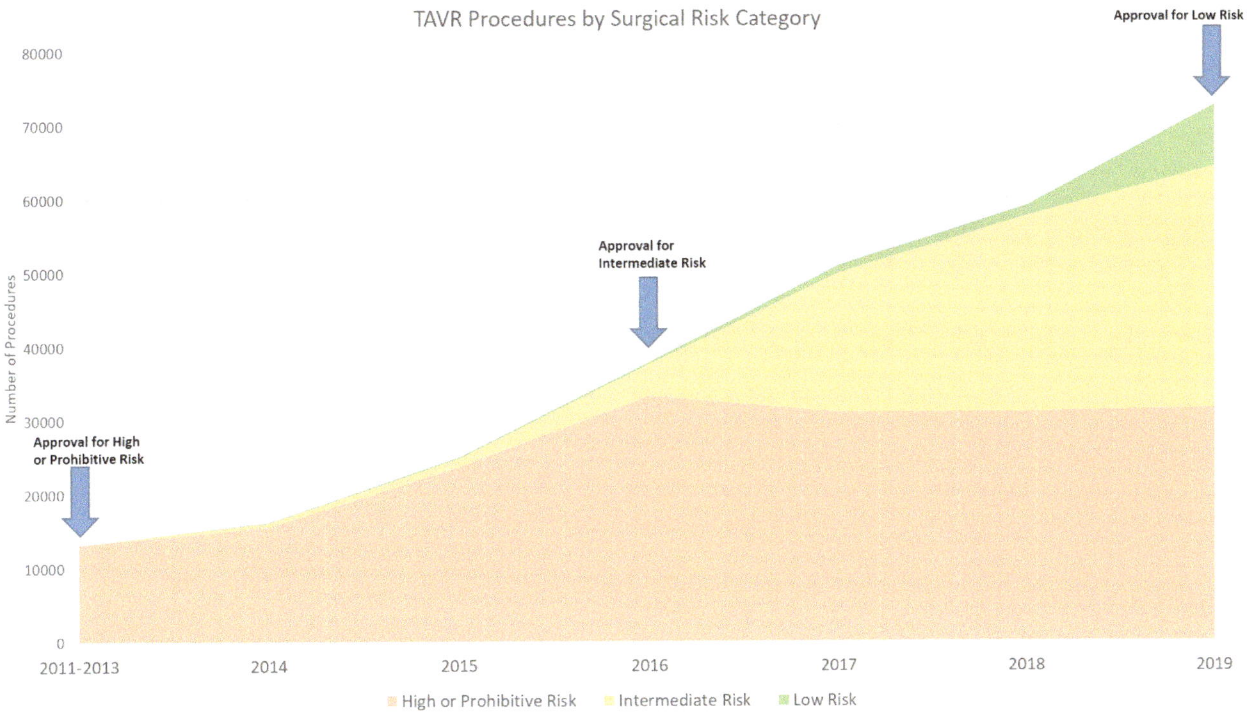

FIGURE 86.3 **Transcatheter aortic valve replacement volume in the United States by year and surgical risk category.**[34] US volume of TAVR valve implantations, with *blue arrows* denoting approval years for new surgical risk groups. These approvals correspond to significant expansions in volume for their respective risk groups.

an indication that none of these catheter lots were in stock, ensuring closed-loop communication regarding inventory search and identification of any affected catheters. Following the Field Safety Notice, a Class I FDA recall was announced for these catheter lots, indicating the possibility of serious health risks from device use.[42] The recall of NC Trek RX and NC Traveler RX lots demonstrates the meticulous organization and resource dedication that manufacturers must provide to support a recall, to ensure that affected products are immediately returned from distribution and further use.

Key takeaways

Postmarket surveillance serves many purposes including, importantly, to improve understanding of device safety and effectiveness after FDA market authorization. This is best achieved when there is frequent and timely communication, as well as cooperation between FDA and device manufacturers across the total product life cycle. Surveillance occurs at multiple levels, including the point of care (e.g., MAUDE), device manufacturers (e.g., product performance reports), and collective clinical experience (e.g., postapproval studies). In all cases, postmarket surveillance is designed with a "least burdensome" approach such that contemporary and future iterations of postmarket surveillance can leverage the power of real-world data combined with advanced computing to reduce the burden of data collection while continuing to protect the public.

Pitfalls to avoid

Not communicating early with the FDA regarding postmarket surveillance orders to develop an appropriate and achievable plan.

Not cooperating with the FDA and devoting resources to adequately address concerns for postmarket surveillance and manufacturer inspections.

Planning around overly optimistic study timelines or required postmarket surveillance.

Get started

Anticipate postmarket surveillance requests from the FDA and develop a systematic method of tracking adverse events.

Assemble a postmarket surveillance team, and consider options for a designated person.

Communicate early with the FDA to develop an adequate plan in response to the receipt of a 522 order. Consider utilizing external consultant resources in the development of the plan.

Resources and further reading

https://www.fda.gov/media/81015/download

FDA guidance on postmarket surveillance: U.S. Department of Health and Human Services. Postmarket Surveillance Under Section 522 of the Federal Food, Drug, and Cosmetic Act: Draft Guidance for Industry and Food and Drug Administration Staff. 2022, https://www.fda.gov/media/81015/download; https://www.fda.gov/media/149346/download.

MAUDE: https://www.accessdata.fda.gov/scripts/cdrh/cfdocs/cfmaude/search.cfm.

FDA guidance on manufacturer inspections: U.S. Food and Drug Administration. Review and Update of Device Establishment Inspection Processes and Standards: Guidance for Industry. 2020, https://www.fda.gov/regulatory-information/search-fda-guidance-documents/review-and-update-device-establishment-inspection-processes-and-standards.

FDA guidance on FDA inspections of device establishments: U.S. Food and Drug Administration. Nonbinding Feedback After Certain FDA Inspections of Device Establishments. 2020, https://www.fda.gov/media/120260/download.

FDA postapproval studies database: https://www.accessdata.fda.gov/scripts/cdrh/cfdocs/cfPMA/pma_pas.cfm.

References

1. Baim DS, Mehran R, Kereiakes DJ, et al. Postmarket surveillance for drug-eluting coronary stents: a comprehensive approach. *Circulation*. 2006;113:891–897.
2. U.S. Food and Drug Administration. *MAUDE - Manufacturer and User Facility Device Experience*; 2021. Published https://www.accessdata.fda.gov/scripts/cdrh/cfdocs/cfmaude/search.cfm. Accessed November 30, 2021. Updated.
3. U.S Food and Drug Administration. *Code of Federal Regulations Title 21 Volume 8: CFR Part 803 Medical Device Reporting, Subpart E - Manufacturer Reporting Requirements*; 2021. Published https://www.accessdata.fda.gov/scripts/cdrh/cfdocs/cfcfr/CFRSearch.cfm?CFRPart=803&showFR=1&subpartNode=21:8.0.1.1.3.5. Accessed October 1, 2021. Updated.
4. Polisena J, Gagliardi A, Clifford T. How can we improve the recognition, reporting and resolution of medical device-related incidents in hospitals? A qualitative study of physicians and registered nurses. *BMC Health Serv Res*. 2015;15:220.
5. Sandberg JM, Gray I, Pearlman A, Terlecki RP. An evaluation of the manufacturer and user facility device experience database that inspired the United States food and drug administration's reclassification of transvaginal mesh. *Investig Clin Urol*. 2018;59:126–132.
6. Scientific B. *Rhythm Managment Product Performance Report*. Q3 ed.; 2021. Published https://www.bostonscientific.com/content/dam/bostonscientific/quality/ppr/2021/q3/Q3%202021%20CRM%20Product%20Performance%20Report%20Rev%20B.pdf. Accessed July 7, 2021. Updated.
7. Abbott. *Cardiac Rhythm Management Product Performance Report*. 1st ed.; 2021. https://www.cardiovascular.abbott/content/dam/bss/divisionalsites/cv/hcp/products/product-performance-reports/documents/Abbott-Product-Performance-Report-2021-First-edition.pdf. Published.
8. Medtronic. *Cardiac Rhythm & Heart Failure Product Performance Report: Important Patient Management Information for Physicians*; 2021. Published https://wwwp.medtronic.com/productperformance-files/Issue%2084%20MDT%20CRHF%20PPR%202021%201st%20Edition%20with%20Micra.pdf. Accessed May 7, 2021. Updated.
9. U.S. Department of Health and Human Services. *Postmarket Surveillance under Section 522 of the Federal Food, Drug, and Cosmetic Act: Guidance for Industry and Food and Drug Administration Staff*; 2022. https://www.fda.gov/regulatory-information/search-fda-guidance-documents/postmarket-surveillance-under-section-522-federal-food-drug-and-cosmetic-act [Accessed 16 May 2024].
10. Deleted in review.
11. U.S. Food and Drug Administration. *Code of Federal Regulations Title 21: CFR Part 822 - Postmarket Surveillance*; 2020. Published https://www.ecfr.gov/current/title-21/chapter-I/subchapter-H/part-822. Accessed June 6, 2020. Updated.
12. U.S. Food and Drug Administration. *Balancing Premarket and Postmarket Data Collection for Devices Subject to Premarket Approval FDA-201-D-0090*; 2015. Published https://www.fda.gov/media/88381/download.
13. Carroll JD, Shuren J, Jensen TS, et al. Transcatheter valve therapy registry is a model for medical device innovation and surveillance. *Health Aff*. 2015;34:328–334.
14. Ross JS, Kesselheim AS. FDA policy and cardiovascular medicine. *Circulation*. 2015;132:1136–1145.
15. Administration USFD. *Code of Federal Regulations Title 21: CFR Part 822 - Postmarket Surveillance, Subpart C Postmarket Surveilance Plan, Sec. 822.9 What Must I Include in My Submission?*; 2020. Published https://www.accessdata.fda.gov/scripts/cdrh/cfdocs/cfCFR/CFRSearch.cfm?fr=822.9. Accessed June 6, 2021. Updated.
16. U.S. Food and Drug Administration. *Code of Federal Regulations Title 21: CFR Part 822 - Postmarket Surveillance, Subpart C Postmarket Surveilance Plan, Sec. 822.9 What Must I Include in My Surveillance Plan?*; 2020. Published https://www.accessdata.fda.gov/scripts/cdrh/cfdocs/cfCFR/CFRSearch.cfm?fr=822.10. Accessed June 6, 2021. Updated.
17. U.S Food and Drug Administration. *Post-Approval Studies Database*. https://www.accessdata.fda.gov/scripts/cdrh/cfdocs/cfPMA/pma_pas.cfm. Accessed 9 January 2022.
18. U.S. Department of Health and Human Services. *Breakthrough Devices Program: Guidance for Industry and Food and Drug Administration Staff*. 2018.
19. Zile MR, Abraham WT, Lindenfeld J, et al. First granted example of novel FDA trial design under expedited access pathway for premarket approval: beat-hf. *Am Heart J*. 2018;204:139–150.
20. Wimmer NJ, Robbins S, Ssemaganda H, et al. Assessing the cost burden of United States FDA-mandated post-approval studies for medical devices. *J Health Care Finance*. 2016;2016.
21. Reynolds IS, Rising JP, Coukell AJ, Paulson KH, Redberg RF. Assessing the safety and effectiveness of devices after us food and drug administration approval: FDA-mandated postapproval studies. *JAMA Intern Med*. 2014;174:1773–1779.
22. U.S. Food and Drug Administration. *Post-Approval Studies (Pas) Database*; 2021. Published https://www.accessdata.fda.gov/scripts/

cdrh/cfdocs/cfpma/pma_pas.cfm?s=t. Accessed December 20, 2021. Updated.

23. *Guidance for the Public, Industry, and Cms Staff: Coverage with Evidence Development*. Center for Medicare and Medicaid Services; November 20, 2014, 2014 https://www.cms.gov/medicare-coverage-database/details/medicare-coverage-document-details.aspx?MCDId=27.
24. Jensen T, Ashby L, Schafer J, Paserchia L, Lotfi R, Hakim R. *Decision Memo for Transcather Mitral Valve Repair (Tmvr) (Vag-00438n)*. 2014.

24a. Zeitler EP, Gilstrap LG, Coylewright M, Slotwiner DJ, Colla CH, Al-Khatib SM. Coverage with evidence development: where are we now? *Am J Manag Care*. 2022;28(8):382–389. https://doi.org/10.37765/ajmc.2022.88870.

25. U.S. Food and Drug Administration. *Review and Update of Device Establishment Inspection Processes and Standards: Guidance for Industry*. 2020.
26. U.S. Food and Drug Administration. *Recalls, Corrections and Removals (Devices)*; 2020. Published https://www.fda.gov/medical-devices/postmarket-requirements-devices/recalls-corrections-and-removals-devices. Accessed September 29, 2020. Updated.
27. Zippel C, Bohnet-Joschko S. Innovation for safe and effective medical devices: contributions from postmarket surveillance. *Ther Innov Regul Sci*. 2017;51:237–245.
28. The Society of Thoracic Surgeons and American College of Cardiology. *STS/ACC Transcatheter Valve Therapy (Tvt) Registry*; 2021. https://www.ncdr.com/WebNCDR/tvt/publicpage. Accessed January 3, 2022. Published.
29. Mack MJ, Brennan JM, Brindis R, et al. Outcomes following transcatheter aortic valve replacement in the United States. *JAMA*. 2013;310:2069–2077.
30. Rogers T, Thourani VH, Waksman R. Transcatheter aortic valve replacement in intermediate- and low-risk patients. *J Am Heart Assoc*. 2018;7.
31. Mack MJ, Leon MB, Thourani VH, et al. Transcatheter aortic-valve replacement with a balloon-expandable valve in low-risk patients. *N Engl J Med*. 2019;380:1695–1705.
32. Popma JJ, Deeb GM, Yakubov SJ, et al. Transcatheter aortic-valve replacement with a self-expanding valve in low-risk patients. *N Engl J Med*. 2019;380:1706–1715.
33. U.S. Food and Drug Administration. *FDA Expands Indication for Several Transcatheter Heart Valves to Patients at Low Risk for Death or Major Complications Associated with Open-Heart Surgery*. FDA News Release; 2019.
34. Carroll JD, Mack MJ, Vemulapalli S, et al. STS-ACC tvt registry of transcatheter aortic valve replacement. *J Am Coll Cardiol*. 2020;76:2492–2516.
35. Rao A, Wright J, Waktare J, Todd D. Sprint fidelis 6949 high-voltage lead appears to be prone to early failure. *Heart Rhythm*. 2008;5:167.
36. Hauser RG, Kallinen LM, Almquist AK, Gornick CC, Katsiyiannis WT. Early failure of a small-diameter high-voltage implantable cardioverter-defibrillator lead. *Heart Rhythm*. 2007;4:892–896.
37. Kallinen LM, Hauser RG, Tang C, et al. Lead integrity alert algorithm decreases inappropriate shocks in patients who have sprint fidelis pace-sense conductor fractures. *Heart Rhythm*. 2010;7:1048–1055.
38. Zeitler EP, Pokorney SD. Fidelis replacement: can we abandon with reckless abandon? *JACC Clin Electrophysiol*. 2019;5:1068–1070.
39. Medtronic. *Crhf Product Performance Esource: 6949 Sprint Fidelis*; 2014. Published https://wwwp.medtronic.com/productperformance/model/6949-sprint-fidelis.html. Accessed December 16, 2021. Updated.
40. Blake K. Postmarket surveillance of medical devices: current capabilities and future opportunities. *J Interv Card Electrophysiol*. 2013;36:119–127.
41. Hauser RG. Here we go again–another failure of postmarketing device surveillance. *N Engl J Med*. 2012;366:873–875.
42. U.S. Food and Drug Administration. *Abbott Vascular Recalls Nc Trek Rx and Nc Traveler Rx Coronary Dilatation Catheters Due to Failure of Balloon (Diameter 4.0mm, 4.5mm and 5.00mm) to Deflate*; 2020. Published https://www.fda.gov/safety/recalls-market-withdrawals-safety-alerts/abbott-initiates-voluntary-recall-specific-lots-two-coronary-catheters. Accessed May 12, 2024. Updated.

Part IX

Clinical implementation

Rohin Moza

Chapter 87

Implementation research

Sherri R. Thumma[1,2] and Disha B. Patel[1]

[1]*Michigan State University College of Osteopathic Medicine, East Lansing, MI, United States;* [2]*Center for Healthcare Outcomes and Policy, University of Michigan, Ann Arbor, MI, United States*

Key points

- Understand the role of implementation research in the translational process
- Identify how behavior change theories and frameworks shape implementation research
- Describe a roadmap to approaching implementation research
- Understand how implementation research can aid in translating evidence-based recommendations to real-world practice in cardiology

Introductory section/why it matters

Cardiovascular disease (CVD) continues to be a leading cause of mortality worldwide with nearly three billion dollars of funding invested annually into its research. As a result, this research yields promising new treatments, effective techniques, and updated guidelines, all aimed at improving the management of CVD and better targeting efforts to reduce its prevalence. Yet, it takes an average of 17 years for research to translate into changes in clinical practice and neither research findings nor guideline publications promise behavioral modifications or adoption of new practices.[1] This 'translation lag' often underscores the persistent challenge of effectively integrating scientific advancements into everyday care of patients. The reasons behind this 'know-do gap' are complex and multifaceted.[2] Given the evidence that implementation of evidence-based practices improves the quality of care and patient outcomes, addressing this gap is more critical than ever.

Implementation research and how it fits into the translational process

Implementation science seeks to bridge the 'know-do-gap' between research and practice by investigating the factors that facilitate or hinder the successful adoption, implementation, and sustainability of interventions, ultimately contributing to the improvement of the quality and outcomes of services provided. It employs the systematic study of methods and strategies aimed at effective integration of evidence-based practices into real-world settings, encompassing a multidisciplinary approach that examines the complex interactions between interventions, individuals, organizations, and the broader socio-cultural context. This ultimately contributes to the improvement of the quality of care delivered and outcomes of services provided.

Implementation research is regarded as an important emerging area of research in translational research so much so that the National Institute of Health has dedicated it its own 'T3 research' stage in the translational process. Preceding T3 are the foundational 'T1 research' stage (scientific question to clinical research) and the 'T2 research' stage (clinical research to clinical practice). The T3 research stage aims to promote and increase uptake and implementation of evidence-based recommendations designed with knowledge from T1 and T2 research. In discussion of T3 research, the term dissemination is often paired with implementation, to create dissemination and implementation research (D&I). D&I and implementation research are often used interchangeably. Dissemination focuses on the process of spreading and communicating knowledge, ensuring its accessibility and usability. Implementation focuses on the actual adoption and integration of these practices into these settings.

Potential pitfalls

There are several common pitfalls to be aware of when conducting implementation research.

- Collaborative decision-making among various stakeholders is crucial to the success of proposed interventions

Translational Cardiology. https://doi.org/10.1016/B978-0-323-91790-2.00013-7

is crucial to the success of their implementation. Engaging everyone early and throughout the process of implementation research will ensure the successful adoption and integration of appropriate interventions.

- Choosing inappropriate theories and frameworks could inhibit evaluation of behavior change techniques and proper implementation of interventions. Efforts for implementation can only yield successful results with proper use of established theories and frameworks.
- Standardizing interventions in a 'one-size-fits-all' generalizing approach, without using an adequate multidisciplinary assessment, can negatively affect the mediators of behavior change and compromise the outcome measures. It's important to consider structural and social determinants of health when designing and implementing interventions to ensure they are effective across diverse populations and settings.

Theoretical basis of implementation research in cardiology

The implementation of various theories and frameworks guide the process of translating scientific knowledge into practical applications that improve patient care. These theories and frameworks help researchers and clinicians elucidate the complex factors involved in implementing new interventions into clinical practice. Three key theories and frameworks are commonly applied to effectively accelerate the implementation of evidence-based practices: Diffusions of Innovations Theory, Reach, Effectiveness, Adoption, Implementation, and Maintenance (RE-AIM) Framework, and the Consolidated Framework for Implementation Research (CFIR). Each provides a unique lens through which to examine and enhance the implementation of new cardiology interventions, ensuring they are not only adopted but also sustained in real-world healthcare settings.

Diffusion of innovations theory

Roger's Diffusion of Innovation Theory explains the adoption of new ideas or innovations based on five attributes: (1) relative advantage, (2) compatibility, (3) complexity, (4) trialability, and (5) observability.[3] Relative advantage pertains to the perceived superiority of an innovation over its predecessor, influencing easier adoption. Compatibility involves how well the innovation aligns with existing values, experiences, and needs, with greater compatibility enhancing adoption likelihood. Complexity refers to the perceived difficulty of understanding and using the innovation, with simplicity being a critical factor in fostering adoption. Roger's theory categorizes innovations on a complexity-simplicity continuum, acknowledging that potential adopters may not fully understand the relevance of the innovation.[3,4] Trialability is the degree to which an innovation can be experimented with on a limited basis, providing an avenue for potential adopters to assess its viability before full implementation. This is particularly crucial, as innovations often require time, energy, and resources. Lastly, observability gauges the visibility of positive outcomes resulting from the innovation's implementation. If potential adopters can see the benefits, the innovation becomes more adaptable.[3,4]

One study explored the adoption of medical innovations among cardiologists using four agent-based computational models. The factors considered include physicians' network connections, leaders' opinions, adopters' categories, autonomy, scientific evidence, patients' pressure, affordability, and promotion. Simulation results indicate that social imitation among local cardiologists significantly influences innovation diffusion, particularly when opinion leaders endorse the innovation.[5] Notably, many cardiologists may accept innovations even without complete scientific evidence, especially if local leaders act as promoters. Patients' pressure is a key influencer on innovation adoption. The models provide insights into physicians' collective behavior in accepting or refusing medical innovations, emphasizing the role of patients' pressure and accessibility to medical coverage in the effective adoption of population healthcare.[5]

Reach, effectiveness, adoption, implementation, and maintenance framework

The RE-AIM Framework, devised over 2 decades ago, emerged as a crucial tool addressing the challenges associated with the translation of scientific evidence into practice and policy.[6] It uniquely underscores the significance of both internal and external validity, placing a particular emphasis on transparency in reporting across its dimensions. RE-AIM has gained substantial traction as a planning and evaluation framework across multiple disciplines, including public health, behavioral science, and implementation science. Its widespread adoption is evidenced by over 700 publications explicitly utilizing RE-AIM for planning, evaluation, or both. This highlights its versatility and applicability across a spectrum of settings, populations, and health issues, extending from clinical and community contexts.[7,8]

In one of the studies conducted at Uganda's largest HIV clinic, the research aimed to assess the effectiveness and implementation of a World Health Organization (WHO) HEARTS intervention integrating hypertension management into HIV care among persons living with HIV (PLHIV). Using a mixed methods approach, the adapted WHO HEARTS intervention included lifestyle counseling, free hypertension medications, treatment protocols, task shifting, and monitoring tools. The study, employing the RE-AIM framework, revealed significant improvements in reach, with 29% of diagnosed hypertensive PLHIV initiating integrated treatment compared to 1% at baseline.[9]

Effectiveness was demonstrated through enhanced hypertension control and improved blood pressure levels. Adoption saw 100% healthcare provider training and acceptance. Implementation was deemed feasible and conducted with fidelity, while maintenance strategies were proposed for long-term sustainability. The WHO HEARTS intervention proved acceptable, feasible, and effective, emphasizing the need for its integration into national guidelines to ensure ongoing success in managing hypertension within the context of HIV care[9]

Consolidated framework for implementation research

In the realm of healthcare implementation, various theories exist to guide the translation of research-based knowledge, such as evidence-based clinical practice guidelines (CPGs), into practical settings. To address the need for a comprehensive framework, Damschroder et al. introduced the CFIR, amalgamating 19 existing implementation theories into five major domains (*intervention characteristics, outer setting, inner setting, characteristics of individuals, and process*) with 39 underlying constructs and subconstructs. The CFIR serves as an explanatory framework, assisting in understanding factors influencing implementation success.[10]

One study suggests pairing the CFIR with a "process" or "action theory" like participatory action research (PAR) for a more complete understanding.[11] The CFIR has been applied in diverse studies, showcasing its utility in explaining and describing research findings across clinical areas. Notably, it has been used to evaluate substance abuse treatments, identify influential contextual constructs, and explore the transportability of interventions. A study applying the CFIR to a fall-prevention CPG implementation project aims to evaluate the framework's comprehensiveness, applicability, and usefulness during an ongoing implementation. This assessment focuses on the constructs, utility, and practical application of the CFIR in guiding implementation processes.[12]

References

1. Rubin R. It takes an average of 17 years for evidence to change practice—the burgeoning field of implementation science seeks to speed things up. *JAMA*. 2023;329(16):1333–1336. https://doi.org/10.1001/jama.2023.4387.
2. Sarkies N, Jones LK, Gidding SS, Watts GF. Improving clinical practice guidelines with implementation science. *Nat Rev Cardiol*. 2022;19(1):3–4. https://doi.org/10.1038/s41569-021-00645-x. https://www.ncbi.nlm.nih.gov/pubmed/34799708.
3. Greenhalgh T, et al. Diffusion of innovations in service organizations: systematic review and recommendations. *Milbank Q*. 2004;82(4):581–629.
4. Scott SD, Plotnikoff RC, Karunamuni N, et al. Factors influencing the adoption of an innovation: an examination of the uptake of the Canadian Heart Health Kit (HHK). *Implement Sci*. 2008;3:41.
5. Borracci RA, Giorgi MA. Agent-based computational models to explore diffusion of medical innovations among cardiologists. *Int J Med Inf*. 2018;112:158–165.
6. Glasgow RE, Vogt TM, Boles SM. Evaluating the public health impact of health promotion interventions: the RE-AIM framework. *Am J Publ Health Res*. 1999;89(9):1322–1327.
7. Glasgow RE, Harden SM, Gaglio B, et al. RE-AIM planning and evaluation framework: adapting to new science and practice with a 20-year review. *Front Public Health*. 2019;7:64.
8. Kwan BM, McGinnes HL, Ory MG, Estabrooks PA, Waxmonsky JA, Glasgow RE. RE-AIM in the real world: use of the RE-AIM framework for program planning and evaluation in clinical and community settings. *Front Public Health*. 2019;7:345.
9. Muddu M, Semitala FC, Kimera ID, et al. Using the RE-AIM framework to evaluate the implementation and effectiveness of a WHO HEARTS-based intervention to integrate the management of hypertension into HIV care in Uganda: a process evaluation. *Implement Sci Commun*. 2023;4:102. https://doi.org/10.1186/s43058-023-00488-2.
10. Damschroder LJ, Aron DC, Keith RE, Kirsh SR, Alexander JA, Lowery JC. Fostering implementation of health services research findings into practice: a consolidated framework for advancing implementation science. *Implement Sci*. 2009;4:50. https://doi.org/10.1186/1748-5908-4-50.
11. Grol RPTM, Bosch MC, Hulscher MEJL, Eccles MP, Wensing M. Planning and studying improvement in patient care: the use of theoretical perspectives. *Milbank Q*. 2007;85:93–138. https://doi.org/10.1111/j.1468-0009.2007.00478.x.
12. Breimaier HE, et al. The consolidated framework for implementation research (CFIR): a useful theoretical framework for guiding and evaluating a guideline implementation process in a hospital-based nursing practice. *BMC Nurs*. 2015;14. https://doi.org/10.1186/s12912-015-0088-4. Accessed January 29, 2024.

Further reading

1. Allen LA, McIlvennan CK. Mis-GUIDED-the importance of negative trials of health care delivery and implementation science. *JAMA Cardiol*. February 1, 2021;6(2):135–136. https://doi.org/10.1001/jamacardio.2020.5778.
2. Chan WV, Pearson TA, Bennett GC, et al. ACC/AHA special report: clinical practice guideline implementation strategies: a summary of systematic reviews by the NHLBI implementation science work group: a report of the American College of cardiology/American Heart association task force on clinical practice guidelines. *Circulation*. February 28, 2017;135(9):e122–e137. https://doi.org/10.1161/CIR.0000000000000481.
3. Fisher ES, Shortell SM, Savitz LA. Implementation science: a potential catalyst for delivery system reform. *JAMA*. 2016;315(4):339–340. https://doi.org/10.1001/jama.2015.17949.

Chapter 88

Design and analysis

Thomaz Alexandre Costa[1], Eduardo Rochitte[2], Carlos Eduardo Rochitte[3] and Bruno B. Lima[4]

[1]*Federal University of Ceara, School of Medicine, Fortaleza, Brazil;* [2]*Centro Universitário FMABC, Faculdade de Medicina, Santo André, São Paulo, Brazil;* [3]*Heart Institute, InCor, University of São Paulo Medical School, São Paulo, Brazil;* [4]*Division of Cardiovascular Medicine, Department of Internal Medicine, Vanderbilt University Medical Center, Nashville, TN, United States*

Get started/objects

This chapter will provide an overview of relevant topics regarding the implementation of evidence-based practices and interventions in daily clinical practice.

Key points

- Overview of the commonly used experimental clinical research methodologies.
- Review implementation research concepts and commonly used methodologies.
- Exemplify measurement challenges, fidelity, external validity, and dissemination strategies through real-world examples.

Introductory section/why it matters

Researchers have historically estimated a prolonged timeline of 17–20 years for integrating new evidence-based practices into the daily standard of care, with less than 50% achieving widespread adoption in clinical settings.[1] The underutilization of clinical innovations imposes substantial costs on our society, as researchers and funding organizations invest significant time and resources in promoting advances in scientific knowledge. However, up to 80% of medical research dollars fail to impact public health for various reasons.[2] Above that, this "chasm" between innovation and implementation has impacted both healthcare systems and patients by hindering advancements in clinical quality of care.[3] Contextual factors, not treatment effectiveness, are suggested as the main determinants of clinical innovation adoption.[1] For example, factors that could affect the adequate implementation of a novel intervention include the need to adapt to a new community setting and target population, the lack of the appropriate knowledge and skills to execute the intervention as intended by the local healthcare providers, or the need to deal with insufficient resources (e.g., in developing nations).[4]

This gap between medical discovery and evidence-based practice implementation triggered the development of the emerging field of "implementation science." This line of research aims to investigate proven evidence-based healthcare interventions and evaluate strategies to promote their adoption, fidelity, and sustained use in routine practice.[5]

This chapter will summarize the most common clinical experimental study designs and provide an overview of implementation research, including its objectives and commonly used methodologies, methodological challenges, and strategies to approach fidelity and external validity. Moreover, we will include real-world published examples of implementation research from the literature on cardiovascular diseases.

Most common experimental clinical research methodologies

Defining scientific research methodologies is crucial to understanding how a study will be designed to answer its proposed research questions. The commonly analytic clinical study designs are divided into two distinct groups: experimental studies and observational studies. Herein, we will further explore experimental clinical studies.

Experimental clinical study types

Experimental studies are designed to facilitate comparisons between distinct groups, emphasizing controlled conditions. While observational studies may involve comparisons, they often primarily serve descriptive purposes. Experimental clinical research aims to test an intervention in a controlled scenario by analyzing prespecified outcomes

Translational Cardiology. https://doi.org/10.1016/B978-0-323-91790-2.00090-3

in the exposed individuals (i.e., intervention group) and comparing them with controls. To conduct safe and successful research and to translate basic studies to clinical utilization of interventions, clinical trials are classically divided into three main phases as the study develops to cover a wider number of scenarios and analyze the data from it.[6]

- Phase 1: This first phase of research intends to evaluate the safety of a drug or procedure and identify its side effects. This phase usually takes over 2–12 months of testing. If the results suggest safety, phase 2 can be started.
- Phase 2: This phase might last about 2 years and involves groups that are tested with different doses of the interventional drug to measure which range of doses are acceptable depending on the aim of the study. Thus, this phase investigates the safety and efficacy (in a limited number of patients) of the proposed intervention with no control group. If the results suggest safety once more, phase 3 can be conducted.
- Phase 3: The volunteers are divided into intervention or control groups, ideally through a randomization process. The intervention is tested on individuals from each group, and data is collected and analyzed over time to determine the efficacy of the implemented innovation. It involves larger populations and might be applied in multiple sites. Moreover, these trials can also be classified based on the applied blinding strategy: (1) open trial: the research team and the volunteers know in which group the participants are, either the control or the intervention group; (2) single-blind randomized trial: only the investigator knows in which group the participants are and the subjects have no information about which group they are included; and (3) double-blind randomized trial: both researcher and volunteer are blinded to which group the volunteers were randomized.[6]

Classically, clinical trials aim either to evaluate an intervention's efficacy or effectiveness. The first tests whether a treatment or intervention produces the expected result under ideal invariant conditions from beginning to end.[7] The primary objective is to establish a cause-and-effect relationship between the intervention and the outcome, focusing on the intervention's internal validity. However, in those trials, the sample, outcome measurements, and intervention are meticulously controlled to achieve this goal, with little regard for general applicability beyond the specific trial set, limiting its external validity. The second type enrolls volunteers in clinical sites similar to those where the intervention is expected to be used to evaluate its effectiveness in a "real world" setting and its external validity (i.e., generalizability to the practice settings in which the clinical innovation will be used). However, as mentioned in the introduction section, proving treatment effectiveness is insufficient to guarantee the adoption of newly discovered evidence-based interventions in routine clinical practice.[5]

In the following sections, we will further introduce, discuss, and exemplify implementation research, which shares characteristics with the above-mentioned study designs but aims to identify the factors that affect the uptake of evidence-based interventions into routine clinical practice.

Implementation science: Objectives and commonly used methodologies

The objective of implementation science is to integrate research findings and other evidence-based practices into regular procedures, aiming to enhance the quality and effectiveness of health services. Implementation studies/trials are often combined with effectiveness trials in hybrid effectiveness-implementation studies to address those questions while also testing the effect of an intervention in clinical outcomes in real-world scenarios.[8]

Numerous implementation strategies with varying levels of complexity have been identified. They are called "discrete implementation strategies" and encompass individual approaches or techniques that might be used in various situations.[9] For example, the distribution of educational materials, communication with local opinion leaders, clinical supervision, or utilizing clinical reminders.[9] Ideally, these discrete strategies should be combined into multifaceted implementation strategies, where two or more approaches are combined to introduce a clinical innovation effectively.[3] The Expert Recommendations for Implementing Change (ERIC) study has meticulously identified 73 such discrete strategies, categorizing them into nine clusters, as outlined and exemplified in Table 88.1.[10] The choice of a specific discrete strategy depends upon the requirements of each setting, the nature of the clinical innovation undergoing implementation, and the characteristics of the individuals who will be utilizing and receiving the innovation, as we illustrate in the following examples.[10]

A variety of tools, guidelines, and methods to properly select and utilize those strategies has been developed, but the four steps usually used in implementation research consist of (1) identify the evidence-based intervention of interest, (2) identify barriers and facilitators for its use, (3) design an implementation strategy to address these barriers/facilitators, and (4) test the implementation strategy.[5] Below, we provide three real-world examples from the published literature that exemplify using some of those strategies and introduce other concepts.

TABLE 88.1 ERIC discrete implementation strategies by thematic cluster with discrete strategies examples.

Cluster	Discrete strategies	
Use evaluative and iterative strategies	Assess for readiness and identify barriers and facilitators	Audit and provide feedback
Provide interactive assistance	Facilitation	Provide local technical assistance
Adapt and tailor to context	Tailor strategies	Promote adaptability
Develop stakeholder interrelationships	Identify and prepare champions	Organize clinician implementation team meetings
Train and educate stakeholders	Conduct ongoing training	Provide ongoing consultation
Support clinicians	Facilitate relay of clinical data to providers	Remind clinicians
Engage consumers	Involve patients/consumers and family members	Intervene with patients/consumers to enhance uptake and adherence
Utilize financial strategies	Fund and contract for the clinical innovation	Access new funding
Change infrastructure	Mandate change	Change record systems

Example #1: A pragmatic effectiveness-implementation trial and implementation fidelity

The term "Implementation fidelity" is defined as the degree to which interventions/programs are implemented as intended by the intervention developers and it is one of the measurements that might predict the success of a new intervention.[11]

The REACH-HF program was tested in a phase 3 multicenter randomized clinical trial and shown to be a clinically effective and cost-effective, home-based, health-care professional-facilitated cardiac rehabilitation program for patients with heart failure, first published in 2019.[12] After its publication, the research group was responsible for developing a pragmatic effectiveness-implementation trial to evaluate its implementation and patient-reported outcome data in four early adopter sites.[13] In this study, they completed two of the four main steps abovementioned. The first step was to identify the evidence-based intervention of interest, in their case, the REACH-HF program. The second step was done by exploring the variations in the implementation process across the four early adopter sites and how such variations impact patients' outcomes. Unfortunately, as with many other new interventions, they found difficulties in implementing its practice in the real world mainly due to challenges in implementation fidelity. They observed lower effectiveness in the early adopter sites, as there were substantial differences between the trial and the sites regarding the patient population and treatment dose/duration proportion. They also encountered exacerbated challenges due to critical service disruptions due to the COVID-19 pandemic, with small sample size, missing data, and the inability to deliver the treatment as it was intended (i.e., face-to-face and in the patient's home).[13] They could not apply the intervention as originally planned and tested in the REACH-HF trial, in other words—failed to achieve implementation fidelity—which affected its success in improving patient outcomes.[11,13] Nevertheless, this study was important as a hypothesis generation study, as future research might be able to investigate stages three and four and properly evaluate the most effective implementation strategy for this intervention while ensuring the fundamental characteristics of the intervention remain intact (fidelity).

Example #2: TAVR and the successful use of "Change infrastructure" cluster strategy to gather evidence of external validity

Transcatheter aortic valve replacement (TAVR) first device approval by the US Food and Drug Administration (FDA) occurred in late 2011 for patients with symptomatic, severe aortic stenosis considered at extreme risk for surgical aortic valve replacement. In the following years, its indications were expanded to patients with high risk (2012), intermediate risk (2016), and low risk (2019) for surgical aortic valve replacement (SAVR).[14,15]

The device's success in being implemented in daily clinical practice is due to the use of multiple strategies since its first approval, such as access to local technical assistance, access to funding, and the involvement of patients/consumers and family members as the technique provided a less invasive procedure, desired by the patients and their families.[9] Nevertheless, one of the mentioned strategies in Table 88.1 was essential in these label expansion approvals:

the "Change infrastructure" cluster strategy. By linking the financial reimbursement to submitting patient-level data for all TAVR patients to a recognized national registry ("Change record systems" strategy), the Centers for Medicare & Medicaid Services (CMS) was essential in providing evidence of the technique's external validity and facilitating its implementation.[1] This was done by establishing a robust observational database comprising data from over a quarter of a million patients and the safety and efficacy analyses derived from this database contributed to implementing this technique in the routine practice of interventional cardiologists.[14]

Example #3: Measuring implementation success through mixed-methods research

Mixed-methods research involves collecting and analyzing both quantitative and qualitative data within the same clinical study.[16] It provides a more comprehensive understanding of research issues than either approach alone, and it has been extensively used as a measurement strategy in implementation research.[17] Qualitative approaches are used to investigate and understand the factors contributing to the success or failure of implementing evidence-based practices. They are also utilized to identify new/different strategies for promoting implementation. On the other hand, quantitative methods are applied to assess and validate hypotheses derived from an established conceptual model. The aim is to achieve a broad understanding of the factors that predict successful implementation.[17]

In Morton et al.[18] researchers employed a mixed method design to examine the implementation of "HOME BP," an effective digital intervention that facilitates healthcare practitioners to initiate and escalate medication when patients' home blood pressure readings rise and that was found to be effective in lowering blood pressure over 12 months.[19] Quantitative intervention usage data and measures regarding adherence were collected from all healthcare practitioners during the trial. Qualitative interviews were conducted with a subsample of practitioners during the trial to investigate their experiences and perceptions of implementing the proposed strategy. By applying this methodology, they gained insights into implementation barriers and facilitators. Consequently, they proposed interventions such as facilitating tracking of patients' readings over time to provide stronger evidence for medication escalation ("Support clinicians" cluster strategy) and permitting more flexibility in decision-making while discouraging clinical inertia due to borderline readings ("Use evaluative and iterative strategies" cluster strategy).

Measurement challenges

There are some challenges concerning measurement in implementation science. These include (1) heterogeneity of descriptors used in publications in the area, (2) the difficulty in registering study protocols due to the iterative nature of implementation strategies, (3) distinction between evidence-based interventions and implementation strategies, (4) assessment of external validity, and (5) variability in defining implementation "success."[20]

Resources that might provide guidance and uniformity during the development of such studies include the use of (1) recommendations for specifying and reporting implementation strategies,[21] (2) frameworks for reporting adaptations and modifications,[22,23] and (3) frameworks to better specify and define implementation outcomes.[24,25]

Dissemination

Dissemination science explores optimal ways to spread evidence-based practices, programs, and policies to a broader societal sector of potential adopters and implementers through communication and education strategies.[1] This definition implies that dissemination encompasses both the goals of external validity, ensuring positive effects can be replicated in diverse settings, and scale-up, replicating positive effects across various conditions and contexts.[26] In contrast, implementation research involves distinct objectives and approaches, although there is often overlap in their purposes and methods.[1]

Potential pitfalls

- Relying solely on clinical treatment efficacy and effectiveness outcomes without measuring implementation outcomes can limit the translation of research into practice.
- The choice of a specific discrete strategy to improve implementation should consider the requirements of each setting, the nature of the intervention undergoing implementation, and the characteristics of the individuals who will be utilizing and receiving the innovation.
- It is important to carefully document implementation strategies and note any changes that occur during the process.
- Utilizing standardized methodologies for specifying and reporting adaptations and modifications to implementation strategies, study protocols, and implementation outcomes is important for ensuring clarity and consistency in research.

Resources

1. Patient Centered Outcomes Research Institute, 2019. PCORI Standards for Studies of Complex Interventions. Retrieved March 16, 2019, from. https://www.pcori.org/research-results/about-our-research/research-methodology/pcori-methodology-standards#Complex.

2. Albrecht, L., Archibald, M., Arseneau, D. et al. Development of a checklist to assess the quality of reporting of knowledge translation interventions using the Workgroup for Intervention Development and Evaluation Research (WIDER) recommendations. Implementation Sci 8, 52 (2013). https://doi.org/10.1186/1748-5908-8-52
3. Implementation Science Journal: https://implementationscience.biomedcentral.com

References

1. Bauer MS, Kirchner J. Implementation science: what is it and why should I care? *Psychiatr Res.* 2020;283:112376. https://doi.org/10.1016/j.psychres.2019.04.025.
2. Chalmers I, Glasziou P. Avoidable waste in the production and reporting of research evidence. *Lancet.* 2009;374(9683):86–89. https://doi.org/10.1016/S0140-6736(09)60329-9.
3. Kirchner JE, Smith JL, Powell BJ, Waltz TJ, Proctor EK. Getting a clinical innovation into practice: an introduction to implementation strategies. *Psychiatr Res.* 2020;283:112467. https://doi.org/10.1016/j.psychres.2019.06.042.
4. Escoffery C, Lebow-Skelley E, Haardoerfer R, et al. A systematic review of adaptations of evidence-based public health interventions globally. *Implement Sci.* 2018;13(1):125. https://doi.org/10.1186/s13012-018-0815-9.
5. Galaviz KI, Barnes GD. Implementation science opportunities in cardiovascular medicine. *Circ Cardiovasc Qual Outcomes.* 2021;14(7). https://doi.org/10.1161/CIRCOUTCOMES.121.008109.
6. Altman DG. Design research: introduction. In: *Practical Statistics for Medical Research.* Chapmann & Hall; 1991.
7. Gartlehner G, Hansen RA, Nissman D, et al. Distinguishing effectiveness from efficacy trials in systematic reviews. In: *Technical Reviews, No. 12.* Agency for Healthcare Research and Quality (US); 2006.
8. Moise N, Cené CW, Tabak RG, et al. Leveraging implementation science for cardiovascular health equity: a scientific statement from the American heart association. *Circulation.* 2022;146(19). https://doi.org/10.1161/CIR.0000000000001096.
9. Powell BJ, Waltz TJ, Chinman MJ, et al. A refined compilation of implementation strategies: results from the Expert Recommendations for Implementing Change (ERIC) project. *Implement Sci.* 2015;10(1):21. https://doi.org/10.1186/s13012-015-0209-1.
10. Waltz TJ, Powell BJ, Matthieu MM, et al. Use of concept mapping to characterize relationships among implementation strategies and assess their feasibility and importance: results from the Expert Recommendations for Implementing Change (ERIC) study. *Implement Sci.* 2015;10(1):109. https://doi.org/10.1186/s13012-015-0295-0.
11. Carroll C, Patterson M, Wood S, Booth A, Rick J, Balain S. A conceptual framework for implementation fidelity. *Implement Sci.* 2007;2(1):40. https://doi.org/10.1186/1748-5908-2-40.
12. Dalal HM, Taylor RS, Jolly K, et al. The effects and costs of home-based rehabilitation for heart failure with reduced ejection fraction: the REACH-HF multicentre randomized controlled trial. *Eur J Prev Cardiol.* 2019;26(3):262–272. https://doi.org/10.1177/2047487318806358.
13. Daw P, Harrison A, Doherty PJ, et al. A pragmatic effectiveness-implementation study comparing trial evidence with routinely collected outcome data for patients receiving the REACH-HF home-based cardiac rehabilitation programme. *BMC Cardiovasc Disord.* 2022;22(1):270. https://doi.org/10.1186/s12872-022-02707-5.
14. Carroll JD, Mack MJ, Vemulapalli S, et al. STS-ACC TVT registry of transcatheter aortic valve replacement. *J Am Coll Cardiol.* 2020;76(21):2492–2516. https://doi.org/10.1016/j.jacc.2020.09.595.
15. Otto CM, Nishimura RA, Bonow RO, et al. 2020 ACC/AHA guideline for the management of patients with valvular heart disease: a report of the American college of cardiology/American heart association joint committee on clinical practice guidelines. *Circulation.* 2021;143(5). https://doi.org/10.1161/CIR.0000000000000923.
16. Tashakkori A, Johnson RB, Teddlie C. *Foundations of Mixed Methods Research: Integrating Quantitative and Qualitative Approaches in the Social and Behavioral Sciences.* Sage Publications; 2020.
17. Palinkas LA, Aarons GA, Horwitz S, Chamberlain P, Hurlburt M, Landsverk J. Mixed method designs in implementation research. *Adm Pol Ment Health.* 2011;38(1):44–53. https://doi.org/10.1007/s10488-010-0314-z.
18. Morton K, Dennison L, Band R, et al. Implementing a digital intervention for managing uncontrolled hypertension in Primary Care: a mixed methods process evaluation. *Implement Sci.* 2021;16(1):57. https://doi.org/10.1186/s13012-021-01123-1.
19. McManus RJ, Little P, Stuart B, et al. Home and Online Management and Evaluation of Blood Pressure (HOME BP) using a digital intervention in poorly controlled hypertension: randomised controlled trial. *BMJ.* 2021:m4858. https://doi.org/10.1136/bmj.m4858. Published online January 19.
20. Chapman A, Rankin NM, Jongebloed H, et al. Overcoming challenges in conducting systematic reviews in implementation science: a methods commentary. *Syst Rev.* 2023;12(1):116. https://doi.org/10.1186/s13643-023-02285-3.
21. Proctor EK, Powell BJ, McMillen JC. Implementation strategies: recommendations for specifying and reporting. *Implement Sci.* 2013;8(1):139. https://doi.org/10.1186/1748-5908-8-139.
22. Wiltsey SS, Baumann AA, Miller CJ. The FRAME: an expanded framework for reporting adaptations and modifications to evidence-based interventions. *Implement Sci.* 2019;14(1):58. https://doi.org/10.1186/s13012-019-0898-y.
23. Miller CJ, Barnett ML, Baumann AA, Gutner CA, Wiltsey-Stirman S. The FRAME-IS: a framework for documenting modifications to implementation strategies in healthcare. *Implement Sci.* 2021;16(1):36. https://doi.org/10.1186/s13012-021-01105-3.
24. Proctor E, Silmere H, Raghavan R, et al. Outcomes for implementation research: conceptual distinctions, measurement challenges, and research agenda. *Adm Pol Ment Health.* 2011;38(2):65–76. https://doi.org/10.1007/s10488-010-0319-7.
25. Glasgow RE, Harden SM, Gaglio B, et al. RE-AIM planning and evaluation framework: adapting to new science and practice with a 20-year review. *Front Public Health.* 2019;7. https://doi.org/10.3389/fpubh.2019.00064.
26. Dearing JW, Kee KF, Quan Peng T. In: Brownson RC, CGA PEK, eds. *Historical Roots of Dissemination and Implementation Science.* 2nd ed. Vol. 55. Oxford Univ Press; 2012.

Chapter 89

Mixed methods research

João P.G. Kasakewitch[1], Henrique Guimarães Barbosa Coelho[2] and Bruno B. Lima[3]

[1]*Beth Israel Deaconess Medical Center, Harvard Medical School, Boston, MA, United States;* [2]*Universidade Federal de Pernambuco, Recife, Brazil;* [3]*Department of Cardiothoracic Imaging, Vanderbilt Heart and Vascular Institute, Vanderbilt University Medical Center, Nashville, TN, United States*

Key points

- Mixed methods research combines quantitative and qualitative approaches to offer a more complete understanding of complex research questions.
- This research approach provides flexibility, allowing researchers to tailor their study design to best address the research question with both quantitative and qualitative methods.
- By integrating diverse data types, mixed methods research enhances the validity of results through cross-verification and provides a more reliable understanding of the topic.

Introductory section/why it matters

In medical research, it is common to see studies focused on generating numerical data. However, qualitative methods, such as semistructured interviews, may give a subjective result of the study. Mixing quantitative and qualitative methods into one study may improve the conclusions and broaden our interpretation of the findings.[1]

Quantitative research is rooted in positivist philosophy, which can be seen in the value it places on concepts such as causality, generalizability, and replicability. Systematic investigation using numerical or statistical data is classified as quantitative research. Therefore, using measurable objects of study. A feature unique to quantitative research is the possibility of testing theories formally through hypotheses and applying statistical analyses.[2]

Qualitative research takes a different path. Instead of collecting numerical data like in quantitative research, it explores and provides deeper insights into real-world problems. Qualitative research gathers participants' experiences, perceptions, and behavior, helping generate hypotheses and understanding quantitative data.[3]

Focused on asking open-ended questions, it searches for the "how" and "why." The positive aspect of qualitative research is the possibility of explaining processes and patterns of human behavior that are not easily quantifiable. As examples, we can highlight factors such as experiences and behaviors. A qualitative approach allows participants to explain how, why, or what they were thinking, feeling, and experiencing at a particular moment or during an event of interest.[3]

Integrating quantitative and qualitative methods, mixed methods research significantly enhances the depth and breadth of findings in healthcare research. This approach is particularly effective in contextualizing patient experiences, a dimension often inadequately captured by quantitative methods alone. Such detailed contextualization is crucial in engaging stakeholders and informs the planning, implementation, assessment, and monitoring of healthcare interventions, making them more responsive to patient needs. Recognizing its comprehensive value, several funding agencies now advocate for including both methodologies, underscoring the importance of a holistic view in developing effective health policies and practices.[1,4]

Examples of ...

Explanatory studies

The mixed methods sequential explanatory design comprises two phases. The first phase involves collecting and analyzing quantitative data, primarily numeric, while the second phase focuses on collecting and analyzing qualitative data, such as texts. This design aims to explain or elaborate on the quantitative results obtained in the first phase. The two phases are connected by an intermediate stage, where data analysis in the first phase guides data collection in the second. Understanding the research problem in more depth is the rationale for subsequently analyzing the quantitative data. This design provides an opportunity to explore quantitative results in detail, which is especially valuable when facing unexpected results from

Translational Cardiology. https://doi.org/10.1016/B978-0-323-91790-2.00070-8

a study. However, it is important to note that the lengthy time and resource availability needed to collect and analyze both data types are limitations of this design.[5]

For instance, Riegel et al. analyzed symptom perception in a sample of patients with chronic Heart Failure who also had an implantable cardioverter defibrillator (ICD) with continuous intrathoracic impedance monitoring capability. This device measured daily thoracic fluid status and was the gold standard for fluid retention.

The study integrated biweekly phone interviews about symptoms, assessing perceived fluid retention, and a symptoms diary, evaluating common HF presentations. Interviews identified symptom interpretation and response patterns, assessing thoughts about signs, symptoms, and causes through open-ended questions. The study compared objective findings, measured by the average fluid index and variability in the fluid index provided by the device, and the subjective data from symptom diaries and interviews. Data integration using triangulation methods identified patterns of symptom interpretation and response, helping demonstrate how patients monitor and interpret symptoms of HF. The study identified problems with symptom detection with evidence of symptom-hemodynamic mismatch, leading to delayed management response and risk of hospitalization.[6]

Exploratory studies

The exploratory sequential design initially explores qualitative data collection and analysis. The findings of the prior will guide the quantitative data collection. This procedure will allow for testing the initial results by designing a quantitative instrument and administering it to a sample from a population.[7]

In a study by Keir et al., research priorities in patients with moderate-to-severe complexity Congenital Heart Disease were assessed. Participants engaged in controlled group activities within their age category to measure between-group and temporal differences in experiences. Through discussions and brainstorming, patients addressed dilemmas that could be potentially improved through research. Using an exploratory sequential design, the authors initially analyzed qualitative data collected during group activities. The data were integrated to develop a list of research priorities, which were then quantitatively assessed for participants' preferences. The overall quantitative results were compared with qualitative findings for generalizability and validity. The study revealed that patients prioritized themes such as physical activity, mental health, cognition, improved communication, and knowledge. Ultimately, the study provides valuable insights to guide future project designs to improve care for individuals with Congenital Heart Disease.[8]

Convergent studies

In a mixed methods convergent (or concurrent) design, the collection and analysis of the quantitative and qualitative data occurs in the same study phase. This approach allows for a dynamic process where ongoing data collection and analysis influence modifications to the data collection procedures. For instance, initial qualitative results may shape the focus and types of quantitative data gathered, or on the contrary. In a more conventional and technically uncomplicated variation, qualitative and quantitative data are collected simultaneously, and the integration analysis begins well after the data collection process has progressed or concluded. Frequently, both types of data are initially separately analyzed before being combined.[7]

In the study by Durante et al., convergent methods were used to understand and predict the burden of heart failure patient's caregivers. The research proceeded with a parallel collection of quantitative and qualitative data, identifying the strengths of these approaches and examining the consistencies across datasets. Data was assessed through semistructured interviews. In that study, the burden was measured using the Caregiver Burden Inventory (CBI); other variants, such as Care dependency, health-related quality of life (HRQL) profile, cognitive assessment in patients, and Social Support, were also evaluated through scales and numerical data. The quantitative data was analyzed using descriptive statistics, while the qualitative data were analyzed using content analysis. The mixed methods analysis was completed through simultaneous integration after individual qualitative and quantitative analysis. The merged results were presented in joint displays depicting the inferences and confirmed, discordant, and expanded meta-inferences.

This approach allowed the authors to identify caregiver-burden-related predictors, such as diminished personal time for care, anticipated bleak future, somatic stress, worry in time, and restructuring life. Furthermore, it explored how burden can affect the capability of informal caregivers to support and care for their relatives with heart failure. Those findings guided the suggestion of further intervention addressing the caregiver burden and, consequently, improving the quality of life of patients and caregivers.[9]

Get started

A mixed methods research project begins with a clear articulation of the research question, as it fundamentally influences the design of the study. The key is to assess whether the research question would benefit from a combination of qualitative and quantitative data. If the question leans toward qualitative aspects, an exploratory sequential design is often suitable. This approach starts with qualitative data to identify key themes and variables, which then inform the subsequent quantitative phase. However, an explanatory sequential design is more appropriate if the question is primarily quantitative, focusing on measuring and analyzing variables. Here, quantitative data collection and analysis precede a qualitative phase, which helps explain the quantitative results in more depth.

Nonetheless, the choice of design also depends on the variables of interest and their measurability. If the variable is known and measurable, an explanatory sequential design is fitting as it allows for a comprehensive analysis followed by a detailed exploration. On the other hand, if the variable is not clearly defined or lacks a reliable measurement method, an exploratory sequential design is advisable to first identify and then measure the variable effectively. Researchers should also consider the time constraints of their study. Sequential designs, while thorough, are time-intensive. If time is a limiting factor, a convergent design, which involves simultaneously collecting and analyzing both qualitative and quantitative data, can be more practical. This approach facilitates a more immediate integration of data types but requires careful planning to ensure effective data synthesis.

Another crucial aspect of initiating mixed methods research is evaluating and assembling the research team. It is essential to have expertise in both qualitative and quantitative methodologies within the team. If the team lacks a qualitative researcher, it is highly beneficial to include one, especially for their skills in data coding and analysis, which are critical in interpreting qualitative data effectively. The right team composition can significantly enhance the study's efficacy, ensuring that both elements of the mixed methods approach are handled with the necessary expertise and attention.[10,11]

Potential pitfalls

In conducting mixed methods research in translational research, it is crucial to avoid the pitfall of underestimating the time required for the study's phases. Mixed methods research inherently involves more time for data collection, especially in sequential designs where the second phase cannot begin until the first phase's results are analyzed. This extended timeframe is further compounded when each phase may require separate submissions to institutional review boards (IRBs). Additionally, qualitative data coding and analysis are time-intensive, and integrating these findings with quantitative data can be challenging. Researchers predominantly trained in quantitative methods may find qualitative analysis particularly demanding, risking both the study's timeline and the quality of the analysis. A strategic approach to mitigate this is the inclusion of a multidisciplinary team with expertise in qualitative methods. However, adding researchers from varied disciplines necessitates a collaborative environment open to diverse perspectives, ensuring the project's smooth progression (Image 89.1 and Table 89.1).[12,13]

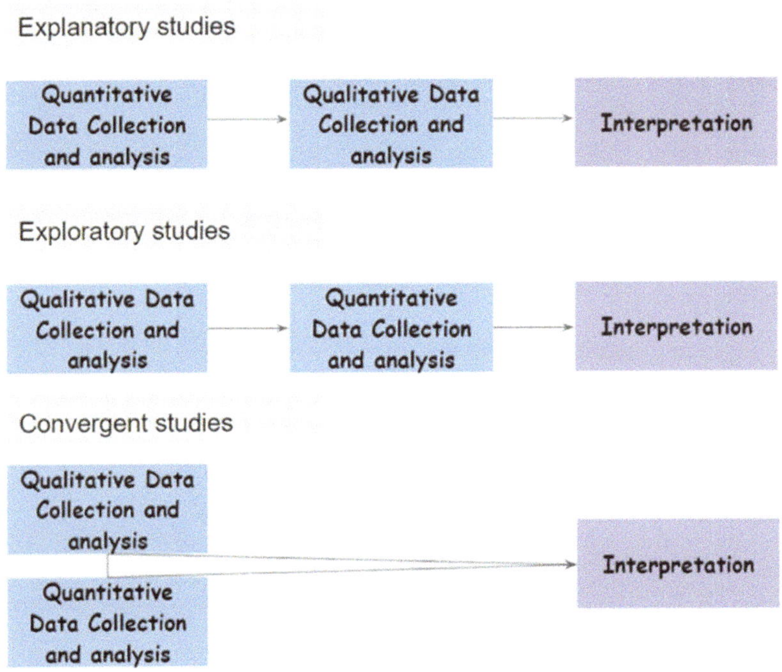

IMAGE 89.1 Basic mixed methods research designs: explanatory sequential, exploratory sequential, and convergent designs.

TABLE 89.1 Mixed methods study models.

Study	Design	Topic	Phase 1	Phase 2	Interpretation
Riegel et al.[6]	Explanatory	Symptom perception in patients with chronic heart failure	Quantitatively measured thoracic fluid status according to the variability and average fluid index provided by the device.	Assessed perceived fluid retention and common HF symptoms through interviews, identified symptom interpretation and response patterns, assessing thoughts about signs, symptoms, and causes through a series of open-ended questions.	The study evidenced issues with symptom detection, detecting the presence of symptom-hemodynamic mismatch, leading to delayed management response and risk of hospitalization.
Keir et al.[8]	Exploratory	Research priorities in patients with Congenital Heart Disease	Patients participated in group discussions and shared their experiences, and brainstormed ways to address care dilemmas through clinical and health services research.	Developed a list of research priorities, quantitatively assessed for participants' preferences	Topics such as physical activity, mental health, cognition, improved communication, and knowledge were revealed to be patients' priorities.
Durante et al.[9]	Convergent	Burden among caregivers of individuals with heart failure	Simultaneous phase Performed semistructured interviews with caregivers, while quantitatively measuring burden through the Caregiver Burden Inventory, care dependency, health-related quality of life profile, cognitive assessment in patients, and social support.		Caregiver-burden-related predictors were identified, including diminished personal time for care, anticipation of a bleak future, somatic stress, concerns about the allocation of time, and the need to restructure life. The study also delved into the ways in which this burden can impact the ability of caregivers to support and care for their relatives with heart failure.

Resources

- Creswell, J.W., Plano Clark, V.L. Designing and conducting mixed methods research, third ed., 2018. Los Angeles, CA: Sage.

References

1. Mixed Methods Research. *Harvard Catalyst.* https://catalyst.harvard.edu/community-engagement/mmr/.
2. Watson R. Quantitative research. *Nurs Stand.* April 1, 2015;29(31):44–48. https://doi.org/10.7748/ns.29.31.44.e8681.
3. Tenny S, Brannan JM, Brannan GD. Qualitative study, 2023 Jan. In: *StatPearls.* Treasure Island (FL): StatPearls Publishing; September 18, 2022. PMID: 29262162.
4. Creswell JW, Klassen AC, Plano Clark VL, Smith KC for the Office of Behavioral and Social Sciences Research. *Best Practices for Mixed Methods Research in the Health Sciences.* National Institutes of Health; August 2011.
5. Ivankova NV, Creswell JW, Stick SL. Using mixed-methods sequential explanatory design: from theory to practice. *Field Method.* 2006;18(1):3–20. https://doi.org/10.1177/1525822x05282260.
6. Riegel B, Dickson VV, Lee CS, et al. A mixed methods study of symptom perception in patients with chronic heart failure. *Heart Lung.* 2018 Mar-Apr;47(2):107–114. https://doi.org/10.1016/j.hrtlng.2017.11.002.
7. Fetters MD, Curry LA, Creswell JW. Achieving integration in mixed methods designs-principles and practices. *Health Serv Res.* December 2013;48(6 Pt 2):2134–2156. https://doi.org/10.1111/1475-6773.12117.
8. Keir M, Tarr C, McFadden C, et al. Determining research priorities with teen and adult congenital heart disease patients: a mixed-methods study. *CJC Pediatr Congenit Heart Dis.* December 13, 2022;2(2):74–81. https://doi.org/10.1016/j.cjcpc.2022.12.002.
9. Durante A, Younas A, Cuoco A, et al. Burden among informal caregivers of individuals with heart failure: a mixed methods study. *PLoS One.* November 17, 2023;18(11):e0292948. https://doi.org/10.1371/journal.pone.0292948.
10. Wasti SP, Simkhada P, van Teijlingen ER, Sathian B, Banerjee I. The growing importance of mixed-methods research in health. *Nepal J Epidemiol.* March 31, 2022;12(1):1175–1178. https://doi.org/10.3126/nje.v12i1.43633.
11. Schoonenboom J, Johnson RB. How to construct a mixed methods research design. *Kolner Z Soz Sozpsychol.* 2017;69(Suppl 2):107–131. https://doi.org/10.1007/s11577-017-0454-1.
12. Deshmukh A, Cornman-Homonoff J. Mixed methods research. In: *Translational Interventional Radiology.* Elsevier; 2023:459–462. https://doi.org/10.1016/b978-0-12-823026-8.00078-x.
13. Tashakkori A, Newman I. Mixed methods. In: *International Encyclopedia of Education.* Elsevier; 2010:514–520. https://doi.org/10.1016/b978-0-08-044894-7.00287-6.

Chapter 90

Population and setting specific implementation

Isaiah V.N. Ware, Amari Evans and Maria Betzabe Diaz
Mercer University School of Medicine, Columbus, GA, United States

Key points

- Women and SGMs have high rates of CVD mortality often due to clinician bias, inconsiderate policies and guidelines, and discriminatory laws.
- Black Americans suffer an increased risk of CVD, higher mortality for CVD, and the highest burden of CVD due to poverty and discrimination.
- Despite efforts to improve rural health, CVD mortality rates continue to be disproportionately higher in rural settings.
- Limited data is available to determine the true causative agent for poorer CVH in certain population specific and setting specific areas.

Introduction

Despite rapid pharmacological and technological advancements in cardiology, cardiovascular disease (CVD) remains a leading cause of death in the United States (US) and impacts certain groups of Americans disproportionately. Lack of representation in clinical trials contributes to persistent health disparities, which hinder clinical understanding of appropriate therapeutic interventions for treating diverse populations. This chapter aims to bring light to these differences and disparities as well as the clinician's responsibility to look at the patient holistically and implement evidence based medicine to improve cardiovascular health (CVH).

Sex specific risk factors for cardiovascular disease

Accounting for 21.8% of deaths in females and 24.2% of deaths in males, CVD is represented almost equally in men and women in the United States.[1] While men and women can experience CVD differently—such as women presenting with more jaw than chest pain during a myocardial infarction[2]—common risk factors include type 2 diabetes, hyperlipidemia, hypertension (HTN), and obesity. However, new evidence indicates women also have sex specific risk factors, including pregnancy- related conditions like gestational diabetes, oral contraceptive use, and reproductive disorders such as polycystic ovary syndrome and early menopause.[1–4] One study found that women with diabetes have an increased relative risk of CVD despite men having a significant absolute risk of mortality. Moreover, the typical cardioprotection of the female sex is reduced in diabetic women.[5]

It is well documented that women have a lower risk of CVD before the onset of menopause.[6] The drastic increase in risk after the age of 55[7] has prompted researchers to investigate the mechanisms underlying these issues. Findings from a study examining the correlation between sex hormones and CVD in postmenopausal women suggested that postmenopausal women were at higher risk of CVD events when they had higher testosterone to estradiol ratio; however, these women had a reduced risk with higher levels of estradiol.[6]

Sexual and gender minority (SGM) populations, which include members of the LGBTQ+ community, are increasing in the United States and must be accounted for by clinicians and researchers. Despite having similar health behaviors, transgender women have higher rates of CVD compared to cisgender men and women; currently, the evidence does not support an increased risk in transgender males.[8–10] However, there is limited data regarding the SGM population due to its smaller footprint, which is estimated to be 0.5% of the United States.[8] One study investigated the association of gender affirming hormone

Translational Cardiology. https://doi.org/10.1016/B978-0-323-91790-2.00022-8

therapy and CVD in gender minority patients, emphasizing the need for clinicians to be vigilant of the longterm effects of these therapies until a better understanding of their impact is achieved.[11]

Gender bias in cardiovascular care

Traditionally, CVD has been thought of as a male dominant disease, which has led to the underrepresentation of women in cardiovascular research trials.[4,12] In 2022, the Journal of the American Medical Association (JAMA) reported that atherosclerotic CVD patients, especially within gender and racial lines, were receiving "suboptimal use of statin therapy, aspirin prescription, and lifestyle counseling" in accordance with the 2013 American College of Cardiology and American Heart Association published guidelines.[13] Further studies have shown that all women, regardless of background, are neglected by their clinicians regarding their CVH. A study by Lee et al. supported this bias, demonstrating that women with premature atherosclerotic CVD were significantly less likely to receive the appropriate treatment or referral to a cardiologist than their male counterparts. One strategy for combating this therapeutic inertia that has become increasingly popular is for clinicians to create sex specific guidelines.[3]

It has been proposed that SGMs experience more CVH disparities and are more likely to die from CVD than most due to marginalization and discrimination, underscoring the need for clinicians to be educated about the norms, culture, and attitudes of SGM individuals.[8,9] Several factors, such as clinician bias or insurance exclusions, contribute to SGMs being less likely to seek healthcare. However, religious exemption laws pose the greatest threat to SGM healthcare. As Fig. 90.1 depicts, nine states have laws that permit medical professionals to decline service to SGMs,[14] a trend that may gain more traction given the Supreme Court's ruling in June 2023. Clinicians must remember their privilege to practice medicine is rooted in the four pillars of medical ethics: autonomy, beneficence, nonmaleficence, and justice.

Race specific risk factors for heart failure

Self identified Black and Hispanic patients have the highest prevalence of heart failure (HF) compared to other racial groups and are twice as likely to be diagnosed at a younger age than White patients.[15] These two patient populations also bear the highest burden of modifiable risk factors, like diabetes, obesity, and HTN, which directly contribute to the increased prevalence of HF. Over the last 18 years, Black Americans have been diagnosed with HF at younger ages and subject to rising mortality rates, with Black men experiencing the highest even when adjusted for age.[15] Despite abundant data on the mortality of CVD in minorities, little exists on the incidence rate, which can demonstrate how quickly a disease spreads through certain racial groups.

The Multi Ethnic Study of Atherosclerosis, a cohort study of 6814 participants, intended to determine the incidence rate of congestive heart failure (CHF) by race since most available data was from White populations. Participants from four racial backgrounds—White (38.5%), Black (27.8%), Hispanic (21.9%), and Chinese American (11.8%)—were contacted every 8 months to report changes to their health. At the end of 4 years, 79 participants developed CHF. Among these cases, Black Americans had the highest incidence rate, followed by Hispanics.

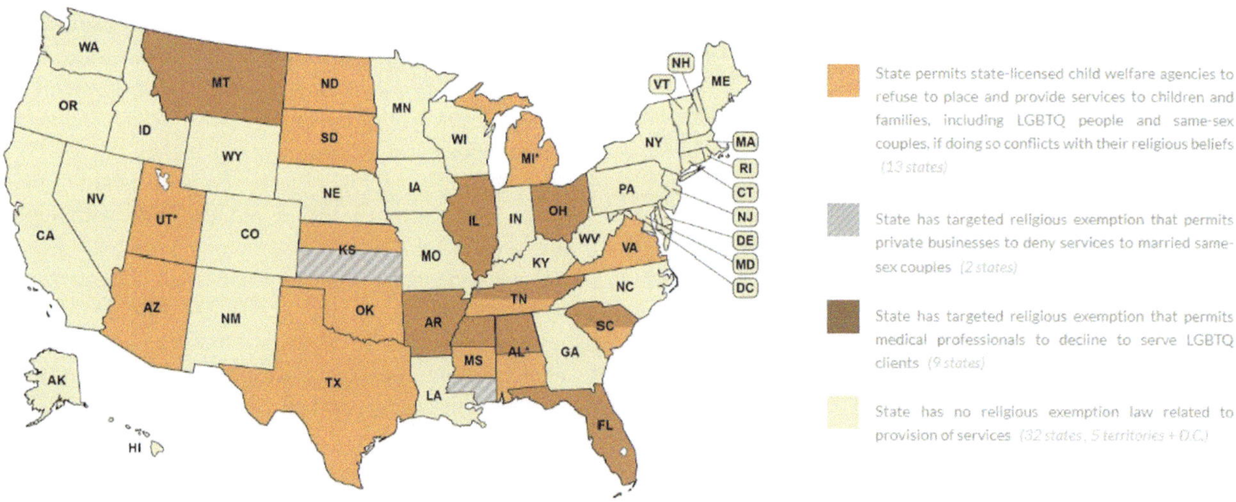

FIGURE 90.1 Religious exemption laws in the United States permit discrimination against members of the LGBTQ. Eighteen states have enacted religious exemption laws permitting the discrimination of LGBTQ members by various business owners and state agencies. Although 32 states (indicated in *yellow*) have no such law, 9 states (indicated in *bronze*) have laws permitting medical professionals to decline service to LGBTQ clients. *Reprinted with permission of Movement Advancement Project.*

Additionally, those diagnosed with CHF carried the highest burden of diabetes, HTN, and obesity.[16]

Racial disparities in cardiovascular research

Evidence based intervention and incorporating social determinants of health are the best tools to decrease racial disparities in healthcare. Yet, practicing evidence based medicine in Black and Hispanic communities may be challenging for clinicians due to the historical shortage of medical research in these populations. In a meta analysis on HF, only 48% of 25 clinical trials reported their outcomes by race. Furthermore, Black Americans only comprise about 5.5% of the treatment groups in landmark HF clinical trials despite their increased burden of HF.[15]

Obtaining data separated by race can aid clinicians in determining how CVD affects groups differently. For example, Black Americans have the highest prevalence of HTN, a known underlying cause of CHF. Lifetime exposure to discrimination increases the risk of developing HTN by 35%.[17] Discrimination is a proposed environmental and social factor that can lead to HTN by exposing individuals to a stressful environment, resulting in the adoption of unhealthy diets and poor physical activity as coping mechanisms. These harmful coping mechanisms are known to lead to comorbidities often seen congruently with CVD, such as diabetes and obesity.[17,18]

Rural–urban disparities in cardiovascular health

Another major topic studied in recent years is the healthcare discrepancies between urban and rural/underserved areas. It has been found that the age adjusted mortality rates of CVD in America from 1999 to 2017 were significantly higher across all ages, sexes, and races/ethnicity in rural areas compared to nonrural. Specifically, the difference in 1999 was 24.0 deaths per 100,000, which nearly doubled to 42.8 in 2017.[19] In another cross sectional study, rural counties in the South had the highest rates of HF related mortality between 2011 and 2018, while urban counties in the Northeast and the Florida peninsula had the lowest rates.[20] Moreover, life expectancy has remained stagnant in urban communities since 2010 and yet declined in rural.[21]

To improve CVH, several unmet social needs in specific populations must be addressed, including reliable public transportation, food security, access to healthcare, financial and housing stability, and education.[22] A data analysis from the Behavioral Risk Factor Surveillance System revealed adults in rural areas are "more likely to be smokers, less likely to maintain normal body weight, and less likely to meet physical activity recommendations compared with adults in urban areas."[23] The association between these social determinants and CVD mortality is consistently stronger in rural communities with a higher proportion of residents receiving food stamps, living in mobile homes, and/or having more than three children in the household.[24]

Despite efforts and policy changes in public health to improve rural health, studies purport that the difference in CVD mortality rates between rural and urban populations continues to increase,[25] with rural areas experiencing disproportionate rates of hospital closures in recent years and an approximate 8.2% decrease in specialists and primary care clinicians.[26] Understanding the true underlying causes of the trends in rural CVH is vital for decreasing healthcare disparities. While it is shown that many social determinants of health play a part, further investigation and research are warranted in hopes to improve healthcare, increase life expectancy, and reduce the increasing risk of CVD in these specific populations.

Socioeconomic factors in cardiovascular health

An individual's socioeconomic status is measured using multiple factors, including income, education, occupation, and one's own perception of their social status. According to the 2020 US Census, the percentage of Americans considered middle class has declined by 2.9%. As a result, more US households are classified as either low or upper class, with the former showing a slightly greater increase.[27] Reviewing data from the 2005–2018 National Health and Nutrition Examination Survey, Shen et al. concluded that a clear and significant relationship exists between individuals in lower income brackets and their increased risk of arteriosclerotic CVD, the main cause of myocardial infarctions and strokes. These individuals were also found to have a higher likelihood of being current smokers, having a low body mass index, and experiencing a higher rate of depression symptoms.[28] A cross sectional analysis of data between 1999 and 2018 also revealed an association between low socioeconomic status and premature cardiovascular mortality between the ages 25 and 64 as well as HTN, obesity, active smoking, and physical inactivity.[29]

The challenges of improving CVH in low and middle income populations are similar to those seen in rural areas—environmental factors and access to quality healthcare, food, education, and proper health insurance. In recent years, the American Heart Association has pushed for improving healthcare in lower socioeconomic populations.[30] It is agreed that proposed interventions should be cost conscious and involve behavioral changes that are focused toward both the individual and community levels. Unfortunately, data regarding such interventions is limited; thus, further study and research are warranted.

Conclusion

It is clear that healthcare inequalities persist in certain populations and settings in the United States. Sex and gender minorities, racial minorities, underserved communities, and those of lower socioeconomic status all suffer a relatively heavier burden of CVD due to factors such as improper, inadequate, and/or inaccessible healthcare; discrimination; social stigma; and limited research. This imperatively reinforces the need for diverse clinical trial populations that accurately capture how CVD affects specific groups differently and racially competent healthcare teams that treat patients holistically. Implanting these measures can help reduce the increasing burden of CVD in America's most vulnerable populations.

Potential pitfalls

- A portion of the studies referenced were meta analyses. These studies require a large portion of personal judgment, making them susceptible to personal biases.
- Evidence used is extracted from a number of primary studies that most likely fail to include a diverse clinical trial population, thus limiting the data.
- Data examined is limited to CVD risks and burdens on populations in the United States.

Examples of implementation

- Awareness of therapeutic inertia and the implementation of sex specific guidelines are two ways to reduce CVD mortality.
- Nine states have laws that permit medical professionals to decline service to SGMs. Despite these laws, clinicians have an ethical obligation to provide equal healthcare to all.
- CVD mortality and the social determinants of health are strongly correlated in rural and low income populations. Although studies are limited, proposed interventions should involve behavioral changes toward the individual and community levels.

Acknowledgments

Special thanks to Anna Dunson, Assistant Director of Skelton Medical Libraries, for your expertise, guidance, and editorial skills. A warm thanks to Nicollette Higgs for reviewing this work. Lastly, we are thankful for the encouragement of Dr. Alice House, Dr. Katey Hughes, and the Mercer University School of Medicine community.

References

1. Freaney PM, Khan SS, Lloyd-Jones DM, et al. The role of sex-specific risk factors in the risk assessment of atherosclerotic cardiovascular disease for primary prevention in women. *Curr Atheroscler Rep.* 2020;22(9):46. https://doi.org/10.1007/s11883-020-00864-6.
2. Shen X, DiMario S, Philip K. Gender disparities in health resource utilization in patients with atherosclerotic cardiovascular disease: a retrospective cross-sectional study. *Adv Ther.* 2019;36(12):3424–3434. https://doi.org/10.1007/S12325-019-01107-0.
3. Lee MT, Mahtta D, Ramsey DJ, et al. Sex-related disparities in cardiovascular health care among patients with premature atherosclerotic cardiovascular disease. *JAMA Cardiol.* 2021;6(7):782–790. https://doi.org/10.1001/JAMACARDIO.2021.0683.
4. Appelman Y, van Rijn BB, ten Haaf ME, et al. Sex differences in cardiovascular risk factors and disease prevention. *Atherosclerosis.* 2015;241(1):211–218. https://doi.org/10.1016/j.atherosclerosis.2015.01.027.
5. Savage K, Williams JS, Garacci E, et al. Association between cardiovascular disease risk factors and mortality in adults with diabetes: a stratified analysis by sex, race, and ethnicity. *Int J Publ Health.* 2022;67. https://doi.org/10.3389/IJPH.2022.1604472.
6. Zhao D, Guallar E, Ouyang P, et al. Endogenous sex hormones and incident cardiovascular disease in post-menopausal women. *J Am Coll Cardiol.* 2018;71(22):2555–2566. https://doi.org/10.1016/j.jacc.2018.01.083.
7. Martinez C, Rikhi R, Haque T, et al. Gender identity, hormone therapy, and cardiovascular disease risk. *Curr Probl Cardiol.* 2020;45(5):100396. https://doi.org/10.1016/j.cpcardiol.2018.09.003.
8. Caceres BA, Jackman KB, Edmondson D, et al. Assessing gender identity differences in cardiovascular disease in US adults: an analysis of data from the 2014–2017 BRFSS. *J Behav Med.* 2020;43(2):329–338. https://doi.org/10.1007/S10865-019-00102-8.
9. Caceres BA, Bynon M, Doan D, et al. Diet, food insecurity, and CVD risk in sexual and gender minority adults. *Curr Atheroscler Rep.* 2022;24(1):41–50. https://doi.org/10.1007/S11883-022-00991-2.
10. Connelly PJ, Freel EM, Perry C, et al. Gender-affirming hormone therapy, vascular health and cardiovascular disease in transgender adults. *Hypertension.* 2019;74(6):1266–1274. https://doi.org/10.1161/HYPERTENSIONAHA.119.13080.
11. Getahun D, Nash R, Flanders WD, et al. Cross-sex hormones and acute cardiovascular events in transgender persons: a cohort study. *Ann Intern Med.* 2018;169(4):205–213. https://doi.org/10.7326/M17-2785/SUPPL_FILE/M17-2785_SUPPLEMENT.PDF.
12. Solola Nussbaum S, Henry S, Yong CM, et al. Sex-specific considerations in the presentation, diagnosis, and management of ischemic heart disease: JACC focus seminar 2/7. *J Am Coll Cardiol.* 2022;79(14):1398–1406. https://doi.org/10.1016/j.jacc.2021.11.065.
13. Mufarreh A, Shah AJ, Vaccarino V, et al. Trends in provision of medications and lifestyle counseling in ambulatory settings by gender and race for patients with atherosclerotic cardiovascular disease, 2006-2016. *JAMA Netw Open.* 2023;6(1):e2251156. https://doi.org/10.1001/jamanetworkopen.2022.51156.
14. Equality Maps: Religious exemption laws: Movement Advancement Project. https://www.lgbtmap.org/equality-maps/religious_exemption_laws. Accessed 29 June 2023.
15. Lewsey SC, Breathett K. Racial and ethnic disparities in heart failure: current state and future directions. *Curr Opin Cardiol.* 2021;36(3):320–328. https://doi.org/10.1097/hco.0000000000000855.
16. Bahrami H, Kronmal R, Bluemke DA, et al. Differences in the incidence of congestive heart failure by ethnicity: the multi-ethnic

study of atherosclerosis. *Arch Intern Med*. 2008;168(19):2138–2145. https://doi.org/10.1001/archinte.168.19.2138.
17. Ogunniyi MO, Commodore-Mensah Y, Ferdinand KC. Race, ethnicity, hypertension, and heart disease: JACC focus seminar 1/9. *J Am Coll Cardiol*. 2021;78(24):2460–2470. https://doi.org/10.1016/j.jacc.2021.06.017.
18. Morris AA, Ko YA, Hutcheson SH, et al. Race/ethnic and sex differences in the association of atherosclerotic cardiovascular disease risk and healthy lifestyle behaviors. *J Am Heart Assoc*. 2018;7(10):e008250. https://doi.org/10.1161/jaha.117.008250.
19. Cross S, Mehra M, Bhatt D, et al. Rural-urban differences in cardiovascular mortality in the US, 1999-2017. *JAMA*. 2020;323(18):1852–1854. https://doi.org/10.1001/jama.2020.2047.
20. Pierce J, Shah N, Petito L, et al. Trends in heart failure-related cardiovascular mortality in rural versus urban United States counties, 2011-2018: a cross-sectional study. *PLoS One*. 2021;16(3). https://doi.org/10.1371/journal/pone.0246813.
21. Abram L, Myrskylä M, Mehta N. The growing rural-urban divide in US life expectancy: contribution of cardiovascular disease and other major causes of death. *Int J Epidemiol*. 2022;50(6):1970–1978. https://doi.org/10.1093/ije/dyab158.
22. *Health, United States Spotlight: Racial and Ethnic Disparities in Heart Disease*. CDC.Gov; April 23, 2019. https://www.cdc.gov/nchs/hus/spotlight/2019-heart-disease-disparities.htm. Accessed June 5, 2023.
23. Mathews K, Croft J, Liu Y, et al. Health-related behaviors by urban-rural county classification - United States, 2013. *MMWR Surveill Summ*. 2017;66(5). https://doi.org/10.15585/mmwr.ss6605a1.
24. Son H, Zhang D, Shen Y, et al. Social determinants of cardiovascular health: a longitudinal analysis of cardiovascular disease mortality in US counties from 2009 to 2018. *J Am Heart Assoc*. 2023;12(2). https://doi.org/10.1161/jaha.122.026940.
25. Loccoh E, Joynt Maddox K, Wang Y, et al. Rural-urban disparities in outcomes of myocardial infarction, heart failure, and stroke in the United States. *J Am Coll Cardiol*. 2022;79(3):267–279. https://doi.org/10.1016/j.jacc.2021.10.045.
26. Germack H, Kandrack R, Martsolf G. When rural hospitals close, the physician workforce goes. *Health Aff*. 2019;38(12):2086–2094. https://doi.org/10.1377/hlthaff.2019.00916.
27. Shrider EA, Kollar M, Chen F, et al. *Income and Poverty in the United States: 2020*. US Census Bureau; 2021:P60–P273. https://www.census.gov/content/dam/Census/library/publications/2021/demo/p60-273.pdf.
28. Shen R, Zhao N, Wang J, et al. Association between socioeconomic status and arteriosclerotic cardiovascular disease risk and cause-specific and all-cause mortality: data from the 2005–2018 National Health and Nutrition Examination Survey. *Front Public Health*. 2022;10. https://doi.org/10.3389/fpubh.2022.1017271/full.
29. Bevan G, Nasir K, Rajagopalan S, et al. Socioeconomic deprivation and premature cardiovascular mortality in the United States. *Mayo Clin Proc*. 2022;97(6):1108–1113. https://doi.org/10.1016/j.mayocp.2022.01.018.
30. Pearson T, Palaniappan L, Artinian N, et al. American Heart Association guide for improving cardiovascular health at the community level, 2013 update. *Circulation*. 2013;127(16):1730–1753. https://doi.org/10.1161/cir.0b013e31828f8a94.

Chapter 91

Guideline development

Karley B. Fischer[1], Darbi Griffith[1] and Damian Valencia[2]

[1]*Boonshoft School of Medicine, Wright State University, Dayton, OH, United States;* [2]*Department of Cardiovascular Medicine, Kettering Health, Kettering, OH, United States*

Key points

- Clinical practice guidelines influence physician decision-making, diagnostic/treatment algorithms, and patient outcomes.
- Development of guidelines is a collaborative, expert-driven, systematic review of scientific evidence with the goal of providing best practice recommendations based on the most recent available data and scientific advancements.
- Effective guidelines must be feasible, measurable, and achievable to allow for successful clinical implementation.

Why it matters

Clinical practice guidelines (CPGs) summarize up-to-date clinical evidence to inform and provide action statements for clinicians. CPGs have the capability to reduce inappropriate variation in clinical practice, efficiently translate research findings into clinical practice, and improve both safety and quality of healthcare. Due to the reduction in care variation, CPGs also inevitably decrease the risk of liability for negligent care and improve the value in the cost ratio of healthcare spending. Quality guidelines should promote evidence-based medicine while being feasible, measurable, and achievable in the targeted patient population. Guideline development requires a stepwise approach shifting from strategic team member selection, to a systematic review of available evidence, and effectively articulating recommendations. This is typically a collaborative process and must frequently be updated based on scientific evidence, including the latest research and advancements in medical technology. Practice of high-quality and evidence-based medicine is increasingly being incorporated into physician evaluation and quality metrics both by employers and healthcare insurance agencies. Therefore, more than ever physicians are looking for quality guideline resources to increase high-quality patient outcomes.

Development of clinical practice guidelines

To be broadly implemented by clinicians, CPGs must be both useful and valid. CPGs are useful for practicing clinicians by providing a summary of high-quality evidence on the topic. Medicine is an evolving field, and CPGs are one reliable source for clinicians to obtain updated information. Therefore, CPGs are a tool for the longitudinal education of a physician. They teach clinicians about scientific advancements, recent research findings, and best-practice medicine, which may differ from their training. CPGs are useful for patients by decreasing inequalities among healthcare practices by providing clinicians with a framework of recommendations applicable to a particular patient population and medical context. This can ultimately optimize patient outcomes.

Validity

Validity refers to the degree that results among study participants reflect true findings among similar individuals in clinical practice. Guideline outcomes should be extrapolated to the targeted patient population. Physicians need to have confidence that they can expect similar outcomes in their own patients. The Conference on Guideline Standardization (COGS) provides a list of 35 components, represented in the following table, deemed necessary to ensure guideline validity (Fig. 91.1).

Process

There are eight fundamental steps to successfully creating unbiased clinical practice guidelines:

Translational Cardiology. https://doi.org/10.1016/B978-0-323-91790-2.00027-7

COGS Recommended Components for Guideline Validity		
Description of evidence collection	Link between recommendations and evidence	Definitions
Criteria for selection of evidence to review	Recommendation strength	Outcomes
Eligibility	Technical report	Developer name
Recommendation	Main focus	Status
Reason	Quantification of harm & benefit	Objectives
Evidence quality	Funding	Number of source documents
Evidence time period	Conflict of interest	Role of patient preference
Evidence quality rating scheme	Specification of harm and benefit	Committee member
Description of evidence combination	Member expertise	Release date
Strength of recommendation rating scheme	Role of sponsor	Alternative strategies
Methods used to reach judgment	Peer review	Care setting
Update plan	Reference	

FIGURE 91.1 COGS recommended components for guideline validity.

Step 1: Establish transparency. Transparency allows clinicians to trust that guidelines are presenting evidence-based medicine without bias, distortions, and conflicts of interest. This is achieved by being able to identify who was involved with developing the guidelines, funding sources, and how recommendations were decided upon. Many organizations also include transparent information regarding anticipated costs, benefits, risks, and harm. Grading Recommendations Assessment, Development and Evaluation (GRADE) is a tool for developers to increase transparency regarding the quality of evidence.

Step 2: Manage conflicts of interest (COI). COI can create bias and bring into question whether a member's actions are motivated by personal gain. COI occurs when a professional obligation, such as creating guideline recommendations, is improperly influenced by a private interest, such as personal financial or intellectual ties. Financial sources of COI include industry funding, and personal payment resulting from consultant work, speaker relationships, stock ownership, and research with the industry. Intellectual COI includes authorship of a publication, participation in research, advocacy, and lobbying of a viewpoint related to the CPG. Committee members should have no or minimal COI. If there is COI, it must be disclosed to each participant and updated at each stage of CPG development.

Step 3: Guideline development group composition and collaboration with relevant organizations and societies. The committee should represent a broad community by selecting experts from a wide spectrum of backgrounds. Diversity of geographic regions, sex, race, ethnicity, intellectual perspectives/biases, and clinical practice settings should be represented. Organizations and professional societies with an interest in the topic should be invited as a collaborating partner as well.

Step 4: Systematic review of the literature and current standards of care. This step begins with a comprehensive search and review of the currently available literature. Evidence should be screened and relevant studies selected for inclusion. This process should be performed by a minimum of two researchers, typically with the involvement of a librarian. Publications from the same study need to be linked to avoid duplication of data.

Step 5: Establishing evidence and rating strength of recommendations: A rating system to evaluate evidence and recommendations must be chosen. GRADE is a widely used evidence and quality assessment tool. There is more information regarding how to incorporate GRADE into guideline development below.

Step 6: Articulation of recommendations: Provide evidence-based recommendations with easy-to-read formatting and the inclusion of flow charts and images that can facilitate the message and increase ease of comprehension. Specific techniques to aid in articulation can be found in the implementability section.

Step 7: External review: An external review committee should consist of methodologists, epidemiologists, clinicians, and biostatisticians who are subject matter experts on the topic. All reviewer comments should be collected and recorded, along with how each comment was addressed with reasoning for inclusion or exclusion in the editions.

Step 8: Periodic updates with emerging data: Guidelines should be updated based on recent study findings. This can be driven by advances in medical therapies and scientific data. At the time of publication, it is best to have a planned timeframe of when CPGs will be updated or at the minimum reviewed for the need to update.

Identifying topics

Identifying topics for potential guideline development requires the ability to recognize areas of clinical importance. Variability in practice patterns, controversy, availability of interventions that have high societal or economic cost, importance for public health, potential impact, disease burden, and availability of high-quality published data are all features that would favor formal guideline development. The PICO (Patient, Intervention, Comparison, Outcome) framework is a useful tool in writing a question for a systematic review of evidence that can lead to guideline development. Start with identifying a target patient population. Then consider the intervention, which includes the therapeutic interventions under investigation. Comparison is the process of including alternative interventions and a control group. And finally, the outcomes detail the comparison between each intervention group.

Action statements

Action statements are the section of CPGs that give the key recommendations supported by the available scientific evidence. These are takeaway messages for the clinician and provide tasks that can be implemented into clinical decision-making and patient care. Action statements should include a verb that reflects the level of obligation based on the strength of the recommendation (i.e., must, should, or may consider). Language must be unambiguous and give a clear indication of the following: when (specific circumstances), who (target clinician), what (precise action), and to whom (targeted patient population). The action statements should be brief but precise to avoid inconsistent interpretations or assumptions by the reader. The explanation of why and how can follow later in the text, alongside the evidence profiles.

Evidence profiles

Evidence profiles appear after the supporting text for each action statement. This includes aggregate evidence regarding quality and certainty, benefit, harm, cost, benefit-harm assessment, value judgments, intentional vagueness, role of patient preferences, and exclusions. Each should be addressed for every action statement in a bulleted list. The evidence profile can then allow readers to quickly understand how and why the action statements were made, increasing overall transparency. GRADEpro, available in the resource section, can provide evidence profile examples to assist in understanding and learning.

Recommendation grades

Recommendation grades represent the strength of a recommendation or action statement. Recommendation grades should always independently assess the strength of the action statement and aggregate quality of evidence. Choosing a rating scale is an important step in guideline development. Several rating scales have been validated, including from the American Academy of Pediatrics (AAP), the United States Preventive Services Task Force (USPSTF), and Grades of Recommendation Assessment Development and Evaluation (GRADE). GRADE is endorsed by more than 100 organizations worldwide, making it the most widely adopted tool to assess the quality of evidence and strength of recommendations. GRADE uses a systematic review to evaluate scientific studies for inclusion in both systematic reviews and clinical practice guidelines (Fig. 91.2).

The systematic approach involves choosing a clinical question, identifying the applicable patient population, identifying two or more alternatives, and identifying the most important outcomes for clinicians. Then, a systematic review provides the risk difference, or effect size for each outcome. At that point, the authors rate the quality of evidence for each outcome. The final GRADE quality of evidence rating is applied by taking the lowest quality of evidence from all the outcomes that are considered critical.

There are four quality categories for evidence[1]: high,[2] moderate,[3] low, and[4] very low. These rankings correlate to the likelihood that the evidence is adequate to support the recommendation or action statement. High quality suggests that there is a high level of confidence that the evidence supports the recommendation. Moderate quality suggests

GRADE Quality Ratings

	Interpretation	Starting Point If:	Indications to decrease	Indications to increase
High	Confident the evidence is adequate to support recommendation	Randomized Trial	Risk of Bias - 1 Serious - 2 Very serious	Large Effect + 1 Large + 2 Very Large
Moderate	Moderately confident the evidence is adequate to support recommendation		Inconsistency - 1 Serious - 2 Very serious	Dose Response + 1 Evidence of a gradient
Low	Limited confidence the evidence is adequate to support recommendation	Observational study	Indirectness - 1 Serious - 2 Very serious	All plausible counfounding + 1 Would reduce a demonstrated effect
Very Low	Likely inadequate evidence to support recommendation		Publication Bias - 1 Likely - 2 Very Likely	+ 1 Would suggest a spurious effect when results show no effect

FIGURE 91.2 Summary of grade quality ratings and modifying components.

the evidence is probably adequate to support the recommendation. Low quality suggests that the evidence might be adequate to support the recommendation. Very low quality suggests the evidence likely cannot support the recommendation.

Confidence in evidence can decrease and can move down the quality ratings due to the risk of bias, inconsistencies, impressions, inconsistency, indirectness, and publication bias. Confidence in the evidence can increase a quality rating based on the magnitude of effect, a dose-response gradient, and a decrease in confounding variables.

A final strength of a recommendation grade or action statement is then provided based on aggregate evidence to represent the desirable effects. The recommendation includes strength (strong or weak/conditional) and directionality (for or against). The spectrum therefore includes a strong recommendation for the intervention, weak or conditional recommendation for the intervention, weak or conditional recommendation against the intervention, or strong recommendation against the intervention (Fig. 91.3).

Implementability describes the effective incorporation of CPGs into clinical practice. It represents the ease at which a practitioner can read, interpret, and apply the information to a patient population. Implementation of CPGs into the clinical practice of medicine is affected by both the content creation process and content communication methods.

The creation of guidelines includes stakeholder involvement, effective evidence synthesis, considered judgment, and feasibility. Stakeholder involvement is crucial; It has been shown that credibility directly correlates with implementation. Therefore, guidelines should be created by topic experts and stated by a respected professional society within that specialty. For example, guidelines for congestive heart failure will have the highest rate of implementation, when authors include cardiologists who specialize in heart failure, and are released by the overseeing respected agency such as the American Heart Association or the American College of Cardiology. Readers can confidently trust the opinions and recommendations provided by these organizations and are more likely to integrate the recommendations into daily patient care activities.

A formal, thorough, and systematic review is necessary. Methods should be clearly defined with an outcome that has strong external validity (reproducibility) to allow a provider to perceive and experience similar outcomes in their own patient population.

Do no harm is a core component of the Hippocratic Oath. Guidelines must address and prioritize probable benefit over the competing possible harm and side effects. This judgment is complex and requires consistent evaluation of evidence and outcomes. In addition, potential variability among patient responses should be considered. The preferences of the patient and physician are also important.

Feasibility determines the ease of incorporation into clinical practice. Potential resource restraints should be minimized when able. Novel recommendations may lag in the community setting due to resource limitations and the comfort of practice patterns.

Guideline communication is the aspect of guideline development with a focus on messaging and format.

Optimization of final messaging increases implementation. Evidence has shown an inverse relationship between the complexity of CPG and the adoption of recommendations. CPGs should be simple, clear, and use persuasive language. Language should be unambiguous and written in an active voice. If evidence shows the recommendation explicitly outweighs the risk, strong language such as "recommend" should be used. If evidence is less clear, language such as "suggest" or "consider" should be used. Sentences should be short with bullet points and flow charts should be included when appropriate. Cognitive load should be reduced to increase understanding and retention

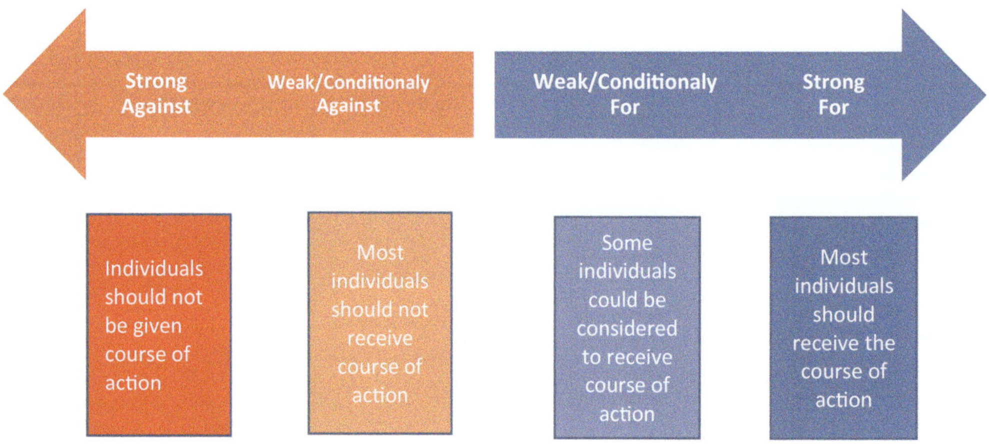

FIGURE 91.3 Strength of recommendation continuum and interpretation.

of the content. This can be achieved by limiting the number of elements, reducing the number of steps within each recommendation, and reducing the number of conditional factors.

Formatting involves deciding on components to include in each version and visual presentation of material to increase perceived ease of use. Document length and structure should enhance the readability of the document, as well as provide section headings for later reference. Proper placement of visual aids, figures, and images should naturally break text, enhance the written text, as well as increase recall when viewed later. Recommendations should always include appropriate patient populations with relevant inclusion and exclusion criteria (Fig. 91.4).

Designing clinical research for guidelines via grade

When designing clinical research studies, researchers should consider the possibility of their findings being included in the evidence-based systematic review for guideline creation. The study design while performing research is critical in determining the quality of evidence that will result from the data. Randomized control trials (RCTs) provide far stronger evidence than that of observational studies (reflected in the starting point quality of evidence in the GRADE assessment). Limitations, however, modify the initial starting point of both RCTs and observational studies.

Current major areas of focus

In recent years there have been numerous advancements in therapy for treating cardiovascular disease. This includes both devices and medical therapies. New therapies are often expensive, therefore before recommending high-cost interventions, clinicians want to be confident in their effectiveness. As a result, guidelines are necessary to guide clinician decision-making. Current areas of major focus include structural heart disease with new options for minimally invasive transcatheter therapies, and advanced heart failure with the emergence of new and effective pharmaceuticals.

Examples of recent guidelines

To further demonstrate the process of guideline development, three real-world examples detailing the process of guideline development are detailed below.

Example 1: Identification of the eight steps used and documented in guideline development, in real practice to create the ACC/AHA/HFSA heart failure guidelines. This reference provides an example of collaboration in action by expert associations, including the American College of Cardiology (ACC), the American Heart Association (AHA), and the Heart Failure Society of America (HFSA) to provide an expert opinion on care guidelines for patients with a diagnosis of heart failure. Note the information included in the guideline development process. ACC and AHA address the 8 standard steps listed above before stating the guidelines:

Title: 2022 AHA/ACC/HFSA Guideline for the Management of Heart Failure: A Report of the American College of Cardiology/American Heart Association Joint Committee on Clinical Practice Guidelines

1. Establish transparency: Addressed in "Preamble." The ACC and AHA state guidelines are based on systematic methods to evaluate evidence with the goal of improving cardiovascular health. No commercial support (or industry funding) was used.
2. Manage conflict of interest (COI): Addressed in "Relationships with Industry and Other Entities." Rigorous policies are in place to ensure there is no bias or

Factors Affecting Clinician Implementation					
Creation				Communication	
Stakeholder	Evidence Synthesis	Considered Judgment	Feasibility	Message	Format
• Credibility • Professional organization • Minimal COI	• Systematic review • Formal • Thorough • High external validity	• Maximize benefit • Minimize harm • Patient preferences • Provider preferences	• Clinician flexibility • Resource availability • Time constraints • Historical practice patterns	• Simple • Clear message • Active Voice • Unambiguous • Persuasive language • Short sentences • Bullet points	• Visual aids • Figures • Images • Length • Complexity

FIGURE 91.4 Implementability factors.

improper influence. The appendix includes disclosure information of all participants.

3. Selection of writing committee members: The committee includes content experts who are representative of the broader cardiovascular community, representing different geographic regions, sexes, races, ethnicities, intellectual perspectives or biases, and clinical practice settings.
4. Clinical practice guideline/systematic review intersection: Addressed in "Evidence Review and Evidence Review Committees." This section states a literature search from May 2020 to December 2020, focusing on RCTs performed, while also including registries, nonrandomized comparative and descriptive studies, case series, cohort studies, systematic reviews, and expert opinion. This particular guideline included literature published in English, derived from MEDLINE, EMBASE, the Cochrane Collaboration, and the Agency for Healthcare Research and Quality.
5. Establishing evidence and rating strength of recommendations: Table 2 addresses the class (strength) of recommendation and level (quality) of evidence used by ACC/AHA in the heart failure guidelines.
6. Articulation of recommendations: The recommendations are organized starting with the contents clearly outlined. 10 take-home messages are clearly stated to provide readers guidance and focus points. Figures are used to enhance understanding and later recall.
7. External review: Addressed in "Document Review and Approval." Stated that the document was reviewed by 2 reviewers nominated by the AHA, 1 reviewer nominated by the ACC, 2 reviewers from HFSA, 1 Joint Committee on Clinical Practice Guideline reviewer, and 32 individual content reviewers. Following the review process, it was approved by the governing bodies of the ACC, AHA, and HFSA.
8. Periodic updating with new emerging evidence: Addressed in methodology that data is reviewed on an ongoing basis with targeted "full revision" and "focused update" will be performed as needed to keep recommendations current.

Example 2: Use of the GRADE quality of evidence assessment.

Note this evaluative process should be performed on each study included in a systematic review. This is just one example of evaluating an individual study with the GRADE quality of evidence assessment.

Question: Should sodium-glucose cotransporter 2 inhibitors (SGLT2i) be used in patients with heart failure and reduced ejection fraction (HFrEF) with the goal of reducing hospitalizations for heart failure and cardiovascular mortality, irrespective of the presence of type 2 diabetes mellitus?

Article title being evaluated: Dapagliflozin in Patients with Heart Failure and Reduced Ejection Fraction.

Quality Assessment					No of Patients (4744)		Effect	Quality	Importance
Study Design	Risk of Bias	Inconsistency	Indirectness	Imprecision	Dapagliflozin	Placebo	Relative (95% CI)		
RCT	Not serious	Not serious	Not serious	Not serious	2371/4744	2373/4744	HR 0.74 (0.65-0.85)	High Quality	Critical

Primary outcome: worsening heart failure or death from cardiovascular causes.

In summary, based on the assessment of evidence, the study is high quality with an outcome suggesting that among patients with HFrEF, the risk of worsening heart failure or death was lower among those who received dapagliflozin (an SGLT2i), regardless of type 2 diabetes mellitus. The study was high quality due to being a randomized control study which places quality at a starting point of high. There were no concerns with the risk of bias, inconsistency, indirectness, or imprecision that would justify reducing quality from the rating of high.

Example 3: Use of the GRADE strength of recommendation assessment.

Recommendation: Utilize SGLT2i in patients with symptomatic HFrEF to reduce hospitalization for heart failure and cardiovascular mortality, irrespective of the presence of type 2 diabetes.

Two categories exist under GRADE for the strength of assessment: strong and weak. Assessing a recommendation as strong should be analyzed with a cautious degree of certainty as one considers the various factors influencing the characterization. When uncertainty outweighs the recommendation, it is best to opt for the characterization as a weak recommendation. It is important to understand implications associated with these guidelines from a patient, clinician, and policy maker's perspective. Qualifying a recommendation as strong is established with confidence that most or all individuals will be best served by the recommendation. In contrast, a weak recommendation has implications that not all individuals will be best served by the course of action. This could also be described as a conditional, discretionary, or qualified recommendation to dispel negative connotations.

Four factors garner the direction and strength of a recommendation. These factors include a balance between favorable and unfavorable outcomes, the quality of evidence for the outcome, the degree of uncertainty in each value, and the cost of the intervention.

Given that evidence from multiple RCTs suggests that most individuals with HFrEF should receive SGLT2i, this is a "strong for" recommendation. ACC/AHA does not use GRADE, so in the guidelines, SGLT2i is a class 2a recommendation. The strength of recommendation is moderate, with evidence suggesting SGLT2i are reasonable to use and can be effective and beneficial.

Budgetary considerations

Guideline development can be a costly process. Randomized clinical trials, if not funded by industry, can be expensive to support. Associated costs include personal, equipment, patient-related expenses, imaging or testing, institutional review board (IRB), statistical support, and therapeutics. Subsequently, this can be a limiting factor in creating high-quality research. Prior clinical trials can therefore be valuable in this regard. However, a thorough systematic review is not entirely free but rather more cost-effective, as it is a labor-intensive process for both research and clinical experts.

How to fund this stage

Funding for guideline development ideally comes from public sources. If industry funding is accepted, strict policies and review processes must be in place to reduce the potential for bias in the final recommendations. The Cochrane Collaboration Steering Group recognizes that for-profit commercial companies (often pharmaceutical and medical device companies) contributing financially could influence the recommendations and undermine credibility. Some organizations do choose to have strict rules against participants with any COI. However, commercial sponsors are often used to assist in the high-cost, labor-intensive research necessary for guideline development. If there is involvement from committee members with financial COI, the degree of COI should be disclosed, including personal investments (employment, consultant, expert testimony, stock, endowments, royalties, patents, and gifts) in relevant pharmaceutical or medical device companies. Most developers feel it is best not to include anyone with a primary income source from industry, but find it acceptable if a member has occasional financial relations with industry. The chair and cochair of the development committee, however, should not have any COI. Funding must be explicitly detailed and publicly available with the CPG.

Get started

After reviewing the information contained in this chapter, the reader is encouraged to take it to the next level and apply the information learned by creating a sample CPG, a process clearly outlined in Fig. 91.5. It is best to select a topic in a personal field of interest, as this will keep the researcher motivated and engaged until the completion of the project. You can reference current CPGs and should start by reading an example thoroughly, taking note of the authors, their credibility, the publishing organization, language choice, and depth of information. Only then, should you move forward with creating an example CPG of your own. Find colleagues or peers that you can collaborate with to form your guideline committee. During your systematic review of current evidence, be sure to use the GRADE method to assess the quality of evidence. The GRADE method should also be used to assess each action statement or recommendation in the sample guideline. Use the resources listed below for further learning and practice on the above topics. A senior mentor can be a source of help and guidance throughout the process.

Pitfalls

CPGs do not come without pitfalls. There are many factors that influence the successful development and use of CPGs; common pitfalls you may encounter are detailed below.

Overreliance on CPGs: Guidelines are intended to recommend practices to meet the needs of patients, in most, but not all circumstances. Therefore, guidelines should not replace clinical judgment. Physician and patient preferences should be incorporated into each decision.

Failure to be transparent and avoid COI: Guidelines can then be viewed as biased and clinicians may lack trust in the recommendations.

Difficult to read: If CPGs do not have clear, precise, easy-to-read action statements, they have the potential to be misleading or improperly interpreted and implemented. Second, if too difficult to read, the CPG may be neglected by the medical community altogether and therefore will not impact patient care or outcomes.

FIGURE 91.5 Getting started with clinical practice guideline development.

Difficult to implement: CPGs must be written with patient context and available resources in mind. If there are too many barriers to the recommendation such as availability or cost, CPGs will not be broadly implemented.

Lack of collaboration: If there are multiple guidelines regarding the same topic, there can be conflicting information and loss of validity. To avoid this, individuals and organizations with a common interest should partner to work together to reach a consensus.

Additional resources and readings

- PubMed Database—provides a search engine for quality research (https://pubmed.ncbi.nlm.nih.gov/)
- ESCSCO—provides a search engine for quality research (https://www.ebsco.com/)
- GRADE Online Learning Modules—provides a series of videos and modules from the World Health Organization (WHO) and Cochrane as an introductory course to GRADE (https://cebgrade.mcmaster.ca/)
- GRADE Handbook—provides extensive detail on how to use the GRADE approach to assess the quality of evidence and strength of recommendations (https://gdt.gradepro.org/app/handbook/handbook.html)
- Database of EtD's and Guidelines—GRADEpro database. Provides GRADE evidence profile and summary of findings table for current evidence. (https://guidelines.gradepro.org/search)

References

1. American Academy of Family Practice. *Clinical Practice Guideline Manual*. AAFP.org; December 2017. Available at: https://www.aafp.org/family-physician/patient-care/clinical-recommendations/cpg-manual.html. Accessed November 3, 2023.
2. American College of Cardiology Foundation, American Heart Association. *Methodology Manual and Policies from the ACCF/AHA Task Force on Practice Guidelines*. American College of Cardiology; 2010;(June:1–88.
3. Afshari A, De Hert S. Pitfalls of clinical practice guidelines in the era of broken science: let's raise the standards. *Eur J Anaesthesiol*. 2018;35(12):903–906. https://doi.org/10.1097/EJA.0000000000000892.
4. Baumann MH, Lewis SZ, Gutterman D. ACCP evidence-based guideline development: a successful and transparent approach addressing conflict of interest, funding, and patient-centered recommendations. *Chest*. 2007;132(3):1015–1024. https://doi.org/10.1378/chest.07-1271.

Further reading

1. Graham R, Mancher M, Miller WD, et al. *Clinical Practice Guidelines We Can Trust, Institute of Medicine Committee on Standards for Developing Trustworthy Clinical Practice Guidelines*. Washington, DC, USA: National Academies Press; 2011.
2. Guyatt G, Oxman AD, Akl EA, et al. GRADE guidelines: 1. Introduction - GRADE evidence profiles and summary of findings tables. *J Clin Epidemiol*. 2011;64(4):383−394. https://doi.org/10.1016/j.jclinepi.2010.04.026.
3. Heidenreich PA, Bozkurt B, Aguilar D, et al. *2022 AHA/ACC/HFSA Guideline for the Management of Heart Failure: A Report of the American College of Cardiology/American Heart Association Joint Committee on Clinical Practice Guidelines*. vol. 145. 2022. https://doi.org/10.1161/CIR.0000000000001063.
4. Kastner M, Bhattacharyya O, Hayden L, et al. Guideline uptake is influenced by six implementability domains for creating and communicating guidelines: a realist review. *J Clin Epidemiol*. 2015;68(5):498−509. https://doi.org/10.1016/j.jclinepi.2014.12.013.
5. Lu MT, Adami EC, Hoffmann U. Practicalities of running a clinical trial. In: *Handbook for Clinical Trials of Imaging and Image-Guided Interventions*. 2016:91−102. https://doi.org/10.1002/9781118849712.ch6.
6. Malmivaara A. Methodological considerations of the GRADE method. *Ann Med*. 2015;47(1):1−5. https://doi.org/10.3109/07853890.2014.969766.
7. McMurray JJV, Solomon SD, Inzucchi SE, et al. Dapagliflozin in patients with heart failure and reduced ejection fraction. *N Engl J Med*. 2019;381(21):1995−2008. https://doi.org/10.1056/NEJMoa1911303.
8. Montgomery P, Movsisyan A, Grant SP, Macdonald G, Rehfuess EA. Considerations of complexity in rating certainty of evidence in systematic reviews: a primer on using the GRADE approach in global health. *BMJ Glob Heal*. 2019;4(suppl 1):e000848. https://doi.org/10.1136/bmjgh-2018-000848.
9. Norris SL, Bero L. GRADE methods for guideline development: time to evolve? *Ann Intern Med*. 2016;165(11):810−811. https://doi.org/10.7326/M16-1254.
10. Rosenfeld RM, Shiffman RN. Clinical practice guideline development manual: a quality-driven approach for translating evidence into action. *Otolaryngol Head Neck Surg*. 2009;140(6 SUPPL):1−43. https://doi.org/10.1016/j.otohns.2009.04.015.
11. Schunemann HJ, Brennan S, Akl EA, et al. The development methods of official GRADE articles and requirements for claiming the use of GRADE - a statement by the GRADE Guidance Group. *J Clin Epidemiol*. May 19, 2023;159:79−84. https://doi.org/10.1016/j.jclinepi.2023.05.010.
12. Schunemann HJ, Brozek J, Guyatt G, Oxman A. GRADE Handbook. *GRADEpro*; October 2013. https://gdt.gradepro.org/app/handbook/handbook.html. Accessed November 1, 2023.
13. Shiffman RN, Shekelle P, Overhage JM, Slutsky J, Grimshaw J, Deshpande AM. Standardized reporting of clinical practice guidelines: a proposal from the conference on guideline standardization. *Ann Intern Med*. 2003;139(6):493−498. https://doi.org/10.7326/0003-4819-139-6-200309160-00013.
14. Siemieniuk R, Guyatt G. What is GRADE? *Br Med J Best Pract*; 2019. https://bestpractice.bmj.com/info/us/toolkit/learn-ebm/what-is-grade/. Accessed November 5, 2023.

Part X

Public health

Clauden Louis

Chapter 92

Public health

Dominique Vervoort[1] and Sameer Hirji[2]

[1]*Institute of Health Policy, Management and Evaluation, University of Toronto, Toronto, ON, Canada;* [2]*Division of Cardiac Surgery, Department of Surgery, Brigham and Women's Hospital, Harvard Medical School, Boston, MA, United States*

Key points

- Cardiovascular diseases are the leading cause of morbidity and mortality worldwide and are rapidly on the rise. A major share of the cardiovascular disease burden could be prevented with adequate and timely population health interventions. This requires public health research to identify, guide, and evaluate healthcare and population health interventions.
- Major contemporary foci of cardiovascular public health research include health promotion, health services research, environmental health, and global health, each striving to work toward an overarching goal of (cardiovascular) health equity for all.
- Cardiovascular public health research should ensure careful program planning and equitable representation in research teams and research populations and recognize potential confounders, taking into account individuals' social determinants of health.

Background: Why it matters

Cardiovascular diseases are the leading causes of morbidity and mortality globally, accounting for approximately 18 million deaths each year.[1] These estimates are expected to reach over 20 million deaths per year by 2025.[1] Despite improvements in acute and chronic management, the pharmacological optimization of secondary prevention, and extensive research into the pathophysiological nature of cardiovascular diseases, the largest share of the cardiovascular disease burden remains preventable. The social determinants of health are the "*conditions in which people are born, grow, work, live, and age, and the wider set of forces and systems shaping the conditions of daily life,*"[2] and may account for up to 60% of individuals' health status and outcomes (Fig. 92.1). Opportunities exist to better understand and address these conditions to promote cardiovascular health equity at the population level.

In this chapter, we discuss the role of cardiovascular research in improving population health through a focus on major and growing areas of public health research. We present practical ways to get started with cardiovascular public health research by describing financial considerations and potential pitfalls that should be avoided.

Population health and the role of public health research in the translational process

Population-level interventions can promote individuals' health by targeting a particular group of individuals (i.e., population health) or society at large (i.e., public health). These interventions can help elucidate healthcare disparities and understand programs' and healthcare's effectiveness in diagnosing, preventing, and treating disease. Thus, by extension, the goal of public health research is to build a bridge between healthcare and the population, to identify drivers of and barriers to population health, and to understand the effectiveness of population-level interventions.

Major areas of focus

Public health research is broad and can pertain to all parts of the care continuum outside the scope of inpatient care and to all social determinants of health. Major contemporary areas of research include but are not limited to the following: health promotion, health services research, environmental health, and global health (Table 92.1).

Health promotion: Population-based cardiovascular disease prevention and management

Promoting healthy lifestyles and optimizing cardiovascular risk factors at the population level are essential to prevent

Translational Cardiology. https://doi.org/10.1016/B978-0-323-91790-2.00031-9

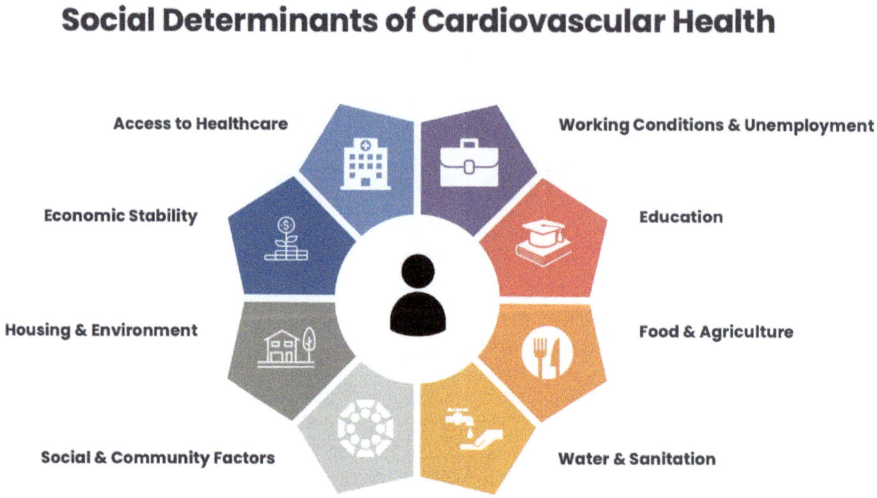

FIGURE 92.1 The social determinants of (cardiovascular) health.

TABLE 92.1 Major current areas of focus in cardiovascular public health research.

Focus area	Description	Select knowledge gaps	Potential data sources	Study examples
Health promotion	Identification and assessment of the effectiveness of population health interventions to improve cardiovascular health	Effectiveness of interventions across populations Role of digital health applications in health promotion	Primary data collection (qualitative, quantitative)	Khetan et al.[3] Kim et al.[4] Marvel et al.[5]
Health services research	Study of access to, costs of, disparities in, and long-term outcomes after cardiovascular care	Variation in care Disparities in care, by procedure, disease, and part of the care continuum	Administrative data; clinical registries; institutional data; survey data	Karamlou et al.[6] Lucas et al.[7] Welke et al.[8]
Environmental health	Understanding of the intersection between the environment, climate change, and cardiovascular health	Projected cardiovascular burdens attributable to climate change Effects of cardiovascular healthcare on the environment	Modeled data; Primary data collection	Grinberg et al.[9] Hayes et al.[10] Yoon et al.[11]
Global health	Cardiovascular research with global implications, in particular with regard to lower-resource settings and remote populations	Unmet cardiovascular care needs Global cardiovascular care capacity Catastrophic health expenditure due to cardiovascular care	Modeled data; primary and secondary data collection	DeWyer et al.[12] Sabatino et al.[13] Vervoort et al.[14]

as well as chronically manage cardiovascular diseases. While lifestyle-related cardiovascular risk factors have long been established, effectively introducing population health programs remains difficult. Moreover, current observational studies and trials remain heterogeneous to the extent of health interventions introduced and the selected populations of interest studied, thereby limiting the generalizability and validity of findings.[15] Opportunities arise in the use of digital health applications, although these add an additional layer of complexity, continuous change, and small per-product sample sizes.[16,17]

In India, Khetan et al.[3] conducted a cluster randomized controlled trial in 1242 Indian adults ages 35–70 who screened positive for at least one cardiovascular risk factor. All individuals were assigned to receive usual community-based care versus home visits with behavioral change communication by community health workers. After 2 years, the intervention group saw a significantly larger reduction in systolic blood pressure, highlighting the importance of population-based risk factor management programs and areas for improvement in future programs.

Health services research: Elucidating and tackling healthcare disparities

Health services research extends beyond traditional clinical research by studying population-level access to care, costs of care, and long-term outcomes.[18] In recent years, there has been an increased interest in exploring the pervasive healthcare disparities facing various global health systems. In the United States, the lack of a universal health coverage system in the setting of racial and socioeconomic disparities often presents challenges associated with equitable healthcare delivery and differential long-term outcomes across different communities and populations.

Welke et al.[8] studied the regional distribution of congenital heart surgery programs across the United States. They found great variations in annual case volumes and case complexity between programs, and a stark concentration of programs with 101 (66%) of 153 hospitals with congenital heart surgery programs being within 25 miles of each other. Despite this, a quarter of patients traveled over 100 miles, including to (often higher-volume) hospitals further away, without differences in outcomes, suggesting the potential for greater regionalization efforts. Similar volume-outcome trends have been observed in other cardiovascular procedures, including but not limited to transcatheter aortic valve replacement,[19] coronary revascularization,[20,21] and thoracic aortic surgery.[22] However, in lower population density areas, travel distance may result in poorer healthcare utilization and/or outcomes,[23] thus requiring further investigation into balancing geographical access with the optimization of cardiovascular outcomes.

Environmental health: Intersection between climate change and cardiovascular medicine

The environments surrounding us affect our daily lives and can contribute to changes in our cardiovascular health. Environmental pollution has been associated with increased cardiovascular disease incidence,[24–26] disproportionally affecting individuals living in poorer neighborhoods, remote areas, or low- and middle-income countries (LMICs). In addition, the physiological and psychological stress from climate change disproportionally affects climate change refugees and individuals living in areas affected by natural disasters. For example, the incidence of myocardial infarctions after Hurricane Sandy in 2012 rose by approximately 22%.[27] Conversely, the management of cardiovascular diseases may contribute to climate change: an eco-audit of cardiac surgical procedures in France showed that the emission from one operation had a global warming equivalent of a 1,080 km single-passenger plane ride.[9]

In South Korea, Yoon et al.[11] used meteorological, governmental-based administrative, and claims data to quantify the disability-adjusted life years attributable to climate change between 2005 and 2008, and forecast it over time. They found that cardiovascular disease accounted for approximately three-quarters of the climate change-related burden of disease in South Korea, projected to double by the end of the century. Such research is vital in informing health policy and practice.

Global health: Leaving no patient behind

There is an increasing understanding of the global burden of cardiovascular disease and unmet care needs. Over 80% of all cardiovascular deaths occur in LMICs,[1] where noncommunicable diseases are rapidly on the rise.[28] Despite major gains in life expectancy and tackling infectious disease, maternal, and child health burdens worldwide, cardiovascular health has trailed in terms of political, academic, and advocacy support.[28–31] For example, six billion people lack access to safe, timely, and affordable cardiac surgical care when needed,[32] resulting in millions of preventable cardiovascular surgical deaths each year.[33]

Vervoort and Cardarelli[14] used Global Burden of Disease data to determine the share of the global burden of congenital heart disease in children that could be considered "excess" morbidity (in disability-adjusted life years) and mortality (deaths) by benchmarking country-level data for different age groups against "best achievers." In this way, the authors found that nearly 200,000 child deaths due to congenital heart disease worldwide may be prevented each year. Global Burden of Disease data is freely accessible (**Additional Resources**) and a helpful resource for public health researchers, giving insights into national and partially subnational disease burdens by age group, sex, year, and other characteristics.

Getting started

Budgetary and funding considerations

Depending on the data needed and the research question at hand, public health research may be done at no cost or may require extensive funding. Certain research (e.g., online surveys, public secondary data) may be conducted for free. The acquisition of data from clinical registries (e.g., societal databases) or administrative databases (e.g., state-level, claims data) can be relatively inexpensive. Population interventions (e.g., mobile health interventions, cardiovascular screening) are generally costly and require careful program planning and budgeting. Various societal and nonsocietal grants are available to support researchers in pursuing public health research (examples in **Additional Resources**).

Pitfalls to avoid

When starting, researchers ought to watch out for various potential pitfalls, including:

1. *Unconscious Biases and Underrepresentation*: Healthcare disparities are, at least in part, driven and perpetuated by conscious and unconscious biases in the healthcare system, as well as a lack of representation by minoritized individuals in the health workforce and research. For example, women and non-White patients have been lacking from randomized clinical trials,[34–37] limiting the generalizability of trial findings to the wider population. This can be addressed by ensuring greater equity and representation in research teams, particularly engaging individuals from the relevant study population (e.g., patient advocates, local researchers, Indigenous Peoples).
2. *Confounding*: Population health intersects with all layers of people's lives, encompassing all social determinants of health. Unlike clinical research, where clinically meaningful variables and outcomes of interest are relatively few, population health is a dynamic and complex function of widely varying aspects of life (e.g., education, environment, social networks, income, beliefs and religion, culture, structural contexts). Thus, conclusions from public health research must be carefully approached in the research methods and analysis, and, when unable to adequately adjust for or explain the effects of potential confounders, be interpreted cautiously.
3. *Disconnect between academia and the public*: The goal of public health research is to directly engage with the public and improve population health. However, there is a risk that public health research remains an "ivory tower" pursuit by academics, whereby the needs of the population are not met or even listened to. Researchers should ensure that research benefits local populations and that the research is conducted as a means to improve population health, rather than being "just another project." This could be done by engaging population representatives in the research process and/or holding focus groups.

Conclusion

Public health research has the potential to better understand and improve cardiovascular population health, despite historically examples being less prioritized compared to clinical and basic science research. Various opportunities arise for researchers from diverse backgrounds to engage in public health research to lay the foundation for cardiovascular health equity.

Additional resources

Books

- Dimick JB, Lubitz CC. *Health Services Research*. Second ed. Springer. 2020.
- Kpodonu J. *Global Cardiac Surgery Capacity Development in Low and Middle Income Countries*. First ed. Springer. 2022

Articles

- Lampropulos JF, Gupta A, Kulkarni VT et al. Most important outcomes research papers on variation in cardiovascular disease. *Circ Cardiovasc Qual Outcomes*. 2013; 6(2):e9–e16. https://doi.org/10.1161/CIRCOUTCOMES.113.000185
- Vervoort D. Moving the Needle: A Guide to Solving the Global Cardiac Surgery Puzzle for Surgeons, Societies, Students, and Researchers. *CTSNet*. February 2020. https://doi.org/10.25373/ctsnet.11872425
- Ebinger J. Health Services Research: Cardiology in the Changing Health Care Landscape. *J Am Coll Cardiol*. 2017; 69(17):2242–2245. https://doi.org/10.1016/j.jacc.2017.03.539

Websites

- Institute for Health Metrics and Evaluation Global Burden of Disease Results Tool: http://ghdx.healthdata.org/gbd-results-tool
- Thoracic Surgery Foundation: https://thoracicsurgeryfoundation.org/
- American Heart Association: https://professional.heart.org/en/research-programs

References

1. Roth GA, Johnson C, Abajobir A, et al. Global, regional, and national burden of cardiovascular diseases for 10 causes, 1990 to 2015. *J Am Coll Cardiol*. 2017;70(1):1–25.
2. World Health Organization. *Social Determinants of Health*; 2021. https://www.who.int/health-topics/social-determinants-of-health. Accessed December 18, 2021.
3. Khetan A, Zullo M, Rani A, et al. Effect of a community health worker-based approach to integrated cardiovascular risk factor control in India: a cluster randomized controlled trial. *Glob Heart*. 2019;14(4):355–365.
4. Kim SR, Choi S, Keum N, Park SM. Combined effects of physical activity and air pollution on cardiovascular disease: a population-based study. *J Am Heart Assoc*. 2020;9(11):e013611.

5. Marvel FA, Spaulding EM, Lee MA, et al. Digital health intervention in acute myocardial infarction. *Circ Cardiovasc Qual Outcome*. 2021;14(7):e007741.
6. Karamlou T, Hawke JL, Zafar F, et al. Widening our focus: characterizing socioeconomic and racial disparities in congenital heart disease. *Ann Thorac Surg*. April 16, 2021. https://doi.org/10.1016/j.athoracsur.2021.04.008.
7. Lucas FL, Sirovich BE, Gallagher PM, Siewers AE, Wennberg DE. Variation in cardiologists' propensity to test and treat: is it associated with regional variation in utilization? *Circ Cardiovasc Qual Outcomes*. 2010;3(3):253–260.
8. Welke KF, Pasquali SK, Lin P, et al. Hospital distribution and patient travel patterns for congenital cardiac surgery in the United States. *Ann Thorac Surg*. 2019;107(2):574–581.
9. Grinberg D, Buzzi R, Pozzi M, et al. Eco-audit of conventional heart surgery procedures. *Eur J Cardio Thorac Surg*. 2021;60(6):1325–1331.
10. Hayes RB, Lim C, Zhang Y, et al. PM2.5 air pollution and cause-specific cardiovascular disease mortality. *Int J Epidemiol*. 2020;49(1):25–35.
11. Yoon S-J, Oh I-H, Seo H-Y, Kim E-J. Measuring the burden of disease due to climate change and developing a forecast model in South Korea. *Publ Health*. 2014;128(8):725–733.
12. DeWyer A, Scheel A, Otim IO, et al. Improving the accuracy of heart failure diagnosis in low-resource settings through task sharing and decentralization. *Glob Health Action*. 2019;12(1):1684070.
13. Sabatino ME, Dennis RJ, Sandoval-Trujillo P, et al. Access to paediatric cardiac surgery in Colombia: a population-based study. *Eur J Cardio Thorac Surg*. September 30, 2021. https://doi.org/10.1093/ejcts/ezab404.
14. Vervoort D, Cardarelli M. The global unmet need of congenital cardiac care: a quantitative analysis of the global burden of disease. *Cardiol Young*. August 20, 2020:1–6.
15. Walton-Moss B, Samuel L, Nguyen TH, Commodore-Mensah Y, Hayat MJ, Szanton SL. Community-based cardiovascular health interventions in vulnerable populations: a systematic review. *J Cardiovasc Nurs*. 2014;29(4):293–307.
16. Vervoort D, Marvel FA, Isakadze N, Kpodonu J, Martin SS. Digital Cardiology: opportunities for disease prevention. *Curr Cardiovasc Risk Rep*. 2020;14(8):10.
17. Vervoort D, Tam DY, Wijeysundera HC. Health technology assessment for cardiovascular digital health technologies and artificial intelligence: why is it different? *Can J Cardiol*. August 28, 2021. https://doi.org/10.1016/j.cjca.2021.08.015.
18. Dimick JB, Lubitz CC. An introduction to health services research. In: Dimick JB, Lubitz CC, eds. *Health Services Research*. Springer International Publishing; 2020:3–9.
19. Bavaria JE. Point: access to transcatheter aortic valve replacement should be limited to high-volume surgical centers. *J Thorac Cardiovasc Surg*. 2013;145(6):1441–1443.
20. Zahn R, Gottwik M, Hochadel M, et al. Volume-outcome relation for contemporary percutaneous coronary interventions (PCI) in daily clinical practice: is it limited to high-risk patients? Results from the Registry of Percutaneous Coronary Interventions of the Arbeitsgemeinschaft Leitende Kardiologische Krankenhausarzte (ALKK). *Heart*. 2008;94(3):329–335.
21. McGrath PD, Wennberg DE, Dickens Jr JD, et al. Relation between operator and hospital volume and outcomes following percutaneous coronary interventions in the era of the coronary stent. *JAMA*. 2000;284(24):3139–3144.
22. Gazoni LM, Speir AM, Kron IL, Fonner E, Crosby IK. Elective thoracic aortic aneurysm surgery: better outcomes from high-volume centers. *J Am Coll Surg*. 2010;210(5):855–859.
23. Damluji AA, Fabbro M, Epstein RH, et al. Transcatheter aortic valve replacement in low-population density areas. *Circ Cardiovasc Qual Outcome*. 2020;13(8):e006245.
24. Burroughs Peña MS, Rollins A. Environmental exposures and cardiovascular disease: a challenge for health and development in low- and middle-income countries. *Cardiol Clin*. 2017;35(1):71–86.
25. Rajagopalan S, Al-Kindi SG, Brook RD. Air pollution and cardiovascular disease: JACC state-of-the-art review. *J Am Coll Cardiol*. 2018;72(17):2054–2070.
26. Peters A, Schneider A. Cardiovascular risks of climate change. *Nat Rev Cardiol*. 2021;18(1):1–2.
27. Swerdel JN, Janevic TM, Cosgrove NM, Kostis JB, Myocardial Infarction Data Acquisition System (MIDAS 24) Study Group. The effect of Hurricane Sandy on cardiovascular events in New Jersey. *J Am Heart Assoc*. 2014;3(6):e001354.
28. Bukhman G, Mocumbi AO, Atun R, et al. The Lancet NCDI Poverty Commission: bridging a gap in universal health coverage for the poorest billion. *Lancet*. 2020;396(10256):991–1044.
29. Chang AY, Cowling K, Micah AE, et al. Past, present, and future of global health financing: a review of development assistance, government, out-of-pocket, and other private spending on health for 195 countries, 1995–2050. *Lancet*. 2019;393(10187):2233–2260.
30. Vervoort D, Parikh UM, Raj A, Swain JD. Global cardiovascular care: an overview of high-level political commitment. *Asian Cardiovasc Thorac Ann*. May 27, 2020;28(5):258–265. https://doi.org/10.1177/0218492320930844.
31. Vervoort D, Genetu A, Kpodonu J. Policy prioritisation to address the global burden of rheumatic heart disease. *Lancet Global Health*. 2021;9(9):e1212.
32. Vervoort D, Swain JD, Pezzella AT, Kpodonu J. Cardiac surgery in low- and middle-income countries: a state-of-the-art review. *Ann Thorac Surg*. August 6, 2020;111(4):1394–1400. https://doi.org/10.1016/j.athoracsur.2020.05.181.
33. Vervoort D, Ghandour H, O'Hara DN, Lin Y, Shrime MG. Global burden of cardiovascular surgical disease: an estimation from the provider perspective. *Bethune Round Table*. 2021.
34. Preventza O, Critsinelis A, Simpson K, et al. Sex, racial, and ethnic disparities in U.S. cardiovascular trials in more than 230,000 patients. *Ann Thorac Surg*. 2021;112(3):726–735.
35. Whitelaw S, Sullivan K, Eliya Y, et al. Trial characteristics associated with under-enrolment of females in randomized controlled trials of heart failure with reduced ejection fraction: a systematic review. *Eur J Heart Fail*. 2021;23(1):15–24.
36. Whitelaw S, Thabane L, Mamas MA, et al. Characteristics of heart failure trials associated with under-representation of women as lead authors. *J Am Coll Cardiol*. 2020;76(17):1919–1930.
37. Ifidon AM, Agbalajobi O, Essien UR. Improving racial and ethnic minority representation in cardiovascular disease trials to advance health equity. *JAMA Cardiol*. 2020;5(5):611.

Chapter 93

Epidemiology

Jef Van den Eynde[1] and Sameer Hirji[2]

[1]Department of Cardiovascular Sciences, KU Leuven, Leuven, Belgium; [2]Division of Cardiac Surgery, Department of Surgery, Brigham and Women's Hospital, Harvard Medical School, Boston, MA, United States

Key points

- The study of epidemiology has led to the establishment of invaluable knowledge about the prevalence, risk factors, and outcomes of various cardiovascular diseases.
- Various types of study design are available within epidemiology, all of which have their own advantages and disadvantages and are applicable in different clinical settings.
- Data generated from epidemiological studies are essential during patient counseling, for the organization of care and in policy making, and it is the responsibility of all stakeholders to ensure that these are being used effectively, ethically, and correctly.

"Why it matters?"

Cardiovascular diseases are the leading cause of death globally. Smoking, excessive alcohol, high blood pressure, unhealthy diet, and physical inactivity increase the risk of developing acquired heart disease and stroke. Early prevention can save lives. We consider all of these truths to be self-evident as if they had been known forever. Nonetheless, this knowledge has only emerged relatively recently as the field of epidemiology started to flourish after World War II. Nowadays, epidemiological studies continue to provide an essential foundation for enhancing our understanding of the prevalence, risk factors, and outcomes of cardiovascular disease. Clinicians, scientists, researchers, and bio- or med-tech entrepreneurs alike should therefore be familiar with some of the central concepts of research in epidemiology to provide meaningful contributions to the literature.

Introduction

According to Last's Dictionary of Epidemiology,[1] "epidemiology is the *study* of the *distribution* and *determinants* of *health-related states and events* in specified *populations*, and the *application* of this study to the control of health problems." In other words, epidemiology is a scientific, systematic, and data-driven approach to what impacts the health of populations. While epidemiology was originally focused on epidemics of infectious diseases, it is now applied more broadly to a wide variety of subjects including environmental exposures, injuries, noncommunicable diseases, natural disasters, and terrorism. The study may describe the frequency (number of events or cases and its relationship to the total size of the population) and pattern (by time, place, and personal characteristics) of health-related states and events. In addition, epidemiology is also concerned about the causes and risk factors that determine or influence the occurrence of health-related states and events. Eventually, these data serve an important role in the planning and evaluation of strategies to effectively prevent and manage these underlying health-related states. Just like the clinical presentation and pathological findings, epidemiology is an integral part of the basic description of a disease state.

While the roots of epidemiology date back to Hippocrates in 400 BCE, the field of epidemiology really started to flourish after World War II.[2] Nowadays, epidemiological studies continue to provide an essential base for our understanding of the prevalence, risk factors, and outcomes of cardiovascular disease. Clinicians, scientists, researchers, and bio- or med-tech entrepreneurs alike should therefore be familiar with some of the central concepts of research in epidemiology. In this chapter, we will discuss these concepts as we go through the main steps of an epidemiological study: defining the question, choosing the optimal study design, and conducting and interpreting the analyses.

Step 1: Defining the question

When faced with an epidemiological issue, one can adopt the 5 Ws to better define the question: What (health-related

Translational Cardiology. https://doi.org/10.1016/B978-0-323-91790-2.00085-X

state or event), Who (person), Where (place), When (time), and Why (causes, risk factors, modes of transmission). Before the distribution and determinants of a certain health-related state or event can be counted, one needs to clarify what is to be considered as a "health-related state or event." For this purpose, case definitions are used, which are sets of standard criteria for classifying whether a person has a particular disease, syndrome, event, or other health condition. These may include symptoms (subjective complaints), signs (objective physical findings), laboratory tests, or other confirmatory examinations.

Some case definitions have been developed and adopted as international standards, which ensures that every case is equivalent and that findings can be compared in a standardized way across different settings. Cognizant of the epidemic of coronary heart disease during the 20th century in most industrialized countries, the World Health Organization in 1959[3] and the American Heart Association in 1964[4] developed consistent and universal case definitions. These definitions fulfilled the need for reliable information on incidence, prevalence, and case-fatality rates of the condition. However, with advances in diagnostic technologies, therapeutic interventions, and changing disease presentations, these definitions have subsequently been revised multiple times commensurate with clinical context. An example of endpoint-focused case definitions is the Valve Academic Research Consortium 3 (VARC-3) definitions that aim to align clinical research in aortic valve disease.[5]

When deciding about the Who, Where, and When, the generalizability of the study results should be carefully considered. A highly selected population from a tertiary referral center might not be representative of the community-dwelling population, and vice versa. Subsequently, the findings of certain associations or treatment effects in one population may not be applicable to the other. Finally, the answer to Why will specify whether one is primarily interested in the distribution (frequency and pattern) of the case or determinants (causes and risk factors) of the case.

Step 2: Choosing the optimal study design

Before we go ahead, it is important to remember the main study designs in clinical cardiovascular research. For this purpose, a summary is provided in Fig. 93.1 and Table 93.1. Representative real-world examples for each of these in cardiovascular epidemiology are provided in Table 93.2. For a more detailed discussion of each study design, the reader is referred to previous chapters in this book.

Designs in cardiovascular epidemiology can be broadly subdivided into experimental or observational studies (Fig. 93.1). In an experimental study, exposures or treatments are assigned by the investigators, whereas an observational study measures a sample without trying to affect it. Furthermore, study designs can be defined as either descriptive or analytical (Fig. 93.1). Descriptive epidemiology focuses on describing disease distribution by characteristics relating to time (when), place (where), and people (who). Examples include surveys, ecological studies, and case reports/series. Although these types of studies can generate hypotheses on risk factors and causes of disease, they cannot test these. The latter is done in analytical epidemiology, which measures the statistical association between a suspected causal exposure and disease occurrence (why) or tests the effectiveness and safety of therapeutic and medical interventions. It also studies the distribution of those exposures and diseases. Examples include case-control, cross-sectional, and cohort studies. Experimental studies are also considered to be part of analytical epidemiology.

The question defined in Step 1 will directly inform the choice of the optimal study design. As shown in Table 93.1, each study design is characterized by different directions, types of measurement, advantages, and limitations. Direction refers to the order in which exposure and outcome are defined and evaluated. If the direction is "Forward," the exposure is determined first and patients are subsequently followed up to register outcomes (interventional trials and prospective cohorts) or the patients' records are reviewed to identify outcomes (retrospective cohort). This is especially useful in the setting of rare exposures but is impractical if the outcome is only rarely observed. The latter are better dealt with in a "Reverse" design such as case-control studies which determine the outcome first and then look for certain exposures that might be associated with that outcome. Other designs such as cross-sectional studies have a directionality that is referred to as "Present," meaning that exposure and outcome are determined at the same point in time. These are easy and inexpensive to conduct but are limited to common exposures and outcomes. Other considerations include cost, time investment, susceptibility to confounding or bias, and ability to conclude about causality.

Of note, the availability of data may inform the choice of study design. Collecting primary data for a specific purpose might be the most solid approach, but is expensive and time consuming. Secondary data, on the other hand, are data that are collected for a different purpose but can nonetheless prove valuable to answer certain epidemiological questions. These include for example birth and death certificates, population census records, patient medical records, disease registries, insurance claim forms and billing records, public health department case reports, or surveys of individuals and households. Two examples in the United States that have been used avidly for cardiovascular research are the National Readmission Database (NRD) and National Inpatient Sample (NIS) databases.

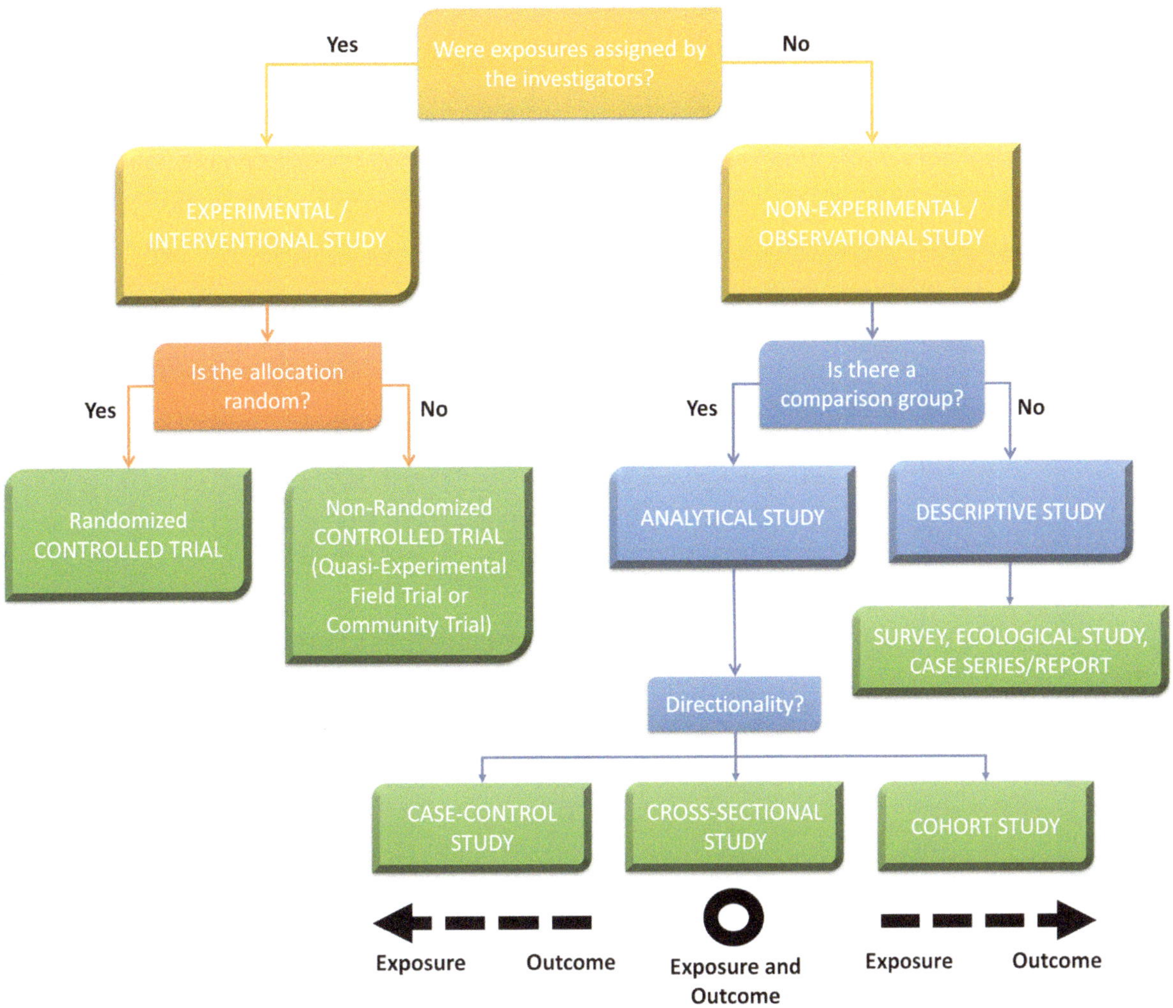

FIGURE 93.1 Flow chart summarizing the main study designs in epidemiological research.

These data are weighted sampled data as part of the Agency for Healthcare Research And Quality. However, secondary data may be limited in terms of accessibility, completeness, depth, accuracy, and timeliness.

Step 3: Conducting and interpreting the analysis

A detailed discourse of the analytical approaches for each of the epidemiological study designs is beyond the scope of this chapter and can be found in previous parts of this book. However, in this section, we would like to highlight three analytical concepts that are commonly encountered in cardiovascular epidemiology: association, causality, and validity (Fig. 93.2).

Associations between exposures and outcomes can be expressed in several ways.[6] First, absolute measures of association include the risk difference (risk of the exposure among cases minus the risk of the exposure among controls) and rate difference (the rate is the number of cases and controls with the exposure divided by the person-years of observation) and are most easily understood. On the other hand, confusion commonly arises around the distinction of risk ratio (RR) and odds ratio (OR), which are both relative measures of association. The RR is the risk of outcomes among cases divided by the risk of the outcome among controls. In contrast, the OR is the odds of exposure (i.e., number of subjects with exposure divided by the number of subjects without exposure) in the cases divided by the odds of exposure in the controls. While cohort studies can determine both RR and OR, the OR is usually the only measure that can be calculated from a case-control study. This results from the fact that the calculation of the RR requires the total number of exposed people, which is not available in a case-control study where subjects have been selected based on outcome status.

"Association does not imply causation" is a common theme in epidemiology. While studies often investigate

TABLE 93.1 Characteristics of the main study designs in epidemiological research.

Study design	Unit of study	Description	Direction	Type of measurement	Advantages	Limitations
Interventional studies/ controlled trials	Individual	Studies and examines an intervention and a control group	Forward	Prevalence, incidence, relative risk, attributable risk	• Strongest evidence for causality of intervention • Control of unknown confounders • Fulfills basic assumptions of statistical hypothesis tests	• Expensive, time consuming • Sometimes generalizability problem
Cohort study	Individual	Studies and examines an exposed and a nonexposed group	Forward	Prevalence, incidence, relative risk, attributable risk	• Useful in setting of rare exposures • Allows to study multiple diseases/outcomes • Can obtain a true measure of risk (relative risk), can determine causality relations	• Not useful in the setting of rare diseases • Prospective cohort studies are time-consuming and costly • Only risk factors measured at the beginning of the study can be measured • Selection bias, loss to follow-up
Case-control study	Individual	Studies and examines a diseased and a nondiseased group	Reverse	Odds ratio	• Useful in setting rare diseases • Allows to study multiple exposures • Relatively inexpensive and efficient	• Can only obtain a relative measure of risk (odds ratio), difficult to establish that "cause" preceded the effect • Selection bias, misclassification bias, recall bias
Cross-sectional study	Individual	Studies a population at a "point in time"	Present	Prevalence, association	• Useful for investigating fixed exposures • Easy and inexpensive	• Not useful for rare diseases or exposures • Selection bias, misclassification bias, recall bias
Survey	Individual	Studies opinions, practices, or patterns in large samples	Present	Prevalence, association	• Collects large amount of data from variety of people in realistic settings • Not constrained by geography	• Dependent on response rate • May include wrong respondents • Bias if inappropriate wording of questions
Ecological study	Population	Compares populations in different places at the same time or in a time series	Present	Correlation	• Allows to study the effects of group-level constructs, based on community group or national level data and trends • Quick and inexpensive	• Ecological fallacy, i.e., associations identified at the population level might not exist at the individual level • Confounding due to lack of comparison
Case series/ report	Individual (1 or more cases)	Describes the characteristics of a (group of) case(s)	Present	Reporting, description	• Quick means of exchanging thoughts with clinical relevance • May be hypothesis-generating	• Inability to determine statistical relations and analysis

For direction, "Forward" refers to a directionality from exposure to the outcome, while "Reverse" refers to a directionality from outcome to exposure and "Present" refers to a design where exposure and outcome are determined at the same point in time.

TABLE 93.2 Examples of the application of each study design in cardiovascular epidemiology.

Study design	Example
Interventional studies/controlled trials	The Salt Substitute and Stroke Study (SSaSS) (NCT02092090)[8] was an open-label, cluster-randomized trial conducted across 600 villages in rural China. The participants were randomly assigned to use a salt substitute (75% sodium chloride and 25% potassium chloride by mass) or regular salt (100% sodium chloride). The primary outcome of stroke was lower with the salt substitute (29.14 events vs. 33.65 events per 1000 person-years; RR 0.86, 95% CI 0.77–0.96, $P = .006$).
Cohort study	The Framingham Heart Study[9] is a population-based, observational cohort study that was initiated by the United States Public Health Service in 1948 to prospectively investigate the epidemiology and risk factors for cardiovascular disease. It is the longest-running cardiovascular epidemiological study, spanning 3 generations of densely phenotyped White persons and 2 cohorts comprised of racial and ethnic minority groups with extensive longitudinal follow-up.
Case-control study	Using data from the Swedish Coronary Angiography and Angioplasty register and linking these other national databases, Tornvall et al.[10] compared risk markers and mortality in patients with Takotsubo stress cardiomyopathy with that of individuals with or without coronary. Their findings provided evidence for the association with beta2-adrenergic agonist agents together with stress related to affective disorders, emphasizing the pathogenic role of sympathetic stimulation.
Cross-sectional study	Gebreab et al.[11] analyzed cross-sectional data from the Jackson Heart Study (JHS), a large population-based cohort of African Americans, to examine the contribution of psychosocial stress to the association of socioeconomic position with cardiovascular risk factors and subclinical atherosclerotic disease.
Survey	The US National Health and Nutrition Examination Surveys (NHANES) is a biennial survey of a nationally representative sample of civilian noninstitutionalized individuals. In one exemplary analysis, Ali et al.[12] determined the prevalence of cardiovascular and renal risks and disease in adults with prediabetes and demonstrated that these had increased over a 25-year period between 1988 and 2014.
Ecological study	Alter et al.[13] linked claims for payment for physicians' services, hospital-discharge abstracts, and vital-status data for 51,591 patients with acute myocardial infarction who were admitted to hospitals in Ontario between April 1994 and March 1997. They found that each 10,000 USD increase in the neighborhood median income was associated with a 10% reduction in the risk of death within 1 year (adjusted HR 0.90; 95% CI 0.86–0.94, $P < .001$).
Case series/report	With the emergence of coronavirus disease 2019 (COVID-19), case reports and series were instrumental in generating awareness about cardiac involvement in affected individuals. As an example, Bangalore et al.[14] reported on 18 patients with COVID-19 who had ST-segment elevation and described their characteristics.

CI, confidence interval; *HR*, hazard ratio; *RR*, relative risk.

associations by isolating different exposures and outcomes, in real life most outcomes are caused by a chain or web consisting of many component causes. Causal inference is the process in which the independent, actual effect of a certain component within this system is determined. The randomized controlled trial is usually considered the gold standard for generating causality of certain interventions. However, various sources of information should ideally be interpreted jointly to determine causality. The Bradford–Hill criteria constitute a set of nine criteria that help determine if observed epidemiologic associations are causal, including strength of association, consistency, specificity, temporality, biological gradient, plausibility, coherence, experiment, and analogy (Table 93.3).[7] It should be noted that different types of causes exist, including probabilistic, sufficient, and necessary causes. In epidemiology, the latter are of major interest as these can be identified and controlled to avoid harmful outcomes.

Because epidemiological data will eventually be used to guide clinical practice and policy making, their validity is an important topic of concern. Validity, or lack of bias, refers to the extent to which a measurement actually reflects the "truth." This is very different from precision, which is a term that describes the extent to which similar information is obtained when the measurement is repeated. Poor precision is usually due to random error and does not shift the effect size if studies are conducted multiple times or in a larger sample. Poor validity, on the other hand, arises from systematic errors such as selection bias, information bias, or confounding. Selection bias occurs when individuals or groups are selected for

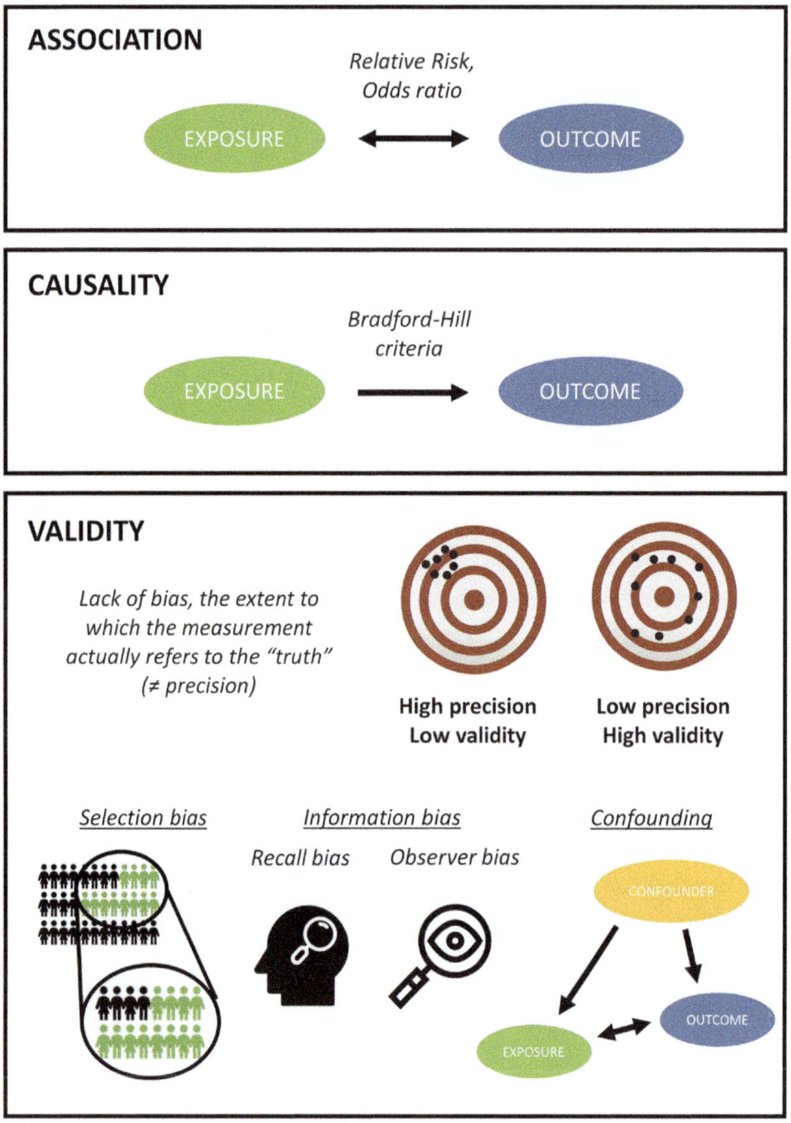

FIGURE 93.2 Association, causality, and validity in epidemiological studies.

inclusion into the study that is systematically different from those in the population of interest. Information bias or measurement bias, occurs due to inaccurate measurement or classification of one of the key study variables such as exposures or outcomes, either due to false recalling of the patient (recall bias) or due to errors by the observer (observer bias). Finally, confounding is a systematic distortion of the measurement due to the mixing of the exposure of interest with another factor that is independently associated with both the exposure and outcome of interest (confounder). Accurately anticipating, identifying, and controlling these sources of bias help ensure that measurements from epidemiological studies are valid.

"Get started"

Using the materials from previous chapters and the concepts discussed in the present chapter, the reader should be able to conduct and critically interpret epidemiological studies, and use the information generated from these to inform clinical practice and policy making. When starting out with any epidemiological project, the research question should be well defined according to the 5 Ws. The question will usually inform the choice of the optimal study design. Furthermore, knowledge about measures of association, causal inference, and validity allows for the correct interpretation and application of epidemiological studies. Eventually, it is the responsibility of all stakeholders, including clinicians,

TABLE 93.3 Bradford–Hill criteria to help determine whether epidemiological associations are causal.

Criterion	Explanation
Strength	A strong association (relative risk or odds ratio >>1 or <<1) is more likely to be causal.
Consistency	An association is more likely to be causal when it is observed in different populations, with different evaluation methods, over time, and across countries.
Specificity	When an exposure is associated with a specific outcome only (e.g., a cancer site or even a particular histological type of this cancer), then it is more likely to be causal. There are exceptions, however, for example, smoking causes several forms of cancers.
Temporality	A cause should not precede the outcome, but also the timing of the exposure should be compatible with any latency or incubation period (mostly in infectious diseases).
Biological gradient	This criterion refers to the dose-response gradient. If the frequency or intensity of the outcome increases when an exposure is more intense or lasts longer, then it is more likely that the association is causal.
Plausibility	An association is more likely to be causal when it is biologically plausible.
Coherence	A cause-and-effect interpretation of an association should not conflict with what is known about the natural history and biology of the disease, or its distribution in time and place.
Experiment	If experimental evidence (e.g., a randomized controlled trial) exists, then the association is more likely to be causal. Such evidence, however, is seldom available in human populations.
Analogy	The existence of an analogy (e.g., if a drug causes birth defects, then another drug could also have the same effect) could strengthen the belief that an association is causal.

scientists, researchers, and bio- or med-tech entrepreneurs alike, to ensure that these are being used effectively and correctly.

References

1. Last JM, ed. *Dictionary of Epidemiology*. 4th ed. New York: Oxford University Press; 2001:61.
2. Thacker SB. Historical development. In: Teutsch SM, Churchill RE, eds. *Principles and Practice of Public Health Surveillance*. 2nd ed. New York: Oxford University Press; 2002:1–16.
3. World Health Organization Regional Office for Europe, Working Group on the Establishment of Ischaemic Heart Disease Registers. *Report of the Fifth Working Group, Copenhagen*. Copenhagen, Denmark: Regional Office for Europe, World Health Organization; 26–29 April 1971, 1971.Report No. Eur 8201(5.
4. Weinstein BJ, Epstein FH, Working Subcommittee on Criteria and Methods, Committee on Epidemiological Studies, American Heart Association. Comparability of criteria and methods in the epidemiology of cardiovascular disease: report of a survey. *Circulation*. 1964;30:643–653.
5. VARC-3 Writing Committee, Généreux P, Piazza N, Alu MC, et al. Valve academic research consortium 3: updated endpoint definitions for aortic valve clinical research. *J Am Coll Cardiol*. June 1, 2021;77(21):2717–2746. https://doi.org/10.1016/j.jacc.2021.02.0386.
6. Knol MJ, Algra A, Groenwold RH. How to deal with measures of association: a short guide for the clinician. *Cerebrovasc Dis*. 2012;33(2):98–103. https://doi.org/10.1159/000334180.
7. Hill AB. The environment and disease: association or causation? *Proc Roy Soc Med*. 1965;58:295–300.
8. Neal B, Wu Y, Feng X, et al. Effect of salt substitution on cardiovascular events and death. *N Engl J Med*. September 16, 2021;385(12):1067–1077. https://doi.org/10.1056/NEJMoa2105675.
9. Andersson C, Nayor M, Tsao CW, Levy D, Vasan RS. Framingham heart study: JACC focus seminar, 1/8. *J Am Coll Cardiol*. June 1, 2021;77(21):2680–2692. https://doi.org/10.1016/j.jacc.2021.01.059.
10. Tornvall P, Collste O, Ehrenborg E, Järnbert-Petterson H. A case-control study of risk markers and mortality in takotsubo stress cardiomyopathy. *J Am Coll Cardiol*. April 26, 2016;67(16):1931–1936. https://doi.org/10.1016/j.jacc.2016.02.029.
11. Gebreab SY, Diez-Roux AV, Hickson DA, et al. The contribution of stress to the social patterning of clinical and subclinical CVD risk factors in African Americans: the Jackson Heart Study. *Soc Sci Med*. November 2012;75(9):1697–1707. https://doi.org/10.1016/j.socscimed.2012.06.003.
12. Ali MK, Bullard KM, Saydah S, Imperatore G, Gregg EW. Cardiovascular and renal burdens of prediabetes in the USA: analysis of data from serial cross-sectional surveys, 1988–2014. *Lancet Diabetes Endocrinol*. May 2018;6(5):392–403. https://doi.org/10.1016/S2213-8587(18)30027-5.
13. Alter DA, Naylor CD, Austin P, Tu JV. Effects of socioeconomic status on access to invasive cardiac procedures and on mortality after acute myocardial infarction. *N Engl J Med*. October 28, 1999; 341(18):1359–1367. https://doi.org/10.1056/NEJM199910283411806.
14. Bangalore S, Sharma A, Slotwiner A, et al. ST-segment elevation in patients with covid-19 - a case series. *N Engl J Med*. June 18, 2020;382(25):2478–2480. https://doi.org/10.1056/NEJMc2009020.

Chapter 94

How to ask the right public health research questions

Clauden Louis

Structural Heart and Transcatheter Endovascular Cardiac Surgery Fellow, Cardiothoracic Surgery, Brigham and Women's Hospital, Boston, MA, United States

Key points

- Qualitative research
- Quantitative research

Introduction

Asking the right public health research questions requires multiple elements to successfully acquire usable information. For the questions to lead to meaningful change, have measurable factors, and have a feasible study design, one must successfully prioritize a research process as well as understand the type of study one seeks to employ to capture the intended study.

Qualitative and quantitative research are two modalities of research that lead to numerous discoveries and answers to questions. One must first consider the nature and goal of a phenomenon one seeks to study to consider which method is appropriate. Whether one learns by immersing oneself in the environment or exploring objective data sources, qualitative and quantitative methods of research serve important roles.

Qualitative research may focus on topics that are debatable. Some of the questions are descriptive "What" and "How" questions seeking to explore topics with a goal to understand. Quantitative research seeks to definitively prove or disprove a concrete matter of interest. The questions are not similar and may be more specific such as "What is the amount" or "How much" with a goal that leads to prediction.

Qualitative research

Qualitative research is research that prioritizes the lived reality of the individuals being researched. "The choice between quantitative and qualitative research methods should be determined by the research question, not by the preference of the researcher."[1] This research embraces complexity with a depth of understanding. Qualitative research makes us examine how we understand science and nature. "Qualitative research can investigate practitioners' and patients' attitudes, beliefs, and preferences, and the whole question of how evidence is turned into practice."—Green and Britten[2]

There are several types of qualitative research, some of which are qualitative interviews, life history interviews, key informant interviews, ethnography, participant observation, focus groups, narrative analysis, and document-text analysis. These various types of research have the same principle however vary as to the methodology of the instrument used to acquire data.

To succeed with qualitative research, there are several epistemological perspectives that should be considered. In research methodology, these positions are objectivism, constructivism, and subjectivism.

Objectivism believes that there is an external reality with meaning residing in the object. In this perspective, the researcher is independent from the object of study. The goals of the objectivist perspective are to reduce bias, generalize studies, and provide reproducibles. Example questions may include What is the ideal INR value of blood thinning for patients with mechanical heart valves to avoid bleeding and clotting complications?

Constructivism subscribes to there being a multitude of realities therefore several stories can be discussed from varying perspectives. The research findings are contextual. Example questions may be: Why don't men get regular colonoscopies on time? How do men in community A define impotence and how do they cope with it? Why do individuals eat diets that are high in carbohydrates and fats?

Translational Cardiology. https://doi.org/10.1016/B978-0-323-91790-2.00015-0

Subjectivism also called interpretivism considers various coming out of an interplay of an imposition of the object on the researcher. This perspective believes knowledge to be internally constructed. Few researchers use this perspective.

Quantitative research

There are numerous examples of quantitative research; some of them include experiments such as trials, surveys, systematic led observations, and secondary research. An experiment seeks to evaluate the influences or change with a hypothesis that is null and one that is alternative. A control is used as a comparison under the same environment to be compared and tested as to whether an intervention led to findings versus random chance.

A survey seeks to acquire answers to questions from of volunteer or paid group of individuals. These questionnaires seek to ascertain an understanding that is key to the cohort(s) questions as their comparisons may serve as a control. Observational research from a quantitative perspective may acquire manifestations of behavior with documented frequencies and severity. Secondary research may be statistical assessments of concepts originally published.

Quantitative research has a systematic process that can be employed for a great majority of investigations. One evaluates the presence of a problem or phenomenon. This problem becomes the source of purpose as one explores potential opportunities for improvement or palliation. A hypothesis question is created. One then explores the current body of literature on the subject in the form of a literature review.

A conceptual framework encompassing this problem, solutions, and relevant knowledge is developed to create an optimal study design. A feasibility study sample may be considered at this time to elicit gross concerns in the study design. Institutional review board (IRB) permissions are sought, and data is collected or accessed per protocol with care to maintain patient confidentiality and HIPPA compliance. Patient identifiers may be removed at this time to deidentify the date and subsequently analyze. Following this result analysis, a discussion is created and penultimately, summary and conclusion of the work at large based on the findings of the study.

For quantitative research, one can consider three types of research questions. For the first type of research question, consider the following: "Between two cohorts of data is there a significant difference?" An example is, is there a significant and statistical difference in test performance on the medical admission test between males and females? This type of question can be analyzed by t-test, ANOVA, ANCOVA, and MANOVA.

For the second type of research question consider the following: "Between two cohorts what is the relationship?" This question seeks to understand correlation, such as positive versus negative, and to what extent of strength is the correlation. An example is, what is the relationship between years of video game playing and surgical precision. This type of question can be analyzed by Pearson r and Spearman Rho.

For the third type of research question, consider the following: "Can one factor of data predict to an extent another factor of data?" An example of this question is "To what extent does socioeconomic status play a role in a patient's health status. This type of question can be analyzed by linear regression, multiple regression, logistic regression, and discriminant analysis.

References

1. Marshall MN. Sampling for qualitative research. *Fam Pract.* 1996;13:522–525.
2. Green J, Britten N. Qualitative research and evidence based medicine. *BMJ.* 1998;316(7139):1230–1232. https://doi.org/10.1136/bmj.316.7139.1230.

Chapter 95

Population- and environmental-specific considerations

Ahmed Alnajar[1] and Sameer Hirji[2]

[1]*The Division of Cardiothoracic Surgery, The Department of Surgery, The University of Miami Miller School of Medicine, Miami, FL, United States;* [2]*Division of Cardiac Surgery, Department of Surgery, Brigham and Women's Hospital, Harvard Medical School, Boston, MA, United States*

Key points

- Environmental factors have a clear role in the development and severity of cardiovascular diseases.
- The measures taken to control the spread of COVID-19 have revealed that health environments can be modified more than was previously believed. This realization, therefore, heightens the potential impact of exploring positive and negative environmental impacts on cardiovascular disease; filling these research gaps can lead to environmental changes that can benefit patients.
- Population exposure to fine particulate matter is mediating and may also lead to cardiovascular events at a significant level, even at concentrations below current regulatory standards.
- Pollution exposure effects occur via mediating pathways, involving changes in blood pressure, lipid profile, and vascular function.
- Policy measures can decrease environmental pollution and therefore reduce the widespread prevalence of exposure affecting both short- and long-term cardiovascular health.

Why it matters?

Despite many notable advances in treatment and management, cardiovascular disease (CVD) remains at the forefront of mortality causes worldwide.[1] While CVD forms from a set of chronic conditions over time, these diseases result from a complex interplay between genetic predisposition, lifestyle factors, and environmental influences that lead to progressive deterioration of the individual. The modifiable nature of CVD is supported by strong evidence that reveals the impact of lifestyle choices[2,3] suggesting that behavior changes can be solutions for high-risk populations. However, the impact of the environment is also critical. While this impact has been known for a long time,[4–7] the COVID-19 pandemic has revealed the modifiable nature of large-scale environmental factors. If environmental changes are a potential modifying factor, exploring and understanding how various environmental factors increase the risk, progression, and severity of CVD is the first step to intervene effectively on the modifiable environmental factors. Meticulous research must be undertaken to ascertain which components of the environment affect CVD risk and to understand the mediating effect of pollution on CVD risk.

Cardiovascular diseases mostly arise from heritable conditions, from complex congenital heart defects to simple dysregulation of blood cholesterol or glucose levels. Researchers have expended enormous effort in understanding the genes and specific DNA sequence variants responsible for this heritability. However, there are also important nongenetic factors that must be considered, such as environmental influences. Environmental influences transform an individual's risk of CVD through factors that introduce mutations or mediate genetic penetrance and phenotypes.[8] Recent evidence points toward the powerful population-level role of social, environmental, and economic factors, referred to as the social determinants of health (SDH), in shaping CVD patterns and outcomes.[9–12] Outcome differences between races, as an example, were not explained by the clinical or graft characteristics in heart transplant recipients.[13] Furthermore, low neighborhood socioeconomic status, as an example of a disparity-driving factor, was associated with higher risks of noncompliance, heart graft rejection, hospitalization, and infection.[14] By considering SDH factors in tandem with medical care, physicians and

Translational Cardiology. https://doi.org/10.1016/B978-0-323-91790-2.00053-8

policymakers can work collaboratively to have a holistic view of individual patients and overall population health impacted by SDH to enhance patient care and promote better disease management and outcomes. SDH is a multidimensional concept that includes political, socio-economic, and cultural construct factors, as well as place-based conditions, including accessible healthcare and education systems, well-designed neighborhoods, availability of healthy food, and key to this discussion, safe environmental conditions.[15–17]

Unique considerations of environmental factors on cardiovascular health, public health, and the economy

The World Health Organization (WHO) identified air pollution as one of the greatest environmental risks to health and acknowledged the importance of implementing policies that reduce outdoor air pollution by developing clean transport, energy-efficient homes, renewable power generation, and green waste management.[18] Environmental pollutants' effect on CVD has also been recognized, and investigations have explored their effect on coronary heart disease (CHD), heart failure (HF), and stroke, as well as preclinical risk factors such as atherosclerosis, endothelial dysfunction, and hypertension. A variety of air, water, and food pollutants may affect cardiovascular health. While the role of air pollution on respiratory diseases has been well recognized,[19] its effect on CVD is increasingly being studied. The effects of ambient air pollution on CVD, however, appear to depend upon individual susceptibility. Individuals with preexisting chronic conditions (such as obesity or diabetes) and smokers seem to be particularly at higher risk to manifest adverse CVD effects of rising temperatures and air pollution.[20] For example, climate change-driven hurricanes have increased risk on patients with specific needs, especially patients with end-stage heart failure[21]; thus, to mitigate the hazardous consequences of environmental factors, it is important to look at the big picture. Sociodemographics, geographic location (rural vs. urban), and economic status, as parts of SDH, are additional factors increasing vulnerable populations' susceptibility to environmental pollutants.[22–25] Legacy and emerging pollutants that we are exposed to through inhalation and consumption (or ingestion) include metals that increase the risk of CHD, HF, and stroke, such as arsenic, cadmium, and lead.[26,27] Noise pollution also plays an important role in major CVD events.[28] Therefore, many individual, social, and natural domains of the human environment may increase CVD risk and lead to unwanted complications; however, the assessment of environmental exposure has been a considerable challenge—when compared to pharmaceutical dosing—since exposure patterns are rarely consistent between individuals. Thus, a population-level study can be important in accounting for variations in behaviors between individuals and measuring the effect of pollutant exposure on a larger scale. Population-wide healthcare costs and productivity losses are subject to the effect of pollution, which could reduce the gross domestic product (GDP) nationally. Air quality improvements may not only reduce deaths from cardiovascular and respiratory diseases but could also yield economic gains. Pollution-related diseases result in healthcare costs that are responsible for 1.7% of annual health spending in high-income countries and for up to 7% of health spending in middle-income countries, which indicates a proportional relationship between increased pollution and an inverse relationship with infrastructure development. There is also evidence of reduced GDP in low- to middle-income countries by up to 2% per year. Productivity losses due to pollution are estimated to amount to $4.6 trillion USD per year: 6.2% of global economic output.[29] The costs attributed to pollution-related disease will likely increase as additional associations between pollution and disease are identified. Thus, environmental risk factors mitigation and prevention can yield large net gains both for individual and global health as well as the national economy. For instance, the 2018 *Lancet commission on pollution and health report* shows that in the United States, there has been an estimated $30 USD (range: $4−88) return for every dollar invested in air pollution control since 1970.[29]

Personal, social, and natural domains of the human environment affecting CVD risk

In addition to the water we drink, the air we breathe, and the food we eat, the natural ecosystem also exerts a powerful influence on human physiology and psychology, through many factors, including the circadian rhythm, the changing seasons, and the neighborhoods we live in. Particularly, with increasing industrialization, some threats were minimized while others persisted. In most Western countries, since World War II, public space has become dominated by private industry and automobiles. Recent years have seen an ambitious push for a healthier environment that is suitable for walking, cycling, and enjoying natural parks in metropolitan areas to promote public health in general and reduce CVD effects in particular. Many policies were motivated to modify the influence of local ecology to promote health and prevent diseases by creating sanitary, climate-friendly, and safe environments where people can live, work, and play. The COVID-19 pandemic has further fueled this challenge, as policies including physical distancing forced the general population to use more space, despite its limited allowance. With the major threats to public health posed by both the spread of infectious disease and the emergence of noncommunicable diseases, the complex role of social environments on health has become a crucial area for study. Thus, the remainder of this chapter will review and summarize current studies of how environmental and social environments will—individually and collectively—affect CVD risk.

Air pollution

In 2015, air pollution was responsible for 21% of all deaths from CVD, 26% of deaths due to CHD, 23% of deaths due to stroke, 51% of deaths due to chronic obstructive pulmonary disease, and 43% of deaths due to lung cancer.[30] WHO estimates air pollution to be associated with 7 million premature deaths per year globally. This includes 1.6 million deaths in China, 1.3 million in India, and 55,000 to 200,000 in the United States.[31,32] Most air pollutants are a mixture of complex aerosols containing both particles and gases. The US Environmental Protection Agency (EPA) set national ambient air quality standards for six principal pollutants that are harming public health and the environment to solve air pollution problems. The Clean Air Act identifies the national ambient air quality standards to provide public health and welfare protection. The public health standards for the six "criteria" pollutants are summarized in Table 95.1. Particulate air pollution consists of particulate matter (PM), which when analyzed for mass fall into two peaks, corresponding to coarse particles (10−2.5 μm in aerodynamic diameter) and fine particles (0.1−2.5 μm [$PM_{2.5}$]).

$PM_{2.5}$, in particular, is associated with 70%−80% of premature deaths due to cardiovascular causes; in fact, aggressive controls of pollution against $PM_{2.5}$ were able to reduce 23% of air pollution-related deaths.[33] Evidence shows that increased exposure to concentrated ambient particles increases atherogenesis, insulin resistance, and thrombosis. Studies have also shown that air pollution induces myocardium apoptosis and then impairs cardiac function and aggravates myocardial infarction via NFκB activation.[34] Furthermore, a recent study by Abohashem et al. adds to the evidence that leucopoietic activity and arterial inflammation could be a biological pathway that independently predicts major adverse cardiovascular events.[35] Other studies showed evidence linking acute exposure to CO, SO_2, NO_2, and $PM_{2.5}$ to increased HF exacerbations among the elderly.[36] However, there is lacking evidence on whether specific lifestyle or pharmacological interventions lower the relative risk associated with exposure to air pollution. Nevertheless, the reduction of pollution exposure will likely come after a series of personal and community actions, as well as public policy measures. In light of the COVID-19 pandemic, we find that community and individual adaptive measures that include staying indoors for prolonged times, using air conditioning, and only going outside for extreme necessity may be helpful but come at a price (negative mental health and increased energy consumption[37]), may not be within feasible for most vulnerable populations. Despite that, the population-based measures intended to halt COVID-19 spread have led to remarkable improvements in air quality, which resulted in fewer deaths from pollution-related diseases. Cleaner air is estimated to have saved 11,000 lives (95% CI: 7000−21,000) in Europe in April 2020, and 77,000 lives in China in January and February. These strategies were focused on the protection of public health and many of them have proven to be cost-effective as they were adopted by cities and countries at every level of income. However, for a lasting effect, the application of the best of these strategies should be carefully planned and

TABLE 95.1 The public health standards for the six "criteria" pollutants.

Pollutant	Averaging time	Level	EPA standard
Carbon monoxide (CO)	8 h	9 ppm	Not to be exceeded more than once per year
Lead (Pb)	Rolling 3-month average	0.15 μg/m^3	Not to be exceeded
Nitrogen dioxide (NO_2)	1 h	100 ppb	98th percentile of 1-h daily maximum concentrations, averaged over 3 years
Ozone (O_3)	8 h	0.070 ppm	Annual fourth-highest daily maximum 8-h concentration, averaged over 3 years
Particle pollution ($PM_{2.5}$)	1 year	12.0 μg/m^3	Annual mean, averaged over 3 years
Particle pollution (PM_{10})	24 h	150 μg/m^3	Not to be exceeded more than once per year on average over 3 years
Sulfur dioxide (SO_2)	1 h	75 ppb	99th percentile of 1-h daily maximum concentrations, averaged over 3 years

supported to enable low-income and middle-income countries to avoid many of the harmful consequences of social distancing while reaping the benefit of pollution and noise reduction and improve the health and wellbeing of the general population. Furthermore, disseminating the health effects and costs of air pollution will continue to require both scientific evidence and public wellness to adopt effective regulatory policies. Such efforts will contribute to the setting and implementation of air quality standards designed to protect not only the general public but also its most vulnerable populations.

Noise pollution

Like air pollution, noise is another important environmental factor that has an important relationship with CVD. Epidemiological studies have shown evidence of traffic noise effect on cardiovascular diseases such as arterial hypertension, myocardial infarction, and stroke.[38] It is estimated by the WHO that in Western Europe a total number of 1,685,000 disability-adjusted life years are lost for ischemic heart disease, cognitive impairment of children, sleep disturbance, tinnitus, and annoyance every year.[39] Noise pollution was also found to be a nonspecific stressor that activates the autonomous nervous system and endocrine signaling. Old evidence that linked noise to CVD was shown as a physiological problem in 1005 German industrial workers who were more pronounced in very noisy industries compared with less noisy industries.[40] A recent study by Dar et al. of 254 individuals observed that chronic exposure to socioeconomic or environmental stressors (e.g., exposure to conditions of lower income, higher crime, or higher noise) has greater stress-associated neurobiological activity (measured as amygdalar metabolic activity).[22] Thus, based on these early observations, there was no doubt that chronic noise exposure may cause and affect CVD.

The interplay of social determinants of health with environmental pollutants

Even though components of the physical environment such as air and noise pollution may influence CVD risk, the most important component of any human environment is not limited by its physical attributes but extends to the neighborhood. SDH is shown to be independently associated with diet quality, adiposity, or weight gain including education, income, race/ethnicity, and neighborhood availability of supermarkets.[15–17,41] In fact, Healthy People 2030—an initiative that sets goals and objectives to improve the health and well-being of people in the United States—have five overarching goals that are specifically related to the fundamental components of SDH (Fig. 95.1). In the United States, exposure to pollutants is significantly different by race and ethnicity due to historical racial segregation in neighborhoods and housing.[42] For example, living within 50 miles of a major roadway is significantly associated with a higher left ventricular mass index without HF may predict the development of both systolic and diastolic HF.[43] Furthermore, the increased exposure to O_3 and $PM_{2.5}$ among black residents—compared to white residents—led to excess attributable risk in cardiovascular admission rates, while Asian/Pacific Islander residents had substantially lower excess attributable risk compared to white residents.[44] Sex disparity has also become evident when women are shown to have an increased CHD due to long-term exposure to $PM_{2.5}$ compared to men. Moreover, women are shown to have higher CVD mortality compared to men due to short-term exposure to PM_{10} and SO_2. Individuals affected by low neighborhood median income, increased total town crime rates, and living in homes with higher noise exposure (>45 dBA) were observing chronic stressors that are significantly associated with major adverse cardiovascular events.[22] Consequently, stressor effects on individuals vary based on their race—most likely due to genetic disadvantage[45]—which evidence suggests that unhealthy coping behaviors for certain demographics combine with negative environmental conditions that contribute to morbidity and mortality.[46] A recent extensive review study has further explained how psychosocial stress and stress conditions are independently associated with CVD, which—in addition to the individual response to a stressor—depends on the degree and duration of stress.[47]

Urban centers in the United States have been shown to have acute rises in arterial blood pressure (BP) due to acute exposure to both $PM_{2.5}$ and PM_{10}, which indicate further evidence of disparity based on urban versus rural areas inhabited by minority groups.[48] Furthermore, rural exposure to air pollution was also linked to increased risk for CVD in high-risk groups.[49] To put this in a bigger context, the burden of pollution-associated disease is most concentrated among children, women, the elderly, and the poor. Thus, environmental factors increase the burden of CVD in vulnerable populations and affect their rights to have equal contributions to their family and societies. The public health community has a critical window of opportunity to invest in research, regulation, and intervention, particularly among the vulnerable population, to reduce the global burden of pollution-attributable CVD.

Lessons learned from the COVID-19 era

While the environmental problem is difficult to solve rapidly, studies seeking to elucidate the mechanisms that promote disease progression contribute to identifying promising novel preventive and therapeutic targets and strategies. Nevertheless, evidence from the COVID-19 era shows promising policies that could be carried out more effectively. A sudden drop in greenhouse gas emissions

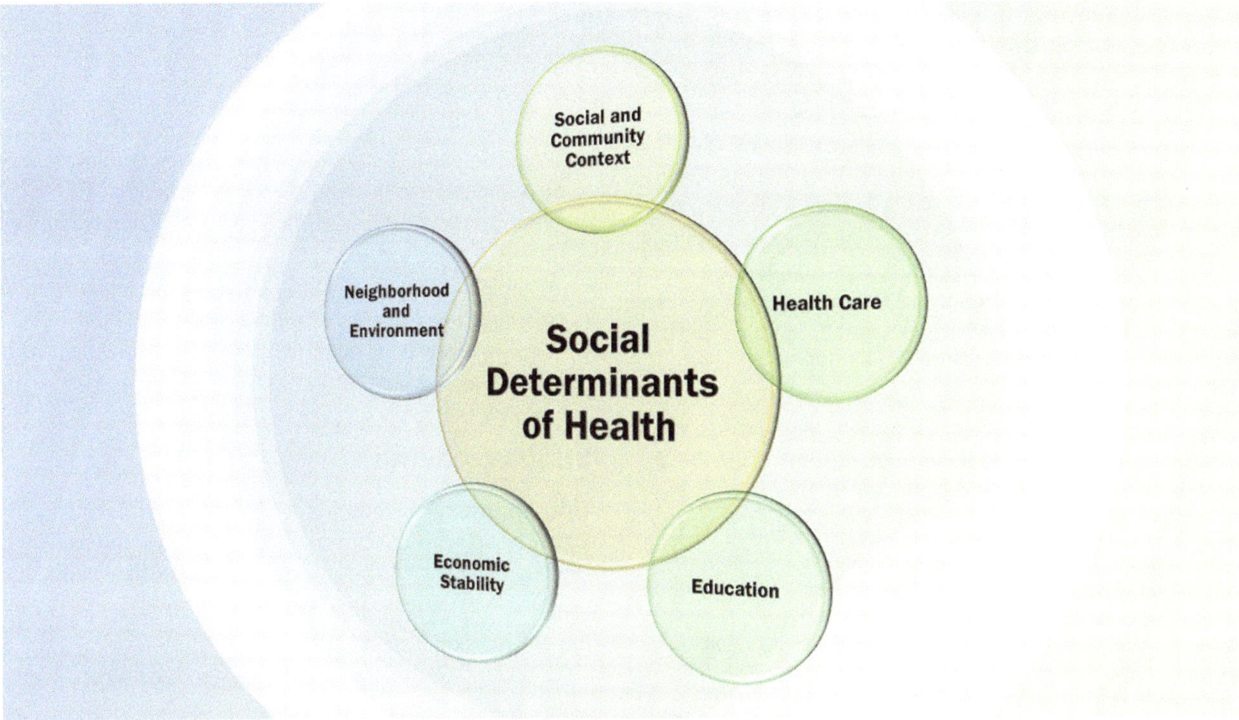

FIGURE 95.1 Essential components of social determinants of health (SDHs): Since SDHs have a major impact on people's health, well-being, and quality of life. Healthy people 2030 have specific data-driven national objectives to improve health and well-being for each of the five domains of SDH: **Economic stability:** Help people earn steady incomes that allow them to meet their health needs. **Education:** Increase educational opportunities and help children and adolescents do well in school. **Health care:** Increase access to comprehensive, high-quality health care services. **Neighborhood and built environment:** Create neighborhoods and environments that promote health and safety. **Social and community context:** Increase social and community support.

(GHGs) was observed as industries, transportation, and companies closed down in response to COVID-19. In New York, the levels of air pollution were reduced by nearly 50% compared to last year because of measures taken to control the pandemic spread.[50] In China, it was estimated that nearly a 50% reduction of N_2O and CO occurred due to the shutdown of heavy industries.[51] Overall, much less consumption of fossil fuels reduced GHG emissions. Oil demand and power generation dropped dramatically. In the first 3 months of 2020, 435,000 oil barrels have dropped globally.[50] Moreover, coal-based power generation was reduced by 26% in India, while it was reduced by 36% in China, the highest coal consumer in the world, compared to the same time of the previous year.[50] Reduction of water pollution was mostly affected by closing the major industrial sources of pollution, which helped to reduce the pollution load.[52] Furthermore, pH level, electric conductivity, and chemical oxygen demand have reduced by almost 1%–10%, 33%–66%, 45%–90%, and 33%–82%, respectively, in different monitoring stations during the lockdown in comparison to the past year.[53] Additionally, the decrease in tourist numbers and water activities led to the Grand Canal in Italy becoming clearer and the return of many aquatic species. A similar phenomenon was observed in the beach areas of Bangladesh, Malaysia, Indonesia, Thailand, and the Maldives.[50]

Successful approaches

The COVID-19 pandemic has induced a global response and made people from different walks of life unite to save themselves, their loved ones, and their planet. However, all the environmental consequences were short-term gains. While personal-level approaches can reduce air pollution exposure and lead to improved health outcomes, population-level strategies that are appropriate for long-term benefit are warranted, especially when considering a long-lasting effect on CVD. Given the broad spectrum of susceptible populations with variant pollution exposure, a framework is needed that incorporates individual- and community-level strategies for harm reduction. For example, to prepare heart transplant candidates receiving care in areas prone to extreme storms and hurricanes, researchers, clinicians, and public health advocates provided an inclusive and detailed mitigation strategy for patients and providers alike, to ensure their safety and uninterrupted access to healthcare resources, while maintaining health care quality before, during, and in the aftermath of a hurricane.[21] Similarly, to mitigate the effect of air pollution, both active and passive exposure mitigation strategies are warranted. Personal efforts may include using home air cleaning tools and humidity modifiers. This will add to an effective personal protection layer that includes utilizing masks or pharmacologic approaches.[54,55] For noise pollution, an effective comprehensive approach will cover road traffic, aircraft, and railroad noises. This includes installing sound-reducing windows in homes and workspaces, as well as changing homes' bedroom orientation toward the quiet side of the residence. Furthermore, installing noise barriers in densely populated areas, applying quiet road surfaces, and roads' speed limits at day-specific times can reduce road traffic noises. For aircraft noises, day-specific time travel bans should be implemented, as well as achieving less vibration during airplane descent to allow for steeper movement with lower throttle settings. For railway noises, in addition to day-specific time travel bans, replacing cast-iron block brakes with composite material and grinding railway tracks should be considered.[56] Eventually, a full circle of evaluation framework will guide and strengthen policy planning and implementation when considering strategic mitigation approaches (Fig. 95.2). This will ensure high-quality evidence, robust and consistent improvement, and an informed decision-making process during the development and implementation of public health policies. Thus, while implementing mitigating strategies, researchers should continue their investigation of key scientific questions, such as the following:

1. What factors may modify the effect of $PM_{2.5}$ and noise on CVD?
2. What distinct effects that different concentrations level of pollutants carry on CVD?
3. What are the biological mechanisms that mediate pollutants' effects on the cardiovascular system?
4. What is the role of SDH in increasing the risk of pollutants negative effects in vulnerable populations?
5. What is the role of SDH in policy implementation and individual engagement?

Pitfalls to avoid

With the increasing number of environmental factors and increasing research opportunities to study the effect of pollutants on the cardiovascular system, some challenges and pitfalls merit careful consideration:

1. Environmental data source: Estimates of pollutant concentration for health studies have been derived from monitors that could be limited in temporal and spatial coverage of the target population. For example, the EPA's $PM_{2.5}$ monitors often only measure every third or sixth day and most counties do not contain regulatory air pollution monitors to provide a good estimate of the

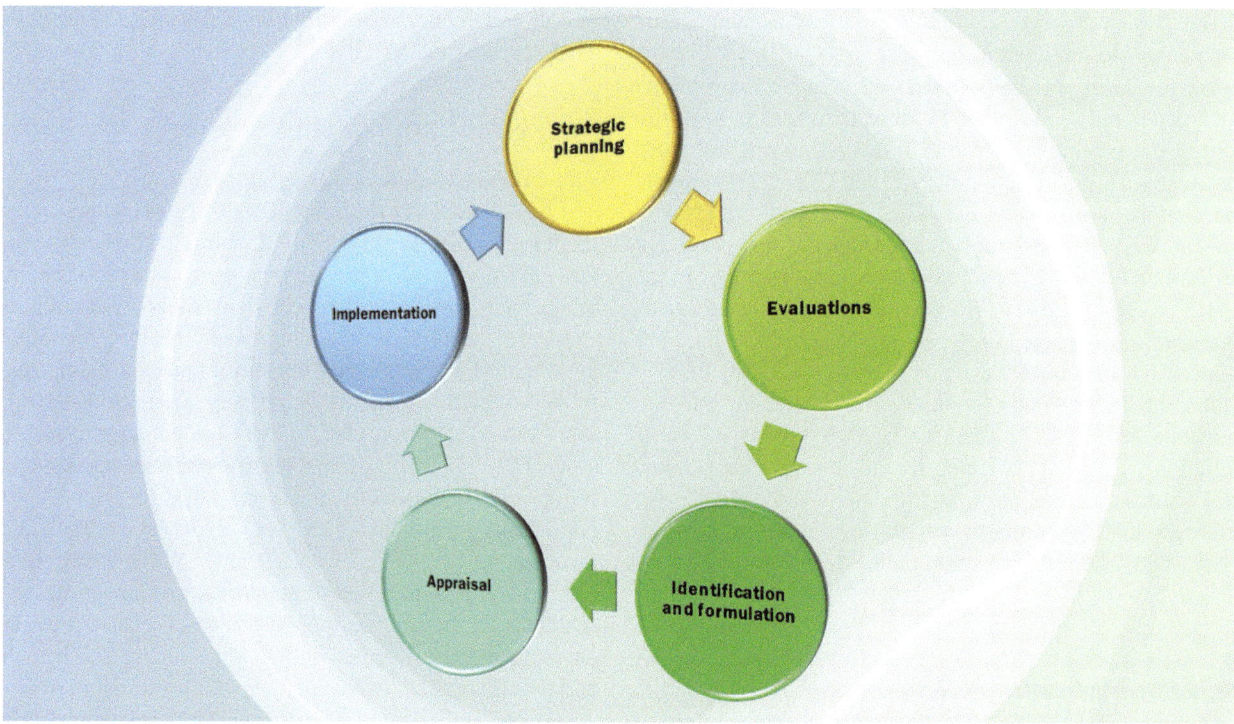

FIGURE 95.2 Evaluation cycle for strategic approaches: For objective and strategic planning, we adopt the infamous project cycle in creating and evaluating mitigation strategies to reduce pollutants' effect on health.

air pollution exposures where every person lives. Thus, researchers should collaborate effectively with environmental scientists and statisticians to estimate pollutant exposures using data from satellites and nationwide geographic databases as well as combine information from different data sources. Various methods of statistical blending should be considered for the analysis, such as spatiotemporal regression kriging, geographically weighted regression, and machine learning methods.

2. Big data research: Enormous databases are used to store the data obtained from long questionnaires and medical records to produce interesting cutting-edge research. The data enables researchers to probe health effects in understudied vulnerable populations. To study subclinical effects on healthy and susceptible populations' exposure to certain pollutants, a large sample is required; however, large databases do not have enough details about certain clinical features such as preexisting disease, immunological markers, and specific disease subtypes that alter responses to pollutants exposure. Thus, researchers are advised to carefully present cause—effect relationships and think clearly when discussing correlations and causation.
3. The role of SDH: When assessing the impact of the natural, social, and personal domains of the individual and his environment, we should consider how these domains influence the effects of other domains and how such interactions collectively affect CVD. Thus, future investigators should integrate the effects of the various components of the environment into a comprehensive model.

References

1. Virani SS, Alonso A, Aparicio HJ, et al. Heart disease and stroke statistics—2021 update. *Circulation*. 2021;143(8):e254—e743.
2. Stampfer MJ, Hu FB, Manson JE, Rimm EB, Willett WC. Primary prevention of coronary heart disease in women through diet and lifestyle. *N Engl J Med*. 2000;343(1):16—22.
3. Worth RM, Kato H, Rhoads GG, Kagan K, Syme SL. Epidemiologic studies of coronary heart disease and stroke in Japanese men living in Japan, Hawaii and California: mortality. *Am J Epidemiol*. 1975;102(6):481—490.
4. Ford ES, Ajani UA, Croft JB, et al. Explaining the decrease in U.S. deaths from coronary disease, 1980—2000. *N Engl J Med*. 2007;356(23):2388—2398.
5. Unal B, Critchley JA, Capewell S. Explaining the decline in coronary heart disease mortality in England and wales between 1981 and 2000. *Circulation*. 2004;109(9):1101—1107.
6. Pekka P, Pirjo P, Ulla U. Influencing public nutrition for non-communicable disease prevention: from community intervention to national programme-experiences from Finland. *Public Health Nutr*. 2002;5(1a):245—251.
7. Critchley J, Liu J, Zhao D, Wei W, Capewell S. Explaining the increase in coronary heart disease mortality in Beijing between 1984 and 1999. *Circulation*. 2004;110(10):1236—1244.
8. Kathiresan S, Srivastava D. Genetics of human cardiovascular disease. *Cell*. 2012;148(6):1242—1257.
9. Erqou S, Echouffo-Tcheugui JB, Kip KE, Aiyer A, Reis SE. Association of cumulative social risk with mortality and adverse cardiovascular disease outcomes. *BMC Cardiovasc Disord*. 2017;17(1):110.
10. Palacio A, Mansi R, Seo D, et al. Social determinants of health score: does it help identify those at higher cardiovascular risk? *Am J Manag Care*. 2020;26(10):e312—e318.
11. Alnajar A. *Identifying Factors Associated With Long-Term Graft Survival Following Heart Transplantation Using Statistics and Machine Learning Methods*. University of Miami; 2021.
12. Alnajar A, Blackstone EH, Reis IM, et al. Abstract 12875: the impact of social determinants of health and race disparities in heart transplant long-term survival. *Circulation*. 2021;144(Suppl_1):A12875.
13. Liu V, Bhattacharya J, Weill D, Hlatky MA. Persistent racial disparities in survival after heart transplantation. *Circulation*. 2011;123(15):1642—1649.
14. Wayda B, Clemons A, Givens RC, et al. Socioeconomic disparities in adherence and outcomes after heart transplant: a UNOS (united network for organ sharing) registry analysis. *Circ Heart Fail*. 2018;11(3):e004173.
15. Kumanyika S, Grier S. Targeting interventions for ethnic minority and low-income populations. *Future Child*. 2006;16(1):187—207.
16. Li F, Harmer PA, Cardinal BJ, et al. Built environment, adiposity, and physical activity in adults aged 50—75. *Am J Prev Med*. 2008;35(1):38—46.
17. Sallis JF, Glanz K. The role of built environments in physical activity, eating, and obesity in childhood. *Future Child*. 2006;16(1):89—108.
18. Organization WH. *Ambient (Outdoor) Air Pollution*; September 22, 2021. Available from: https://www.who.int/news-room/fact-sheets/detail/ambient-(outdoor)-air-quality-and-health.
19. Chen H, Goldberg MS, Villeneuve PJ. A systematic review of the relation between long-term exposure to ambient air pollution and chronic diseases. *Rev Environ Health*. 2008;23(4):243—297.
20. Gold DR, Mittleman MA. New insights into pollution and the cardiovascular system: 2010 to 2012. *Circulation*. 2013;127(18):1903—1913.
21. Alnajar A, Bud Frazier OH, Elgalad A, Smith PA, Shultz JM. Preparing end-stage heart failure patients and care providers in the era of climate change-driven hurricanes. *J Card Surg*. 2021;36(10):3491—3493.
22. Dar T, Osborne MT, Abohashem S, et al. Greater neurobiological resilience to chronic socioeconomic or environmental stressors associates with lower risk for cardiovascular disease events. *Circ Cardiovasc Imag*. 2020;13(8):e010337.
23. Tibuakuu M, Michos ED, Navas-Acien A, Jones MR. Air pollution and cardiovascular disease: a focus on vulnerable populations worldwide. *Curr Epidemiol Rep*. 2018;5(4):370—378.
24. Buekers J, Colles A, Cornelis C, Morrens B, Govarts E, Schoeters G. Socio-economic status and health: evaluation of human biomonitored chemical exposure to per- and polyfluorinated substances across status. *Int J Environ Res Publ Health*. 2018;15(12).
25. Tyrrell J, Melzer D, Henley W, Galloway TS, Osborne NJ. Associations between socioeconomic status and environmental toxicant

concentrations in adults in the USA: NHANES 2001–2010. *Environ Int*. 2013;59:328–335.

26. Cosselman KE, Navas-Acien A, Kaufman JD. Environmental factors in cardiovascular disease. *Nat Rev Cardiol*. 2015;12(11):627–642.
27. Wu W, Jiang S, Zhao Q, et al. Environmental exposure to metals and the risk of hypertension: a cross-sectional study in China. *Environ Pollut*. 2018;233:670–678.
28. Osborne MT, Radfar A, Hassan MZO, et al. A neurobiological mechanism linking transportation noise to cardiovascular disease in humans. *Eur Heart J*. 2020;41(6):772–782.
29. Landrigan PJ, Fuller R, Acosta NJR, et al. The Lancet Commission on pollution and health. *Lancet*. 2018;391(10119):462–512.
30. GBD 2015 Risk Factors Collaborators. Global, regional, and national comparative risk assessment of 79 behavioural, environmental and occupational, and metabolic risks or clusters of risks, 1990–2015: a systematic analysis for the Global Burden of Disease Study 2015. *Lancet*. 2016;388(10053):1659–1724.
31. Bhatnagar A. Environmental determinants of cardiovascular disease. *Circ Res*. 2017;121(2):162–180.
32. Caiazzo F, Ashok A, Waitz IA, Yim SH, Barrett SR. Air pollution and early deaths in the United States. Part I: quantifying the impact of major sectors in 2005. *Atmos Environ*. 2013;79:198–208.
33. Group GMW. Burden of disease attributable to coal-burning and other air pollution sources in China. *Spec Rep*. 2016;20.
34. Li X, Geng J, Chen Y, et al. Exposure to particulate matter induces cardiomyocytes apoptosis after myocardial infarction through NFκB activation. *Biochem Biophys Res Commun*. 2017;488(1):224–231.
35. Abohashem S, Osborne MT, Dar T, et al. A leucopoietic-arterial axis underlying the link between ambient air pollution and cardiovascular disease in humans. *Eur Heart J*. 2021;42(7):761–772.
36. Shah AS, Langrish JP, Nair H, et al. Global association of air pollution and heart failure: a systematic review and meta-analysis. *Lancet*. 2013;382(9897):1039–1048.
37. Marroquín B, Vine V, Morgan R. Mental health during the COVID-19 pandemic: effects of stay-at-home policies, social distancing behavior, and social resources. *Psychiatr Res*. 2020;293:113419.
38. The adverse effects of environmental noise exposure on oxidative stress and cardiovascular risk. *Antioxid Redox Signal*. 2018;28(9):873–908.
39. Fritschi L, Brown A, Kim A, Schwela D, Kephalopoulos S. *WHO and JRC Report: Burden of Disease from Environmental Noise*. 2011.
40. Jansen G. Effects of noise on health. *Ger Med Mon*. 1968;13(9):446–448.
41. Rummo PE, Guilkey DK, Ng SW, Popkin BM, Evenson KR, Gordon-Larsen P. Beyond supermarkets: food outlet location selection in four U.S. Cities over time. *Am J Prev Med*. 2017;52(3):300–310.
42. Morello-Frosch R, Lopez R. The riskscape and the color line: examining the role of segregation in environmental health disparities. *Environ Res*. 2006;102(2):181–196.
43. Van Hee VC, Adar SD, Szpiro AA, et al. Exposure to traffic and left ventricular mass and function: the multi-ethnic study of atherosclerosis. *Am J Respir Crit Care Med*. 2009;179(9):827–834.
44. Hackbarth AD, Romley JA, Goldman DP. Racial and ethnic disparities in hospital care resulting from air pollution in excess of federal standards. *Soc Sci Med*. 2011;73(8):1163–1168.
45. Abohashem S, Osborne M, Choi KW, et al. Abstract 16270: genetic susceptibility to stress associates with higher amygdalar activity and greater myocardial infarction risk. *Circulation*. 2020;142(Suppl_3):A16270.
46. Jackson JS, Knight KM, Rafferty JA. Race and unhealthy behaviors: chronic stress, the HPA axis, and physical and mental health disparities over the life course. *Am J Public Health*. 2010;100(5):933–939.
47. Dar T, Radfar A, Abohashem S, Pitman RK, Tawakol A, Osborne MT. Psychosocial stress and cardiovascular disease. *Curr Treat Option Cardiovasc Med*. 2019;21(5):23.
48. Dvonch JT, Kannan S, Schulz AJ, et al. Acute effects of ambient particulate matter on blood pressure: differential effects across urban communities. *Hypertension*. 2009;53(5):853–859.
49. Baumgartner J, Zhang Y, Schauer JJ, Huang W, Wang Y, Ezzati M. Highway proximity and black carbon from cookstoves as a risk factor for higher blood pressure in rural China. *Proc Natl Acad Sci U S A*. 2014;111(36):13229–13234.
50. Rume T, Islam SMD. Environmental effects of COVID-19 pandemic and potential strategies of sustainability. *Heliyon*. 2020;6(9):e04965.
51. Caine P. Environmental impact of COVID-19 lockdowns seen from space. *Sci Nat*. 2020;2.
52. Yunus AP, Masago Y, Hijioka Y. COVID-19 and surface water quality: improved lake water quality during the lockdown. *Sci Total Environ*. 2020;731:139012.
53. Arif M, Kumar R. *Reduction in Water Pollution in Yamuna River Due to Lockdown under COVID-19 Pandemic*. 2020.
54. Al-Kindi SG, Brook RD, Biswal S, Rajagopalan S. Environmental determinants of cardiovascular disease: lessons learned from air pollution. *Nat Rev Cardiol*. 2020;17(10):656–672.
55. Langrish JP, Li X, Wang S, et al. Reducing personal exposure to particulate air pollution improves cardiovascular health in patients with coronary heart disease. *Environ Health Perspect*. 2012;120(3):367–372.
56. Münzel T, Miller MR, Sørensen M, Lelieveld J, Daiber A, Rajagopalan S. Reduction of environmental pollutants for prevention of cardiovascular disease: it's time to act. *Eur Heart J*. 2020;41(41):3989–3997.

Chapter 96

Law, policy, ethics

Clauden Louis

Structural Heart and Transcatheter Endovascular Cardiac Surgery Fellow, Cardiothoracic Surgery, Brigham and Women's Hospital, Boston, MA, United States

Key points

- Law
- Policy
- Ethics

Introduction

Laws, policies, and ethics have been used to help generate, establish, and reinforce the very foundations and organizations that safeguard the nation's well-being and health while creating opportunities for persons to enhance their position in society. Multiple examples of these three constructs working together have led to safer work environments, increased vaccination rates, and decreased tobacco exposure. Ethics, in particular, moves beyond regulations but consists of moral principles, and when used in this setting they set the tone and foundation. Libraries exist covering these three topics individually. I must admit any methodology to cover these topics briefly is ambitious.

Law

> *The Commission on Human Rights has attached, and is continuing to attach, the highest importance to the implementation of the International Covenant on Civil and Political Rights, particularly to its provisions from which no derogation is permissible*

Siracusa Principles[1]

Siracusa Principles is a famous legal document discussing measures in which a state party can take measures derogating from its obligations and justified in infringing on human rights such as is seen during times of quarantine or times of war in draft. Most importantly, this document reveals nonderogable rights where there is never a justification to remove human rights. From the *Journal of Law, Medicine, and Ethics*, authors Childress et al. discuss five conditions as a rationale for infringing individual rights: efficacy, proportionality, necessity, least infringement method, and public justification.[2]

Laws are obligatory regulations placed to protect the safety and welfare of society. Constitutional authority comes from the power of the federal and state governments. They too have the ability to enforce the laws as well.[3]

Policy

Since most people support the goals of public health, so understood we should take care not to infuse the theory and practice of public health with contentious moral and political commitments.

Public health ethics 2011, Jonathan Anomaly

Policy is defined as the "principles that govern action directed toward given ends."[4] Several types of policies exist with one of the more notable being public policy. Public policy is formed by the government for example as is seen by health warning labels on goods being sold. Social policy seeks to promote public welfare such as limiting access to alcoholic beverages to individuals of age. Health policy seeks to promote the state of wealth as well as maintain prevention programs. Institutional policies serve specific businesses and organizational policies via associations and programming.

Public policy may refer more broadly to federal, state, or local legislation and regulations. Common policy challenges include cost control, physician organization and payment, and the organization of health coverage and services.[5]

The Council on Education for Public Health (CEPH), which serves as the accreditation body for Public Health

Translational Cardiology. https://doi.org/10.1016/B978-0-323-91790-2.00034-4

students, has updated their standards and goals requiring that training programs teach, and students learn, to "advocate for political, social or economic policies and programs that will improve health in diverse populations.[6]"

With public health having stronger implications recently, it is imperative that policies are in place that are efficient, supported, and advocated to help ratify sound policy into regulation and several methods must be in place that require both time and insight.

Ethics

> *"... the sense of justice and injustice is not derived from nature, but rises artificially, though necessarily from education, and human conventions." ... "The convention is not of the nature of a promise: For even promises themselves ... arise from human conventions. It is only a general sense of common interest; which sense all the members of the society express to one another, and which induces them to regulate their conduct by certain rules."*
>
> *- A treatise of Human Nature, P. 375–379, David Hume 1711–1776*

The above represents a key statement in meta-philosophy and ethics (meta-ethics) pertaining to the meaning of judgment with two other large categories of ethics being normative ethics and applied ethics. Normative ethics seeks to understand how one aligns moral behavior. Applied ethics seeks to understand the application of ethics to real-world scenarios. Morality is concerned with good and bad behaviors. Ethics is different from morality in that it serves to be a critical analysis of morality. Bioethics is the moral concerns and problems that arise due to modern medicine and research.

As it pertains to public health there are principles that guide decisions and justification of public health interventions. Per Upshur et al., the harm principle, originally from John Stuart Mill, allows one to do anything desired, however to limit things that cause harm. The principles of least restrictive and coercive, reciprocity, and transparency require that every public health discussion and intervention should have rationale available to the public for criticism and understanding.[7]

There are many sectors of ethics. For example, academic ethics, which correlates well with matters pertaining to academic integrity, has a focus on deciphering authorship, appropriate credit, and responsibilities associated with ever-increasing creations in the setting of collaboration, discovery, and innovation. Deviation from the ideals of this concept may lead to academic dishonesty and similar relevant moral code violations. These concepts serve as the written and unwritten policies as they pertain to issues surrounding attribution. Another large area of ethics is medical ethics. This branch of ethics encompasses values pertaining to the treatment and care of patients' bodies, their electronic data, and even their preferences when a choice cannot be rendered by them. Medical ethics may serve as the obligations of doctors to serve and serve well others engrained deep within the inviolable Hippocratic oath. As areas become gray and protecting, one ideal may compromise others and one may pursue hospital-appointed ethics boards for assistance in weaving complex perspectives. Research ethics seeks to create and maintain scientific integrity among individuals and institutions at large by creating personal and professional responsibilities for actions and maintaining an environment. Research ethics involves issues of several categories including publication, conflicts of interest, conflicts of interest, mentorship and collaboration, human subject research, and data management.

Key views of ethics as it pertains to patients are concepts surrounding autonomy. Autonomy as defined by German philosopher Immanuel Kant had several criteria that must be met. One must have the ability to understand moral concepts and be able to have and exercise judgment as a free will. Kant also believed that should variable "x" be a means to an end, then "x" is not an autonomy agent. Therefore, a means is a tool and an end is a project. Kant believed anything lacking these qualities was simply a tool. Critics of Kant's view supply a sufficient condition for mattering morally. They believe he is wrong to feel that if you lack such qualities one no longer matters morally and should not be a necessary condition.

Utilitarianism seeks to identify a method for decision-making. Although complexities exist, at the crux, one can measure the pain and pleasure of A and measure the pain and pleasure of B. The best choice is to pick the choice that maximizes pleasure to the least amount of pain as Utilitarianism seeks to promote actions that have utility and benefit the majority. Originally founded by Jeremy Bentham and supported by John Stuart Mill and Peter Singer, utilitarianism at large seeks to maximize happiness and well-being although some argue as to the role of legislature and hedonism.

Several types of justice are of importance and should be discussed as it pertains to law, policy, and ethics. There are two types of distributive justice, one egalitarian and the other non-egalitarian. Egalitarian serves as equality and fairness of resources and opportunities such as procedural/type versus substantive/amount, equality in capability, equality of wealth, and equality of welfare. It is more widely accepted across philosophical views. Non-egalitarian: justice based on utility; justice is to optimize welfare. Optimize the greatest amount of beneficence or welfare. Another justice is retributive versus corrective. Retributive for instance seeks to give back what is stolen, whereas corrective implements a punishment such as going to jail and losing freedom.

References

1. UN Commission on Human Rights. *The Siracusa Principles on the Limitation and Derogation Provisions in the International Covenant on Civil and Political Rights*; September 28, 1984. E/CN.4/1985/4, available at: https://www.refworld.org/docid/4672bc122.html. Accessed December 30, 2021.
2. Childress J, Faden R, Gaare R, et al. Public health ethics: mapping the terrain. *J Law Med Ethics*. 2002;30(2):170–178.
3. Lewis MA, Tamparo CD. *Medical Law, Ethics, and Bioethics for Ambulatory Care*. F.A. Davis; 2002.
4. Titmuss RM. *Social Policy: An Introduction*. London: Allen & Unwin; 1974:33–38.
5. Marchildon GP, Cafaro CS, Brown A. Myths, misperceptions, and policy learning: comparing healthcare in the United States and Canada. *J Law Med Ethics*. 2018;46(4):833–837. https://doi.org/10.1177/1073110518821975.
6. Council on Education for Public Health. Accreditation criteria for schools of public health & public health programs. *available at*: https://storage.googleapis.com/media.ceph.org/wp_assets/2016.Criteria.pdf; October 2016 (last visited March 14, 2019).
7. Upshur RE. Principles for the justification of public health intervention. *Can J Public Health*. 2002;93.

Chapter 97

Healthcare institutions and system

Faith Kanneh

Johns Hopkins Bloomberg School of Public Health, Baltimore, MD, United States

Introduction/why it matters

Healthcare industry is an essential aspect of society that varies across the globe. Healthcare institutions and systems vary greatly, with the United States health system being a mix of public and private, for-profit, and not-for-profit insurers. In the United States, private insurance provides the majority of the coverage.1 In 2018, the national health expenditure was 3.6 trillion and is projected to reach 6.2 trillion by 2028.2,3 Among other high-income countries such as the United Kingdom, Norway, and Australia, the United States ranks last on access to care, administrative efficiency, equity, and healthcare outcomes, despite spending more of its gross domestic product on healthcare.

Get started

- Healthcare industry is an essential aspect of society that varies across the globe.
- Healthcare institution is a broad term that can be defined as any hospital, health clinic, or health maintenance organization dedicated to promoting, restoring, and maintaining one's health
- Stakeholders have very intricate interrelationships within the healthcare system.
- Affordable healthcare coverage is rare due to rising premiums and strict requirements that impede many from obtaining the care they need.
- Healthcare providers often face the difficult task of balancing the role of gatekeeper for insurance companies and advocacy for patients.
- Patients are at the center of the healthcare system.
- The US healthcare system is an imperfect market that warrants regulation by private or public entities at federal, state, and local county and city levels.

Key points

- Healthcare institutions and systems vary greatly, with the United States health system being a mix of public and private, for-profit, and not-for-profit insurers.
- Healthcare providers often face the difficult task of balancing the role of gatekeeper for insurance companies and advocacy for patients.
- Patients are at the center of the healthcare system.

Potential pitfalls

- Among other high-income countries such as the United Kingdom, Norway, and Australia, the United States ranks last on access to care, administrative efficiency, equity, and healthcare outcomes, despite spending more of its gross domestic product on healthcare.
- Pharmaceutical companies' primary goal of existence may not necessarily be to promote, maintain, or restore consumers' health but rather to maximize wealth.
- Healthcare providers often function under a "risk factor" paradigm centered around behavioral modification for high-risk populations as the primary method of disease prevention via advising patients to quit smoking, improve diet, and increase physical activity. Despite continued use, this approach is not preventative and thus is not successful in reducing disease burdens early such as intervening in coronary heart disease before pathologic states.

Examples of ...

- Beckfield J, Olafsdottir S, Sosnaud B. Healthcare systems in comparative perspective: classification, convergence, institutions, inequalities, and five missed turns. *Annu Rev Sociol*. 2013;39:127–146.

Translational Cardiology. https://doi.org/10.1016/B978-0-323-91790-2.00005-8

- Hyde JK, Shortell SM. The structure and organization of local and state public health agencies in the US: a systematic review. *Am J Prev Med*. 2012;42(5):S29–S41.
- Kumar S, Ghildayal NS, Shah RN. Examining quality and efficiency of the US healthcare system. *Int J Health Care Qual Assur*. 2011;24(5):366–388.

Resources

- Alcaraz KI, Wiedt TL, Daniels EC, Yabroff KR, Guerra CE, Wender RC. Understanding and addressing social determinants to advance cancer health equity in the United States: a blueprint for practice, research, and policy. *CA Cancer J Clin*. 2020;70(1):31–46.
- Closser S, Mendenhall E, Brown P, Neill R, Justice J. The anthropology of health systems: a history and review. *Soc Sci Med*. 2022;300:114314.
- Thibault GE. The future of health professions education: emerging trends in the United States. *FASEB Bio-Adv*. 2020;2(12):685.

Healthcare industry is an essential aspect of society that varies across the globe. Healthcare institutions and systems vary greatly, with the United States health system being a mix of public and private, for-profit, and not-for-profit insurers. In the United States, private insurance provides the majority of the coverage.[1] In 2018, the national health expenditure was 3.6 trillion and is projected to reach 6.2 trillion by 2028.[2,3] Among other high-income countries such as the United Kingdom, Norway, and Australia, the United States ranks last on access to care, administrative efficiency, equity, and healthcare outcomes, despite spending more of its gross domestic product on healthcare.[4]

Healthcare institution is a broad term that can be defined as any hospital, health clinic, or health maintenance organization dedicated to promoting, restoring, and maintaining one's health.[5,6] Various stakeholders are responsible for delivering these goals in each healthcare organization, with consequences arising when responsibilities are disregarded. Key stakeholders that inform healthcare systems include insurance providers, pharmaceutical companies, patients, the government, and healthcare providers.[7] Fig. 97.1 displays the relationship of these entities supporting one another in delivering healthcare goals.

Pharmaceutical and insurance companies

Stakeholders have very intricate interrelationships within the healthcare system. They are beholden to delivery while also responsible for being sustainable and providing quality of care. Pharmaceutical companies are a telling example to highlight this point as their primary goal of existence may not necessarily be to promote, maintain, or restore consumers' health but rather to maximize wealth.[8] Their economically driven goals are achieved through methods such as stock exchange and patent holding. These intellectual properties allow pharmaceutical companies to set higher prices and increase profits when no other company can offer a drug alternative at a lower price. Arguments in favor of increasing drug costs are to offset the cost of research, cost of intellectual properties, drug testing, and marketing.

To support the true primary goals of some entities within healthcare, political advocacy in the form of lobbying is aggressively pursued. Having legislative influence is a modality necessary for some such as pharmaceutical firms to maintain power. Big pharma, or the "global pharmaceutical industry" spends more on lobbying than any other industry, with goals to counteract regulatory initiatives such as lowering drug costs.[9]

Insurance

The insurance industry remains profit-driven despite the need to provide healthcare plans and services to patients. Affordable healthcare coverage is rare due to rising premiums and strict requirements that impede many from obtaining the care they need. Elements such as quarterly reports from stockholders redirect companies to focus on profit management instead of equal access to their insurance. In return, insurance companies implement rigid regulations that benefit healthy individuals but encumber patients with preexisting conditions that need coverage the most. This pattern is unethical as it perpetrates healthcare as a profit-centered industry and creates more barriers to affordable care. Many insurance companies need to find an equilibrium between their responsibilities of adequate patient plans and profit stability with shareholders.

Healthcare providers

Healthcare providers often face the difficult task of balancing the role of gatekeeper for insurance companies and advocacy for patients. Many physicians experience the concern of ever-increasing patient loads while lacking the adequate time necessary to treat them, so some may administer more tests and imaging to reduce liability and risk. This way of managing the demand also contributes to increasing healthcare costs. Healthcare providers are ethically inclined to keep patients' best interests ahead however difficulties arise as patients may choose providers that may perform care and therapeutics specific to the interest of the patient as opposed to clinical concerns. These patients may choose to continue care or establish new care with providers amenable to the pursued interests which is also known as "doctor shopping." Some reimbursement methods reimburse set charges for services such as the fee-for-service method.[10] This modality seeks to reduce the

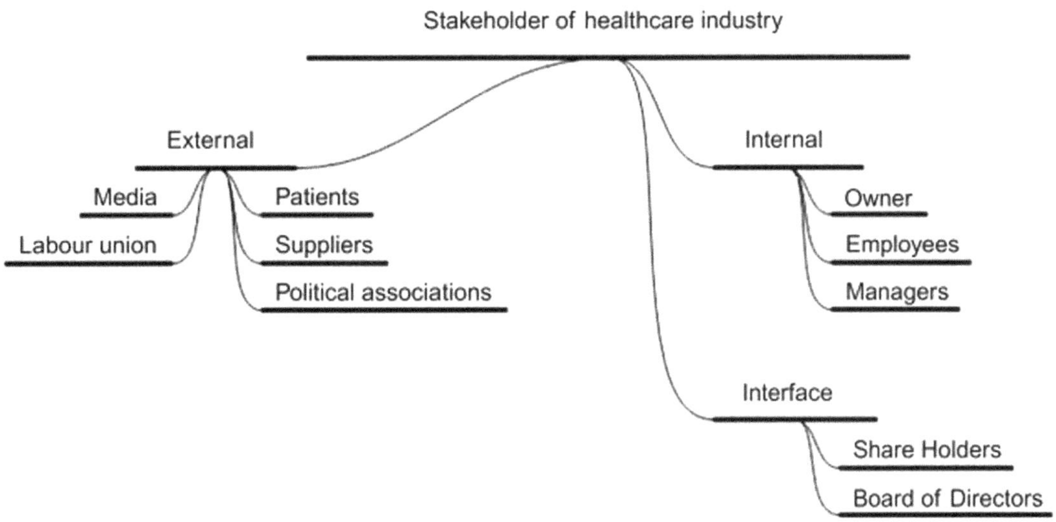

FIGURE 97.1 Stakeholder of healthcare industry. *From Sapra R.* Stakeholder of Healthcare Industry. *Dhaliwal P, ed.; 2021.*

financial incentive of needless additional testing and workup as the hospital receives the same amount of reimbursement regardless of hospital stay or additional workup. Despite the many challenges and competing interests of healthcare systems, providers must be cognizant of patients' unique needs and navigate a plan to deliver the best care for the patient.

Patients

Patients are at the center of the healthcare system. Each patient unless lacking competence is autonomous to make decisions that impact their health. However, sufficient evidence supports that social determinants of health contribute more to one's health status than lifestyle choices. For example, numerous studies suggest that social determinants of health (SDOH) account for 30%−55% of health outcomes.[11] Healthcare institutions and systems aim to provide the public with information and treatment to achieve and maintain a healthy freestyle. Yet not all institutions and systems consider the strong impact that SDOH has on one's internal and external ability to manage an ideal healthy lifestyle. Healthcare providers often function under a "risk factor" paradigm centered around behavioral modification for high-risk populations as the primary method of disease prevention via advising patients to quit smoking, improve diet, and increase physical activity.[12] Despite continued use, this approach is not preventative and thus is not successful in reducing disease burdens early such as intervening in coronary heart disease before pathologic states. Beneath the thresholds of awareness, before the onset of symptoms, the general population may worsen clinical health status with continued inappropriate health-related choices and behavior.[13] Another example is that low-income communities have lower concentrations of food sources that promote healthy eating,[14] but higher a concentration of convenience stores that have been linked to subsequent comorbidities such as substantial tobacco and alcohol use. Correspondingly, the sparse availability of fresh produce, joined with concentrated fast-food chains and limited recreational facilities, results in narrow opportunities to access nutritional foods and safe physical activity.[15]

Government

Although many would agree that the government has a principal influence on the healthcare institutions and system, the role of government is very specific regarding introducing, organizing, financing, policy, and delivering healthcare initiatives. Government-funded clinics and hospitals especially in low-income communities that offer adequate and affordable care is an example of one way the government provides health care. An extension of this is also purchasing healthcare for millions of disabled individuals and elderly populations via Medicare; government employees, retirees, and children enrolled in the State Child Health Insurance also receive this benefit.[16] The US healthcare system is an imperfect market that warrants regulation by private or public entities at federal, state, and local county and city levels.[17] Most of the regulations concerning quality of care rely on systems within state and local governments. One example of a state-level regulation is the licensure of healthcare providers, professionals, and institutions. This is primarily facilitated by state departments of health, departments of public health, board of registrations, or similar agencies. The Food and Drug Administration (FDA) is an example of a federal-level regulation. The FDA has regulatory authority over a broad range of categories including medical devices, vaccines, medical products such as prescription and nonprescription drugs, and electronic products that give off radiation.

References

1. Tikkanen R, Osborn R, Mossialos E, Djordjevic A, Wharton GA. *United States*. Commonwealth Fund; June 5, 2020. https://www.commonwealthfund.org/international-health-policy-center/countries/united-states.
2. Centers of Disease and Prevention. *FastStats - Health Expenditures*. Centers for Disease Control and Prevention; 2019. https://www.cdc.gov/nchs/fastats/health-expenditures.htm.
3. Centers for Medicare and Medicaid Services. *NHE Fact Sheet*. CMS; December 16, 2020. https://www.cms.gov/research-statistics-data-and-systems/statistics-trends-and-reports/nationalhealthexpenddata/nhe-fact-sheet.
4. The Common Wealth Fund. *Mirror, Mirror 2021: Reflecting Poorly*. Commonwealth Fund; August 4, 2021. https://www.commonwealthfund.org/publications/fund-reports/2021/aug/mirror-mirror-2021-reflecting-poorly.
5. Cornell Legal Information Institute. *29 U.S. Code § 152 - Definitions. LII/Legal Information Institute*; 1992. https://www.law.cornell.edu/uscode/text/29/152#14. Accessed December 30, 2021.
6. WHO. *Everybody's Business Strengthening Health Systems to Improve Health Outcomes*; 2007. https://www.who.int/healthsystems/strategy/everybodys_business.pdf.
7. Kumar P, Kumar Y, Tawhid MA. *Machine Learning, Big Data, and IoT for Medical Informatics*. London San Diego, Ca Cambridge: Ma Oxford Elsevier, Academic Press; 2021:327–347.
8. Perkins L. Pharmaceutical companies must make decisions based on profit. *West J Med*. 2001;175(6):422–423. https://www.ncbi.nlm.nih.gov/pmc/articles/PMC1275981/. Accessed November 25, 2021.
9. Wouters OJ. Lobbying expenditures and campaign contributions by the pharmaceutical and health product industry in the United States, 1999–2018. *JAMA Intern Med*. 2020;180(5). https://doi.org/10.1001/jamainternmed.2020.0146.
10. Teitelbaum D. An Incentive Perspective on U.S. Healthcare. *N.Y.Univ J Legis Publ Pol*; 2021. https://nyujlpp.org/quorum/teitelbaum-an-incentive-perspective-on-us-healthcare/. Accessed December 30, 2021.
11. World Health Organisation. *Social Determinants of Health*. World Health Organisation; 2021. https://www.who.int/health-topics/social-determinants-of-health#tab=tab_1.
12. Rose G. Sick individuals and sick populations. *Int J Epidemiol*. 1985;14(1):32–38. https://doi.org/10.1093/ije/14.1.32.
13. Ebrahim S, Taylor F, Ward K, Beswick A, Burke M, Davey Smith G. Multiple risk factor interventions for primary prevention of coronary heart disease. *Cochrane Database Syst Rev*. 2011;(1). https://doi.org/10.1002/14651858.cd001561.pub3.
14. Hilmers A, Hilmers DC, Dave J. Neighborhood disparities in access to healthy foods and their effects on environmental justice. *Am J Public Health*. 2012;102(9):1644–1654. https://doi.org/10.2105/ajph.2012.300865.
15. Braveman P, Gottlieb L. The social determinants of health: it's time to consider the causes of the causes. *Publ Health Rep*. 2014;129(2):19–31. https://doi.org/10.1177/00333549141291s206.
16. Tang N, Eisenberg JM, Meyer GS. The roles of government in improving health care quality and safety. *Joint Comm J Qual Saf*. 2004;30(1):47–55. https://doi.org/10.1016/s1549-3741(04)30006-7.
17. World Health Organization. *United States of America Region of Americas*; 2014. https://www.who.int/health-laws/countries/usa-en.pdf?ua=1.

Chapter 98

Public health institutions and systems

Faith Kanneh
Johns Hopkins Bloomberg School of Public Health, Baltimore, MD, United States

Introduction/why it matters

Healthcare and public health systems have several activities that work interchangeably. As discussed in the previous chapter, the government plays a vital role in healthcare institutions and systems. The role of the government is a portion of public health institutions and systems, but it is an imperative component. In 1994, the United States Public Health Service delivered the "Public Health in America Statement," which defined the goals and services government and public health agencies are responsible for to maintain population health.1 These goals still inform public health practice decades later.

Get started

- Healthcare and public health systems have several activities that work interchangeably.
- The Institute of Medicine's (IOM) report, "The Future of Public Health" in 1988 catalyzed the Public Health in America Statement.
- The first core function: assessment, systematically collects, analyzes, and obtains data that concerns of health the overall population or particular groups.
- Policy development includes using scientific knowledge and health policy analysis to guide decision-making and policy implementation.
- The assurance core function ensures fundamental components of an effective health system are in place.

Key points

- Healthcare and public health systems have several activities that work interchangeably.
- In 1994, the United States Public Health Service delivered the "Public Health in America Statement," which defined the goals and services government and public health agencies are responsible for to maintain population health.
- The Institute of Medicine's (IOM) report, "The Future of Public Health" in 1988 catalyzed the Public Health in America Statement.
- This IOM report also defined three core functions of public health—assessment, policy development, and assurance.
- The Ten Essential Public Health Services (EPHS) were developed in 1994 and revised in 2020.3 The center of the wheel in the 2020 version is focused on equity, whereas, in 1994, it was on research and systems.

Potential pitfalls

- The role of the government is a portion of public health institutions and systems, but it is an imperative component.
- At the time the IOM report came out, there were few issues in terms of society and health, so people were neglecting the need for public health. The report was galvanizing for the field of public health because it revealed that the United States had lost sight of the public health goals and that the system of public health had fallen into disarray.
- The matter of public health is a state's responsibility due to its lack of reference in the US Constitution.

Examples of ...

- Black A, MacCannell DR, Sibley TR, Bedford T. Ten recommendations for supporting open pathogen genomic analysis in public health. *Nat. Med.* 2020;26(6): 832–841.
- Jarvis T, Scott F, El-Jardali F, Alvarez E. Defining and classifying public health systems: a critical interpretive synthesis. *Health Res. Policy Syst.* 2020;18:1–12.
- Mendez DD, Scott J, Adodoadji L, Toval C, McNeil M, Sindhu, M. Racism as public health crisis: assessment and review of municipal declarations and resolutions across the United States. *Front Public Health* 2021;9:686807.

Translational Cardiology. https://doi.org/10.1016/B978-0-323-91790-2.00062-9

Resources

- Eisen RJ, Paddock CD. Tick and tickborne pathogen surveillance as a public health tool in the United States. *J. Med. Entomol.* 2021;58(4):1490–1502.
- Kooli C. COVID-19: public health issues and ethical dilemmas. *Ethics Med. Public Health* 2021;17:100635.
- Levin MB, Bowie JV, Ragsdale SK, Gawad AL, Cooper LA, Sharfstein JM. Enhancing community engagement by schools and programs of public health in the United States. *Annu. Rev. Public Health* 2021;42:405–421.
- Underwood JM. Overview and methods for the youth risk behavior surveillance system—United States, 2019. *MMWR Suppl.* 2020;69.

Healthcare and public health systems have several activities that work interchangeably. As discussed in the previous chapter, the government plays a vital role in healthcare institutions and systems. The role of the government is a portion of public health institutions and systems, but it is an imperative component. In 1994, the United States Public Health Service delivered the "Public Health in America Statement," which defined the goals and services government and public health agencies are responsible for to maintain population health.[1] These goals still inform public health practice decades later. They are as follows:

- Prevent epidemics and the spread of disease
- Protect against environmental hazards
- Prevent injuries
- Promote and encourage healthy behaviors
- Respond to disasters and assist communities in recovery
- Assure the quality and accessibility of health services

The Institute of Medicine's (IOM) report, "The Future of Public Health" in 1988 catalyzed the Public Health in America Statement.[2] At the time the IOM report came out, there were few issues in terms of society and health, so people were neglecting the need for public health. The report was galvanizing for the field of public health because it revealed that the United States had lost sight of the public health goals and that the system of public health had fallen into disarray.[2] This IOM report also defined three core functions of public health—assessment, policy development, and assurance. The core functions proposed a method to examine problems, organize responses, and implement solutions. It is important to note that these core functions are an evolving process rather than a step-by-step guide.

The first core function: assessment, systematically collects, analyzes, and obtains data that concerns the of health the overall population or particular groups. If we apply this core function to the COVID-19 pandemic, we would answer questions such as how many cases are in the United States or specific states? Policy development includes using scientific knowledge and health policy analysis to guide decision-making and policy implementation. State stay-at-home mandates or declarations of emergency around COVID are relevant examples of the policy development function. Finally, the assurance core function ensures fundamental components of an effective health system are in place. The government can examine this core function although implementation is often performed by others. This function would pertain to evaluating the impact of a COVID lockdown and determining if it decreased the number of cases and if it is reasonable to lift restrictions. The three core functions are a valuable intellectual framework for government and public health agencies to reference to ensure they deliver their responsibilities. However, the core functions lack tangible descriptions of the work of public health agencies. The Public Health in America Statement, informed by IOM's core functions, defined a series of Ten Essential Public Health Services that facilitate daily responsibilities for local state and federal government public health systems.

Fig. 98.1 The Ten Essential Public Health Services (EPHS) were developed in 1994 and revised in 2020.[3] The center of the wheel in the 2020 version is focused on equity, whereas, in 1994, it was on research and systems. The purpose of the revision was to integrate modern and future public health practices into the framework. EPHS provides a blueprint that governmental, nongovernmental systems and communities can follow to what is needed to protect the public's health. These public health services are delivered through a synthesis of local/federal government agencies and nongovernmental organizations.

Governmental agencies

The matter of public health is a state's responsibility due to its lack of reference in the US Constitution. States can maintain authority or receive aid from the federal government—this is referred to as the "home rule."[4] States can also delegate their authority to local agencies at the city, county, or other local levels. Some responsibilities of local public health agencies include licensing of health professionals, administering nutrition programs, and regulating health facilities, such as nursing homes. The federal government often works with local agencies to facilitate the core functions of public health. The Department of Health and Human Services is the central public health agency of the federal government. The Centers for Disease Control and Prevention (CDC) and the National Institute of Health (NIH) are among the essential federal health agencies of the Department of Health and Human Services. The CDC is the leading agency for epidemic investigation and prevention, health data collection, and public health measures aimed at disease control and prevention. The CDC's role in connecting and informing the public of COVID-19 protocols

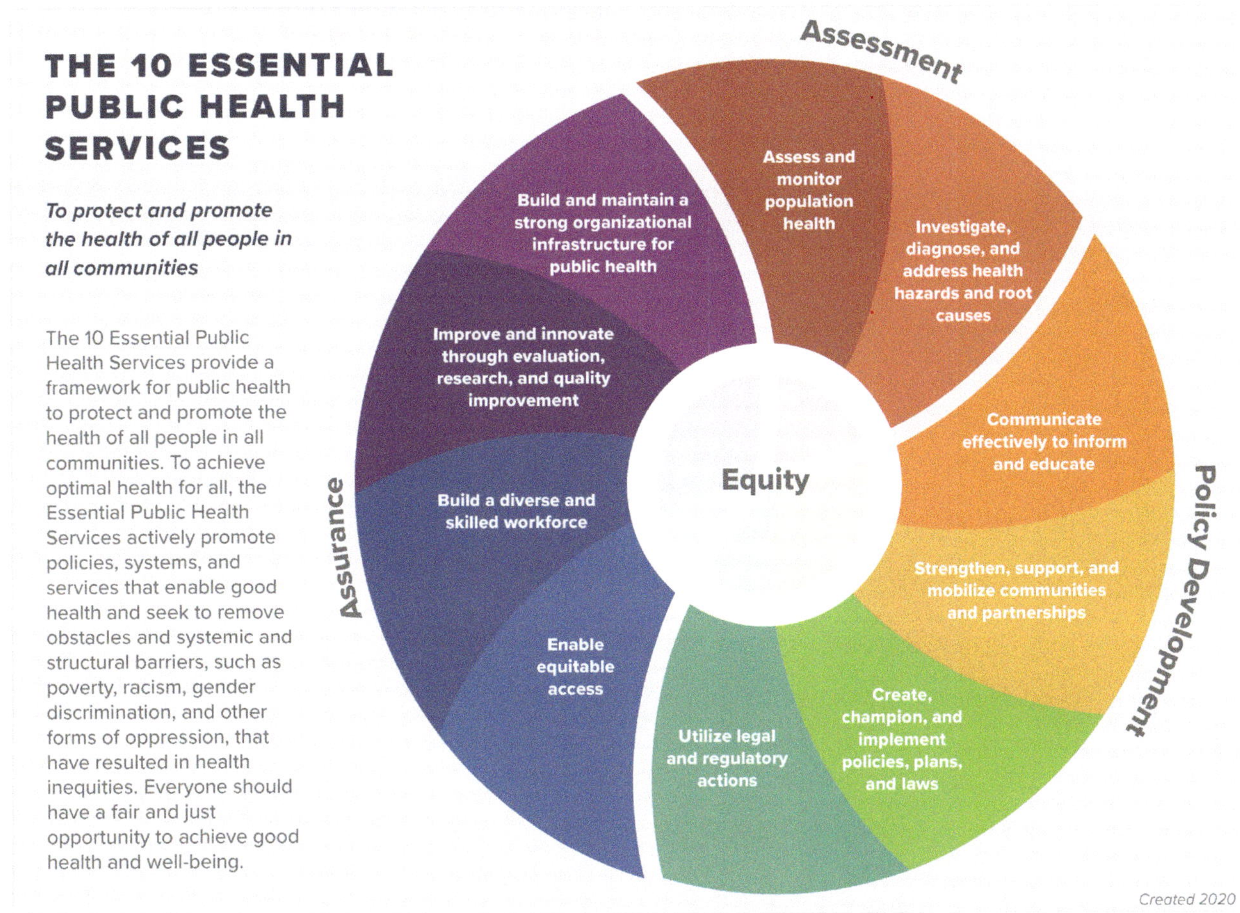

FIGURE 98.1 The Centers for Disease Control and Prevention (CDC).

continues to be an essential contribution to mitigating cases. The NIH is the leading research agency, with most of its efforts devoted to translating scientific research into practice.

Nongovernmental agencies

Nongovernmental agencies increasingly serve in public health roles in the United States and around the world. The American Red Cross is a fitting example of a private organization advocating for public health causes, delivering services, and supporting public health efforts. This organization plays a central role in obtaining volunteers for blood donations and ensuring the safety and effectiveness of blood supply, with the help of the US Food and Drug Administration.[5] Mothers Against Drunk Driving (MADD) is an exemplary organization that advocates for a specific public health intervention. This nongovernmental agency group has had a substantial influence on the passage of enforcement of drunk driving laws. CARE and OXFAM are global nongovernmental organizations (NGOs) that provide services germane to global health-related crises.

References

1. Harrell J, Baker E. *The Essential Services of Public Health*. In: *The Public Health National Center for Innovations*; 1994. Published 1994 https://phnci.org/uploads/resource-files/The-Essential-Services-of-Public-Health-Harrell-Baker-1994.pdf.
2. Institute of Medicine (US). Committee on assuring the health of the public in the 21st century. In: *The Future of the Public's Health in the 21st Century*. National Academies Press (US); 2002. https://www.ncbi.nlm.nih.gov/books/NBK221223/. Accessed December 16, 2021.
3. Public Health CDC. *System and the 10 Essential Public Health Services*. OSTLTS: Centers for Disease Control and Prevention.
4. McCarty KL, Nelson GD, Hodge JG, Gebbie KM. Major components and themes of local public health laws in select U.S. Jurisdictions. *Publ Health Rep*. 2009;124(3):458–462. https://www.ncbi.nlm.nih.gov/pmc/articles/PMC2663884/. Accessed December 30, 2021.
5. American Red Cross. *Nation's Largest Blood Centers Conducting Innovative Research That Could Help Lead to Additional Changes in Donor Eligibility for Men Who Have Sex With Men*. https://www.redcross.org/about-us/news-and-events/press-release/2021/nation-s-largest-blood-centers-conducting-innovative-research.

Part XI

Practical resources

Amr Salama

Chapter 99

Presenting data

Amr Salama[1] and Jeremiah P. Depta[2]

[1]*Interventional Cardiology Department, University of Alabama at Birmingham, Birmingham, AL, United States;* [2]*Sands-Constellation Heart Institute, Rochester Regional Health, Rochester, NY, United States*

Manuscript

Presenting data is one of the challenging steps in the process of writing manuscripts. It is as important as the phase of data production. It will guide how competitive the paper will be. The golden rule is to reflect an organized flow of ideas throughout the different sections of the paper, to convey a clear message in a storytelling manner, with no gaps or disconnections. Readers should be curious to read the following section. Simplicity and clarity are essentials. In general, it might be confusing to report all research findings; it is more important to focus on relevant information necessary to assess and explain the findings.

Text plays a central role in presenting data; however, visual representation is a whole additional key to success. They complement each other in providing clear findings. Readers should quickly answer three key questions: what the research question is, how is it done, what are the relevant results, and how does it affect the daily clinical practice. Moving from a "general to specific data" theme usually serves the goal. It is crucial to determine the "target journal" and be aware of the author's instructions ahead of time before starting the draft. It is always helpful to review some of the recently published manuscripts in the journal to gauge what is acceptable for publication.

Having a structure for presenting the data makes it easy to present the manuscript effectively. The IMRAD style is one of the most commonly used structures, required by the majority of medical journals: Introduction (Question asked?), Methods (How was it studied?), Results (What was found?), Moreover, Discussion (What do the findings mean?). It is important to remember that many people will read only the title and the abstract, while few will go over the entire text. So, time should be spent to pick a "catchy" title and an abstract representing the work comprehensively and stimulating readers' curiosity.

Here are a few tips for writing the sections.

Introduction

It is believed that one of the most common reasons for manuscript rejections by editors is the lack of straightforward research questions. The introduction section should summarize the knowledge gap that led to the research question. The two most common reasons to pursue research work, either it is the next logical step or prior studies, were deficient in answering clinical questions. Authors should justify, highlight the research question, and rationalize the need for their research work. A brief literature review of related 2−4 studies is reasonable beforehand. It is preferred to avoid referring to papers using author names, year of publication, and journal name as they tend to disturb the pace of the writing. Instead, refer to the clinical conclusion or statistical results, with a number at the end of the sentence. By the end of this section, readers should answer the purpose of the study. Worth mentioning that social media creates a channel for interaction with world top scientists and invites peer-to-peer discussion, which overall helps spread new data and aims for quicker implementation of new meaningful research into clinical practice.

Methods

It should state the patient population, methods selection, and statistical analysis used. Authors should organize information under subheadings if the section is lengthy.

If the study reports negative results, a comment about sample size and power calculations may be helpful.

Results

This is the "number section" All relevant results should be reported objectively as facts, without any interpretation of the findings. Past tense is preferred in describing the results. Close attention to the units; they should match throughout

Translational Cardiology. https://doi.org/10.1016/B978-0-323-91790-2.00009-5

the represented data. A table of the baseline characteristics of the study population followed by the relevant representative data in graphical figures is the usual pattern to include most of the data. Descriptive statistics should be reported (means, medians, SDs). Repetition should be avoided; if authors decide to share results visually in a table or figure, the text should not describe the findings again.

Graphs

Graphs give the researcher the advantage of visual emphasis on the data and a quicker grasp of the results while minimizing lengthy text descriptions.

Tables are the most used tools to summarize descriptive statistics effectively. Other figures should illustrate the major imaging findings. Arrows on figures are encouraged to be used for further highlighting.

Tables or any included graphic should have a title, figure legend, and footnotes along with the figure itself, where the title explains what the figure is about as well as the purpose. Having multiple graphs in the same figure is one of the tricks to deliver the outcome. In the ISCHEMIA trial, Fig. 99.1 is composed of four side-by-side line charts of the primary and secondary outcomes.

The footnotes give a chance to expand and add more data without affecting the graph itself. Careful use of the signs and symbols should be identical in different graphs. In the ADAPTABLE trial, authors summarize findings using three tables and one figure. The figure included time-to-event curves (i.e., line chart) for primary and safety outcomes. The footnote of T was a great example of the utilization of the footnote section, to minimize any crowded data in the table.

Discussion

One of the most critical sections reports the data and relays the authors' clinical perspective. It is usually one-third of the manuscript. A suggested flow summarizes the major findings at the beginning, which will address the questions posed in the introduction. If several major points were found, presenting the result in a new paper should be considered to minimize the dilution of the content. Then, comparing the results with the most recent trials should follow, whether the results supported or declined the current thinking.

Authors should explain the reasons for the discrepant results, if any, followed by the study's possible clinical and scientific implications. In this section, numbers and facts should be translated into clinical implications. Finally, a suggestion for further work (Fig. 99.2). If negative results, self-criticism of the work should be made, but it is essential to avoid undermining the methods.

Conclusion

The conclusion should be limited to one or two sentences and should reflect the purpose and findings of the research.

Presenting data on social media

There is currently a lack of data about the benefits and limitations of sharing data on social media. Historically, it is believed it takes more than 1 decade to disseminate new knowledge in practice. Social media is now changing the nature and speed of health themes. It brings a new dimension to the research world by offering a quick medium used by researchers to promote their work. It creates a channel for interaction with world top scientists and invites peer-to-peer discussion, which overall helps spread new data and aims for quicker implementation of new meaningful research into clinical practice. So proper use of these tools is a must in this era. Commonly used websites for sharing new data include Twitter, LinkedIn, Facebook, and ResearchGate.

Worth mentioning that it remains informal and unregulated, so careful consideration of data privacy and confidentiality should be maintained while sharing the data online. Furthermore, most traditional publishing models prohibit publishing articles except for paid subscribers. However, the open-access publishing model is becoming more frequent now, where authors can share preprint/postprint copies, and the article is accessible to everyone for free.

Categorical Data	Nominal Data	Relationship between values	Relationship between volumes	Trends or patterns
•Bar Chart •Pie Chart	•Bar Chart •Histogram if discrete •Line Chart if continuous	•Line Chart •Scatter Plot	•Area Chart •Bubble Chart	•Line Chart •Scatter Plot

FIGURE 99.1 Most commonly used charts for different types of data.

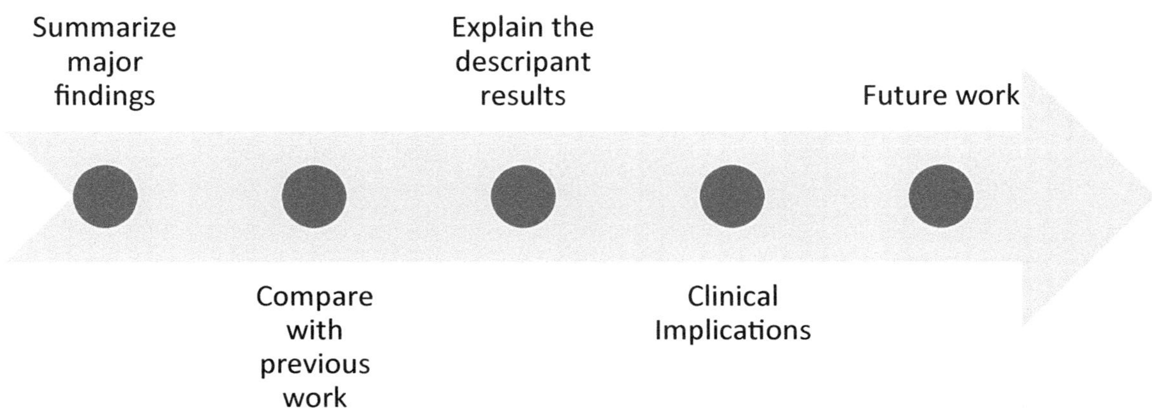

FIGURE 99.2 Proposed flow of the "discussion" section.

Common pitfalls to avoid

- Excessive use of qualitative words, which are subjective and away from being precise and objective, for example, extremely, remarkably.
- Repetition of the results or highlighting the statistical analysis in the discussion section.
- Extensive literature review in the discussion section, without a focus on the relevant data.
- Large number of reference citations.

Resources: Additional reading, consultants, contractors

- Reviewing (How to Write a Paper third edition Edited by George M Hall)
- Writing It Up: A Step-by-Step Guide to Publication for Beginning Investigators* Mark A. Kliewer
- Review A New Dimension of Health Care: Systematic Review of the Uses, Benefits, and Limitations of Social Media for Health Communication
- Presenting data in tables and charts* Rodrigo Pereira Duquia
- HOW TO PRESENT RESEARCH DATA? Tong Seng Fah MMed
- https://www.ahrq.gov/talkingquality/translate/display/index.html

Chapter 100

Manuscript preparation

Christine Shen
Division of Interventional Cardiology, Scripps Clinic, La Jolla, CA, United States

Key points

- Manuscript writing is the natural continuation of scientific work, the means to effectively communicate the findings of research.
- Information should be introduced in the proper context first, then with communication of results, followed by an explanation of its importance.
- Language must be clear and concise.

"Why it matters?"

Writing is the means to effectively communicate research once it is complete. This chapter will describe the different types of manuscripts, the anatomy of the manuscript, and formatting. The final goal is a publication in a peer-reviewed journal.

There are a few types of common research papers accepted by most peer-reviewed journals. Types include full, original research papers, review papers, and letters or short communications (Table 100.2). Letters and short communications are also original research, but they are often smaller studies that do now warrant a full research paper. Sometimes, a letter or communication references already published work but provides new commentary to add to the literature. Because researchers often read many already published papers on a topic when planning a project, they can summarize that knowledge in a review article, which can also highlight the gaps that set the stage for their own research.

The format of a manuscript is dependent on the journal itself, so it is important to first decide on the intended journal for publication.[1] There are also reporting guidelines available to assist researchers when reporting their research based on the specific type of study performed.[2–7] Often, peer reviewers and editors also refer to these guidelines when critically appraising a paper.

The general template of a scientific manuscript includes the following: Abstract, Introduction, Methods, Results, Discussion, Tables, and Figures.[8] Table 100.1 describes each structured element in depth. Each paper also needs a title, which should be concise to clearly and accurately convey the results of the research. For clarity, abbreviations should be avoided. Journals may have specific guidelines regarding formatting for the title. Some journals require keywords, which are a simple way of allowing the general audience to easily search for or discover a specific article.

The abstract is easiest to write last, when the trajectory of the paper has already been set. Journals usually have

TABLE 100.1 Anatomy of a manuscript.

Abstract	Part of the journal article first read by the audience, and it is a short, usually structured paragraph that summarizes the main points of the article. It should provide a short introduction, methodology, results, and conclusion, so that readers can easily grasp it.
Introduction	Include 2–3 paragraphs outlying the current context and demonstrating a knowledge gap, setting the scene for the results described in the study.
Methods	Should provide details so that others can critically evaluate the robustness of the study and reproduce the research.
Results	Should be presented in a clear fashion, utilizing critical tables and figures. This section should contain only data from the study itself.
Discussion	Needs to put the study in a global context, explaining why the paper is applicable. Therefore, a thorough understanding of the surrounding context through a comprehensive literature search is vital to apply the research.
References	Cite publications in this section according to the journal's style.

Translational Cardiology. https://doi.org/10.1016/B978-0-323-91790-2.00021-6

TABLE 100.2 Types of manuscripts.

Original research	Involves research studies and are typically longer in length, about 2000–5000 words, with 5 figures, and 30–50 references
Review	Longer paper summarizing the state of research for a particular topic, often by invitation
Letter or short communication	Shorter research paper with findings typically too small for an original research paper

specific instructions for structured abstracts including a maximum word count, which should be followed.

Next, the introduction should provide context for the research and explain gaps in the current literature that need to be filled. Mensch and Kording described the "context-content-conclusion scheme," recommending starting by setting of the broader context, then proceeding with the content of the research, before finishing with a conclusion of the importance of that content. This scheme allows readers to be prepared to receive the actual research results.[9] One example of context is a manuscript by Witberg et al. in the New England Journal of Medicine regarding COVID-19 vaccination and myocarditis. The short introduction details the specific dates of the timeline of vaccinations in Israel, a statement from the CDC, and a report from the Israeli Ministry of Health, which gives context to readers, preparing them to understand the importance and relevance of the study.[10]

The methods section is often the easiest section to start first because it is a straightforward description of the research protocol itself. Planning of population groups should be explained with explicit inclusion and exclusion criteria. Primary and secondary endpoints should be clearly stated. Part of the methods should have a detailed description of the statistics performed, and consulting with a statistician can be helpful. It is recommended to specify the power analysis in this section. Studies with human patients should include a statement regarding approval from the Institutional Review Board.

The results section should parallel the methods, starting with a detailed description of patient groups, included and excluded patients, as well as the duration of follow-up and loss to follow-up. Any significant differences in patient demographics should be identified. Then, the primary and secondary outcomes should be stated, often accompanied by a table or figure. When writing, clarity is a key goal, particularly any time there is a large set of results. The language serves to make the results clear, even to those outside the particular area of expertise.[11] If it is not the most important data, consider placing those results in the supplementary section. However, the key results supporting the main purpose of the paper should be in the main body of the text.

The discussion section plays two main roles. First, it defends the methodology of the research, including the protocol and the data analysis. Second, it explains how the results fit into the context of current research and fill in the gaps explained in the introduction. Because no study is perfect, each discussion should have a few lines explaining the limitations of the research. Finally, the discussion should explain how readers should apply the research findings in their own practices or what future research is needed.

When choosing tables and figures, they should be arranged to best tell the story, not necessarily the order in which results were obtained. Often, the first figure shows the protocol design, with included and excluded patients. The first table is often for demographics showing any significant differences between groups. They should be presented simply to communicate results, ideally self-explanatory even without the main text. For example, Allard et al. published in the Journal of the American College of Cardiology regarding exercising training for skeletal muscles in long-term statin users, utilizing the central illustration to depict the main points of the article using easily digestible pictorials.[12] While there is no perfect number of figures and tables, often journals have a maximum that can be included in the main manuscript.

All of the authors on the manuscript are responsible for the work, so it is important to circulate the paper to ensure agreement from all co-authors. This is also a valuable time to obtain feedback regarding the paper, and it is common to make significant revisions based on this feedback. Often, journals request a detailed description of the contribution of each author.

There are common reasons for a journal to reject a manuscript for publication (Table 100.3).[1,13] Often, there are issues with the research itself, particularly the methods, but there are also mistakes in writing that can lead to rejection. It is not enough to present the results if the appropriate context and applicability are not conveyed. Often, a journal requests a revision. Reviewers list comments, and a revision should be accompanied by a point-by-point response to each comment, explaining what

TABLE 100.3 Common reasons for manuscript rejection.

Formatting	Contents in the wrong sections, not formatted correctly for the journal, incomplete references
Methods	Small sample size, type I error, short follow-up
Results	Data previously published elsewhere
Discussion	Poor description of the results

changes were made to the manuscript. Sometimes, comments may also call for additional experimentation or data analysis, which may be time consuming but will improve the overall research and manuscript.

While the process of preparing a manuscript can be daunting, there are discrete steps to get started. Guidance from experienced writers is valuable. With persistence, following these steps can result in the publication of a peer-reviewed manuscript.

"Get started"

1. Choose the journal for your manuscript submission and read the journal's format and requirements.
2. Write an outline.
3. Draft the manuscript and obtain feedback.

Pitfalls to avoid

1. Do not use incorrect formatting for the journal.
2. Avoid subjective, defensive commentary.
3. Cut out overcomplicated language.

Resources

- Balch CM, McMasters KM, Klimberg VS et al. Steps to Getting Your Manuscript Published in a High-Quality Medical Journal. *Ann Surg Oncol.* 2018; 25(4):850-855. https://doi.org/10.1245/s10434-017-6320-6
- Mensh B, Kording K. Ten simple rules for structuring papers. Markel S, ed. *PLoS Comput Biol.* 2017; 13(9):e1005619. https://doi.org/10.1371/journal.pcbi.1005619

References

1. Balch CM, McMasters KM, Klimberg VS, et al. Steps to getting your manuscript published in a high-quality medical journal. *Ann Surg Oncol.* 2018;25(4):850–855. https://doi.org/10.1245/s10434-017-6320-6.
2. Liu X, Cruz Rivera S, Moher D, et al. Reporting guidelines for clinical trial reports for interventions involving artificial intelligence: the CONSORT-AI extension. *Nat Med.* 2020;26(9):1364–1374. https://doi.org/10.1038/s41591-020-1034-x.
3. Cohen JF, Korevaar DA, Altman DG, et al. STARD 2015 guidelines for reporting diagnostic accuracy studies: explanation and elaboration. *BMJ Open.* 2016;6(11):e012799. https://doi.org/10.1136/bmjopen-2016-012799.
4. von Elm E, Altman DG, Egger M, Pocock SJ, Gøtzsche PC, Vandenbroucke JP. The strengthening the reporting of observational studies in epidemiology (STROBE) statement: guidelines for reporting observational studies. *Int J Surg.* 2014;12(12):1495–1499. https://doi.org/10.1016/j.ijsu.2014.07.013.
5. Percie du Sert N, Ahluwalia A, Alam S, et al. Reporting animal research: explanation and elaboration for the ARRIVE guidelines 2.0. *PLoS Biol.* 2020;18(7):e3000411. https://doi.org/10.1371/journal.pbio.3000411.
6. Gagnier JJ, Kienle G, Altman DG, et al. The CARE guidelines: consensus-based clinical case reporting guideline development. *Case Rep.* 2013;2013(oct23 1). https://doi.org/10.1136/bcr-2013-201554. bcr2013201554-bcr2013201554.
7. Moher D, Hopewell S, Schulz KF, et al. CONSORT 2010 explanation and elaboration: updated guidelines for reporting parallel group randomised trials. *Int J Surg.* 2012;10(1):28–55. https://doi.org/10.1016/j.ijsu.2011.10.001.
8. Ibrahim AM, Dimick JB. Writing for impact. In: *Medical and Scientific Publishing.* Elsevier; 2018:81–92. https://doi.org/10.1016/B978-0-12-809969-8.00009-7.
9. Mensh B, Kording K. Ten simple rules for structuring papers. *PLoS Comput Biol.* 2017;13(9):e1005619. https://doi.org/10.1371/journal.pcbi.1005619.
10. Witberg G, Barda N, Hoss S, et al. Myocarditis after covid-19 vaccination in a large Health care organization. *N Engl J Med.* 2021; 385(23):2132–2139. https://doi.org/10.1056/NEJMoa2110737.
11. Gewin V. How to write a first-class paper. *Nature.* 2018;555(7694): 129–130. https://doi.org/10.1038/d41586-018-02404-4.
12. Allard NAE, Janssen L, Aussieker T, et al. Moderate intensity exercise training improves skeletal muscle performance in symptomatic and asymptomatic statin users. *J Am Coll Cardiol.* 2021;78(21): 2023–2037. https://doi.org/10.1016/j.jacc.2021.08.075.
13. Audisio RA, Stahel RA, Aapro MS, Costa A, Pandey M, Pavlidis N. Successful publishing: how to get your paper accepted. *Surg Oncol.* 2009;18(4):350–356. https://doi.org/10.1016/j.suronc.2008.09.001.

Chapter 101

Promoting research

Adam Vohra and Harsimran S. Singh
Division of Cardiology, Department of Medicine, Weill Cornell Medicine - New York Presbyterian Hospital, New York, NY, United States

Key points

- Multiple databases exist to query clinical research, including PubMed and Google Scholar.
- h-Index is a single number criterion to measure research impact by determining the number of publications with *h* citations.
- Social media plays an increasingly important role in disseminating research.

Why it matters?

There has been an exponential growth in scientific publications over the past 2 decades. The amount of available information has made it difficult for researchers and providers to keep abreast of current and past research.[1–3] Promoting and accessing research is crucial for researchers to build upon work that is being done alongside their own. It is also crucial for providers to utilize and incorporate the most up-to-date science. This chapter will focus on (1) utilizing research databases to find research articles, (2) understanding measures of research impact, and (3) exploring the role of social media in promoting research.

Utilizing research databases

There are a variety of research databases that are utilized in clinical practice (see Table 101.1). PubMed is traditionally one of the most commonly used databases for clinical searches. The database has over 33 million citations and abstracts primarily from MEDLINE, a database run through the United States National Library of Medicine. Publications in MEDLINE are indexed using Medical Subject Headings (MeSH), which are manually curated.[4] PubMed algorithm searches through titles, abstracts, and MeSH terms for indexed articles.

In recent years, Google Scholar has become an increasingly utilized database using a natural-language processed search. Google Scholar differentiates from traditional Google search by using weights on citation counts to emphasize academic articles.[5] In addition to peer-reviewed literature, Google Scholar indexes articles from preprint servers and conference proceedings.[6] Limited research has suggested that while Google Scholar often provides a broader array of articles, PubMed may provide higher-quality and more relevant results.[5,7]

Measures of research impact

Journal-level measures of research impact

Citation analysis has been an important discipline in determining the relative importance and reach of individual journals and scientists. Traditionally, these indicators were used by libraries to determine which journals to carry, but have become increasingly used in the assessment of researchers in grant proposals.[8]

The most widely used journal-level bibliometric indicator is the journal impact factor. It was developed by Eugene Garfield in the 1970s to better map an individual journal's network.[9] The impact factor is calculated by taking the number of citations in the current year from journal articles published in the prior 2 years, divided by all the articles published in that journal for two prior years. For example, for 2021, this would equal the total number of citations in 2021 for articles published in 2019 and 2020 divided by the total number of articles published in 2019 and 2020.[8] These are published on a yearly basis in Journal Citation Reports from aggregated publication and citation information from Web of Science.

There have been various critiques of the journal impact factor. First, the journal impact factor is a mean value. Since the citations for most journals are not normally skewed, a median has been suggested as a more appropriate measure. Second, journal impact factors can be inflated since citations from all sources are included in the numerator, while only citable items (articles and reviews) are included in the denominator. Items such as abstracts,

Translational Cardiology. https://doi.org/10.1016/B978-0-323-91790-2.00087-3

TABLE 101.1 Research databases for accessing information.

Database	Producer	Details
PubMed/MEDLINE[a]	National Library of Medicine	Free database. Contains 33 million citations or abstracts, primarily through the MEDLINE database. Citations are indexed using Medical Subject Headings (MeSH)
EMBASE[b]	Elsevier	Subscription database. Includes 37 million records, including those in MEDLINE. Indexed using Emtree search terms.
Scopus[b]	Elsevier	Subscription database. Searches abstracts and citations in Elsevier database, but does not use Emtree search terms.
Web of Science[c]	Clarivate	Subscription database. Originally produced by the Institute for Scientific Information. Has 171 million records from different disciplines.
Google Scholar[d]	Google	Free database. Index full texts of articles and websites using a web robot algorithm.

[a] *PubMed Overview. National Library of Medicine. 2021. https://pubmed.ncbi.nlm.nih.gov/about/.*
[b] *Embase resources, Elsevier. 2021. https://www.elsevier.com/solutions/embase-biomedical-research/learn-and-support.*
[c] *Web of Science. Clarivate. 2021. https://clarivate.com/webofsciencegroup/solutions/web-of-science/.*
[d] *About. Google Scholar. 2021. https://scholar.google.com/intl/en/scholar/about.html.*

letters, editorials, corrections, or retractions are not included as citable items.[8]

Article-level measures of research impact

There are a number of article-level measures of research impact. These include a number of citations, article views and downloads, and Altmetric Attention Score.

- Number of citations: The number of citations is traditionally a very common method to determine an article's overall research impact. It is an indicator that the article is being read and that it is contributing to and influencing further research. However, this can only assess the impact at a slow pace as articles will not be cited for likely months, if not years, after publication.
- Article views/downloads: In a digital age, measuring article views and downloads is a method to quickly assess the impact of an article, even only hours to days after being published. However, given various access and venues for people to read articles, it is a crude method and does not fully account for an article's overall reach.
- Altmetric Attention Score: This is a weighted count of all the attention an article has received and is based on three factors: volume, sources, and authors. For volume, the more an article is mentioned, the higher the score rises. For sources, each category of mention (i.e., newspaper, tweet, etc.) contributes a different base score. For authors, the score is modified based on the type of author. For example, a journal Twitter account mentioning an article will count for less than an unconnected researcher mentioning the article. These factors are weighted and a score is assigned.[10] In a cardiovascular disease specifically, research has shown that Altmetric scores have a weak but significant relationship with a number of citations.[11,12] However, there is concern that Altmetric scores may be biased by personal opinions and not truly represent the scientific value of an article.[13]

Individual measures of academic productivity

In addition to the measures of research impact, various indicators have been created to determine the overall academic productivity and impact for individual scientists. Traditionally, these have focused on measures that look at the scientific output for researchers. These include total number of papers, total number of citations, citations per paper, or citations per paper However, each of these methods has advantages and disadvantages (see Table 101.2), with a tradeoff between evaluating total productivity versus total impact.[14]

The h-index was created by Jorge E. Hirsch in 2005 to better characterize the scientific output of a researcher incorporating both productivity and impact in a single-number criteria. He defined the h-index as the number of papers with citation number $\geq$ h. For example, if a scientist has 20 papers with at least 20 citations, his or her h-index would equal 20. This can be visualized in Fig. 101.1. To correct for scientific age, he also proposed the m-index which divided the h-index by years of productivity (i.e., a scientist with an h-index of 20 who has been productive for 20 years would have an m-index of 1).[14]

Promoting research through social media

Over the past 2 decades, social media has created further opportunities to promote research. Universal access to platforms such as Twitter, Facebook, LinkedIn, and others has created a democratization of discourse that has allowed researchers at all ranks from trainees to full professors to promote, disseminate, and comment on their research work.[15]

Social media has become an important tool in disseminating information and research. Traditional organizations such as professional societies (i.e., American College of Cardiology or American Heart Association) can utilize social media to disseminate information and knowledge quickly to large-scale audiences.[16] Similarly, journals can promote research articles on Twitter immediately after release instead of relying on delays in printing and mailing. Perhaps most importantly, individual researchers can promote and disseminate their own work in ways that were not possible before with the ability to connect with far-reaching and widespread audiences.

Social media also plays an important role in critical appraisal and peer review of research in addition to traditional methods.[17] In the past, much of the debates and discussions surrounding research and literature existed at regional and national meetings and were concentrated among a group of academic experts. Social media has allowed for real-time feedback of literature as it is published with discussion from a wide array of people, from clinical trialists to healthcare practitioners.

Get started

- Utilize PubMed and Google Scholar in searching for research articles.
- Calculate your personal h-index to get a sense of your overall research impact.
- Incorporate social media into promoting your own and learning about others' research

TABLE 101.2 Indicators of research impact, researcher level.
Commonly used single-number criterion to evaluate impact for researchers.

Metric	Definition	Advantages	Disadvantages
Number of papers	Total number of papers	Measures total productivity	Does not capture the total impact of work
Number of citations	Total number of citations for all papers	Captures the total impact of work	Can be skewed by a small number of papers with high citations
Citations per paper	Average number of citations per paper	Allows comparison of researchers at different scientific ages	Can reward those with low productivity and penalize those with high productivity
Number of significant papers	Number of papers with more than y citations	Can measure the broad sustained impact	Y is arbitrary and can differ between those of different scientific age
Number of citations for each of q most cited papers	Number of citations for top q cited papers	Can measure the broad sustained impact	Q is an arbitrary number. Criterion is not a single number and is difficult to compare

Adapted from Hirsch JE. An index to quantify an individual's scientific research output. *Proc Natl Acad Sci U S A.* 2005;102(46):16569–16572.

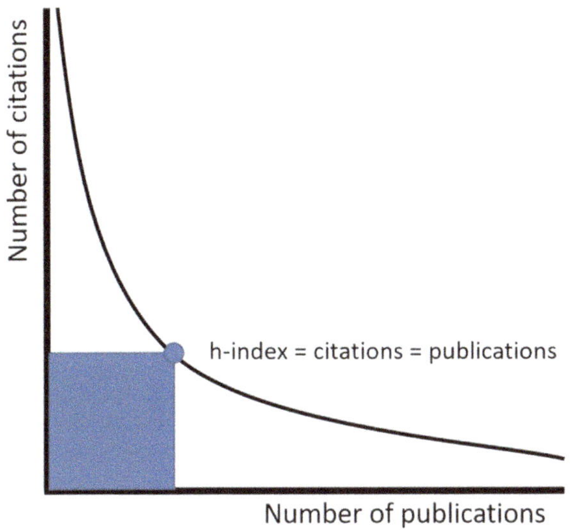

FIGURE 101.1 **h-Index.** h-Index is equal to the number of papers with an h number of citations. For example, a researcher with 20 papers with at least 20 citations would have an h-index of 20.

References

1. Gusenbauer M, Haddaway NR. Which academic search systems are suitable for systematic reviews or meta-analyses? Evaluating retrieval qualities of Google Scholar, PubMed, and 26 other resources. *Res Synth Methods*. March 2020;11(2):181–217. https://doi.org/10.1002/jrsm.1378.
2. Larsen PO, von Ins M. The rate of growth in scientific publication and the decline in coverage provided by Science Citation Index. *Scientometrics*. September 2010;84(3):575–603. https://doi.org/10.1007/s11192-010-0202-z.
3. Opthof T, Coronel R. Productivity in science: more more and more? *Cardiovasc Res*. November 2002;56(2):175–177. https://doi.org/10.1016/s0008-6363(02)00651-x.
4. *PubMed Overview*. National Library of Medicine; 2021. https://pubmed.ncbi.nlm.nih.gov/about/.
5. Morshed T, Hayden S. Google versus PubMed: comparison of Google and PubMed's search tools for answering clinical questions in the emergency department. *Ann Emerg Med*. 2020;75(3):408–415. https://doi.org/10.1016/j.annemergmed.2019.07.003.
6. Giles J. Science in the web age: start your engines. *Nature*. December 01, 2005;438(7068):554–555. https://doi.org/10.1038/438554a.
7. Shariff SZ, Bejaimal SA, Sontrop JM, et al. Retrieving clinical evidence: a comparison of PubMed and Google Scholar for quick clinical searches. *J Med Internet Res*. August 15, 2013;15(8):e164. https://doi.org/10.2196/jmir.2624.
8. Opthof T. Comparison of the impact factors of the most-cited cardiovascular journals. *Circ Res*. 2019;124(12):1718–1724. https://doi.org/10.1161/CIRCRESAHA.119.315249.
9. Garfield E. Citation analysis as a tool in journal evaluation. *Science*. November 03, 1972;178(4060):471–479. https://doi.org/10.1126/science.178.4060.471.
10. *How Is the Altmetric Attention Score Calculated?* Almetric; 2021.
11. Patel RB, Vaduganathan M, Bhatt DL, Bonow RO. Characterizing high-performing articles by altmetric score in major cardiovascular journals. *JAMA Cardiol*. 2018;3(12):1249–1251. https://doi.org/10.1001/jamacardio.2018.3823.
12. Parwani P, Martin GP, Mohamed MO, et al. Relationship of altmetric attention score to overall citations and downloads for papers published in JACC. *J Am Coll Cardiol*. 2020;76(6):757–759. https://doi.org/10.1016/j.jacc.2020.06.025.
13. Tomasoni D, Coats AJS. The Altmetric Attention Score: how science tries to meet social media. *Eur J Heart Fail*. 2021;23(5):693–697. https://doi.org/10.1002/ejhf.2136.
14. Hirsch JE. An index to quantify an individual's scientific research output. *Proc Natl Acad Sci U S A*. November 15, 2005;102(46):16569–16572. https://doi.org/10.1073/pnas.0507655102.
15. Yeh RW. Academic cardiology and social media: navigating the wisdom and madness of the crowd. *Circ Cardiovasc Qual Outcomes*. 2018;11(4):e004736. https://doi.org/10.1161/CIRCOUTCOMES.118.004736.
16. Walsh MN. Social media and cardiology. *J Am Coll Cardiol*. 2018;71(9):1044–1047. https://doi.org/10.1016/j.jacc.2018.01.037.
17. Mandrola J, Futyma P. The role of social media in cardiology. *Trends Cardiovasc Med*. 2020;30(1):32–35. https://doi.org/10.1016/j.tcm.2019.01.009.

Chapter 102

Research versus quality improvement

Shaw-Ree Chen
Oak Orchard Community Health Centers, Brockport, NY, United States

Key points

- Regulatory requirements of research and quality improvement are different—know the difference and when to utilize an Institutional Review Board.
- Research and quality improvement use the same approach and framework. Both should utilize a robust data collection and analysis process.
- Identifying quality improvement activities starts with knowing your key performance indicators.

Why it matters

You may be wondering what a single chapter about quality is doing in a book on clinical and translational research. As a researcher turned quality improvement specialist, I initially asked the same question, as there are significant differences in research and quality improvement work. And yet, I was compelled to participate in this venture because of the number of times I've been told by physicians and healthcare workers that they don't know anything about quality improvement. I could not disagree more with that statement—if you are familiar with the scientific method, you are familiar with the basics of quality improvement. As healthcare evolves, the practice of continuous quality improvement is becoming central to practicing medicine.Â Accrediting bodies expect that if you are asked, "What quality improvement projects are you working on and how is it going" you should be able to easily reference one or two projects. While you may not always be participating in research, if you are in the healthcare field, you should absolutely be participating in quality improvement. This chapter is for individuals in the position to lead those quality improvement initiatives and to help you ensure your staff can always answer the question of "What quality improvement projects are you working on?"

Is it research or quality improvement? This is a critical question to answer before starting your work. As you have read previously, any research that involves patients should involve a human subjects review. Quality improvement, however, does not require the same level of review. Therefore, it is vital to distinguish the two early on. The Department of Health and Human Services (DHHS) indicates that research is designed to develop new knowledge, not to implement knowledge. In contrast, DHHS says this about quality improvement (QI):

> *QI is an integral part of the ongoing management of the system for delivering clinical care, not an independent, knowledge-seeking enterprise. QI practitioners design QI activities to bring about immediate improvements in care, relying on theory and evidence from research and practical experience to identify changes that are very likely to be beneficial.*[1]

For example, research was conducted to show that annual fecal immunochemical tests (FITs) were as good at detecting colon cancer as every 10-year colonoscopies.[2] At the time the research was conducted, it was not yet established that annual FIT tests would provide the same benefit as every 10-year colonoscopies. In this instance, there was significant uncertainty about whether or not FIT testing, as the intervention, would be beneficial, and the study was designed to produce generalizable knowledge about the intervention, not immediately improved care. Once it was established as a best practice,[3] offering FIT testing in addition to colonoscopy referrals as a way of increasing the percentage of patients screened for colon cancer is QI. There is no longer a question that annual FIT testing would be beneficial, and the rationale for the change is to increase the number of patients screened. Note that this is not to say that in implementing FIT testing, no new knowledge can be obtained and shared. The method by which annual FIT testing is implemented will undoubtedly result in knowledge gained as a clinic identifies ideal workflows, documentation, and education

Translational Cardiology. https://doi.org/10.1016/B978-0-323-91790-2.00103-9

that can, and should, be shared through publications. See this resource for information about ethical considerations of using QI methods.[4]

Performance metrics and balancing metrics: Ensuring quality improvement benefits patients

Once one has determined that one is about to engage in quality improvement and not research, one is not required to go through the same process of planning a human subjects study and involving an institutional review board. The rationale for excluding quality improvement from human subjects research is that quality improvement should be a continuous process ... requiring that clinics and providers engage IRBs for every change in practice would become onerous and costly, making it difficult to continually strive to do better for one's patients. However, this does not mean that care should not be taken to ensure that the quality improvement initiative does not inadvertently cause negative outcomes for patients or their families. We do this by identifying performance and balancing metrics.

Performance Metrics are defined broadly as any type of measurable data that tracks processes and outcomes of a business. When we think about performance metrics, we think about those metrics that tell us we are doing a good job, and doing right by our patients. Healthcare in particular is inundated with performance metrics. The Center for Medicaid Services (CMS) and the Agency for Healthcare Research and Quality (AHRQ) are well-known resources to identify performance metrics for different types of care and specialties of care. CMS has a measures inventory tool at cmit.cms.gov. Performance metrics is a term that can be used interchangeably with quality metrics, outcomes metrics, or key performance indicators. We will use the term performance metrics here to emphasize the use of metrics to measure the overall performance of clinical care.

Before initiating any kind of quality improvement, it is very important to identify your performance metric. This metric should answer the question, "what am I trying to achieve, and how will I know when I get there?" For example, in the case of implementing FIT testing, we can go to the CMS measure inventory tool and find a metric for colon cancer screening: the percentage of patients 50–75 years old who had appropriate screening for colorectal cancer. In the definition, inclusion and exclusion criteria are clearly identified. Our hypothetical clinic might have a screening rate of 24% and desire to increase it to 30%. As this book is focused on clinical work, I won't go into performance metrics that are more operational in nature and might not have as clearly defined metrics. When seeking to improve clinical care, performance metrics are typically well defined, with established inclusion and exclusion criteria, and QI practitioners would do well to identify and use those performance metrics.

Balancing metrics are ones that are selected to ensure that an improvement in one area does not negatively impact another. A classic example of metrics that balance each other in the acute care setting is readmissions and 30-day survival rates. With the implementation of financial penalties in 2012 associated Hospital Readmission Reduction Program (HRRP), there was concern that efforts to ensure that patients were not readmitted would increase or decrease 30-day survival rates (patients surviving 30 days after discharge). One could hypothesize that a hospital might enact an intervention to divert an admission that would occur within 30 days after discharge, and this might result in the death of the patient. A 2018 study found an association of the HRRP program with reduced 30 and 1-year readmission rates and increased 30 and 1-year mortality rates.[5] While this relationship has yet to be shown as causal, it highlights to need to ensure that balancing metrics are chosen carefully to ensure that an intervention does not have unwanted negative consequences to overall quality of care.

Having spoken generally about the selection of outcomes and balancing metrics as a way of addressing the potential positive and negative impact of quality improvement projects on patient outcomes, now we will get into the practical steps of implementing a quality improvement project.

Quality versus process versus performance improvement: Words matter

Before we begin, I'd like to address the semantic difference between three terms that can be perceived to have very different, but overlapping definitions:

- Quality improvement—refers to activities that bring about immediate improvements in patient care and outcomes.
- Process improvement—refers to activities that are focused on process versus people.
- Performance improvement—sometimes perceived to be focused on the performance of people, but can also be defined as an activity that brings about immediate improvements in performance metrics (Which can be quality metrics)

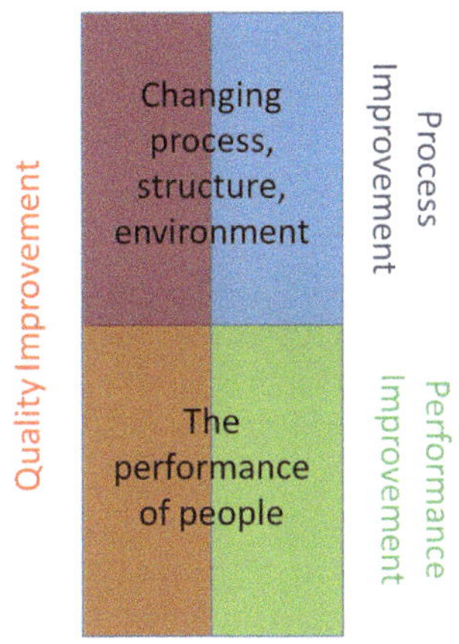

"Quality improvement" is the least exclusionary of all the terms and is the one that will be used here. However, it should be noted that process improvement should be the primary method by which quality improvement is achieved. As noted in the seminal work "Crossing the Quality Chasm,[6]" quality problems typically occur "not because of a failure of goodwill, knowledge, effort, or resources devoted to healthcare, but because of fundamental shortcomings in the ways care is organized. The nation's current health care system often lacks the environment, the processes, and the capabilities needed to ensure that services are safe, effective, patient-centered, timely, efficient and equitable." Process improvement should lead to performance improvement because it will be easier for people to do the right thing, which leads to the best outcomes.

Given that process improvement is the fundamental goal, let's examine two hypothetical quality improvement projects conducted by two clinics serving similar patient populations.

Clinic A is 2 months away from submitting its end-of-year quality performance review. It notices that colon cancer screening rates are well below target … only 22% of eligible patients have been screened for colon cancer. They decide to implement a competitive push to increase colon cancer screening rates before the end of the year to 35%. They split the eligible patient population that has not yet been screened into four groups, assign each of the groups to a nurse and a physician, and challenge the four dyads to engage as many patients as possible in colon cancer screening. The dyads will do this by educating patients on the benefits of colonoscopy and offering to schedule their appointment in the next month. They announce that the winning dyad will receive three additional paid time off days. They decide to use a balancing metric of staff satisfaction to ensure their project is not overburdening staff.

Clinic B is 2 months away from submitting its end-of-year quality performance review. It notices that colon cancer screening rates are well below target … only 22% of eligible patients have been screened for colon cancer. They ask patients who have not yet been screened why they haven't been screened. The top three reasons for not getting it done are fear of the colonoscopy and lack of time to schedule the colonoscopy. The clinic decides to offer fecal immunochemical testing as an alternative to colonoscopy. When providers note that they cannot speak to the pros and cons of the two different types of screening, the clinic implements an in-service and plans on repeating the in-service yearly in collaboration with the local American Cancer Society representative. They ensure that the order for the FIT and patient educational material comparing FIT to colonoscopy is linked to the advisory that pops up when a patient is due for colon cancer screening, and set a target of 35% or more screened for the next year, implementing a monthly monitoring process. They decided to use a balancing metric of staff satisfaction to ensure their project is not overburdening staff.

Of the two cases above, which clinic focused on the performance of people and which focused on the process? Which do you think will be most successful in engaging patients to complete colon cancer screening in the long term? Why?

While Clinic A is more likely to show better performance than Clinic B in its end-of-year quality performance review, the long-term prognosis for better performance is more likely to lie with Clinic B. Clinic A has focused on provider performance … how many patients can they engage? Clinic B has focused on process change … how and what manner of screening is provided and how it fits into their workflow. Clinic B has also taken the time to understand the reasons for low patient engagement in screening … which brings us to discuss the process of quality improvement.

PDSA versus PDCA versus DMAIC: Frameworks matter

The two most common techniques for quality improvement are PDS/CA and DMAIC. PDSA and PDCA are often used interchangeably, but the C versus S does represent a fundamental difference such that PDSA is the more current and recommended technique. Dr. Edward Deming considered the "father of the quality movement" is typically attributed as the founder of the PDCA and PDSA cycles, although a very interesting history of the evolution of the PDSA cycle speaks to how the Deming cycle became the PDCA and eventually PDSA cycles.[7] The difference in the two is semantic. Plan, Do Check/Study, Act is a framework that describes the steps of quality improvement. Check implies that all the practitioner is doing is checking to ensure that the desired outcome has been achieved. As long as the desired outcome is achieved, then one can proceed to Act. Study implies a more analytical process, in which one determines if the outcome has been achieved, but also asks what can be learned from the results, presumably for the purpose of continuous improvement.

Let's take the example of Clinic A, where an incentive program was implemented to increase colon cancer screening rates. Assume that a month after the incentive program begins, screening rates have increased from 22% to 36%. Following the PDCA cycle, a check would indicate that the incentive program was successful and that they should move into Act—implement a continuous incentive program to ensure that screening rates are always above 35% and set ever-increasing targets. However, assume Clinic A does not stop at a check and that they ask the clinical teams how they are feeling about the process and why it's working. They might find that the most successful team has been staying late several days out of the week to make phone calls after their regularly scheduled visits are complete. Perhaps that team is also harboring resentment of their colleagues, who do not stay late to ensure their patients are well informed and this is creating tension in the office. A less successful team might tell the story of calling patients who tell them that they can't get transportation or time off to attend a lengthy screening appointment and are frustrated because they have no alternatives. The team might conclude that, while the program is successful at quickly increasing screening rates, it's an unsustainable process that is creating frustration within the team.

While many organizations use PDCA without adhering to the concept of "just check, don't analyze" the difference between the two terms does raise the issue of understanding why frameworks are used, and which ones will work best in a clinical environment. Frameworks are used so that we don't miss steps and so that we know that there's a path forward that needs to be taken. Imagine if Clinic B spent an entire year planning their quality improvement project—knowing that PDSA is a process, and that plan is just one part of an entire project can help to set expectations ... at some point, planning has to end, and doing must begin. Likewise, organizations can forget to study ... they implement an intervention and assume all is well unless someone tells them otherwise. Because frameworks are used to ensure that all the right steps are taken, it is my preference to use DMAIC. DMAIC, PDSA, and even the scientific method all describe the same steps. DMAIC spells them out more discreetly, and for that reason, it's my preferred method of engaging teams in quality improvement. As you can see from below the Figure, the "Plan" portion of PDSA contains three sections of DMAIC and the scientific method. It's not that teams that use PDSA will forget to measure performance, do research, or understand the root cause of variation, it's just that the framework does not explicitly call it out. Because it's not explicitly stated, it's easy to skip when doing a PDSA.

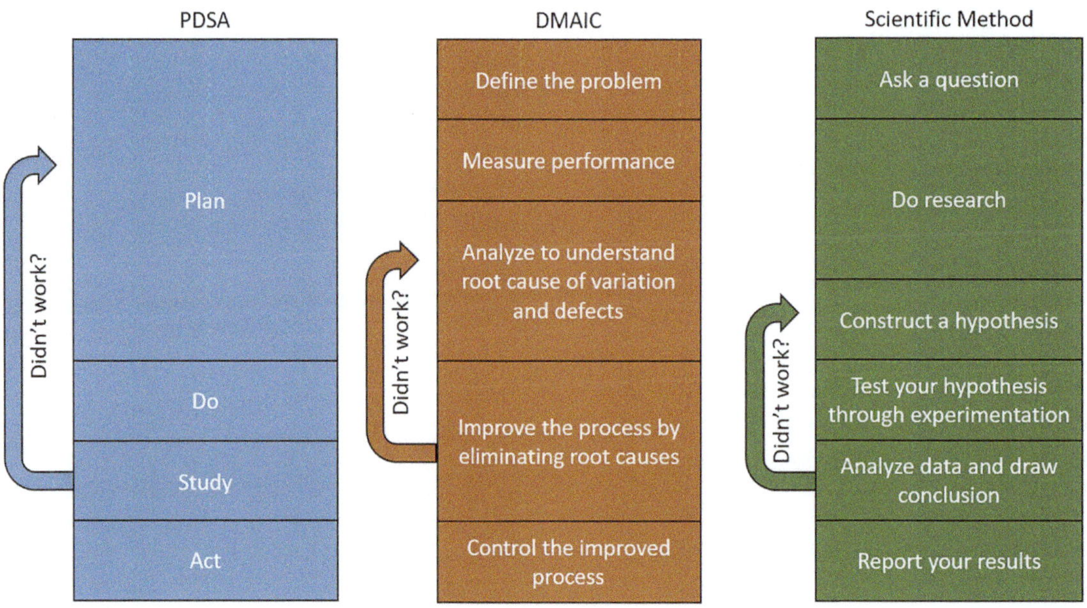

Let's take an example of Clinic X in a hypothetical city in the USA. The practice manager of Clinic X attends a seminar on the value of Fit testing to increase colon cancer screening rates and thinks it's a great idea. The practice manager returns, with a toolkit, templates for the electronic health record, and plans with the clinical team on how to implement. Three months later, they move from Plan to Do, and their new Fit testing process has been implemented, at significant resource costs to the organization, as time and money has been spent to implement a new workflow. Clinic X's team is thrilled to implement this new process and looks forward to seeing improvements. "How much will we improve?" asks a technician. "10%?" says the practice manager. The technician shakes his head, "But … we're already at 95% … and the only reason the 5% aren't showing up as having completed testing is because we haven't entered their data into our electronic health record yet … it's data entry staff we really need"

Now that's an unlikely hypothetical example (I'd like to learn from the clinic that has a 95% colon cancer screening rate), but I've seen quality improvement projects get started with planning, only to have someone say "so how big is our problem, really?" and then find out that the problem wasn't as big as they thought. This is why the steps of DMAIC are so valuable. But in reality, there are unique aspects to each of the three frameworks that add value to any clinical quality improvement. If I could combine them all, the steps would look like this:

(1) Define the problem
(2) Measure performance
(3) Analyze to understand the root cause of variation and defects
(4) Research to learn how others have addressed the issue and what best practice is
(5) Identify process changes that would address root causes and implement best practice and hypothesize how this will affect performance and balancing metrics
(6) Test your hypothesis by implementing your process change—use process metrics
(7) Study: Analyze data and draw conclusions
(8) Control the improved process
(9) Report your results

A DIY framework for quality improvement: DMARITSCR

It will never win any awards for the catchy acronym of the year, but the nine-step process that combines the best of PDSA, DMAIC, and the scientific method does capture critical elements of quality improvement that really should not be skipped. Let's go back to Clinic B and how they implemented fit testing.

Define the problem

Clinic B knows that the percentage of eligible patients screened for colon cancer isn't meeting the target. They need to get at least 35% of their patients screened.

Measure performance

Here, Clinic B defined their problem using their performance metric of % patients screened. This isn't always the case … sometimes a team has a gut feeling that something isn't right. They may feel like wait times are long, but not know for sure because they haven't measured it.

Analyze to understand the root cause of variation and defects

Clinic B knew that only 22% of their eligible patients had been screened. At the time, they may not have known why the other 78% were not screened. To understand the root cause of variation, they would need to ask patients who had not been screened why they had not been screened. They could do this during regular visits where colon cancer screening is discussed, or if they wanted to get the information faster they could hold a focus group[8,9] or send out a survey to eligible patients who have not completed colon cancer screening. Once they've received back data, they would want to analyze their data to understand what some of the root causes of not getting colon cancer screening are in their patient population.

Research to learn how others have addressed the issue and what best practice is

Once you have some idea of the root cause, it's a good idea to do research on how this issue may have been addressed elsewhere. If fear of colonoscopy and the prep involved is a barrier, a quick internet search of alternatives to colonoscopy will reveal viable alternatives that have been shown to be as effective as colonoscopy when done regularly.[3] You might want to go to sites like the American Cancer Society to see how they've implemented workflows to encourage regular Fit testing and find the FluFIT program, that integrates FIT testing with Flu vaccines to ensure that FIT tests are done annually.[10]

Identify process changes that would address root causes and implement best practice and hypothesize how this will affect performance and balancing metrics

Depending on what you've found during your research, you may choose to use an existing toolkit or develop your own, depending on your patient population. In the case of Clinic B, they implemented a process change to offer FIT testing but chose not to implement the FluFIT program, as most of their eligible patients had already gotten flu vaccines. They decided to select staff satisfaction as a balancing metric because they were worried about the burden that the new workflow would put on staff. They did not feel that patient satisfaction was likely to be impacted. Perhaps during their focus groups, they found that patients would appreciate having an alternative to colonoscopy.

Test your hypothesis by implementing your process change—Use process metrics

Once you've decided on the process change, now is the time to implement. One oft-forgotten step of implementation is selecting a process metric. Process metrics are metrics that are used to define if a process is working as it should. A process metric should answer the question "how will we know we are doing what we said we would?" Do not be fooled into thinking that your outcome metric can double as your process metric. If colon cancer screening rates increase, does this mean that your process is being implemented as you said you would? Not necessarily. Perhaps a highly visible celebrity passed away from colon cancer and spurred patients who were on the fence about screening to get their screens done. You'd be surprised how much time you can put into developing a new process and then when the team is ready to put it into place, the demands of the day and stressors of patient care can cause teams to slip back into business as usual. Then weeks later, when your outcome metric hasn't changed, you're left wondering if it's because the new process isn't having the impact it should, or if you're not implementing the new process at all. A good process metric might be the percentage of times a physician offered a patient fit testing if he declined the colonoscopy.

Study: Analyze data and draw conclusions

Don't wait too long to analyze data. Analyze your process metrics early and often to ensure that your new process is being implemented as it should. Understand how long it might take for outcome metrics to change. In the case of FIT testing, a kit has to be given to the patient, and then the patient has to return the completed FIT test, which then needs to be sent to a lab. It might take two to 3 weeks from when the patient is given the FIT test to have a result back in the electronic health record. As it's the result that triggers the patient to be marked as completing screening, from the kick-off of the new project, to when you might see changes in outcome metrics could take 1–2 months. As the timeline for expected changes draws near, look at your data and draw conclusions.

Control the improved process

Now that you've identified a process that gives you the results you want, it's time to think about controlling that process. How do you make the process permanent and monitor whether or not it's taking place? It's easy to assume that a process that's working will continue to work indefinitely. But what if staff turns over and new staff don't know about FIT testing as a viable alternative to colonoscopy? The control phase is about ensuring that the new, better process is sustainable and error proof. In the case of FIT testing, changing the electronic medical record to provide a best practice advisory that suggests every 10-year colonoscopy or annual fit testing and includes a link to additional information for the patient and clinician might be a way of hardwiring the FIT testing process.

Report your results

It's easy to forget this step. Once you've gone through the steps above, isn't it enough to rest on your laurels and pat yourself on the back for a job well done? Why bother taking the time to write up your study and share it with others? The answer to this question lies in what you did above when you did research to understand best practices. It's always helpful to learn from the experiences of others. The more we can share our quality improvement work with others, the easier it becomes for those after us to get the best outcomes for their patients.

References

1. Department of Health and Human Services. *Code of Federal Regulations: Protection of Human Subjects*; June 23, 2005. http://www.hhs.gov/ohrp/humansubjects/guidance/45cfr46.htm#46. Accessed December 5, 2021, 102, section 46.102(d).
2. Chen H, Lu M, Liu C, et al. Comparative evaluation of participation and diagnostic yield of colonoscopy vs fecal immunochemical test vs risk-adapted screening in colorectal cancer screening: interim analysis of a multicenter randomized controlled trial (TARGET-C). *Am J Gastroenterol*. August 2020;115(8):1264–1274. https://doi.org/10.14309/ajg.0000000000000624. .
3. Rex DK, Boland CR, Dominitz JA, et al. Colorectal cancer screening: recommendations for physicians and patients from the U.S. Multi-society task force on colorectal cancer. *Am J Gastroenterol*. July 2017;112(7):1016–1030. https://doi.org/10.1038/ajg.2017.174. .
4. Baily MA, Bottrell M, Lynn J, Jennings B, Hastings Center. The ethics of using QI methods to improve health care quality and safety. *Hastings Cent Rep*. 2006;36(4):S1–S40. https://doi.org/10.1353/hcr.2006.0054.
5. Gupta A, Allen LA, Bhatt DL, et al. Association of the hospital readmissions reduction program implementation with readmission and mortality outcomes in heart failure. *JAMA Cardiol*. 2018;3(1):44–53. https://doi.org/10.1001/jamacardio.2017.4265.
6. Institute of Medicine (US) Committee on Quality of Health Care in America. *Crossing the Quality Chasm: A New Health System for the 21st Century*. Washington (DC): National Academies Press (US); 2001, 1, A New Health System for the 21st Century. Available from: https://www.ncbi.nlm.nih.gov/books/NBK222273/.
7. Moen R. *Foundation and History of the PDSA Cycle*; June 2020. https://deming.org/wp-content/uploads/2020/06/PDSA_History_Ron_Moen.pdf. Accessed December 21, 2021.

8. Cancer Research UK, *Using Focus Groups to Gather Patient Insight*, https://www.cancerresearchuk.org/funding-for-researchers/patient-involvement-toolkit-for-researchers/planning-your-patient-involvement/choosing-your-patient-involvement-method/using-focus-groups-to-gather-patient-insight#focus_group_plan2, Accessed 22 December 2021.
9. The University of Mississippi. *Office of Institutional Research, Effectiveness, and Planning, Guidelines for Conducting a Focus Group*; 2005. https://irep.olemiss.edu/wp-content/uploads/sites/98/2016/05/Trinity_Duke_How_to_Conduct_a_Focus_Group.pdf. Accessed December 22, 2021.
10. American Cancer Society. *FluFIT Implementation Guide*; 2014. https://www.aachc.org/wp-content/uploads/2019/03/american-cancer-society-flufobt-program-implementation-guide-for-primary-care-practices.pdf. Accessed December 22, 2021.

Chapter 103

The importance of teams

Chen Shaw-Ree
Oak Orchard Community Health Center, Brockport, NY, United States

Key point

Engaging in any large body of work requires planning, communication and collaboration between stakeholders. The principles of project management organizes a body of work into a clear plan. Assessments such as DISC and Myers Briggs inform a team how to adapt their behaviors with others.

Why it matters

There is unlikely to be any situation within clinical translational research in which an entire project will only require the collaboration of a few, like-minded people. Utilizing project and people management tools can keep a large team focused on project goals.

It is impossible to complete the kind of complex work that clinical translational research requires alone, or even with a group of people within a single discipline. From concept to research to application, the skills, knowledge, and abilities required for each step require specialized training and experience. As important as this work is to achieve positive outcomes in population health, training and education provided to researchers and clinicians on teamwork and leadership is minimal and of questionable effectiveness.

While the scale of translational work required to get from concept to application is immense, that does not mean the work that a team sets out to do must be that way. While this chapter is about teams, it must also be about project and people management, because the tools of project management facilitate seeing a project through to completion while maintaining collegial and respectful relationships.

Project management

Project management is a collection of techniques for proposing, planning, implementing, managing, and evaluating projects, while at the same time, managing people. It has been noted that translational teamwork is different from traditional teamwork such as those in corporate, information technology, or manufacturing because translational teamwork has permeable boundaries—the final milestone may be to improve population health, but how that comes about can evolve as the project evolves because the work exists at the boundaries of what is known.[1] For that reason, it may seem that managing a translational project will not be as concrete or straightforward as a construction or software deployment project. However, projects are defined as unique, temporary endeavors that are undertaken to achieve planned objectives. Planned objectives can be outputs, outcomes, or benefits. While project management is most often used for those projects where the output or outcome consists of a known series of concrete steps, the tools of project management were designed to be used whether or not the path to the outcome is straightforward. In fact, the less well known the path, the more the tools of project management aid in keeping the team moving forward, avoiding the pitfalls of dysfunctional teams.[2]

It should be noted that the science and methodology of project management is a specialty unto itself. Ideally, translational teams should avail themselves of a project manager with experience, a portfolio of work that demonstrates success in working with teams of disparate specialties, and who can demonstrate the emotional intelligence needed to manage what will undoubtedly be a group of highly intelligent individuals with diverse perspectives. There is a certification in project management called the Project Management Professional (PMP) that requires passing a 4-h exam covering project management skills and abilities. However, more important to translational teams is the portfolio of work. Passing the PMP requires significant resources devoted to studying for and passing—not all successful practitioners of project management may be able to obtain certification, but still have a significant amount of experience.

Translational Cardiology. https://doi.org/10.1016/B978-0-323-91790-2.00010-1

In the event that your translational team is unable to engage a project management professional, aspects of project management can still be applied by any one or several leaders of the project team and are described below based on the phases of a project.

Project initiation

During this phase of the project, the team answers the questions "Why are we doing this work?" and "What do we want to achieve?" While a popular method for setting goals is to develop SMART (Specific, Measurable, Attainable, Realistic, and Timely) goals, in translational work, SMART goals are more difficult to define, as new knowledge can require a project to evolve. In this case, CLEAR goals may be more appropriate (Collaborative, Limited, Emotional, Appreciable, and Refinable). The team should also assess if the goals are feasible, ensuring that feasibility testing is completed if unsure. Ensure that all of your stakeholders are in agreement that the objectives are feasible before proceeding.

If the project stakeholders give the green light for the project to proceed, the development of a project charter is a critical first step. There are multiple templates for project charters, the CDC utilizes a template that can be found here: https://view.officeapps.live.com/op/view.aspx?src=https%3A%2F%2Fwww2a.cdc.gov%2Fcdcup%2Flibrary%2Ftemplates%2FCDC_UP_Project_Charter_Template.doc.

Project planning

During this phase of the project, the team develops the roadmap for the project by developing the project plan, schedule, and constraints. Project plans can be developed using the following steps:

(1) Understand the scope of the project by establishing the milestones or deliverables that should be met to achieve the goals

(2) Develop a work breakdown structure by defining the tasks that are needed to reach the milestone or produce the deliverable and ensuring that an individual or team is assigned to the task.

(3) Establish the project schedule by reconciling any existing deadlines against reasonable amounts of time that will be necessary to perform the work. Build in task dependencies as necessary. These connect tasks that cannot be done unless another task is finished. This step requires input from all team members and subject matter experts.

Development of the project plan is a collaborative task. The project plan should exist in a place where it can be reviewed at any time by any team member. There is also a balance between a project plan with enough work breakdown structure vs. one with too much work breakdown, where the plan accounts for tasks at too detailed of a level. A good rule of thumb is that all project members should be able to see the project plan.

Project execution/monitoring

With a well-done project charter and plan, the stage is set for the team to be accountable and attentive to results. But as Mike Tyson famously said, "Everybody has a plan until they get punched in the mouth." How the team reacts to the punches determines whether or not the team will be successful.

As a team embarks on a project plan they are converting a theoretical discussion about work into a practical implementation of their plan. During this time, issues can arise or new information can make a previous assumption irrelevant or plain wrong. Frequent check-ins can ensure that the team follows some critical principles of project execution

(1) Following processes—regular review of the project plan ensures that agreed-upon processes are followed.

(2) Managing people—highlight when milestones are met and celebrate incremental victories.

(3) Distributing information—check-ins should address when work is not going according to plan, or when punches to the mouth have occurred or are feared. It is important to foster an environment where disagreement is encouraged. Healthy disagreement is valuable to assessing risk. Facilitators of the check-in should make space and encourage disagreement, as long as it stays at the level of process and does not become personal.

When the three principles of execution are followed in a respectful manner, teams can avoid mistrust, conflict, and losing sight of why they committed to the project in the first place.

Project closure

This phase occurs when the project is complete. At this time, contractors hired to work on the project are terminated, and the project manager typically conducts a "postmortem" to evaluate what went well and identify project failures. This can be helpful to document lessons learned so that improvements can be made for future projects. At this time, it is also important to note any tasks that were not accomplished and to make plans for completing them in a follow-up project, or to note why they were not considered necessary to this project. All project documents and deliverables should also be collected and stored in a single location for posterity.

Team dynamics

Above, you may have noticed that I mentioned that when the three principles of execution are followed in a respectful manner, teams can avoid mistrust and conflict. One of the challenges of working with diverse groups of people is that "respectful" means different things to different people. One person might think respect means not questioning the decision of a subject matter expert. Another may feel that it is respectful to get clarity through asking questions. Two common tools to manage team dynamics include DiSC[3] and Myers Briggs.[4] These require payment to use, although there are free versions of each of these that can be accessed online.

DiSC

DiSC originated with William Moulton Marston, who published the Marston Model of the Four Dimensions of Behavior. These evolved into the four dimensions of DiSC: Dominance, Influence, Steadiness, and Conscientious.

- Dominance: People with D personalities tend to be confident and place emphasis on accomplishing bottom-line results
- Influence: People with I personalities tend to be more open and place an emphasis on relationships and influencing or persuading others
- Steadiness: People with S personalities tend to be dependable and place an emphasis on cooperation and sincerity
- Conscientious: People with C personalities tend to place emphasis on quality, accuracy, expertise, and competency

DiSC has been used by organizations to improve teamwork, communication, and productivity. Knowing how one's own DiSC style interacts with another team member's DiSC style can help team members understand how they respond to conflict, what causes stress, and how they best solve problems and communicate with others.

Myers Briggs Type Indicator

Myers Briggs Type Indicator (MBTI) was constructed by Katharine Cooke Briggs and Isabel Briggs Myers in the early 1900s and is based on Carl Jung's theory that people experience the world using four principal psychological functions, sensation, intuition, feeling and thinking, and that one of these four is dominant for a person most of the time.[5] The development of MBTI emphasizes that individuals have preferences for how the experience the world and that these preferences underlie a person's interests, needs, values, and motivation.[6]

The MBTI looks at these four key areas and like DiSC, has been used by organizations to improve teamwork, communication, and productivity. Knowing how teammates get energy, take in information, and organize their time and environment can increase understanding of why team members say or do certain things.

- How individuals get their energy (extraversion vs. introversion)
- How individuals take in information and learn (sensing vs. intuition)
- How individuals make decisions (thinking vs. feeling)
- How individuals organize time and environment (judging vs. perceiving)

Get started

Set clear goals and timelines, then start defining the work and resources that will be needed to meet both.

Potential pitfalls

Projects can fail for many reasons, but some common pitfalls include.

(1) Not adequately engaging stakeholders (which can derail a plan because they introduce objections to the plan late in the game),
(2) Setting unrealistic deadlines and finding yourself needing to continually push your deadlines back, which can cause your resources run into scheduling problems,
(3) "Scope creep" where your project seems to continually expand and take on new goals—leading to a project that never ends.

Resources

CDC example template for a project charter: https://view.officeapps.live.com/op/view.aspx?src=https%3A%2F%2Fwww2a.cdc.gov%2Fcdcup%2Flibrary%2Ftemplates%2FCDC_UP_Project_Charter_Template.doc.

Project management tools: Smartsheets is an excellent product for project management. While opinions may differ, it is relatively easy to use compared to Microsoft Project, and allows for collaboration with teammates.

References

1. Begerowski SR, Traylor AM, Shuffler ML, Salas E. An integrative review and practical guide to team development interventions for translational science teams: one size does not fit all. *J Clin Transl Sci.* 2021;5(1):e198. https://doi.org/10.1017/cts.2021.832.
2. Sutton L, Berdan LG, Bolte J, et al. Facilitating translational team science: the project leader model. *J Clin Transl Sci.* 2019;3(4):140–146. https://doi.org/10.1017/cts.2019.398.

3. Personality Profile Solutions. LLC; https://www.discprofile.com/. Accessed 28 December 2021.
4. The Meyers-Briggs Company. https://www.mbtionline.com/. Accessed 28 December 2021.
5. Jung CG. In: *Psychological Types*. Vol. 6. Princeton University Press; 1971.
6. Kaplan RM, Saccuzzo DP. *Psychological Testing: Principles, Applications, and Issues*. 5th ed. Wadsworth/Thomson Learning; 2001.

Chapter 104

Patent basics

Kelley N. Benck[1], Alex Smith[1], Joseph Lamelas[2] and Ahmed Alnajar[3]

[1]The University of Miami Miller School of Medicine, Miami, FL, United States; [2]Department of Cardiac Surgery, The University of Miami Miller School of Medicine, Miami, FL, United States; [3]The Division of Cardiothoracic Surgery, The Department of Surgery, The University of Miami Miller School of Medicine, Miami, FL, United States

Key points

- Protections for intellectual property have different types that researchers should be familiar with.
- Building a strong portfolio when preparing a patent application can serve as a defensive tool to protect the invention and become a source of revenue in the future through license or sale.
- To streamline the process, researchers should understand the best practices for communication and submission of a patent application.

Why it matters?

Patents are important, especially in a rapidly evolving field, as they help safeguard our inventions. While a patent can protect any product, medical devices and tool designs should meet certain specifications according to their originality, practicality, suitability, and utility. In most cases, a patent can protect an invention for up to 20 years. In the current era of minimally invasive procedures, researchers and inventors should be familiar with the process that can protect their rights and streamline their inventions from ideas to products used in a competitive market.

Introduction

The protection of intellectual property is a major driver of innovation within the United States and around the world.[1] One subtype of protection, patents, has been especially influential in biomedical fields[2] such as cardiology. Started in 1790 and established through the US Constitution, the patent system was in part designed to incentivize the sharing of knowledge by offering protections to inventors and their work for a window of time. The concurrence of invention disclosure to the public and acknowledging ownership of the inventor's work increases the publication of useful, novel information to further discovery and innovation.[3]

Patents have significant utility in specialties such as cardiology due to the benefits they provide patients. Innovators hoping to improve patient outcomes with their inventions need the ability to produce on a large enough scale to reach a high volume of patients, and commercialization is one of the best methods to achieve this goal. However, commercialization is often not feasible without the help of investors, and many investors consider unprotected inventions to be uninvestable due to the risk of a competitor replicating the technology before the investor can capitalize on their investment. Therefore, inventors aspiring to utilize their inventions to better the lives of cardiology patients should consider the use of a patent.

Within the United States, patents are granted and regulated through the United States Patent and Trademark Office (USPTO). This body not only ensures the legitimacy and fairness of potential patents, but also requires novelty and utility for the betterment of the innovative process.[4] Applying for a patent can be complex, requiring multiple entities such as registered attorneys and agents, and involving many components such as an understanding of the types of patents available, conducting thorough patent searches, and planning for timelines and budgets. The goal of this chapter is to provide an explanation of the patent system and the process involved in applying for a patent.

Options to protect intellectual property

Though the focus of this chapter is patents, it is first important to understand the various options and categories available to protect one's intellectual property. Table 104.1 has the basic descriptions of the three main types of protections—patents, trademarks, and copyrights.

Translational Cardiology. https://doi.org/10.1016/B978-0-323-91790-2.00049-6

TABLE 104.1 Other forms of intellectual property.

	Trademark/Servicemark	Copyright	Trade Secret
Protection	Word, phrase, name, symbol, design, service—or a combination of those—which denotes a particular product or brand *Note: provides ownership of the association between the work, name, or phrase and the product or service*	Authorship or original literary, musical, theatrical and/or other forms of intellectual and artistic works that are either published or unpublished	When an invention is not disclosed in a patent or copyright, rather it is held by a company or individual as a secret
Benefit	Distinguish one product or service from another, but do not prevent the production or sale of similar products and services as long as they are under a different trademark or servicemark	Assume or claim ownership that allow filing a copyright infringement lawsuit.	There is no time limit (unlike 20-year period of protection with a patent) *Note: if secret become public, it is no longer owned by the company, also someone else may patent to same invention before it is made public*

Patent

A patent is used for the protection of technical inventions. Once one is granted, it excludes "others from making, suing, offering for sale, or selling" such item within the United States or through importation to the United States.[5] Enforcement of the patent is under the responsibility of the patentee, not the USPTO. Terms usually last 20 years from the application date and patents granted by USPTO are only effective within the United States and its territories and possessions.

There are three subtypes of patents available[6]:

1. *Utility patents*: protect functional or structural features of inventions such as processes (of a technical or industrial background), compositions of matter (e.g., chemical compounds), machines, and products of manufacture.
2. *Design patents*: design is defined as the "surface ornamentation" of the invention, so this patent subtype only protects the appearance of the invention.
3. *Plant patents*: inventions or discoveries of a new, distinct variation of the plant; the plan must be reproduced asexually.

On average, the total filing process for a patent takes between 18 and 24 months. The USPTO website offers updated averages on the length of total pendency (the time it takes from application filing to final disposition).

Intellectual property protection in cardiology

Within the field of cardiology, most inventors are seeking protections for devices or drugs, making patents—often utility patents—the most frequent protection. A preliminary search of patents using the terms "cardiology OR cardiac OR cardiovascular" yields 1.6 million results. Even limiting the scope to the top 10,000 patents in the field shows significant activity in both the device space and pharmaceutical space (Fig. 104.1). This may be overwhelming for physicians new to medical device innovation. However, despite the vast number of patents already in the space, there is often room to protect a novel combination of features. Take intracardiac catheters for example. Early intracardiac catheter patents are simplistic and broad; however, modern cardiac catheter patents have much narrower claims to reach the legal requirements of a patent (e.g., novelty) (Fig. 104.2). Similar evolutions can be seen with the development of the prosthetic aortic valve, where early patents often contain a single claim,[9] compared to prosthetic valve patents of the 21st century often containing over 20 claims.[10] Breaking into such saturated fields as cardiology can be daunting for new innovators. The subsequent sections describe the framework needed to build a sufficient patent.

Patents

Building a portfolio

Once an inventor has determined a patent is the appropriate category of protection for their intellectual property, an important next step involves compiling a portfolio. USPTO requires patents to have utility, be novel, have unobviousness, and include a proper disclosure of the invention,[4] and a portfolio can help inventors ensure such criteria are met. Below are three key components to be included to prepare a strong patent.

Prior art search

A prior art search is useful in helping inventors better understand the field of their patent. For example, an inventor

FIGURE 104.1 Preliminary search of cardiology-related patents correlating box size with proportion of patent topics.

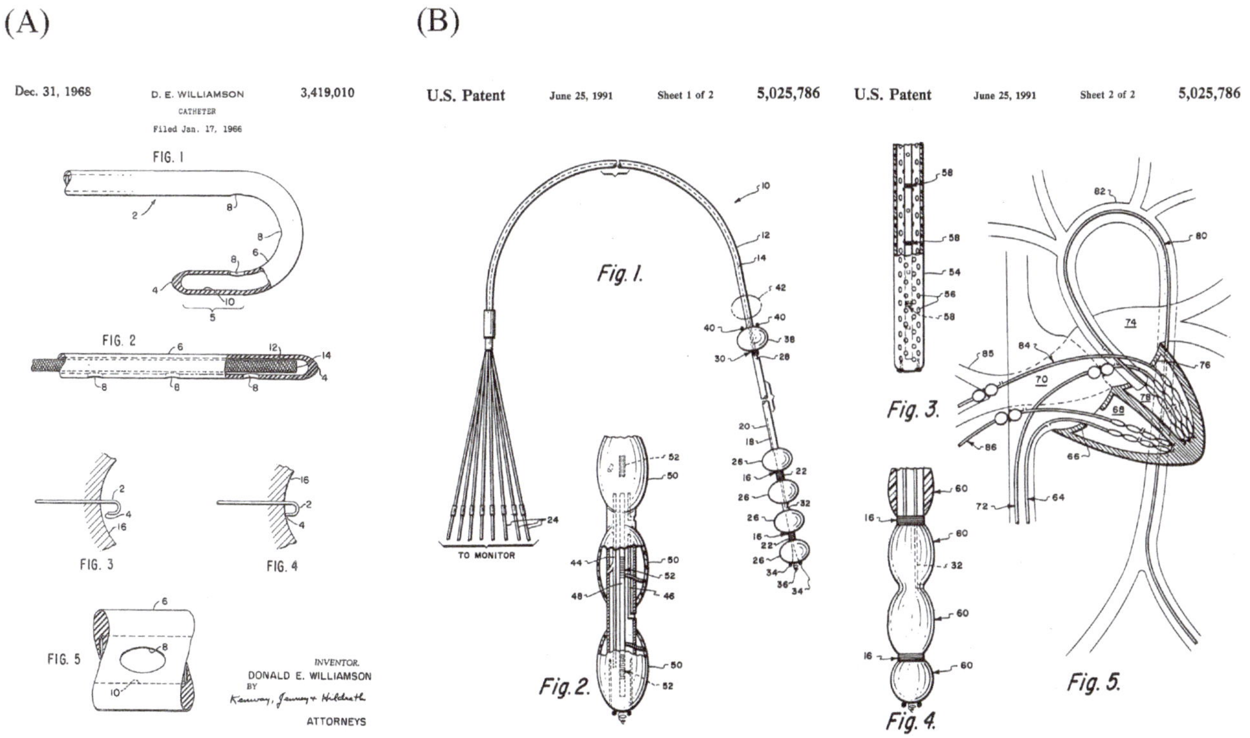

FIGURE 104.2 Evolution of patentability requirements of intracardiac catheters. (A) Patent drawing for intracardiac catheter in the 1960s.[7] (B) Patent drawing for intracardiac catheter created in 1990s.[8]

looking to patent a new cardiac pacemaker device would be able to understand how saturated the field is with similar devices. In addition, a thorough investigation into previous patents can help an inventor further stratify their invention from existing products. Lastly, this search ensures the novelty criteria are being met as an invention should not be previously known or used.[1] Prior art is defined as information accessible through mediums such as previous patent applications, journal publications, social media, websites, or conference presentations.[11] This search should first be done by the inventor and/or their team but will also be completed by a licensed patent official for a fee.

Patentability assessment

A patent search is an important preliminary step to help inventors develop an understanding of the field; however, it is not fully reliable. Patentability assessments take the patent search a step further by completing a detailed analysis of uncovered materials, accompanied by a written report on the invention's patentability.[12] The assessment provides a deeper analysis of the perception of the invention related to previous works, including whether the potential patent would meet more involved criteria such as unobviousness and usefulness. Patentability assessments are more expensive than a prior art search and often require a patent attorney; however, the information gathered may be crucial, especially in more saturated fields such as minimally invasive cardiology devices.

Freedom to operate

A freedom-to-operate (FTO) patent search helps inventors understand their risk of infringing on another individual's intellectual property. FTOs are defined as the ability to produce or commercialize an invention without infringing on previous work or products. The concept of an FTO can be challenging as someone can hold a patent but not be granted an FTO.[13] A simplified example of this can be demonstrated with intraluminal devices. An innovator may be granted a patent for a device that repairs an aortic aneurysm utilizing a stabilizing intraluminal stent. Though this device is patentable—fulfilling the requirements of novelty, utility, and nonobviousness—if a patent already exists describing a metal, cylindrical intraluminal device maintaining vessel patency (i.e., coronary stent) then the innovator is unable to commercialize their product due to patent infringement. One way to think about this is as though two patents are required to build a device, but the two patents are held by different companies. A common approach to solving this before patenting is to craft the aims to avoid competing patents. If an inventor is already granted a patent and later finds they do not have FTO, they can negotiate a licensing agreement to gain access to the patent preventing their FTO. FTOs are often completed by patent attorneys and can help mitigate legal risk and help refine patent applicants to have better chances of later production and commercialization. FTO patent searches usually cost more than patentability assessments, so additional budget considerations may be required with their use.

Path to protection

Medicine-related patents typically go through a more vast protection process within the Patent Cooperation Treaty (PCT) system managed by the World Intellectual Property Organization (WIPO).[14] The PCT is an international system that helps inventors seek protections within multiple countries with a single application filing. There are three main phases to this process (Fig. 104.3), which are described in more detail below.[15]

Initial patent filing

The first step in the PCT process is to file an application. This is often completed by the submission of a provision patent application, which is granted through the USPTO, setting a priority date and protecting an invention temporarily, over a 1-year period, before it is converted into a PCT application. A provisional patent can be thought of as a placeholder for inventors with qualifying inventions and can be especially useful in fast-paced, competitive industries due to the first-to-file rule.[1] The provisional patent application still requires disclosure of information about the invention, but not to the same degree as will be required in the utility patent application.[16]

Provisional applications differ from regular patent applications in a few key ways. First, the registration date for the invention created through the provisional application is earlier than the eventual registration date issued through the regular application. In addition, the provisional patent cannot be searched and will not be published without later being granted and references through a regular patent application.[16]

Though this process can be beneficial, especially if intending to file through the PCT system, there are some considerations to be mindful of. The fees for the provisional patent are less than for a regular patent, which may be beneficial for an inventor needing more time to raise an adequate budget. However, each application contains separate fees, which may not be feasible for all inventor's overall, long-term budget. In addition, time is a consideration as examination of the patent only begins once the regular application is filed, not when the provisional patent application is submitted. Amendments have been made to allow provisional applications to be upgraded to utility patent applications.

In preparing and completing a provisional patent, a provisional patent template can be used to aid in that process. It can provide the foundation for the patent as an

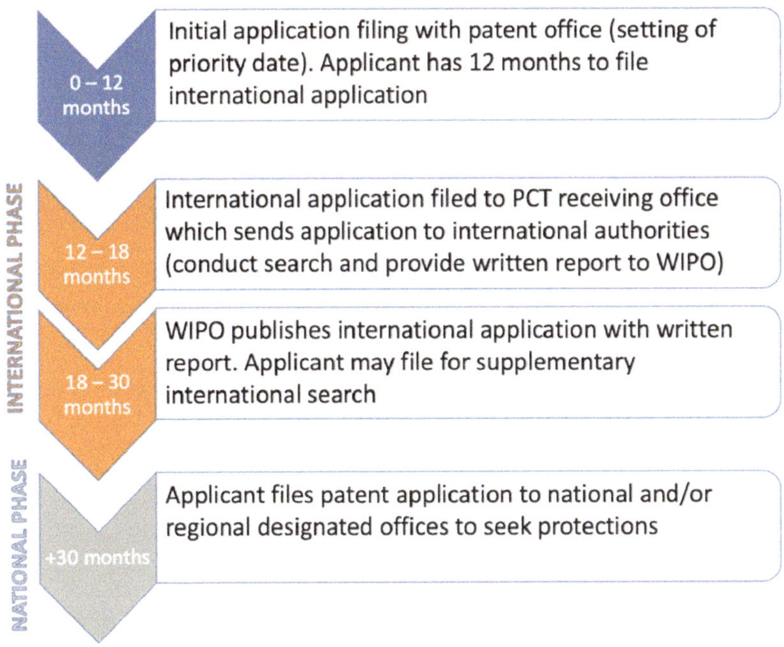

FIGURE 104.3 General timeline (showing months from priority date) for patent filing under the Patent Cooperation Treaty (PCT) system with the World Intellectual Property Organization (WIPO).[15]

inventor prepares for the regular patent application, helping organize required information and ensuring a sufficient provisional application is submitted. The USPTO does not require one specific template but expects a minimum of the following information[16]:

- Completion of the "Provisional Application for Patent Cover Sheet" from the USPTO website
- Inventor name
- Invention title
- Field of invention
- If applicable, references to related applications
- Comprehensive description of the invention
- Background information on the invention, including description of related art
- Drawings and list of figures with brief descriptions
- Minimum of one example of invention's use

Provisional patent templates can be a great resource for helping an inventor organize the components of their work and prepare for the final patent process. However, it is not foolproof or a substitute for adequate research into the complex patent process. It is recommended that inventors either use the assistance of a patent attorney or be well versed in the literature and rules surrounding nonprovisional patent applications.[17]

International phase

As mentioned above, an international application under the PCT system should be filed within 1 year of submitting the provisional patent application. The international application should be submitted to a receiving office, which includes national or regional patent offices or the International Bureau of WIPO.[17] The receiving office then sends the application to the International Search Authority (ISA) and the International Bureau of WIPO. ISA is the body responsible for conducting the international search, similar to a prior art search, but on a much larger and more comprehensive scale. Once the search is conducted, the ISA submits a report, the International Search Report (ISR), to the Internal Bureau about the patentability of the invention. The ISR must be sent within 3 months of the ISA receiving the application or 9 months from the priority date established from the provisional patent application (whichever deadline comes first). Typically, around 18 months after the initial filing, the International Bureau publishes the patent application, allowing for third-party analysis. During this 10-month window, any third party files disputes on the invention, such as observations against the novelty of the patent.[18]

National phase

At around 30 months from the initial filing date or priority date, the application transitions from the International Phase to the National Phase. This stage is when inventors can receive grants from the national or regional patent offices in countries of interest.[17]

Budgetary considerations

Fees for patents vary based on the type of patent being filed, the manner in which one presents the invention in the application, and the experience/characteristics of the law firm or attorney being used (i.e., generally patents for medical devices are best advised by attorneys with a background in engineering). On average, the cost to the USPTO office ranges from $10,000 to $25,000. The USPTO website offers updated fee schedules, which should be conferred before application submission. Required fees involve items such as application filing (nonrefundable), USPTO searches, examination, postallowance (including the "issue fees" if the patent is granted), and maintenance (paid at 3.5, 7.5, and 11.5 years after the patent is granted to maintain legal protection).[15] Optional fees may include extensions of time, trial and appeal, and petition.

Communication of intellectual property

Communication and collaboration are important components to the development of one's intellectual property, from its early stages of creation all the way through the granting of a patent.

The patent process is quite strict, so it is important for inventors to be aware of the rules surrounding documents and sharing their intellectual property. Lab notebooks are a standard documentation tool that can act as a preliminary security for inventors. In addition, in the context of collaboration, more confidential methods of communication (e.g., nondisclosure agreements) are less risky for inventors in the context of the first-to-file rule and allow a more relaxed timeline than communicating through nonconfidential methods. In addition, the steps from patent application to production can be improved by understanding available public and private resources.

Lab notebooks

Lab notebooks are an important tool for record keeping throughout the entire patent process. Its use spans from brainstorming to tracking the evolution of ideas to recording processes. Most importantly, if constructed properly, it has the ability to act as legal documentation of one's intellectual property. In theory, notebooks should record the invention's development so comprehensively that any reader could replicate the process.[19] Below are some methods to compose a valid notebook:

- Notebook setup
  - Use of a notebook that will physically show if any pages are torn out
  - Cover should indicate author's name, institution, contact information (phone number and email), project name, supervisor (if applicable), and department; this information should be copied onto the inside front cover or first page.[20]
  - Initial pages start with a table of contents with subsequent pages being numbered.
  - Spaces reserved after each entry to sign and date work.
- Writing and content
  - Write with a pen (a nonerasable utensil)
  - Dedicate a new page for each new test or experiment with a clearly labeled title and date
  - Data should be entered in real time with a comprehensive list of materials and tools used
  - Persons involved in each entry or experiment should be recorded
  - Clearly label and associate any external references (e.g., online resources, diagrams, pictures) with corresponding pages/entries

Lab notebooks may be completed manually (i.e., pen and paper) or through electronic formats. There are many resources to aid in the electronic creation and maintenance of a notebook to ensure it retains its legal validity.

Nondisclosure agreements

A nondisclosure agreement (NDA), or confidential discloser agreement (CDA), is a legally binding contract where the recipient agrees not to disclose information deemed confidential by the owner. NDAs have many uses, one of which promotes a collaborative environment for researchers hoping to share information while protecting their intellectual property. Although a patent may be a better long-term solution, an NDA can be a useful alternative in the interim for the sharing of unpublished intellectual property. Below are some important components and protections included within an NDA.

- No warranty: protects the owner in the case of any errors in the confidential information if the recipient were to use that information and incur damages. This can be thought of as the owner not warranting the accuracy of the information within the NDA
- Risk of accepting disclosure: in addition to no warranty, this ensures the recipient acknowledges the risk of utilizing any provided information
- Limited license: confines the recipient to only accessing the confidential information for review for future collaboration and not using the information for any unspecified purposes
- General provisions: clearly lay out components of the NDA such as descriptions of the information, purpose for disclosure, termination date, return policy, and more[17]

Public disclosure

Public disclosure is a nonconfidential method of conveying information. Examples include conference abstracts or presentations, publications, websites, product sales, advertising, verbal conversations outside of the inventor's team, or communication conducted outside of an NDA. Public disclosure is important if an inventor is hoping to file a patent in the future, as per US laws, there is only a 1-year period to submit a patent application starting with the first disclosure of information to the public.[21,22]

The use of confidential documents such as NDAs can be beneficial to avoid being constricted to the 1-year timeline. If an NDA is not possible, it is important to be mindful of the amount of information an inventor discloses if hoping to avoid a public disclosure.

Technology transfer

Inventors from the public sector looking to transition their inventions to large-scale production may benefit from the technology transfer system. It is a process to transfer technological information to industry from federal or university research institutions for the purpose of commercial production. This process was first incentivized under the Stevenson-Wydler Technology Innovation Act of 1980 and the Bayh Dole Act, to support the increasing output of technology products and services.[23] It is especially useful for encouraging private industry to accept federally funded work for commercialization. For inventors working through university labs, the use of this system can often be facilitated by the university's patent office.

Getting started

Establishing ownership of work

There are circumstances where multiple parties may be involved in the patent process, so it is important to clearly define roles early on. The inventor is initially the presumed owner of the patent, although it is possible to reassign ownership or establish joint ownership. This distinction of ownership is important as the patent owner, not necessarily the inventor or a collaborator, reaps the benefits of the patent title, rights, and monetary interest. Awareness of ownership is especially important for inventors working at an institution, which may own the full or partial rights of any intellectual property produced by its employees or faculty.[19] Use of a Personnel Intellectual Property Assignment Agreement (PIPAA) can help clarify and establish ownership for all involved parties.

University technology transfer offices

For innovators working in academic settings, it may be beneficial to understand the roles universities play in the patent process. These university departments may go by different names such as Technology Transfer Office (TTO), Technology Licensing Office (TLO), and Office of Technology Commercialization. Each institution runs its office differently, so inventors should be in communication to understand expectations around areas such as filing and licensing autonomy. These departments help the commercialization of patents by facilitating licensing.

Pursuing commercialization

In cardiology, commercialization is the key component to delivering an invention to the bedside. As mentioned above, universities help facilitate this process through the use of a TTO. Many physician innovators are employees of academic hospitals, and as noted, the employer (e.g., hospital or university) will own the filed IP as the assignee. If the physician chooses to pursue commercialization, it is likely that they will need to form a company to license from the university. Negotiating with a university, especially a nonprofit institution, is not like negotiating with other businesses. Creating alignment in the goals and motivations of the university and innovator is imperative to successful product production. As most academic institutions are nonprofits, they are not incentivized to invest in processes of production. However, universities are interested in promoting biomedical innovations produced by their faculty and staff. Therefore, these institutions value licensing. Alignment in the goals of the institution and innovator is important during negotiations, especially in the context of the inclusion of antishelving clauses. These clauses prevent the patent owner (i.e., the university) from licensing a patent to a company with the sole purpose of shelving the product to prevent competition, rather than drive innovation of new products.

Commercialization in academic settings can be a more involved process, so a thorough understanding of roles and expectations increases the chances of successful commercialization.

Patent attorney

The application process can be quite complex, including the need to conduct a thorough patent search. Though clinicians and scientists are competent in the processes of literature searches and composing research manuscripts, the patent process presents different needs. Patent attorneys and agents specialize in this system and can be beneficial in navigating the application process and understanding the language surrounding claims. The inventor is well versed in their respective field, but such legal support can aid in drafting applications, providing feedback to better optimize chances of being granted a patent, managing communication with USPTO examiners, and handling an appeal if the patent application receives a rebuttal. Though the use of an attorney is not necessary and will increase the budget, legal aid may be worth the cost for more complex patents or crowded industries.

Pitfalls to avoid

Late or no use of a lab notebook

To reiterate, lab notebooks are an important record-keeping tool for inventors. Not utilizing this resource early enough in the invention process, or not at all, could lead to missed documentation of intellectual property. This is especially important as lab notebooks can serve a legal purpose for inventors.

The fine line of filing

The first-to-file rule enacted in the United States in 2013 can put pressure on inventors to file quickly. There is validity in making sure another inventor does not submit an application before you; however, there is also a risk of losing the quality of the application. Although it is important to work efficiently and not unnecessarily delay an invention's progress, inventors should avoid filing with only an idea. It is more defensible and more cost-effective to wait until there is data or at least a descriptive drawing of the invention.

The breadth of your claim

In theory, most inventors desire broad claims to possess more vast rights to the invention and future variants. However, most patent offices work to narrow an application's claims, and the patentability of broader claims is more likely to be refused. There is a balance for inventors to not file too narrow of a claim and miss potential rights to their work and to not file too broad of a claim and risk refusal of their invention. In competitive, highly patented markets such as cardiology, finding that balance is important for inventors trying to establish the novelty of their independent claim. One of the most effective ways to defend the breadth of that claim is with the use of data, which can also help narrow the range of the claim, increasing its defensibility. It may be advisable for inventors to utilize the provisional patent period to include and finalize their claim data. Once the provisional application is filed, the time constraints of the 12-month window for regular patent filing may speed up the need for data production at a rate that can cost quality.

References

1. Patino R. Intellectual property rights and research disclosure in the university environment: preserving the commercialization option and optimizing market interest. *J Am Assoc Lab Anim Sci.* 2009;48(2):138–143.
2. Cockburn I, Long G. The importance of patents to innovation: updated cross-industry comparisons with biopharmaceuticals. *Expert Opin Ther Pat.* 2015;25(7):739–742.
3. Tulasi GK, Rao BS. A detailed study of patent system for protection of inventions. *Indian J Pharm Sci.* 2008;70(5):547–554.
4. Elliott G. Basics of US patents and the patent system. *AAPS J.* 2007;9(3):E317–E324.
5. Alexandria V. *General Information Concerning Patents.* United States Patent and Trade Office; 2022. https://www.uspto.gov/patents/basics/general-information-patents. Accessed March 12, 2022. Published.
6. United States Patent and Trademark Office. *1502.01 Distinction Between Design and Utility Patents [R-07.2015]. Patent Laws, Regulations, Policies & Procedures Web site.* https://www.uspto.gov/web/offices/pac/mpep/s1502.html. Accessed 20 February 2022.
7. Williamson DE, Inventor. *Catheter.* 1966.
8. Siegel SB, Inventor. *Intracardiac Catheter and Method for Detecting and Diagnosing Myocardial Ischemia.* 1988.
9. Cromie HW, Inventor. *Prosthetic Sutureless Heart Valve.* 1963.
10. Mcguckin JF, Hinchliffe PWJ, Inventors. *Percutaneous Aortic Valve.* 2001.
11. United States Patent and Trademark Office. *901 Prior Art [R-08.2012]*; 2020. https://www.uspto.gov/web/offices/pac/mpep/s901.html. Accessed March 7, 2022. Published.
12. United States Patent and Trademark Office. *2103 Patent Examination Process [R-10.2019]*; 2020. https://www.uspto.gov/web/offices/pac/mpep/s2103.html. Accessed March 12, 2022. Published.
13. World Intellectual Property Organization. *IP and Business: Launching a New Product: Freedom to Operate.* WIPO Magazine; 2005. https://www.wipo.int/wipo_magazine/en/2005/05/article_0006.html. Accessed March 19, 2022. Published.
14. Berks AH. Patent information in biotechnology. *Trend Biotechnol.* 1994;12(9):352–364.
15. World Intellectual Property Organization. Overview of the PCT system. In: *Protecting Your Inventions Abroad: Frequently Asked Questions About the Patent Cooperation Treaty (PCT).* WIPO; 2020.
16. United States Patent and Trademark Office. *Provisional Application for Patent*; 2022. Published 2021 https://www.uspto.gov/patents/basics/types-patent-applications/provisional-application-patent.
17. Tidwell JL, Liotta LA. Inventions and patents: a practical tutorial. *Method Mol Biol.* 2012;823:391–408.
18. Office USPaT. *1842 Basic Flow Under the PCT [R-10.2019]*; 2020. https://www.uspto.gov/web/offices/pac/mpep/s1842.html. Published.
19. Merges RP. The law and economics of employee inventions. *Harv J Law Technol.* 1999;13(1):1.
20. UM Ventures. *Lab Notebooks: Protecting Your Intellectual Property.* University of Maryland; 2014. https://www.umventures.org/for-inventors/inventor-resources/lab-notebooks-protecting-your-intellectual-property. Accessed March 3, 2022. Published.
21. United States Patent and Trademark Office. *2153.02 Prior Art Exception under AIA 35 U.S.C. 102(b)(1)(B) to AIA 35 U.S.C. 102(a)(1) (Inventor or Inventor-Originated Prior Public Disclosure Exception) [R-10.2019]*; 2020. https://www.uspto.gov/web/offices/pac/mpep/s2153.html#ch2100_d20033_25935_125. Accessed March 7, 2022. Published.
22. Lipkus NB, Mackie JE, Singer PA. Guidance for reconciling patent rights and disclosure of findings at scientific meetings. *Health Res Policy Syst.* 2010;8:15.
23. United States Patent and Trademark Office. *Technology Transfer*; 2021. https://www.uspto.gov/ip-policy/patent-policy/technology-transfer. Accessed March 14, 2022. Published.

Chapter 105

Small Business Innovative Research (SBIR) and Small Business Technology Transfer (STTR)

Syedah Aleena Haider[1] and Jonathan Teng Fai Loke[2]

[1]Department of Internal Medicine, Glangwili Hospital, Carmarthen, Wales, United Kingdom; [2]Department of Medicine, Trinity College Dublin, Dublin, Ireland

Key points

- The National Institutes of Health (NIH) is the world's largest funder of biomedical research, including grants such as Small Business Innovative Research (SBIR) and Small Business Technology Transfer (STTR) which encourage private sector commercialization of technology.
- Unlike SBIR, STTR encourages partnership between businesses and research institutions (RIs), the implications of which should be considered when deciding on the appropriate grant application.
- There are multiple grant eligibility and registration requirements which must be taken into account prior to submitting an application on the 'Application Submission System and Interface for Submission Tracking'.

What is the National Institutes of Health?

The NIH is a part of the United States Department of Health and Human Services, functioning as the largest biomedical agency in the world.[1] It comprises 27 different institutes and centers (Fig. 105.1). Each one has its own specific budget and research agenda, often specializing in a particular disease or bodily system.

NIH grants: An introduction

The NIH is also the world's largest funder of biomedical research, investing approximately $32 billion annually with the aim to 'enhance health, lengthen life, and reduce illness and disability'.[2] In order to achieve these goals, the NIH provides financial support for cooperative agreements, contracts, research grants, and other research-related activities. These activities include, but are not limited to, training, fellowships, scientific conferences, and other aspects of career development.[3]

The need for NIH grants

Historically, early-stage biomedical innovators have found it challenging to attract potential investors due to the high risk and unpredictability of research outcomes.[4] This absence of private investment makes a strong case for public intervention via grants from organizations such as the NIH.

Therefore, the NIH has established the SBIR and STTR grants. Such grants provide federal funding to overcome this relative absence of private investment in the early stages of biomedical research.

In fact, $1.2 billion of the NIH's annual research and developmental (R&D) budget is set aside under the SBIR and STTR programs with the goal of boosting private sector commercialization of technology to increase small business participation in federally funded R&D.[5]

The SBIR programs support Small Business Concerns (SBCs) in the early stages of translational innovation, while the STTR supports the same innovation by partnerships between SBCs and research institutions.

Small Business Innovative Research

The main goal of the SBIR grant is to support R&D in for-profit institutions with ideas that have the potential for commercialization, which will stimulate technological

Translational Cardiology. https://doi.org/10.1016/B978-0-323-91790-2.00078-2

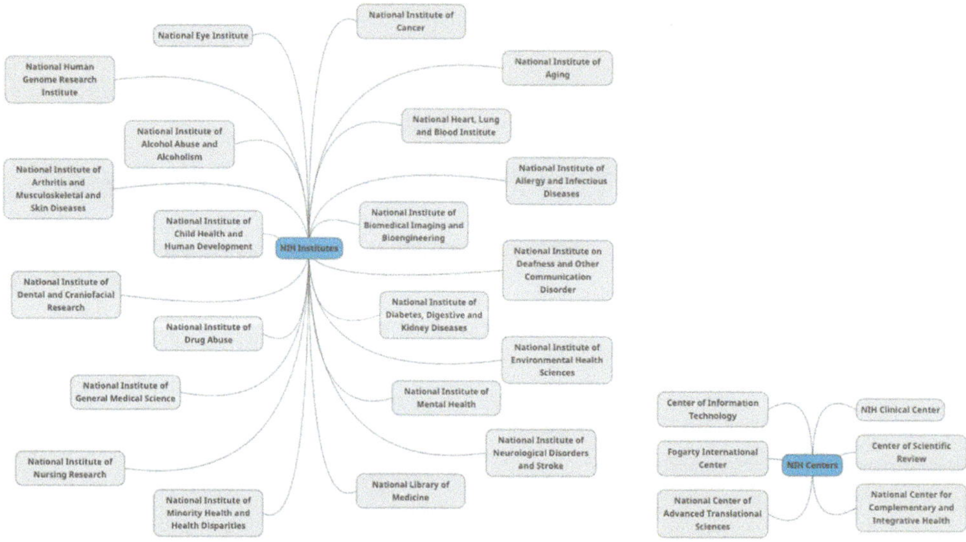

FIGURE 105.1 National Institutes of Health Organization Chart, 1a (*left*): NIH institutes, 1b (*right*): NIH centers.

innovation in the private sector. Eligibility of the SBIR is limited to SBCs in the United States, and the primary employment of the project lead must be with the SBC. All institutes and centers utilize the SBIR except the Fogarty International Center.

Small Business Technology Transfer

The main goal of the STTR grant is to stimulate scientific and technological innovation through encouraging collaborative R&D between SBCs and RIs.[6] As with SBIR, the grant supports for-profit institutions that have the potential for commercialization of innovative technologies. Eligibility of the STTR is also limited to SBCs in the United States, but the primary employment of the project lead can either be with the SBC or with the participating nonprofit research institution, provided they have a formal appointment to the applicant SBC. Again, all institutes and centers utilize the STTR except the Fogarty International Center.

Get started: Phases of the programs

Both SBIR and STTR programs are three-phased programs,[7] summarized in Table 105.1. The phases are detailed as follows:

1. Phase I: This phase determines the technical feasibility and merit of the proposed research and ascertains the quality of performance of the applicant before any further federal support is provided in Phase II. According to the statutory guidelines, Phase I has a maximum budget of $275,766 for a timeline between 6 months and 2 years.[7]
2. Phase II: The key objective of Phase II is to continue the research or R&D efforts that were previously initiated in Phase I. Funding is based on the results of Phase I and evaluation of the scientific, technical merit, and commercial potential of the Phase II application. Award periods for both SBIR and STTR may not exceed 3 years and $1,838,436 for Phase II.[7]
3. Phase III: The key objective of Phase III, if appropriate, is to pursue commercialization based on the results in Phase I and II. Note that Phase III does not allow for the use of SBIR or STTR funds.

Key differences between SBIR and STTR grants, as adapted from the NIH website[7]

The first key difference between SBIR and STTR grants pertains to research partnership. The SBIR program allows for subcontracting and encourages research partnership. This is however not a requirement, and the SBC may conduct the entire project without any external collaboration.

On the other hand, collaboration with a nonprofit research institution for both phases of the STTR program is mandatory. Examples of such institutions include universities, nonprofit hospitals, nonprofit research organizations, and federally funded research and development centers. NIH also denotes a specific proportion of work to be done by each partner, with at least 40% of the research to be conducted by the SBC and at least 30% to be conducted by the single, nonprofit research institution. The STTR grant is then awarded directly to the SBC, which will then be tasked with disbursement of the funding to the research institution.

TABLE 105.1 Phases of the SBIR and STTR programs.

Phase 1
Discovery and feasibility checks: Focus on the commercial potential and wider benefits of the innovation.
Phase II
Development and full R&D: Continuation of R&D efforts from phase I.
Fast-track
This allows a single application for both phase I and phase II in order to prevent stagnation of funding. However, full details of phase II must be outlined in advance.
Direct to phase II (only applicable to SBIR)
A project can bypass phase I to progress to phase II if it has already proven sufficient feasibility.
Phase IIB
Some NIH institutes and centers provide phase IIB awards if a phase II project requires further time beyond the standard allocated time for phase II.
Commercialization Readiness Pilot (CRP) program
CRP offers technical assistance and funding if it is required for late-stage development of phase II small businesses.

The next main distinguishing factor is the employment status of the Principal Director or Principal Investigator (PD/PI). The SBIR program requires the PD/PI to have primary employment with the small business during the duration of the entire project and at the time of the award.

The STTR program however allows the PD/PI to have primary employment with either the small business or the partnering research institution. The PD/PI must also devote at least 10% of their time to the STTR program.

Table 105.2 highlights the main differences between work requirements between the two programs. SBIR allows outsourcing of 33% of Phase I research and 50% of Phase II research, while STTR has minimal work requirements divided between the small business and research institution partner.

Examples of when to use SBIR[8]

- When there is no plan to include a nonprofit research entity (RI) in the project, or when there is a plan to subcontract less than 33% of a Phase I research effort to a nonprofit research entity.
- When potential investors could prefer not to involve nonprofit RI due to implications in allocation of intellectual property between the involved parties.
- When personnel at the RI prefer to contribute to the project in a manner that is independent to the RI, i.e., when they do not wish to be subcontracted as part of the nonprofit RI.

Examples of when to use STTR

- When further subcontracting and collaboration may be required/advantageous, for example, to attract investors or due to the nature of the project. An STTR allows up to 60% of the R&D to be conducted by the nonprofit RI. In fact, the nonprofit RI must contribute a minimum of 30%, which means that a potential of 30% of the remaining work could be conducted by the same nonprofit RI or subcontracted to another institution.

Eligibility requirements

Only SBCs within the United States concerns are eligible to submit applications, provided that they meet all of the following criteria at the time of award for Phase I and II:[9]

1. **Profit**: Organized for profit with a place of business located in the United States, operates primarily within the United States or makes a significant contribution to the United States economy through taxes or use of products, materials, or labor.
2. **Business**: In the legal form of an individual proprietorship, partnership, limited liability company, corporation, joint venture, association, trust, or cooperative.
3. **Ownership**: More than 50% directly owned and controlled by one or more individuals (who are citizens or permanent resident aliens of the United States).
4. **Size**: No more than 500 employees.

TABLE 105.2 Key differences between SBIR and STTR.

	SBIR	STTR
Partnership requirement	Encourages partnership	Requirements: a nonprofit research institution partner
Work requirement	Allows outsourcing	Minimum work requirement: 40% small business, 30% nonprofit research institution partner
Employment requirement	PD/PI must have primary employment with the small business	PD/PI may have primary employment with either the research institution partner or small business

Application requirements

Upon fulfillment of the aforementioned eligibility requirements, the grant application must be submitted electronically via Application Submission System and Interface for Submission Tracking on electronic Research Administration (eRA) Commons or Grants.gov workspace.[10]

Below are the required registrations prior to application:

1. **DUNS number (Company)**: All applicants must be issued a Dun and Bradstreet Universal Numbering System (DUNS) number, a unique global business identification system to identify and validate businesses worldwide.
2. **System for Award Management (SAM)**: All applicants have to complete and maintain an active SAM registration; this requires a minimum annual renewal. SAM is an official system required to participate in business with the U.S. government and update, renew, or check the status of an application.
3. **Grants.gov (Company)**: Upon receiving an active DUNS number and SAM registration, applicants can complete the Grants.gov registration, a requirement for application of a federal grant.
4. **eRA Commons (Company and all PD/PIs)**: Applicants require an active DUNS number and SAM registration for the eRA Commons registration. The eRA Commons is the NIH's electronic system for grants and administrations that allows PD/PIs to track the status of application.
5. **SBA Company Registry at** SBIR.gov: Small Business Administration (SBA) is required to register the business as a distinct legal entity.

Resources

1. Office of The Director, National Institutes of Health. (2022). Notice of Change to Award Budget for PHS 2021-02 Omnibus Solicitation of the NIH, CDC, and FDA for Small Business Innovation Research (SBIR) and Small Business Technology Transfer (STTR) Grant Applications. Available from: https://grants.nih.gov/grants/guide/notice-files/NOT-OD-22-043.html [accessed 20 January 2022].

References

1. Packalen M, Bhattacharya J. NIH funding and the pursuit of edge science. *Proc Natl Acad Sci U S A*. 2020;117(22):12011–12016. https://doi.org/10.1073/pnas.1910160117.
2. National Institutes of Health. *Mission and Goals*; 2017. Available from: https://www.nih.gov/about-nih/what-we-do/mission-goals. Accessed January 20, 2022.
3. Ben-Menachem G, Ferguson S, Balakrishnan K. Beyond patents and royalties: the perception and reality of doing business with the NIH. *Bioentrepreneur*; 2005. https://doi.org/10.1038/bioent895. Accessed January 20, 2022.
4. Meslin E, Blasimme A, Cambon-Thomsen A. Mapping the translational science policy 'valley of death. *Clin Transl Med*. 2013;2(1):14. https://doi.org/10.1186/2001-1326-2-14. Accessed January 20, 2022.
5. Audretsch D, Banucci G, Bond TJ, et al. Committee on capitalizing on science, technology, and innovation: an assessment of the small business innovation research program—phase II; board on science, technology, and economic policy; policy and global affairs; national academies of sciences, engineering, and medicine. In: *SBIR/STTR at the National Institutes of Health*. Washington (DC): National Academies Press (US); 2015. Available from: https://www.ncbi.nlm.nih.gov/books/NBK338158/. Accessed January 20, 2022.
6. Sun R, Kamat I, Byju A, et al. Advancing pediatric medical device development via non-dilutive NIH SBIR/STTR grant funding. *J Pediatr Surg*. 2021;56(11):p2118–p2123. https://doi.org/10.1016/j.jpedsurg.2021.01.025. Accessed January 20, 2022.
7. National Institutes of Health. *Understanding SBIR and STTR*; 2022. Available from: https://seed.nih.gov/small-business-funding/small-business-program-basics/understanding-sbir-sttr. Accessed January 20, 2022.
8. Small Business Innovation Research. *SBIR or STTR? Which One is Right for me*; 2022. Available from: https://www.sbir.gov/tutorials/program-basics/tutorial-3. Accessed January 20, 2022.
9. National Institutes of Health. *NIH Grants Policy Statement*; 2021. Available from: https://grants.nih.gov/grants/policy/nihgps/html5/section_18/18.5.2_eligibility.htm. Accessed January 20, 2022.
10. U.S. Small Business Administration. *Guide to SBIR/STTR Program Eligibility*; 2013. Available from: https://www.sbir.gov/sites/default/files/elig_size_compliance_guide.pdf. Accessed January 20, 2022.

Glossary

2D/3D Two-dimensions/Three dimensions

510(k) Premarket Notification

513(g) Form to determine FDA classification of the medical device

95% CI 95% confidence interval (a range of values that contains the true mean of the population by 95% confidence level)

A cost-effective study, or cost-effectiveness analysis (CEA), is an economic evaluation method used to compare the relative costs and outcomes of different interventions or courses of action within healthcare or other fields.

AA Pathway FDA established Accelerated Approval (AA) Program in 1992 to allow for earlier approval of orphan drugs.

Accumulation effect It occurs when patients must be exposed to a risk factor for a long period to develop an event.

Accuracy The proportion of all correct predictions (both true positives and true negatives) out of the total number of predictions made.

ACS Acute coronary syndrome

Adaptive Designs Dynamic methodologies in clinical trials that allow for real-time adjustments to trial parameters based on accumulating data, enhancing trial efficiency and responsiveness.

Adaptive Randomization A method of randomizing participants in a clinical trial where the likelihood of assignment to different treatment arms is adjusted based on accumulating data.

ADME Absorption, distribution, metabolism, and excretion

Adulterated The contamination of a substance with inferior or harmful elements, reducing its quality or safety. This term often applies to various products, including food, drugs, and beverages.

Adverse Effect An unwanted negative effect that occurs when a patient is administered a certain drug. Adverse effects of drugs can also be classified as side effects of the drug.

AF Atrial fibrillation

AI Artificial Intelligence

Altmetric score Article-specific bibliometric that is a weighted count of attention an article weighing volume, authors, and sources of mentions.

Analytical pitfalls Common errors and misconceptions encountered during the data analysis process in translational cardiology research.

Animal welfare Humane care and well-being of all research animals. It includes both the physical and the mental state of an animal with regard to the conditions in which it lives and dies. In a research setting, suffering should always be avoided.

ANOVA Analysis of Variance

Antishelving clause Permits a licensee the commercial use of a patented invention within a set period of time, and the licensee must notify the patentee if it no longer sells or manufactures the invention

APOB Apolipoprotein B

ARRIVE Animal Research Reporting of In Vitro Experiments

ASD Atrial septal defect

Assay Sensitivity The trial's ability to distinguish between an effective therapy and one that is not effective.

Assurance Assurance is focused on ensuring the findings gathered through assessment and policy-related research are effectively implemented.

At risk Participants in a survival study are "at risk" at any time point after time zero if they are alive and under observation and the event has not yet occurred at that time.

Auditing Periodic independent review of core trial process and documents

Barriers Obstacles or challenges that hinder the successful adoption and integration of evidence-based practices into routine healthcare settings. These barriers can encompass a range of factors such as organizational challenges, practitioner resistance, resource constraints, and other impediments that inhibit the smooth translation of research findings into everyday clinical practice.

BDD Breakthrough Device Designation

Best Practice A standard or set of guidelines that has been generally accepted to be superior to known alternatives based on superior outcomes.

Bias A systematic error, deviation in results, or inferences from the truth.

Bioavailability The fraction of the initial dosage of a therapeutic agent that reaches the binding site, promoting the expected action.

Biocreep The cumulative decline in effectiveness that can occur when a series of noninferiority/equivalence trials are performed, resulting in a new standard of care that is not superior to placebo.

Biologic License Application An application, processed through the FDA, that biologic sponsors (usually pharmaceutical or biotech companies) formally propose that the FDA approve a new biologic for sale and marketing in the United States.

Blinding The practice of concealing the allocation of participants into treatment or control groups from either the participants, the researchers, or both, to prevent bias in the conduct or interpretation of the study.

BP Blood pressure

CAD Coronary artery disease

Case An instance of a particular disease, injury, or other health conditions that meets selected criteria (i.e., case definition).

Case Series Descriptive studies that observe and report characteristics, clinical courses, and outcomes of a group of patients with a specific condition or treatment.

Case-control study Observational study conducted on retrospective data.

Causal Inference The process of determining whether a cause-and-effect relationship exists between variables.

CBER Center for Biologics Evaluation and Research

CDER Center for Drugs Evaluation and Research

CDRH Center for Device and Radiological Health

CDS Clinical Decision Support

CEBM Center for Evidence-Based Medicine

Censoring A feature of time-to-event data in biomedical studies which indicates that the event of interest does not occur for all individuals while they are under observation.

CFR US Code of Federal Regulations

CHD Coronary heart disease

CKD Chronic kidney disease

Claim States what a patent covers to establishes the boundaries of an invention's protection.

Clinical Development A blanket term defining the process of bringing a new drug, biologic, or device to the market.

Clinical Practice Guidelines Statements that include recommendations intended to optimize patient care that are informed by a systematic review of evidence and an assessment of the benefits/harms of alternative care options.

Clinical Relevance Determine what the test is intended to diagnose or predict. The objective should be clinically relevant, addressing a specific diagnostic question or gap in medical practice.

Clinical research A comprehensive study of the safety and effectiveness of the most promising advances in patient care.

Clinical trial Experiment that involves human subjects to analyze different types of therapeutic approaches, including drugs, procedures, and devices.

Clinical Trials Phases Phase 1: Safety evaluation and side effect identification (2–12 months). Phase 2: Dose range testing for safety and efficacy (about 2 years). Phase 3: Efficacy testing with larger populations and multiple sites.

Cluster randomization The process of randomizing a group of patients to a treatment under the same study arm as opposed to randomizing the individual.

CMC Chemistry, Manufacturing, and Controls

CO Carbon monoxide

Cohort A group of individuals representing the larger population pool, who are followed over time in a longitudinal study.

Cohort Studies Follow a group of individuals over time to assess how well the test predicts outcomes.

Collective Case Series Studies that compile multiple cases to enable cross-comparisons, identifying patterns and variances across cases.

Composite Endpoints A determined end point that contains two or more distinct component endpoints.

Comprising clinicians, statisticians, and other experienced professionals in clinical trials, this committee is appointed by the study sponsor. Crucially, it remains independent from the study investigators, distancing itself from the trial's day-to-day operations and thereby avoiding influence from investigators when making recommendations.

Computational Models Mathematical representations of biological systems, processes, and phenomena using algorithms to simulate and predict behaviors, interactions, and results under different circumstances.

Confidence interval (CI) A range of values that is likely to contain the true effect size.

Confounders Variables that are associated with both the exposure and outcome but do not directly explain the causality of the outcome by the exposure.

Confounding A situation where the observed association between the exposure and outcome is distorted by the presence of another variable that is related to both.

Confounding bias It occurs when a variable is associated with both the exposure and outcome.

Confounding Variables A variable that influences both the dependent and independent variables and can produce erroneous statistical associations.

Confusion Matrix A table used in classification problems to assess the performance of a model by comparing predicted and actual outcomes.

Convergent Design Simultaneous collection and analysis of quantitative and qualitative data, with integration occurring after data collection.

Copyright Legal right to the creator or an assignee of a print, performance, film, artwork, or musical material to copy, sell, or market such products.

Correlated data The value of a data point at a certain time point correlates with its value at a future point in time.

Correlation Statistical measure that quantifies the degree of association or relationship between two variables.

Covariance The linear relationship between two variables. If it is zero, that means no linear relationship exists.

COVID-19 Coronavirus disease 2019

Crossover studies are longitudinal studies where each participant is assigned to a sequence of two or more interventions or treatments rather than one intervention or treatment

Cross-sectional study A type of observational study that analyzes data from a population at a specific point in time.

CSV (Comma-Separated Values) A plain text file format commonly used to store tabular data, where each line represents a row, and the values within each row are separated by commas. CSV files are often used for importing and exporting data between different software applications.

CTCAE Common Terminology Criteria for Adverse Events

CVD Cardiovascular disease

Data Analysis It is the process of systematically applying statistical and/or logical techniques to describe and illustrate, condense and recap, and evaluate data.

Data Collection and Analysis With some rare exceptions, new tests, while developed, often go through several stages of trials.

Data management plan (DMP) A document that outlines how research teams handle data during and after a project, including details about data acquisition, organization, safety and quality assurance, and plans on how to use and share the data.

dBA decibels (measurement of Noise level)

DeNovo DeNovo Request for Device Classification

Dependent Variable A variable that depends on the values of other variables.

Descriptive studies Observational studies which describe the patterns of disease occurrence in relation to variables such as person, place, and time.

Design bias It occurs when the control group is significantly different from the intervention group.

Designated person An individual who conducts or supervises the conduct of postmarket surveillance. If the postmarket surveillance plan includes a team of investigators, as defined below, the designated person is the responsible leader of that team.

Determinant Any factor, whether event, characteristic, or other definable entity, that brings about a change in a health condition or other defined characteristic. This term generally includes the causes (including agents), risk factors (including exposure to sources), and modes of transmission, but does not include the resulting public health action.

Device recall Removal of a device from the market, either voluntarily or as ordered by FDA. Recall may be temporary (i.e., to update or fix the device and return it to market), or permanent.

DHT Digital Health Technology

Diagnostic Accuracy Measures These include sensitivity (ability to detect the condition when it is present), specificity (ability to exclude the condition when it is not present), positive predictive value (PPV), and negative predictive value (NPV).

Diagnostic test A clinical tool used to test for the presence or absence of a disease.

Disease model An animal model in which some specific pathological processes or characteristics of a disease are induced to mimic human disease.

Drug repurposing The active campaign to redefine the usage of a specific drug, with a previously established indication.

Drug Safety Refers to the process of continuous monitoring, collection, and analysis of data on the effects and potential toxicity of drugs to prevent adverse effects and events.

Drug testing To analyze therapeutic agents used to treat different medical conditions, as well as to evaluate known drugs that serve as basis for further studies.

DSMP Data and Safety Monitoring Plan

EBP Evidence-Based Practice

Effect size A quantitative measure of the strength of a phenomenon.

Effectiveness The measure of an intervention under normal, real-world conditions.

Effectiveness Trials Studies testing treatment outcomes under real-world conditions for generalizability.

Efficacy The measure of an intervention under ideal and controlled conditions.

Epidemic An epidemic is an unexpected increase in the number of disease cases in a specific geographical area. Yellow fever, smallpox, measles, and polio are prime examples of epidemics. An epidemic disease does not necessarily have to be contagious. Epidemics can refer to a disease or other specific health-related behavior (e.g., smoking) with rates that are clearly above the expected occurrence in a community or region.

Epidemiology The study of the distribution and determinants of health-related states and events in specified populations, and the application of this study to the control of health problems.

Equivalence Trials Clinical trial aiming to demonstrate that two treatments are equivalent (similar enough)

Equivalence/ Non-Inferiority Margin A predefined value that is considered smaller than a clinically significant effect that is used to establish equivalence or noninferiority between drugs.

ERG Human ether-a-go-go related gene

Error Failure of a planned action to be completed as intended or the use of a wrong, inappropriate, or incorrect plan to achieve an aim.

Ethical Considerations Ensure the study design follows ethical guidelines, including informed consent, confidentiality, and the right to withdraw.

Event The study end-outcome based on the research question. Potential events are (1) death from any cause, (2) disease progression, or (3) diagnosis with a specific disease.

Evidence-Based Medicine The use of the scientific method to organize and apply current data to improve healthcare decisions.

Evidence-Based Interventions Proven practices, strategies, or approaches grounded in empirical evidence, systematically designed to enhance effectiveness in yielding optimal outcomes.

Evidence-based research Research that answers questions related to patient care that is based on valid, efficient, and transparent methods.

Evidence-to-Practice Gaps Disparities between established scientific evidence, often derived from clinical research, and the actual implementation of that evidence into routine healthcare practices. These gaps highlight instances where medical interventions supported by robust evidence are not consistently or effectively applied in real-world clinical settings, leading to variations in patient care and potentially suboptimal outcomes.

Excel A popular spreadsheet software developed by Microsoft, commonly used for data analysis, calculation, graphing tools, and pivot tables. Excel files typically have the extension ".xlsx" and can handle large datasets with various functionalities.

Experimental Studies Research testing an intervention in a controlled scenario, comparing intervention and control groups.

Experimentation Phase The processes of performing proceduralized testing to determine if a certain hypothesis is true; a process by which something new is discovered.

Expert Opinion A belief or opinion given by a qualified expert in the subject matter.

Explanatory Sequential Design Collects and analyzes quantitative data first, followed by qualitative data to explain the quantitative results.

Explanatory trials Trials aimed at evaluating efficaciousness and internal validity of an intervention under ideal parameters with rigid implementation of protocols aimed at reducing confounders, biases and establishing causal relationship

Exploratory Sequential Design Begins with qualitative data collection and analysis, guiding subsequent quantitative data collection.

Exposure Contact with a disease-causing factor or receipt of an intervention.

External Validity The extent to which the findings of a study can be generalized to contexts outside of the study conditions.

Facilitators Factors or elements that support and enhance the successful adoption and integration of evidence-based practices into routine healthcare settings. Facilitators can include organizational support, clinician engagement, effective communication strategies, sufficient resources, and other positive influences that promote the smooth translation of research findings into everyday clinical practice.

FAIR Guiding Principles Guiding principles of data management designed by representatives from funding agencies, academia, and the health industry to enhance both manual and automated use and sharing of data.

FD&C Act Food, Drug, and Cosmetic Act of 1938

FDA US Food and Drug Administration

FDA regulatory class Medical devices are classified into three categories based on risk, which reflects the level of controls necessary to assure device safety and effectiveness. This largely correlates with device-related risk. Class I devices present minimal risk, while class II devices pose moderate risk, and class III devices are highest risk.

Feasibility The extent to which a new innovation, treatment, or recommendation can be successfully implemented within a target patient population.

FINER Feasible, Interesting, Novel, Ethical, and Relevant.

First-in-Human studies Initial experiments that establish the safety range for dosage of new therapeutic agents.

Freedom to operate The ability to produce or commercialize an invention without infringing on previous work or products

GCP Good Clinical Practice

Gender minority Individuals who identify as a gender other than their assigned sex at birth.

Genome-wide association studies (GWAS) Quantification of the association between genetic variation and observed phenotype.

Gold Standard The gold standard is the best proven method for diagnosing the condition of interest. The new test's performance is measured against this standard to assess its accuracy and reliability.

Good Laboratory Practices Precepts that promote a minimum structure and criteria for carrying out experiments and managing research data.

Google Scholar Database maintained by Google that uses natural-language algorithm to query academic articles.

Group Sequential Designs An adaptive design that permits interim analyses at predefined intervals during a clinical trial, enabling early stopping for efficacy or futility.

Guideline Adherence Consistent application of established and evidence-based clinical guidelines, ensuring that healthcare practices align with recommended standards to enhance patient care and outcomes.

H-index Researcher-specific bibliometric that takes the number of publications with citation number $\geq$ h.

Half-life The period that a drug requires to decrease in half its initial concentration.

Harm Any negative effect on a patient's health or well-being, whether or not it is evident to the patient.

Harmonization Working on those areas which are complementary in order to have the plans working together for the achievement of an overall strategic objective

Hawthorne effect A modification in the behavior of an individual as a function of observation

HDE Humanitarian Device Exemption

Health authority A government body's (usually a health or public health department or agency) legal ability to further public health by using tools such as community engagement, data collection, scientific research, laws and regulations, enforcement, and the many methods of sharing information and guidance with those who need it most.

Health disparity Preventable differences in healthcare an individual receives due to conscious or unconscious bias that can lead to poor health outcomes.

Health Insurance Portability and Accountability Act of 1996 (HIPAA) A federal law passed by the Congress in 1996 regulating health care in several spheres, including the standards for protecting and confidential handling of sensitive health information.

Health-related quality of life The way health is empirically estimated to affect the quality of life.

Healthcare providers A healthcare provider is an individual health professional or a health facility organization licensed to provide healthcare diagnosis and treatment services including medication, surgery, and medical devices.

Heterogeneity The variation in study outcomes between studies included in a meta-analysis.

HF Heart failure

Homeostasis The tendency of a biological system to maintain a stable equilibrium.

HTS High-throughput screening

HUD Humanitarian Use Designation

Hypothesis A testable conjecture, prediction, or educated guess to what the answer may be to a scientific question.

Hypothetico-deductive Question A question that arises from prior knowledge.

Hypothetico-inductive Question A question that arises from observation.

IBM (International Business Machines Corporation) A multinational technology and consulting corporation known for producing hardware, software, and providing hosting and consulting services. IBM has a significant presence in various industries, including computer hardware, software development, and artificial intelligence.

ICF Informed Consent Form

ICH International Conference on Harmonization

IDE Investigational Device Exemption

IDEAL Framework The IDEAL (Idea, Development, Exploration, Assessment, Long-term Study) Framework is a structured approach to evaluating surgical innovations. It provides guidelines and recommendations for the stages and processes involved in assessing and introducing new surgical techniques and devices.

Implementation Fidelity Degree to which interventions are implemented as intended by developers.

Implementation Research The systematic study of methods and strategies aimed at effective integration of evidence-based practices into real-world settings, encompassing a multidisciplinary approach that examines the complex interactions between interventions, individuals, organizations, and the broader socio-cultural context. It seeks to bridge the gap between research and practice by investigating the factors that facilitate or hinder the successful adoption, implementation, and sustainability of interventions, ultimately contributing to the improvement of the quality and outcomes of services provided.

Implementation Science Study of methods to integrate research findings into routine healthcare to improve quality and effectiveness.

In a factorial design, increasing the levels or numbers of factors will require a higher sample size to have a high probability of detecting the specified effect size (power). The power of the factorial design depends on the level of each factor, the number of factors, the number of possible combinations, and the significance level (Type I error).

In Chemico testing An approach that replaces animal (in vivo) testing for abiotic chemical procedure.

In clinical studies treatment or intervention can be called a factor. Factorial design is a study design that allows testing multiple treatments or interventions on the same subjects, which permits an assessment of potential interactions among the treatments. The simplest form of factorial design is 2×2 design. In this design, two treatments are evaluated. If we study drugs A and B, we will have four groups. The groups will receive the following treatment: (Active drug A and B), (active drug A and placebo B), (placebo A and active B), and (placebo A and B).

In Silico Testing This method analyzes mathematically the relations between all components of a system through computed models.

In vitro diagnostic devices Laboratory tools tested and studied by the Food and Drug Administration that have been designed to test for the presence or absence of a disease

Incidence The number of new cases of a condition or disease in a population over a specific period of time.

Incidence rate Also called person-time rate, this is a measure of incidence that incorporates time directly into the denominator. It is the ratio of the number of cases to the total time the population is at risk of disease.

IND Investigational New Drug

Independent Variable A variable whose variation does not depend on the values of other variables.

Instrumental Case Series Studies that use typical examples of a condition to understand broader phenomena or general principles related to the condition.

Insurance companies An insurance company, also known as an insurer, provider, or underwriter, is a business that offers insurance policies to individuals or organizations. These policies protect against financial losses due to unexpected events, such as accidents, illnesses, or natural disasters.

Intellectual property An invention that one has the right to apply for a protection (e.g., patent, copyright, trademark), of that product.

Interim Analyses Periodic analyses of accumulating trial data conducted before the completion of the study to assess safety, efficacy, or futility.

Internal Validity The degree to which the study design, conduct, and analysis avoid bias and thus provide confidence in the causal relationships identified within the study population.

Interpretative challenges Challenges and complexities encountered in the accurate comprehension and contextualization of research results within the spectrum of translational cardiology.

Intervention Development Steps that aim to create substantial evidence-based interventions that are maximally effective, feasible, and relevant to a target population.

Intrinsic Case Series Studies focusing on unique or rare cases to provide detailed insights into their presentation, management, and prognosis.

Investigational Device Exemption (IDE) Designation by the FDA that a medical device can be used in the context of a clinical study prior to market authorization, or outside the existing device labeling. This utilizes various levels of oversight and controls incorporated into study design, including, but not limited to, human subjects protections and reporting to the FDA.

Investigational Drug A pharmaceutical substance under study in clinical trials or research studies that has not yet received marketing approval from regulatory authorities.

IRB Institutional Review Board

IRDiRC International Rare Diseases Research Consortium officially launched in 2011 for the treatment and care of people living with a rare disease.

JASP JASP is a free and open-source statistical software program designed for both beginners and advanced users. It provides a user-friendly interface for conducting statistical analyses and supports a wide range of statistical methods and techniques.

JMP JMP is a statistical analysis software package developed by SAS Institute. It is widely used for data visualization, exploration, and analysis, especially in fields such as engineering, science, and business.

Journal impact factor Journal-specific bibliometric that is calculated by taking the number of citations in the current year from journal articles published in prior 2 years, divided by the all the articles published in that journal for two prior years

Key Stakeholders Key stakeholders are people or groups that have a significant impact on the success of a company or project.

LDLR Low-density lipoprotein receptor gene

Lead Optimization The process of improving potency, selectivity, and safety after identifying a potential molecule. This process usually involves adding chemical components to optimize drug safety and tolerability.

Lead time bias It occurs when the screened group has a longer duration of disease compared to the nonscreened group because of early detection.

Life-supporting or life-sustaining device used outside a device user facility A device essential to, or yielding information essential to, the restoration of a bodily function important to the continuation of human life that is used outside a hospital, nursing home, ambulatory surgical facility, or diagnostic or outpatient treatment facility. A physician's office is not a device user facility.

Linux Linux is an open-source operating system kernel initially developed by Linus Torvalds. It is widely used in servers, mainframes, and other computing devices. Linux distributions, such as Ubuntu, Fedora, and Debian, offer various graphical user interfaces and software packages.

Longitudinal data Repeated observations recorded periodically in a group of individuals.

LVADs Left ventricular assist devices

M-index Researcher-specific bibliometric that takes h-index divided by researchers' scientific age.

Mac OS Mac OS (formerly Macintosh Operating System) is the operating system developed by Apple Inc. for its line of Macintosh computers. It is known for its user-friendly interface, stability, and integration with other Apple products and services.

Macros Macros are sequences of instructions or commands that automate tasks within software applications. In the context of spreadsheet software like Excel, macros can be used to record and replay user actions, enabling repetitive tasks to be performed more efficiently.

Major procedure study A study in which major anatomical regions (e.g., thoracic, abdominal) are accessed. The normal anatomy may be permanently altered, or a disease model may be induced. Major procedures are more likely to have complications and involve a difficult recovery than minor procedures.

Market authorization A procedure for approval of a medical product for marketing after it has undergone a process of evaluation to determine the safety, efficacy and quality of the product and the appropriateness of the product information

Matching The process of selecting cases and controls so that they are similar in certain characteristics to reduce confounding.

Maximum tolerable dose The highest dosage of the test drug to which subjects in the experiments are submitted that are predicted to not affect subjects' course of life from adverse effects.

MDUFMA Medical Device User Fee and Modernization Act

Mechanism of Action The specific biochemical interaction through which a pharmaceutical or biologic exerts its pharmacological effect.

Meta-analysis A statistical technique for combining the findings from independent studies.

Metadata Structured data that describe and help to locate research data (e.g., equipment details, software versions).

Method Phase A stepwise approach used in the experimentation phase of the scientific method to ensure reproducibility of data collected.

Minor procedure study This type of study is usually shorter than a major procedure study. The procedure is restricted to superficial tissues, requires either minimal or local anesthesia, and is expected to have minimal complications.

Misbranded Products that are inaccurately labeled or advertised, failing to comply with legal requirements or correctly represent the product's ingredients, uses, or characteristics.

Mixed-effects model Regression model used to predict a mean response trajectory by accounting for intrasubject and intersubject sources of variation.

Mixed-Methods Approach The use of both qualitative (e.g., interviews, case histories) and quantitative (e.g., numerical data analysis) research methods to gather and analyze data.

Mixed-Methods Research Combines quantitative and qualitative data for comprehensive understanding.

ML Machine Learning
Monitoring The process of surveilling the procedures of a clinical trial.
Monotonic relationship As one variable increases, the other variable either consistently increases or decreases, but not necessarily at a constant rate.
MPI Myocardial perfusion imaging
mRNA messenger Ribonucleic Acid
Multiplicity The issue of making multiple statistical comparisons in a trial, requiring adjustments to control the overall risk of making at least one Type I error.
N_2O Nitrous oxide
National Institutes of Health (NIH) The primary federal agency for conducting and supporting medical research.
NCI National Cancer Institute
Negative Predictive Value The likelihood that a negative test result is a true negative.
New Drug Application (NDA) An application, processed through the FDA, that drug sponsors (usually pharmaceutical or biotech companies) formally propose that the FDA approve a new pharmaceutical for sale and marketing in the United States.
Next-Generation Sequencing (NGS) Fragmentation of DNA into small segments and subsequent sequencing of each nucleotide base.
NFκB Nuclear Factor Kappa B
NIS Nationwide Inpatient Sample database
NO_2 Nitrogen dioxide
Non-Inferiority Trials Clinical trials aiming to prove that a treatment is not worse than a predefined margin when compared to another. The new treatment may be equivalent or superior.
Nonresponse bias A bias that occurs when the individuals who choose not to participate in the study differ significantly from those who do participate.
Nonsurvival study A study in which the animal will not recover from the procedure, and euthanasia will be performed under anesthesia immediately after the study endpoints have been met. Anesthesia is maintained because only sedation, and not recovery, is required. Also referred to as an acute study.
NSTEMI Non-ST-elevation myocardial infarction.
Null Hypothesis A prediction that there is no significant association between the observed phenomenon and the factors being evaluated.
O_3 Ozone
Observational Study A study design in which no intervention is performed. Study subjects are only observed to collect data for analysis.
Observer Bias Bias introduced when the researcher's expectations or knowledge influence the data collection or interpretation.
OCP Office of Combination Products
Odds ratio (OR) A measure of association between an exposure and an outcome.
OMERACT Outcome Measures in Rheumatology
Operational Outcomes Consider test feasibility, cost, time, ethical aspects, medicolegal aspects, and patient acceptability
Orphan drug A drug that has been designated with the goal of treating a rare disease.
Orphan Drug Act Passed in 1983, provided guidelines and incentives for orphan drug discovery.
Orphan Drug Designation The product or drug must be intended to treat rare or neglected diseases.
Orphan Drug Legislation Orphan Drug Legislation is practiced by many countries to enhance the development, research, and marketing of orphan drugs.
Outcome Typically refers to the development of a disease.
P-Value A probability that measures the evidence against a null hypothesis. A small p-value (e.g., less than 0.05) suggests strong evidence against the null hypothesis.
PAH Pulmonary arterial hypertension.
Parameter Number that describes a specific feature of a population.
Patent A legal protection of an invention for a set period of time, providing the right to exclude other from producing, using, or selling the product
Patent search Used to distinguish patent and nonpatent literature to help establish an invention's patentability
Patentability assessment Written report on an invention's patentability using combination of a patent search and an analysis of references uncovered
Patient safety The protection of patients from harm during medical procedures, including surgical interventions, minimizing risks, ensuring appropriate care, and monitoring and managing complications.
Patient-Centric Design A focus on adapting trial parameters to prioritize the needs and experiences of patients, ensuring a more tailored and responsive approach to study design in adaptive trials
PCSK9 Proprotein convertase subtilisin/kexin type 9
PDUFA Prescription Drug User Fee Act
Pearson's rank correlation coefficient Statistical measure that quantifies the strength and direction of a linear relationship between two continuous variables, ranging from -1 to $+1$.
PECO Population, Exposure, Comparison, and Outcome

PET Positron emission tomography

PGHD Patient Generated Health Data

Pharmaceutical companies An industry in medicine that discovers, develops, produces, and markets pharmaceutical drugs for use as medications to be administered to patients, intending to cure and prevent diseases, or alleviate symptoms.

Pharmacovigilance The continuous monitoring, identification, evaluation, and mitigation of adverse effects and other drug-related issues from preclinical trials through postmarket surveillance. This process aims to protect public health and improve medication safety.

Phocomelia A congenital condition characterized by underdeveloped or absent limbs, resulting in shortened or flipper-like appendages directly attached to the trunk.

PI Principal Investigator

PICO Population, Intervention, Comparison, and Outcome

PICOS Population, Intervention, Comparison, Outcome and Study design

PICOT Population, Intervention, Comparison, Outcome and Time

Pilot study A study with a small sample, conducted to assess the feasibility of a new research idea, or test a prototype as a proof of concept. Its results, if sufficiently promising, are used as the basis of a larger, more extensive protocol.

Platform Trials Adaptive clinical trial designs that evaluate multiple treatments simultaneously, allowing for the seamless addition or removal of treatment arms based on emerging evidence.

PM2.5 Particulate matter 2.5

PMA Premarket approval

PMA Premarket approval application

PMN Premarketing notification

PMOA Primary Mode of Action

Policy development Policy development in public health is the process of advancing and implementing public health laws, regulations, or voluntary practices to improve health. These practices can influence organizational change, systems development, and individual behavior.

Positive Predictive Value The likelihood that a positive test result is a true positive.

Post-Approval Study Study mandated by the FDA as a condition of device approval to provide further characterization of device performance. A PAS is designed to meet outstanding questions of safety and effectiveness so can take on a variety of forms ranging from bench or animal studies to registries, to traditional clinical trials.

Post-Hoc Analysis Post hoc in Latin means "after this"; it refers to a statistical analysis specified after a study has been concluded and the data collected.

Postmarket surveillance It is the monitoring of a substance or device after it has been released for public use to ensure that it is up to standard and that it does not have any harmful effect on its consumers.

Post-review, the committee can make recommendations to the study sponsor. These could involve continuing the trial as planned, suggesting modifications, or recommending termination. Additionally, the committee may propose the collection of further data or the execution of additional analyses.

Practical Applications These involve the use and implementation of theoretical concepts or methods to address specific problems or challenges. In various fields, practical applications refer to utilizing knowledge and tools to yield real-world solutions or outcomes.

Pragmatic trials Trials aimed at evaluating effectiveness and external validity of an intervention under real-world parameters with flexibility in protocol implementation allowing for practitioner autonomy and usually studied with concrete outcomes; it is aimed at maximizing intervention effect.

PRECIS-2 Pragmatic-Explanatory Continuum Indicator Summary; a tool developed by Thorpe et al. in 2009 to help assess a study for the impact that its results will have on clinical decisions and applicability.

Precision The proportion of true positives out of all predicted positives. Often used interchangeably with Positive Predictive Value.

Preclinical Considerations Factors that need to be taken into account during the early stages of developing and testing interventions. These studies are done on animal models before application to human subjects is approved

Preclinical studies Use microorganisms and animals to formulate data about the test drug.

Predictive enrichment techniques are used to select participants who have the highest chance to benefit from the treatment. Previous studies which showed a higher response in specific subgroups could be beneficial. Non-compliant patients, or those reporting adverse events, are not considered for study enrolment

Prevalence The total number of cases of a condition or disease in a population at a given time.

Prior art Used to determine an invention's patentability, specifically by establishing novelty and nonobviousness.

PRO Patient-reported outcome

Product performance report A publicly available device-specific report of device complications and failures, as reported to and compiled by the device manufacturer.

Project management A collection of techniques for proposing, planning, implementing, managing, and evaluating projects and people.

Project Management Professional Certification in project management

PTP Pretest probability

Public health Public health is the science of protecting and improving the health of people and their communities. This work is achieved by promoting healthy lifestyles, researching disease and injury prevention, and detecting, preventing, and responding to infectious diseases.

Public health systems Public health systems are commonly defined as 'all public, private, and voluntary entities that contribute to the delivery of essential public health services within a jurisdiction.

Publication bias The tendency for journals to publish positive findings over non-significant or negative results.

PubMed Database maintained by the National Library of Medicine that is indexed by Medical Subject Headings (MeSH). Commonly used for clinical searches.

Python Python is a high-level, interpreted programming language known for its simplicity and readability. It supports multiple programming paradigms and is widely used for web development, data analysis, artificial intelligence, scientific computing, and more.

QoL Quality of Life

Qualitative Research Explores real-world problems by gathering participants' experiences, perceptions, and behaviors, focusing on explaining processes and patterns not easily quantifiable.

Quality Healthcare The degree to which healthcare services for individuals or populations increase the likelihood of the desired health outcome and are consistent with current professional knowledge.

Quality of life The individuals' perceptions of their position in life in the context of the culture and value systems in which they live and in relation to their goals, expectations, standards, and concerns.

Quantitative Research Involves systematic investigation using numerical or statistical data, focusing on causality, generalizability, and replicability.

Quasiexperimental trials Empirical interventional study that estimates the causal impact of an intervention on a target population without random assignment

R R is a programming language and environment specifically designed for statistical computing and graphics. It is widely used among statisticians and data miners for data analysis, visualization, and the development of statistical models.

Race Social construct that lacks genetic basis but allows individuals to be grouped based on perceived physical traits.

Rare disease Defined by the U.S. Food and Drug Administration (FDA) as a disease that affects less than 200,000 people.

RCTs Randomized controlled trial(s) are prospective studies that measure the effectiveness of a new intervention or treatment between controlled groups with random assignment.

Recall Another term for sensitivity; it is the proportion of actual positives correctly identified by the test.

Recall bias It occurs when participants do not recall events accurately.

Reductionist Breaks down complex principles into smaller subsets and tackle those.

Registries Prospective databases or records that collect and store information about patients who undergo surgical procedures, medical devices, or interventions. Registries are used to track outcomes, safety, and trends in real-world clinical practice.

Regression is a statistical method that evaluates the relationship between a dependent variable (also known as outcome) and independent variables (also known as predictors). The regression models are broadly divided into linear and nonlinear. However, the distinction between linear and nonlinear models is not as straightforward as expected. Linear regression is by far the most common type. The main difference is that the outcome variable is dependent on the sum of the regression coefficients for linear models.

Regulatory affairs A profession developed from the desire of governments to protect public health by controlling the safety and efficacy of products in areas including pharmaceuticals, veterinary medicines, medical devices, pesticides, agrochemicals, cosmetics and complementary medicines, and by the companies responsible for the discovery, testing, manufacture and marketing of these products wanting to ensure that they supply products that are safe and make a worthwhile contribution to public health and welfare.

Regulatory Considerations The process of navigating and adhering to regulatory requirements and guidelines when implementing adaptive designs in clinical trials.

Regulatory reliance The act whereby the National Regulatory Authority in one jurisdiction may take into account and give significant weight to assessments performed by another National Regulatory Authority or trusted institution, or to any other authoritative information in reaching its own decision

Relative Risk The ratio of the probability of an event occurring in an exposed group to the probability of it occurring in a nonexposed group.

Relative Risk/Hazard Ratio The risk of developing the outcome in the exposed group compared to the nonexposed group.

Reliability studies Research designed to test the consistency of a measurement instrument or method by determining whether repeated measurements provide the same results across different occasions, different observers, or different items within the same test.

Research data All information collected and generated during a research project.

Research Methods The process by which data are collected to answer a research problem.

Research Problem The issue identified in a given population to which a solution is sought.

Respondent Validation A process where researchers verify their study conclusions with the participants to ensure the accuracy and reliability of the interpretations.

RFD Request for Designation

Risk Modeling It is a mathematical representation of a system, commonly incorporating probability distributions. It specifically involves using statistical and mathematical techniques to estimate the likelihood of various outcomes and their potential impact.

Risk Ratio Another term for Relative Risk; it compares the risk of an event occurring in two groups.

RPM Remote Patient Monitoring

Sample size (N) Number of observations in a sample.

Sample Size Calculation Calculate the necessary sample size based on the expected performance of the test, the prevalence of the condition in the target population, and the desired precision of the results.

Sample Size Re-estimation An adaptive design approach that allows for modification of the sample size during an ongoing trial based on interim data, optimizing statistical power.

Sampling The process of selecting a subset of individuals from a population to represent the entire population.

SAS (Statistical Analysis System) SAS is a software suite used for advanced analytics, business intelligence, data management, and predictive analytics. It is widely used in various industries for data analysis and decision-making.

SBA Small Business Administration

SBIA Small Business and Industry Assistance

SBIR Small Business Innovation Research

Scientific Method A set of steps which offer a procedural approach to answer a specific question objectively based on reproducibility. The steps are (1) Observation/Question, (2) Research, (3) Hypothesis, (4) Experimentation, (5) Analysis, and (6) Conclusion.

SDH Social determinants of health

SE Substantially equivalent

Selection Bias A distortion of the study results due to the method of selecting participants.

Selection Criteria The predefined conditions that determine which cases are included or excluded from a study to prevent information overload and selection bias.

Sensitivity The probability that a person with the disease will have a positive test result

Sentinel Initiative An initiative by the FDA to establish a database called Sentinel, to collect information from consumers, healthcare providers, and manufacturers alike, regarding drug safety. It allows the FDA to facilitate the continuous monitoring of medical products accessible on the markets.

Serious adverse health consequence Any significant adverse experience related to a device, including device-related events that are life-threatening, or events that involve long-term morbidity.

Sexual minority Individuals whose sexual orientation differs from normative sexual behavior.

Shared Decision-Making A collaborative approach between healthcare providers and patients in making decisions about their care. It often involves exchanging information, discussing treatment options, and considering patient preferences and values.

Small Business Concern (SBC) An independently owned and operated business that does not dominate its field of operation and complies with specific industry criteria.

Small Business Innovative Research (SBIR) A federal grant or funding agreement that enables small business concerns to receive support during the early stages of translational innovation, with no obligation for partnership with research institutions.

Small Business Technology Transfer (STTR) A federal grant or funding agreement that enables small business concerns to receive support during the early stages of translational innovation, with an obligation to form partnerships with nonprofit research institutions.

Small Molecule Drugs An organic molecule with low molecular weight which may or may not have an advantage in therapeutics.

SO_2 Sulfur dioxide

Social determinants of health Factors and conditions in which individuals are born, live, work, play, and grow that could potentially influence their health outcome

Spearman's rank correlation coefficient Statistical measure assessing the strength and direction of monotonic association between two variables by comparing their ranked values, providing a nonparametric measure of correlation.

Specificity The likelihood that a person without the disease will have a negative test result.

SPSS (Statistical Package for the Social Sciences) SPSS is a software package used for statistical analysis. It provides tools for data management, advanced statistical analysis, and reporting. SPSS is commonly used in social science research and business analytics.

Stakeholders The term "stakeholders" includes national primary care and specialty clinicians, patients, advocacy groups, and other organizations with expertise/interest in the topic being discussed.

Standard deviation (σ) Measurement of dispersion of data points in a given data set. Square root of variance.

Standardization To bring into conformity with a standard especially in order to assure consistency and regularity.

Standardization of Interventions The process of developing and implementing uniform and consistent approaches to healthcare practices, or interventions. This involves establishing clear and standardized protocols, procedures, or guidelines for delivering specific interventions.

Start time "Time zero" is known as "the time origin," which is the time at which participants are considered at-risk for the outcome of interest. In many studies, time at risk is measured from the start of the study (i.e., at enrollment).

Stata Stata is a statistical software package used for data analysis and visualization. It provides a wide range of statistical techniques and tools for managing, analyzing, and visualizing data, particularly in fields such as economics, sociology, political science, and epidemiology.

Statistic Number that describes a specific feature of a sample.

Statistical Distributions They are parameterized mathematical models that give the probabilities of different outcomes for a random variable. There are discrete and continuous distributions depending on the random value they model. Common examples include the normal, binomial, Poisson, and the exponential distribution.

Statistical power ($1-\beta$) The probability of observing a statistically significant effect, given sample size, alpha level (i.e., probability of type I error), and a hypothetical effect size of interest.

Statistical Principles These refer to the fundamental concepts and rules that direct the orderly collection, analysis, interpretation, and presentation of data in a systematic way. These principles ensure the accuracy and reliability of statistical methods, thereby assisting in drawing valid conclusions from data.

Statistics Discipline that involves the collection, analysis, interpretation, presentation, and organization of data to extract meaningful patterns, makes informed decisions, and draws inferences about populations.

Study design limitations Constraints inherent in the structuring of cardiology research that may impact the applicability and interpretation of study outcomes. Limitations may arise from factors such as sample size, the choice of control groups, as well as data collection methods.

Surgical innovation The development and introduction of new surgical techniques, devices, or procedures that aim to improve patient outcomes, enhance surgical processes, or address unmet clinical needs.

Surgical Innovator An individual who leads the development and introduction of novel surgical techniques, devices, or procedures aimed at enhancing patient care and surgical practices.

Surgical Operators An individual who has adopted a novel surgical technique, device or procedure which was created by the surgical innovator.

Surgical technique The specific method, procedure, or approach used by surgeons during a surgical operation to achieve a desired outcome, which may involve precise skills and tools.

Surveillance bias It occurs when certain outcomes are detected in patients who are more closely monitored.

Survival bias A form of selection bias in cross-sectional studies when the individuals who are surveyed or included are those who have survived a condition up to the time of the study, overlooking those who may have had the condition but did not survive until the time of the study.

Survival study In the preclinical setting, a survival study involves a procedure in which the animal is expected to recover and survive the postoperative period. It is also referred to as a chronic study.

Systematic review A type of literature review that collects and critically analyzes multiple research studies or papers.

Systems Biology A study of the interaction of complex biological networks.

T-score (T) A statistic used in sample size calculation in scenarios where sample size is small, and the standard deviation of population is unknown.

t-test A t-test is a statistical hypothesis test used to determine if there is a significant difference between the means of two groups. It is commonly used in experimental and observational studies to compare means and assess the significance of differences.

Target Identification The process of identifying molecules/compounds for the use of meeting a clinical and commercial need while being safe and efficacious.

Target Population Identify the population in which the test will be used, including age, race, gender, disease prevalence, and specific risk factors.

Target Validation The process of validating molecules/compound through whole animal disease models, molecular signaling pathways, and disease/molecule association.

TAVR Transcatheter Aortic Valve Replacement

Test article In a preclinical setting, this term refers to the investigational drug or device to be tested.

The committee conducts regular meetings to assess trial data, focusing on intervention safety, effectiveness, and trial progress. It may also review data quality and the investigators' adherence to the trial protocol.

The committee's role is instrumental in promoting participant safety and well-being in clinical trials. Its recommendations can contribute to the safe and ethical conduct of the trial, ensuring the generation of accurate and reliable results.

The historically controlled study is a study design in which a new treatment or intervention is compared to historically adequately documented active treatment or intervention in comparable patients.

The monitoring committee, also known as a data monitoring committee (DMC) or data and safety monitoring board (DSMB), is an autonomous team of experts tasked with scrutinizing clinical trial data to safeguard participants' safety and well-being.

Theoretical Sampling The process of selecting cases based on specific theories or hypotheses, aiming to explore a wide range of conditions or treatment effects.

Therapeutic inertia A phenomenon where a clinician fails to initiate or increase a therapy when treatment goals are not met.

Total product life cycle (TPLC) The stages of a medical device from concept to approval, production and dissemination, and finally through postmarket surveillance.

Trademark Legally registered protection of a word(s), phrase, or design allowing exclusive rights.

Translation The process of turning observations in the laboratory, clinic, and community into interventions that improve the health of individuals and the public—from diagnostics and therapeutics to medical procedures and behavioral changes.

Translational Cardiology A branch of Translational Medicine targeting the conversion of scientific research into clinical usage for cardiovascular diagnoses and treatment.

Translational Medicine A branch of medicine or medical research targeting the elimination of the barriers between laboratory and clinical research.

Translational Research Seeks to produce more meaningful, applicable results that directly benefit human health.

Type I error (α) The probability of rejecting the null hypothesis when, in fact, it is true.

Type I Error Rates The probability of incorrectly rejecting a null hypothesis when it is true; a critical consideration in adaptive designs to control the risk of false-positive findings.

Type II error (β) The probability of not rejecting the null hypothesis when it, in fact, is false.

U.S United States

VAERS Vaccine Adverse Event Reporting System

Validation studies Research that assesses the accuracy of a measurement instrument or method against a reference standard (gold standard) to confirm that the tool or method in question accurately measures the concept or condition it is intended to measure.

Variance The term variance refers to a statistical measurement of the spread between numbers in a data set.

Windows Windows is a series of operating systems developed by Microsoft. It is the most widely used operating system for personal computers, known for its graphical user interface and support for a wide range of software applications and hardware devices.

Withdrawal trials are clinical trials where all participants receive active treatment for a specific time (open-label phase). Then responders are randomized to receive either active treatment or placebo (withdrawal of active treatment)

Z-score Statistical measure that quantifies the distance between a data point and the mean of a dataset. Measured in standard Deviations.

Index

'*Note:* Page numbers followed by "f" indicate figures and "t" indicate tables.'

H

I

J

K

L

M

N

O

P

S